ASTROPHYSICAL MASERS AND THEIR ENVIRONMENTS

IAU SYMPOSIUM No. 242

COVER IMAGE:

Alice Springs is located in the geographical centre of Australia and is almost equidistant between Adelaide in the south and Darwin in the north. The town lies within the East and West MacDonnell ranges which extend over hundreds of kilometres and form a landscape of great variety and striking landforms. This photograph shows the red sandstone rocks at Standley Chasm in the western ranges.

Photo credit: Andrej Sobolev, Ural State University, Russia

INTERNATIONAL ASTRONOMICAL UNION
UNION ASTRONOMIQUE INTERNATIONALE

International Astronomical Union

IAU

ASTROPHYSICAL MASERS AND THEIR ENVIRONMENTS

PROCEEDINGS OF THE 242th SYMPOSIUM OF THE INTERNATIONAL ASTRONOMICAL UNION HELD IN ALICE SPRINGS, AUSTRALIA MARCH 12–16, 2007

Edited by

JESSICA M. CHAPMAN
CSIRO Australia Telescope National Facility, Australia

and

WILLEM A. BAAN
ASTRON, the Netherlands

CAMBRIDGE
UNIVERSITY PRESS

CAMBRIDGE UNIVERSITY PRESS
The Edinburgh Building, Cambridge CB2 2RU, United Kingdom
32 Avenue of the Americas, New York, NY 10013–2473, USA
477 Williamstown Road, Port Melbourne, VIC 3207, Australia
Ruiz de Alarcón 13, 28014 Madrid, spain
Dock House, The Waterfront, Cape Town 8001, South Africa

First published 2008

Printed in the United Kingdom at the University Press, Cambridge

Typeset in System LaTeX 2ε

A catalogue record for this book is available from the British Library

Library of Congress Cataloguing in Publication data

ISBN 9780521874649 hardback
ISSN 1743-9213

Table of Contents

Session 6. Galactic structure and the Galactic Centre
Chair: Luis Rodríguez

Session 7. Masers in AGN environments
Chair: Moshe Elitzur

Session 10. New millimeter and sub-millimeter masers

Chair: Indra Bains

Session 11. Future facilities and conference summary

Chairs: Elizabeth Humphreys, Karl Menten

Preface

IAU Symposium 242, *Astrophysical Masers and Their Environments*, was held in Alice Springs from 12–16 March 2007. Alice Springs is located in the centre of Australia, in an area of outstanding beauty next to the MacDonnell Ranges and Simpson Desert. The town is surrounded by a red sand sea the size of Europe. It was a tremendous experience to bring together 125 astronomers, from 18 different countries, to the Australian 'outback' for an intense week of presentations, discussions and social events. Many participants took time to explore the region which includes natural wonders such as Uluru (Ayers Rock) and Kings Canyon.

Since the initial discovery of 'mysterium' in the early 1960s, later identified as maser emission from the OH molecule, maser emission has been identified from many different molecules. Masers are observed in the interstellar medium and star-formation regions, in the expanding winds from evolved stars, in the compressed shells of supernovae remnants, in comets and in the nuclei of galaxies. Systematic surveys have led to the detections of thousands of maser sources. As well as providing strong signals that can be used to investigate a diverse range of astrophysical environments, masers also provide ideal tools for studying the structure and dynamics of our own Galaxy, while masers in other galaxies are now used for cosmological studies and studies of galactic nuclei.

This meeting was the third international symposium on astronomical masers. The first was held in March 1992 in Virginia US (*Lecture Notes in Physics: Astrophysical Masers, 1993, eds. Clegg & Nedoluha*). Considerable progress was made during the 1990s in understanding and observing the complexities of the regions in which the masers are found and these were were discussed at IAU Symposium 206 in March 2001, held in Mangaratiba, Brazil (*IAU 206: Cosmic Masers: From Protostars to Black Holes, eds. Migenes & Reid*). At the Brazil meeting it was proposed that the next maser meeting be held in Australia.

This meeting in Alice Springs brought the community together to review the observational and theoretical progress over the last six years, to show recent results from new facilities such as the sub-millimeter array (SMA), and to look ahead to what may be possible with future facilities such as the Atacama Large Millimeter Array (ALMA), and the Square Kilometre Array (SKA). The meeting had a notably strong contribution from younger scientists: 20 of the participants were PhD students and seven of these gave excellent talks. The meeting was scheduled with 11 science topics, each with a review talk, two or more invited talks and contributed talks. Two poster sessions allowed discussion of the 75 posters which were on display all week.

We are grateful to the sponsors for the meeting. We thank the CSIRO Australia Telescope National Facility for their support and encouragement over many months, Qantas for providing flights within Australia, and the Northern Territory Tourist Commission for providing beautiful photographs used to promote the meeting. We also thank businesses in Alice Springs for their excellent services, especially the Alice Springs Convention Centre where the meeting was held, and the Alice Springs Desert Park where we had a memorable conference banquet and disco. We look forward to the next international maser meeting on another continent!

Jessica M. Chapman (Australia Telescope National Facility, Australia)
Philip J. Diamond (Jodrell Bank Observatory, UK),
Willem A. Baan (ASTRON, the Netherlands)

ORGANIZING COMMITTEES

Scientific

J.M. Chapman (co-Chair, Australia)
W.A. Baan (Netherlands)
E.M.L. Humphreys (USA)
A.J. Kemball (USA)
L.F. Rodriguez (Mexico)
J. Yang (China)
P.J. Diamond (co-Chair, UK)
M. Elitzur (USA)
H. Imai (Japan)
K.M. Menten (Germany)
S. Slysh (Russia)

Local

J.M. Chapman (chair)
A.J. Green
V. Drazenovic
J.L. Caswell
H. Green
M. Wardle

Acknowledgements

The symposium was sponsored and supported by IAU Division X and Commission 40 (Radio Astronomy).

Funding and support from
the International Astronomical Union
CSIRO Australia Telecope National Facility
Qantas
and
the Northern Territory Tourist Commission
are gratefully acknowledged.

Photo credit: Shaun Amy, CSIRO

Participants

Shaun **Amy**, Australia Telescope National Facility, Australia	shaun.amy@csiro.au
Esteban **Araya**, New Mexico Tech & National Radio Astronomy Observatory, USA	earaya@nmt.edu
Megan **Argo**, Jodrell Bank Observatory, UK	mkargo@jb.man.ac.uk
Yoshiharu **Asaki**, Institute of Space and Astronautical Science, Japan	asaki@vsop.isas.jaxa.jp
Willem **Baan**, ASTRON, the Netherlands	baan@astron.nl
Indra **Bains**, Swinburne University of Technology, Australia	ibains@astro.swin.edu.au
Leszek **Blaszkiewicz**, University of Warmia and Mazury, Poland	leszekb@matman.uwm.edu.pl
Roy **Booth**, Hartebeesthoek Radio Astronomy Observatory, South Africa	roy@hartrao.ac.za
James **Braatz**, National Radio Astronomy Observatory, USA	jbraatz@nrao.edu
Jan **Brand**, INAF-Istituto di Radioastronomia, Italy	brand@ira.inaf.it
Shari **Breen**, University of Tasmania, Australia	slbreen@utas.edu.au
Crystal **Brogan**, National Radio Astronomy Observatory , USA	cbrogan@cv.nrao.edu
Valentin **Bujarrabal**, Observatorio Astronmico Nacional, Spain	v.bujarrabal@oan.es
Michael **Burton**, University of NSW, Australia	m.burton@unsw.edu.au
Paola **Castangia**, INAF-Osservatorio Astronomico di Cagliari, Italy	pcastang@ca.astro.it
James **Caswell**, Australia Telescope National Facility, Australia	james.caswell@csiro.au
Jessica **Chapman**, Australia Telescope National Facility, Australia	jessica.chapman@csiro.au
Xi **Chen**, Shanghai Astronomical Observatory, China	chenxi@ntnu.edu.tw
Richard **Crutcher**, University of Illinois, USA	crutcher@uiuc.edu
Jeremy **Darling**, University of Colorado, USA	jdarling@origins.colorado.edu
James **de Buizer**, Gemini Observatory, Chile	jdebuizer@gemini.edu
Itziar **de Gregorio-Monsalvo**, European Southern Observatory, Chile	idegrego@eso.org
Alet **De Witt**, University of South Africa, South Africa	aletdewitt@webmail.co.za
Shuji **Deguchi**, Nobeyama Radio Observatory, NAOJ, Japan	deguchi@nro.nao.ac.jp
Jean-Francois **Desmurs**, Observatorio Astronomico Nacional, Madrid, Spain	desmurs@oan.es
Phil **Diamond**, Jodrell Bank Observatory, UK	pdiamond@jb.man.ac.uk
Akihiro **Doi**, Yamaguchi University, Japan	doi@vsop.isas.jaxa.jp
Vicki **Drazenovic**, Australia Telescope National Facility, LOC, Australia	vicki.drazenovic@csiro.au
Phil **Edwards**, Australia Telescope National Facility, Australia	philip.edwards@csiro.au
Moshe **Elitzur**, University of Kentucky, USA	moshe@pa.uky.edu
Simon **Ellingsen**, University of Tasmania, Australia	simon.ellingsen@utas.edu.au
Dieter **Engels**, Hamburger Sternwarte, Germany	dengels@hs.uni-hamburg.de
Jarken **Esimbek**, Urumqi Observatory, China	jarken@ms.xjb.ac.cn
Vincent **Fish**, NRAO, Socorro, USA	vfish@nrao.edu
Richard **Forster**, UC Berkeley, Hat Creek Radio Observatory, USA	rick@hcro.berkeley.edu
Michael **Gaylard**, Hartebeesthoek Radio Astronomy Observatory, South Africa	mike@hartrao.ac.za
Ciriaco **Goddi**, Harvard-Smithsonian Center for Astrophysics, USA	cgoddi@cfa.harvard.edu
Sharmila **Goedhart**, Hartebeesthoek Radio Astronomy Observatory, South Africa	sharmila@hartrao.ac.za
Yolanda **Gomez**, Centro de Radioastronoma y Astrofsica, Mexico	y.gomez@astrosmo.unam.mx
Carlos **Gonzalez Fernandez**, Instituto de Astrofsica de Canarias, Spain	carlos.gonzalez@iac.es
Miller **Goss**, NRAO, USA	mgoss@nrao.edu
Malcolm **Gray**, University of Manchester, UK	malcolm.gray@manchester.ac.uk
Anne **Green**, University of Sydney, Australia	agreen@physics.usyd.edu.au
Harvey **Green**, LOC, Australia	hagreen@bigpond.net.au
James **Green**, Jodrell Bank Observatory, UK	james.green@postgrad.man.ac.uk
Lincoln **Greenhill**, Harvard-Smithsonian Center for Astrophysics, USA	greenhill@cfa.harvard.edu
Harm **Habing**, Leiden Observatory, the Netherlands	habing@strw.leidenuniv.nl
JinLin **Han**, National Astronomical Observatories, China	hjl@bao.ac.cn
Lisa **Harvey-Smith**, Joint Institute for VLBI in Europe, the Netherlands	harvey@jive.nl
Jinhua **He**, Insitute of Astronomy & Astrophysics, Academia Sinica, Taiwan	jhhe@asiaa.sinica.edu.tw
Christian **Henkel**, Max-Planck-Institut fuer Radioastronomie, Germany	chenkel@mpifr-bonn.mpg.de
Jack **Hewitt**, Northwestern University, USA	j-hewitt@northwestern.edu
Tracey **Hill**, Leiden Observatory, the Netherlands	thill@strw.leidenuniv.nl
Tomoya **Hirota**, National Astronomical Observatory of Japan, Japan	tomoya.hirota@nao.ac.jp
Peter **Hofner**, New Mexico Tech, USA	phofner@nrao.edu
Mareki **Honma**, National Astronomical Observatory of Japan, Japan	honmamr@cc.nao.ac.jp
Shinji **Horiuchi**, Canberra Deep Space Communication Complex, Australia	shoriuchi@astro.swin.edu.au
Liz **Humphreys**, Harvard Center for Astrophysics, USA	ehumphreys@cfa.harvard.edu
Roberta **Humphreys**, University of Minnesota, USA	roberta@umn.edu
Todd **Hunter**, NRAO, USA	thunter@nrao.edu
Busaba **Hutawarakorn Kramer**, National Astronomical Research Institute of Thailand, Thailand	busaba@nari.or.th
Hiroshi **Imai**, Kagoshima University, Japan	imai@astro.sci.kagoshima-u.ac.jp
Terry **Jones**, University of Minnesota, USA	tjj@astro.umn.edu
Seiji **Kameno**, Kagoshima University, Japan	kameno@sci.kagoshima-u.ac.jp
Ryuichi **Kamohara**, Mizusawa VERA Observatory, NAOJ, Japan	ryuichi.kamohara@nao.ac.jp
Athol **Kemball**, University of Illinois, USA	akemball@uiuc.edu
Hideyuki **Kobayashi**, National Astronomical Observatory of Japan, Japan	hideyuki.kobayashi@nao.ac.jp
Edo **Loenen**, Kapteyn Astronomical Institute, the Netherlands	loenen@astro.rug.nl
Steven **Longmore**, University of New South Wales, Australia	snl@phys.unsw.edu.au
Colin **Lonsdale**, MIT Haystack Observatory, USA	cjl@haystack.mit.edu
Kim **McAlpine**, Rhodes University, South Africa	kim.mcalpine@gmail.com
Korinne **McDonnell**, Macquarie University, Australia	korinnem@ics.mq.edu.au
Eduardo **Mendoza**, Instituto Nacional de Astrofisica Optica y Electronica, Mexico	mend@inaoep.mx
Karl **Menten**, Max -Planck-Institut fuer Radioastronomie, Germany	kmenten@mpifr-bonn.mpg.de
Victor **Migenes**, Universidad de Guanajuato, Mexico	vmigenes@astro.ugto.mx
Nanako **Mochizuki Suzuyama**, Japan Aerospace Exploration Agency, Japan	nanakom@vsop.isas.jaxa.jp
George **Moellenbrock**, National Radio Astronomy Observatory, USA	gmoellen@aoc.nrao.edu
Jim **Moran**, Harvard Smithsonian Center for Astrophysics, USA	moran@cfa.harvard.edu
Masaki **Morimoto**, Nature Education Center, Japan	hello@ojisan.jp
Luca **Moscadelli**, INAF - Osservatorio Astrofisico di Arcetri, Italy	mosca@arcetri.astro.it
Akihura **Nakagawa**, Kagoshima University, Japan	nakagawa@astro.sci.kagoshima-u.ac.jp
Jun-ichi **Nakashima**, ASIAA, Taiwan	junichi@asiaa.sinica.edu.tw
Jagadheep **Pandian**, Max Planck Institute for Radio Astronomy, Germany	jagadheep@astro.cornell.edu
Rodrigo **Parra**, Onsala Space Observatory, Sweden	rodrigo@oso.chalmers.se
Nimesh **Patel**, Harvard-Smithsonian Center for Astrophysics, USA	npatel@cfa.harvard.edu
Michele **Pestalozzi**, University of Herfordshire, UK	michele.pestalozzi@gmail.com
Chris **Phillips**, ATNF, Australia	chris.phillips@csiro.au
Ylva **Pihlstrom**, University of New Mexico, USA	ylva@unm.edu

Lyshia **Quinn**, University of Manchester, UK lquinn@jb.man.ac.uk
Mark **Reid**, Harvard Smithsonian Center for Astrophysics, USA reid@cfa.harvard.edu
Anita **Richards**, Jodrell Bank Observatory, UK amsr@jb.man.ac.uk
Laura **Richter**, Rhodes University, South Africa llrichter@gmail.com
Tim **Robishaw**, University of California, Berkeley, USA robishaw@astro.berkeley.edu
Luis **Rodriguez**, Universidad Nacional Autonoma de Mexico, Mexico l.rodriguez@astrosmo.unam.mx
Georgij **Rudnitskij**, Sternberg Astronomical Institute, Russia gmr@sai.msu.ru
Mayumi **Sato**, National Astronomical Observaotry of Japan, Japan mayumi.sato@nao.ac.jp
Satoko **Sawada-Satoh**, ASIAA, Taiwan satoko@asiaa.sinica.edu.tw
Derck **Smits**, University of South Africa, South Africa smitsdp@unisa.ac.za
Andrej **Sobolev**, Ural State University, Russia andrej.Sobolev@usu.ru
Rebeca **Soria Ruiz**, Joint Insitute for VLBI in Europe, the Netherlands soria@jive.nl
Vladimir **Strelnitski**, Maria Mitchell Observatory, USA vladimir@mmo.org
Koichiro **Sugiyama**, Yamaguchi University, Japan j011vc@yamaguchi-u.ac.jp
Kazuyoshi **Sunada**, Nobeyama Radio Observatory, Japan sunada@nro.nao.ac.jp
Marian **Szymczak**, Torun Centre for Astronomy, Poland msz@astro.uni.torun.pl
Daniel **Tafoya**, Harvard-Smithsonian Center for Astrophysics, USA dtafoya@cfa.harvard.edu
Andrea **Tarchi**, INAF - Osservatorio Astronomico di Cagliari, Italy a.tarchi@ira.inaf.it
Kalle **Torstensson**, Leiden Observatory/JIVE, the Netherlands kalle@strw.leidenuniv.nl
Miguel Angel **Trinidad**, Departamento de Astronomia, Universidad de Guanajuato, Mexico trinidad@astro.ugto.mx
Tomofumi **Umemoto**, Nobeyama Radio Observatory, Japan umemoto@nro.nao.ac.jp
Lucero **Uscanga**, Centro de Radioastronomia y Astrofisica - UNAM, Mexico l.uscanga@astrosmo.unam.mx
Johan **van der Walt**, Unit for Space Physics, North-West University, South Africa johan.vanderwalt@nwu.ac.za
Wouter **Vlemmings**, Jodrell Bank Observatory, UK wouter@jb.man.ac.uk
Maxim **Voronkov**, Australia Telescope National Facility, Australia maxim.voronkov@csiro.au
Andrew **Walsh**, James Cook University, Australia andrew.walsh@jcu.edu.au
Mark **Wardle**, Macquarie University, Australia wardle@ics.mq.edu.au
Jonathon **Weintroub**, Harvard Smithsonian Center for Astrophysics, USA jweintroub@cfa.harvard.edu
Markus **Wittkowski**, European Southern Observatory, Germany mwittkow@eso.org
Daniel **Wong-McSweeney**, University of Manchester, UK d.wong-mcsweeney@postgrad.manchester.ac.uk
Alwyn **Wootten**, National Radio Astronomy Observatory/ALMA, USA awootten@nrao.edu
Ye **Xu**, Max Planck Institut fuer Radioastronomie, Germany xuye@mpifr-bonn.mpg.de
Aya **Yamauchi**, University of Tsukuba, Japan yamauchi@physics.px.tsukuba.ac.jp
Farhad **Yusef-Zadeh**, Northwestern University, USA zadeh@northwestern.edu
Jiang-shui **Zhang**, Center for astrophysics, Guangzhou University, China jszhang365@yahoo.com.cn
Xingwu **Zheng**, Nanjing University, China xwzheng@nju.edu.cn
Jianjun **Zhou**, Urumqi Observatory, China zhoujj@ms.xjb.ac.cn

Tribute to Raymond James Cohen

The proceedings of this meeting are dedicated to the memory of Jim Cohen. Jim died on 1 November 2006 at the age of fifty-eight, after a brain aneurism. He was a friend and colleague to many of us and would have attended this meeting as an invited speaker. He is a co-author of six of the papers in these proceedings.

Jim's research career was based at Jodrell Bank. He arrived in Britain from Australia in 1970 and obtained his PhD on the distribution of hydrogen near the Galactic Centre, using observations taken with the Lovell telescope (then called the Mk1). His first paper was published in 1974, co-authored with Rod Davies and Andrew Wilson. Since then there have been some 190 papers in a variety of scientific and popular journals and publications. These first papers on hydrogen in the Galactic Centre inspired Jessica Chapman to become Jim's first PhD student in 1981, and do maser research with MERLIN at the onset of its first high-resolution images.

Jim was a much appreciated colleague at Jodrell Bank. He was a walking encyclopedia on star-formation and other astronomical subjects. One colleague said that Jim in his quiet and gentle manner played a large role in the success of Jodrell. He was full of ideas and acted on them.

Jim's maser research has made significant contributions to just about all aspects of observational maser astronomy including comets, star formation, evolved stars, Galactic structure, polarization studies, and megamaser galaxies. With his students and collaborators he used MERLIN to obtain many beautiful sub-arcsec images of OH and water masers.

As a supervisor Jim was always willing to look at an exciting new result or talk through a problem. He was an extremely clear-headed thinker and brilliant at explaining things in a simple way with great insight into the underlying physics. He had the gift of teaching and ably communicated his insights, not only at the telescope but also in the classroom. Five of his eleven PhD students attended this meeting (Jessica Chapman, Anita Richards, Busaba Hutawarakorn Kramer, Lisa Harvey-Smith and Jimi Green).

Jim also maintained strong links with his native Australia. Most recently he was the leading member of the consortium of researchers from Great Britain and Australia to build the multi-beam receivers and execute the ambitious Parkes methanol survey of the southern skies. After his death, John Reynolds, who is in charge at the Parkes Observatory, said that "Jim was one of our best loved visitors in recent times and had the respect and friendship of all here who knew him."

Jim's professional career included many years of active involvement in activities related to protecting radio astronomy from interference and light pollution. He was a member and the Chair of IAU Commission 50 and a member of the Scientific Committee on the Allocation of Frequencies (ICSU, International Council for Science). With his open attitude and his quiet and stable demeanor, Jim was an effective participant whose substantial input over a long period of time made him a leader in this group. A special memory for Willem Baan was the time, in 1993, of working together with Jim and John Ponsonby to negotiate the agreement between the radio astronomers and the generals of the Russian Space Forces, about the GLONASS satellite system. Jim's contribution was of great import in resolving the conflict. Fittingly, the pub that serves as a meeting place before dinner for the "defenders of the passive radio services" during ITU meetings in Geneva is the *The Lord Jim* on Rue de Lausanne.

Some of Jim's friends will also remember his love of music. Musically inclined since childhood, he was an accomplished player of the recorder, the whistle and the flute, the instrument he played in his local amateur orchestra at Alderley Edge,and in his own ceilidh bands, *Merlin* (named after the array of telescopes based at Jodrell Bank) and *Hartigan's Fancy*.

Jim was a family man, devoted to his wife Pat and his daughters Beth and Helen. He lived a life of service. The Biblical scripture that best describes Jim's character is the command from Ephesians "Be ye kind one to another, tender-hearted, forgiving one another". As a man of deep faith in God, yet a researcher of meticulous scientific integrity, Jim was a rarity in today's world.

Jim will be missed greatly at Jodrell, by the maser research community, by the spectrum protection community, and at the Rose Cottage in Rudyard.

A 2006 visit to the construction site of the Sardinia Radio Telescope during a meeting of the CRAF spectrum committee. Continuing the tradition of placing something personal or valuable under the foundation of the new structure, Jim just buried his business card.

Jessica Chapman
Willem Baan

Astrophysical Masers and their Environments
Proceedings IAU Symposium No. 242, 2007
J.M. Chapman & W.A. Baan, eds.

doi:10.1017/S1743921307012483

Alice's astronomical ancestry

Philip Edwards

CSIRO ATNF, Paul Wild Observatory, Narrabri NSW 2390, Australia
email: philip.edwards@csiro.au

Abstract. IAU242 was held in Alice Springs, a town of 28,000 in central Australia. In this paper the Alice after whom the town is named is identified, and her astronomical connections described.

Keywords. history and philosophy of astronomy, time

1. Introduction

Alice Springs was established as one of eleven telegraph relay stations between Adelaide and Darwin on the Overland Telegraph. The Overland Telegraph, completed in 1872, connected to the undersea cable from Java, providing the first telegraphic link from Australia to the rest of the world. Alice Springs was named after Alice Todd, whose husband, Charles, was Government Astronomer and Superintendent of Telegraphs for the colony of South Australia and who oversaw the planning and construction of the Overland Telegraph.

The township was originally named Stuart, with Alice Springs being the name of the telegraph station, but to end confusion between the two (and a reflection of the importance of the Overland Telegraph) the town of Stuart was officially renamed Alice Springs in August 1933.

Although there is no professional astronomical observatory in central Australia, a wide range of astronomical observations have been conducted from sites near the town, including balloon-borne studies at X-ray and gamma-ray energies (Leventhal *et al.* 1978, Pietsch *et al.* 1980, Zanrosso *et al.* 1981, Sood *et al.* 1988a, O'Neill *et al.* 1989, Covault *et al.* 1992, Chapuis *et al.* 1993, Greenhill *et al.* 1993, Leventhal *et al.* 1993, Barthelmy *et al.* 1994), studies of meteors using radar (Thomas *et al.* 1986), optical astronomy (Nozawa *et al.* 2000), and, at radio wavelengths, Very Long Baseline Interferometry (Preston *et al.* 1989). The Arrernte (also spelt Aranda, Arunda, Arunta) traditional owners of the area, like other aboriginal peoples, have a cultural tradition which incorporates astronomical legends and beliefs (see, e.g., Chapter 1 of Haynes *et al.* 1996). In this paper, however, we will briefly recount some of the endeavours and contributions to astronomy of Alice's husband, Charles Todd, her son-in-law, William Henry Bragg, and her grandson, William Lawrence Bragg.

2. Charles Todd

Charles Todd was born in London in 1826 and educated at Greenwich. He entered the workforce at the age of 15 as Astronomical Computer at the Royal Observatory, Greenwich, and later became Assistant Astronomer at the Cambridge Observatory, before returning to Greenwich in 1854 to take charge of the new galvanic department.

Todd was appointed, on the recommendation of the Astronomer Royal, G.B. Airy, as Astronomical Observer and Superintendent of Telegraphs for the young colony of South

Australia. South Australia was founded in 1836, forty-eight years after the arrival of the First Fleet at Port Jackson (Sydney). Before sailing from England, Todd married Alice Bell, ten years his junior. The family story of their "courtship" is recounted in Thomson (1999). They arrived in Adelaide in November 1855, and are pictured in Figure 1.

3. The overland telegraph

Todd's first telegraph line, which connected the city with Port Adelaide, opened in February 1856 and used the magnetic telegraph system of Wheatstone and Cooke. The next line, between Adelaide and Melbourne, was completed in July 1858. The Superintendent of Telegraphs for the colony of Victoria was Samuel McGowan, who had been a student of Samuel Morse's in the USA before moving to Australia, and so not surprisingly Morse code was adopted for the transmission on this (and all future Australian) telegraph lines. Sydney and Melbourne were connected in October 1858, providing a link between Adelaide and Sydney; a direct line was completed between Adelaide and Sydney in 1867.

Todd first proposed a telegraph line crossing Australia from south to north in 1859, but it was not until John McDouall Stuart's crossing of the continent in 1861–62 that the idea started to receive support. In 1870, the South Australian Government committed to build a telegraph line across Australia to connect with an undersea cable being laid from Java (Indonesia).

Todd's plan divided the construction of the line into three parts, with the southern and northern sections of the line assigned by contract, and the longest and most difficult central section constructed by Todd's department. The department's work was finished ahead of schedule, and the department went on to finish the work in the northern section when the contractors experienced severe difficulties with conditions imposed by the tropical wet season.

The line was finally completed in August 1872. The honour of sending the first message was given to Todd, who cabled "We have this day, within two years, completed a line of communications two thousand miles long through the very centre of Australia, until a few years ago a terra incognita believed to be a desert." Taylor (1980) summarises the achievement: "In spite of almost overwhelming odds... [Todd] built a line more than three thousand kilometres long through country that had been crossed only once, that was for the most part uninhabited by white people, and had one of the most inhospitable climates in the world. Not only was it an outstanding achievement by any standard, but it had been accomplished in the remarkable time of one year and eleven months." Detailed descriptions of the planning and construction of the Overland Telegraph are given by Clune (1955), Taylor (1980), Moyal (1984), and Harcourt (1987). A more recent account was given by Paul Davies on the ABC radio Science Show, the transcript of which is available at www.abc.net.au/rn/scienceshow/stories/2005/1515660.htm.

Following the construction of the Overland Telegraph, communication with England took a matter of hours, compared with the several months it had previously taken for mail delivered by ship – a factor of $\sim$500 improvement – with obvious benefits to Australian industry, trade and government. In his book "The Victorian Internet," Standage (1998) draws interesting parallels between the information age brought about by the rapid spread of the telegraph over 120 years ago, and that brought about by computer networks over the last 20 years.

There is an ATNF connection to the Overland Telegraph: Ron Ekers' (the former Director of the Australia Telescope National Facility) great-grandfather worked on its construction! And the husband of Ron's great-great-Aunt worked at Barrow Creek, the first relay station north of Alice Springs, losing his life there in 1874 in an incident known as the Barrow Creek massacre.

The Australia colonies were all finally linked in 1877 with the completion of the line linking Adelaide to the Western Australian capital, Perth. That year, South Australia became the first colony to join the International Telegraph Union, and in 1885 Todd attended the Union's international conference in Berlin.

4. Astronomy

Although the construction and operation of telegraphs, and later the postal service, took much of Todd's time, he also made significant contributions to astronomy and meteorology (Edwards 1993). Todd established the Adelaide Observatory, which provided the time service for the colony, and which conducted a variety of astronomical observations. Todd and his assistant Alexander Ringwood (who had previously worked as a surveyor during the construction of the Overland telegraph) started a long series of observations of the planet Jupiter, and of eclipses, occultations and transits of the Galilean satellites. They noted occasions where satellites were visible through Jupiter's limb for a short time after their occultation, which they interpreted as evidence for an extensive gaseous atmosphere on Jupiter.

Todd also participated in observations of the transits of Venus in 1874 and 1882 (Edwards 2004), which it had been hoped would enable the measurement of the Astronomical Unit to be refined. The 1874 transit was observed at the Adelaide Observatory with a new eight inch (20.3 cm) Cooke & Son equatorial. Todd and Ringwood both noted during the egress of Venus from the sun's disk that the portion of the planet outside the solar disk remained visible and clearly defined. Like many other observers who observed this phenomenon, they were surprised by this evidence of the dense Venusian atmosphere, although there had been reports of it during the previous transits in 1761 and 1769.

Todd observed the 1882 transit from Wentworth in south-western NSW, with an 4.5 inch (11.4 cm) equatorial that had previously belonged to Charles Babbage (Edwards 2004). The following year, the Overland Telegraph line was used to transmit time signals from Singapore in order to more accurately determine the longitudes of Australian sites for the analysis of results from the 1882 transit of Venus.

5. Time zones

Conference participants cannot have failed to notice that the Northern Territory, like South Australia, is in an unusual half-hour time zone. Todd was instrumental in the introduction of time zones in Australia, initially arguing for a simple implementation with a single time zone, 9 hours ahead of Greenwich Mean Time (GMT), for the whole of Australia (see, e.g., Edwards 1996). The eastern states did not accept this proposal, electing to move to GMT + 10 hrs, but when time zones were introduced in 1895, South Australia (which until 1911 included the Northern Territory) adopted the zone GMT − 9 hrs. However, the South Australian Chamber of Commerce claimed that local commerce was suffering from the fact that their eastern states competitors had one hour to respond to the cables received overnight from England before local merchants received them in the morning. As a result of their lobbying, parliament amended the standard time bill so that from May 1st, 1899, South Australia was 9 hours 30 minutes ahead of Greenwich — corresponding to a longitude that does not even run through the state! The Astronomical Society of South Australia unanimously passed a resolution that "... if in the commercial

Figure 1. Alice (1836–1898) and Charles (1826–1910) Todd (reproduced from The Chronicle, February 4, 1905)

interest of South Australia any change is desirable, it should be in the alteration of the nominal hours of business rather than in the present standard time ...", but in vain.

6. Todd and Bragg

In February 1886, William Henry Bragg arrived in Adelaide to take up the position of Professor of Mathematics and Physics at the University of Adelaide. Bragg met Todd on his first full day in Adelaide and became a regular visitor at the Observatory, marrying Alice and Charles' daughter Gwendoline in 1889 (Jenkin 1986). The Braggs had two sons, William Lawrence, born in 1890, and Robert Charles, born in 1892.

The demonstrations by Hertz in 1887–88 of electromagnetic radiation were replicated in laboratories around the world. Todd and Bragg became interested in the possibility of wireless telegraphy, and Bragg used a letter of introduction from Todd to meet William Preece and Guglielmo Marconi while in London on study leave in 1898 (Jenkin 1986). On returning to Adelaide in 1898, Bragg and his assistant, Arthur Rogers, succeeded in transmitting radio messages over increasing distances, and ultimately the 8 km from the Observatory to Henley Beach.

These experiments with radio ended in 1990 as other interests took priority, Bragg working with others in Adelaide to follow up Röntgen's discovery of X-rays. Lawrence Bragg broke his arm at the age of five, and the X-ray taken in his father's laboratory at the University is said to be the first surgical use of X-rays in Australia.

The Braggs left Adelaide in 1909, with WH Bragg becoming Cavendish Professor of Physics at the University of Leeds. WL Bragg, who had completed his undergraduate degree at the University of Adelaide with first class Honours in mathematics, enrolled at Cambridge.

7. WL Bragg and radio astronomy

WH Bragg and WL Bragg were awarded the Nobel Prize for Physics in 1915 "for their services in the analysis of crystal structure by means of X-rays." From 1919 to 1937, WL Bragg held the position of Professor of Physics at the Victoria University of Manchester. After the second World War, he returned to Cambridge where, following the belief that "the ideal research unit is one of six to twelve scientists and a few assistants," he split the Cavendish Laboratory into research groups.

Molecular biologist Max Perutz wrote "Bragg's period of office in Cambridge coincides with the decline of the Cavendish Laboratory as the world's leading centre of atomic physics. This was an inevitable consequence of the war and the transformation of atomic physics to 'Big Science', to which the tradition and structure of Cambridge University were ill-adapted. Rather than fight a rearguard action, Bragg decided to back two new applications of physics in which the Cavendish was again to lead the world." One of these was molecular biology, the field in which Perutz shared the Nobel Prize for Chemistry in 1962. "The other new application of physics which Bragg helped to initiate at the Cavendish Laboratory was Radio Astronomy"
(see http://www.phy.cam.ac.uk/cavendish/history/years/molbiol.php).

Martin Ryle recalled the efforts of establishing the radio astronomy group at Cambridge, "During these early months, and for many years afterwards, both Ratcliffe and Sir Lawrence Bragg, then Cavendish Professor, gave enormous support and encouragement to me. Bragg's own work on X-ray crystallography involved techniques very similar to those we were developing for "aperture synthesis", and he always showed a delighted interest in the way our work progressed"
(see http://nobelprize.org/nobel_prizes/physics/laureates/1974/ryle-autobio.html).

Bragg's choice of radio astronomy as one of the new fields to develop was undoubtedly made carefully and rationally, but it is interesting to speculate whether his grandfather's work may have had some subtle influence on his decision. The Adelaide Observatory, that Todd established, closed in 1952 (Edwards 1994). On learning of this, WL Bragg wrote "I was very grieved to hear that the Adelaide Observatory has ceased to function. So many of my most vivid early memories are bound up with the old building and my grandfather's work there: and I deeply regret that nothing has been created to take its place and carry on its magnificent early pioneer work in the Southern Hemisphere." The collaboration and competition between Cambridge and the CSIRO Division of Radiophysics was certainly instrumental in spurring on the "magnificent early pioneer work" in the new field of radio astronomy.

8. In conclusion

Charles Todd, best known for his association with the Overland Telegraph, carried out a range of astronomical observations over his career. Alice Springs is named for Todd's wife, while the Todd River and Todd Mall bear his name. A number of other roadways in the town bear the names of people who worked on the Overland Telegraph, such as Patterson Crescent, Babbage St, Harvey St, Knuckey Ave, McMinn St, Mills St, and the Ross Highway. Fittingly, as Todd also served as Postmaster General, the Overland Telegraph is commemorated on several Australian philatelic items: the centenary of the Overland Telegraph was marked by the issue of a 7c stamp in 1972, the Alice Springs telegraph station is pictured on a 27c stamp issued in 1982, and Todd is pictured on the 45c pre-stamped envelope issued in 1997. It is known that Charles Todd travelled along the overland telegraph route. It appears, however, that Alice Todd never had the chance to visit the town named after her.

In summary, Todd was an active astronomer, studied astronomical atmospheres (albeit planetary rather than stellar), contributed to the development of the cosmological distance ladder, experimented with radio waves, employed the most rapid methods of communication for astronomical research, encouraged an interest in astronomy in his children and grand-children, and travelled to international conferences. He would have felt right at home at IAU Symposium 242!

Acknowledgements

I would like to thank Jessica Chapman, Glenys & Glen Edwards, Ron Ekers, and Miller Goss for their encouragement and contributions to this paper.

References

Barthelmy, S. D., *et al.* 1994, *ApJ*, 427, 519
Chapuis, C. G. L., *et al.* 1993, *ApJ*, 403, 332
Clune, F. 1955, *Overland Telegraph* (Angus & Robertson, Sydney)
Covault, C. E., Grindlay, J. E., & Manandhar, R. P. 1992, *ApJ*, 388, L65
Dodwell, G. 1911, *MNRAS*, 71, 272
Edwards, P. G. 1993, *Proc. ASA*, 10, 349
Edwards, P. G. 1994, *Proc. ASA*, 11, 206
Edwards, P. G. 1996, *Sky and Space*, 9, no. 4, 30
Edwards, P. G. 2004, *J. Ast. Hist. Heritage*, 7, 1
Greenhill, J. G., Sharma, D. P., Dieters, S. W. B., Sood, R. K., Waldron, L., & Storey, M. C. 1993, MNRAS, 260, 21
Harcourt, E. 1987, *Taming the Tyrant* (Allen & Unwin, Sydney)
Haynes, R., Haynes, R., Malin, D., & McGee, R. 1996, *Explorers of the Southern Sky* (Cambridge University Press)
Jenkin, J. 1986, *The Bragg Family in Adelaide: A Pictorial Celebration* (University of Adelaide Foundation, Adelaide)
Leventhal, M., MacCallum, C. J., & Stang, P. D. 1978, *ApJ*, 225, L11
Leventhal, M., Barthelmy, S. D., Gehrels, N., Teegarden, B. J., Tueller, J., & Bartlett, L. M. 1993, *ApJ*, 405, L25
Moyal, A. 1984, *Clear Across Australia* (Thomas Nelson, Melbourne)
Nozawa, H., Misawa, H., Takahashi, S., Morioka, A., Okano, S., Sood, R., & Yumoto, K. 2000, *Adv. Sp. Res.*, 26, 1525
O'Neill, T. J., Kerrick, A. D., Ait-Ouamer, F., Tumay Tumer, O., Zych, A. D., & White, R. S. 1989, *Science*, 244, 451
Pietsch, W., Voges, W., Reppin, C., Truemper, J., Kendziorra, E., & Staubert, R. 1980, *ApJ*, 237, 964
Preston, R. A., Jauncey, D. L., Meier, D. L., Tzioumis, A. K., Ables, J., *et al.* 1989, AJ, 98, 1
Sood, R. K., *et al.* 1988, *MNRAS*, 234, 73P
Standage, T. 1998, *The Victorian Internet* (Weidenfeld & Nicolson)
Taylor, P. 1980, *An End to Silence* (Methuen, Sydney)
Thomas, R. M., Whitham, P. S., & Elford, W. G. 1986, *Proc. ASA*, 6, 303
Thomson, A. 1999, *The Singing Line* (Random House)
Zanrosso, E., Long, J. L., Zych, A. D., & White, R. S. 1981, *Adv. Sp. Res.*, 1, 235

Astrophysical Masers and their Environments
Proceedings IAU Symposium No. 242, 2007
J.M. Chapman & W.A. Baan, eds.

doi:10.1017/S1743921307012495

Recent developments in maser theory

Moshe Elitzur

Physics & Astronomy Dept., University of Kentucky, Lexington, KY 40506-0055, USA
email: moshe@pa.uky.edu

Abstract. This review covers selected developments in maser theory since the previous meeting, *Cosmic Masers: From Proto-Stars to Black Holes* (Migenes & Reid 2002). Topics included are time variability of fundamental constants, pumping of OH megamasers and indicators for differentiating disks from bi-directional outflows.

Keywords. masers, polarization, radiative transfer, accretion disks, cosmology: miscellaneous

1. Fun with masers

Rather then maser theory, the first couple of items covered here can be more properly described as "fun with masers". These items were selected for discussion because they do not belong under the title of any other talk in this meeting. Both involve weak OH signals that lead to spectacular results.

1.1. *How do we know masers exist?*

There are numerous arguments why the radiation from various astronomical sources involves maser amplification (e.g., Elitzur 1992). But the only direct proof can come from measurements of excitation temperatures that yield negative values. The well known technique for measuring a line excitation temperature, T_x, requires a molecular cloud in front of a distance continuum source that provides background radiation with measurable brightness temperature T_b at the transition frequency. The measured brightness temperatures in the "off" and "on" positions, away from and toward the background source, obey

$$T^{\rm off} = T_x\left(1 - e^{-\tau}\right), \qquad T^{\rm on} = (T_x - T_e)\left(1 - e^{-\tau}\right)$$

where τ is the line optical depth of the cloud. The two measurements yield two equations for the two unknowns T_x and τ. In the classic experiment by Rieu *et al.* (1976), the OH main lines from a cloud in front of the extragalactic radio source 3C123 switched from emission in the off-source position to on-source absorption. The excitation temperature of both lines is $T_x \simeq 5$–7 K. But in the case of the 1720 MHz line, off-source emission turned into even stronger emission in the on-source position, indicating that the cloud amplifies the background radiation with $\tau \simeq -0.1$. The corresponding excitation temperature is $T_x(1720) \simeq -10$ K.

Almost 30 years later, Weisberg *et al.* (2005) came up with a beautiful variant of this approach in OH spectral measurements toward 18 pulsars. One member of the sample, pulsar B1641–45, emits every 0.455 seconds a pulse of radio radiation that passes through an OH cloud on its way toward Earth. The entire system is positioned in front of an HII region that provides additional, steady background radiation. When the pulsar is off, spectra of the OH ground-state lines detected from the cloud display absorption features in the 1612, 1665 and 1667 MHz lines and an emission feature at 1720 MHz. When the pulsar is on, passage of its radiation through the cloud deepens the absorption features. But at 1720 MHz the emission feature becomes stronger, showing that at this frequency the pulsar radiation is amplified by passing through the intervening screen. This is the equivalent of an object amplifying a background light instead of casting a shadow. It is the same effect as observed earlier by Rieu *et al.*, but it does not involve any change in the telescope pointing. The pulsar method also increases confidence in the results, thanks to the repeated detection of amplified signals that arrive more than twice every second. The amplification again is modest, τ is only ~ -0.05. Although the input signals are amplified by only a few percent in both cases, these weak masers are the only unambiguous direct evidence for amplification.

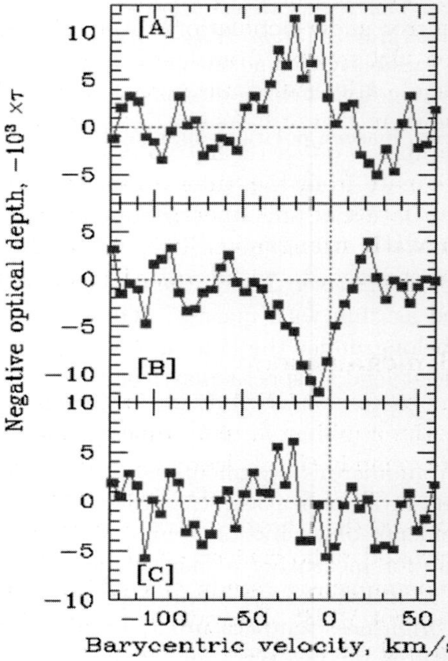

Figure 1. OH satellite line spectra toward PMN J01340931 plotted against barycentric velocity relative to the source redshift, $z = 0.76355$. (A) 1720 MHz transition; (B) 1612 MHz transition; (C) sum of the 1612 and 1720 MHz spectra; this spectrum is consistent with noise. (Kanekar *et al.* 2005)

1.2. *Are fundamental constants truly constant?*

A cosmological implication of many variants of string theories is that fundamental constants such as the fine structure constant $\alpha = e^2/\hbar c$, the electron-proton mass ratio m_e/m_p, etc. should vary with time. Indeed, from measurements of QSO absorption lines it has been suggested that a variation $\Delta\alpha/\alpha \sim 10^{-5}$ has possibly occurred between the $z = 3.5$ epoch and the present (Murphy *et al.* 2001). However, the uncertainty in this suggestion is substantial because measurements based on atomic lines are faced with fundamental difficulties. Comparing the frequencies of different lines of the same atom does not yield information on α because the entire spectrum of a given species has the same α-dependence, therefore frequency shifts that would result from a change in α cannot be distinguished from the effects of redshift. And comparison of lines of different atoms suffers from the uncertainty arising from the possibly different locations, and different z, of the species.

Darling (2003) pointed out that the complexity of the OH spectrum makes it most suitable for getting around this problem. The centimeter OH transitions are a combination of hyperfine splitting and Λ-doubling, and the two effects depend differently on α. As a result, by comparing the frequencies of different OH radio transitions one can constrain the cosmic evolution of α from a single species, avoiding systematic errors in α

measurements from multiple species which may have relative velocity offsets. The problem of different locations still exists, though, because different OH lines may still arise from different locations because of preferential excitations in different environments. This problem was solved by Kanekar *et al.* (2005) who noted that many weak signals from the OH ground-state satellite lines appear in conjugate pairs of absorption and emission. The two halves of the Λ-doublet consist of pairs of hyperfine states. Because of the different overall spins of hyperfine states, their coupling to different excited states makes it easy to transfer population between them without affecting the overall population of each half of the Λ-doublet. The result is an overpopulation, or even inversion, of one satellite line and a mirror under-population of the other (see, e.g., Elitzur 1992).

Taking OH spectra from the $z \sim 0.765$ gravitational lens toward PMN J01340931, Kanekar *et al.* identified a conjugate pair of 1612 MHz absorption and 1720 MHz emission, shown in Figure 1. As is evident from the figure's bottom panel, the sum of the two satellite lines produces a flat spectrum consistent with noise, a strong indication that they originate from the same region. The deduced line frequencies place tight bounds on the variation of fundamental constants over the past ~ 6.5 G yr: $\Delta\alpha/\alpha < 6.7 \times 10^{-6}$ and a similarly strong upper limit of 1.4×10^{-5} for $m_{\mathrm{e}}/m_{\mathrm{p}}$. These fundamental constants have not changed, to within this tight tolerance, for half the age of the Universe.

2. OH megamasers

OH megamasers (OHMM) are extremely luminous extragalactic sources with isotropic luminosities a million or more times greater than their Galactic OH maser counterparts. They are found in the nuclear regions of luminous infrared galaxies where intense star formation is occurring. The "standard model" for OHMM was introduced by Baan (1985), who proposed that the maser emission is produced by low gain unsaturated amplification of background radio continuum. This proposal was confirmed by the Henkel & Wilson (1990) comprehensive study of OHMM. However, compact maser emission discovered in subsequent VLBI observations produced conflicts with the standard model, conflicts that reached their climax around the time of the previous maser meeting (Lonsdale 2002). Strong compact masers without correspondingly bright continuum background implied amplification factors in excess of 800. In addition, the 1667/1665 line ratios of the compact masers greatly exceeded those of the diffuse OHMM emission. An additional surprise of these observations was that linewidths remained large (tens of $\mathrm{km\,s^{-1}}$) even on the smallest observed angular scales. This has led to the suggestion that there are two different modes of OHMM operation — low gain, unsaturated amplification responsible for the diffuse maser emission and high gain, saturated emission producing the compact sources. It was also suggested that the two classes may have different pumping mechanisms (Lonsdale 2002).

A major recent development is the realization that a two-component picture is not necessary. The compact masers can be explained within the standard model as chance alignment of low-gain maser clumps. The key was the interferometric observations of IIIZw35 by Pihlström *et al.* (2001) and the modeling by the same authors and later, in greater detail, by Parra *et al.* (2005). The model explains both the diffuse and compact emission from IIIZw35 by assuming a clumpy maser medium. Each cloud generates low gain, unsaturated emission, and strong compact emission occurs when the line of sight intersects more than one cloud. A similar model has subsequently been used to explain the megamaser emission from IRAS 17208-0014 (Momjian *et al.* 2006). The clumpy maser model also explains the observation that compact masers are always found embedded in the diffuse emission and do not occur in isolation (Lonsdale 2002). Thus the only additional ingredient necessary to supplement Baan's "standard model" is clumpiness — the OHMM phenomenon is explained as the amplification of background radio continuum by low gain ($\tau_{1667} \lesssim 1$) clumps.

Lockett & Elitzur (2007) performed pumping calculations to explain the OHMM inversion process. Pumping models should explain the outstanding properties of OHMM emission: (1) The line ratio of observed fluxes always obeys $F(1667)/F(1665) > 1$ and the satellite lines are always weaker than the main lines. By comparison, in evolved stars the main lines emission region displays a similar behavior — 1667 MHz is the strongest maser, and the satellite line emission from that region is weaker than either main line. In contrast, Galactic star forming regions show a different, more complex emission pattern. Generally, the main lines are stronger than the satellite lines and display $F(1667)/F(1665) < 1$, the opposite of EHM and evolved stars. But there are sources in which the strongest line is the 1665 MHz is and some where it is a satellite line.

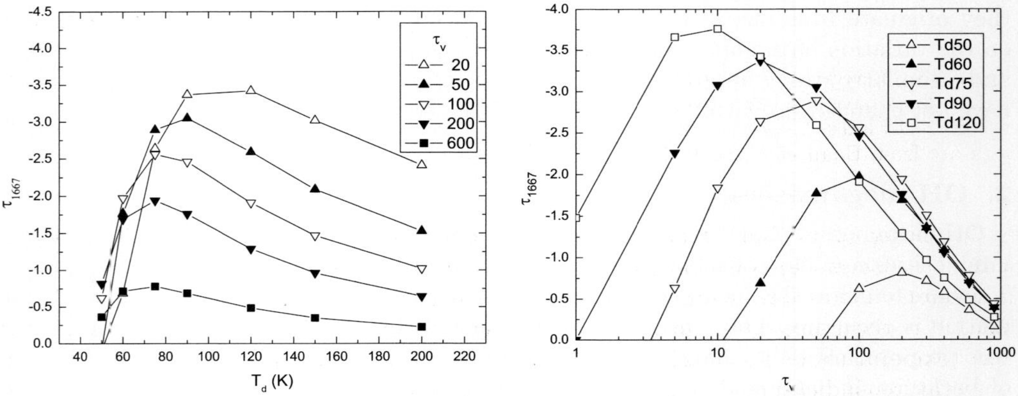

Figure 2. Dependence of the 1667 MHz maser optical depth, τ_{1667}, on the parameters of the IR pumping flux (see eq. 2.1). *Left:* τ_{1667} versus dust temperature for varying dust optical depth. *Right:* τ_{1667} versus dust opacity for different dust temperatures (Lockett & Elitzur 2007).

In the Lockett & Elitzur models, the inversion arises from radiative excitations by far-IR radiation (FIR). The FIR radiation field is described by the simple expression

$$F_\nu = B_\nu(T_d)\left(1 - e^{-\tau_\nu}\right), \tag{2.1}$$

corresponding to emission from dust with a uniform temperature T_d and optical depth τ_ν. The frequency dependence of τ_ν follows that of standard interstellar dust so the only free parameters are T_d and τ_V, the dust optical depth at visual. Combinations of parameters that produce main lines inversion are shown in figure 2. The left panel shows the 1667 MHz optical depth versus dust temperature for a range of dust opacities. A minimum dust temperature of ~45 K is needed for maser production. Increasing the dust temperature increases the maser strength up to a point, after which it decreases. Maximum maser inversion occurs for dust temperatures between ~70–120 K. The right panel shows the 1667 MHz optical depth versus dust opacity for a range of dust temperatures. Two important results can be seen in this figure. First, when the opacity becomes large the dust radiation approaches that of a black-body and the maser begins to thermalize. At $\tau_V \gtrsim 200$ there is little dependence of maser strength on dust temperature. Second, for each dust temperature there is a minimum dust opacity needed for inversion. This implies that a minimum FIR pumping flux is needed for maser production. Further understanding can be found by examining the maser dependence on the FIR intensity

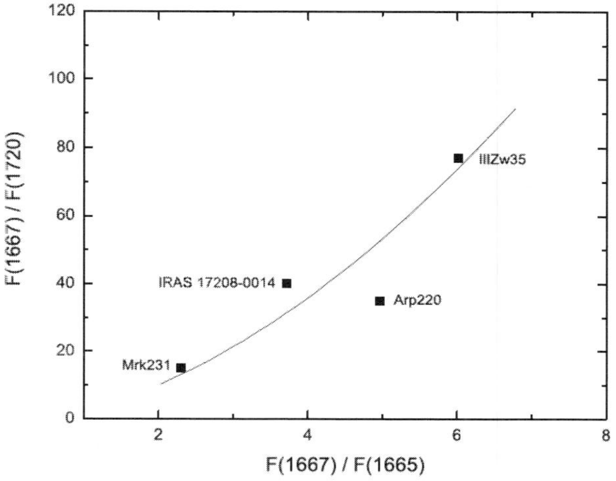

Figure 3. "Color-color" diagram of OH maser line ratios. The line shows the variation when all four lines have the same excitation temperature, which is the situation when the linewidth exceeds ~ 10 km s^{-1}, as τ_{1667} increases from 0.5 at the curve's lower end to 4 at its upper end. Data are from Baan *et al.* 1989, 1992 and Martin *et al* 1989 (Lockett & Elitzur 2007).

in the maser pump line at 53 μm. The pumping flux is determined by the product of the two factors in equation 2.1. The first is the Planck function, which depends on dust temperature, the second is the self-absorption term, which depends on the dust opacity at the pumping frequency. The FIR flux is increased by increasing either the dust temperature or its opacity. As a result, a larger flux is needed for a lower dust temperature, indicating that the FIR spectral shape is important for maser pumping. Thus a smaller IR luminosity can support inversion as long as the dust temperature is sufficiently high — in agreement with observations (Baan 1989, Darling & Giovanelli 2002). Similarly, inversion requires a minimum dust temperature. Below $\sim$50 K, the Planck function at 53 μm decreases very rapidly with dust temperature. The flux can be increased by raising the dust opacity but cannot exceed the black-body limit, which is insufficient for inversion at dust temperatures below about 45 K.

An important correlation is the increase of $L_{\rm OH}$ with $L_{\rm IR}$. Early OHMM observations revealed an approximately quadratic relation between the two (Baan 1989), but subsequent studies established the flatter dependence $L_{\rm OH} \propto {L_{\rm IR}}^{1.2}$ (Darling & Giovanelli 2002). Low gain amplification of background radio continuum implies that $L_{\rm OH}$ is proportional to the radio luminosity, which in turn is proportional to $L_{\rm IR}$ over the relevant luminosity range. Since the observed correlation is steeper than linear, Baan (1989) conjectured that maser optical depth $\tau_{\rm maser}$ also varies linearly with the pump luminosity $L_{\rm IR}$. However, the Lockett & Eitzur model results show that, depending on the different parameters, $\tau_{\rm maser}$ in fact can either increase or decrease with the IR luminosity. The increase of $L_{\rm OH}$ in excess of linear proportion to $L_{\rm IR}$ may be explained by the results of the Gao & Solomon (2004) of HCN emission from luminous IR galaxies. They find a tight proportionality between IR and HCN luminosities, and conclude that the amount of dense molecular gas, which is traced by the HCN, is linearly proportional to $L_{\rm IR}$. It is thus reasonable to assume that, together with the increase of dense molecular gas, the number of OH maser clouds is also increasing with $L_{\rm IR}$.

Line overlap is a major component of the OHMM inversion and the calculations show that the linewidth, Δv, is the most important factor in determining the 1667/1665 MHz line ratio. Widths in excess of ~ 2 km s^{-1} cause the 1667/1665 ratio to be consistently larger than one, while widths $\lesssim 2$ km s^{-1} can cause this ratio to be greater than or less than one, depending on the actual Δv. Large linewidths ($\gtrsim 10$ km s^{-1}) produce ground state hyperfine transitions that are all inverted with approximately the same (negative) excitation temperature. Detailed comparison with observations is difficult because the satellite lines are very weak and have only been detected for relatively nearby sources. Even in those cases the signal to noise ratio is small, resulting in large uncertainties in line strength ratios. Figure 3 plots the 1667/1720 MHz line ratio versus the 1667/1665 MHz line ratio for four relatively nearby OHMM. The agreement is surprisingly good considering the large uncertainties in the line ratios.

The variation of line ratios with Δv suggests that FIR might be a common pump of all OH main line masers, Galactic as well as megamasers, with the magnitude of velocity controlling line overlap the factor explaining the observed differences. The maser patterns in Galactic star forming regions stand apart because of the relatively small values of Δv in these sources. The similarity between OHMM and evolved stars may reflect the large velocities controlling line overlap in both environments.

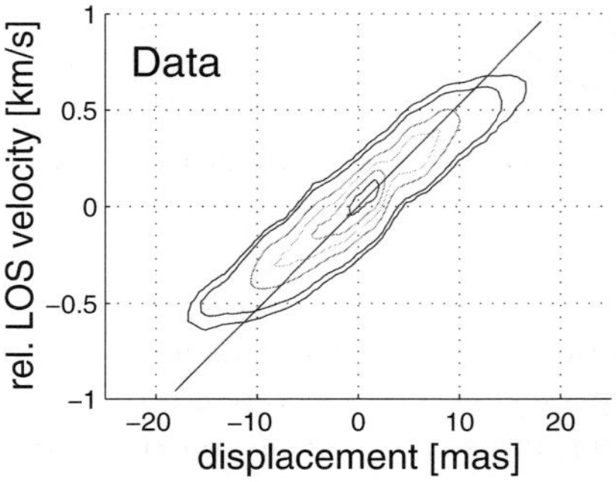

Figure 4. Position-velocity diagram of 12 GHz methanol maser amplification contours in NGC 7538 IRS 1N (Pestalozzi *et al.* 2004).

3. Disks and outflows

Class II methanol maser emission often shows linear structures in both spatial maps and in position–velocity (p–v) diagrams. It has been suggested that these structures delineate protostellar edge-on disks (Norris *et al.* 1998). However, similar structures can be produced also in bi-directional outflows, a geometry that actually seems more appropriate in a large fraction of methanol masers (De Buizer 2003). The fact that such disparate maser geometries can produce such similar signatures raises a fundamental question: Short of proper motion measurements, is there a reliable way to distinguish between the two? The methanol maser source towards IRS 1N in NGC 7538 may offer an answer. Its unique emission contours display a centrally-peaked large, elongated structure in the p–v diagram, shown in figure 4; in contrast, most other sources typically display in such diagrams a string of maser features. The structure in NGC 7538 was successfully modeled in great detail by Pestalozzi *et al.* (2004) with an edge-on Keplerian disk. Although this interpretation was questioned by De Buizer & Minier (2005), Kraus *et al.* (2006) have demonstrated the existence of jet precession in this source, settling the issue in favor of the disk geometry.

It turns out that the p–v structure seen in figure 4 implies fairly tight constraints that significantly narrow the range of possible geometries (Pestalozzi *et al.* 2007). Consider first a bipolar outflow. When the outflow axis is in the plane of the sky, the emission

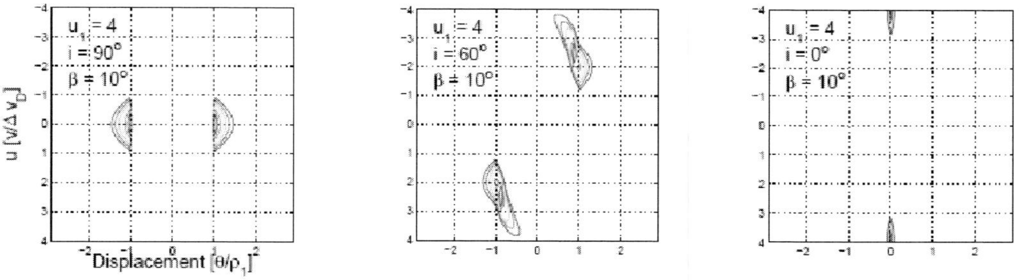

Figure 5. Position-velocity diagram of maser amplification contours for bipolar outflow with representative properties. The outflow axis is in the plane of the sky in the left panel, aligned with the line-of-sight in the right panel and 60° off the line-of-sight in the center one (Pestalozzi *et al.* 2007).

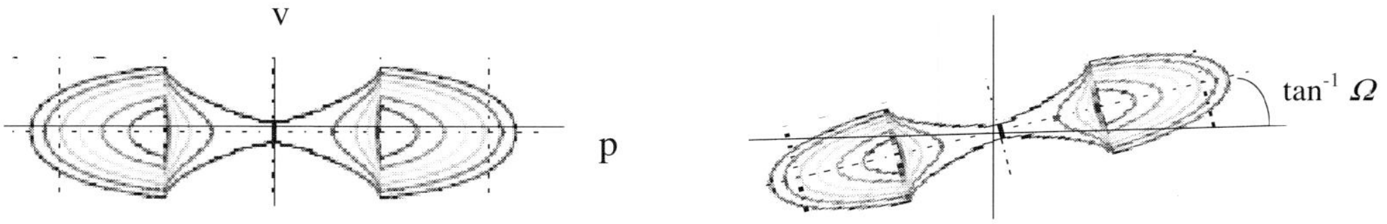

Figure 6. Position-velocity diagram for maser amplification in an edge-on disk. *Left*: A quiescent disk, only thermal velocities. *Right*: The same disk subjected to solid-body rotation with angular velocity Ω (Pestalozzi *et al.* 2007).

from each lobe is centered on velocity $v = 0$. Because the density declines rapidly with radial distance, the emission is peaking at two regions displaced symmetrically from the outflow origin. The resulting structure in the p–v diagram is two blobs that straddle the origin at $v = 0$, as seen in the left panel of figure 5. When the outflow axis is along the line-of-sight, the emission from both cones occurs at the same position, but at different velocities — blue and red shifted by the same amount from the systemic velocity. The p–v diagram, shown in the right panel of figure 5, again displays two blobs, only this time they are in the same position but at different velocities. In essence, rotating the outflow axis by 90° effects a similar rotation of the emission contours in the p–v diagram. It is not surprising that an intermediate orientation of the outflow axis likewise produces an intermediate rotation in the p–v diagram, as can be seen from the center panel of figure 5. Whatever the inclination of its axis, a bipolar outflow cannot produce the single centrally peaked structure seen in figure 4. A smooth outflow will produce two mirror features displaced symmetrically from the center. Fragmentation of the outflow may introduce an asymmetry between the two halves but will always avoid the center, never producing a centrally peaked feature.

Consider now the case of an edge-on disk. When the disk does not rotate, its optical depth obeys $\tau(p,v) = \tau(p)\phi(v)$; that is, the positional variation of the amplification is controlled solely by the length of the amplifying column because the velocity profile ϕ is the same at every position. For uniform inversion, $\tau(p)$ is simply proportional to the pathlength through the disk, which is maximal at the two tangents to the disk inner radius. The p–v diagram of the amplification contours in this case is shown in the left panel of figure 6. Similar to the outflow geometry, the contours show two distinct peaks, displaced symmetrically from the center. The center does show amplification in this case, but it is a local minimum. When the disk is set in rotation, the Doppler velocity of the emission from a point at radius r and impact parameter p obeys $v = \Omega(r)p$, where $\Omega(r)$

is the local angular velocity. Therefore, in solid body rotation the material along each impact parameter remains fully velocity coherent, only the velocity profile is centered on $v = \Omega p$; that is, the optical depth obeys $\tau(p,v) = \tau(p)\phi(v - \Omega p)$. As a result, the structure of amplification contours in the p–v diagram remains the same, only rotated by the angle $\tan^{-1}\Omega$ (and slightly stretched to maintain the peak positions at the two tangents to the inner radius), as shown in the right panel of figure 6. Similar to bipolar outflow, an edge-on disk rotating as a solid body will never produce a centrally peaked structure as observed in NGC 7538 (figure 4). Only differential rotation, such as Keplerian motion, can produce such structure. Furthermore, even in that case, turning the local minimum at the center of the p–v diagram into a local maximum requires rotation rates in excess of a certain threshold (Pestalozzi *et al.* 2007). Detailed analysis of the p–v diagram is, evidently, a powerful tool that deserves more study.

4. Open problem — maser polarization

Maser polarization behaves differently according to whether the Zeeman shift ν_B is larger or smaller than the linewidth $\Delta\nu_D$. The case $\nu_B \gg \Delta\nu_D$ is well understood. Spontaneous decays occur in pure Δm transitions, producing in each transition radiation that is fully polarized and centered on a different Zeeman frequency. Under these circumstances the amplification process preserves the polarization, therefore thermal and maser polarizations are the same. The only difference between the two cases is the disparity between the π and σ maser intensities, reflecting their different growth rates (Elitzur 1996).

In the opposite limit, $\nu_B \gg \Delta\nu_D$, the Zeeman components overlap and the amplification mixes the different polarizations. The pumping processes produce radiation in three different modes of polarization with respect to the magnetic field ($\Delta m = 0, \pm 1$), but only two independent combinations propagate in any direction because the electric field must be perpendicular to the wave vector. Elimination of the longitudinal component implies specific phase relations among the three pump-generated electric fields; only waves launched with these phase relations produce superpositions that are purely transverse so that they can be amplified by propagation in the inverted medium. The phase relations translate to the following polarization for maser radiation propagating at angle θ from the magnetic axis:

$$\frac{Q}{I} = -1 + \frac{2}{3\sin^2\theta}, \qquad \frac{V}{I} = \frac{16xx_B}{3\cos\theta}, \tag{4.1}$$

where $x = (\nu - \nu_0)/\Delta\nu_D$ and $x_B = \nu_B/\Delta\nu_D$. This solution, which was first derived by Goldreich, Keeley & Kwan (1973) in the limit $x_B = 0$ and extended by Elitzur (1996) to finite $x_B < 1$, is obtained also from the requirement that the four Stokes parameters produce fractional polarizations that remain unaffected by the amplification process.

How the unpolarized radiation produced in spontaneous decays evolves into this stationary polarization solution remains an open problem. Numerical calculations that follow the growth of the Stokes parameters do not reach the stationary solution for any reasonable amount of amplification, and this slow growth can be understood from simple analytic considerations. However, the Stokes parameters are intensities, quadratic in the electric fields of the waves, and in formulating their growth equations all information about the phases of the electric vectors is lost. Since the stationary solution reflects specific phase relations between the π and σ components, the formalism employed for calculating the growth of the Stokes parameters negates the very effect it seeks to describe. Eliminating the phase information, the interactions selecting the stationary polarization are removed from the calculation. What is still missing is a full simulation of the ensemble evolution of interacting particles and electric fields. Such simulations are preformed in studies of laboratory plasma but have not yet been attempted for astronomical maser radiation.

5. Coupled Escape Probability (CEP)

Pumping models require reliable calculations of the population distribution in multi-level systems. The radiative terms in the statistical rate equations require the local radiation field, which must be determined from a separate solution of the radiative transfer equation for each and every line. Because of the complexity and computational demands of exact solution methods, most codes bypass altogether the radiative transfer equations, employing instead the escape probability approximation. In this approach only the level populations are considered, calculated from rate equations that include photon escape factors which are meant to account approximately for the effects of radiative transfer. This approach amounts to an uncontrolled approximation without internal error estimates because it is founded on a plausibility assumption right from the start. Nevertheless, this inherent shortcoming is often tolerated because of the simplicity and usefulness of the escape probability approach.

Elitzur & Asensio Ramos (2006) have recently developed a new exact method, the Coupled Escape Probability (CEP), that retains all the advantages of the naive escape probability approach. In this new technique the source is divided into zones, and formal level population equations are derived rigorously from first principles. Different zones are coupled through terms resembling standard escape probability expressions, resulting in a set of coupled algebraic equations for level populations that are self-consistent with the line radiation they generate. Any desired accuracy can be achieved by increasing the number of zones. In comparative studies of a number of standard problems, the CEP method outperformed the current leading techniques by substantial margins. While the new method holds a great performance edge, its greatest advantage is its simplicity and ease of implementation. The CEP method employs a set of algebraic equations that are already incorporated in all codes based on the escape probability approximation. All that is required for an exact solution is to augment the escape probability in such codes with a zone-coupling sum. With this simple modification, the multi-level line transfer problem is solved exactly and efficiently.

For my own pumping calculations I developed over the years the escape probability code MOLPOP that can handle any arbitrary atom or molecule for which atomic data exist. For any species, the energy levels, collision rates and Einstein A-coefficients are tabulated in ordinary ASCII text files and MOLPOP selects from its database the appropriate files according to input instructions. We are in the process of implementing the CEP technique into MOLPOP, creating a general purpose tool for atomic and molecular line analysis that will become publicly available. In addition, the atomic and molecular data team at Meudon Observatory has embarked on the development of another tool that accesses through the web their central database and creates on the user's local computer atomic and molecular data files in the format required by MOLPOP. This auxiliary tool will become part of the MOLPOP/CEP package and will provide users the most current atomic data at all times. The complete package will enable exact pump modeling for every maser molecule. We hope to have a first release by the end of 2007.

Acknowledgements

The author's maser research is supported by NSF through grant AST-0507421.

References

Baan, W. 1985, Nature, 315, 26

Baan, W. 1989, ApJ, 388, 804

Baan, W. A., Haschick, A., & Henkel, C. 1989, ApJ, 346, 680

Baan, W. A., Haschick, A., & Henkel, C. 1992, AJ, 103, 728

Darling, J. 2003, Physical Review Letters, 91, 011301

Darling, J. & Giovanelli, R. 2002, AJ, 124,100

De Buizer, J. 2003, MNRAS, 341, 277

De Buizer, J. M., & Minier, V. 2005, ApJ, 628, L151

Elitzur, M. 1992, Astronomical Masers (Kluwer Academic Publishers)

Elitzur, M. 1996, ApJ 457, 415

Elitzur, M. & Asensio Ramos, A. 2006, MNRAS 365, 779

Gao, Y., & Solomon, P. M. 2004, ApJ, 606, 271

Goldreich, P., Keeley, D. A. & Kwan, J. Y. 1973, ApJ, 179 111

Henkel, C., & Wilson, T. L. 1990, A&A, 229, 431

Kanekar, N., *et al.* 2005, Physical Review Letters, 95, 261301

Kraus, S., *et al.* 2006, A&A, 455, 521

Lockett, P. & Elitzur, M. 2007, in preparation

Lonsdale, C. J. 2002, Cosmic Masers: From Proto-Stars to Black Holes, IAU Symposium #206, 413

Martin, J., Bottinelli, Dennedeld, M., Gougenheim, L., & Lesqueren, A., 1989, A&A, 208,39

Migenes, V., & Reid, M. J. 2002, Cosmic Masers: From Proto-Stars to Black Holes, IAU Symposium #206 (San Francisco: ASP)

Momjian, E., Romney, J. D., Carilli, C. L., & Troland, T. H. 2006, ApJ, 653, 1172

Murphy, M. T., Webb, J. K., Flambaum, V. V., Dzuba, V. A., Churchill, C. W., Prochaska, J. X., Barrow, J. D., & Wolfe, A. M. 2001, MNRAS, 327, 1208

Parra, R., Conway, J.E., Elitzur, M., & Pihlstrom. Y.M. 2005, A&A, 443, 383

Pestalozzi, M. R., Elitzur, M., Conway, J. E., & Booth, R. S. 2004, ApJ 603, L113; Erratum: 2004, ApJ, 606, L173

Pestalozzi, M. R., Elitzur, M. & Conway, J. E., 2007, in preparation

Pihlström, Y. M., Conway, J. E., Booth, R. S., Diamond, P. J., & Polatidis, A. G. 2001, A&A, 377, 413

Norris, R., Byleveld, S., & Diamond, P. J. 1998, ApJ, 508, 275

Rieu, N. Q., Winnberg, A., Guibert, J., Lepine, J. R. D., Johansson, L. E. B. & Goss, W. M. 1976, A&A, 46, 413

Weisberg, J. M., Johnston, S., Koribalski, B. & Stanimirovic, S. 2005, Science 309, 106

Astrophysical Masers and their Environments
Proceedings IAU Symposium No. 242, 2007
J.M. Chapman & W.A. Baan, eds.

doi:10.1017/S1743921307012501

Tracing pumping routes in OH

Malcolm D. Gray[1] D. A. Howe[1] and B. M. Lewis[2]

[1]Astrophysics Group, School of Physics & Astronomy, University of Manchester, PO Box 88, Manchester M60 1QD, United Kingdom
email: Malcolm.Gray@manchester.ac.uk

[2]Arecibo Observatory, HC3, Box 53995, Arecibo, PR 00612, U.S.A.

Abstract. We introduce a general method of restoring detailed information about the population flow in molecules under non-local-thermodynamic-equilibrium (NLTE) conditions. This information is usually discarded in numerical algorithms which generate only a solution. We apply the method to tracing the pumping schemes for OH in models that represent three common astrophysical maser environments: the envelopes of asymptotic-giant-branch (AGB) stars, the envelopes of red supergiants, and Galactic star-forming regions. In all three of these cases, we show that a large fraction, typically 0.8 or more, of the maser inversion can be recovered from a set of routes that depend on a much smaller fraction (considerably less than 0.1) of the total number of input coefficients to the model. Therefore, these cases display underlying simplicity in the pumping scheme.

Keywords. masers, molecular data, molecular processes, radiative transfer, methods: numerical, stars: AGB and post-AGB, stars: formation

1. Introduction

Our current knowledge of maser pumping schemes is patchy. In the case of OH, conspicuous successes include the 1612-MHz line in the circumstellar envelopes of AGB and OH/IR stars. This pump was recovered by analytic estimates, backed up by numerical calculations, and was found to be based on the absorption of 35-μm and 53-μm radiation (Elitzur, Goldreich & Scoville (1976); Elitzur (1981); Dickinson (1987)). This absorption is followed by a radiative cascade down the $^2\Pi_{1/2}$ ladder of levels, and then a final cross-ladder transition to the $^2\Pi_{3/2}, J = 3/2$ rotational ground state. A large inversion results if this last transition is optically thick. Observational support for these schemes is so far partial (He & Chen (2004); He *et al.* (2004)), and further observations are required with greater sensitvity and spectral resolution than the LWS instrument aboard the Infrared Satellite Observatory (ISO). A pumping scheme in similar detail for the 1720-MHz line of OH was developed by Elitzur (1976). In contrast to the 1612-MHz scheme, this 1720-MHz pump is driven primarily by collisions with H_2.

By contrast, information about pumping of the OH main-line masers in star-forming regions has been much harder to glean from a variety of observations and models. The models have been employed mainly in parameter-space searches, and these yield only broad inferences from the initial conditions fed into them. For example, a trend that shows increasing maser inversion as a function of dust temperature strongly suggests that far-infrared (FIR) radiation forms part of the pump, but the details are lost because each run of the model, corresponding to a point in the parameter space, provides only an NLTE solution. No details of the population flow are stored. One feature which is expected to be important in main-line pumping is FIR line overlap (for example Bujarrabal *et al.* 1980). Another is a strong continuum FIR field, probably from dust. The combination of this continuum and the FIR line overlap have been shown to be effective in several models, for example Gray, Field & Doel (1992), Pavlakis & Kylafis (1996). However, most of these models used the large velocity gradient (LVG) approximation, which makes it virtually impossible to study the region of parameter space where velocity shifts across the model are small compared to the local sound speed, and where thermal FIR line overlap only is present. As a large amount of velocity-driven FIR line overlap is a fundamental part of any LVG OH model, it is, in a way, unsurprising that such models find it essential.

To recover the detail of pumping schemes from a numerical model, a record of the algebraic operations which are performed on the system of equations must be stored. This allows coefficients of a partially-solved system to be decomposed into those of a less-processed form of the equations, building up a network of population flow cycles involved in pumping. The identification of such cycles was attempted by Sobolev (1986). These cycles are somewhat analogous to electrical circuits, obeying rules similar to Kirchoff's laws Sobolev & Deguchi (1994). A Monte-Carlo computer code based on this method has been applied to water masers Sobolev (1989) and to methanol masers Sobolev & Deguchi (1994). In both cases, the indications are that it may not be possible to identify simple pumping schemes in these molecules. The case of OH is discussed below. The tracing method is not Monte-Carlo, but computer-algebra based. The method is implemented in the computer code TRACER, and details appear in Gray, Howe & Lewis and Gray (2007).

2. Population tracing

2.1. *The parameter-space model*

The numerical solution used as the basis for population tracing is still drawn from a parameter-space search. However, the analysis, in terms of the pumping scheme, is completely different: instead of drawing an inference from the physical conditions supplied to a large number of models, a single zone from one model is examined in detail with TRACER. For the environments studied in the current work, the parameter-space models were all of the accelerated lambda-iteration (ALI) type, using LU-factorization as the numerical method. Local and non-local FIR line overlap were included.

2.2. *The* TRACER *model*

There are two parts to this model: one is written into a special subroutine of the ALI solver, which runs only once as an extra iteration when convergence has already been achived by the ALI method. In this final iteration, a naive method of solving the matrix is used for ease of human interpretation, and a log is maintained of the operations carried out on the matrix. In the naive method the all-process rate coefficient for the transition from level x to level y, with z eliminations to go, may be written

$$k^z_{x,y} = k^{z+1}_{x,y} + k^{z+1}_{x,z} k^{z+1}_{z,y} / k^{z+1}_{z,z} \tag{2.1}$$

The first term on the right represents the coefficient from the previous elimination, whilst the second term represents transfer between level x and level y via level z. The log of operations compares the size of these two terms and, according to the accuracy of the trace, determines whether both must be kept, or one can be omitted.

The operation log described above forms the input to the tracer code itself. This begins with an analytical expression for the inversion, derived from a small, highly processed, matrix with $z = 3$ or $z = 4$. Expressions for all four ground-state inversions appear in Section 3. Tracer algebraically expands the expression for a given inversion, considering antagonistic pairs of expressions that represent the forward and backward parts of each route. Every time an expression like the second term on the right-hand side of 2.1 must be included, a branch to a new route is introduced, leading to increased complexity. The trace is complete when all coefficients are the unprocessed originals that were fed into the matrix as input data. These have $z = N + 1$, where N is the number of energy levels in the model. There is no guarantee of underlying simplicity, and expressions could become completely unmanageable for values of z considerably smaller than $N + 1$.

3. Ground-state inversions

We present here the remaining ground-state inversions, as analytical solutions of either a 4×4 or 3×3 matrix. Obviously, the forms for the transitions which include level 4 are more complicated, but contain one layer of possible complexity fewer when TRACER is run. Values of $\mathcal{N}$ and X are given in Gray (2007).

It is convenient to break these expressions up into 'direct' and 'indirect' parts. The former involve no other uneliminated level, whilst the latter do: for example in the 1612-MHz inversion, 3.1, the 'direct' part involves only the maser levels, 2 and 3, whilst the indirect part also involves the other uneliminated level, level 1. Of course both the 'direct' and 'indirect' parts can involve, when expanded by TRACER, a large number of previously eliminated levels.

$$\Delta\rho_{32} = \frac{\mathcal{N} k^4_{1,2}}{D}\left\{(1+\eta)\left(\frac{k^4_{2,3}}{3} - \frac{k^4_{3,2}}{5}\right) + \frac{\eta k^4_{2,1}}{3} - \frac{k^4_{3,1}}{5}\right\} \tag{3.1}$$

$$\Delta\rho_{31} = \frac{\mathcal{N} k^4_{2,2}}{3D}\left\{k^4_{1,3} - k^4_{3,1} + \left(\frac{k^4_{1,2}k^4_{2,3} - k^4_{3,2}k^4_{2,1}}{k^4_{2,2}}\right)\right\} \tag{3.2}$$

$$\begin{aligned}\Delta\rho_{42} = \frac{\mathcal{N}'}{5}\Bigg\{ & k^5_{2,4} - k^5_{4,2} + \frac{k^5_{1,1}}{X}\left(k^5_{2,3}k^5_{3,4} - k^5_{4,3}k^5_{3,2}\right) + \frac{k^5_{3,3}}{X}\left(k^5_{2,1}k^5_{1,4} - k^5_{4,1}k^5_{1,2}\right) \\ & + \frac{k^5_{2,1}k^5_{1,3}k^5_{3,4} - k^5_{4,3}k^5_{3,1}k^5_{1,2}}{X} + \frac{k^5_{2,3}k^5_{3,1}k^5_{1,4} - k^5_{4,1}k^5_{1,3}k^5_{3,2}}{X}\Bigg\}\end{aligned} \tag{3.3}$$

$$\begin{aligned}\Delta\rho_{41} = \mathcal{N}'\Bigg\{ & \frac{k^5_{1,4}}{5} - \frac{k^5_{4,1}}{3} + \frac{k^5_{3,3}}{X}\left(\frac{k^5_{1,2}k^5_{2,4}}{5} - \frac{k^5_{4,2}k^5_{2,1}}{3}\right) + \frac{k^5_{2,2}}{X}\left(\frac{k^5_{1,3}k^5_{3,4}}{5} - \frac{k^5_{4,3}k^5_{3,1}}{3}\right) \\ & + \frac{k^5_{1,2}k^5_{2,3}k^5_{3,4}}{5X} - \frac{k^5_{4,3}k^5_{3,2}k^5_{2,1}}{3X} + \frac{k^5_{1,3}k^5_{3,2}k^5_{2,4}}{5X} - \frac{k^5_{4,2}k^5_{2,3}k^5_{3,1}}{3X}\Bigg\}\end{aligned} \tag{3.4}$$

4. Example pump traces

4.1. *1612 MHz in a detached Mira envelope*

This study considered the behaviour of masers in the envelope of an AGB star at the end of a phase of rapid mass loss: as it finishes, the envelope detaches from the star, leaving a hollow shell. The pumping scheme at shell detachment was compared with the scheme 40 yr later, when the maser emission at 1612 MHz had become very weak. Further details appear in Gray, Howe & Lewis (2005).

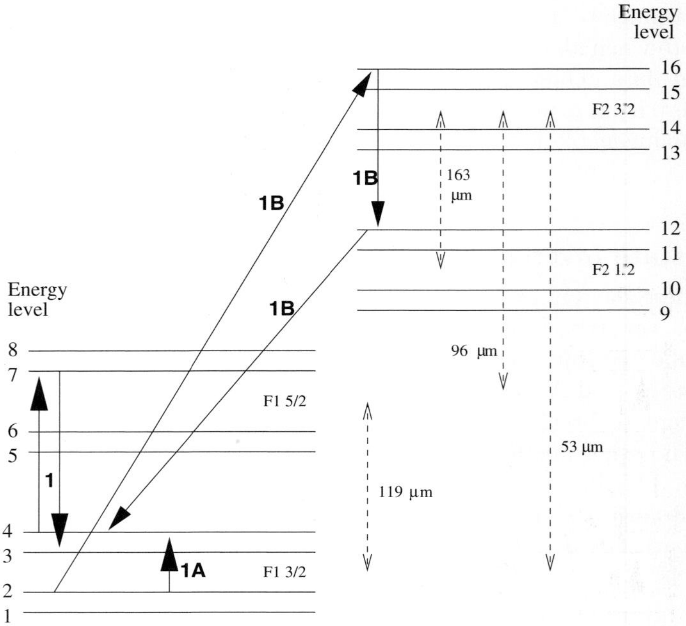

Figure 1. The 'direct' part of the 1612-MHz pump at shell detachment. OH levels are numbered by increasing energy. Rotational, lambda-doubling and hyperfine splittings are shown, but not to scale. Approximate rotational transition wavelengths are marked in μm. Transitions shared by more than one route are shown only once. Reverse transitions are omitted.

We traced the 'direct' part of the pump, a positive-definite multiplier times $k_{2,3}^4/3 - k_{3,2}^4/5$, and the 'indirect' part, which can be written as a multiplier times the expression,

$$\frac{1}{k_{1,2}^4}\left[\left(\frac{k_{2,1}^4}{3}\right)k_{1,3}^4 - k_{3,1}^4\left(\frac{k_{1,2}^4}{5}\right)\right].$$

At shell detachment, the 'direct' part supplied approximately 2/3 of the inversion. The most significant change over 40 yr was the complete loss of the 'indirect' routes, which by that time contributed significant absorption. The 'direct' and 'indirect' parts of the pump are shown respectively in figure 1 and figure 2. The notation used for the routes employs numbers for the primary branches, capital letters for secondary branches, and roman numerals for tertiary branches. As expected, the most important radiative pump employs 53 μm radiation, but the 'indirect' routes are the more dependent. Therefore, as the hollow shell expands and cools the energy density of the 53 μm radiation falls, leading to preferential collapse in the part of the pump which operates via level 1 (the 'indirect' part).

4.2. *1612 and 1667 MHz in a supergiant envelope*

The pumping schemes for OH masers in a larger, warmer envelope were traced. The parameters of the envelope were loosely based on the red supergiant, S Perseii (spectral type M4.5). The most important routes (solid lines) are displayed in figure 3 for the 1665-MHz main line. The 1612-MHz pump has much in common with figure 3, and both are based on initial absorption of 35 μm radiation. In the 1612-MHz pump, the most important route is completed by the expected decay to the rotational ground state Elitzur, Goldreich & Scoville (1976), but for 1665 MHz, we see from figure 3 that the

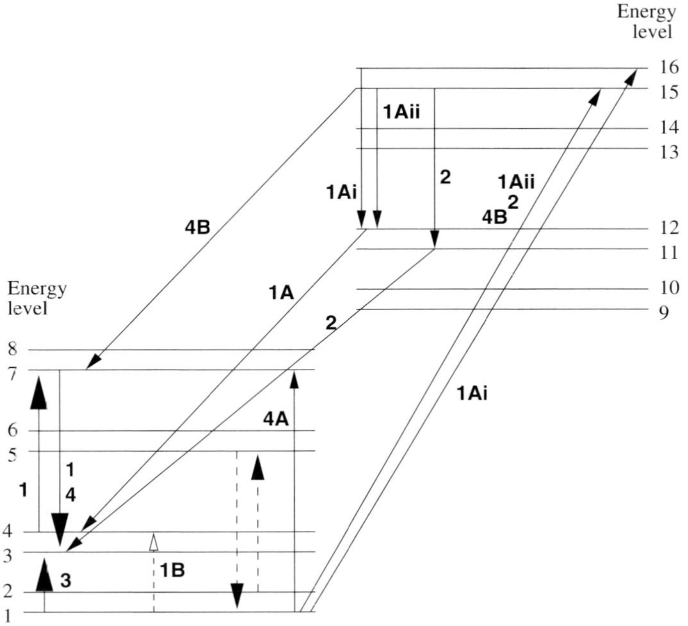

Figure 2. As figure 1, but for the 'indirect' part of the pump, which operates via level 1.

radiative cascade does not complete, and three additional absorptions result in transfer to the $^2\Pi_{3/2}$ ladder, followed by radiative decays through that stack of levels to the upper level of the maser.

4.3. *Main lines in a star-forming region*

The ALI solution was drawn from a region of a parameter space search with very high main-line ground-state inversions, yielding unsaturated integrated gains of 20.039 at 1665 MHz and 35.661 at 1667 MHz. The ALI slab had no velocity shift or microturbulence and constant physical conditions: $n_{H_2} = 10^7$ cm^{-3}; [OH] = 0.2 ppm; T_K = 30 K and T_d = 70 K. Eighty logarithmically-spaced model slabs covered a depth of 2×10^{13} m.

The trace of the 1665-MHz pump appears in figure 4, showing the five most important routes, contributing 82% of the total inversion. These were expanded from both the 'direct' and 'indirect' parts of the pump. As pump rates, the five routes contribute, (i) $(k^6_{1,5}k^6_{5,3} - k^6_{3,5}k^6_{5,1})/k^6_{5,5} = 6.01\times 10^{-5}$ s^{-1}; (ii) $(k^6_{1,5}k^6_{5,2}k^6_{2,4}k^6_{4,3} - k^6_{3,4}k^6_{4,2}k^6_{2,5}k^6_{5,1})/(k^4_{2,2}k^5_{4,4}k^6_{5,5}) = 4.86\times 10^{-5}$ s^{-1}; (iii) $(k^5_{1,4}k^5_{4,3} - k^5_{3,4}k^5_{4,1})/k^5_{4,4} = 2.72\times 10^{-5}$ s^{-1} and two more minor terms totalling 2.46×10^{-5} s^{-1}. Route (i) is especially simple, having no additional complexity on expansion back to original coefficients. 73% of route (ii) is formed from three sub-routes, A (via levels 4, 6 and 7), B (via level 20) and C (via levels 10 and 14). Route (iii) is even more complex with only 50% of its strength coming from the most important three subroutes: A (via levels 9, 10, 14 and 20), B (via levels 10 and 14) and C (direct collisional transfer from level 4 to level 3).

The traced pump routes for 1667 MHz are shown in figure 5. Both main-line pump routes lie predominantly in the $^2\Pi_{3/2}$ ladder. Most transitions are radiative, but there are several important collisional links. The strongest route, route (i), in both main lines is very similar, with a radiative upward pump, and a collisional decay (see the bold, solid routes in figure 4 and figure 5. The route is $1 \rightarrow 5$, $5 \rightarrow 3$ for 1665 MHz, and one level higher at 1667 MHz. Further details appear in Gray (2007).

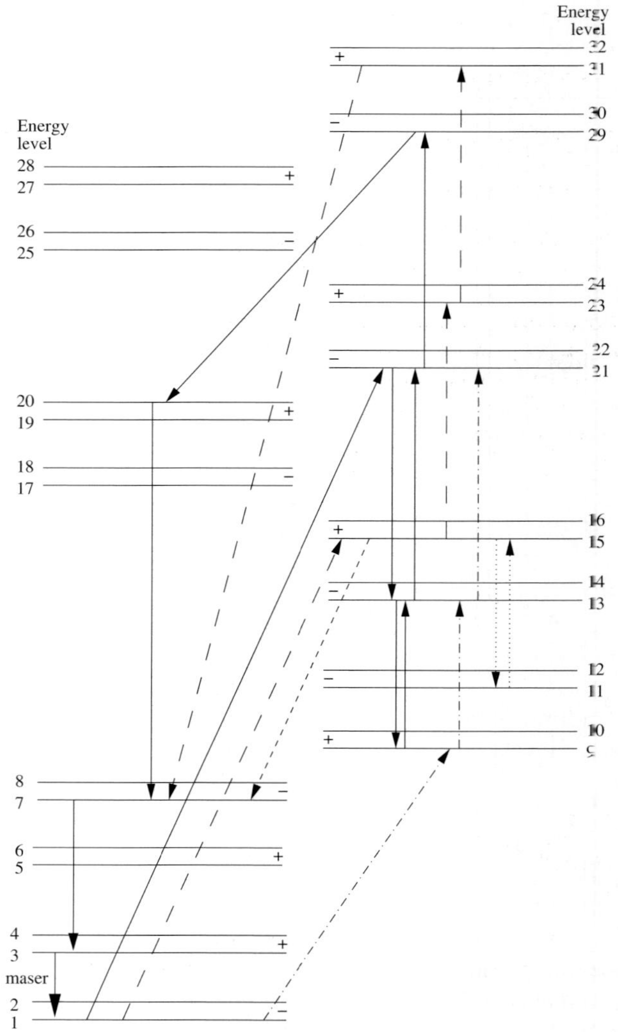

Figure 3. Pump routes for 1665 MHz in an M-supergiant. The most important routes are shown as solid lines, followed in order of inverting effect by those shown in the long dashed, chained, short-dashed and dotted styles.

A more detailed analysis of the strongest routes only, breaking the coefficients down into their individual processes, shows a hierarchy of processes with the Einstein A-values dominating over the radiative pumps of type BJ, and these in turn dominating over the collisions. Values in Hz for the 1665-MHz line are: $C_{5,3} = 3.75(-4); C_{3,5} = 1.13(-5)$ $A_{5,1} = 0.1244; B_{5,1}\bar{J}_{1,5} = 2.160(-2); C_{5,1} = 3.16(-4)$ $B_{1,5}\bar{J}_{1,5} = 3.60(-2); C_{1,5} = 9.48(-5)$. Ignoring products of collisional terms, and $C_{3,5}$ in comparison to $C_{5,3}$, we find the route (i) expression is $k_{1,5}k_{5,3} - k_{3,5}k_{5,1} \sim B_{1,5}\bar{J}_{1,5}C_{5,3} - A_{5,1}C_{3,5}$. In terms of downward coefficients only, and as $h\nu/(kT_k) \gg 1$ for the $5 \rightarrow 1$ transition, this reduces to $k_{1,5}k_{5,3} - k_{3,5}k_{5,1} \sim K((\bar{J}_{1,5}/B_\nu(T_K) - 1)$ where K is a positive definite expression which cannot decide the sign of the inversion. Therefore, the inversion depends on two things: firstly the rotational energy gap is large compared with T_K, favouring $C_{5,3}$ over $C_{3,5}$; secondly, the dust continuum, at $T_d = 70\,$K, is hotter than the kinetic temperature, allowing mean intensity greater than a black-body at T_K. An

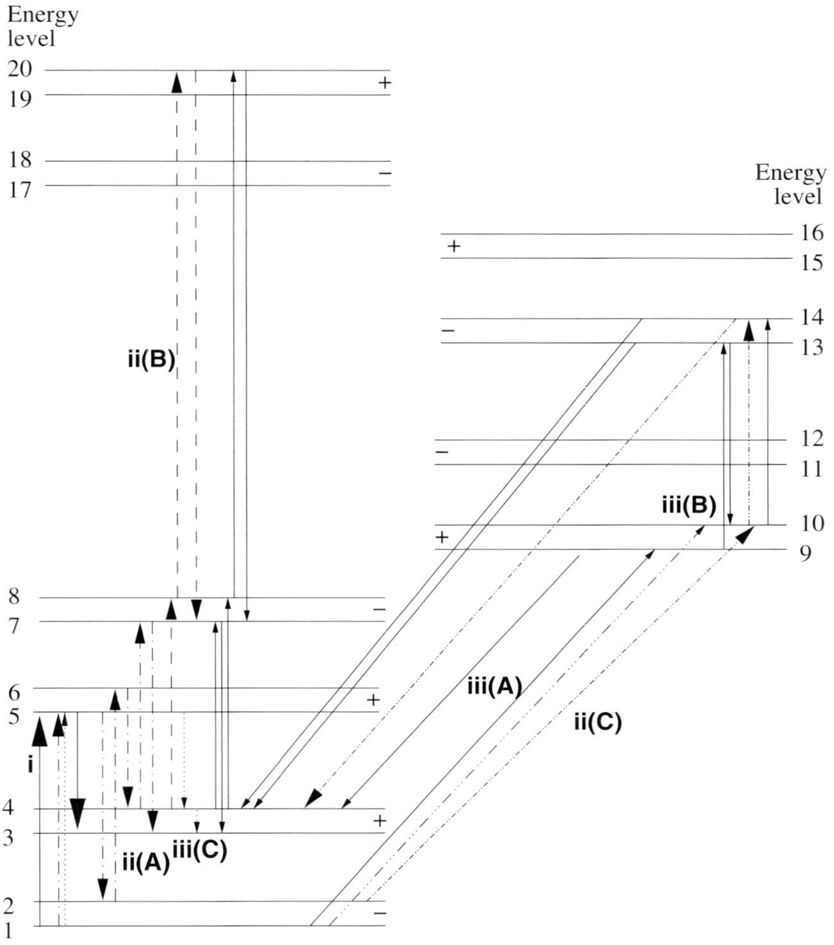

Figure 4. Strongest pump routes at 1665 MHz: transitions appearing in more than one route are repeated, but reverse routes are not shown for clarity.

analysis of the ALI slabs near the inner boundary in terms of radiation diffusion shows that the source of the dust radiation is the optically thick boundary itself. Radiation in the pumping 5, 1 pumping line penetrates because it is effectively scattered and not thermalised.

Mirror routes, an example being the a transfer from level 3 to level 5 via level 7 (instead of 5) in the 1665-MHz case, do not cancel the inverting effect of the pump because of parity propensity in the collisions of OH with para-H_2. This propensity makes $C_{7,1}$ significantly smaller than $C_{5,3}$ at values of T_K below 100 K, where the para-species of H_2 dominates.

5. Conclusions

It is possible to obtain detailed pumping routes generating the great majority of the inversion from a small fraction of the input coefficients. AGB and supergiant traces broadly support earlier theory. Significant contributions to main-line pumps have just two steps: radiative 'up' and collisional 'down'.

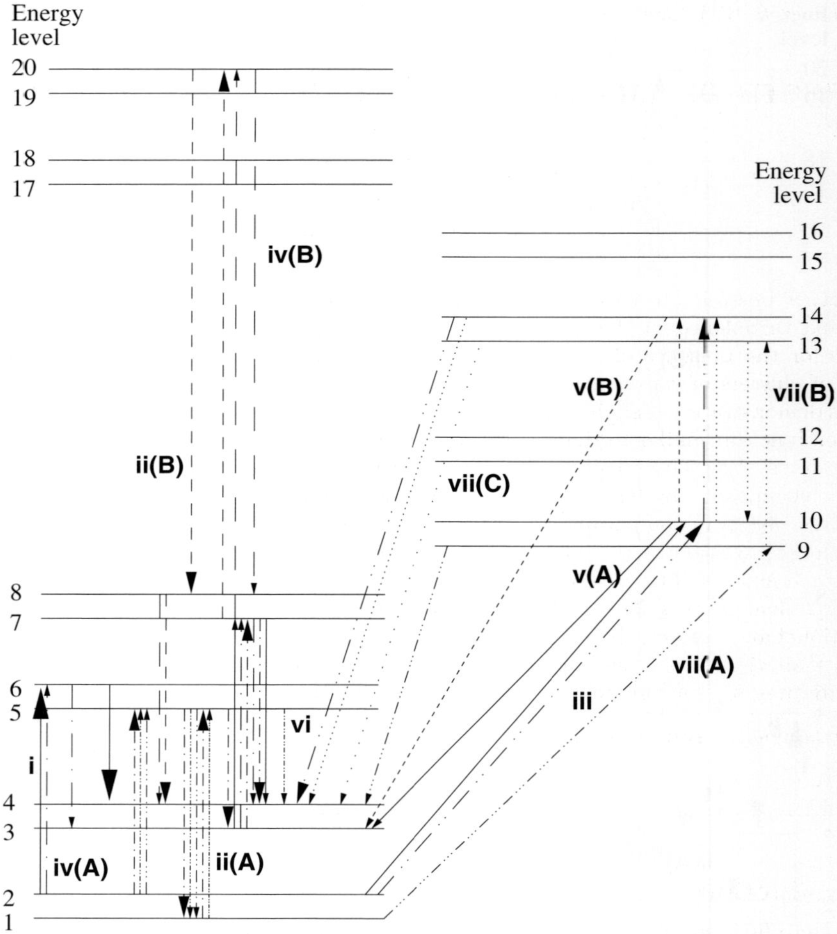

Figure 5. As figure 4 but for 1667 MHz. The seven routes shown have relative strengths 1.0, 0.413, 0.290, 0.225, 0.228, 0.203, 0.112, with route i being the strongest.

References

Bujarrabal, V., Guibert, J., & Nguyen-Q-Rieu 1980, *A&A* 84, 311

Dickinson, D. F. 1987, *ApJ* 313, 408

Elitzur, M. 1976, *ApJ* 203, 124

Elitzur, M. 1981, in: I. Iben & A. Renzini (eds.), *Physical Processes in Red Giants*, (Dordrecht: Reidel), p. 363

Elitzur, M., Goldreich, P., & Scoville, N. 1976, *ApJ* 205, 384

Gray, M. D. 2007, *MNRAS* 375, 477

Gray, M. D., Field, D., & Doel, R. C. 1992, *A&A* 262, 555

Gray, M. D., Howe, D. A., & Lewis, B. M. 2005, *MNRAS* 364, 783

He, J. H., & Chen, P. S. 2004, *New Astronomy* 9, 54

He, J. H., Szczerba, R., Chen, P. S., & Sobolev, A. M. 2005, *A&A* 434, 201

Pavlakis, K. G., & Kylafis, N. D. 1996, *ApJ* 467, 309

Sobolev, A. M. 1986, *Sov. Ast.* 30, 399

Sobolev, A. M. 1989, *Astronomiche Nachrichten* 310, 343

Sobolev, A. M., & Deguchi, S. 1994, *ApJ* 433, 719

Astrophysical Masers and their Environments
Proceedings IAU Symposium No. 242, 2007
J.M. Chapman & W.A. Baan eds.

doi:10.1017/S1743921307012513

Masers as probes of supersonic turbulence

Vladimir Strelnitski[1]

[1]Maria Mitchell Observatory, Nantucket, MA
email: vladimir@mmo.org

Abstract. A possible intimate connection between astrophysical masers in regions of star formation and turbulence has been a subject of increasing interest during the last two decades. Evidence for the presence of a residual turbulent component in the observed expansion and rotation of clusters of water masers was shown by multi-epoch VLBI maps. The water maser hot spots demonstrate self-similar (fractal) spatial clustering and a power-law two-point velocity correlation function similar to that of incompressible turbulence – with the power index close to "Kolmogorov's" 1/3. The possibility of using maser sources for studying supersonic turbulence critically depends on whether the observed hot spots are an integral effect of radiative transfer over a large distance, comparable to the size of the whole maser source, or whether they are compact local physical objects, such as small random shocks, in which the mechanical energy of turbulence dissipates. If the latter hypothesis is correct, the compact and bright maser hot spots may be excellent local probes of the spatial and kinematic structure of supersonic turbulence. Observational and theoretical arguments for and against these hypotheses are discussed and the first quantitative results about supersonic turbulence obtained by statistical analysis of maser sources are presented in this review.

Keywords. masers, turbulence, stars:formation, stars:circumstellar matter, ISM:kinematics and dynamics

---

1. Introduction

More than 50 years ago, von Weizsäcker (1951) pointed out that with its typical scales, temperatures and densities, the ISM is characterized by very high Reynolds numbers and thus should be turbulent. Since the macroscopic velocities of the interstellar gas are, typically, much higher than the local speed of sound, it is clear that interstellar turbulence must also be highly supersonic – the type of turbulence that does not naturally occur on earth and is exceedingly difficult to reproduce in the lab. Will the properties of supersonic turbulence revealed by astronomical observations be similar to or different from those of incompressible turbulence known from experimental and theoretical studies on earth?

A fundamental property of incompressible turbulence is the power-law character of the two-point velocity correlation function, with the classical ("Kolmogorov") value of the power index = 1/3. This value of the power index is a sign of the dissipationless cascade of mechanical energy from the largest ("outer") scale, where the energy is pumped into the medium, through the intermediate scales, down to the smallest ("inner") scale, where the energy dissipates via viscosity.

Much effort has been put into the study of various manifestations of interstellar turbulence since the seminal Weizsäcker's paper [see, for example, reviews by Scalo (1987) and Elmegreen (1993)]. The methods of study included the statistics of interstellar clouds (Larson 1981); the analysis of the absorption line profiles in molecular clouds (Falgarone & Phillips 1990); the comparison of the observed line profiles with the results of computer simulations of supersonic turbulence (Falgarone *et al.* 1994); and studies of scintillations of compact remote radio sources due to the scattering on inhomogeneities of the ionized turbulent component of ISM (e.g. Ricket 1990).

The two major results obtained by these methods can be summarized as follows: (1) The ubiquity of a Kolmogorov, or somewhat steeper, velocity correlation function (power index = 0.3 → 0.5). (2) Signs of intermittency (spatial inhomogeneity, spottiness) of turbulence.

The ionized ISM, of course, affects the radiation of masers propagating through it as well. However, in this review we will leave aside the studies of such observed or possible effects of maser propagation throught the general ISM as broadening of the sources (e.g. Moran *et al.* 1973; Gwinn *et al.* 1988; Hansen *et al.* 1992), their "wandering" (Gwinn *et al.* 1988), or scintillations (Clegg & Cordes 1991). Our topic is different: Masers as possible probes of turbulence *in situ*, the turbulence with which they seem to coexist and, at least in some cases, by which they may be ignited.

2. Are astrophysical masers "spooks" or "things"?

Turbulence on earth is investigated experimentally using probes that are placed directly in a turbulent flow and measure local fluctuations of the velocity field. *A priori*, it seems that in the case of interstellar turbulence only integral characteristics of the velocity field can be obtained from the observed spectral lines, which are a result of radiative transfer through the turbulent medium. However, the data obtained during the last two decades provide hope that, in fact, we may be able to use some masers, at least some water masers, as *local* probes of turbulence, not dissimilar from those used to study turbulence on earth. This hope is based on the assumption (not shared by all the investigators) that an observed water maser hot spot is not an integral effect of radiative transfer over a large distance, comparable with the size of the whole maser source, but rather a compact *physical object*, singled out from its environment by, for example, its enhanced density and/or temperature.

The two approaches to the interpretation of the maser hot spots – real compact physical objects ("things") or a mere optical effect ("spooks") – appeared soon after the discovery of the first cosmic masers. This dilemma is still not completely resolved. It is possible that both cases are represented by different classes of masers.

Radhakrishnan, Goss and Bhandary (1975) were the first to propose that the broad spectra of H_2O masers in several sources [see, for example, a recent broad W49-H_2O spectrum in McGrath *et al.* (2004)] are not due to the Doppler effect of rapidly moving physical condensations but to an optical effect of stimulated Raman scattering. More recently, this hypothesis was explored by Deguchi (1994) and Gangadhara *et al.* (1999) to explain the extremely broad (± 2000 km s^{-1}) water maser spectrum in the galaxy NGC4258.

A crucial experiment that could decide between the Raman scattering and real kinematic motions in the galactic H_2O sources was suggested earlier by Strelnitski & Sunyaev (1973) who argued that the observed H_2O spectrum in W49 is so broad that, if the maser hot spots are real gas condensations, their cluster cannot be gravitationally bound and must expand with the typical observable speed of a few milliarcseconds per year. This prediction was later confirmed by multi-epoch VLBI experiments, first for Ori-KL (Genzel *et al.* 1981) and then for W49 (Gwinn *et al.* 1992). For more recent results on proper motions of water masers see e.g. Imai *et al.* (2002). The observed proper motions of maser spots exclude the Raman scattering as an explanation for the large breadths of H_2O spectra.

The exceedingly high brightness temperatures of H_2O masers brought forward another kind of a "spook" hypothesis, namely, that the hot spots do not correspond to single local gas condensations, but rather to widely separated pairs of condensations that happen to lie along the same line of sight and have, by chance, equal radial velocities (Deguchi & Watson, 1989). Exponential unsaturated amplification of the radiation from one condensation by the other would then create a very narrow beam of bright emission and free the theoretical models from the necessity of an exceedingly powerful pumping mechanism, which is needed if the radiation is more or less isotropic (e.g. Strelnitski 1984). The distant pair hypothesis may work for the exceptionally bright hot spots, but it would hardly be possible to account for the regular proper motions revealed by the VLBI experiments (expansion and rotation of maser clusters), if all, or even most of the hot spots were due to distant pairs of amplifying blobs.

Theoretical models that connect astrophysical masers with turbulence use both the "Spook"and the "Thing" hypotheses to explain the observed hot spots. We will review them by turn in the next two sections.

3. Masers as an optical effect of propagation through a homogeneous medium with a random velocity field

Apparently, Gail *et al.* (1975) were the first to demonstrate a high sensitivity of an unsaturated maser to local velocity gradients. Deguchi (1982) performed the first numerical simulation of maser propagation through an (expanding) homogeneous medium with a random velocity field and demonstrated the possibility of explaining, in this way, the spectra (multitude of narrow velocity components) and maps (patchiness) of the 1612 MHz OH masers in the expanding envelopes of red supergiants. Examples of more recent models of this type are those by Wallin *et al.* (1998) or Sobolev *et al.* (1998). In these models, a random velocity field is numerically simulated in an N^3 grid point box by sampling from a Kolmogorov-type power spectrum, and an unsaturated (exponential) maser amplification along many lines parallel to the line of sight is calculated. The qualitative properties of the bright H_2O maser spectrum in the galaxy NGC 4258 (Wallin *et al.* 1998, 1999), as well as the spectra and maps of some methanol masers (Sobolev *et al.* 1998, 2003; Minier *et al.* 2002) are satisfactorily described by these models (see also Deguchi & Watson 1989; Wiebe & Watson 1998; Watson & Wiebe 2001; Watson *et al.* 2004).

The major conclusion of all these models is that the observed maser hot spots may be a purely optical effect – no local density or temperature enhancements are needed to create bright, isolated hot spots at the output of an exponentially amplifying medium with a random velocity field.

The success of these models in qualitatively accounting for the observed properties of the spectra and maps of corresponding sources provides indirect evidence of a turbulent velocity field in the source. However, since, according to this approach, each observed hot spot is a result of an integration along a considerable part of the medium, the observations of the sources described by these models are not supposed to give direct information about the local properties of turbulence.

4. Masers as local probes of turbulence

The first indications that water masers may be connected with turbulence appeared in the 1980s. Walker (1984), using the VLBI maps of the H_2O source in W49N obtained earlier by Walker *et al.* (1982), demonstrated that both the two-point velocity and spatial correlation functions showed power-law dependences on maser spot separation. This behavior is typical of a turbulent flow, although Walker did not favor the turbulence interpretation.

Another piece of evidence in support of turbulence in H_2O maser sources came from the modeling of the results of the multi-epoch VLBI of the H_2O source in SgrB2(N) (Reid *et al.* 1988). The modeling revealed a superposition of two regular motions in the source – expansion and rotation. However, besides these regular components, the presence of a residual *random* component was indicated, with a typical dispersion of total velocity vector $\approx 20 \rightarrow 30$ km s^{-1}.

Gwinn (1994) carried out a statistical analysis similar to that of Walker (1984) but using better VLBI results for W49N from Gwinn *et al.* (1992). He confirmed the power-law dependency of velocity increments and spatial density of masing spots on point separation and interpreted this as possibly being due to turbulence produced by an outflow from a young star into the surrounding gas.

Strelnitski *et al.* (1998; 2002), Imai *et al.* (2002) and Ripman (2007) carried out detailed statistical analyses of several water maser sources. They concluded that the slope of the two-point velocity correlation function is, typically, quite close to Kolmogorov's 1/3. Imai *et al.* (2002) measured a slope of 0.29 ± 0.03 for W3(IRS5) and Ripman obtained the average value of 0.29 ± 0.04 for the overall (covering all the available values of point separations) fit slope for five sources. However, for small values of point separations, he found the slope to be steeper, with an average of 0.50 ± 0.06.

The 1/3 value for the power index of the velocity correlation function corresponds to the conservation of mechanical energy in the energy cascade along the scales and thus it confirms that highly supersonic turbulence may have an "inertial subrange" of scales, similar to that of incompressible turbulence. The steeper slope for the small separations can be due to the increasing dissipation of turbulent energy at the smaller scales, approaching the inner scale, where most of the mechanical energy dissipates. Strelnitski *et al.* (2002) hypothesized that supersonic turbulence may have a "shock wave" inner scale, which is much larger than the viscous inner scale of incompressible turbulence. Similar expressions for the size of the putative shock-wave inner scale were obtained by Stelnitski *et al.* (2002) and by Boldyrev (2002). H_2O masers need high density and temperature for their collisional pumping and, since the 1970s (Shmeld *et al.* 1976), they have been supposed to be produced behind shocks. The hypothesis of their genetic connection with the random shocks dissipating turbulent energy seems therefore quite natural.

Another important result of these papers is the possibility of representing the two-point spatial correlation function of the H_2O hot spots by a power law. This demonstrates a self-similar (fractal) structure of the cluster of spots and gives another indirect indication that the masers are associated with turbulence. Indeed, incompressible turbulence on earth is known to be intermittent, with the active regions typically distributed on a fractal set with a dimension of ≈ 2.7 (Mandelbrot 1982). The fractal dimension for the clusters of water masers is, however, considerably lower. Strelnitski *et al.* (2002) cite values between 0.2 and 1.0 for several sources. Ripman (2007) obtained an average of 0.87 ± 0.05 for five sources.

If the hypothesis of von Weizsäcker, that many structures in the Universe are created by the dissipation of turbulence, is correct, then it seems significant that the fractal dimension determined for such various cases as the distribution of galaxies (1.2, de Vaucouleurs 1970); distribution of the gas in interstellar clouds and starlight distribution in spiral galaxies (1.4; Elemegreen & Elemegreen 2001); star forming complexes in dwarf galaxies (0.5-1.7, Odekon 2006); clusters of young stars in Taurus (1.4, Larson, 1995); and H_2O masers ($\leqslant 1$, Strelnitski *et al.* 2002) are all considerably lower than the fractal dimension of incompressible turbulence (2.7, Mandelbrot 1982). Is this an indication that highly supersonic turbulence is much more intermittent (less space filling) than incompressible turbulence?

Strong support for the hypothesis of an intimate connection of water masers with highly intermittent turbulence is also provided by the analysis of their Probability Distribution Function (PDF). All the five H_2O sources investigated by Strelnitski *et al.* (2002) demonstrate strong deviations from the Gaussian PDF – an excess of large deviations from the mean values of the two-point velocity differences at all the investigated scales.

5. Prospects

The discrimination between the non-local and local hypotheses for the origin of maser hot spots can be made by comparison of the results of appropriate computer simulations with the observed maps and spectra of the masers. Holder *et al.* (2007) have recently simulated maser propagation in a turbulent medium considering both unsaturated and saturated amplification and a homogeneous and fractal distributions of active regions within the medium. Comparing the synthetic spectra and maps with the observed ones, they conclude that strong H_2O masers are probably saturated and that they are compact local physical objects distributed on a fractal set of low dimension. This supports the hypothesis that the water masers may be connected with the ultimate dissipation of turbulent energy in random shocks and that they may serve as good local probes of the kinematics and structure of turbulence.

A problem with the computer simulations of turbulence is the need for still higher processor speeds. To compare the results of the simulations with the VLBI observations of H_2O masers (which typically cover 3-4 orders of magnitude in spatial scale), more than 1024^3 grid points are needed, which is still a challenge.

Statistical studies of more H_2O sources, as well as other maser sources (OH, CH_3OH, SiO) may help to better understand the putative connection of masers with turbulence. The first attempt of a two-point correlation analysis of a few methanol masers by Truong (2007) showed quite different results from those for water masers. The fractal dimension of the spot distribution is still low – the average for the three measured Class I sources was 1.02 ± 0.25, and for one measured Class II source it was 1.53 ± 0.40. However, the power index of the two-point velocity correlation function was found to be considerably higher for methanol sources than for water sources – 0.65 ± 0.08 for the three Class I sources, and 0.54 ± 0.07 for the Class II source. It is not clear whether the difference is due to the physical difference between the turbulent flows associated with the water and methanol sources, or to the fact that methanol maser hot spots are *not* compact local "things" within the turbulent flow but rather "spooks" at the output surface of the vast amplifying medium. More statistical studies of sources of various classes are needed.

6. Conclusions

1. The possibility for masers to serve as probes of turbulence depends on whether the observed maser hot spots are real compact physical objects ("things") within the turbulent flow, or a mere optical effect ("spooks") at the output of the amplifying turbulent medium.

2. Methanol masers may be "spooks" and, in this case, they can only confirm the fact of their association with a turbulent velocity field, but cannot provide such quantitative characteristics of turbulence as the index of the power spectrum or the fractal dimension of the spatial set on which turbulence is active.

3. Strong Galactic water masers seem to be "things"; they point to (1) a high degree of intermittency of supersonic turbulence, (2) a lack of energy dissipation at larger scales,

and (3) possible genetic connection of maser condensations with the putative shock-wave dissipation scale of turbulence.

4. More statistical studies of maser hot spots, including those of OH, CH_3OH and SiO, are needed, as well as computer simulations of maser propagation in turbulent media with greater numbers of grid points.

Acknowledgements

The author is grateful to J. Alexander, S. Gezari, B. Holder, N. Nezhdanova, J. Moran, M. Reid, B. Ripman, V. Shishov, and P. Truong – the students and collegues with whom he collaborated in several projects to obtain the results mentioned in this review. These projects were supported by the NSF grants AST-9820555, AST-0097694 and AST-0354056 and by the Nantucket Maria Mitchell Association.

References

Boldyrev, S. 2002, *ApJ* 569, 841

Clegg, A. W. & Cordes, J. M. 1991, *ApJ* 374, 150

Cordes, J. M. 1993, in: A. W. Clegg & G. E. Nedoluha (eds.) *Astrophysical Masers* (Berlin-Heidelberg: Springer-Verlag), p. 251

Deguchi, S. 1994, *ApJ* 420, 551

Deguchi, S. & Watson, W. D. 1989, *ApJ* 340, L17

Elmegreen, B. G. 1993, in: E. H. Levy & J. I. Lunine (eds.) *Protostars and Planets III* (Tucson: Univ. Arizona Press), p. 97

Elemegreen, B. G. & Elemegreen, D. M. 2001, *AJ* 121, 1507

Falgarone, E. & Phillips, T. G. 1990, *ApJ* 359, 344

Falgarone, E., Lis, D. C., Phillips, T. G., Pouquet, A., Porter, D. H. & Woodward, P. R. 1990, *ApJ* 436, 728

Gail, H.-P., Kegel, W. H. & Sedlmayr, E. 1975, *A&A* 42, 81

Gangadhara, R. T., Deguchi, S. & Lesch, H. 1999, *Physics of Plasmas* 6, 4088

Genzel, R. *et al.* 1981, *ApJ* 247, 1039

Gwinn, C. R. 1994, *ApJ* 429, 241

Gwinn, C. R., Moran, J. M., Reid, M. J., & Schneps, M. H. 1981, *ApJ* 330, 817

Gwinn, C. R., Moran, J. M., & Reid, M. J. 1992, *ApJ* 393, 149

Hansen, J., Booth, R. S., Dennison, B. & Diamond, P. J. 1993, in: A. W. Clegg & G. E. Nedoluha (eds.) *Astrophysical Masers* (Berlin-Heidelberg: Springer-Verlag), p. 255

Holder, B. P., Nezhdanova, N., Shishov, V. I. & Strelnitski, V. 2007, *in preparation*

Imai, H., Deguchi, S. & Sasai, T. 2002, *ApJ* 567, 971

Larson, R. B. 1981, *MNRAS* 194, 809

Larson, R. B. 1995, *MNRAS* 272, 213

Mandelbrot, B. B. 1982, *The Fractal Geometry of Nature* (New York: Freeman)

McGrath, E. J., Goss, W. M. & DePree, C. G. 2004, *ApJS* 155, 577

Minier, V., Booth, R. S. & Conway, J. E. 2002, *A&A* 383, 614

Moran, J. M., Papadopoulos, G. D., Burke, B. F. *et al.* 1973, *ApJ* 185, 535

Odekon, M. C. 2006, *AJ* 132, 1834

Radhakrishnan, V., Goss, W. M. & Bhandari, R. 1975, *Pramana* 5, 49

Reid, M. J., Schneps, M. H., Moran, J. M., Gwinn, C. R., Genzel, R., Downes, D. & Rönnäng, B. 1988, *ApJ* 330, 809

Ricket, B. J. 1990, *ARAA* 28, 561

Ripman, B. H. 2007, *Journal of Young Investigators*, accepted (October 2007 issue)

Scalo, J. M. 1987, in: D. J. Hollenbach & H. A. Thronson, Jr. (eds.) *Interstellar Processes* (Dordrecht:Reidel), p. 349

Shmeld, I. K., Strelnitski, V. & Muzylev, V. V. 1976, *Soviet Astron.* 20, 411

Sobolev, A. M., Wallin, B. K. & Watson, W. D. 1998, *ApJ* 498, 763

Sobolev, A. M., Watson, W. D. & Okorokov, V. A. 2003, *ApJ* 590, 333
Strelnitski, V. 1984, *MNRAS* 207, 339
Strelnitski, V. & Sunyaev, R. A. 1973, *Soviet Astron.* 16, 579
Strelnitski, V. S., Alexander, J., Moran, J. M. & Reid, M. J. 1998, in: J. A. Zensus, G. B. Taylor & J. M. Wrobel (eds.) *Radio Emission from Galactic and Extragalactic Compact Sources*, IAU Coll. 164, ASP Conf. Series, 144, p. 369
Strelnitski, V., Alexander, J., Gezari, S., Holder, B. P., Moran, J. M. & Reid, M. J. 2002, *ApJ* 581, 1180
Truong, P. 2007, *Journal of Young Investigators*, accepted (October 2007 issue)
von Weizsäcker, C. F. 1951, *ApJ.* 114, 165
Walker, R. C. 1984, *ApJ.* 280, 618
Walker, R. C., Matsakis, D. N. & Garcia-Barreto, J. A. 1982, *ApJ.* 255, 128
Wallin, B. K., Watson, W. D. & Wyld, H. W. 1998, *ApJ.* 495, 774
Wallin, B. K., Watson, W. D. & Wyld, H. W. 1999, *ApJ.* 517, 682
Watson, W. D. & Wiebe, D. S. 2001, *ApJ.* 557, 967
Watson, W. D.; Wiebe, D. S.; McKinney, J. C. & Gammie, C. F. 2004, *ApJ.* 604, 707
Wiebe, D. S. & Watson, W. D. 1998, *ApJ.* 503, L71

Astrophysical Masers and their Environments
Proceedings IAU Symposium No. 242, 2007
J.M. Chapman & W.A. Baan, eds.

doi:10.1017/S1743921307012525

Influence of turbulence on the shape of a spectral line: the analytical approach

N. A. Silant'ev[1,2], E. E. Lekht[1], J. E. Mendoza-Torres[1] and G. M. Rudnitskij[3]

[1]Instituto Nacional de Astrofísica, Óptica y Electrónica, Luis Enrique Erro No. 1, Apdo Postal 51 y 216, 72840 Tonantzintla, Puebla, México
email: silant@inaoep.mx

[2]Main Astronomical Observatory, Russian Academy of Sciences, Pulkovo, St.-Petersburg, 196140 Russia

[3]Sternberg Astronomical Institute, Moscow State University, 13 Universitetskij prospekt, Moscow, 119992 Russia
email: gmr@sai.msu.ru

Abstract. We consider the propagation of spectral-line radiation in a correlated turbulent atmosphere. The ensembles of turbulent velocities $\mathbf{u}(\mathbf{r}, t)$ and optical depths, τ_ν, are assumed to be Gaussian. We investigate the explicit analytical solution of the stochastic radiative transfer equation for the intensity I_ν of radiation. The scattering term is not taken into account. It is shown that, in addition to the usual Doppler broadening of the spectral line, correlated turbulent motions of atoms and molecules give rise to considerable changes in the shape of a spectral line. We find that the mean intensity $I_\nu^{(0)}$ ($I_\nu = I_\nu^{(0)} + I'_\nu$, $\langle I'_\nu \rangle = 0$) obeys the usual radiative transfer equation with renormalized extinction factor $\alpha_\nu^{\rm eff}$ if the correlation length R_0 of the turbulence is small as compared to a photon free path. A simple analytical expression for $\alpha_\nu^{\rm eff}$ is given. This expression integrally depends on the two-point correlation function of the turbulent velocity field.

Keywords. turbulence, molecular processes, radio lines: ISM, radio lines: stars, masers: circumstellar matter

---

In a turbulent layer the effective absorption coefficient $\alpha_\nu^{\rm eff}$ does not coincide with the mean absorption coefficient $\alpha_\nu^{(0)} \equiv \langle \alpha_\nu \rangle$. The transition $\alpha_\nu^{\rm eff} \to \alpha_\nu^{(0)}$ takes place if the optical depth at the line center τ_1 of correlation length R_0 tends to zero (the case of short-correlated turbulence), or if the characteristic thermal velocity $u_{\rm Th}$ is equal to or greater than characteristic turbulent velocity u_k (parameter $\xi = u_{\rm Th}/u_k \geqslant 1$).

It is found that $\alpha_\nu^{\rm eff} \leqslant \alpha_\nu^{(0)}$. Statistically, a turbulent layer with a finite correlation length is more transparent than a layer of short-correlated turbulence for all the frequencies of a spectral line. It can be estimated even from the simple example of the averaging of intensities corresponding to two realizations with optical lengths $\tau_\nu^{(0)} + \tau'_\nu$ and $\tau_\nu^{(0)} - \tau'_\nu$.

The averaged spectral line intensity $I_\nu^{(0)}$ is narrower than that described by the radiative transfer equation with an averaged extinction factor $\alpha_\nu^{(0)}$ (see Fig. 1). This means that turbulence with a finite correlation length diminishes the equivalent width of an absorption line as compared with short-correlated turbulence.

The distortion of the shape increases with increasing optical depth τ_0 of the turbulent layer at the line center. An increase in the parameter τ_1 and decrease in ξ also give rise to an increase in the distortion.

If the turbulent layer contains many ($\tau_0/\tau_1 \geqslant 1$) turbulent cells (rotating curls), then the effects of finite correlation length turbulence disappear. If the number of turbulent

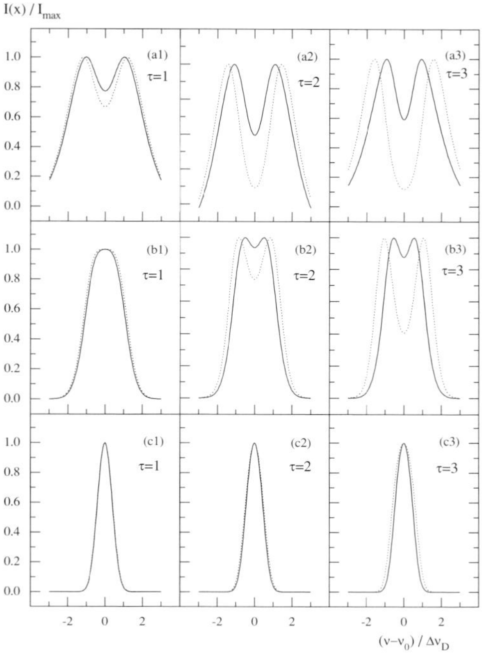

Figure 1. Normalized shape $I_\nu(x)/I_{\nu\max}$ of the maser line with a Gaussian initial profile after the passage of a turbulent cloud (solid lines). The dotted lines correspond to the usual model of short-correlated turbulence. The cases (an), (bn) and (cn) represent $\Delta\nu_D/\Delta\nu_S = 0.5, 1$ and 2, respectively; $\Delta\nu_D$ and $\Delta\nu_S$ are the Doppler widths in the cloud and maser source. The optical depth of the cloud τ at the line center takes the values $1, 2$ and 3. Turbulence correlation length R_0 is taken to be one-fifth of the layer's depth, and parameter $\xi = u_{\rm Th}/u_k = 0.1$; $u_{\rm Th}$ and u_k-values are characteristic thermal and turbulent velocities in the cloud.

cells is not large, but $u_{Th} \geqslant u_k$, the finite correlation length effects also tends to disappear. In the first case the situation looks like a cloud of many chaotically moving absorbing atoms. The second case corresponds to a very small shift of the center of the α_ν – profile for either correlated or uncorrelated turbulence.

These effects were mentioned earlier in various papers as the results of purely numerical calculations (see, for example, Gail *et al.* 1974, Levshakov *et al.* 1997, Magnan *et al.* 1976), without a simple physical explanation. The simple analytical formulae derived in our paper (Silant'ev *et al.* 2006) allow us to understand the expected effects of turbulence without complicated numerical calculations.

In our paper (Silant'ev *et al.* 2006) we discuss the problem of how to obtain the main turbulent parameters from the analysis of the shape of an absorption line or maser line propagating through a turbulent interstellar cloud. It should be emphasized that in the case of finite correlation length turbulence one can estimate separately the characteristic thermal velocity u_{Th} and turbulent velocity u_k. This is impossible if one uses the short-correlated model of turbulence.

As an illustrative example, we have estimated the possible parameters of the H_2O maser source in S252A (Berulis *et al.* 1996). We hope that the obtained analytical formulae will be useful for a detailed explanation of various observational data.

References

Berulis, I. I., Lekht, E. E., Mendoza-Torres, J. E., 1996, *Astron. Rep.* 40, 329

Gail, H.-P., Hundt, E., Kegel, W. H., *et al.* 1974, *Astron. Astrophys.* , 32, 65

Levshakov, S. A., & Kegel, W. H. 1997, *MNRAS* 288, 787

Magnan, C. 1976, *J. Quant. Spectrosc. Radiat. Transfer* 16, 281

Silant'ev, N. A., Lekht, E. E., Mendoza-Torres, J. E., Rudnitskij, G. M., 2006, *Astron. and Astrophys.* 453, 989

Astrophysical Masers and their Environments
Proceedings IAU Symposium No. 242, 2007
J.M. Chapman & W.A. Baan, eds.

doi:10.1017/S1743921307012537

Short-term variations in Class I methanol masers?

Preethi Pratap[1], Vladimir Strelnitski[2], Samantha Hoffman[1] and Jenna Lemonias[2]

[1]MIT Haystack Observatory, Westford, MA 01886
email: ppratap@haystack.mit.edu

[2]Maria Mitchell Observatory, Nantucket, MA
email: vladimir@mmo.org

Abstract. Class I methanol masers are found near regions of high-mass star formation and are pumped by collisions. Some indication of long-term variability in these masers has previously been reported (Kurtz et al. 2004). Here we will present evidence for variations in the intensity of the Class I methanol masers at 44 GHz on much shorter timescales. We present preliminary results of the observation of the 44 GHz Class I methanol transition toward sources in the DR21 region that indicates variability on timescales of days to possibly hours.

Keywords. masers, stars:formation

1. Introduction

Class I methanol masers are found around young, high mass star forming regions. They are most often associated with outflows and are often found offset from the star forming site and sometimes in regions with no obvious signs of star formation. In most of these isolated cases, signs of star formation are eventually detected. The masers are thus thought to probe a very early stage in the star formation process.

Methanol masers, until recently, have not been the subject of variability studies. Recent observations of the strong 6.7 GHz Class II maser by Goedhart *et al.* (2004) found that these masers vary significantly with timescales on the order of days to years. Several sources were also found to have periodic variability. Different variability timescales need differing theories to explain the observations, such as small scale turbulence, shock activity resulting from outflow, intrinsic variation in the exciting sources and also possibly interstellar scintillation. In the case of Class I masers, long term variability has been reported in a few sources (Kurtz et al. 2004; Kalenskii et al. 2005) but no systematic measurements have been made. Systematic studies of the lower frequency transitions would provide added information regarding the nature and environment of these masers.

2. Observations and results

Data were taken with the Haystack 37-m telescope over several observing periods - 2006 January, 2006 October and 2006 December. The observations in January were part of a mapping project and were not optimized to study the variability. However, over a period of a week there were indications of variability in DR21N, a newly discovered Class I methanol source, and DR21OH. The October 2006 observations were conducted more systematically, but some issues with pointing reduced the amount of usable data. Both the January and October data showed evidence of variability of the order of $> 20\%$

on timescales of days. The variability was determined from the ratios of strong maser components since this minimizes systematic errors.

The ratios of the intensities of two maser features in DR21OH were examined as a function of time over a period of several days in December 2006. The sources were observed from rise to set, except for about 3 hours at transit. The results show that the ratios change by about 20% in a period of $\tilde{6}$ hours. These particular maser features are situated within about 10 arcseconds of each other (Kurtz et al. 2004). This is important when studying the variation since the distribution of Class I masers in DR21OH is over an area of about 30″. Variability detected in the ratios of other maser features that are separated by larger fractions of the Haystack antenna main beam (45″ at 44 GHz) can be mimicked by pointing drifts.

3. Discussion

Possible short term variability in Class I methanol masers can put constraints on the excitation mechanism for these masers and on the source of the variability. One can rule out variability in the central radiative pumping source since the different velocity components vary differently. The variability could possibly arise in the small-scale turbulent medium in which the masers exist. Methanol masers appear to co-exist with H_2O masers which also exhibit variability with timescales of 10's of days (Felli et al. 2006 and references therein). Some linear polarization that is detected in our data has also been seen in DR21-W at 133 GHz (Wiesemeyer et al. 2004). Variability with a timescale t can be used to set an upper limit on the size of the emitting area: $D < cta^{1/2}$, where a is the amplitude of the variation. Variations on timescales of a few hours results in an upper limit on the size of about 2×10^{15} cm, which is somewhat compatible with measured characteristic sizes of hot spots (Kogan and Slysh 1998). High angular resolution observations are needed to confirm the size limits.

The Class I methanol maser variability can be used to set constraints on the modeling of the maser excitation and the maser environment. These masers appear to be associated with a very early stage of star formation - one in which no near-IR or radio continuum sources are yet detected. Follow up, high resolution observations of the masers, the molecular gas and the continuum sources in the maser vicinity are needed to understand the mechanism for the variability. Observations of other sources are also needed to examine whether the variability is ubiquitous.

Acknowledgements

Support for undergraduate research with the 37-m telescope at MIT Haystack Observatory comes from the National Science Foundation (NSF). The REU programs at the Maria Mitchell Observatory and MIT Haystack Observatory are supported by grants from the NSF.

References

Felli, M., Massi, F., Robberto, M., & Cesaroni, R. *A&A* 453, 911

Goedhart, S., Gaylard, M. J., & van der Walt, D. J. *M.N.R.A.S*, 355, 553

Kalenskii, S. V., Promyslov, V. G., Slysh, V. I., Bergman, P. & Winnberg, A. 2005, *Astronomy Reports* 50, 289

Kogan, L. & Slysh, V. 1998, *Ap.J.* 497, 800

Kurtz, S., Hofner, P. & Alvarez, C. V. 2004, *Ap.J.S.* 155, 149

Wiesemeyer, H., Thum, C., & Walmsley, C. M. *A&A* 428, 479

Photo: Shaun Amy

PhD students at IAU 242

From left to right: James Green, Carlos Gonzalez-Fernadez, Edo Loenen, Steven Longmore, Daniel Wong-McSweeney, Kalle Torstensson, Lyshia Quinn, Tim Robishaw, Estaban Araya, Mayumi Sato, Korinne McDonnell, Shari Breen, Laura Richter, Alet de Witt, Kim McAlpine, Jack Hewitt, Rodrigo Parra, Koichiro Sugiyama (not shown: Daniel Tafoya, Paola Castangia)

Astrophysical Masers and their Environments
Proceedings IAU Symposium No. 242, 2007
J.M. Chapman & W.A. Baan, eds.

doi:10.1017/S1743921307012549

A review of maser polarization and magnetic fields

W. H. T. Vlemmings[1,2]

[1]Argelander Institute for Astronomy, University of Bonn, Auf dem Hügel 71, 53121 Bonn, Germany

[2]Jodrell Bank Observatory, Univ. of Manchester, Macclesfield, Cheshire SK11 9DL, U.K.
email: Wouter.Vlemmings@manchester.ac.uk

Abstract. Through polarization observations masers are unique probes of the magnetic field in a variety of different astronomical objects, with the different maser species tracing different physical conditions. In recent years maser polarization observations have provided insights in the magnetic field strength and morphology in, among others, the envelopes around evolved stars, Planetary Nebulae (PNe), massive star forming regions and supernova remnants. More recently, maser observations have even been used to determine the magnetic field in megamaser galaxies. This review will present an overview of maser polarization observations and magnetic field determinations of the last several years and discuss the implications of the magnetic field measurements for several important fields of study, such as aspherical PNe creation and massive star formation.

Keywords. masers, polarization, magnetic fields

1. Introduction

Because of their compactness, their high brightness and the fact that they occur in a wide variety of astrophysical environments, masers are excellent astrophysical probes. As the masers are often highly linearly and circularly polarized, polarization observations add fundamental information on both the masing process (such as pumping and level of maser saturation) and the physical conditions in the masing gas. With a detailed theory of maser polarization propagation, the observations can yield the strength of the magnetic field along the maser line of sight and the two-dimensional (or even three-dimensional) field structure. Consequently, they allow for a determination of the dynamical importance of the magnetic field. And, since masers are specifically suited for high angular resolution observations with interferometry instruments, the polarization observations can probe the magnetic field properties at unprecedented small scales.

In the period since the previous IAU maser conference (Migenes & Reid 2002), the number of maser polarization observations have increased significantly. They have been used to determine the strength and structure of magnetic field in the circumstellar envelopes (CSE) of Asymptotic Giant Branch (AGB) stars, high-mass star forming regions and supernova remnants (SNRs), while observations of the magnetic field strength in megamasers have also become possible. This review details specifically the results on magnetic field measurements derived from maser polarization observations since 2001, focusing on the maser transitions in the radio-wavelength regime. In §2, the theoretical background of maser polarization and analysis considerations are discussed briefly. Recent magnetic field measurements and their consequences for related astrophysical problems involving evolved stars and high-mass star forming regions are presented in §3 and §4 respectively. §5 details observations of SNR and megamaser magnetic fields and finally §6 presents perspectives for future instruments and observations.

2. Background and considerations

There exists extensive literature on the theory of maser polarization, as polarization during maser amplification differs from the regular thermal emission case due to the stimulated emission process and a range of other properties of the masing process that can influence the radiation polarization characteristics. The main theoretical problem is framed by constructing the density matrix evolution and radiative transfer equations for the maser emission including the Zeeman terms (Goldreich *et al.* 1973). The Zeeman effect occurs when the degeneracy of magnetic substates is broken under the influence of a magnetic field. The magnitude of the Zeeman effect is significantly different for paramagnetic (e.g. OH) and non-paramagnetic molecules (e.g. SiO, H_2O and methanol), due to the ratio between the Bohr magneton ($\mu_B = e\hbar/2m_e c$) and the nuclear magneton ($\mu_N = e\hbar/2m_n c$). In these expressions e is the electron charge, $\hbar$ is the Planck constant and c the speed of light. The ratio of the two ($\mu_B/\mu_N \approx 10^3$) is determined by the ratio of the electron mass (m_e) and the nucleon mass (m_N) and implies three orders of magnitudes larger Zeeman splitting for a similar magnetic field strength in the case of paramagnetic molecules compared to the non-paramagnetic ones.

A fundamental difference in the treatment of polarized maser emission exists between cases when the magnetic transitions overlap in frequency or when they are well separated. This can be defined by the splitting ratio $r_Z = \Delta\nu_Z/\Delta\nu_D$, where $\Delta\nu_Z$ is the Zeeman splitting and $\Delta\nu_D$ the Doppler line width. Typically, $r_Z \gtrsim 1$ for the paramagnetic molecule OH, while $r_Z < 1$ for the other, non-paramagnetic maser species. In the case of $r_Z > 1$ there are no theoretical ambiguities and the Zeeman components are well separated and resolved. The magnetic transitions $\Delta m_F = \pm 1$ give rise to the $\sigma^{\pm}$ components, circularly polarized perpendicular to the magnetic field B. The transition $\Delta m_F = 0$ gives rise to the π component, linearly polarized along B. For an arbitrary angle between B and the maser propagation direction θ, the resultant components are elliptically polarized for $\theta < \pi/2$, and linearly polarized for $\theta = \pi/2$. The observed splitting of the Zeeman components directly gives the magnetic field strength $B\cos\theta$.

In the case of $r_Z < 1$, the Zeeman components overlap. Theoretical work has been done both analytically and numerically with different implications for the derived magnetic field strengths (e.g.Watson 1994, Elitzur 1996, and references therein). A comparison of the different polarization theories is given by Gray (2003), who finds the numerical models can be used to accurately describe the maser polarization. This is further supported by H_2O maser polarization observations that are best described by the numerical models by Nedoluha & Watson (1992). When $r_Z < 1$ the relation between the linear polarization angle and the magnetic field direction is complex. The magnetic field direction is either parallel or perpendicular to the magnetic field, depending on θ. When $\theta < \theta_{\rm crit} \approx 55°$, the polarization vectors are parallel to the magnetic field and when $\theta > \theta_{\rm crit}$ they are perpendicular. However, this relation is only valid when the Zeeman frequency shift under the influence of the magnetic field $g\Omega$ is much larger than the rate of stimulated emission R. Otherwise, the relation between polarization angle and magnetic field direction is dependent on the maser intensity. However, this is unlikely to affect any except the strongest of maser features. The predicted relation between θ and the polarization angle from H_2O maser theory is supported by the observations of a 90° polarization angle flip when θ crosses the critical angle in combination with the predicted decrease in linear polarization fraction (Vlemmings & Diamond 2006).

The derived B strengths when $r_Z < 1$ do not only depend on the circular polarization fraction but also depend on the maser saturation level. Especially for masers that are saturated, simple assumptions with a fixed proportionality between circular polarization and magnetic field strength can lead to B being overestimated by up to a factor of 4. On the other hand, velocity gradients along the maser path can cause the true field strength to be underestimated by a factor of two (Vlemmings 2006). Also the blending of maser features in lower angular resolution observations typically causes another factor of two underestimation of B (Sarma et al. 2001).

In both cases ($r_Z < 1$ and $r_Z > 1$) there are several other properties of the maser and its surrounding medium that need to be taken into account when interpreting polarization observations. Especially at the low frequencies, Faraday rotation can make a direct connection between the polarization angle and the B-field uncertain. External Faraday rotation can cause significant vector rotation originating along the line of sight to the maser source. For instance at 1.6 GHz, a typical interstellar electron density and magnetic field strength can cause up to a full $\sim 180°$ rotation towards the W3(OH) star forming region. Additionally, internal Faraday rotation can alter the polarization characteristics of individual maser features in a source in defferent ways, possibly destroying any large scale structure in the linear polarization measurements (e.g. Fish & Reid 2006).

3. Evolved stars and planetary nebulae

Maser polarization observations are the predominant source of information about the role of magnetic fields during the late stages of stellar evolution. Most observations have focused on the masers in the CSEs of AGB stars, as OH, H_2O and SiO masers are fairly common in these sources. However, polarization observations of masers around post-AGB stars and (Proto-)PNe are becoming more common as more such sources with maser emission are found (see Gómez, 2007, these proceedings).

3.1. *AGB stars*

While polarization observations of CSE 1.6 GHz OH masers are fairly commonplace, recent years have seen an increase in 22 GHz H_2O and 43 GHz SiO maser observations of mainly Mira variables and supergiants. As the different maser species typically occur in different regions of the CSE, combining observations of all three species allows us to form a more complete picture of the magnetic field throughout the entire envelope. Close to the central star, SiO maser linear polarization reveals an ordered B-field with a linear polarization fraction ranging from $m_l \sim 15\%$ to $m_l \sim 65\%$ (e.g. Kemball & Diamond 1997). Strikingly, radially elongated jet-like SiO maser structures ,that have been observed around for example o Ceti and R Aql, are apparently aligned with the B-field (Cotton *et al.* 2006). A recent large single dish survey of SiO maser polarization revealed an average field strength of 3.5 G when assuming a regular Zeeman origin of the polarization, indicating a dynamically important B-field (Herpin *et al.* 2006). The observations find no specific support for other (non-Zeeman) interpretations of the polarization.

Further out in the envelope, also 22 GHz H_2O maser measurements reveal significant B-fields, both around Miras and supergiants (Vlemmings *et al.* 2005, and references therein). The measured field strength is typically of the order of $\sim 100-300$ mG but can be up to several Gauss. The strongest field strengths are found around Mira variables, consistent with the H_2O masers occurring closer to the star. As no linear polarization has been detected thus far, describing the magnetic field shape is difficult. However, for the Supergiant VX Sgr, the complex maser structure reveals an ordered field reversal across the maser region consistent with a dipole B-field. Interestingly, the orientation of the field determined from the H_2O maser polarization is similar to the orientation of the

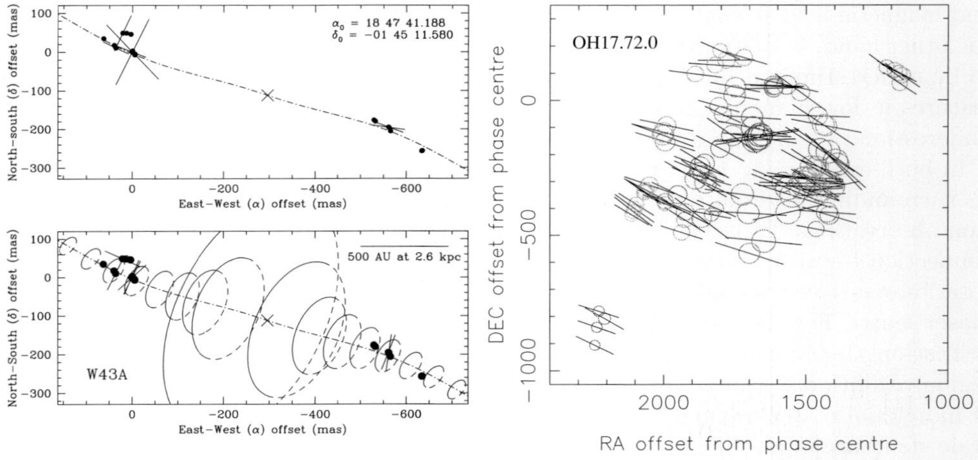

Figure 1. (left; top) The H_2O masers in the precessing jet (dashed-dotted line) of W43A (indicated by the cross). The maser features with the determined linear polarization vectors scaled linearly according to the fractional linear polarization. The polarization vectors lie predominantly along the jet with a median angle of $\chi = 63 \pm 12°$ east of north. (left; bottom) The toroidal B-field of W43A. The vectors indicate the determined direction of B, perpendicular to the polarization vectors, at the location of the H_2O masers. The ellipses indicate the toroidal field along the jet, scaled with $B \propto r^{-1}$ (Vlemmings *et al.* 2006a). (right) The spatial distribution of the 1612 MHz linearly polarized maser components around the P-PNe OH17.7-2.0. Symbol size is proportional to Stokes I flux density; vector size is proportional to logarithmically scaled linearly polarized flux density. More blue-shifted components are represented by darker (blue) symbols and more red-shifted are lighter (red). The clear, overall regular magnetic field structure is consistent with a stretched dipole field (Bains *et al.* 2004)

outflow determined from other H_2O maser observations as well as the orientation of a dipole field determine from OH maser polarization (Vlemmings et al. 2005).

Finally, at typically even larger distances, 1.6 GHz OH maser polarization are fairly common. Often strongly linearly polarized, the masers reveal an ordered B-field at several thousands AU from the star. The field strength at these distances is typically of the order of a few mG (e.g. Deacon *et al.* 2007, in prep.; Etoka & Diamond 2004).

3.2. *Proto-Planetary nebulae*

Most of the polarization work of P-PNe is done on OH masers. Similar to the magnetic field strengths around their progenitor stars, the P-PNe fields are $\sim$ 1 mG in the OH maser region. Single dish surveys reveal linear and circular polarization in respectively $\sim$ 50% and $\sim$ 75% of the sources, dependent on the OH maser line (polarization is more common in the 1612 MHz OH satellite line than in the 1665 and 1667 main line masers). The polarization fraction is typically less than 15% (Szymczak & Gérard 2004). The OH maser polarization has also been mapped using MERLIN. As shown in Fig. 1(right), the linear polarization vectors reveal a highly ordered B-field (Bains *et al.* 2003 & 2004).

A very small fraction of the Post-AGB/P-PNe maser stars show highly collimated H_2O maser jets (see Imai 2007; these proceedings). These so-called water-fountain sources are likely the progenitors of bipolar PNe and there are indications that they evolve from fairly high-mass AGB stars. The archetype of this class is W43A (Imai et al. 2002) and polarization observations have recently revealed that the maser jet is magnetically collimated (Fig.1 and 2; Vlemmings et al. 2006a). This lends strong support to the theories of magnetic shaping of PNe (e.g. García-Segura et al. 2005). In addition to the

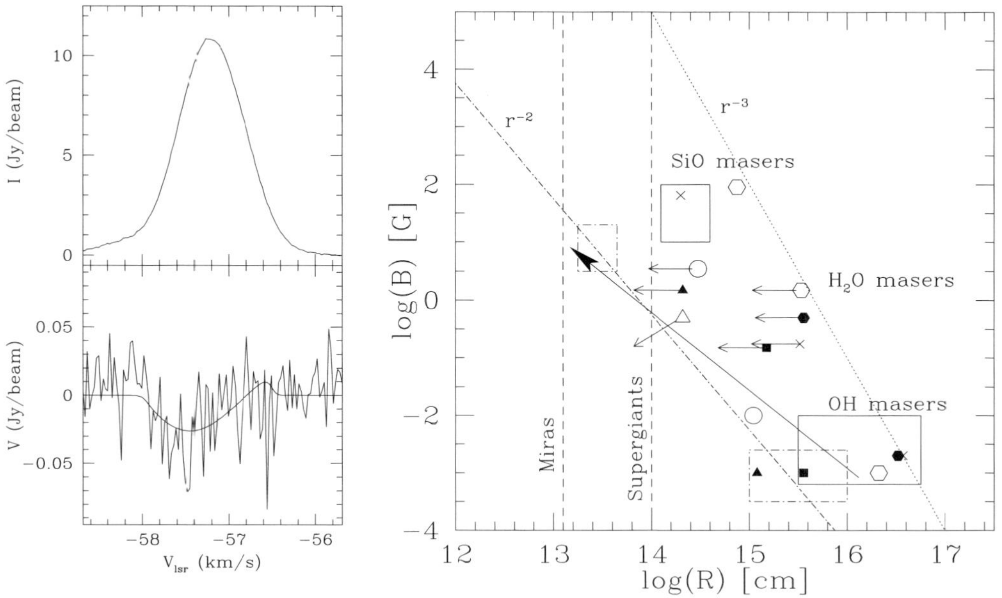

Figure 2. (left) The total intensity and circular polarization spectrum of the H_2O maser feature of W43A for which circular polarization was detected corresponding to a magnetic field strength of 85 mG (Vlemmings et al. 2006a). (right) The figure reproduced from Vlemmings et al. (2005) of measured B-fields on the masers in the CSEs of evolved stars. The dashed-dotted boxes indicate the range of magnetic field strengths measured on the SiO and OH masers of Mira stars and the solid boxes those of Supergiant stars. The thin arrow indicate the H_2O maser B-fields measured in Vlemmings et al. (2002) and Vlemmings et al. (2005), where the length of the arrows indicate the thickness of the H_2O maser shell with the symbols drawn on the outer edge. The symbols without arrows are the measurements on SiO and OH masers from the literature on the same sample of stars. The thick solid arrow indicates B in the jet of W43A.

observations of W43A, recent Australia Telescope Compact Array (ATCA) observations of the likely water-fountain source IRAS 15445-5449 (Vlemmings & Chapman 2007, in prep.) also indicate a magnetic H_2O maser jet.

3.3. *Planetary Nebulae*

There are only a handful of PNe known which show maser emission, and even less of these have masers that are strong enough to provide B-field measurements from polarization observations. One of the sources that shows both OH and H_2O maser emission is the very young PNe K3-35. In this source, the OH masers indicate a B-field of a few mG at 800 AU from the central object (Gómez et al. 2006).

3.4. *Summary*

Fig. 2(right) gives a summary of the current magnetic field measurements in CSEs of both Mira and supergiants. Although the exact relation between the magnetic field strength and distance to the central star remains uncertain, the field strengths are obvious strong enough to dynamically influence the shaping of the outflow and help shape asymmetric PNe. This is further supported by recent observations of strong B-fields on the surface of the central stars in several PNe (Jordan *et al.* 2005) as well as recent PNe dust polarization observations (e.g. Sabin et al. 2007). The magnetic field could also the missing component in the stellar mass-loss mechanism, as recent models indicate the pulsation

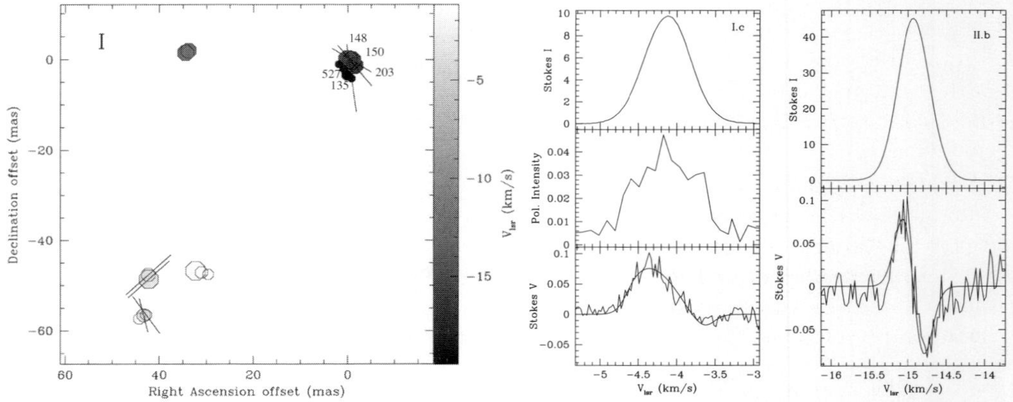

Figure 3. From Vlemmings et al. (2006a): (left) The polarization of the H_2O masers that are argued to be excited in a expanding shockwave through a rotated protostellar disk in Cepheus A HW2 (Gallimore *et al.* 2003). The linear polarization vectors, scaled logarithmically according to polarization fraction P_l, are over-plotted. For the maser features where the Zeeman splitting was detected the magnetic field strength is indicated in mG. (right) Total power (I) and V-spectra for selected maser features in Cepheus A. Additionally, the linear polarized flux density, $\sqrt{(Q^2 + U^2)}$, is shown when detected. The flux densities are given in Jy beam^{-1}. The thick solid line in the bottom panel shows the best non-LTE model fit to the circular polarization V. The V-spectrum is adjusted by removing a scaled down version of the total power spectrum and are clearly not necessarily symmetric.

and radiation pressure alone might not be enough (Woitke 2006). However, the question of the origin of the magnetic field remains and will likely have to be sought in the interaction with a heavy planet, a binary companion or a circumstellar disk, although several models also claim a magnetic dynamo can be maintained by a single star.

4. High-mass star formation

Star forming regions, and especially those forming high-mass stars, often contain a wide variety of maser species tracing many different density and temperature regimes. As was the case for evolved stars, most information on the small scale magnetic fields comes from maser polarization observations, dominated by OH maser measurements but also with an increasing number of H_2O and methanol maser observations. As SiO masers are uncommon during star formation, polarization observations are understandably rare, however, observations of the 86 GHz SiO masers near the massive protostar Orion IRc2 reveal a high level of fractional polarization and a magnetic field along the disk (Plambeck *et al.* 2003).

4.1. *OH masers*

There exists a large number of OH maser polarization measurements, probing densities from $\sim 10^5 - 10^8$ cm^{-3}. OH masers are often strongly polarized and as the Zeeman splitting is large, observations of the separate σ^+ and σ^- components directly yields a magnetic field strength. Observations of the 100% linearly polarized π-component are extremely rare however, likely due to magnetic beaming and the overlap of several differently polarized masers along the line of sight (Fish & Reid 2006). The measured field strengths are typically around $\sim$ 1 mG although recent observation have also revealed individual maser spots with field strengths up to $\sim$ 40 mG (Fish & Reid 2007). The field direction observed from OH masers seems to maintain the direction of the ambient B-field as the observations show a uniform direction of the field with at most one reversal across regions as large as several arcmin (e.g. Bartkiewicz *et al.* 2005). While linear polarization measurements can sometimes be used to constrain the three-dimensional B-field configuration (e.g. Hutawarakorn *et al.* 2002, Gray et al. 2003), internal and external Faraday rotation make a direct interpretation of the 1.6 GHz OH maser linear polarization difficult.

Magnetic field measurements have also been made using several of the excited OH maser transitions at 6 and 13 GHz. These observations suffer less from Faraday rotation while tracing mostly the same regions of the star forming region. The polarization and B-field values are fully consistent with the lower excitation measurements (e.g. Baudry & Diamond 98, Caswell 2003, Desmurs *et al.* 1998, Etoka *et al.* 2005).

4.2. *H_2O masers*

After the first discovery of interstellar H_2O maser Zeeman splitting by Fiebig & Güsten (1989) using single dish observations, there have been an increasing number of higher spatial resolution circular polarization observations confirming the earlier results (e.g. Sarma *et al.* 2001, 2002). These observations typically reveal B-field strengths between 15 and 150 mG at densities of $n_{H_2} = 10^8 - 10^{11}$ cm^{-3}. The largest field was recently found in a proposed protostellar disk around one of the several protostars in the Cepheus A HW2 region, with a field strength of ~ 650 mG, as shown in Fig. 3 (Vlemmings et al. 2006b). Such high B-field strength implies a nearby source enhancing the magnetic field. Considering the size of the disk this implies a B-field strength of ~ 2.5 G near the embedded protostar. Alternatively, the observations indicate much higher densities then current H_2O maser theory allows in the shockwave where the masers are excited (Elitzur et al. 1989, and references therein).

In addition to the circular polarization, low levels of linear polarization (typically $\lesssim 2\%$) are also observed in star forming regions. While often structure in the B-field direction is detected, the observations show rapid changes of direction over small scales (see Fig. 3).

4.3. *Methanol masers*

Although the 6.7 and 12 GHz methanol masers are some of the most abundant masers in high-mass star forming regions, polarization observations have been rare. Ellingsen (2002) presented ATCA linear polarization measurements of the 6.7 GHz masers, with a typical polarization fraction $m_l \sim 1.5\%$, while Koo *et al.* (1988) presented similar polarization fractions for the 12.2 GHz maser. Finally, Wiesemeyer *et al.* (2004) claim up to 40% linear polarization of the millimeter methanol masers. Only recently has the first 6.7 GHz methanol maser polarization map been made of the star forming region W3(OH), as shown in Fig. 4(left) (Vlemmings et al. 2006c). These observations show that the B-field is aligned with the large scale methanol maser filament and are consistent with previous OH maser observations. Importantly, the observations show that because they are less influenced by both internal and external Faraday rotation, methanol masers are better probes of the overall B-field structure than OH masers. Since the Zeeman splitting of methanol is small and circular polarization measurements are often hindered by dynamic range problems, the observations of Vlemmings *et al.* (2006c) only provide an upper limit to the B-field strength. However, the first tentative detection of a B-field strength from methanol observations has recently been made (Green *et al.* 2007, this proceedings).

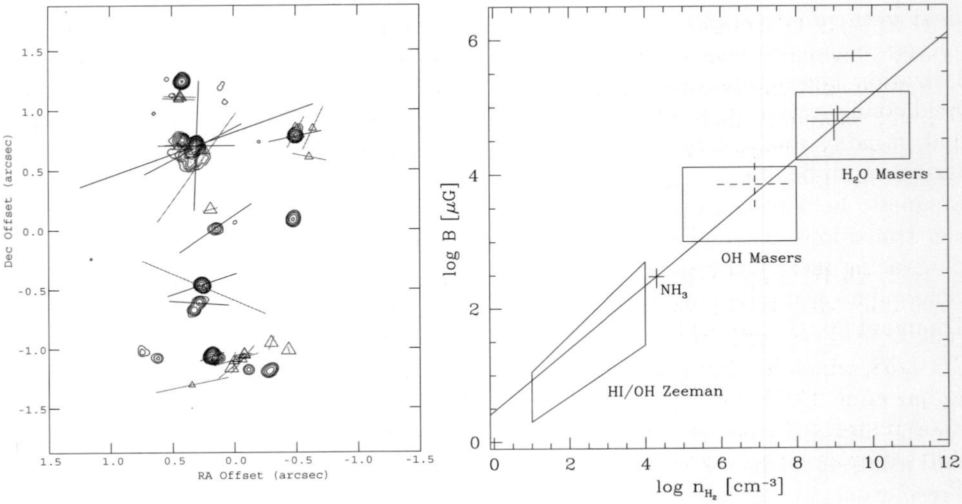

Figure 4. (left) The methanol masers of W3(OH) (contours) including the polarization vectors (black) scaled linearly according the fractional linear polarization (Vlemmings et al. 2006c). The positions in the map are indicated with respect to R.A.(J200)= $02^h 27^m 03.7743^s$, Dec(J2000)= $61°52'24.549''$. The (blue) triangles denote the main line OH masers from Wright et al. 2004b for which polarized intensity was detected at 5σ significance (polarized flux > 75 mJy), and the grey (red) vectors are their linearly scaled polarization vectors. The main line OH maser polarization vectors lengths are scaled down by a factor of 5 with respect to the lengths of the methanol maser polarization vectors. (right) The magnetic field strength B in massive star forming regions measured from Zeeman measurements as a function of n_{H_2}, the number density of neutral hydrogen. The boxes indicate the literature values for HI/non-masing OH, OH maser and H_2O maser Zeeman splitting observations. Also indicated are the H_2O maser measurements (thick solid crosses) and OH maser measurements (thick dashed cross) in the star forming region Cepheus A (Vlemmings *et al.* 2006b, Bartkiewicz et al. 2005) as well as an NH_3 measurement for the same source from Garay et al. (1996). The solid line is the empirical relation $B \propto n^{0.49}$ determined by Crutcher (1999) fixed to the NH_3 magnetic field measurement.

4.4. *Summary*

Fig. 4 (right) shows the B-field measurements of both masers and non-maser observations as a function of number density. The figure seems to indicate that (with the exception of regions such as the above mentioned protostellar disk), the B-field follows the density scaling law over an enormous range of densities, implying that the magnetic field remains partly coupled to the gas up to the highest number density (see Crutcher 2007, these proceedings). However, the shock excited H_2O masers are short-lived (with a typical lifetime $\tau_m \sim 10^8$ s) compared to the typical adiabatic diffusion timescale at the highest densities ($\tau_d \sim 10^9$ s), implying that in the non-masing gas of similar densities, magnetic field strengths are likely lower due to the adiabatic diffusion. Still, the maser B measurements strongly imply a dynamical importance of magnetic fields during the high-mass star formation process, especially in shaping outflows and jets.

5. Other sites of maser emission

Besides the evolved stars and star forming regions, masers are found in several other types of sources. Of those, both supernova remnant (SNR) masers and megamasers have had polarization measurements reported in the last few years.

5.1. *Supernova remnants*

The 1720 MHz OH masers in SNRs have been the target of several polarization observations and recently high resolution observations have for the first time directly resolved the Zeeman splitting between right- and left-circularly polarized spectra (Hoffman et al. 2005, and references therein). Linear polarization reveals the post-SNR shock B-field is well ordered and consistent with other magnetic field tracers and typical derived B-field strengths are $|B| \sim 0.5 - 1.0$ mG (see Brogan 2007, these proceedings).

5.2. *Extragalactic megamasers*

Finally, both H_2O and OH megamasers have been targeted for polarization studies. As the accretion disk H_2O masers are weak, only upper limits have been determined for the magnetic field strength in the masing region. A limit of 30 mG was found in the disk of NGC 4258, which has been used to determine a black hole accretion rate $< 10^{-3.7}$ M./yr (Modjaz *et al.* 2005). The tightest limit found to date is ~ 11 mG in the disk of NGC 3079 at 0.64 pc from the central black hole (Vlemmings et al. 2007), while a further limit of ~ 150 mG was found for the B-field in the Circinus H_2O maser disk (McCallum *et al.* 2007).

The first actual measurements of surprisingly strong extragalactic maser magnetic field strengths were recently made by Robishaw et al. (2007, these proceedings) on the 1667 MHz OH masers in a number of ultra-luminous infrared galaxies (ULIRGs).

6. Future perspectives

Since the previous IAU maser conference (Migenes & Reid 2002) there has been a huge increase in maser polarization measurements. H_2O maser polarization observations have become much more common, the first extragalactic maser polarization detections have been made and methanol maser polarization observations have shown their enormous potential. With the advent of new instruments such as the SMA and ALMA, high resolution dust polarization observations will soon be able to bridge the gap between the small scale maser magnetic field measurements and the very large scale single dish dust polarization observations. This should finally be able to convincingly show the importance of maser polarization observations in star forming regions for magnetic field studies, where especially methanol masers will be able to play a hugely important role. Further in the future, the SKA should be able to bring H_2O maser observations to a new level, fully imaging the polarization of the H_2O envelopes of evolved stars, PNe and even AGN accretion disks. But already in the near future, existing and soon to be upgraded instruments such as (e)MERLIN, the (e)VLA and the VLBA will be able to answer the questions about the role of magnetic fields during star formation and late stellar evolution by further increasing the size of the source samples.

References

Bains, I., Gledhill, T. M., Yates, J. A., & Richards, A. M. S. 2003, *MNRAS*, 338, 287

Bains, I., Richards, A. M. S., Gledhill, T. M., & Yates, J. A. 2004, *MNRAS*, 354, 529

Bartkiewicz, A., Szymczak, M., Cohen, R. J., & Richards, A. M. S. 2005, *MNRAS*, 361, 623

Baudry A., & Diamond, P. J. 1998, *A&A*, 331, 697

Caswell, J. L. 2003, *MNRAS*, 341, 551

Cotton, W. D., *et al.* 2006, *A&A*, 456, 339

Crutcher, R. M. 1999, *ApJ*, 520, 706

Desmurs, J. F., Baudry, A., Wilson, T. L., Cohen, R. J., & Tofani, G., 1998, *A&A*, 334, 1085

Elitzur, M., Hollenbach, D. J., & McKee, C. F. 1989, *ApJ*, 346, 983
Elitzur, M. 1996, *ApJ*, 457, 415
Ellingsen, S. P. 2002, IAU Symposium, 206, 151
Etoka, S., Cohen, R. J., & Gray, M. D., 2005, *MNRAS*, 360, 1162
Etoka, S. & Diamond, P. 2004, *MNRAS*, 348, 34
Fiebig, D. & Güsten, R. 1989, *A&A*, 214, 333
Fish, V. L., & Reid, M. J. 2006, *ApJS*, 164, 99
Fish, V. L., & Reid, M. J. 2007, *ApJ*, 656, 952
Gallimore, J. F., Cool, R. J., Thornley, M. D., & McMullin, J. 2003, *ApJ*, 586, 306
García-Segura, G., López, J. A., & Franco, J. 2005, *ApJ*, 618, 919
Garay, G., Ramirez, S., Rodriguez, L. F., Curiel, S., & Torrelles, J. M. 1996, *ApJ*, 459, 193
Goldreich, P., Keeley, D. A., & Kwan, J. Y. 1973, *ApJ*, 179, 111
Gómez, Y., Tafoya, D., Anglada, G., Franco-Hernandez, R., Torrelles, J. M., & Miranda, L. F. 2006, *Rev. Mexicana AyA*, 26, 22
Gray, M. D. 2003, *MNRAS*, 343, L33
Gray, M. D., Hutawarakorn, B., & Cohen, R. J. 2003, *MNRAS*, 343, 1067
Herpin, F., Baudry, A., Thum, C., Morris, D., & Wiesemeyer, H. 2006, *A&A*, 450, 667
Hoffman, I. M., Goss, W. M., Brogan, C. L., & Claussen, M. J. 2005, *ApJ*, 627, 803
Hutawarakorn, B., Cohen, R. J., & Brebner, G. C. 2002, *MNRAS*, 330, 349
Imai, H., Obara, K., Diamond, P. J., Omodaka, T., & Sasao, T. 2002, *Nature*, 417, 829
Kemball, A. J. & Diamond, P. J. 1997, *Ap. Lett.*, 481, L111
Koo, B.-C., Williams, D. R. D., Heiles, C., & Backer, D. C. 1988, *ApJ*, 326, 931
Migenes, V., & Reid, M. J. 2002, IAU Symposium, 206
McCallum, J. N., Ellingsen, S. P., & Lovell, J. E. J. 2007, *MNRAS*, 103
Modjaz, M., Moran, J. M., Kondratko, P. T., & Greenhill, L. J. 2005, *ApJ*, 626, 104
Nedoluha, G. E. & Watson, W. D. 1992, *ApJ*, 384, 185 (NW92)
Plambeck, R.L., Wright, M. C. H., & Rao, R. 2003, *ApJ*, 594, 911
Sabin, L., Zijlstra, A. A., & Greaves, J. S. 2007, *MNRAS*, 376, 378
Sarma, A. P., Troland, T. H., Crutcher, R. M., & Roberts, D. A. 2002, *ApJ*, 580, 928
Sarma, A. P., Troland, T. H., & Romney, J. D. 2001, *ApJ*, 554, L217
Szymczak, M., & Gérard, E. 2004, *A&A*, 423, 209
Vlemmings, W. H. T. 2006, *A&A*, 445, 1031
Vlemmings, W. H. T., Bignall, H.E., & Diamond, P. J. 2007, *ApJ*, 656, 198
Vlemmings, W. H. T., & Diamond, P. J. 2006, *ApJ*, 648, L59
Vlemmings, W. H. T., Diamond, P. J., & Imai, H. 2006a, *Nature*, 440, 58
Vlemmings, W. H. T., Diamond, P. J., & van Langevelde, H. J. 2002, *A&A*, 394, 589
Vlemmings, W. H. T., Diamond, P. J., van Langevelde, H. J., & Torrelles, J.M. 2006b, *A&A*, 448, 597
Vlemmings, W. H. T., Harvey-Smith, L., & Cohen, R. J. 2006c, *MNRAS*, 371, L26
Vlemmings, W. H. T., van Langevelde, H. J., & Diamond, P. J. 2005, *A&A*, 434, 1029
Watson, W. D. 1994, *ApJ*, 424, L37
Wiesemeyer, H., Thum, C., & Walmsley, C. M. 2004, *A&A*, 428, 479
Woitke, P. 2006, *A&A*, 460, L9
Wright, M. M., Gray, M. D., & Diamond, P. J., 2004, *MNRAS*, 350, 1272

Astrophysical Masers and their Environments
Proceedings IAU Symposium No. 242, 2007
J.M. Chapman & W.A. Baan, eds.

doi:10.1017/S1743921307012550

Magnetic fields in the non-masing ISM

Richard M. Crutcher

Department of Astronomy, University of Illinois, Urbana, IL 61801, USA
email: crutcher@uiuc.edu

Abstract. Observations of the Zeeman effect in OH and H_2O masers provide valuable information about magnetic field strength and direction, but only for the very high density gas in which such masers are found. In order to understand the role of magnetic fields in the evolution of the interstellar medium and in the star formation process, it is essential to consider the maser results in the broader context of magnetic fields in lower density gas. This contribution will (very briefly) summarize the state of observational knowledge of magnetic fields in the non-masing gas. Magnetic fields in H I and molecular clouds may be observed via the Zeeman effect, linear polarization of dust emission, and linear polarization of spectral-line emission. Useful parameters that can be inferred from observations are the mass-to-flux ratio and the scaling of field strength with density. The former tells us whether magnetic fields exert sufficient pressure to provide support against gravitational contraction; the latter tells whether or not magnetic fields are sufficiently strong to determine the nature (spherical or disk geometry) of the contraction. Existing observations will be reviewed. Results are that the strength of interstellar magnetic fields remains roughly invariant at 5-10 microgauss between densities of $0.1\,\mathrm{cm}^{-3} < n(H) < 1,000\ \mathrm{cm}^{-3}$ but increases proportional to approximately the square root of density at higher densities. Moreover, the mass-to-flux ratio is significantly subcritical (strong magnetic support with respect to gravity) in diffuse H I clouds that are not self-gravitating, but becomes approximately critical in high-density molecular cloud cores. This suggests that MCs and GMCs form primarily by accumulation of matter along magnetic field lines, a process that will increase density but not magnetic field strength. How clumps in GMCs evolve will then depend crucially on the mass-to-flux ratio in each clump. Present data suggest that magnetic fields play a very significant role in the evolution of molecular clouds and in the star formation process.

Keywords. ISM: magnetic fields, techniques: polarimetric, stars: formation

---

1. Introduction

Observing polarization of electromagnetic radiation from the cosmos is the principal and most direct way to learn about cosmic magnetic fields. It has become increasingly clear that cosmic magnetic fields are pervasive, ubiquitous, and likely important in the properties and evolution of almost everything in the Universe, from planets to quasars (*e.g.*, Wielebinski & Beck 2005). One area where the role of magnetic fields is far from being understood is star formation – an outstanding challenge of modern astrophysics. In spite of significant progress in recent years, there remain unanswered fundamental questions about the basic physics of star formation. In particular, what drives the star formation process? The prevailing view for most of the past 30 years has been that self-gravitating dense clouds are supported against collapse by magnetic fields (*e.g.*, Mouschovias & Ciolek 1999). However, magnetic fields are frozen only into the ionized gas and dust, while the neutral material (by far the majority of the mass) can contract gravitationally unaffected directly by the magnetic field. Since neutrals will collide with ions in this process, there will be support against gravity for the neutrals as well as the ions. But there will be a drift of neutrals into the core without a significant increase in the magnetic field strength in the core; this is ambipolar diffusion. Eventually the core mass will become sufficiently large that the magnetic field can no longer support the core, and dynamical collapse and star formation can proceed. The other extreme from the magnetically dominated star formation scenario is that molecular clouds are intermittent phenomena in an interstellar medium dominated by turbulence (*e.g.*, MacLow & Klessen 2004), and the problem of cloud support for long time periods is irrelevant. In this picture, clouds form and disperse by the operation of compressible supersonic turbulence, with clumps sometimes achieving sufficient mass to become self-gravitating. Even if the turbulent cascade has resulted in turbulence support, turbulence then dissipates rapidly, and the cores collapse to form stars. Hence, there are two competing models for driving the star formation process. The issue of what drives star formation is far from settled, on either observational or theoretical grounds.

In this paper we discuss how observations of magnetic fields in molecular clouds can distinguish between these models. In §2 we describe the predictions of the two models that may be tested via observations of magnetic fields. In §3 we briefly review polarization observational techniques for observing magnetic fields in molecular clouds and describe new observational results. In §4 we apply the tests and discuss the results. In §5 we summarize the conclusions about the role of magnetic fields in star formation, and finally in §6 suggest new observations that may answer definitively the question – what drives star formation?

2. Star formation theory – predictions and observational tests

The ambipolar diffusion and turbulence models for driving star formation have different predictions for magnetic field strength, which form the basis for tests of the two models using observations of magnetic fields. Of course, it is clear that there are both magnetic fields and turbulence in real clouds. In order to sharpen the distinctions between the two models, we will consider only turbulence models in which magnetic fields are negligibly weak and magnetic support/ambipolar diffusion models without turbulence. Here, we will discuss two tests of the two extreme-case models.

2.1. *Test 1: mass-to-flux ratio*

The ratio of the mass in a magnetic flux tube to the magnitude of the magnetic flux is a crucial parameter for the magnetic support/ambipolar diffusion model. The critical value for the mass in a disk with uniform density that can be supported by magnetic flux Φ is $M_{Bcrit} = \Phi/2\pi\sqrt{G}$ (Nakano & Nakamura 1978); the precise value of the numerical coefficient is slightly model dependant (*e.g.*, Mouschovias & Spitzer 1976, who calculated the result for a more realistic density stratified disk model). It is convenient to state observed M/Φ in units of the critical value, and to define $\lambda \equiv (M/\Phi)_{obs}/(M/\Phi)_{crit}$. Inferring λ from observations is possible if the column density N and the magnetic field strength B are measured:

$$\lambda = \frac{(M/\Phi)_{obs}}{(M/\Phi)_{crit}} = \frac{mNA/BA}{1/2\pi\sqrt{G}} = 7.6 \times 10^{-21} \frac{N(H_2)}{B} \tag{2.1}$$

where $m = 2.8m_H$ allowing for He, A is the area of a cloud over which measurements are made, $N(H_2)$ is in cm^{-2}, and B is in μG.

Ambipolar diffusion model: Clouds are initially subcritical, $\lambda < 1$. Ambipolar diffusion is fastest in shielded, high-density cores, so cores become supercritical, and rapid collapse ensues. The envelope continues to be supported by the magnetic field. Hence, the prediction is that λ must be < 1 in cloud envelopes (models typically have $\lambda \sim 0.3-0.8$), while in collapsing cores λ becomes slightly > 1. Hence, this model tightly constrains λ.

Turbulence model: The turbulent model imposes no direct constraints on λ, although strong magnetic fields would resist the formation of gravitationally bound clouds by compressible turbulence. Also, if magnetic support is to be insufficient to prevent collapse of self-gravitating clumps that are formed by compressible turbulence, the field must be supercritical, $\lambda > 1$. λ may take any value > 1, although of course for the turbulence model with very weak magnetic fields that we are considering, clouds and cores will be highly supercritical, $\lambda >> 1$.

2.2. *Test 2: scaling of B with ρ and σ*

The relationship between magnetic field strength and the volume density of gas is usually parameterized as $B \propto \rho^{\kappa}$. However, a dependence on the velocity dispersion σ may also be significant if turbulence plays a role in cloud support.

Consider first flux-freezing possibilities. If compression of the gas occurs parallel to the magnetic field, there will be no change in B, so $\kappa = 0$. If compression is orthogonal to the field, both the gas density and the field strength increase together, and $\kappa = 1$. Mestel (1966) showed that if the field is small enough to be dynamically unimportant, so gravitational collapse will be spherical, then $\kappa = 2/3$. Finally, if a cloud is supported parallel to the field by kinetic motions (thermal and/or turbulent) and orthogonal to the field primarily by the magnetic field, $\kappa = 1/2$ (Mouschovias & Ciolek 1999).

Ambipolar diffusion model: Clouds are subcritical before the formation of cores, with little change in B as ambipolar diffusion gradually increases the mass to magnetic flux ratio in the cloud center, forming a high-density core. Therefore κ for the cloud is small, $\sim 0.1 - 0.2$, in the early stages of evolution. Once the core becomes supercritical and rapid collapse ensues, the magnetic field is dragged inward with the collapsing gas, and κ gradually approaches 0.5; a typical value during the core collapse phase is $\kappa \approx 0.4$ (*c.f.*, Ciolek & Mouschovias 1994). The specific signature of ambipolar diffusion (as opposed to flux freezing in ideal MHD) is κ slightly less than 1/2. Little change in B takes place in the envelope during the entire process, however, so there is essentially no predicted correlation of B with ρ in the envelope.

Turbulence model: The correlation of field strength with other parameters has been discussed by several authors (*e.g.*, Mestel 1966; Padoan & Nordlund 1999; Basu 2000; Ballesteros-Paredes & MacLow 2002; Passot & Vazquez-Semadeni 2003). In the low-density case before gravity becomes important no relation is expected between the magnetic field strength and other parameters such as density. Hence, the prediction is similar to the ambipolar diffusion model, $\kappa \approx 0$. Once a clump becomes self-gravitating, neither magnetic fields (which are by assumption weak) or turbulence (which dissipates on roughly a free-fall time scale) can prevent approximately free-fall gravitational collapse; since magnetic fields are too weak to impose a preferred geometry, a spherical contraction is predicted. Flux freezing ($M \propto \Phi$, or $\rho R^3 \propto BR^2$) gives $R \propto B/\rho$, and a spherical collapse of a constant mass ($M \propto \rho R^3$) gives $R \propto \rho^{-1/3}$. Then $B \propto \rho^{2/3}$, so $\kappa = 2/3$. For the more general case of turbulent and magnetic field support, a rough equipartition of energies is achieved. Although clouds will generally not be in equilibrium, we can extract the form of the correlations if we assume virial equilibrium between gravity and turbulence ($3GM^2/5R = 3M\sigma^2/2$); then $\rho R^2 \propto \sigma^2$. Together with flux freezing, we find $B \propto \sigma\rho^{1/2}$. This is the same $\kappa = 1/2$ dependance discussed above with the inclusion of velocity dispersion term σ. In general σ varies much less than $n(H)$ among molecular clouds, so the dominant dependance would be expected to be the $B \propto \rho^{1/2}$ (but see Basu 2000).

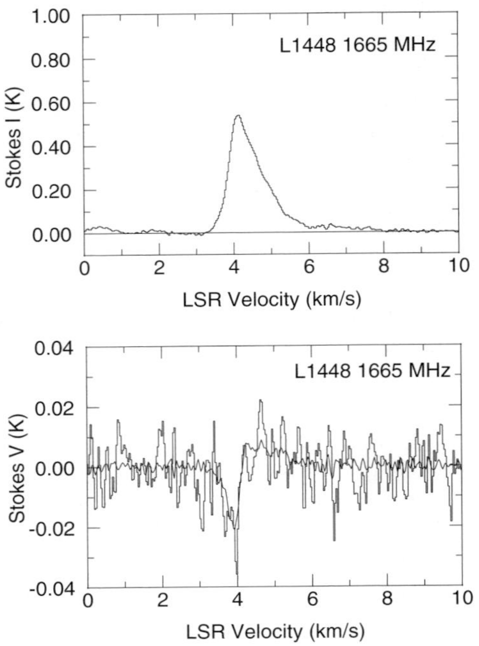

Figure 1. Arecibo Stokes I and V spectra of the 1665 and 1667 MHz lines of OH toward L1448. Observed data are histogram plots; fits to Stokes V are the dark lines. The combined result $\overline{B}_{los} = -26\ \mu$G, together with $N(H_2) \approx 5 \times 10^{21}$ cm^{-2} inferred from the OH lines, yields a mass-to-flux ratio with respect to critical of $\lambda \approx 1.5$ (before any geometrical correction), which is nominally slightly supercritical.

3. Observational techniques and new results

There are three main techniques for measuring magnetic fields that are applicable to molecular clouds: the Zeeman effect, linear polarization of thermal radiation from dust, and linear polarization of spectral-line emission due to the Goldreich-Kylafis (1981) effect. Space precludes detailed discussion; see Heiles & Crutcher (2005). But because maser studies of magnetic fields rely exclusively on the Zeeman effect, some brief comments about the Zeeman effect are worthwhile here. First, the normal Zeeman splitting term $\delta\nu_z$ is proportional to B_{total}, the total magnetic field strength, with the proportionality constant depending on the specific spectral line being observed.. An essential point for the Zeeman effect is that if the Zeeman splitting $\delta\nu_z < \Delta\nu_{sl}$, the width of the spectral line, only the line-of-sight component B_{los} of **B** can be determined. This is because a radio-telescope receiver sensitive (say) to left circularly polarized radiation will detect both the left elliptically polarized Zeeman σ component (which will be shifted by $\delta\nu_z$ from the rest frequency), half of the linearly polarized π (unshifted in frequency) component, and half of the linearly polarized part of the right elliptically polarized σ component (shifted in the opposite sense from the other σ component). This will "pull" the observed frequency of the "left" circularly polarized line toward the unshifted frequency; similarly for right circular polarization. The result is that the observed separation of the left and right circularly polarized lines will be proportional to B_{los} and not B_{total}. This is true whether the line is thermally excited or masing. See Crutcher *et al.* (1993) for details. On the other hand, if $\delta\nu_z > \Delta\nu_{sl}$, which can occur in some masers, such as OH, then the

σ and π Zeeman components are resolved, the observed splitting is directly the Zeeman splitting $\delta\nu_z$, and B_{total} is measured.

It is possible to correct statistically for the fact that only $B_{los} = |\mathbf{B}|\cos\theta$ is what is usually measured. Averaged over an ensemble of clouds with random **B** orientation with respect to the line of sight, $\overline{B}_{los} = B_{total}/2$. If **B** is strong, clouds will have a disk morphology with **B** along the minor axis. To properly measure λ, one needs B and N along a flux tube, *i.e.*, parallel to the minor axis. For this case, $\overline{M/\Phi} = (M/\Phi)_{obs}/3$, and we define $\lambda_C = \lambda/3$. See Heiles & Crutcher (2005) for details.

Heiles & Crutcher (2005) reviewed observational results in the diffuse and molecular interstellar medium. New results have come from an extensive survey (Troland & Crutcher 2007) of the OH Zeeman effect toward 33 dark cloud core positions; there were 10 new detections. Fig. 1 shows an example of the Stokes I and V profiles, toward L1448.

4. Test results

In applying here the tests of star formation theory, I will use the following sources of data: the "Millennium" Arecibo H I, survey (Heiles & Troland 2005), the Vela cloud optical polarization survey (Pereyra & Magalhaes 2007, the Crutcher compilation (Crutcher *et al.* 1999), and the Arecibo OH cores survey (Troland & Crutcher 2007). The details of this analysis are preliminary, and additional data are available and will be added in the final analysis. Moreover, the exact values for the volume densities in the abscissa of the next two figures are preliminary estimates and will likely be modified for the final journal publication of this analysis. However, these modifications will not be systematic and will not affect the trends discussed below or the overall results and conclusions.

4.1. *Test 1 – M/Φ*

Figure 2 shows a plot of λ_C versus n(H), the volume density of protons. Error bars are not shown because they would overwhelm the figure. The n(H) have been estimated in various ways, in some cases are preliminary, and often have large uncertainties for individual points, but are not believed to be systematically in error. Although the statistical correction of 1/3 for geometrical bias has been applied to each point, so that statistically this plot should be valid, for any individual point the true λ could be higher or lower than the plotted λ_C. A large part of the scatter in figure 2 is therefore be due to geometrical projection effects. Moreover, all of the masses (column densities) have been estimated from trace constituents (dust, OH, CO, etc.); an uncertainly of ~ 2 probably exists in the mass estimates.

The results show two clear trends. Below $n(H) \sim 10^3$ cm^{-3}, the mass-to-flux ratio is clearly subcritical, with an apparent trend with density from highly subcritical to roughly critical. Because molecular clouds must form from more diffuse gas, study of λ in diffuse gas ($n < 10^3$ cm^{-3}) is relevant to star formation. Hartmann *et al.* (2001) pointed out that for the typical very diffuse ISM values $n(H) \approx 1$ cm^{-3} and $B \approx 5$ μG (which would be highly subcritical), if the formation of clouds involves mass accumulation along field lines over distances greater than ~ 430 pc, eq. 2.1 implies that the resulting cloud will be supercritical. Accumulation over a flux tube with $R \approx 100$ pc and $L \approx 430$ pc would result in $\lambda = 1$ and $M \approx 3 \times 10^5$ $M_\odot$ – a typical GMC mass. An intermediate step would be the formation of H I clouds, with $n(H\ I) \sim 50$ cm^{-3}. The Heiles & Troland (2005) "Millinium" survey results are for a mean $n(H) \approx 50$ cm^{-2}, and are subcritical by a factor ~ 10.

The results for $n > 10^3$ cm^{-3} are for molecular clouds. The data suggest that $\overline{\lambda} \approx 1$, or approximately critical. This agrees with the expectation for the strong magnetic field

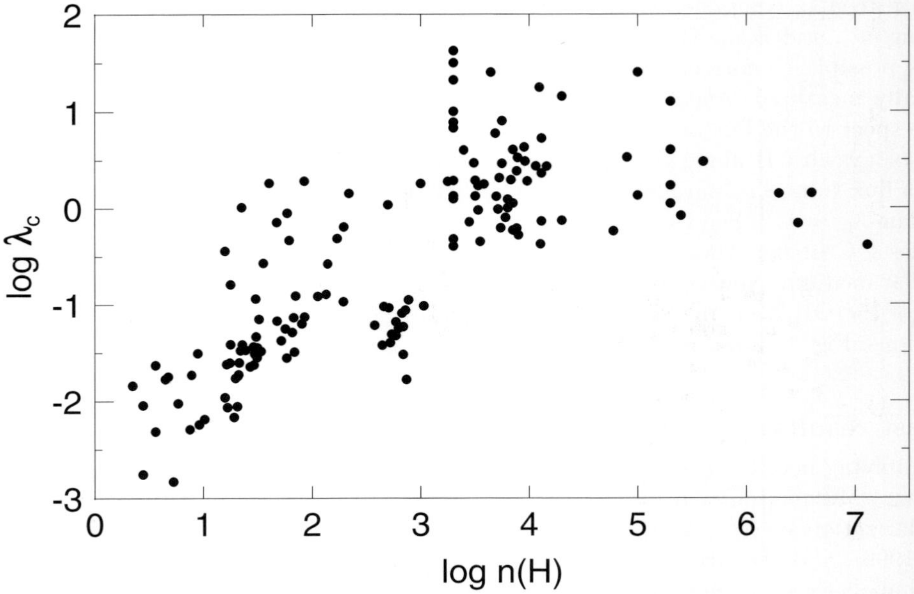

Figure 2. Observed, statistically corrected mass-to-flux ratios versus density.

model. While perhaps not fatal for the turbulence model, which predicts $\lambda > 1$ (strictly, $\lambda >> 1$), it certainly implies that magnetic fields are sufficiently strong that they cannot be neglected in models and simulations.

4.2. *Test 2 – B vs. ρ and σ*

At low densities $n \sim 0.1 - 100$ cm^{-3}, it has been clear for some time that there is no correlation of B with ρ (Troland & Heiles 1986); $\overline{B} \approx 5 - 10\ \mu$G. At higher densities the earlier analysis (Crutcher 1999) showed that $B \propto \rho^{\kappa}, \kappa = 0.47 \pm 0.08$.

Figure 3 shows the total magnetic field strengths B_{total} versus the volume density of protons, n(H). Both the turbulence and magnetic support models predict little or no correlation between B and ρ at low densities, as observed. At higher densities, a weighted fit to the data yields $\kappa = 0.48 \pm 0.06$, consistent with the signature ($\kappa = 0.5$) of flux freezing for a cloud support by kinetic pressure parallel to the magnetic field and kinetic and magnetic pressure perpendicular to the field. The observations do not support the weak magnetic field, free-fall collapse prediction $\kappa = 2/3$. Breaking of flux freezing for the magnetically supported model by ambipolar diffusion would produce κ slightly less than 0.5; the present data are not sufficient to show whether this is the case or not.

5. Conclusions

The fact that the magnetic field strength is essentially constant ($B \approx 5 - 10\ \mu$G) from the lowest densities in the interstellar medium up to the densities where molecular clouds are self gravitating provides a very significant clue about the formation of molecular clouds. If densities increased perpendicular to magnetic field lines, field strengths would increase linearly with density. Hence, molecular clouds must form primarily by

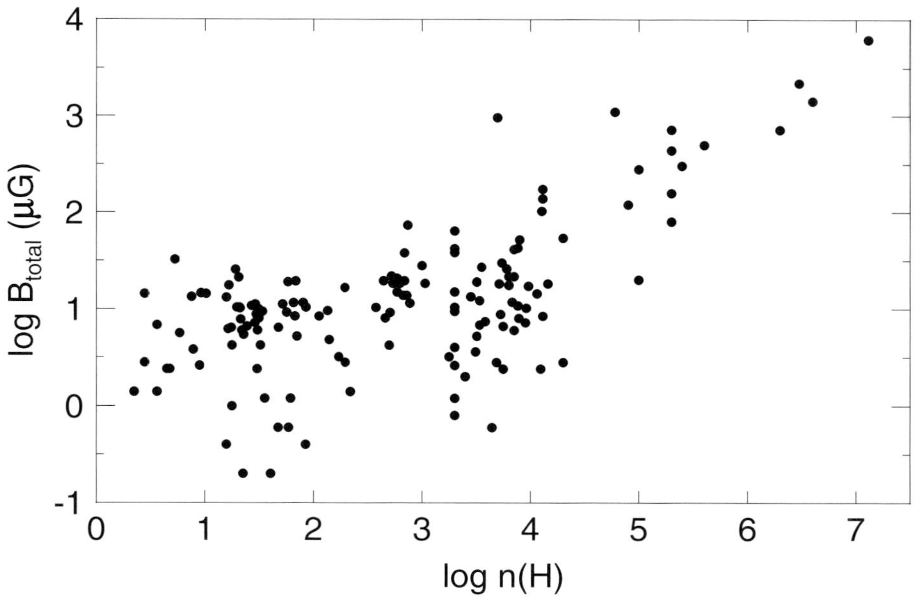

Figure 3. Observed, statistically corrected total magnetic field strengths versus density.

accumulation of matter along field lines. This process would increase densities but not field strengths.

Diffuse clouds with n(H I) ~ 50 cm^{-3} are significantly subcritical ($\overline{\lambda} \approx 0.03$) but not self-gravitating. Molecular clouds are approximately critical, $\overline{\lambda} \approx 1$ (see figure 2). The change in λ from subcritical values in diffuse clouds to critical ones in molecular clouds probably takes place during the molecular cloud formation process, by material accumulating along flux tubes to form dense clouds (*e.g.*, Hartmann *et al.* 2001). Although this would not actually increase the mass-to-flux ratio in a flux tube, observers of individual H I clouds in the flux tube would infer a lower λ than would be found after H I clouds aggregate to form a single dense molecular cloud. A combination of accumulation of matter within flux tubes, turbulence-driven ambipolar diffusion (Heitsch *et al.* 2004), and gravity-driven ambipolar diffusion may all be important at different stages in molecular cloud formation and collapse.

The data show that M/Φ is subcritical in H I clouds and approximately critical in molecular clouds, in agreement with ambipolar diffusion. Also, the fact the κ seems to be inconsistent with 2/3 contradicts the weak field model. The available data clearly favor the strong magnetic field model of star formation.

6. The future

The tests described above are limited by the fact that only one component of the three-component vector **B** can be measured, requiring statistical analysis that may not be convincing. However, there is a prediction of the ambipolar diffusion theory that is subject to a direct test, object by object. The theory absolutely requires that M/Φ *increase* from the envelope of a cloud to its core. On the other hand, "observations" of cores formed in converging turbulent flow simulations (Dib 2006) appear to show that M/Φ *decreases* with density. Hence, observing the *differential* M/Φ between envelope and core should provide a definitive test. Even though only B_{los} can be measured via Zeeman observations, the angle between the regular magnetic field and the line of sight will be essentially the same between envelope and core. Moreover, if one uses the same species (such as OH) to measure B_{los} between envelope and core, the problem of knowing the abundance ratio [X/H] between the Zeeman species X and H is eliminated. One might suggest that [OH/H] would decrease from envelope to core due to astrochemistry, but this would not negate a test of ambipolar diffusion. A smaller [OH/H] between envelope and core would reduce the differential M/Φ between envelope and core – the opposite of the ambipolar diffusion effect. Hence, detection of an increase in the differential M/Φ in a selection of molecular clouds with cores would provide very strong support for that theory. Such an observational program has been approved. Crutcher & Troland will use the GBT to measure $N(OH)/B_{los}$ toward cores with Arecibo detections of B_{los} (Troland & Crutcher 2007) by observing at the four cardinal positions surrounding but excluding the cores. Together with the Arecibo results for the cores, the GBT results for the envelopes will yield the change in mass to flux ratio from envelope to core. Failure to detect the differential M/Φ predicted by ambipolar diffusion will be difficult for advocates of that theory to dismiss. On the other hand, success would provide powerful evidence for ambipolar diffusion.

Acknowledgements

This work was partially supported by NSF grants AST 02-05810 and 06-06822.

References

Ballesteros-Paredes, J. & MacLow, M.-M. 2002, ApJ, 570, 734
Basu, S. 2000, ApJ, 540, L103
Ciolek, G. E. & Mouschovias, T. Ch. 1994, ApJ, 425, 142
Crutcher, R. M., Troland, T. H., Goodman, A. A., Heiles, C., Kazès, I., & Myers, P. C. 1993, ApJ, 407, 175
Crutcher, R. M. 1999, ApJ, 520, 706
Dib 2006, private communication
Goldreich, P. & Kylafis, N. D. 1981, ApJ, 243, L75
Hartmann, L., Ballesteros-Paredes, J., & Bergin, E. A. 2001, ApJ, 562, 852
Heiles, C. & Crutcher, R. 2005, in *Cosmic Magnetic Fields* Lecture notes in Physics 664, ed. R. Wielebinski & R. Beck, (Springer: Berlin), p. 137
Heiles, C. & Troland, T. 2005 ApJ, 624, 773
Heitsch, F., Zweibel, E. G., Slyz, A. D., & Devriendt, J. E. G. 2004, ApSS, 292, 45
MacLow M.-M. & Klessen R. S. 2004, Rev. Mod. Phys., 76, 125
Mestel, L. 1966, MNRAS, 133, 265
Mouschovias, T. Ch. & Spitzer, L. 1976, ApJ, 210, 326
Mouschovias, T. Ch. & Ciolek, G. E. 1999, in *The Origin of Stars and Planetary Systems*, eds C. J. Lada & N. D. Kylafis (Kluwer), p. 305
Nakano, T. & Nakamura, T. 1978, PASJ, 30, 681
Padoan, P. & Nordlund, Å. 1999, ApJ, 526, 279
Passot, T. & Vázquez-Semadeni, E. 2003, A&A, 398, 845
Pereyra, A. & Magalhaes, A. M. 2007, astro-ph/0702550
Troland, T. H. & Heiles, C. 1986, ApJ, 301, 339
Troland, T. H. & Crutcher, R. M. 2007, in preparation
Wielebinski, R. & Beck, R. 2005, *Cosmic Magnetic Fields* Lecture notes in physics 664, (Springer: Berlin)

Astrophysical Masers and their Environments
Proceedings IAU Symposium No. 242, 2007
J.M. Chapman & W.A. Baan, eds.

doi:10.1017/S1743921307012562

Magnetic fields in our Galaxy on large and small scales

J. L. Han

National Astronomical Observatories, Chinese Academy of Sciences
Jia-20 DaTun Road, Chaoyang District, Beijing 100012, China
email: hjl@bao.ac.cn

Abstract. Magnetic fields have been observed on all scales in our Galaxy, from AU to kpc. With pulsar dispersion measures and rotation measures, we can directly measure the magnetic fields in a very large region of the Galactic disk. The results show that the large-scale magnetic fields are aligned with the spiral arms but reverse their directions many times from the innermost arm (Norma) to the outer arm (Perseus). The Zeeman splitting measurements of masers in HII regions or star-formation regions not only show the structured fields inside clouds, but also have a clear pattern in the global Galactic distribution of all measured clouds which indicates the possible connection of the large-scale and small-scale magnetic fields.

Keywords. ISM: magnetic fields, pulsars: general, HII regions, radio lines: ISM, ISM: structure

1. Introduction

Inside our Galaxy, the diffuse interstellar medium (ISM), molecular clouds, and very dense cloud core or HII regions, are all permeated by magnetic fields. Many physical processes in the ISM are related to magnetic fields. For example, magnetic fields in the diffuse medium contribute to the hydrostatic balance and stability of the ISM (Boulares & Cox 1990); magnetic fields are the main agent for transporting charged cosmic-rays(e.g. Tinyakov & Tkachev 2002, Prouza & Smída 2003), while magnetic fields in molecular clouds obviously play an important role in the star formation process (Heiles & Crutcher 2005). However, our knowledge of the Galactic magnetic fields is far from complete.

It is possible that the fields in clouds are enhanced when interstellar gas contracts to form a cloud or the cloud core, so that the observed field strength increases with gas density (Crutcher 1999). Here, I review the observational results of large-scale magnetic fields in our Galaxy mainly revealed by pulsar rotation measures, which are related to the diffuse gas and the spiral structure. Small-scale magnetic fields have been detected by other means, such as continuum polarization surveys of the Galactic plane. I will show that the Galactic distribution of the small scale (AU) magnetic fields detected by maser-lines from clouds are probably related to the large-scale Galactic magnetic field structure. In this paper *large scale* means a scale larger than the separation of spiral arms, i.e. a scale of 1 kpc, while *small scale* means a smaller scale.

There are five observational tracers for the Galactic magnetic fields: Zeeman splitting, polarized thermal emission from the dusts in clouds, polarization of starlight, synchrotron radio emission, and Faraday rotation of polarized sources. Observations of Zeeman splitting of spectral lines, and of polarized thermal emission from dust at mm, sub-mm or infrared wavelengths, have been used to detect respectively the line-of-sight strength and the transverse orientation of magnetic fields in molecular clouds (e.g. Crutcher 1999, Novak *et al.* 2003, Fish *et al.* 2003). Starlight polarization data show that the local field is parallel to the Galactic plane and follows the local spiral arms (see Mathewson & Ford

1970, Andreasyan & Makarov 1989, Heiles 1996a). Since we live near the edge of the Galactic disk, we cannot have a face-on view of the global magnetic field structure in our Galaxy through polarized synchrotron emission, as is possible for nearby spiral galaxies (see reviews by Beck *et al.* 1996, Beck 2005). Polarization observations of synchrotron continuum radiation from the Galactic disk give the transverse orientation of the field in the emission region and some indication of its strength. Large-angular-scale features are seen emerging from the Galactic disk, for example, the North Polar Spur (e.g. Junkes *et al.* 1987, Duncan *et al.* 1997, Duncan *et al.* 1999, Reich *et al.* 2002), and the vertical filaments near the Galactic Center (Haynes *et al.* 1992, Duncan *et al.* 1998). Faraday rotation of linearly polarized radiation from pulsars and extragalactic radio sources is a powerful probe of the diffuse magnetic field in the Galaxy (e.g. Simard-Normandin & Kronberg 1980, Han & Qiao 1994, Han *et al.* 2006, Brown *et al.* 2007). Galactic magnetic fields both on large scales and small scales can be revealed by these tracers.

2. Galactic magnetic fields on large scales

Our Galaxy is a spiral galaxy. Spiral galaxies have large-scale magnetic fields (Beck 2005). In the last decade, significant progress has been made on revealing the large-scale magnetic fields of our Galaxy in the Galactic central region, in the Galactic halo and the Galactic disk.

Near the Galactic Center, non-thermal gaseous filaments have been discovered (LaRosa *et al.* 2004, Nord *et al.* 2004, Yusef-Zadeh *et al.* 2004). The majority of bright non-thermal filaments are perpendicular to the Galactic plane, indicating predominantly poloidal fields of maybe mG strength, but some filaments are not, indicating a more complicated field structure than just the poloidal field. LaRosa et al. (2005) detected the diffuse radio emission and argued for a weak pervasive field of tens of μG near the Galactic Center. The new discovery of an infrared 'double helix' nebula (Morris *et al.* 2006) reinforces the presence of strong poloidal magnetic fields merging from the rotated circumnuclear gas disk near the Galactic Center (Yusef-Zadeh & Morris 1987). With the development of polarimetry at mm, submm or infrared wavelengths, toroidal fields have been observed in the molecular cloud zone near the Galactic Center (Novak *et al.* 2003, Chuss *et al.* 2003), complimenting the poloidal fields shown by the vertical filaments. The large RMs of radio sources near the Galactic Center (Roy *et al.* 2005) may indicate toroidal field structure.

From the RM distribution in the sky, Han *et al.* (1997, 1999) identified the striking antisymmetry in the inner Galaxy with respect to the Galactic coordinates after removing the RM "outliers" compared to their neighborhoods. Such an antisymmetry should result from azimuthal magnetic fields in the Galactic halo with reversed field directions below and above the Galactic plane. Such a field can be produced by an 'A0' mode of dynamo. The observed filaments near the Galactic Center should result from the dipole field in this dynamo scenario. The local vertical field component of 0.2 μG (Han & Qiao 1994, Han *et al.* 1999) may be a part of this dipole field in the solar vicinity. At present, we have observed another 1700 radio sources in the Northern sky by the Effelsberg 100 m telescope (Han, Reich *et al.* 2007, in preparation), and wish to do more in the Southern sky with Parkes, so that the rotation measure (RM) sky can be described quantitatively.

Magnetic fields in a large-part of the Galactic disk have been delineated by Faraday rotation data of pulsars, which gives a measure of the line-of-sight component of the magnetic field. Extragalactic sources have the advantage of large numbers but pulsars have the advantage of being spread through the Galaxy at approximately known distances, allowing direct three-dimensional mapping of the magnetic field. Pulsars also give a very

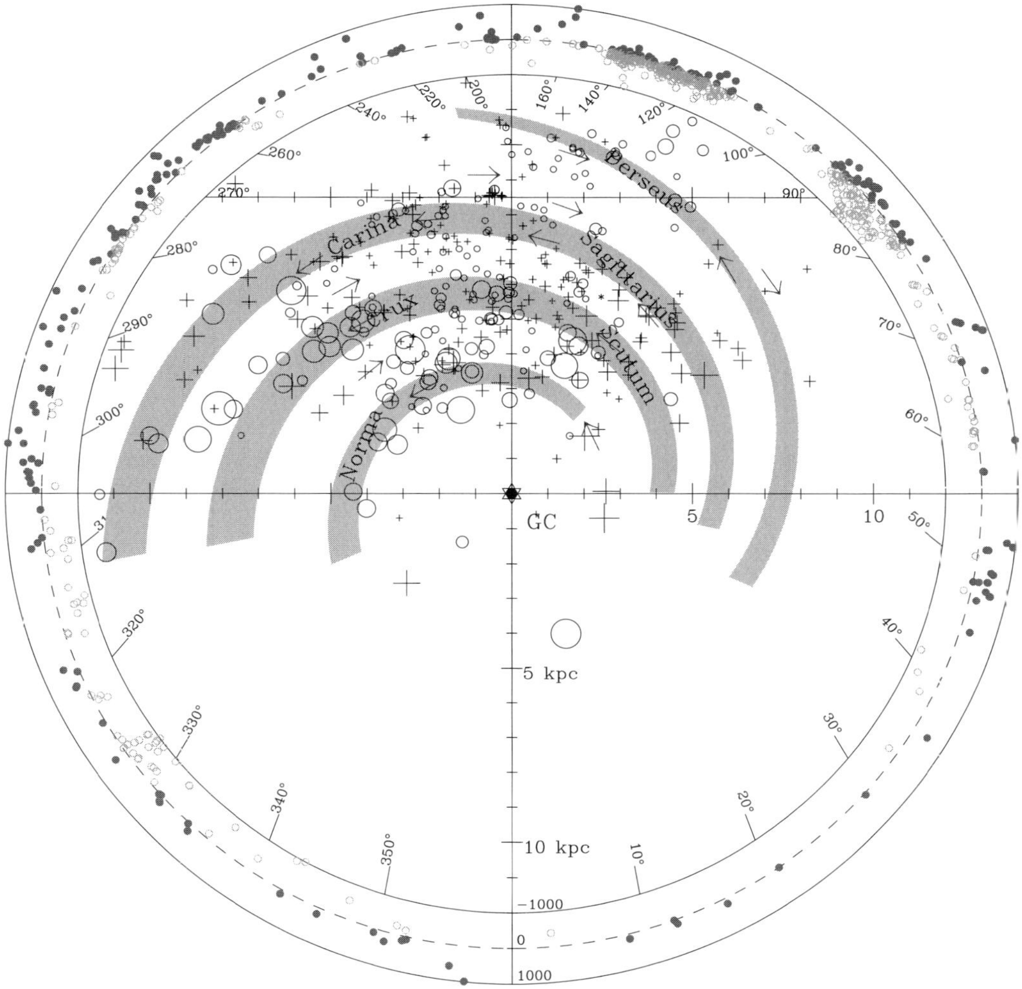

Figure 1. The RM distribution of 374 pulsars with $|b| < 8$ deg, projected onto the Galactic plane. The linear sizes of the symbols are proportional to the square root of the RM values. The crosses represent positive RMs, and the open circles represent negative RMs. The approximate locations of four spiral arms are indicated. The large-scale structure of magnetic fields derived from pulsar RMs are indicated by thick arrows. See Han *et al.* (2006) for details. RMs of extragalactic radio sources of $|b| < 8$ deg (data from Clegg *et al.* 1992, Gaensler *et al.* 2001, Brown *et al.* 2003, Brown *et al.* 2007) are also displayed in the outskirt ring. Positive RMs are shown by filled circles and negative RMs by open circles. The RM limits of ± 1000 rad m^{-2} are set at the outer and inner edges of the ring. As one can see from this plot, we have not many measurements for the magnetic fields for one half of the Galactic disk. The RMs of extragalactic radio sources become scarcer and scarcer in the region of $|l| < 45$ deg.

direct estimate of the strength of the field through normalization by the dispersion measure (DM). The RM is defined by $\phi = \mathrm{RM}\,\lambda^2$, where ϕ is the position angle in radians of linearly polarized radiation relative to its infinite-frequency ($\lambda = 0$) value and λ is its wavelength (in m). For a pulsar at distance D (in pc), the RM (in radians m^{-2}) is given by $\mathrm{RM} = 0.810 \int_0^D n_e \mathbf{B} \cdot d\mathbf{l}$, where n_e is the electron density in cm^{-3}, $\mathbf{B}$ is the vector magnetic field in μG and $d\mathbf{l}$ is an elemental vector along the line of sight toward us (positive RMs correspond to fields directed toward us) in pc. With the DM $= \int_0^D n_e dl$, we obtain a direct estimate of the field strength weighted by the local free electron density

$$\langle B_{||} \rangle = \frac{\int_0^D n_e \mathbf{B} \cdot d\mathbf{l}}{\int_0^D n_e dl} = 1.232 \, \frac{\text{RM}}{\text{DM}}. \tag{2.1}$$

where RM and DM are in their usual units of rad m^{-2} and cm^{-3} pc and $B_{||}$ is in μG.

Previous analysis of pulsar RM data has often used the model-fitting method (Han & Qiao 1994, Indrani & Deshpande 1999), i.e., to model magnetic field structures in all of the paths from the pulsars to us (observer) and fit them, together with the electron density model, to the observed RM data. *Significant improvement* can be obtained when both RM and DM data are available for many pulsars in a given region with similar lines of sight. Measuring the gradient of RM with distance or DM is the most powerful method of determining both the direction and magnitude of the large-scale field local in that particular region of the Galaxy (Lyne & Smith 1989, Han *et al.* 2002, Weisberg *et al.* 2004, Han *et al.* 2006). Field strengths in the region can be *directly derived* (instead of *modeled*) from the slope of trends in plots of RM versus DM. Based on Equation 2.1, we got

$$\langle B_{||} \rangle_{d1-d0} = 1.232 \frac{\Delta\text{RM}}{\Delta\text{DM}} \tag{2.2}$$

where $\langle B_{||} \rangle_{d1-d0}$ is the mean line-of-sight field component in μG for the region between distances $d0$ and $d1$, $\Delta\text{RM} = \text{RM}_{d1} - \text{RM}_{d0}$ and $\Delta\text{DM} = \text{DM}_{d1} - \text{DM}_{d0}$.

So far, RMs of 550 pulsars have been observed (Hamilton & Lyne 1987, Qiao *et al.* 1995, Han *et al.* 1999, Weisberg *et al.* 2004, Han *et al.* 2006). Most of the new measurements are for the fourth and first Galactic quadrants where sources are relatively distant. This enables us to investigate the structure of the Galactic magnetic field over a much larger region than was previously possible. We detected counterclockwise magnetic fields in the most inner arm, the Norma arm (Han *et al.* 2002). A more complete analysis for the fields near the tangential regions of most probable spiral of our Galaxy (Han *et al.* 2006) gives a similar picture for the coherent large-scale fields aligned with the spiral-arm structure in the Galactic disk, as shown in Fig.1: magnetic fields in all inner spiral arms are counterclockwise when viewed from the North Galactic pole. On the other hand, at least in the local region and in the inner Galaxy in the fourth quadrant, there is good evidence that the fields in interarm regions are similarly coherent, but clockwise in orientation. There are at least two or three reversals in the inner Galaxy, probably occurring near the boundary of the spiral arms. The magnetic field in the Perseus arm cannot be determined well. The negative RMs for distant pulsars and extragalactic sources (see Fig. 1) in fact suggest the interarm fields both between the Sagittarius and Perseus arms and beyond the Perseus arm are predominantly clockwise.

The fluctuations in the RM distribution of extragalactic radio sources (Clegg *et al.* 1992, Gaensler *et al.* 2001, Brown *et al.* 2003, Brown *et al.* 2007) with Galactic longitude, especially these of the fourth Galactic quadrant, are consistent with magnetic field directions derived from the tangential regions of the arms (see Fig. 1). This implies that the dominant contribution of RMs of extragalactic radio sources behind the Galactic disk comes from the interstellar medium mainly in tangential regions. However, modeling the averaged RM distribution of scattered RM data of background radio sources behind the Galactic disk requires fewer field reversals (Brown *et al.* 2007) maybe due to the smearing effect in the box-car averaging process over 9° bins along the Galactic longitude, larger than the separation between the inner arms.

With much more pulsar RM data now available, Han *et al.* (2006) were able to *measure*, rather than *model*, the regular field strength near the tangential regions in the 1st and 4th Galactic quadrants, and then plot the dependence of regular field strength on the Galacto-radii. Although the "uncertainties", which in fact reflect the random fields, are large, the tendency is clear that fields get stronger at smaller Galactocentric radii and weaker in interarm regions. To parameterize the radial variation, an exponential function was used. This was chosen to give the smallest χ^2 value and to avoid a singularity at $R = 0$ (for $1/R$) and unphysical values at large R (for the linear gradient). The function is, $B_{\rm reg}(R) = B_0 \exp\left[\frac{-(R-R_\odot)}{R_{\rm B}}\right]$, with the strength of the large-scale or regular field at the Sun, $B_0 = 2.1 \pm 0.3$ μG and the scale radius $R_{\rm B} = 8.5 \pm 4.7$ kpc.

3. Galactic magnetic fields on small scales

Small-scale magnetic fields can be revealed by polarization surveys of diffuse radio emission of the Galactic plane, and by polarization observations of molecular clouds and supernova remnants by either linear polarization mapping or Zeeman splitting of line emission or maser emission. The statistics of these observations and the statistics of RMs of pulsars or extragalactic radio sources give the overall properties of the small-scale magnetic fields.

Polarization surveys of the Galactic plane have been comprehensively reviewed by Reich (2007). See references therein. Here I would like to mention that the observed polarized emission of the Galactic plane often has a scale size of about tens or hundreds of pc, and is the sum of all contributions coming from various regions along a line of sight with different polarization properties (i.e. polarization angle and polarization percentage) at different distance. Emission from more distant regions suffers from more Faraday effect produced by foreground interstellar medium. If the emission brightness at various distances are more or less similar, then the observed polarized emission is predominantly from local regions. Accumulating polarized emission from different regions should "depolarize" each other. This is more obvious in lower frequencies. Note that disordered polarized structures smaller than an observational beam could be smeared out by the so-called beam depolarization. Therefore, observations with a smaller beam would detect more polarized structures. Observations at higher frequencies should show structures at larger distances. These polarized structures are closely related to the magnetic field structure where the emission is generated. The Sino-German 6 cm polarization survey (Sun *et al.* 2007) of the Galactic plane using the Urumqi 25m telescope is currently been carried out, and we have detected some magnetic structures from Faraday screens and supernova remnants, some of which would not be detectable at lower frequencies (Sun *et al.* 2007, Xu *et al.* 2007).

In recent years, with development of instruments and backend technology, a lot of molecular clouds have been directly mapped for polarized emission at mm, submm or infrared bands, some by single dish, some by interferometers (see review by Heiles & Crutcher 2005). These maps always show field orientation on a scale of the cloud size, i.e. a scale of pc to tens of pc. I do not have to cover this topic as readers can find comprehensive information from the reviews given by Crutcher (2007) and Vlemmings (2007) in this volume.

Some maser spots have a very small scale size, of AU rather than pc. From star formation regions or HII regions, the Zeeman splitting of the maser lines can directly gives the field strength *in situ* as well as the field direction in the line of sight. In recent years, there have been many new measurements, e.g. VLBA measurements by Fish *et al.*

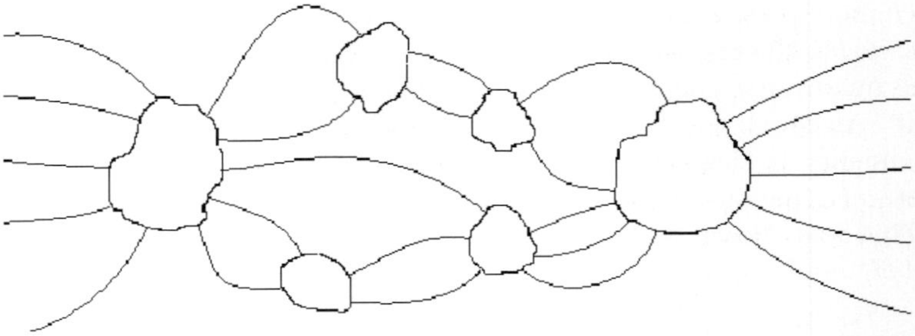

Figure 2. The magnetic fields on small scales, e.g. in clouds, are possibly related to the large-scale fields. See also Beck *et al.* (1991).

(2006) or other measurements by Hutawarakorn & Cohen (2005). See Han & Zhang (2007) for a list of all previous measurements. It is amazing that magnetic fields on such a small scale are possibly related to the large-scale Galactic magnetic fields (see below).

One important tool for studying small-scale magnetic fields is to use statistics. Detailed statistics for polarization survey data of the Galactic plane have been rare, but would be very useful. Statistics of the mean deviations of "polarization vectors" for clouds have been used to estimate the field strength in the cloud (see e.g. Crutcher *et al.* 2004). Minter & Spangler (1996) have worked out the structure function of a group of RMs of extragalactic radio sources in a mid-latitude region, showing that the magnetic fields probably follow the Kolmogorov spectrum for a scale less than a few pc, while it becomes flatter above a few pc up to 80 pc, maybe due to the thin shape of the Galactic disk. Sun & Han (2003) found that the structure function for RMs at the two Galactic poles are flat, indicating the random RM distribution, but at lower latitudes it becomes inclined with different slopes at different Galactic longitudes. Haverkorn *et al.* (2006) found that the structure functions of RMs in the arm tangential directions have much larger slopes than those in the interarm directions, indicating that the arm regions are more turbulent. The outer scale is as small as only 17 pc. Han et al. (2004) used statistics of pulsar RMs in a very large region of the two third of the Galactic disk, and obtained a flat magnetic energy spectrum for scales between 500 pc and 15 kpc, which is different from the Kolmogorov spectrum at small-scales and should constrain the theoretical simulations (e.g. Balsara & Kim 2005) on how the Galactic magnetic fields were generated and maintained.

4. The connection of magnetic fields on large and small scales

On one hand, molecular clouds were formed by contraction of diffuse gas in the interstellar medium, and the magnetic fields are so enhanced that they have the same energy as the kinetic energy (Crutcher 1999). Magnetic fields of clouds have been observed to have an hourglass shape, which is an indication of field direction conservation in the ISM during the contraction (see Fig. 2 and Beck *et al.* 1991). On the other hand, magnetic fields in the shell of supernova remnants are also enhanced in ISM due to supernova explosions, as can be seen from the polarized emission of remnants. Magnetic fields at this scale, i.e. the energy injection scale into ISM from the kinetic, should be the strongest. On average, random magnetic fields are stronger than the large-scale fields (Rand & Kulkarni 1989, Han & Qiao 1994, Heiles 1996b, Han *et al.* 2004).

However, evidence emerges for a possible relationship between the field directions of large-scale and small-scale magnetic fields. Recently, the observations (Li *et al.* 2006) have shown that the majority *JinLin do you mean 'majority' or something else?* of magnetic field orientations of molecular clouds seem to be preferentially parallel to the Galactic plane. From the radio observations of supernova remnants, the bilateral remnants always tend to have the longer axis parallel to the Galactic plane (Gaensler 1998). This could be understandable given the fact that the Galactic magnetic fields are confined to the Galactic plane, and the strongest component of the Galactic magnetic fields is the azimuthal component, much stronger than the radial and vertical components (Han & Qiao 1994, Han *et al.* 1999).

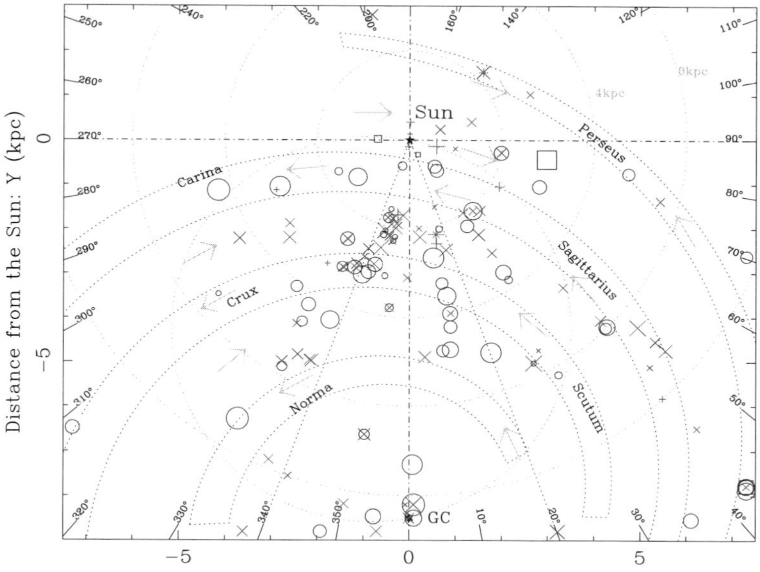

Figure 3. The medians of field measurements from Zeeman splitting of OH masers (cross and circles) in 137 objects or HI or OH lines of 17 molecular clouds (plus and squares) projected onto the Galactic plane, with the rough indication of spiral arms and the magnetic field directions (arrows) derived from pulsar RM data. The linear sizes of the symbols are proportional to the square root of the field-strength values, with limits of 0.5 and 9 mG for fields from the median maser fields and of 15 μG and 270 μG for median cloud fields. The crosses or pluses on the right (0 deg $< l <$ 180 deg) represent positive B, i.e. the field direction going away from observer, and circles or squares going towards us. The symbols on the left (180 deg $< l <$ 360 deg) are reversed, so that all crosses and pluses are consistent with the clockwise fields viewed from the Northern Galactic pole, and all circles and squares with counterclockwise fields. See Han & Zhang (2007) for details.

If the magnetic fields in molecular clouds are preserved from the permeated magnetic fields in the interstellar medium during cloud formation, then the question arises as to whether or not the magnetic fields in molecular clouds can still "remember" the large-scale magnetic fields in the interstellar medium? Are they sufficiently strong that their correlation with the large-scale fields was not destroyed by turbulence in clouds? If the answer is yes, then observations of cloud fields could be an independent approach to reveal the large-scale structure of the Galactic magnetic fields.

Han & Zhang (2007) collected measurements of the magnetic fields in molecular clouds, including Zeeman splitting data of OH masers in clouds of HII regions and OH or HI absorption or emission lines of clouds themselves. The data show clear structures in the sign distribution of the median of line-of-sight components of the magnetic field (see Fig. 3). Compared to the large-scale Galactic magnetic fields derived from pulsar RMs, the sign distribution of the Zeeman data shows similar large-scale field reversals. We conclude that the magnetic fields in the clouds may still "remember" the directions of magnetic fields in the Galactic ISM to some extent, and could be used as complementary tracers of the large-scale Galactic magnetic structure.

How can such coherent magnetic field directions occur from low density ($\sim 1\mathrm{cm}^{-3}$) to higher density ($\sim 10^3\mathrm{cm}^{-3}$) ISM clouds, even to the highest density maser regions ($\sim 10^7\mathrm{cm}^{-3}$), after a density compression of about 3, or even 10, orders of magnitude? One implication of this result is that the clouds probably do not rotate much after they are formed as otherwise the field directions of clouds we measured would be random. During the process of star formation, the clouds seem to be too heavy to be rotated, although there are jets or disks from newly formed stars which may have some dynamic effects. Furthermore, the fields in the molecular clouds are strong enough after the contraction that the turbulence in the clouds cannot significantly alter the magnetic field status.

5. Conclusions

Magnetic fields have been observed on all scales in our Galaxy. However, our knowledge of Galactic magnetic fields and their impact on ISM physics is far from complete. Large-scale magnetic fields in some regions have been delineated by enriched pulsar RM data(Han *et al.* 2006). However, large-scale magnetic fields in many regions remain to be measured with more pulsar RM data or probed by extragalactic radio sources (Brown *et al.* 2007). Small-scale fields have been measured in several ways: from polarization surveys of the Galactic plane (tens pc or hundreds pc), polarization mapping of clouds and supernova remnants (pc or tens pc), using the structure function of RMs (pc or tens pc), and Zeeman splitting observations of line emissions (AU to pc). Although small-scale magnetic fields appear as random or "noise" on the larger-scale, and are stronger than the large-scale magnetic fields, observational evidence already shows some physical connections between the small scale magnetic fields and the large-scale magnetic fields.

Acknowledgements

I am very grateful to colleagues who have collaborated with me: Dr. R.N. Manchester from Australia Telescope National Facility, CSIRO, Prof. G.J. Qiao from Peking University (China), Prof. Andrew Lyne from Jodrell Bank Observatory (UK), and Dr. Katia Ferriére from Observatory of Midi-Pyrénées (France), Dr. JiangShui Zhang from GuangZhou University (China), Dr. Willem van Straten from Swinborne University (AU). The author is supported by the National Natural Science Foundation of China (10521001 and 10473015).

References

Andreasyan, R. R. & Makarov, A. N. 1989, Afz, 31, 247

Balsara, D. & Kim, J. 2005, ApJ 634, 390

Beck R. 2005, In: Magnetic Fields in the Universe: From Laboratory and Stars to Primordial Structures, AIP 784, p.343

Beck, R., Berkhuijsen, E. M., & Bajaja, E., 1991, IAU Sypm. 146, p.209

Beck R., Brandenburg A., Moss D., Shukurov A. M., & Sokoloff D. D., 1996, ARA&A, 34, 155

Boulares, A. & Cox, D. 1990, ApJ, 365, 544

Brown, J. C., Taylor, A. R., & Jackel, B. J. 2003, ApJS, 145, 213

Brown, J. C., Haverkorn, M., Gaensler, B. M., & *et al.* 2007, ApJ, in press

Chuss, D. T., Davidson, J. A., Dotson, J. L., & *et al.* 2003, ApJ, 599, 1116

Clegg, A. W., Cordes, J. M., Simonetti, J. M., Kulkarni, S. R. 1992, ApJ, 386, 143

Crutcher, R. M. 1999, ApJ, 520, 706
Crutcher, R. M. 2007, IAU Symp.242, this volumn.
Crutcher, R. M., Nutter, D. J., Ward-Thompson, D., & Kirk, J. M. 2004, ApJ, 600, 279
Duncan, A. R., Haynes, R. F., Jones, K. L., & Stewart, R. T. 1997, MNRAS, 291, 279
Duncan, A. R., Haynes, R. F., Reich, W., Reich, P., & Gray, A. D. 1998, MNRAS, 299, 942
Duncan, A. R., Reich, P., Reich, W., & Fürst, E. 1999, A&A, 350, 447
Fish, V. L., Reid, M. J., Agron, A. L. & Menten, K. M., 2003, ApJ, 596, 328
Gaensler, B. M. 1998, ApJ, 493, 781
Gaensler, B. M., Dickey, J. M., McClure-Griffiths, N. M., & *et al.* 2001, ApJ, 549, 959
Hamilton, P. A. & Lyne, A. G. 1987, MNRAS, 224, 1073
Han, J. L. & Qiao, G. J. 1994, A&A, 288, 759
Han, J. L., Ferriere, K., & Manchester, R. N. 2004, ApJ, 610, 820
Han, J. L., Manchester, R. N., Berkhuijsen, E. M., & Beck, R. 1997, A&A, 322, 98
Han, J. L., Manchester, R. N., & Qiao, G. J. 1999, MNRAS, 306, 371
Han, J. L., Manchester, R. N., Lyne, A. G., & Qiao, G. J. 2002, ApJ, 570, L17
Han, J. L., Manchester, R. N., Lyne, A. G., Qiao, G. J., & van Straten, W. 2006, ApJ, 642, 868.
Han, J. L., & Zhang, J. S. 2007, A&A, 464, 609
Haverkorn, M., Gaensler, B. M., Brown, J. C., *et al.* 2006, ApJ 637, L33
Heiles, C. 1996a, ApJ, 462, 316
Heiles, C. 1996b, in ASP Conf. Ser. 97: Polarimetry of the Interstellar Medium, 457
Heiles, C, & Crutcher, R. 2005, In: Cosmic Magnetic Fields, LNP 664, 137
Haynes, R. F., Stewart, R. T., Gray, A. D., & *et al.* 1992, A&A, 264, 500
Hutawarakorn, B. & Cohen, R. J., 2005, MNRAS, 357, 338
Indrani, C. & Deshpande, A. A. 1999, New Astronomy, 4, 33
Junkes, N., Fürst, E., & Reich, W. 1987, A&AS, 69, 451
LaRosa, T. N., Nord, M. E., Lazio, T. J. W., & Kassim N. E. 2004, ApJ, 607, 302
LaRosa, T. N., Brogan C. L., Shore S. N., *et al.* 2005, ApJ, 626, L23
Li, H., Griffin, G. S., Krejny, M., & *et al.* 2006, ApJ, 648, 340
Lyne, A. G. & Smith, F. G. 1989, MNRAS, 237, 533
Mathewson, D. S. & Ford, V. L. 1970a, MemRAS, 74, 139.
Minter, A. H. & Spangler, S. R. 1996, ApJ, 458, 194
Morris, M., Uchida, K.,& Do, T. 2006, Nature, 440, 308
Nord, M. E., Lazio, T. J. W., Kassim, N. E., *et al.* 2004, AJ, 128, 1646
Novak, G., Chuss, D. T., Renbarger, T., & *et al.* 2003, ApJ, 583, L83
Prouza, M. & Smída, R. 2003, A&A 410, 1
Qiao, G. J., Manchester, R. N., Lyne, A. G., & Gould, D. M. 1995, MNRAS, 274, 572
Rand, R. J. & Kulkarni, S. R. 1989, ApJ, 343, 760
Reich, W. 2007, in: Cosmic Polarization, (astro-ph/0603465), p. 91
Reich, W., Fürst, E., Reich, P., *et al.* 2002, in AIP Conf. Proc. 609: p. 3
Roy, S., Rao, A. P., Subrahmanyan, R. 2005, MNRAS 360, 1305
Simard-Normandin, M. & Kronberg, P. P. 1980, ApJ, 242, 74
Sun, X. H. & Han, J. L. 2004, in: Magnetized Interstellar Medium, p. 25
Sun, X. H., Han, J. L., Reich, W., & *et al.* 2007, A&A 463, 993
Tinyakov, P. G. & Tkachev, I. I. 2002, Astropart. Ph. 18, 165
Vlemming, W. 2007, IAU Symp. 242, this volume
Weisberg, J. M., Cordes, J. M., Kuan, *et al.* 2004, ApJS, 150, 317
Xu, J. W., Han, J. L., Reich, W., & *et al.* 2007, A&A, in press
Yusef-Zadeh, F., & Morris, M. 1987, ApJ, 320, 545
Yusef-Zadeh, F., Hewitt, J. M., & Cotton, W. 2004, ApJS, 155, 421

Astrophysical Masers and their Environments
Proceedings IAU Symposium No. 242, 2007
J.M. Chapman & W.A. Baan, eds.

doi:10.1017/S1743921307012574

VLBI OH maser polarimetry with the Australian Long Baseline Array: the star-forming region G340.054–0.244

I. Bains[1,2], J. Caswell[3], A. M. S. Richards[4], C. Phillips[3], S. Tingay[1], B. Hutawarakorn Kramer[5], R. J. Cohen[4] and M. Cunningham[2]

[1]Centre for Astrophysics and Supercomputing, Swinburne University of Technology, PO Box 218, Hawthorn, Victoria 3122, Australia
email: ibains@astro.swin.edu.au
[2]School of Physics, University of New South Wales, Sydney, NSW 2052, Australia
[3]Australia Telescope National Facility, CSIRO, PO Box 76, Epping, NSW 1710, Australia
[4]Jodrell Bank Observatory, Department of Physics and Astronomy, University of Manchester, Macclesfield, Cheshire, SK11 9DL, UK
[5]National Astronomical Reseach Institute of Thailand, Physics Building, Chiang Mai University 502000, Thailand

Abstract. Until recently, high spatial resolution full Stokes maser polarimetry was the sole domain of northern interferometers and a wealth of sources in the far south remained unexplored due to a lack of suitable instrumentation having both high spatial and high velocity resolution. The Australia Telescope Long Baseline Array (LBA) has now switched to disk-based software correlation, permitting full Stokes observing in spectral line mode with velocity channels which are sufficiently narrow to sample usefully the polarization structure. To illustrate the utility of this valuable addition to radio astronomy, we present preliminary results of the first such polarimetric observation, the subject of which are the OH masers in the star-forming region G340.054–0.244.

Keywords. masers – polarization – techniques: interferometers – techniques: high angular resolution – stars: formation – ISM: individual (G340.054–0.244) – ISM: magnetic fields – ISM: molecules – radio lines: ISM

1. Introduction

Emission from mainline 1665- and 1667-MHz OH maser transitions commonly occurs in regions associated with high mass star formation (e.g. Caswell 1998). Maser polarization is thought to arise due to Zeeman splitting of the energy levels by a magnetic field (e.g. Elitzur 1996). Measuring the magnetic field strength in regions of star formation enables a determination of the available magnetic energy that can provide support against gravitational collapse (the 'mass-to-magnetic flux ratio'; Crutcher 1999). In order to study magnetic fields with maser polarimetry, high spatial resolution, such as that provided by VLBI observations, is required to verify that the apparent signatures of Zeeman splits in Stokes V spectral profiles do indeed arise from a single maser feature.

2. Observations

These observations utilized four stations of the LBA: Parkes, Mopra, Hobart and the ATCA in tied array mode. The maximum baseline length of 1400 km provided an angular resolution of 35 mas at the observing wavelength of 18 cm. The Swinburne software correlator (Deller *et al* 2007) was configured with 8192 channels over a 4 MHz bandwidth and the integration time was 2 s.

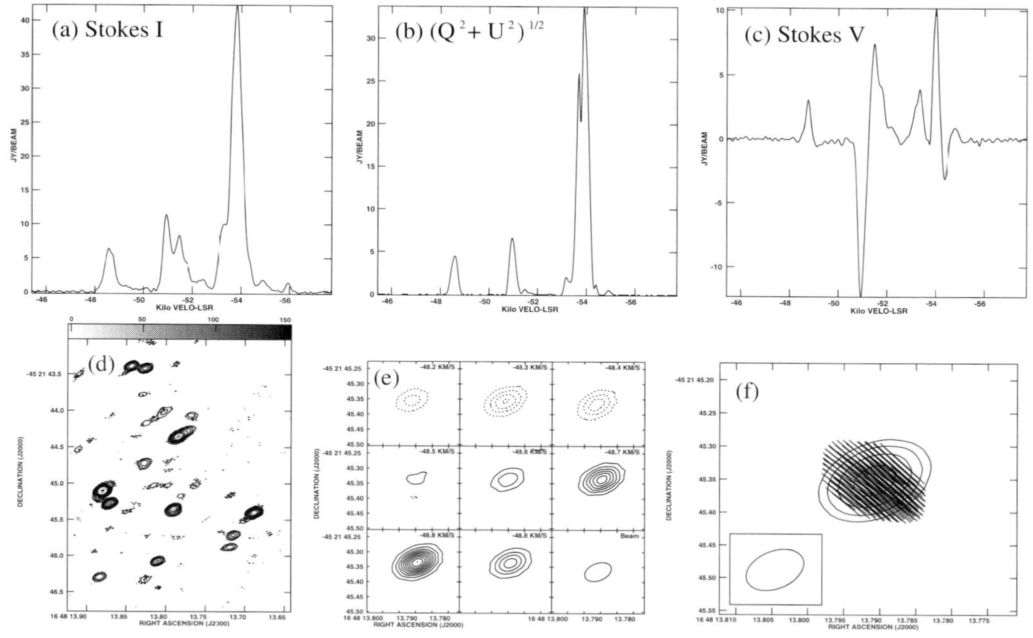

Figure 1.

3. Results

Figs. 1a-c show the 1665-Mz velocity profiles of the total intensity, linearly-polarized and circularly-polarized emission (with velocity increasing to the left). Figs. 1a & 1b show that linear polarization exceeding 50% is present in several strong features. The high resolution imagery has revealed that the two most prominent apparent signatures of Zeeman splitting within the Stokes V spectrum (given by the smooth S-shaped profiles) are in fact due to separate +V and -V components that align in velocity but not in space. Fig. 1d shows the total intensity image summed over the full velocity extent of the 1665-MHz emission. At the near kinematic distance of 4.6 kpc, the $3''$ angular extent of the maser region corresponds to 14000 AU. Fig. 1e shows the Stokes V contoured channel maps of a 1665-MHz Zeeman σ-component pair, which is associated with a single Stokes I feature. Dashed contours indicate negative flux density. The velocity split corresponds to a magnetic field strength of -1 mG, which is directed towards the Earth. Fig. 1f shows an example of a linearly polarized maser spot. Linear polarization vectors are overlaid on the total intensity contours; vector length is proportional to linearly polarized flux density, with a peak of 1 Jy beam^{-1}. The magnetic field direction in the sky plane is perpendicular to the polarization angle.

These observations represent the first full-polarimetric spectral line experiment performed with the LBA and serve to illustrate that interferometric full Stokes maser polarimetry is now possible for far southern sources.

References

Caswell, J. 1998, *MNRAS* 297, 215
Crutcher, R. 1999, *ApJ* 520, 706
Deller, A. T., Tingay, S. J., Bailes, M., & West, C. 2007, *PASP* 119, 318
Elitzur, M. 1996, *ApJ* 457, 415

Astrophysical Masers and their Environments
Proceedings IAU Symposium No. 242, 2007
J.M. Chapman & W.A. Baan, eds.

doi:10.1017/S1743921307012586

Possible CO clouds interacting with the SNRs G21.8-0.6 and G32.8-0.1

Jianjun Zhou[1], Jarken Esimbek[1], Xizheng Zhang[2] and Hongbo Zhang[2]

[1]National Astronomical Observatories/Urumqi Observatory, CAS, China, 40-5 South Beijing Road, Urumqi Xinjiang 830011, China; [2] National Astronomical Observatories, CAS, China, 20A, Datun Road, Chaoyang Dis, Beijing 100012.

Abstract. We have obtained the first CO(1-0) maps towards the SNRs G21.8-0.6 and G32.8-0.1, which are associated with OH 1720 MHz masers. Based on the morphological correspondence between the radio remnant and the CO cloud, and the velocity and position agreement between the OH maser and CO clouds, we tentatively suggest the clouds may be interacting with G21.8-0.6 and G32.8-0.1. However, the shock-excited line broadening was not obvious for these two SNRs. We discuss the possible reasons for this.

Keywords. ISM:clouds — ISM: individual (G21.8-0.6, G32.8-0.1) — ISM:molecules — masers — supernova remnants

---

OH 1720 MHz masers are a known signature of SNR-molecular cloud interactions. The excitation of OH 1720 MHz maser emission requires very dense gas ($\sim 10^5 cm^{-3}$) with a temperature of 50−125K. A C-type shock caused by expanding supernova remnant will create such conditions (Elitzur 1976; Lockett *et al.* 1999). This has been confirmed by many molecular line observations (Koralesky *et al.* 1998; Reach & Rho 1998; Frail & Mitchell 1998; Reynoso *et al.* 2000). We took CO(1-0) observations towards the SNRs G21.8-0.6 and G32.8-0.1, both of which are associated with 1720 MHz OH masers. The sensitivity of the observation is 0.26 K (rms) for an integration time of 60 seconds. The results indicate that ^{13}CO(1-0) emission of these two SNRs is very weak, while no C^{18}O(1-0) was detected. The CO clouds around these two SNRs are displayed in Figure 1 and Figure 2.

Based on the morphological signature of CO clouds, and position and velocity agreement between the OH maser and CO clouds, we tentatively suggest that cloud C of G21.8-0.6 and cloud F of G32.8-0.1 may be physically related to the SNR. However, we did not detect line broadening in cloud F, and line broadening is not very strong in cloud C. A possible reason for this is that the shocked molecular gas may be highly excited, so the emission from the low-lying CO(1-0) transition is weak (Wilner *et al.* 1998). Moreover, the background cloud emission could mask such weak features. Taking into consideration the complex CO spectra, we think that confusion due to different velocity components along the line of sight plays an important role here.

Acknowledgements

This work was funded by the National Natural Science Foundation of Chinese through Grant 10203003 and the Program of the Light in China's Western Region LCRW under No.RCPY200605 and ZZBS200601.

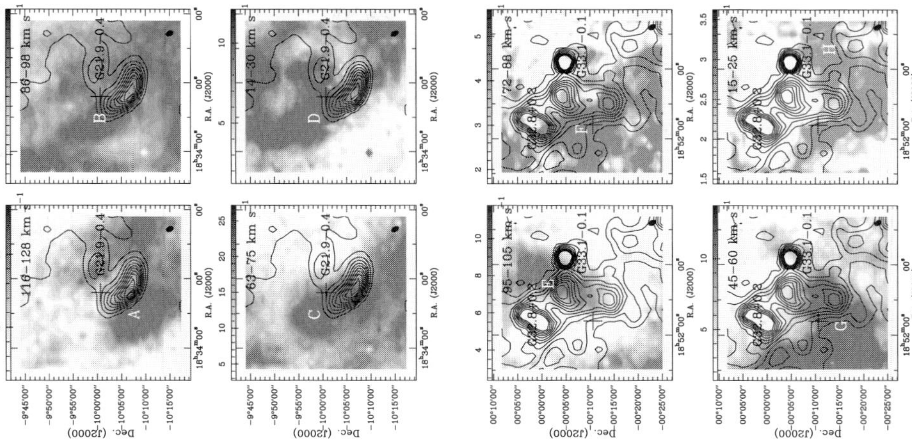

Figure 1. Left: Set of four channel maps of ^{12}CO(1-0) emission integrated over velocity ranges 116-128, 86-98, 63-75 and 14-30 km s^{-1} toward G21.8-0.6. The grey scale is in K km s^{-1} and is shown on top of the image. The 2σ noise level of these maps are 1.25, 1.65, 2.18, and 1.29 K km s^{-1} respectively. The beam, $106'' \times 70''$, is indicated by black filled ellipse in the bottom right corner. The contour map was obtained from the radio continuum survey of the Galactic plane at 11 cm wavelength(Reich *et al.* 1984). G21.9-0.4 is a HII region. The plus sign shows the position of OH 1720 MHz maser.

Right: Set of four channel maps of ^{12}CO(1-0) emission integrated over the velocity ranges 95-105, 72-88, 45-60 and 15-25 km s^{-1} toward G32.8-0.1. The grey scale is in K km s^{-1} and is shown on top of the images. The 2σ noise level of these maps are 1.46, 1.11, 1.53 and 0.99 K km s^{-1} respectively. The beam, $106'' \times 70''$, is indicated by black filled ellipse in the bottom right corner. the continuum contour of G32.8-0.1 was also obtained from the radio continuum survey of the Galactic plane at 11 cm wavelength (Reich *et al.* 1984), which including two HII regions, G32.8+0.2 and G33.1-0.1. The plus sign indicates the position of OH maser.

References

Elitzur M., 1976, ApJ, 203, 124
Frail D.A., Mitchell G.F., 1998, ApJ, 508, 690
Koralesky B., Frail D.A., Goss W.M., Claussen M.J., & Green A.J., 1998, AJ, 116, 1323
Lockett T.L., Gauthier E., Elitzur M., 1999, ApJ, 511, 235
Reach W.T., Rho J., 1998, ApJ, 507, L93
Reich W., Fürst E., Steffen P., Reif K., Haslam C.G., 1984, A&AS, 58, 197
Reynoso E.M., Mangum J.G., 2000, ApJ, 545, 874
Wilner D.J., Reynoso S.P., Moffett D.A., 1998, ApJ, 115, 247

Astrophysical Masers and their Environments
Proceedings IAU Symposium No. 242, 2007
J.M. Chapman & W.A. Baan, eds.

doi:10.1017/S1743921307012598

Maser science at Tidbinbilla

Shinji Horiuchi[1] **and Jim Lovell**[2]†

[1]Canberra Deep Space Communication Complex, Raytheon Australia, PO Box 1035, Tuggeranong, ACT 2901, Australia
email: shoriuchi@cdscc.nasa.gov

[2]CSIRO Australia Telescope National Facility and CSIRO Industrial Physics, NSW, Australia
email: jim.lovell@utas.edu.au

Abstract. The 70 m antenna at Canberra Deep Space Communication Complex (Tidbinbilla) is the largest and most sensitive dish in the southern hemisphere, working at 1, 3, 13 and 18 cm, and as such is in high demand from astronomers both in Australia and overseas. In this paper we present the current status of the single dish spectroscopy system and highlight some recent results in maser science.

1. 12 mm system and spectrometer

The Canberra Deep Space Communications Complex (CDSCC) at Tidbinbilla is part of NASA's Deep Space Network (DSN) and is home to the largest and most sensitive single dish in the southern hemisphere. This 70-m antenna is primarily used for spacecraft tracking but, as part of the Host Country agreement with NASA, a fraction of time on the antenna is allocated for independent scientific activities sponsored by the Australian Government. This time allocation, typically 300 hours per year, is divided into single-dish and VLBI observing. Access to this time is obtained via proposal through the ATNF and is open to anyone.

Of particular interest is the 12 mm (18–26 GHz) system. With a SEFD of 60 Jy it is 10 times more sensitive than the current Parkes system and when the Parkes 12 mm receiver is upgraded will still be 3.5 times more sensitive. The 12 mm system at Tidbinbilla currently operates with an on-axis feed and provides dual-polarisation output with an instantaneous available bandwidth of 600 MHz. An upgrade is currently underway to provide a simultaneous off-axis beam to allow simultaneous on- and off-source spectra to be taken, effectively halving the required observing time for pointed observations.

Currently Tidbinbilla is equipped with a Parkes Multibeam correlator block which allows two IFs to be correlated at a maximum bandwidth of 64 MHz. For more information see http://www.atnf.csiro.au/observers/tidbinbilla. It is hoped that a new spectrometer will soon be obtained to provide sufficient bandwidth and channelisation to handle all four IFs from the upgraded front-end.

2. Galactic masers

Ellingsen *et al.* (2004) searched for 19.9 GHz methanol masers in star-forming regions using the 70 m telescope and found six new sources. The results indicate a strong correlation between 19.9 GHz methanol masers and the Class II methanol masers associated with 6035 MHz OH masers.

† Present address: Department of Maths and Physics, University of Tasmania, Private Bag 21, Hobart, Tasmania 7001, Australia

Valdettaro *et al.* (2007) performed a high sensitivity survey of water masers in bright rimmed clouds compressed by ionization fronts from nearly massive stars. They found masers from seven bright rimmed interstellar clouds out of 44 such regions. The low detection rate suggests that low-mass star formation is the most natural outcome of such external compression.

Deacon *et al.* (2007) observed a sample of 85 post-Asymptotic Giant Branch (post-AGB) candidate stars, selected on the basis of their OH 1612 MHz and far-infrared properties. New detections of water masers were identified for 21 stars. The source b292 is a young post-AGB star that is highly likely to be a 'water fountain' source with highly collimated, high velocity (up to 210 km/s) water masers. (See Chapman *et al.* in these proceedings.)

3. Extragalactic masers

Greenhill *et al.* (2003a) discovered water maser emission in seven AGN using a custom-built 4096 channel spectrometer with a bandwidth of 5300 km/s. A continuation of the search was made, resulting in new detection in eight more AGN (Kondratko *et al.* 2006). Greenhill *et al.* (2003a) also discovered high velocity components in the Seyfert nucleus of the Circinus galaxy.

Tarchi *et al.* (2007, in preparation) have been searching for water maser emission toward a sample comprised of all galaxies with Dec. < 30 deg with IRAS 100 micron fluxes > 50 Jy in order to confirm the relation between FIR flux and water maser detection rate found in the northern hemisphere (Henkel *et al.* 2005). While this survey is still on going, so far they have detected water masers in one of the sources in the sample.

Greenhill *et al.* (2003b) presented water maser VLBI maps of the the Circinus galaxy obtained by the Australian Long Baseline Array (LBA) in which the Tidbinbilla 70m played a key role as the most sensitive antenna in the array. The outflowing masers are irregularly distributed above and below a warped, edge-on accretion disk (maximum rotation speed 260 km/s), with relative outflow velocities up to $\pm$ 160 km/s, projected along the line of sight.

Greenhill *et al.* (1997) have discovered that the water maser lines in the Circinus galaxy vary on time scales as short as a few minutes. Follow up observations were made at Tidbinbilla by McCallum *et al.* (2005) and the same phenomena were confirmed. McCallum *et al.* (2005) showed that this rapid variability can be explained by interstellar scintillation, similar to that seen in a significant number of flat-spectrum AGN.

References

Deacon, R. M., Chapman J. M., Green, A. J., Sevenster, M.N 2007, *ApJ* 658, 1096

Ellingsen, S. P., Cragg, D. M., Lovell, J. E. J., Sobolev, A. M., Ramsdale, P. D., Godfery, P. D. 2004, *MNRAS* 354, 401

Greenhill, L. J., Ellingsen, S. P., Norris, R. P., Gough, R. G., Sinclair, M. W., Moran, J. M., Mushotzky, R. 1997, *ApJ* 464, L103

Greenhill, L. J., Kondratko, J. E., Lovell, J. E. J., Kuiper, T. B. H., Moran, J. M., Jauncey, D. L., Baines, G. P. 2003, *ApJ* 582, L11

Greenhill, L. J., Booth, R. S., Ellingsen, S. P., Herrnstein, J. R., Jauncey, D. L., McCulloch, P. M., Moran, J. M., Norris, R. P., Reynolds, J. E., Tzioums, A. K. 2003, *ApJ* 590, L162

Kondratko, P. T., Greenhill, L. J., Moran, J. M., Lovell, J. E. J., Kuiper, T. B. H., Jauncey, D. L., Cameron, L. B., Gómez, J. F., García-Miró, C., Moll, E., de Gregorio-Monsalvo, I., Jiménez-Bailón, E. 2006, *ApJ* 638,100

McCallum, J. N., Ellingsen, S. P., Jauncey, D. L., Lovell, J. E. J., Greenhill, L. J. 2005, *AJ* 129, 1231

Valdettaro, R., Chapman, J. M., Lovell, J. E. J., Palla, F. 2007, *A&A* 466, 247

Photos (clockwise from top left): James Moran, James Moran, Andrej Sobolev, Miller Goss

Top row: Professor Xingwu Zheng with Associate Professor Zhaofen Ling (left)
Vicki Drazenovic, Todd Hunter and Crystal Brogan (right)
Bottom row: James and Sheena Caswell (left),
Michael Gaylard and Leszek Blaszkiewicz (right)

Astrophysical Masers and their Environments
Proceedings IAU Symposium No. 242, 2007
J.M. Chapman & W.A. Baan, eds.

doi:10.1017/S1743921307012604

Masers and star formation

Vincent L. Fish

Jansky Fellow, National Radio Astronomy Observatory, 1003 Lopezville Rd., Socorro, NM 87801, USA
email: vfish@nrao.edu

Abstract. Recent observational and theoretical advances concerning astronomical masers in star forming regions are reviewed. Major masing species are considered individually and in combination. Key results are summarized with emphasis on present science and future prospects.

Keywords. masers — stars: formation — radio lines: ISM — ISM: molecules — ISM: jets and outflows — ISM: kinematics and dynamics

1. Introduction

This review summarizes maser results pertinent to star formation appearing in the literature since the last maser meeting (IAU Symposium 206). References are drawn from recent literature when possible.

2. Masing species

2.1. *Water (H_2O)*

The 22.235 GHz water line is the predominant water maser line. Masers in this transition are very bright, easily observable, and inverted under a wide range of conditions (e.g., Babkovskaia & Poutanen 2004). Several millimeter and submillimeter transitions of water are also seen as masers. Discussion of these transitions can be found in the section of these proceedings devoted to millimeter and submillimeter masers.

Water masers are frequently seen in outflows from both high-mass and low-mass YSOs (Honma *et al.* 2005; Moscadelli, Cesaroni, & Rioja 2005; Goddi & Moscadelli 2006; Moscadelli *et al.* 2006). These jets are seen in deceleration (Imai *et al.* 2002) and often have substructure on AU scales (Torrelles *et al.* 2003; Furuya *et al.* 2005; Uscanga *et al.* 2005).

Water masers are sometimes believed to trace disks as well as outflows (Seth, Greenhill, & Holder 2002; Gallimore *et al.* 2003), possibly excited by an expanding shock wave. Indeed, shocks likely are responsible for arc-like maser distributions (Honma *et al.* 2004) and may excite masers in accreting material as well (Menten & van der Tak 2004). Water masers appear in Bok globules (Gómez *et al.* 2006), likely tracing bipolar molecular outflows (de Gregorio-Monsalvo *et al.* 2006), as well as bright rimmed clouds (Valdettaro *et al.* 2005), again likely associated with outflows (Urquhart *et al.* 2006). The location of water masers near the ionization front of large H II regions provides evidence in support of triggered star formation (Healy, Hester, & Claussen 2004). The common thread of all these environments is the existence of energetic shocks, which fits with conventional wisdom regarding water maser pumping.

Despite the small Zeeman splitting coefficient of water, line-of-sight magnetic fields of tens to hundreds of milligauss have been measured in star forming regions via water maser Zeeman splitting (Sarma *et al.* 2002; Vlemmings *et al.* 2006), although direct interpretation of the Stokes V profile as a magnetic field strength may be in error by up to a factor of two depending on local velocity and magnetic field gradients (Vlemmings 2006). Linear polarization observations of water masers can provide information on the orientation of the magnetic field in the plane of the sky. The hourglass morphology of the magnetic field in W3 IRS 5 appears to be due to the processes of collapse rather than a result of the outflow traced by water masers (Imai *et al.* 2003). Further interpretation of water maser polarization can be found in the review by Wouter Vlemmings.

Water masers have a fractal distribution over 4 orders of magnitude in spatial scale, possibly indicating that they appear at the turbulent dissipation scale (Strelnitski *et al.* 2002; Ripman & Strelnitski 2006; Strelnitski *et al.* 2007). Turbulence may also be responsible for variability, including changes in the line-of-sight velocity of individual components, of the water masers in some sources (Lekht *et al.* 2006a). Larger-scale variations may contribute as well, such as changes in outflow parameters or cyclic variability of the central star (Pashchenko & Lekht 2005; Lekht *et al.* 2006b). An ordered structure is inferred in W31(2) from successive flaring of features at different velocities (Lekht, Munitsyn, & Tolmachev 2005).

2.2. *Methanol (CH_3OH)*

Methanol masers divide into two categories, known as class I and class II, based on their propensity for certain transitions to produce masers while others are seen in absorption. Traditionally, class I and class II masers do not mix (e.g., Ellingsen 2005); however, fine-tuned conditions may rarely excite lines from both classes simultaneously, as appears to be the case in OMC-1 (Voronkov *et al.* 2005). For purposes of this review, class I and class II methanol masers will be treated separately. Improved laboratory data for rest frequencies have been obtained for many methanol transitions in both classes (Müller, Menten, & Mäder 2004).

2.2.1. *Class I*

Class I masers are primarily collisionally pumped. They are typically found in younger sources that are Class II masers and may trace distant parts of outflows interacting with dense molecular gas (Beuther *et al.* 2005; Ellingsen 2006). Comparison of interferometric maps of the 9.9 and 104.3 GHz transitions with H_2 data confirms the outflow association in IRAS 16547−4247 (Voronkov *et al.* 2006). Linear polarization suggests that Class I masers may appear in oblique shocks parallel to the outflow axis and perpendicular to the magnetic field in OMC-2 (Wiesemeyer, Thum, & Walmsley 2004), although interpretation of methanol polarization may be complicated (e.g., Elitzur 2002).

Many different Class I transitions have been observed. Weak masers at 84.5 and 95.2 GHz to the southwest of W3(OH) provide strong evidence in support of collisional pumping and allow for physical conditions to be inferred (Sutton *et al.* 2004). The latter transition is commonly seen as a maser in both Class I and Class II sources (Minier & Booth 2002). Strong 36 GHz maser emission is believed to be an indicator of an early evolutionary stage, as may line ratios of other transitions (Gillis, Pratap, & Strelnitski 2005; Hoffmann, Pratap, & Strelnitski 2006). The intensity ratio of highly-excited 146.6 and 156.8 GHz methanol masers may be a sensitive probe of density and temperature in masing regions (Lemonias, Strelnitski, & Pratap 2006). Short-timescale variability is seen in the 44 and 146.6 GHz transitions (Pratap, Hoffmann, & Strelnitski 2006).

2.2.2. *Class II*

The key Class II maser transitions are at 6668 and 12178 MHz. Class II masers are often found in an earlier evolutionary stage than ultracompact (UC) H II regions (e.g., Minier *et al.* 2005), but observations of methanol masers cospatial with both millimeter and centimeter continuum emission indicate that Class II masers appear over a wide range of evolutionary stages (Pestalozzi *et al.* 2006). While the lower stellar mass limit for methanol masers is still a subject of research, 6.7 GHz masers do not appear below approximately 3 solar masses (Minier *et al.* 2003).

Linear structures of masers with organized velocity structures have led some to conclude that methanol masers often trace edge-on disks (Norris *et al.* 1993, 1998). Observations of maser proper motions (Minier *et al.* 2000), shocked H_2 (De Buizer 2003), and SiO (De Buizer *et al.* 2006) indicate that, in the majority of cases at least, methanol masers are aligned with an outflow, not a disk. These structures may be explained by propagation of a shock front into a region with large-scale velocity structure, such as rotation (Dodson, Ojha, & Ellingsen 2004). Disk candidate sources remain (Slysh *et al.* 2002b; Pestalozzi *et al.* 2004; Pillai *et al.* 2006), although these, too, may turn out to be associated with outflows when studied at high resolution in the mid infrared (e.g., De Buizer & Minier 2005). The conclusion to be drawn is that a linear distribution of masers with a velocity gradient does not by itself present convincing evidence that the masers trace an edge-on disk. An intriguing variant is the possibility of methanol masers tracing a face-on disk in G23.657−0.127 (Bartkiewicz, Szymczak, & van Langevelde 2005).

There is evidence to support the hypothesis that most methanol masers are tracing shocked regions, often in the presence of outflows. Methanol masers appear preferentially near radio sources with a spectral index indicative of an outflow (Zapata *et al.* 2006). Mid-infrared images of some sources indicate that masers are found along the shocked material on the surface of an outflow cavity (De Buizer 2006, 2007). In some sources, methanol masers appear near but offset from UCH II regions, suggesting that they appear in the shocked molecular gas outside the ionization front, similar to hydroxyl masers (Phillips & van Langevelde 2005).

The 6.7 GHz transition is a popular line for Galactic maser surveys. Several surveys were reported on during IAU Symposium 206. Details of the Arecibo and Parkes multi-beam 6.7 GHz surveys can be found in the section of these proceedings devoted to Galactic maser surveys. Unsurprisingly, their distribution correlates well with Galactic structure (Pestalozzi, Minier, & Booth 2005; Pestalozzi *et al.* 2007). The 12.2 GHz line, when it occurs, is almost always weaker than 6.7 GHz emission (Błaszkiewicz & Kus 2004). Both lines have also been the subject of monitoring studies (e.g., Goedhart, Gaylard, & van der Walt 2005), which find variability in a large fraction of sources including periodic variability and a time delay between features possibly due to light travel time (Goedhart, Gaylard, & van der Walt 2005; Goedhart *et al.* 2005). Based on comparison of spectra over a period of a decade, the lifetime of an individual 6.7 GHz maser feature is about 150 years (Ellingsen 2007), while the lifetime of the 6.7 GHz maser phase in a source is a few $\times\ 10^4$ years (Codella *et al.* 2004; van der Walt 2005), similar to the lifetime of the OH maser phase (e.g., Fish & Reid 2006).

Numerous other Class II transitions have been observed. New maser sources have been found in rare transitions at 85.5, 86.6, and 107.0 GHz (Minier & Booth 2002; Ellingsen *et al.* 2003) and a torsionally-excited line at 44.9 GHz (Voronkov, Austin, & Sobolev 2002). Several weak maser lines near 165 GHz have also been detected (Salii & Sobolev 2006). Emission in the 19.9 GHz transition is usually weak and correlates well with 6035 MHz OH masers (Ellingsen *et al.* 2004). A search for 23.1 GHz emission resulted in no new detections beyond the previously known maser in NGC 6334F (Cragg *et al.* 2004). Observations of these less common methanol maser transitions can help constrain physical parameters in maser models. Improved collisional rate data has also allowed refinement of methanol models, which slightly affects predicted excitation conditions and brightness temperatures but not which transitions are expected to produce detectable Class II masers (Cragg, Sobolev, & Godfrey 2005).

2.3. *Hydroxyl (OH)*

Hydroxyl masers are usually studied in sources with associated UCH II regions (e.g., Fish *et al.* 2005) but are also found toward less evolved massive protostellar objects (Edris, Fuller, & Cohen 2007). A different class of OH masers is seen at the ends of the jet in the W3 TW object (Argon, Reid, & Menten 2003).

Hydroxyl masers around more evolved sources are usually seen in expansion (sometimes very rapid; see Stark *et al.* 2007) ahead of the ionization front of a UCH II region (Fish & Reid 2006). Sometimes the masers appear to trace a molecular disk or torus (Slysh *et al.* 2002a; Hutawarakorn & Cohen 2005; Edris *et al.* 2005; Nammahachak *et al.* 2006). Masers are often seen along arcs or filaments (Cohen *et al.* 2006), with extended filamentary emission especially common at 4.7 GHz (Palmer, Goss, & Devine 2003).

Multitransition overlaps are of special interest because of their ability to constrain physical conditions in models. The 4765 MHz line is observed to be the strongest line of the 6 cm triplet but is usually only weakly inverted and often spatially extended (e.g., Palmer, Goss, & Whiteoak 2004; Harvey-Smith & Cohen 2005). A histogram of 18 cm emission resembles 4.7 GHz lineshapes in W49A, suggesting that the high-gain 18 cm emission and low-gain 6 cm emission have similar velocity distributions, even if the 4660 MHz emission is spatially separate (Palmer & Goss 2005). Much is made of overlaps between 4765 and 1720 MHz masers (Palmer *et al.* 2003; Niezurawska *et al.* 2004, 2005). It should be noted that 6035 MHz maser emission correlates more strongly with 4765 MHz than does 1720 MHz, even if the velocities do not always agree (Dodson & Ellingsen 2002; Smits 2003). Masers in the 1720 MHz transition also appear to correlate with 1665 and 6035 MHz OH masers and 6.7 GHz methanol masers, at least to arcsecond accuracy (Caswell 2004a). Masers in the 6030 MHz transition are almost always accompanied by stronger emission at 6035 MHz (Caswell 2003), with excellent spatial coincidence and agreement of magnetic field strengths with each other and with 1665 MHz masers (Etoka, Cohen, & Gray 2005). Masers in the highly-excited 13441 MHz transition are rare but are always accompanied by 6035 MHz masers at the same velocity, although the intensities in the two transitions do not show a high degree of correlation (Baudry & Desmurs 2002; Caswell 2004c). The ground-state, satellite line transitions at 1612 and 1720 MHz are usually conjugate with respect to absorption and maser emission (Szymczak & Gérard 2004). Sources do exist in which both transitions are inverted, though not in direct spatial overlap (e.g., Wright, Gray, & Diamond 2004).

Magnetic fields as strong as 40 mG are seen in OH masers (Slysh & Migenes 2006; Fish & Reid 2007). Magnetic fields are highly ordered in star forming regions (Fish & Reid 2006) and support pictures in which the processes of star formation do not tangle field lines significantly. While magnetic field strengths are usually stable from epoch to epoch, monotonic decay of the field in a Zeeman group in Cep A continues to be observed (Bartkiewicz *et al.* 2005). High spectral resolution observations support the conventional assumption that the Zeeman splitting coefficient appropriate for $\sigma^{\pm 1}$ components should be assumed when measuring magnetic fields at 1612 and 1720 MHz (Fish, Brisken, & Sjouwerman 2006c). Linear polarization is of limited usefulness in determining the full, three-dimensional orientation of the magnetic field, likely due to a combination of Faraday rotation and anisotropic magnetohydronamic turbulence (Watson *et al.* 2004; Fish & Reid 2006).

Extreme variability is occasionally seen in OH. The 1665 MHz maser in W75 N briefly flared to nearly 1 kJy to become the brightest OH maser in the sky (Alakoz *et al.* 2005).

The 4765 MHz transition is highly time-variable (Palmer *et al.* 2004): the maser in Mon R2 flared to nearly 80 Jy before disappearing (Smits 2003) and reappearing (Fish *et al.* 2006b). Short-timescale variability is seen in the ground-state lines (Ramachandran, Deshpande, & Goss 2006; Goss *et al.* 2007).

2.4. *Formaldehyde* (H_2CO)

Formaldehyde masers are seen near a handful of several massive YSOs, with several new detections in recent years (Araya *et al.* 2005, 2006). They have both a compact and an extended component with velocity gradients (Hoffman *et al.* 2003; Hoffman, Goss, & Palmer 2007). A short-duration flare has been detected toward one source (Araya *et al.* 2007a). Further details can be found in the review on formaldehyde masers (Araya *et al.* 2007b).

2.5. *Silicon monoxide (SiO)*

While SiO masers are commonly seen in evolved stars, they are rare in star forming regions. They are seen in bipolar outflows in W51 IRS 2 and Orion KL source I and appear much closer to the central source than do water masers (Eisner *et al.* 2002; Greenhill *et al.* 2004). As is the case in evolved stars, the maser species SiO, water, and OH occur at progressively larger distances from source I (Cohen *et al.* 2006). While OH masers are ubiquitous throughout Orion, there is a "zone of avoidance" associated with source I in which they do not appear but SiO and water masers do. Interestingly, the $v = 1$, $J = 2 \rightarrow 1$ masers are found closer to the protostar in source I than are the $J = 1 \rightarrow 0$ masers, a finding that is difficult to understand in the context of SiO maser pumping models (Doeleman *et al.* 2004).

2.6. *Other species*

Few other new maser species or transitions have been reported in the literature since the last meeting. The first $(J, K) = (6, 6)$ ammonia (NH_3) maser has been detected centered on a millimeter peak in NGC 6334 I (Beuther *et al.* 2007). Weakly inverted acetaldehyde (CH_3CHO) has been detected in the $1_{11} \rightarrow 1_{10}$ transition at 1065.075 MHz toward Sgr B2 (Chengalur & Kanekar 2003).

3. Multi-species associations

Methanol, OH, and water masers are frequently in the same source, although water and methanol masers usually originate in different regions (Beuther *et al.* 2002; Caswell 2004b; Edris *et al.* 2005; Szymczak, Pillai, & Menten 2005). Most 6.7 GHz methanol maser sources have associated OH masers, almost always at 1665 MHz and frequently at 1667 MHz as well (Szymczak & Gérard 2004), while the correlation between 6.7 GHz methanol masers and 22 GHz water masers is less strong (e.g., Breen *et al.* 2007). The distributions of 6.7 GHz methanol masers and 6.0 GHz OH masers in W3(OH) are very similar, although direct overlap of the two species is rare (Etoka *et al.* 2005). Similar phenomena are also seen between 6.7 GHz methanol and 1.6/4.7 GHz OH masers (Harvey-Smith & Cohen 2006).

All three species correlate more strongly with mid infrared emission than centimeter or near infrared emission, and all are frequently found in linear groupings (De Buizer *et al.* 2005). However, the luminosity of water masers correlates less strongly with far infrared luminosity than is the case for methanol and OH masers, likely because water masers are not predominantly pumped by infrared photons (Szymczak *et al.* 2005), although they may not be pumped entirely by collisions either (Liu, Forster, & Sun 2005; Liu *et al.* 2007). In any case, the existence of either masers (water and methanol) or outflows towards a UCH II region is an excellent predictor of the other (Codella *et al.* 2004), indicating that both masers and outflows are usually detectable in the UCH II phase.

4. Observational advances

4.1. *Proper motions and geometric distances*

Evidence in support of the kinematic interpretation of maser motions continues to pile up, with reports of the persistence of spot shapes in methanol (Moscadelli *et al.* 2002) and water masers (Goddi *et al.* 2006) and the inferred average overdensity of masers as compared to non-masing material (Fish, Reid, & Menten 2005; Fish *et al.* 2006a). In addition to tracing internal source motions, maser proper motions can be used to obtain geometric parallax distances and measurements of Galactic rotation. Recent years have seen this technique used for both methanol and water masers in W3(OH) to obtain distances accurate to a few percent (Xu *et al.* 2006; Hachisuka *et al.* 2006). Further details can be found in proceedings in the Galactic structure session.

4.2. *Spectral resolution*

Very high spectral line observations at VLBI (very long baseline interferometry) spatial resolution have been obtained toward masers in several species in star forming regions. Detailed line profile analyses of water masers conclude that the near-Gaussian lineshapes indicate that they occur in hot ($\sim$ 1200 K) gas with small beaming angles (Watson, Sarma, & Singleton 2002). Similar observations of 12.2 GHz methanol masers (Moscadelli *et al.* 2003) and the ground-state quartet of OH in W3(OH) (Fish *et al.* 2006c) find similarly Gaussian spectral profiles as well as maser spot positional gradients as a function of velocity (equivalently, velocity gradients). The positional gradients show no clear large-scale spatial organization but have similar magnitudes in methanol and three of the four OH transitions. It is possible that these positional gradients represent turbulent motions on very small spatial scales. If so, it is important to understand the characteristics of the turbulence, since observed maser properties, including variability, can be highly sensitive to turbulence in the masing region (Böger, Kegel, & Hegmann 2003; Sobolev, Watson, & Okorokov 2003; Silant'ev *et al.* 2006). Similar polarization characteristics in Class II methanol features at different velocities may indicate that velocity gradients induce velocity redistribution (Wiesemeyer *et al.* 2004), which may play a critical role in preventing saturated rebroadening (e.g., Nedoluha & Watson 1988).

4.3. *Infrared pumping lines*

Molecular infrared transitions observations, particularly of OH, are essential for measuring maser pump efficiencies and may help place observational constraints on the radiative pump cycles of some models (e.g., Gray 2007). Archival Infrared Space Observatory data have been searched for the 34.6 and 53.3 μm pumping lines of OH with limited success, due to the low spectral resolution of the instruments (He & Chen 2004; He 2005). Herschel and SOFIA will have the spectral resolution and frequency coverage required to observe pumping lines of OH as well as the critical 560 μm line of methylidyne (CH).

5. Further remarks

Two quotes from De Buizer *et al.* (2005) serve to summarize the common themes of the observations over the past five years. The first is that "maser emission in general can trace a variety of phenomena associated with massive stars including shocks, outflows, infall, and circumstellar disks. No one maser species is linked exclusively to one particular process or phenomenon." Indeed, while certain maser species may preferentially turn up in a particular context (possibly a result of observational biases), the set of all observations of any one maser species resist being pigeonholed into a particular phenomenon. The second quote is that water, OH, and methanol masers "do not seem to be associated with different early evolutionary stages of massive stars. Instead it appears that they all trace a variety of stellar phenomena throughout many early stages of massive stellar evolution." As is clear from §3, different maser species are commonly found together, independent of the evolutionary stage of the source. While proposed sequences in which certain masers turn on before others may be useful for statistical evaluation of evolutionary phases, important exceptions to such sequences exist. Those oddball masers that do not seem to fit present standard paradigms should be studied in especial detail, since we cannot predict beforehand what will be thereby learned about their environment or about maser processes in general.

It was only a few years ago that Ellingsen (2004) referred to masers as "the Bart Simpson of star formation research," noting that they are "under-achievers" in comparison with masers in other environments due to the lack of sensitive, high resolution observations at complementary wavelengths. While this may once have been true, recent maser observations have made great advancements in probing a wide range of dynamic structures relevant to star formation. Maser VLBI allows observations of small- and large-scale morphologies, magnetic fields, and motions on AU scales and is showing great promise as a tool to trace Galactic structure. Maser models for some species are becoming sufficiently refined to provide good constraints on physical conditions. The community is beginning to appreciate the role of turbulence and the ability to probe its properties using maser observations. Synergies with mid infrared instruments have clarified many of the mysteries of linear structures with velocity gradients. We have entered the era of greatly improved far infrared instrumentation, and the ALMA (Atacama Large Millimeter Array) era, with unprecedented sensitivity and angular resolution at submillimeter wavelengths, will begin in a few years. Further advancements in radio instrumentation, including new space VLBI missions and the SKA (Square Kilometre Array), will provide even greater insights. It is perhaps more correct to state that maser observations are at the vanguard of star formation research: yesterday's observations can be explained by complementary data and theory today, and today's observations lay the groundwork for the breakthroughs that will be achieved in the context of tomorrow.

Acknowledgements

The National Radio Astronomy Observatory is a facility of the National Science Foundation operated under cooperative agreement by Associated Universities, Inc.

References

Alakoz, A. V., Slysh, V. I., Popov, M. V., & Val'tts, I. E. 2005, *Astron. Lett.*, 31, 375

Araya, E., Hofner, P., Goss, W. M., Kurtz, S., Linz, H., & Olmi, L. 2006, *ApJ*, 643, L33

Araya, E., Hofner, P., Kurtz, S., Linz, H., Olmi, L., Sewilo, M., Watson, C., & Churchwell, E. 2005, *ApJ*, 618, 339

Araya, E., Hofner, P., Sewilo, M., Linz, H., Kurtz, S., Olmi, L., Watson, C., & Churchwell, E. 2007a, *ApJ*, 654, L95

Araya, E., *et al.*, 2007b, (these proceedings)

Argon, A. L., Reid, M. J., & Menten, K. M. 2003, *ApJ*, 593, 925

Babkovskaia, N., & Poutanen, J. 2004, *A&A*, 418, 117

Bartkiewicz, A., Szymczak, M., Cohen, R. J., & Richards, A. M. S. 2005, *MNRAS*, 361, 623

Bartkiewicz, A., Szymczak, M., & van Langevelde, H. J. 2005, *A&A*, 442, L61
Baudry, A., & Desmurs, J. F. 2002, *A&A*, 394, 107
Beuther, H., Thorwith, S., Zhang, Q., Hunter, T. R., Megeath, S. T., Walsh, A. J., & Menten, K. M. 2005, *ApJ*, 627, 834
Beuther, H., Walsh, A., Schilke, P., Sridharan, T. K., Menten, K. M., & Wyrowski, F. 2002, *A&A*, 390, 289
Beuther, H., Walsh, A. J., Thorwith, S., Zhang, Q., Hunter, T. R., Megeath, S. T., & Menten, K. M. 2007, *A&A*, in press, astro-ph/0702190
Błaszkiewicz, L., & Kus, A. J. 2004, *A&A*, 413, 233
Böger, R., Kegel, W. H., & Hegmann, M. 2003, *A&A*, 406, 23
Breen, S. L., *et al.* 2007, *MNRAS*, in print, astro-ph/0702673
Caswell, J. L. 2003, *MNRAS*, 341, 551
Caswell, J. L. 2004a, *MNRAS*, 349, 99
Caswell, J. L. 2004b, *MNRAS*, 351, 279
Caswell, J. L. 2004c, *MNRAS*, 352, 101
Chengalur, J. N., & Kanekar, N. 2003, *A&A*, 403, L43
Codella, C., Lorenzani, A., Gallego, A. T., Cesaroni, R., & Moscadelli, L. 2004, *A&A*, 417, 615
Cohen, R. J., Gasiprong, N., Meaburn, J., & Graham, M. F. 2006, *MNRAS*, 367, 541
Cragg, D. M., Sobolev, A. M., Caswell, J. L., Ellingsen, S. P., & Godfrey, P. D. 2004, *MNRAS*, 351, 1327
Cragg, D. M., Sobolev, A. M., & Godfrey, P. D. 2005, *MNRAS*, 360, 533
De Buizer, J. M. 2003, *MNRAS*, 341, 277
De Buizer, J. M. 2006, *ApJ*, 642, L57
De Buizer, J. M. 2007, *ApJ*, 654, L147
De Buizer, J. M., & Minier, V. 2005, *ApJ*, 628, L151
De Buizer, J. M., Radomski, J. T., Telesco, C. M., & Piña, R. K. 2005, *ApJS*, 156, 179
De Buizer, J. M., Redman, R., Feldman, P., Longmore, S., & Caswell, J. 2006, *BAAS*, 38, 1059
de Gregorio-Monsalvo, I., Gómez, J. F., Suárez, O., Kuiper, T. B. H., Anglada, G., Patel, N. A., & Torrelles, J. M. 2006, *AJ*, 132, 2584
Dodson, R. G., & Ellingsen, S. P. 2002, *MNRAS*, 333, 307
Dodson, R., Ojha, R., & Ellingsen, S. P. 2004, *MNRAS*, 351, 779
Doeleman, S. S., Lonsdale, C. J., Kondratko, P. T., & Predmore, C. R. 2004, *ApJ*, 607, 361
Edris, K. A., Fuller, G. A., & Cohen, R. J. 2007, *A&A*, 465, 865
Edris, K. A., Fuller, G. A., Cohen, R. J., & Etoka, S. 2005, *A&A*, 434, 213
Eisner, J. A., Greenhill, L. J., Herrnstein, J. R., Moran, J. M., & Menten, K. M. 2002, *ApJ*, 569, 334
Elitzur, M. 2002, *Astrophysical Spectropolarimetry*, 225
Ellingsen, S. P. 2004, *IAU Symposium*, 221, 133
Ellingsen, S. P. 2005, *MNRAS*, 359, 1498
Ellingsen, S. P. 2006, *ApJ*, 638, 241
Ellingsen, S. P. 2007, *MNRAS*, in press, astro-ph/0702506
Ellingsen, S. P., Cragg, D. M., Lovell, J. E. J., Sobolev, A. M., Ramsdale, P. D., & Godfrey, P. D. 2004, *MNRAS*, 354, 401
Ellingsen, S. P., Cragg, D. M., Minier, V., Muller, E., & Godfrey, P. D. 2003, *MNRAS*, 344, 73
Etoka, S., Cohen, R. J., & Gray, M. D. 2005, *MNRAS*, 360, 1162
Fish, V. L., Brisken, W. F., & Sjouwerman, L. O. 2006c, *ApJ*, 647, 418
Fish, V. L., & Reid, M. J. 2006, *ApJS*, 164, 99
Fish, V. L., & Reid, M. J. 2007, *ApJ*, 656, 943
Fish, V. L., Reid, M. J., Argon, A. L., & Zheng, X.-W. 2005, *ApJS*, 160, 220
Fish, V. L., Reid, M. J., & Menten, K. M. 2005, *ApJ*, 623, 269
Fish, V. L., Reid, M. J., Menten, K. M., & Pillai, T. 2006a, *A&A*, 458, 485
Fish, V. L., Zschaechner, L. K., Sjouwerman, L. O., Pihlström, Y. M., & Claussen, M. J. 2006b, *ApJ*, 653, L45
Furuya, R. S., Kitamura, Y., Wootten, A., Claussen, M. J., & Kawabe, R. 2005, *A&A*, 438, 571
Gallimore, J. F., Cool, R. J., Thornley, M. D., & McMullin, J. 2003, *ApJ*, 586, 306

Gillis, R. G., Pratap, P., & Strelnitski, V. 2005, *BAAS*, 37, 1472

Goddi, C., & Moscadelli, L. 2006, *A&A*, 447, 577

Goddi, C., Moscadelli, L., Torrelles, J. M., Uscanga, L., & Cesaroni, R. 2006, *A&A*, 447, L9

Goedhart, S., Gaylard, M. J., & van der Walt, D. J. 2003, *MNRAS*, 339, L33

Goedhart, S., Gaylard, M. J., & van der Walt, D. J. 2005, *Ap&SS*, 295, 197

Goedhart, S., Minier, V., Gaylard, M. J., & van der Walt, D. J. 2005, *MNRAS*, 356, 839

Gómez, de Gregorio-Monsalvo, I., Suárez, O., & Kuiper, T. B. H. 2006, *AJ*, 132, 1322

Goss, W. M., *et al.*, 2007, (these proceedings)

Gray, M. D. 2007, *MNRAS*, 375, 477

Greenhill, L. J., Reid, M. J., Chandler, C. J., Diamon, P. J., & Elitzur, M. 2004, *IAU Symp.* 221, 155

Hachisuka, K., Brunthaler, A., Menten, K. M., Reid, M. J., Imai, H., Hagiwara, Y., Miyoshi, M., Horiuchi, S., & Sasao, T. 2006, *ApJ*, 645, 337

Harvey-Smith, L., & Cohen, R. J. 2005, *MNRAS*, 356, 637

Harvey-Smith, L., & Cohen, R. J. 2006, *MNRAS*, 371, 1550

He, J. H. 2005, *New Astron.*, 10, 283

He, J. H., & Chen, P. S. 2004, *New Astron.*, 9, 545

Healy, K. R., Hester, J. J., & Claussen, M. J. 2004, *ApJ*, 610, 835

Hoffman, I. M., Goss, W. M., & Palmer, P. 2007, *ApJ*, 654, 971

Hoffman, I. M., Goss, W. M., Palmer, P., & Richards, A. M. S. 2003, *ApJ*, 598, 1061

Hoffmann, S., Pratap, P., & Strelnitski, V. 2006, *BAAS*, 38, 947

Honma, M., Bushimata, T., & Choi, Y. K., *et al.* 2005, *PASJ*, 57, 595

Honma, M., Choi, Y. K., & Bushimata, T., *et al.* 2004, *PASJ*, 56, L15

Hutawarakorn, B., & Cohen, R. J. 2005, *MNRAS*, 357, 338

Imai, H., Horiuchi, S., Deguchi, S., & Kameya, O. 2003, *ApJ*, 595, 285

Imai, H., Watanabe, T., Omodaka, T., Nishio, M., Kameya, O., Miyaji, T., & Nakajima, J. 2002, *PASJ*, 54, 741

Lekht, E. E., Munitsyn, V. A., & Tolmachev, A. M. 2005, *Astron. Lett.*, 31, 315

Lekht, E. E., Silant'ev, N. A., Krasnov, V. V., & Munitsyn, V. A. 2006, *Astron. Rep.*, 50, 638

Lekht, E. E., Trinidad, M. A., Mendoza-Torres, J. E., Rudnitskij, G. M., & Tolmachev, A. M. 2006, *A&A*, 456, 145

Lemonias, J. J., Strelnitski, V., & Pratap, P. 2006, *BAAS*, 38, 947

Liu, H., Forster, J. R., Liu, Y., & Sun, J. 2007, *ApSS*, in press

Liu, H.-P., Forster, J. R., & Sun, J. 2005, *Chin. J. Astron. Astrophys.*, 5, 175

Menten, K. M., & van der Tak, F. F. S. 2004, *A&A*, 414, 289

Minier, V., & Booth, R. S. 2002, *A&A*, 387, 179

Minier, V., Booth, R. S., Ellingsen, S. P., Conway, J. E., & Pestalozzi, M. R. 2000, *EVN Symposium*, Proceedings of the 5th European VLBI Network Symposium, 179

Minier, V., Burton, M. G., Hill, T., Pestalozzi, M. R., Purcell, C. R., Garay, G., Walsh, A. J., & Longmore, S. 2005, *A&A*, 429, 945

Minier, V., Ellingsen, S. P., Norris, R. P., & Booth, R. S. 2003, *A&A*, 403, 1095

Moscadelli, L., Cesaroni, R., & Rioja, M. J. 2005, *A&A*, 438, 889

Moscadelli, L., Menten, K. M., Walmsley, C. M., & Reid, M. J. 2002, *ApJ*, 564, 813

Moscadelli, L., Menten, K. M., Walmsley, C. M., & Reid, M. J. 2003, *ApJ*, 583, 776

Moscadelli, L., Testi, L., Furuya, R. S., Goddi, C., Claussen, M., Kitamura, Y., & Wootten, A. 2006, *A&A*, 446, 985

Müller, H. S. P., Menten, K. M., & Mäder, H. 2004, *A&A*, 428, 1019

Norris, R. P., *et al.* 1998, *ApJ*, 508, 275

Nammahachak, S., Asanok, K., Hutawarakorn Kramer, B., Cohen, R. J., Muanwong, O., & Gasiprong, N. 2006, *MNRAS*, 371, 619

Nedoluha, G. E., & Watson, W. D. 1988, *ApJ*, 335, L19

Niezurawska, A., Szymczak, M., Cohen, R. J., & Richards, A. M. S. 2004, *MNRAS*, 350, 1409

Niezurawska, A., Szymczak, M., Richards, A. M. S., & Cohen, R. J. 2005, *Ap&SS*, 295, 37

Norris, R. P., Whiteoak, J. B., Caswell, J. L., Wieringa, M. H., & Gough, R. G. 1993, *ApJ*, 412, 222

Palmer, P., & Goss, W. M. 2005, *MNRAS*, 360, 993
Palmer, P., Goss, W. M., & Devine, K. E. 2003, *ApJ*, 599, 324
Palmer, P., Goss, W. M., & Whiteoak, J. B. 2004, *MNRAS*, 347, 1164
Pashchenko, M. I., & Lekht, E. E. 2005, *Astron. Rep.*, 49, 624
Pestalozzi, M. R., Chrysostomou, A., Collett, J. L., Minier, V., Conway, J., & Booth, R. S. 2007, *A&A*, 463, 1009
Pestalozzi, M. R., Elitzur, M., Conway, J. E., & Booth, R. S. 2004, *ApJ*, 603, L113
Pestalozzi, M. R., Minier, V., & Booth, R. S. 2005, *A&A*, 432, 737
Pestalozzi, M. R., Minier, V., Motte, F., & Conway, J. E. 2006, *A&A*, 448, L57
Phillips, C. J., & van Langevelde, H. J. 2005, *Ap&SS*, 295, 225
Pillai, T., Wyrowski, F., Menten, K. M., & Krügel, E. 2006, *A&A*, 447, 929
Pratap, P., Hoffmann, S., & Strelnitski, V. 2006, *BAAS*, 38, 948
Ramachandran, R., Deshpande, A. A., & Goss, W. M. 2006, *ApJ*, 653, 1314
Ripman, B. H., & Strelnitski, V. 2006, *BAAS*, 38, 1053
Salii, S. V., & Sobolev, A. M. 2006, *Astron. Rep.*, 50, 965
Sarma, A. P., Troland, T. H., Crutcher, R. M., & Roberts, D. A. 2002, *ApJ*, 580, 928
Seth, A. C., Greenhill, L. J., & Holder, B. P. 2002, *ApJ*, 581, 325
Silant'ev, N. A., Lekht, E. E., Mendoza-Torres, J. E., & Rudnitskij, G. M. 2006, *A&A*, 453, 989
Slysh, V. I., & Migenes, V. 2006, *MNRAS*, 369, 1497
Slysh, V. I., Migenes, V., Val'tts, I. E., Lyubchenko, S.Yu., Horiuchi, S., Altunin, V. I., Fomalont, E. B., & Inoue, M. 2002, *ApJ*, 564, 317
Slysh, V. I., Voronkov, M. A., Val'tts, I. E., & Migenes, V. 2002, *Astron. Rep.*, 46, 969
Smits, D. P. 2003, *MNRAS*, 339, 1
Sobolev, A. M., Watson, W. D., & Okorokov, V. A. 2003, *ApJ*, 590, 333
Stark, D. P., Goss, W. M., Churchwell, E., Fish, V. L., & Hoffman, I. M. 2007, *ApJ*, 656, 943
Strelnitski, V., *et al.*, 2007, (these proceedings)
Strelnitski, V., Alexander, J., Gezari, S., Holder, B. P., Moran, J. M., & Reid, M. J. 2002, *ApJ*, 581, 1180
Sutton, E. C., Sobolev, A. M., Salii, S. V., Malyshev, A. V., Ostrovskii, A. B., & Zinchenko, I. I. 2004, *ApJ*, 609, 231
Szymczak, M., & Gérard, E. 2004, *A&A*, 414, 235
Szymczak, M., Pillai, T., & Menten, K. M. 2005, *A&A*, 434, 613
Torrelles, J. M., Patel, N. A., Anglada, G., *et al.* 2003, *ApJ*, 598, L115
Urquhart, J. S., Thompson, M. A., Morgan, L. K., & White, G. J. 2006, *A&A*, 450, 625
Uscanga, L., Cantó, J., Curiel, S., Anglada, G., Torrelles, J. M., Patel, N. A., Gómez, J. F., & Raga, A. C. 2005, *ApJ*, 634, 468
Valdettaro, R., Palla, F., Brand, J., & Cesaroni, R. 2005, *A&A*, 535, 540
van der Walt, J. 2005, *MNRAS*, 360, 153
Vlemmings, W. H. T. 2006, *A&A*, 445, 1031
Vlemmings, W. H. T., Diamond, P. J., van Langevelde, H. J., & Torrelles, J. M. 2006, *A&A*, 448, 597
Voronkov, M. A., Austin, M. C., & Sobolev, A. M. 2002, *A&A*, 387, 310
Voronkov, M. A., Brooks, K. J., Sobolev, A. M., Ellingsen, S. P., Ostrovskii, A. B., & Caswell, J. L. 2006, *MNRAS*, 373, 411
Voronkov, M. A., Sobolev, A. M., Ellingsen, S. P., & Ostrovskii, A. B. 2005, *MNRAS*, 362, 995
Watson, W. D., Sarma, A. P., & Singleton, M. S. 2002, *ApJ*, 570, L37
Watson, W. D., Wiebe, D. S., McKinney, J. C., & Gammie, C. F. 2004, *ApJ*, 604, 707
Wiesemeyer, A., Thum, C., & Walmsley, C. M. 2004, *A&A*, 428, 479
Wright, M. M., Gray, M. D., & Diamond, P. J. 2004, *MNRAS*, 350, 1272
Xu, Y., Reid, M. J., Zheng, X. W., & Menten, K. M. 2006, *Science*, 311, 54
Zapata, L. A., Rodríguez, L. F., Ho, P. T. P., Beuther, H., & Zhang, Q. 2006, *AJ*, 131, 939

Astrophysical Masers and their Environments
Proceedings IAU Symposium No. 242, 2007
J.M. Chapman & W.A. Baan, eds.

doi:10.1017/S1743921307012616

How do methanol masers manage to appear in the youngest star vicinities and isolated molecular clumps?

A. M. Sobolev[1], D. M. Cragg[2], S. P. Ellingsen[3], M. J. Gaylard[4], S. Goedhart[4], C. Henkel[5], M. S. Kirsanova[6], A. B. Ostrovskii[1], N. V. Pankratova[1], O. V. Shelemei[1], D. J. van der Walt[7], T. S. Vasyunina[8] and M. A. Voronkov[9]

[1]Ural State University, Ekaterinburg, 620083, Russia, email: Andrej.Sobolev@usu.ru
[2]Monash University, Clayton, Australia, email: Dinah.Cragg@monash.edu.au
[3]University of Tasmania, Hobart, Australia, email: Simon.Ellingsen@utas.edu.au
[4]Hartebeesthoek Radio Astronomy Observatory, South Africa, email: sharmila@hartrao.ac.za
[5]MPIfR, Bonn, Germany, email: p220hen@mpifr-bonn.mpg.de
[6]Institute for Astronomy, Moscow, Russia, email: kirsanova@inasan.ru
[7]North-West University, South Africa, email: johan.vanderwalt@nwu.ac.za
[8]MPIA, Heidelberg, Germany, email: Vasyunin@mpia-hd.mpg.de
[9]ATNF CSIRO, Sydney, Australia, email: Maxim.Voronkov@csiro.au

Abstract. General characteristics of methanol (CH_3OH) maser emission are summarized. It is shown that methanol maser sources are concentrated in the spiral arms. Most of the methanol maser sources from the Perseus arm are associated with embedded stellar clusters and a considerable portion is situated close to compact HII regions. Almost 1/3 of the Perseus Arm sources lie at the edges of optically identified HII regions which means that massive star formation in the Perseus Arm is to a great extent triggered by local phenomena. A multiline analysis of the methanol masers allows us to determine the physical parameters in the regions of maser formation. Maser modelling shows that class II methanol masers can be pumped by the radiation of the warm dust as well as by free-free emission of a hypercompact region (hcHII) with a turnover frequency exceeding 100 GHz. Methanol masers of both classes can reside in the vicinity of hcHIIs. Modelling shows that periodic changes of maser fluxes can be reproduced by variations of the dust temperature by a few percent which may be caused by variations in the brightness of the central young stellar object reflecting the character of the accretion process. Sensitive observations have shown that the masers with low flux densities can still have considerable amplification factors. The analysis of class I maser surveys allows us to identify four distinct regimes that differ by the series of their brightest lines.

Keywords. masers; catalogs; surveys; stars: formation; ISM: clouds, evolution, kinematics and dynamics, structure; Galaxy: structure; radio lines: ISM

1. Introduction

Methanol is a complex molecule — it consists of six atoms, and two of them are heavy. It is surely the 1st prize winner by complexity among the molecules producing cosmic masers. It is really surprising that such a molecule gives birth to numerous and extremely bright masers that are only matched by the 1.7 GHz hydroxyl (OH) and 22 GHz water vapor (H_2O) lines.

This paper describes the methanol maser phenomenon on different spatial scales emphasizing the physical conditions that characterize methanol maser sites. Several aspects of this topic are covered in papers of other participants of this conference. We concentrate on results obtained by researchers from the Ural State University mostly in cooperation with scientists from other institutes.

2. General characteristics

Methanol masers are traditionally divided into two classes according to the sets of transitions in which they emit. Historically, the first methanol masers (later attributed to class I) were discovered in Orion in the $J_2 - J_1$ E series of lines at about 25 GHz (Barrett *et al.* 1971). The existence of the second class of methanol masers (eventually named class II) was justified in the paper by Batrla *et al.* (1987) after the discovery of strong masers in the $2_0 - 3_{-1}$ E line at 12.1 GHz. The existence of an intermediate class of methanol masers is predicted but these masers are not likely to attain fluxes comparable to those of the strongest lines of the two major classes (Voronkov *et al.* 2005). Methanol masers of both classes often co-exist in the same star-forming region, though there are bright and notorious examples of class-I-only (OMC-1) and class-II-only (W3(OH)) sources.

Methanol maser lines can achieve extremely high brightness: class I maser lines at 44 GHz are reported to be brighter than 10^8 K (Kogan & Slysh 1998) and class II lines at 6.7 GHz achieve even higher brightness exceeding 10^{12} K (Menten *et al.* 1992). Spectra of the bright methanol masers usually contain numerous components (see, e.g., Caswell *et al.* 1995) spread in a range smaller than 20 km/s around the systemic velocity of the respective molecular core (Malyshev & Sobolev 2003). Maser components are usually narrow (about 0.5 km/s) and the narrowest ever reported line (<0.1 km/s) belongs to a class I methanol maser (Voronkov *et al.* 2006). Methanol masers display time variability of various kinds including periodic, chaotic and episodic flaring (Goedhart *et al.* 2004).

3. Methanol masers on a Galactic scale

Studies of the distribution of methanol masers in the Galaxy can be reduced to class II maser sources because most of the class I maser sites are found close to class II ones (see, e.g., Slysh *et al.*. 1994 and Ellingsen 2005). At present there exist three unified catalogues of class II methanol masers each containing information on more than 500 objects (Malyshev & Sobolev 2003; Xu, Zheng & Jiang 2003 and Pestalozzi *et al.* 2005). The distinguishing feature of the Malyshev & Sobolev (2003) catalogue is that it contains data on several maser transitions and cross-references to the data on molecular shock tracers. This catalogue is updated on a monthly basis and is available at http://astroserver.astro.usu.ru.

Most of the class II maser sources were found toward positions of OH masers and IRAS sources with specific spectral characteristics. A new way of targeting based on GLIMPSE data is suggested in Ellingsen (2007). Blind surveys, including a recent Arecibo survey by Pandian *et al.* (2007) led to the detection of many new maser sources in selected areas. A real breakthrough is expected after the completion of the blind survey within the MMB (Methanol Multibeam) project (see the paper by J. Green *et al.* in this volume).

Information on the Galactic distribution of class II methanol masers was presented in several papers which established that these masers mostly reside in the molecular ring of our Galaxy (for one of the most recent articles, see Pestalozzi *et al.* 2005). At present, class II masers were found only in high mass star-forming regions (Minier *et al.* 2003) and it is natural to expect that these sources should be concentrated in the spiral arms.

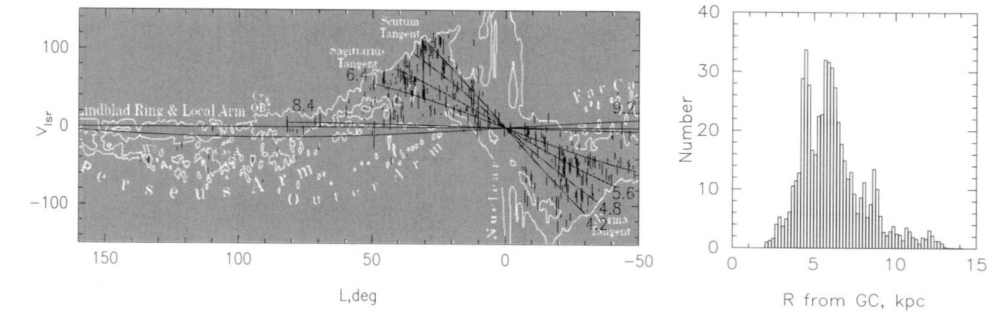

Figure 1. Class II methanol masers in the $V_{LSR} - l$ diagram (lines of equal Galactocentric distances are shown, background: CO data from Dame *et al.* 2001) and distribution of the maser sources as a function of Galactocentric distance.

For the outer Galaxy, the concentration of methanol maser sources to the Perseus spiral arm is quite certain: the velocity-Galactic latitude (l) diagram shows that the masers mostly reside in spiral arms which are well outlined by CO emission (see figure 1). Evidence for a concentration of the methanol masers in the spiral arms is more difficult to prove in the tangled regions of the inner Galaxy and was first provided in Sobolev *et al.* (2005). New maser sources were reported since then, mainly in Pandian *et al.* (2007) and Ellingsen (2007). Figure 1 shows a $V_{LSR} - l$ diagram for the masers from the inner part of the Galaxy along with lines of equal Galactocentric distance, R, and the distribution of the maser sources as a function of R. The updated Malyshev & Sobolev (2003) catalogue and the Brand & Blitz (1993) rotation curve were used. The concentration of masers to certain ranges of Galactocentric distance is apparent. The values of R correlate with the nearest spiral arm sections of the inner Galaxy. It is, however, not possible to conclude that all class II masers reside in the spiral arms. For the bulk of the sources there are only kinematic distances which are not accurate enough. Moreover, most of the distance estimates are based on velocities of 6.7 GHz maser peaks which are significantly scattered with respect to the systemic velocities of the clouds (Malyshev & Sobolev 2003 and van der Walt, Sobolev & Butner 2007). So a sophisticated analysis of the Galactic methanol maser distribution requires a drastic improvement in the precision of distance estimates.

4. Masers in the Perseus Arm

In this section we will discuss the association of methanol masers with the large scale constituents of star forming complexes in the Perseus arm. We confine ourselves to the Perseus Arm because these objects have relatively simple structure, most of them have no confusing and strong back- or foreground and the objects are relatively near. We realize that the Perseus Arm complexes are somewhat different from those residing in the other parts of the Galaxy but it is easier to start with the least complicated case. 28 sources were selected from the Malyshev & Sobolev (2003) catalogue on the basis of their position in the $V_{LSR} - l$ diagram. Using the SkyView virtual observatory (http://skyview.gsfc.nasa.gov/) we compiled an atlas of images of these sources displaying their association with developed HII regions seen at optical wavelengths (R colour images), embedded star clusters revealed in near infrared (K_s colour images) and compact HII regions detected in the NRAO VLA Sky Survey (NVSS). The numbers of maser sources associated with these types of objects are shown in figure 2. It is found that 57 % of the sources are associated with embedded clusters which means that these maser sites can be affected by several young stellar objects (YSOs). This is especially true for class I masers which are found apart from the YSOs. The fraction of cluster associations can be

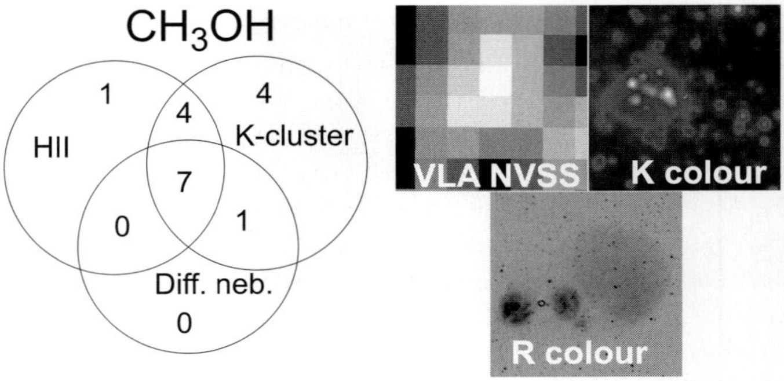

Figure 2. Statistics of the association of class II maser sources from the Perseus Arm with optically visible HII regions, embedded stellar clusters, and VLA NVSS sources.

substantially higher since K_s colour images do not show very deeply embedded clusters revealed at submm and mm wavelengths. For 43 % of the targets we find an association with VLA sources which means that a considerable fraction of maser sources is associated with rather developed compact HII regions around luminous YSOs. 29 % of the sources appear to be associated with optical HII regions. We conclude that a considerable fraction of the methanol masers is triggered by the expansion of evolved HII regions. Since the time duration of the methanol masing phase is quite short (see, e.g., van der Walt *et al.* 2005) this means that contemporary massive star formation in the Perseus Arm is to a great extent controlled by local phenomena. A significant fraction of the sources (~40 %, mostly the distant ones) did not show a clear association with the objects of our atlas.

5. Class II Methanol maser environments and pumping

The environment of class II methanol masers was discussed in many papers (see, e.g., Ellingsen 2006, Minier *et al.* 2005 and Walsh *et al.* 2001) and talks at this symposium (review by V. Fish, talks by J. de Buizer, S. Ellingsen, T. Hill, S. Longmore, L. Moscadelli and others). Here we will instead emphasize two facts which are decisive for our understanding of the maser pumping: it is established that almost all of the maser sources are associated with mid-infrared (MIR) emission (Walsh *et al.* 2001) and some of them are associated with submm sources without cm - IR counterparts (Hunter *et al.* 2006).

5.1. *Class II methanol maser pumping regime*

Both observations and modelling of class II masers show that the 6.7 GHz transition always manifests the highest brightness temperature (Malyshev & Sobolev 2003; Sobolev *et al.* 1997). Observations have not shown significant differences in the positions of the maser spots seen in different transitions (see, e.g., Menten *et al.* 1992; Sutton *et al.* 2001). So, there is basically only one known pumping regime of class II masers. However, ratios of the brightnesses of different maser lines show considerable dependence on the values of the physical parameters. The hypothesis that the masers are pumped by MIR emission of the warm dust allows us to explain observed brightness temperatures and to determine physical parameters of several sources on the basis of multi-transitional observations (e.g., Sutton *et al.* 2001, Cragg, Sobolev & Godfrey 2005). A model of the "common" class II methanol maser source was constructed on the basis of extensive surveys (e.g., Cragg *et al.* 2004 and Ellingsen *et al.* 2004). The results of such studies are summarized in Sobolev *et al.* (2002) and Cragg, Sobolev & Godfrey (2005) and will not be presented here.

5.2. *Class II methanol maser variability*

The response of the maser fluxes to changes in the physical parameters allows us to determine these parameters and their variation in time via detailed modelling. For example, interferometric studies of variability in the source G9.62+0.20 shows that different parts of the source are synchronized, suggesting a radiative origin for the variability (Goedhart *et al.* 2005). Monitoring observations have shown that this source is almost strictly periodical (see the papers by Goedhart & Gaylard and Gaylard & Goedhart in this volume). Periodicity is revealed in the variations of G9.62+0.20 in three maser transitions at 6.7, 12.2 and 107 GHz (van der Walt *et al.* in preparation). A possible cause for the synchronized variability might be a variation of the dust temperature in the region where maser is formed. Model calculations show that corresponding changes of fluxes can be reproduced by variations of the dust temperature by a few per cent which may follow variations in the brightness of the central YSO reflecting the character of the accretion process.

5.3. *Class II methanol maser pumping in hypercompact* HII *region environments*

Recent observations have shown that the masers which were previously "hanging in nowhere" (e.g., the brightest 6.7 GHz masers in NGC6334F, Ellingsen *et al.* 1996) are associated with submm emission (Hunter *et al.* 2006) and have high chances to be formed in the vicinity of hypercompact HII regions (hcHIIs). These objects manifest strong free-free emission which in some cases dominates the dust emission up to frequencies as high as 300 GHz (Kurtz 2005). We have undertaken model calculations in order to explore whether methanol masers can be pumped by free-free emission with high turnover frequency (i.e., high emission measure) without considerable dust content. Historically, first calculations of this kind were presented in the paper by (Slysh *et al.* 2002) but these authors did not account for very powerful torsional transitions which are crucial for pumping, at least in terms of intensities. Our calculations are based on the most recent transition rates that were used in the model calculations presented in Cragg, Sobolev & Godfrey (2005). The model parameters are close to those presented by Cragg, Sobolev & Godfrey (2005) with the major difference that the warm dust emission is substituted by the free-free emission from the hcHII. We have chosen the following set of parameters: gas temperature $T_k = 150$ K, hydrogen number density $n_H = 10^7$ cm^{-3}, methanol specific column density $N_m/dV = 10^{12}$ cm^{-3} s, beaming factor $1/\epsilon = 10$, hcHII electron temperature $T_e = 12\,000$ K, turnover frequency $f_e = 250$ GHz and hcHII dilution factor $W_{Hii} = 0.5$. The calculations show that bright class II masers are indeed produced.

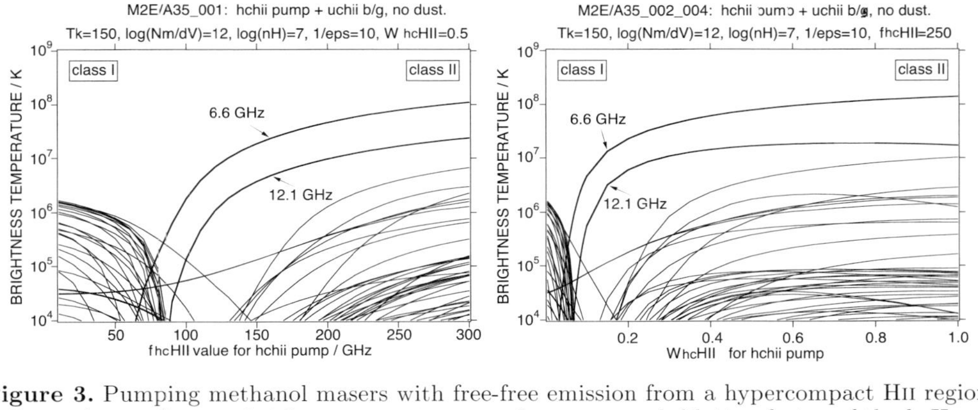

Figure 3. Pumping methanol masers with free-free emission from a hypercompact HII region. Dependence of maser brightnesses on turnover frequency and dilution factor of the hcHII.

Figure 3 presents results from two series of models. The left panel reveals variations of brightness temperature when modifying the turnover frequency. It is clearly seen that to pump bright class II masers free-free emission of hcHII should have $f_e > 100$ GHz, i.e. it should posess a very high emission measure. The right panel demonstrates that the pumping of class II methanol masers takes place quite close to the hcHII where the free-free emission is not greatly diluted ($W_{Hii} > 0.1$). It is noteworthy that, when the distance from the hcHII increases ($W_{Hii} \ll 0.1$), some dense, warm and elongated clumps created by interaction of the hcHII with the ambient medium can give birth to rather strong class I methanol masers. This is one of the possible explanations of the fact that masers of different classes can reside close to each other (see, e.g., Ellingsen 2005).

5.4. *Are weak masers really masing?*

It was shown that important constraints on the models can be obtained through an evaluation of upper limits for the fluxes in particular maser lines. Within this context it is very important to know whether the weak lines are masers or whether a given transition is thermalized. In order to find a solution to this problem we conducted a very sensitive search for emission in the $9_2 - 10_1 A^+$ and $10_1 - 9_2 A^-$ lines at 23.1 and 23.4 GHz using the 100-m telescope at Effelsberg. The transitions at 23.1 and 23.4 GHz highly differ in their response to factors producing departures from LTE: e.g., W3(OH) is strongly masing at 23.1 GHz and shows absorption at 23.4 GHz (Menten *et al.* 1986).

In contrast to previous surveys at Parkes and Onsala the detection rate in the maser line at 23.1 GHz was rather high and we detected emission in 10 sources out of 20. The 23.1 GHz line in the sources W51, W75N, DR21(OH), Cep A and G12.91–0.26 showed linewidths comparable to those of corresponding "quasi-thermal" lines from these objects. In these sources, the 23.4 GHz line shows very similar intensities and line shapes (see the W51 spectra in figure 4) which means that the transitions are thermalized.

Toward the sources G9.62+0.19, G23.01+0.41 and G188.95+0.89 a clear detection of the 23.1 GHz line was not accompanied by 23.4 GHz emission. The detected lines are considerably narrower than "quasi-thermal" lines from these objects (see figure 4). Line

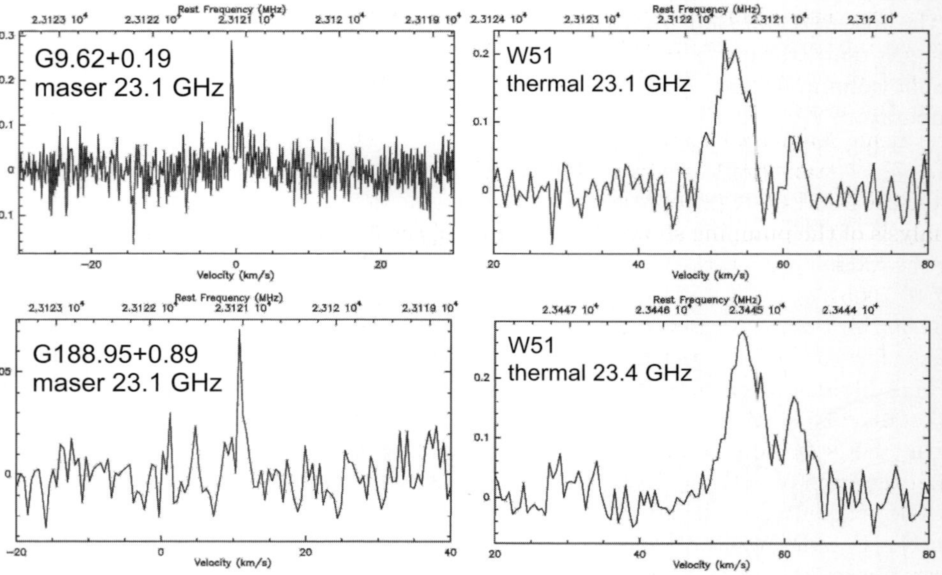

Figure 4. Maser emission at 23.1 GHz and "quasi-thermal" emission at 23.1 and 23.4 GHz.

narrowing is a property of unsaturated masers with considerable optical depth. So, we can state that we detected three new weak but real masers and that the weakness of masers does not necessarily mean that their amplification factors are low.

6. Class I methanol maser environments and pumping

The environment of class I masers is highly turbulent. Lines of the strong masers usually contain several spectral components and their spots form clusters. General characteristics of the maser images and spectra can be reproduced by modelling the turbulent velocity field (Sobolev *et al.* 1998). The existence of the interferometric data on the images and time variability of the $J_2 - J_1E$ maser line series at about 25 GHz in OMC-1 (Johnston *et al.* 1992) allow us to construct models constraining the characteristics of the turbulence in this source (Sobolev *et al.* 1998; Sobolev *et al.* 2003).

Class I methanol masers are less strong and widespread than class II ones. Most of them were found in the vicinity of class II maser sources (see, e.g., Slysh *et al.* 1994 and Ellingsen *et al.* 2005). There were extensive searches (e.g., Slysh *et al.* 1994 and Val'tts *et al.* 2000) for class I sources but no unified catalogues of these objects are published in the journals and no successful blind surveys were performed yet. There are several lines of evidence that class I masers trace dense parts of molecular outflows, probably locations where new stars might form (e.g., Voronkov *et al.* 2006 and Sutton *et al.* 2004) but generally the environment of class I sources is not well established yet.

Recently, sensitive and extensive surveys of the southern class I masers were conducted utilizing the unique capabilities of the Australia Telescope Compact Array (see the paper by Voronkov *et al.* in these proceedings). They allow us to draw some general conclusions on the class I maser pumping and identify four distinct regimes differing by the series of the brightest (in terms of brightness temperature) line.

The most frequently encountered maser regime shows the brightest lines which belong to the $J_{-1} - (J\text{-}1)_0E$ series. The $4_{-1} - 3_0E$ and $5_{-1} - 4_0E$ transitions at 36.1 and 84.5 GHz are weak masers under normal conditions of massive star forming regions. Usually, the maser nature of the lines in this regime is difficult to prove observationally. Nevertheless, there are cases where the line profiles contain narrow spikes and the maser nature is proved interferometrically. The sources Sgr B2, G30.8-0.1, and G1.6-0.025 can be considered as representatives of this maser regime (Sobolev 1996).

In the second class I maser regime the lines of the $J_0 - (J\text{-}1)_1A^+$ series become prevalent. Numerous sources show definitely maser lines arising in the $7_0 - 6_1A^+$ and $8_0 - 7_1A^+$ transitions at 44.1 and 95.2 GHz, respectively. Masers in the sources DR21W, NGC2264 and OMC-2 represent this regime (Menten (1991)). A preliminary theoretical analysis of the pumping shows that the lines of the $J_0 - (J\text{-}1)_1A^+$ series become brightest in the models with rather high beaming (> 20) and moderate column densities (Sobolev *et al.* 2005).

The third maser regime is less widespread and is represented by the sources where the lines of the $J_2 - J_1E$ series at about 25 GHz are the brightest. OMC-1 is the prototypical source (Johnston *et al.* 1992) and some additional bright sources were found recently (Voronkov *et al.* these proceedings). Lines of this series become brightest in the models with high specific column densities ($lg(N_m/dV\ [\mathrm{cm}^{-3}\cdot\mathrm{s}]) \geqslant 12$) and require relatively high temperatures ($T_k = 75-100$ K) and densities ($lg(n_H\ [\mathrm{cm}^{-3}]) = 5-7$).

The existence of the forth maser regime with the brightest lines from the $J_{-2} - (J\text{-}1)_{-1}E$ series was previously uncertain. Recent ATCA observations confirmed however, the existence of such a regime in the sources W33-Met and G343.12-0.06. Preliminary modelling shows that the $9_{-2} - 8_{-1}E$ line at 9.9 GHz becomes brightest in models with specific column densities which are as high as required for the previously discussed regime. However, a smaller beaming angle and either lower densities or higher temperatures ($T_k > 100$ K) are also required.

References

Barrett, A. H., Schwartz, P. R., & Waters, J. W. 1971, *ApJ* 168, L101
Batrla, W., Matthews, H. E., Menten, K. M., & Walmsley, C. M. 1987, *Nature* 326, 49
Brand, J. & Blitz, L. 1993, *A&A* 275, 67
Caswell, J. L., Vaile, R. A., Ellingsen, S. P. & Norris, R. P. 1995, *MNRAS* 274, 1126
Cragg, D. M., Sobolev, A. M., Caswell, J. L., *et al.* 2004, *MNRAS* 351, 1327
Cragg, D. M., Sobolev, A. M. & Godfrey, P. D. 2005, *MNRAS* 360, 533
Dame, T. M., Hartmann, D. & Thaddeus P. 2001, *ApJ* 547, 792
Ellingsen, S. P., Norris, R. P. & McCulloch, P. M. 1996, *MNRAS* 279, 101
Ellingsen, S. P., Cragg, D. M., Lovell, J. E. J., Sobolev, A. M., *et al.* 2004, *MNRAS* 354, 401
Ellingsen S. P. 2005, *MNRAS* 359, 1498
Ellingsen S. P. 2006, *ApJ* 638, 241
Ellingsen S. P. 2007, *MNRAS* 377, 571
Goedhart, S., Gaylard, M. J. & van der Walt, D. J. 2005, *MNRAS*, 355, 553
Goedhart, S., Minier, V., Gaylard, M. J. & van der Walt, D. J. 2005, *MNRAS*, 356, 839
Hunter, T. R., Brogan, C. L., Megeath, S. T. *et al.* 2006, *ApJ* 649, 888
Johnston, K. J., Gaume, R., Stolovy, S., *et al.* 1992, *ApJ* 385, 232
Kogan, L. & Slysh, V. 1998, *ApJ* 497, 800
Kurtz S. 2005, *ProcIAU* 1, 111
Malyshev A. V. & Sobolev A. M. 2003, *A&ATr* 22, 1
Menten, K. M., Walmsley, C. M., Henkel, C., Wilson, T. L. *et al.* 1986, A&A 169, 271
Menten K. M. 1991, *ApJ* (Letters) 380, L75
Menten, K. M., Reid, M. J., Pratap, P., *et al.* 1992, ApJ (Letters) 401, L39
Minier, V., Ellingsen, S. P., Norris, R. P. & Booth, R. S. 2003, *A&A* 403, 1095
Minier, V., Burton, M. G., Hill, T. Pestalozzi, M. R. *et al.* 2005, *A&A* 429, 945
Pandian, J. D., Goldsmith, P. F. & Deshpande, A. A. 2007, *ApJ* 656, 255
Pestalozzi M., Minier V. & Booth R. 2005, *A&A* 432, 737
Slysh, V. I., Kalenskii S. V., Val'tts I. E. & Otrupcek, R. 1994, *MNRAS* 268, 464
Slysh, V. I., Kalenskii S. V. & Val'tts I. E. 2002, *Astron.Rep.* 46, 49
Sobolev, A. M., Ostrovskii, A. B., Malyshev, A. V., *et al.* 2002, *ASP Conf.Ser.* 206, 179
Sobolev, A. M. 1996, *ASPC* 102, 68
Sobolev, A. M., Cragg, D. M. & Godfrey, P. D. 1997, *MNRAS* 288, 39
Sobolev, A. M., Wallin, B. K. & Watson, W. D. 1998, *ApJ* 498, 763
Sobolev, A. M., Watson, W. D. & Okorokov, V. A. 2003, *ApJ* 590, 333
Sobolev A. M., Ostrovskii A. B., Kirsanova M. S., *et al.* 2005, *ProcIAU* 1, 174
Sutton, E. C., Sobolev, A. M., Ellingsen, S. P., *et al.* 2001, *ApJ* 554, 173
Sutton, E. C., Sobolev, A. M., Salii, S. V., *et al.* 2004, *ApJ* 609, 231
Val'tts I. E., Ellingsen, S. P., Slysh, V. I., *et al.* 2000, *MNRAS* 317, 315
Voronkov, M. A., Sobolev, A. M., Ellingsen, S. P., *et al.* 2005, *Ap&SS* 295, 217
Voronkov, M. A., Brooks, K. J., Sobolev, A. M., *et al.* 2006, *MNRAS* 373, 411
Walsh, A. J., Bertoldi, F., Burton, M. G. & Nikola, T. 2001, *MNRAS* 326, 36
van der Walt, D. J. 2005, *MNRAS* 360, 153
van der Walt, D. J., Sobolev, A. M. & Butner, H. 2007, *A&A* 464, 1015
Xu Y., Zheng X.-W. & Jiang D.-R. 2003, *ChJAA* 3, 49

Astrophysical Masers and their Environments
Proceedings IAU Symposium No. 242, 2007
J.M. Chapman & W.A. Baan, eds.

doi:10.1017/S1743921307012628

Methanol masers as tools to study high-mass star formation

Michele R. Pestalozzi[1]

[1]Centre for Astrophysics Research, University of Hertfordshire, AL10 9AB Hatfield, UK
email: michele.pestalozzi@gmail.com

Abstract. In this contribution I will attempt to show that the study of Galactic 6.7 and 12.2 GHz methanol masers themselves, as opposed to the use of methanol masers as signposts, can yield important conclusions contributing to the understanding of high-mass star formation. Due to their exclusive association with star formation, methanol masers are the best tools to do this, and their large number allows us to probe the entire Galaxy. In particular I will focus on the determination of the luminosity function of methanol masers and on the determination of an unambiguous signature for a circumstellar masing disc seen edge-on. Finally I will try to point out some future fields of research in the study of methanol masers.

Keywords. masers, methanol, star formation, high-mass stars

1. Introduction

The Galactic maser emission of methanol appears to be exclusively associated with sites of high-mass star formation (Minier *et al.* 2003). In particular, the *class II* masers at 6.7 and 12.2 GHz are among the brightest masers in the Milky Way. Since their discovery, a number of extended searches have been put into action with the aim of gathering a potentially large number of objects to be studied in depth. A good list of these searches is presented in Pestalozzi *et al.* (2005). Because of their high brightness, methanol masers are obvious targets for interferometric observations that reveal the finest dynamical and morphological details at large distances (1 AU at 1 kpc).

The census of 6.7 GHz masers in the Milky Way counts to date some 520 sources, all potentially locating a high-mass protostar (Pestalozzi *et al.* 2005). This number is subject to increase as the Methanol MultiBeam (MMB) Survey (see contribution by J. Green in these proceedings) covers more regions of the Galactic Plane. The distribution of the known methanol masers in the Galaxy is shown in Fig. 1.

Research connected with methanol masers can be summarised and organised as shown in Table 1. In the squares, I put some examples of the type of investigation that were undertaken to study methanol masers. Methanol maser studies can be divided into two main categories: studies having the maser emission itself as target of their investigation on the one hand, and studies having the source hosting the maser emission as their target on the other. Both these paths give rise to two subclasses: studies of the maser sources as a group (statistical or global analysis) on the one hand, and deep studies on specific sources on the other. The outcomes from these studies can also be distributed into a similar diagram. The goal is to make the outcomes fall in the most general possible category (bottom left square in Table 1), i.e. to understand high-mass star formation.

The common property of the early searches for 6.7 GHz (and 12.2 GHz) methanol masers was that they concentrated their efforts in the characterisation of the host of the masers (bottom left square of Table 1). This was in fact a natural consequence

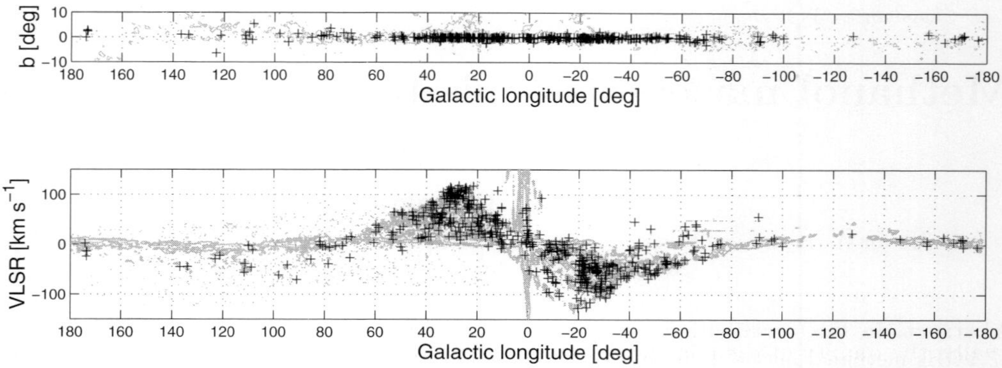

Figure 1. Distribution of methanol masers in Galaxy, superposed on the CO contours from Dame *et al.* 1987, in space (top) and LOS velocity (bottom). The methanol masers seem to accurately follow the overall structure of the Galaxy, both in space and LOS velocity. Particularly visible in the bottom panel is the fact that methanol masers are tracing the spiral arms ($150^\circ > l > 80^\circ$ and $-40^\circ > l > -90^\circ$) and the high rotational velocity of the nuclear ring ($l \approx 0^\circ$).

of the strategy adopted to find new sources, i.e. targetting IRAS sources according to selection criteria corresponding to particular objects, as Ultra Compact (UC) HII regions with more or less red colours (e.g. Schutte *et al.* 1993,Szymczak *et al.* 2000), or known star formation regions hosting OH masers (e.g. Caswell *et al.* 1995). The relatively low detection rate of these surveys revealed the inaccuracy of the definition of methanol maser hosts and called for untargetted searches, where large areas of sky were covered with observations (e.g. Ellingsen *et al.* 1996, Szymczak *et al.* 2002, Pestalozzi *et al.* 2002b, Pandian *et al.* 2007). The main outcome of these studies was that a large fraction of methanol masers at 6.7 GHz were found not to be associated with known bright IR emitters but with very deeply embedded and therefore probably young objects (e.g. Pestalozzi *et al.* 2002a). This conclusion relies on accurate astrometric information both for the masers and for the IR objects, which is only recently starting to become available at a large number of wavelengths and for a large number of targets. More recently, large follow-up campaigns in the millimetre and sub-millimetre bands in continuum and spectral line emission have given further insight into the nature of the hosts of 6.7 GHz methanol masers (Purcell *et al.* 2006, Hill *et al.* 2005). The main outcome from these studies is that methanol masers are probably the first signal for the existence of a high-mass protostar when it is still embedded in its thick primordial dust cocoon.

Studies having the masers themselves as object are more challenging. This is mainly due to the fact that the maser emission is non-thermal in nature and therefore the extraction of information on the physical conditions of the emitting medium is not trivial. Theoretical modelling of the maser emission of methanol faces the challenge of including a larger number of molecular transitions in the calculations, potentially all interacting and producing a large number of masing transitions (e.g. Sobolev *et al.* 1997, Cragg *et al.* 2005). These studies are tightly linked to multi transitional observations of maser-bearing objects, that are starting to include larger samples of targets (see Pestalozzi *et al.* 2005 for a list). In the near future these studies will greatly benefit from the clear improvement of backends that will allow simultaneous observation of different frequency bands covering large regions of the spectrum at once. Polarisation measurements appear to be technically challenging, and from the levels of polarisation obtained, it is not always easy to discriminate between pumping effects and scattering effects (Wiesemeyer *et al.* 2004).

Table 1. Conceptual division of the study of methanol masers.

	Global studies	Particular studies
Masers	maser luminosity structure of the Galaxy maser theory polarisation	NGC 7538: disc signature Location of new sources proper motions (variability)
Hosts of masers	Associations SEDs follow-up observations	IRAS 20126, S255 Protoclusters

Finally, the quality of the conclusions from variability studies is critically dependent on long time monitoring, and has only recently started to produce interesting results. Some sources show a remarkable regularity, and represent a challenge for modelling (see e.g. Goedhart *et al.* 2004).

In this contribution I will concentrate on the upper row of Table 1 and present one example for each square. In particular I will concentrate on the extraction of the luminosity function of the masers and on the recognition of a unique signature for an edge-on rotating disc marked by maser emission.

2. A global study: the luminosity function of the masers

The distribution of the known methanol masers in the Milky Way is shown in Fig. 1. The majority of the masers are located in a ring, covering Galactic longitudes of $\pm 50^\circ$. What is interesting is that a comparison of the Galactic distribution of methanol masers with OB associations (Bronfman *et al.* 2000) shows excellent agreement. This fact reinforces the association of methanol masers with high-mass star formation.

The study of the luminosity function of astronomical objects has to first solve the problem of the distance determination. In the Milky Way, kinematic distances can be attributed to sources observed in spectral line emission (as e.g. masers). Kinematic distances for sources orbiting the Galactic Centre on orbits internal to the Sun's orbit suffer from an ambiguity which can be resolved only on a case-to-case basis. Several methods have been used to solve this ambiguity with varying success (see e.g. Busfield *et al.* 2006 for the HI self absorption method, and van der Walt 2005 for a probabilistic approach). Here I present an approach to the study of the luminosity function of methanol masers that removes the problem of the heliocentric distance determination.

What is needed is to disentangle the spatial distribution of the sources under investigation from their intrinsic luminosity distribution. In fact, the *observed* luminosity distribution of masers (diamonds and circles in Fig. 2) is the convolution of a spatial distribution with an *intrinsic* luminosity distribution after applying an observational sensitivity cutoff. Knowing the spatial distribution of the masers and the sensitivity cutoff leaves the intrinsic luminosity function to be the only unknown in the problem.

The starting point is the distribution of masers in Galactocentric distance. The latter is not affected by any ambiguity, and only assumes a rotation curve of the Galaxy. The inaccuracies of the rotation curve of the Galaxy do not significantly affect the final result, because, as seen in Fig. 2, most of the methanol masers are distributed in a ring with Galactocentric radius ~5 kpc: at that radius, even the influence of the central bar is not strong enough to produce significant deviations from a circular motion (see e.g. Fux 1999). The distribution of masers in Fig. 2 is assumed to represent with a high level of confidence *the shape* of the true, azimuthally averaged, spatial distribution of methanol masers in the

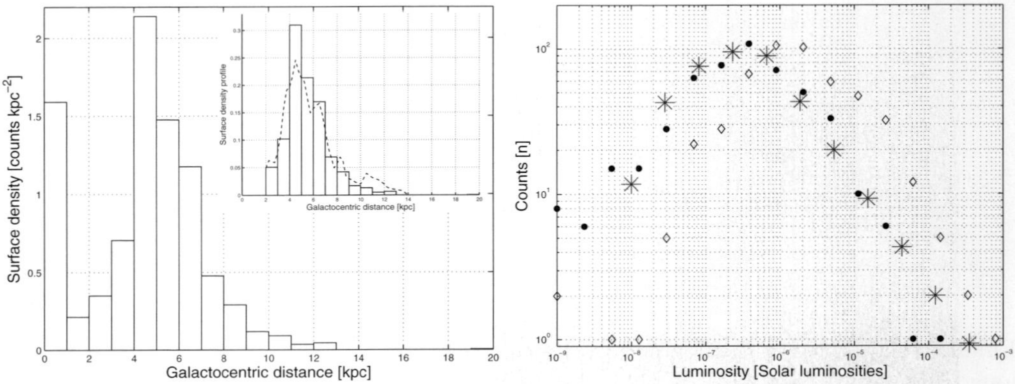

Figure 2. Left: Surface density of the known methanol masers in Galactocentric distance. The distribution shows a peak at the same radius as OB associations. The inset shows a comparison of the surface density of the masers with the molecular gas in the Milky Way. Right: Luminosity distribution of methanol masers in the Galaxy. Diamonds and circles plot all the (ambiguous) sources at the near- and far heliocentric distance, respectively. Stars show a synthetic luminosity distribution obtained by observing with a 2 Jy sensitivity a population of 5000 masers spatially distributed according to a Gaussian and having an intrinsic luminosity distribution as a power-law between cutoffs of index -1.7.

Galaxy, $F(R)$. This will be convolved with the intrinsic luminosity distribution $G(L)$ and "observed" with a certain sensitivity to fit the Galactocentric distribution of observed masers shown in Fig. 2.

The free parameters in the fit are the real total number of masers in the Galaxy (the integral of $F(R)$) and the function $G(L)$. To keep the number of unknowns to a minimum $G(L)$ was expressed as a single power-law between two cutoffs. These cutoffs were fixed by taking a range of sensible luminosities from the literature, so that the only two free parameters in the fit were the index of the luminosity function and the total number of methanol masers in the Galaxy. The two parameters cannot be found simultaneously but a plausible range can be estimated. On the basis of previous estimates of the total number of sources in the Galaxy (van der Walt 2005), the index for the luminosity function was found to lie between -1.5 and -2.0. In Fig. 2 an example of observed synthetic luminosity distribution is shown (stars). The model consists of a total number of 5000 sources distributed in luminosity according to a power-law with index -1.7 and observed with a detection threshold of 2 Jy. The details of this work are presented in Pestalozzi *et al.* 2007a.

The significance of the intrinsic luminosity distribution of the masers is still not clear, both in terms of high-mass star formation and in terms of maser modelling. A major contribution will be given by the completion of the MMB Survey, that is expected to determine with a very high level of confidence how many methanol masers there are in the Milky Way and thus limiting one more parameter in the model presented here.

3. Particular study: the disc signature

NGC7538 is a very well known and deeply studied high-mass star forming region in in the Perseus arm. It contains some 11 catalogued IR sources spanning over a wide range of ages. The centre of the figure is occupied by an UC HII region (IRS 1) that hosts bright 6.7 and 12.2 GHz methanol masers (see the 6.7 GHz spectrum in Fig. 3).

All spectral features have been observed at high angular resolution using MERLIN and the EVN. In particular, positions were recently obtained for features F and G and

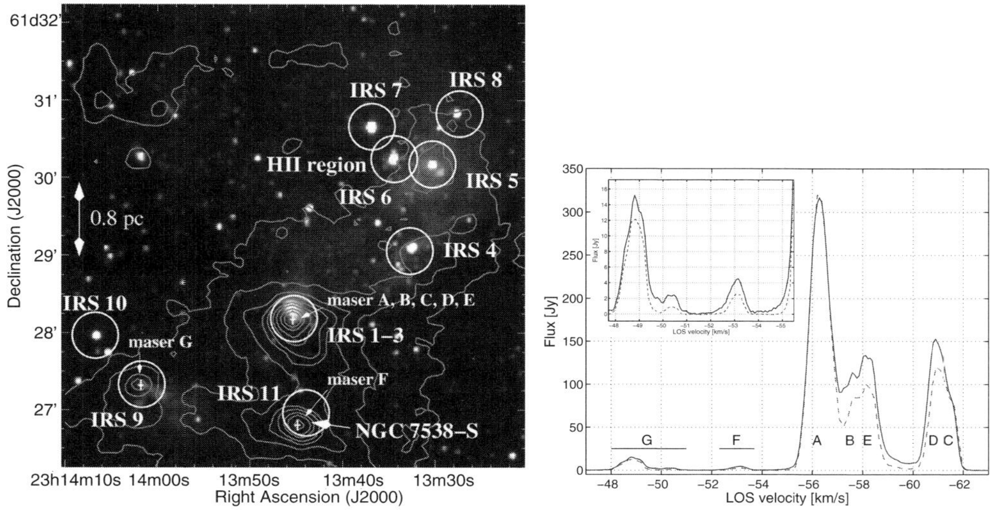

Figure 3. Left: General view of the high-mass star forming region NGC7538. Grey contours of the 1.2 mm dust continuum are overlaid on a JHK 2-mass image. Circles indicate the catalogued IR sources, crosses the methanol masers. The methanol masers appear in association with the youngest and most massive objects. Right: Spectrum of the 6.7 GHz maser emission toward NGC7538. Features A, B, C, D, E are cospatial with IRS 1. The maser disc is feature A. The inset is a zoom of the features F and G. The 12.2 GHz spectrum is similar to the 6.7 GHz one.

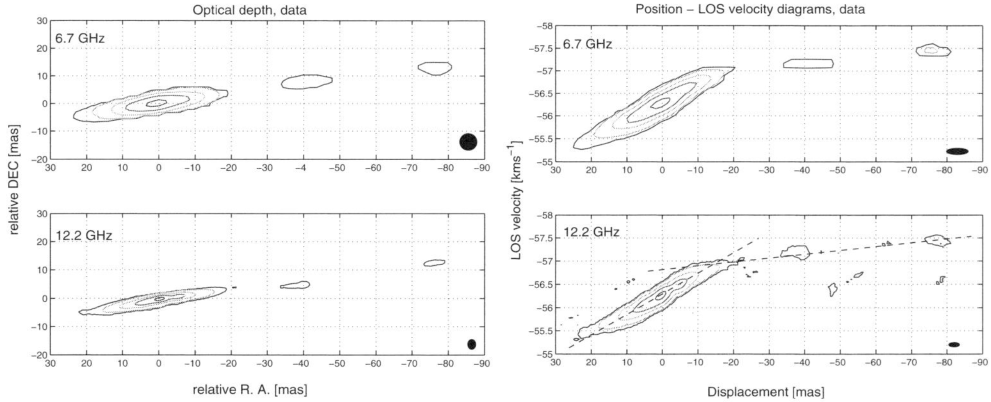

Figure 4. Maps (left) and velocity-displacement diagrams (right) of the main spectral feature of the 6.7 and 12.2 GHz methanol masers toward NGC7538 IRS 1. The beam size is shown in the bottom right corner in all panels. The contours are 1, 5, 10, 30, 50, 70, 90% of the peak in all panels. Dashed lines indicate the change in gradient between the main feature and the outliers.

appear to be cospatial with the youngest and most massive sources in the region, namely NGC7538 S and IRS 9, respectively (Pestalozzi *et al.* 2006). By determining the position of the masers with an accuracy of better than 30 mas, the authors claim to have accurately determined the position of the protostars.

The main spectral feature (A in Fig. 3) reveals a single, smooth spatial feature 6-20 beams across and shows a mainly linear LOS velocity gradient (Fig. 4).

The linear structure in both space and LOS velocity was naturally interpreted as arising from a (Keplerian) rotating disc seen edge-on amplifying a background continuum, and where the *maser spots* are orbiting a central object at a fixed radius (Minier *et al.* 1998). The maser emission was subsequently modelled as coming from a uniformly distributed material in a rotating disc seen edge-on and confined between an inner and outer radii. This avoided the assumption of a single radius and produced astonishingly accurate fits to the data (Pestalozzi *et al.* 2004).

Assuming amplification of a background continuum I_B, the optical depth of the maser emission at displacement θ_x and θ_y from the centre of each frequency plane with LOS velocity v is expressed by $\tau(\theta_x, \theta_y, v) = \ln(I(\theta_x, \theta_y, v)/I_B)$. In the case of a thin disc, one spatial dimension can be dropped and the modelling reduces to the study of the position-LOS velocity diagram of the normalised optical depth $F(\theta_x, v) = \tau(\theta_x, v)/\tau_0$, where τ_0 is the optical depth at displacement 0. The function F is then expressed as:

$$F(\theta_x, v) = \int \eta(\rho) \exp\left[-\frac{1}{2}\left(\frac{v - \Omega(\rho)\,\theta_x}{\Delta v_D}\right)^2\right] \frac{d\rho}{\sqrt{1 - (\theta_x/\rho)^2}} \tag{3.1}$$

where $\Omega(\rho)$ is the angular velocity at every radius and the product $\Omega(\rho)\,\theta_x$ is the LOS component of the rotation velocity. Δv_D is the linewidth at $\theta_x = 0$ and η is the gain of the maser that also defines the limits of integration. Pestalozzi *et al.* (2004) modelled η as a power-law with index 0.5 and with sharp cutoffs at inner and outer radii ρ_{in} and ρ_{out}.

The aim is to reproduce the right panels in Fig. 4. Two features are of extreme importance in this matter: the fact that most of the emission is symmetric around a peak (that we will consider to be the central LOS across the disc) and that there is a significant drop of the overall gradient at displacements larger than 10-15 mas. These two facts greatly constrain the dynamics in the modelling.

The study of F is best summarised by the study of the *spine*, defined by the maximum of F at every LOS, $\partial F/\partial v = 0$. In the case $\Omega(\rho) = constant$, i.e. solid body rotation, the study of the spine becomes very much simplified, as the exponential in eq. 3.1 comes out of the integral. The location of the spine is defined by $v - \Omega\,\theta_x = 0$ and the optical depth on the spine by the integral. The spine is then a straight line with constant slope (Ω). F on the spine will have maxima at displacements tangent to the inner edge of the disc and a local minimum at $\theta_x = 0$. The spectrum will have two features at $v = \pm\Omega\theta_{in}$, where θ_{in} is the LOS tangent to the inner edge. In the special case of $\Omega = 0$ (quiescent disc) the spine will be parallel to the displacement axis, F on the spine will be the same as the previous case and the spectrum will have only one feature, centered at $v = 0$ and of width Δv_D. The above considerations unambiguously show that solid body rotation (and a quiescent disc) cannot reproduce the data in Fig. 4, differential rotation is required.

The study of the spine for $\Omega \neq const$ gives a constraint on the rotation velocity: in order for F to have a maximum at $\theta_x = 0$, the rotation velocity at the inner edge has to be at least some 3–4 times higher than Δv_D (Pestalozzi *et al.* 2007b). Also, in order to account for the drop in gradient (dashed lines in the lower right panel of Fig. 4), the spine analysis indicates that the dynamics has to be Keplerian, i.e. $\Omega \sim \rho^{-3/2}$. Finally, it is important to notice that if the rotation velocity does not meet the above constraint, both spectrum and F-profiles (F on the spine) will have three peaks of emission, in the same way that was predicted by e.g. Ponomarev *et al.* (1994).

The general conclusion at this point is that a maximum of emission in the centre of both map and velocity-displacement diagrams is a strong hint for (fast) differential rotation seen edge-on. This is also supported by the existence of a clear bend of the gradient in the velocity-displacement diagram. These two facts unambiguously exclude solid body rotation (and a quiescent disc) as these would produce linear features both in space and velocity but no maximum in the centre, and no bend in the velocity diagram.

4. Summary and future views

In this paper I have tried to show how the study of methanol masers themselves (as opposed to the study of the environment of methanol masers) can lead to important conclusions about the maser emission (luminosity function) and the hosts of the masers (disc signature and accurate localisation of high-mass protostars).

The luminosity function of the maser will be refined in the near future thanks to the outcomes of the MMB Survey, which will also yield the ultimate census of the population of methanol masers in the Galaxy. The study of the accurate luminosity function of the masers will give insights into some characteristics of the maser mechanism, instrumental for extracting the physical conditions of the maser environment from the maser emission itself.

By accurately modelling the maser emission in one source, we were able to unambiguously determine what the signature for a circumstellar (differentially) rotating disc is as opposed to e.g. a bipolar outflow. Such model will be applied to other sources in order to test the existence of thin discs around high-mass protostars. The finding (or not!) of further discs around high-mass protostars will contribute in refining the high-mass star formation scenario.

Furthermore, there is a strong call for studies of the Galactic structure of the Milky Way, as methanol masers seem to represent fairly equally all Galactic longitudes (and latitudes) and carry an important dynamical information. This is unprecedented, as previous studies of the structure of our Galaxy were relying on the observation and positioning of HII regions.

One aspect that will be crucial for the future studies of methanol masers will be the gathering of data on the extended maser emission. As has been observed by Minier *et al.* (2002) and Harvey-Smith & Cohen (2006), it appears that maser emission can probe large surroundings of a high-mass star forming region. The mechanisms of this low brightness maser emission will have to be understood in order to be able to correctly interpret the data and increase our understanding of the physical conditions around high-mass protostars.

Finally it is exciting to keep in mind that the next few years will be characterised by large scale Galactic legacy surveys at a number of wavelengths. In particular, the SCUBA-2 surveys (SASSy and JPS) will yield the most complete catalogue of star forming clouds in the Milky Way detected at sub millimetre wavelengths. The cross correlation of these catalogues with the final MMB catalogue will enable the definition of an evolutionary sequence for the initial stages of the formation of high-mass stars. More in the future, the Herschel survey of the Galactic Plane will give the long waited view of the Galaxy at previously unavailable frequencies. Again, this will allow an accurate characterisation of methanol maser hosts and ultimately the definition of an accurate evolutionary sequence for high-mass star formation.

Acknowledgements

The present paper is a summary of the work done in collaboration with an extended group of people to whom I am deeply thankful: A. Chrysostomou and J. Collett (Univ. of Hertfordshire, UK), M. Elitzur (Univ. of Kentucky, USA), V. Minier (Saclay, Paris, France), J. Conway (Onsala Space Observatory, Sweden), R. Booth (HartRAO,

South Africa), and the NGC7538 collaboration (see the website http://star-www.herts.ac.uk/~michele/website_ngc/index.htm). I also thank D. Nutter for careful reading of the final manuscript.

References

Bronfman, L., Casassus, S., May, J. & Nyman, L. -Å. 2000 *A&A*, 358, 521

Busfield, A. L., Purcell, C. R., Hoare, M. G., Lumsden, S. L., Moore, T. J. T. & Oudmaijer, R. D. 2006 *MNRAS*, 366, 1096

Caswell, J. L., Vaile, R. A., Ellingsen, S. P., Whiteoak, J. B. & Norris, R. P. 1995, *MNRAS*, 272, 96

Cragg, D. M., Sobolev, A. M. & Godfrey, P. D. *MNRAS*, 360, 533

Dame, T. M., Ungerechts, H., Cohen, R. S., Geus, E. J., Grenier, I. A., May, J., Murphy, D. C., Nyman, L. -Å. & Thaddeus, P. 1987, *ApJ*, 322, 1706

DeBuizer, J. M. 2003, *MNRAS*, 341, 277

De Buizer, J. M. & Minier, V. 2005, *ApJ*, 628, L151

Ellingsen, S. P., von Bibra, M. L., McCulloch, P. M., Norris, R. P., Deshpande, A. A. & Phillips, C. J. 1996, *MNRAS*, 280, 378

Fux, R. 1999 *A&A*, 345, 787

Goedhart, S., Gaylard, M. J. & van der Walt, D. J., 2004 *MNRAS*, 355, 553

Hill, T., Burton, M. G., Minier, V., Thompson, M. A., Walsh, A. J., Hunt-Cunningham, M. & Garay, G. 2005, *MNRAS*, 363, 405

Harvey-Smith, L. & Cohen, R. J. 2006, *MNRAS*, 371, 1550

Minier, V., Booth, R. S., & Conway, J. E. 1998, *A&A*, 336, L5

Minier, V., Booth, R. S. & Conway, J. E. 2002, *A&A*, 383, 614

Minier, V., Ellingsen, S. P., Norris, R. P. & Booth, R. S. 2003, *A&A*, 403, 1095

Pandian, J. D., Goldsmith, P. F. & Deshpande, A. A. 2007, *ApJ*, 656, 255

Pestalozzi, M. R., Humphreys, E. M. L. & Booth, R. S. 2002a, *A&A*, 384, L15

Pestalozzi, M. R., Humphreys, E. M. L. & Booth, R. S. 2002b, in: Migenes, V. and Reid, M. (eds.), *Cosmic Masers: from Proto-Stars to Black Holes*, IAU 206 (ASP Conference Series), p. 139

Pestalozzi, M. R., Elitzur, M., Conway, J. E., & Booth, R. S. 2004, *ApJL*, 603, L113, erratum: 2004, ApJ, 606, L173

Pestalozzi, M. R., Minier, V., & Booth, R. S. 2005, *A&A*, 432, 737

Pestalozzi, M. R., Minier, V., Motte, F., & Conway, J. E. 2006, *A&A*, 448, L57

Pestalozzi, M. R., Chrysostomou, A., Collett, J. Minier, V., Conway, J. E. & Booth, R. S. 2007a, *A&A*, 463, 1009

Pestalozzi, M. R. and Elitzur, M. and Conway, J. E. 2007b, *in prep. for ApJ*

Ponomarev, V. O. and Smith, H. A. and Strelnitski, V. S. 1994, *ApJ*, 424, 976

Purcell, C. and Balasubramanyam, R. and Burton, M. G. *et al.* 2006, *MNRAS*, 367, 553

Schutte, A. J., van der Walt, D. J., Gaylard, M. L. & MacLeod, G. C. 1993, *MNRAS*, 261, 783

Sobolev, A. M., Cragg, D. M. & Godfrey, P. D. 1997, *MNRAS*, 288, L39

Szymczak, M., Hrynek, G. & Kus, A. J. 2000, *A&AS*, 143, 269

Szymczak, M., Kus, A. J., Hrynek, G., Kepa, A. & Pazdereski, E. 2002, *A&A*, 392, 277

Walsh, A. J., Burton, M. G., Hylard, A. R., & Robinson, G. 1998, *MNRAS*, 301, 640

van der Walt, J. 2005 *MNRAS*, 360, 153

Wiesemeyer, H., Thum, C. & Walmsley, C. M. 2004 *A&A*, 428, 479

Discussion

Y. Rodriguez: What is the nature of the outliers in your maps of the NGC7538 maser emission at displacements >30 mas?

Pestalozzi: These are considered to be local enhancements of the optical depth that do not fall on the spine. They helped in constraining the drop of gradient that finally indicated Keplerian rotation.

Astrophysical Masers and their Environments
Proceedings IAU Symposium No. 242, 2007
J.M. Chapman & W.A. Baan, eds.

doi:10.1017/S174392130701263X

Periodic variations in 6.7-GHz methanol masers

Sharmila Goedhart[1], Michael J. Gaylard[1] and D. Johan van der Walt[2]
[1]Hartebeesthoek Radio Astronomy Observatory, PO Box 443, Krugersdorp 1740, South Africa
email: sharmila@hartrao.ac.za
[2] Unit for Space Physics, North-west University, South Africa

Abstract. An intensive monitoring program of 54 6.7-GHz methanol maser sources was carried out at the Hartebeesthoek Radio Astronomy Observatory from January 1999 to April 2003. The monitoring program was subsequently continued on 19 sources of interest. Analysis of the resulting time-series stretching over eight years shows that six of the sources are periodic, with periods ranging from 133 days to 504 days. The waveforms in individual sources range from sinusoidal fluctuations to sharp flares and there can be other long term trends in the time-series. The amplitudes of the variations can also change from cycle to cycle. The time-series of the periodic masers will be presented, and possible causes of the variability discussed.

Keywords. masers, radio lines: ISM, stars: formation

1. Introduction

The existence of variability in 6.7-GHz masers was established by Caswell, Vaile & Ellingsen (1995) and MacLeod & Gaylard (1996). However, the nature of the variability had not been fully characterised until the study of Goedhart, Gaylard & van der Walt (2004) who monitored 54 sources for four years. The results of this study showed a range of variability behaviour, including monotonic, quasi-periodic and what appeared to be periodic variations. The monitoring programme has been continued on the most variable sources, enabling the confirmation of periodicity in six sources, despite considerable upgrades to the telescope during this time. The time-series of the periodic candidates are presented here.

2. Observations

The observations were done using the 26-m telescope at the Hartebeesthoek Radio Astronomy Observatory. The observations from January 1999 to April 2003 were of left-circular polarisation using a 256-channel spectrometer. Observations resumed in September 2003 with the cryogenic receiver upgraded from single to dual polarisation, a 2x1024 channel spectrometer and a new telescope control system. The telescope surface was replaced by solid panels, with the final alignment taking place in September 2004.

Amplitude calibrations were based on monitoring of Virgo A, Hydra A and 3C123. The calibrations were checked against the relatively quiescent source G351.42+0.64.

3. Results

Periodicity analysis was done on time-series extracted from velocity channels at the maser peaks. Fourier analysis was done using the program PERIOD04 developed by Lenz

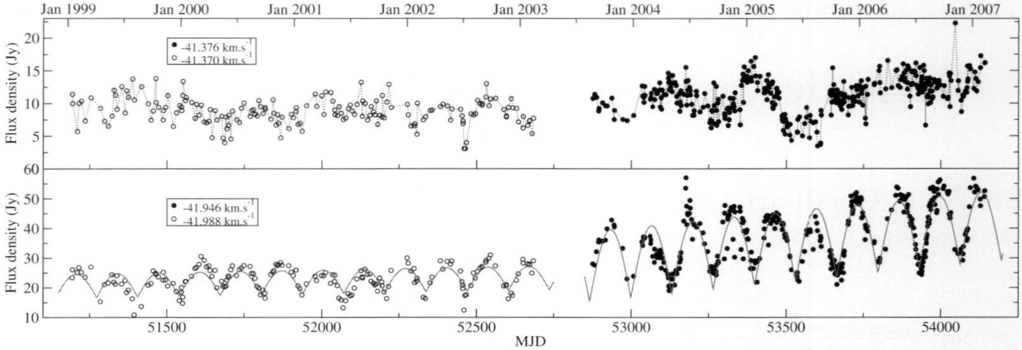

Figure 1. Time-series for selected peak velocity channels in G338.93-0.06. The solid curves show the result of fitting a first-order polynomial plus |sin|.

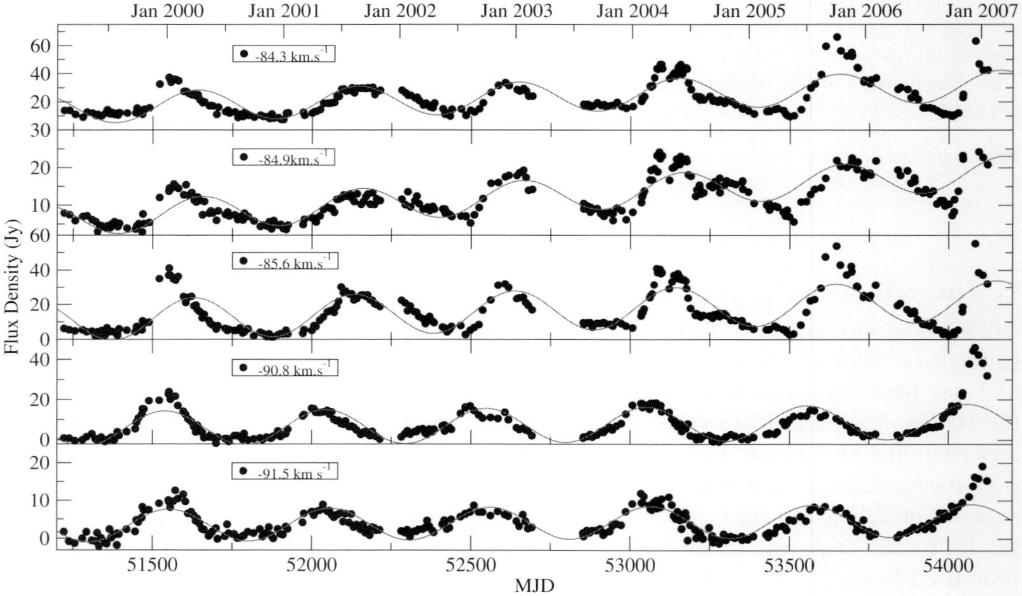

Figure 2. Time-series for selected peak velocity channels in G331.13-0.24. The solid curves show the results of fitting a sinusoid plus a first order polynomial.

& Breger (2005). Since many of the time-series are non-sinusoidal, the periods were confirmed by epoch-folding using the test statistic developed by Davies, 1990.

Time-series at velocity channels of interest will be presented. Figure 1 shows the time-series for the two strongest spectral features in G338.93-0.06. The intensity variations in a single velocity channel are shown. The variations in the two features are uncorrelated, with the feature at -41.9 km s^{-1} showing regular variations with a period of 133 days. Nineteen cycles have been observed to date in this source, with the overall intensity of the periodic feature steadily increasing. The amplitude of the variations have also been increasing. The shape of the light-curve can be fitted by a function of the form |sin|.

G331.13-0.24 shows variations with a period of 504 days (Figure 2). The maser features in the velocity range -92 to -89 km s^{-1} show extremely regular variations which follow a sinusoidal form for part of the cycle. The group at -87 to -83.5 km s^{-1} show regular variations with a time-delay from the start of the flares at the first group. The durations

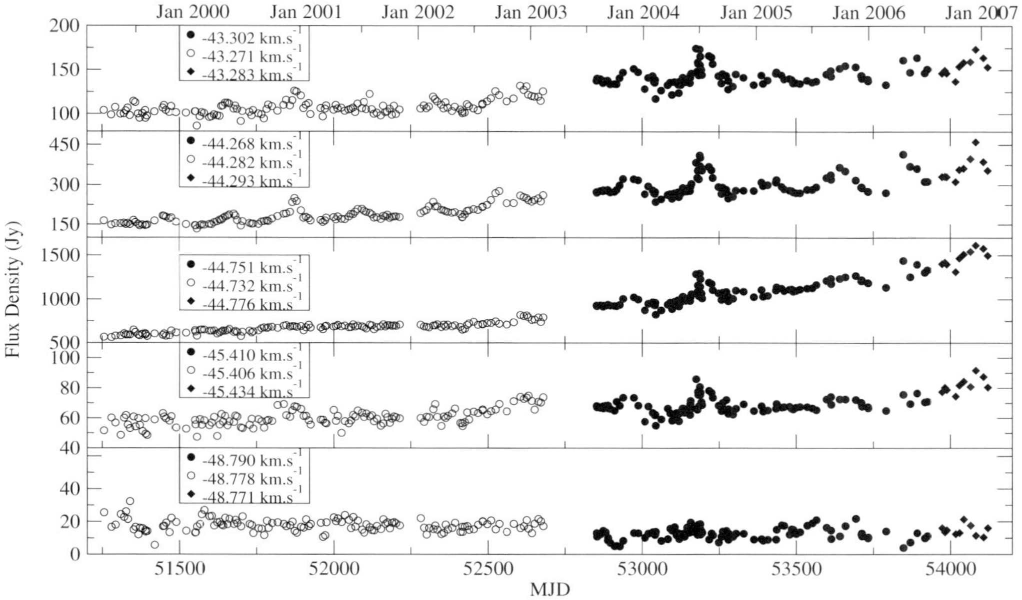

Figure 3. Time-series for selected peak velocity channels in G328.24-0.55.

of these flares vary, as well as the shape of the flares but the start times of the flares appear to be periodic.

G328.24-0.55 shows clear periodicity with a period of 220 days in only one spectral feature (Figure 3) at -44.3 km s^{-1}. Weak fluctuations with the same period are seen in the feature at -43.3 km s^{-1}. Interestingly, a large flare in June 2004 was seen at all features in the velocity range -46 to -43 km s^{-1}.

G339.62-0.12 shows periodic variations with a period of 201 days in all the spectral features (Figure 4). The peak at -35.7 km s^{-1} generally has low amplitude variations but occasionally shows very strong flares.

G9.62+0.20 shows periodic flaring with a period of 244 days. This source is known to show simultaneous flaring at 12.2 GHz with an even greater amplitude (Goedhart, Gaylard & van der Walt, 2003; Gaylard & Goedhart, this volume).

G188.95+0.89 showed low amplitude sinusoidal variations with a period of 404 days (Figure 6). However, the behaviour of this source has been changing since mid-2003, with the spectral feature at -11.4 km s^{-1} decreasing steadily in intensity. Linear detrending of the time-series reveals that the sinusoidal variations persisted until ~mid-2006. There was a sharp increase in intensity in August 2006, marking the time at which the time-series no longer show sinusoidal behaviour. This source may be at the end of the periodic phase in its lifetime or the cycle pattern is changing. Further monitoring will be necessary to see what happens next.

4. Discussion

The environment in which methanol masers are found is complex, with a number of factors that could affect the maser intensity. The radiation for pumping the masers originates at the central star but will have been considerably reprocessed by the time it arrives at the maser region. The central (proto)star may be surrounded by a rotating, inhomogeneous disk which may also be feeding an outflow. The outflows may be sporadic

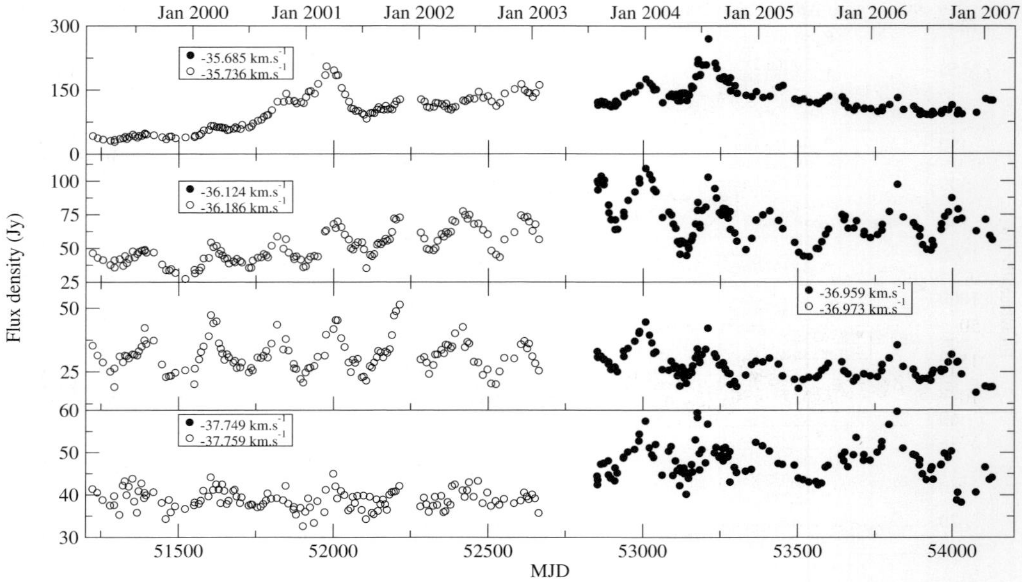

Figure 4. Time-series for selected peak velocity channels in G339.62-0.12.

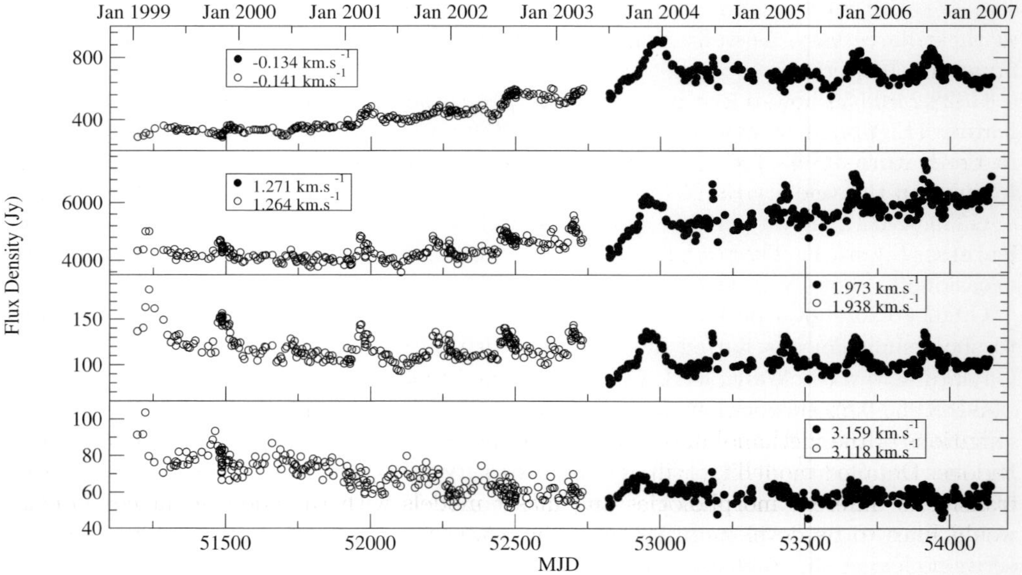

Figure 5. Time-series for selected peak velocity channels in G9.62+0.20.

or there may be precessing jets. The central object may even be a close binary which could still have associated accretion disks, leading to complex dynamical interactions. The source of the pump photons, believed to be the HII region, could also be variable. Interpretation of the variability is also limited by the uncertainty of where the masers are situated in the circumstellar environment. Current observational evidence is unclear but the masers may be tracing circumstellar disks, outflow cavities or weak shocks. Periodic mechanisms can be invoked for any of these scenarios.

The variations in the masers shown here appear to have a periodic component with periods in the range 133 to 504 days. This must be pointing to a specific mechanism. The

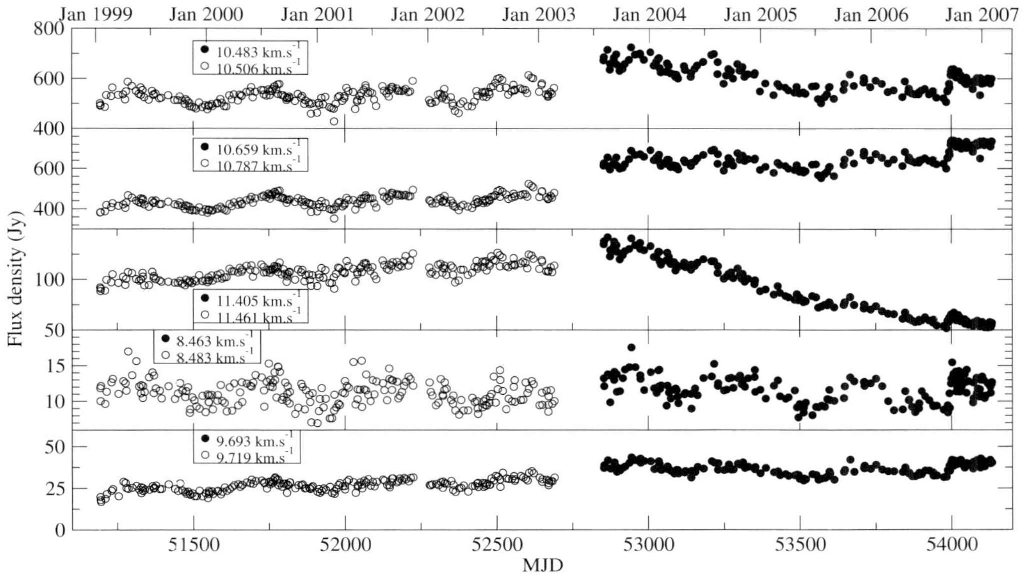

Figure 6. Time-series for the peak velocity channels in G188.95+0.89.

typical periods are much larger than those seen in pulsating stars, with the exception of long-period variables such as Miras. However, these are evolved stars and unlikely to be found in the vicinity of the methanol masers. O and B type stars have extremely rapid rotation rates, around 0.5 to 2 days, which are far too short to be related to the observed periodicity. Keplerian orbits within 1 to 10 AU of the central star have periods in the observed range. However, it is not clear what the orbiting objects would be or how this would lead to a modulation in the masers. Since there is a high incidence of binarity or multiplicity in optically visible stars, it is not inconceivable that a binary system could be present. On the other hand, we may be seeing a process related to an accretion disk.

5. Conclusion

Periodicity was found in $\sim 10\%$ of the original sample of 54 sources monitored. No clear cause of the variations can be found at this stage. High-resolution, multi-wavelength observations of the methanol maser sources are necessary to understand conditions in these regions. Detailed modelling will also be necessary to understand the effects of radiative transfer in different morphologies and maser models with time-dependent components would have to be developed. This may be a way to gain greater insight into processes occurring in regions that can not be directly observed.

References

Caswell, J. L., Vaile, R. A., & Ellingsen, S. P., 1995, *Publ. Astron. Soc. Aust.* 12, 37

Davies, S. R., 1990, *MNRAS* 244, 93

Goedhart, S., Gaylard, M. J., & van der Walt, D. J., 2003, *MNRAS* 339, L36

Goedhart, S., Gaylard, M. J., & van der Walt, D. J., 2004, *MNRAS* 355, 553

Lenz, P., & Breger, M, 2005, *Commun. Astroseismology* 146, 53

MacLeod, G. C., & Gaylard, M. J., 1996, *MNRAS* 280, 868

Astrophysical Masers and their Environments
Proceedings IAU Symposium No. 242, 2007
J.M. Chapman & W.A. Baan, eds.

doi:10.1017/S1743921307012641

The infrared environments of masers associated with star formation

James De Buizer
Gemini Observatory, Casilla 603, La Serena, Chile
email: jdebuizer@gemini.edu

Abstract. The near infrared (1–2 μm) and the thermal infrared (3–25 μm) trace many of the environments in which masers are thought to reside, including shocks, outflows, accretion disks, and the dense medium near protostars. After a number of recent surveys it has been found that there is a higher detection rate of mid-IR emission towards masers than cm radio continuum emission from UC HII regions, and that the mid-IR emission is actually more closely cospatial to the maser locations. A high percentage of water and methanol masers that are not coincident with the UC HII regions in massive star forming regions are likely to be tracing outflows and extremely young high mass stars before the onset of the UC HII region phase. After a decade of groundwork supporting the hypothesis that linearly distributed class II methanol masers may generally trace accretion disks around young massive stars, compelling evidence is mounting that these masers may generally be associated with outflows instead. Substantiation of this claim comes from recent outflow surveys and high angular resolution mid-IR imaging of the maser environments.

Keywords. masers, accretion disks, ISM: jets and outflows, stars: formation, stars: early-type, infrared: ISM

In the proceedings of the last maser meeting in Brazil in 2001, I discussed the largely untapped potential of the infrared spectral regime as a tool for understanding the nature of the circumstellar environments of masers associated with star formation (De Buizer 2002a). In that article, I described the early stages of the work I had started using the infrared (IR) to study maser environments, and gave several possible paths for future work in the field. In the intervening years since that meeting and this one, I have explored several of those paths in detail, and will present in this article some of the highlights of that work.

1. The close association of masers and IR emission

In the Brazil proceedings article I described how the near infrared (1–2 μm) and the thermal infrared (3–25 μm) trace everything from the direct photospheric emission from young stellar sources to the relatively cool dusty environments that can be distributed far from their parent stars. The thermal infrared in fact has been shown to trace dusty circumstellar environments from temperatures of ~1500 K at 3 μm, which is very close to a star and near the dust sublimation temperature, to cool dust tens of thousands of AU away at wavelengths near 25 μm. Because of this, it was suggested that the IR would be a good tracer of the emission from many of the environments in which masers were thought to reside: shocks fronts, outflow shocks, and accretion disks.

But how well does the infrared trace the environments of maser emission? There now exist a number of studies at several different wavelengths to help address this question. Since it is easiest to obtain radio continuum observations of maser regions, the most common studies of maser environments have traditionally involved searches for ionized gas emission in the form of ultracompact (UC) HII regions and partially ionized outflows. Well before the Brazil meeting it was known that the percentage of maser sources directly associated with cm radio continuum emission was not very high, which was remarkable for a phenomenon that is suppose to trace young massive star formation.

From studies with reasonable detection limits at a variety of wavelengths, one can piece together the dominant wavelengths at which maser environments can be seen and studied. As mentioned above, the cm radio continuum detection rate towards methanol and water masers is low, and has been found to be about 20% from the recent surveys of Walsh *et al.* (1998) and Beuther *et al.* (2002). Whereas in the surveys of De Buizer *et al.* (2000), Walsh *et al.* (2001), and De Buizer *et al.* (2005) it was found that the detection rate of mid-IR (typically 10 and 18 μm) emission towards class II methanol masers is ~70%, and towards water masers ~80%. Coarser spatial resolution observations in the sub-mm and mm are showing detection rates approaching 100% towards both water and methanol masers (Walsh *et al.* 2003; Beuther *et al.* 2002). Hence, all masers appear to be more closely associated with regions of hot to warm (300-30K) thermal dust emission than ionized gas emission.

Furthermore, De Buizer *et al.* (2005) showed that mid-IR emission when detected in a maser region was more closely associated with actual maser locations than the radio continuum emission detected towards those same maser regions (i.e., Figure 1b). The survey of Hofner & Churchwell (1996) showed that the median separation between water maser and the radio UC HII regions they detected is ~18800 AU. The median separation between water masers and mid-IR sources is ~8700 AU (De Buizer *et al.* 2005). Therefore, not only is there a higher detection rate of mid-IR emission towards masers than cm radio continuum emission, mid-IR emission is actually more closely cospatial to the maser locations.

2. Using masers to pin-point extremely young massive stars

A large subset of water and class II methanol masers are found in regions of massive star formation, but are not coincident with the UC HII regions found there. What are these masers tracing?

In the case of the G29.96-0.02 region, the methanol and water masers are offset from the cometary UC HII region and were instead found to be coincident with a hot molecular core discovered in the ammonia line images by Cesaroni *et al.* (1994). Many hot molecular cores are high mass protostellar objects (HMPOs) which are at an extremely early stage of massive stellar birth. A HMPO consists of a massive protostar surrounded by a thick envelope of accreting dust and gas. They are compact sources seen in radio-wavelength ammonia (or molecular line) images but are so young that they have not had time to ionize their surroundings, and hence typically have little to no detectable radio continuum emission of their own. Other cases similar to G29.96-0.02, where masers were found associated with hot molecular cores (HMCs) and offset from UC HII regions, lead to the idea that some maser species may generally trace these very young sources at the earliest stages of massive star formation.

Some of the first models of the HMPO phase of massive star formation that were constructed (e.g. Osorio *et al.* 1999) showed that by changing parameters such as spectral type, accretion rates, and core radius, one can construct a variety of model spectral energy distributions (SEDs) for HMPOs. Throughout an array of physically reasonable parameter space, the SEDs peak around 80 μm on average. On the Rayleigh-Jeans side of the SED (>80 μm), there is little change in the shape of the SED when parameters are changed. However, on the Wien side of the SED (<80 μm), there is considerable change in

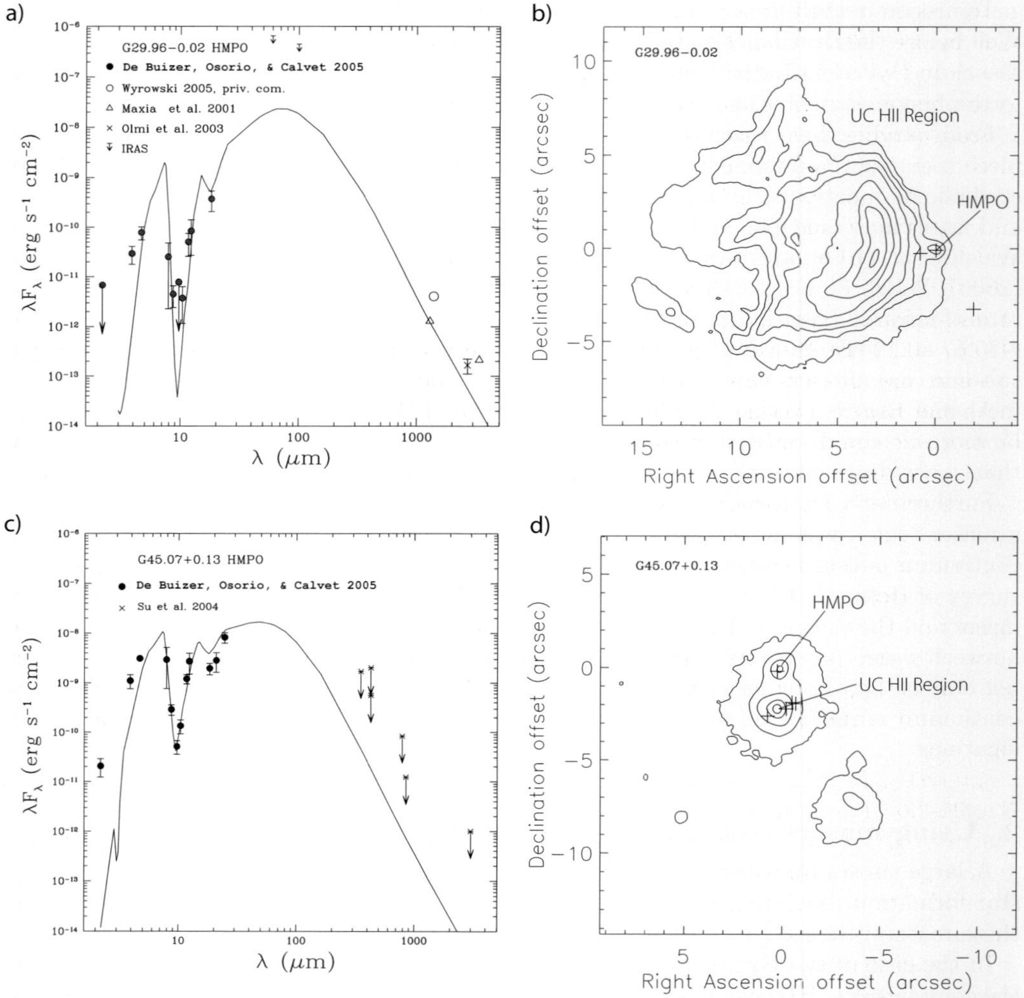

Figure 1. **(a)** Observed and model SEDs for the G29.96-0.02 HMPO. The different symbols represent the observed values of the flux densities from the references indicated. Error bars are 3–σ for the IR data. Upper limits are represented by arrows. The solid line represents the best fit model obtained with $L_\star = 1.8 \times 10^4$ $L_\odot$, the equivalent luminosity of a B0 star. **(b)** A contour plot of the 11.7 μm image of the G29.96-0.02 field with the extended mid-IR emission from the UC HII region and compact mid-IR emission of the HMPO indicated. Plus signs represent the location of water masers from Hofner & Churchwell (1996). The masers are not associated with the UC HII region, but some are associated with the HMPO. **(c)** Same as in (a) but for the G45.07+0.13 HMPO. The solid line represents the best fit model obtained with $L_\star = 2.5 \times 10^4$ $L_\odot$. **(d)** Same as in (b) but for G45.07+0.13. In this case, there are some water masers associated with the UC HII region, however the clump of masers to the north are associated with a HMPO.

SED shape with the variation of each parameter, especially if one could observe the mid-IR emission of these sources (see Figure 1a and 1c). The most noticeable changes in the HMPO SEDs are between 3 and 30 μm, and in the depth and shape of the 10 μm silicate feature. This modeling showed that the mid-infrared could be key to understanding the properties of the youngest massive stars, if we could detect the mid-IR emission from such embedded and distant objects.

To see if we could indeed detect these HMPOs in the mid-IR and test the hypothesis that masers offset from UC HII regions may be associated with HMPOs, I began my search in the mid-IR with the prototypical source, G29.96-0.02. These observations led to the first confirmed direct detection of a HMC at mid-IR wavelengths (De Buizer *et al.* 2002a), proving that indeed some HMPOs are bright enough to observe in the mid-IR. To further test the hypothesis that non-radio continuum emitting sources of maser emission could be associated with HMPOs, a mid-infrared study of several fields from the survey of water maser and UC HII regions of Hofner & Churchwell (1996) was performed in an attempt to find more mid-infrared bright HMPOs. Concentrating on the fields that have UC HII regions with water masers well offset, this survey led to the detection of mid-infrared emission from the locations of two HMPO candidates, G11.94-0.62 and G45.07+0.13 (De Buizer *et al.* 2003). These observations seemed to support the idea that in some cases masers can indeed be used to find the locations of HMPOs.

At the Brazil meeting I claimed that these mid-IR-bright sources held the promise that, if one could perhaps create a well-sampled SED from observations of a HMPO at many wavelengths in the mid-infrared, the data could then be fit with the new HMPO models to derive accurate physical parameters, such as mass, luminosity and accretion rate for these youngest massive stars.

This was attempted by De Buizer, Osorio, & Clavet (2005), for the three above mentioned sources: G29.96-0.02, G11.94-0.62, and G45.07+0.13. Though the modeling performed on these sources led to relatively accurate luminosities, degeneracies and a large number of free parameters prevent the accurate assessment of physical parameters from SED modeling alone and can only constrain such parameters. More recent and more detailed SED modeling of YSOs by Robitaille *et al.* (2007) have come to the same conclusion.

Interestingly, of the three sources modeled in De Buizer, Osorio, & Calvet (2005), G11.94-0.62 has a luminosity to small to be a true HMPO. Though this source lies near an UC HII region, is cospatial with water maser emission, and has a deep silicate feature in the mid-infrared indicative of a highly embedded star, it is likely a young and embedded intermediate mass YSO. Hence, there exist intermediate mass YSOs that have similar observational characteristics to HMPOs that may contaminate HMPO studies.

From all of this, one can conclude that some masers offset from UC HII regions can indeed be associated with HMPOs. However, in De Buizer *et al.* (2005) it is shown that some of these masers are likely offset from known YSOs because they are associated with outflows, and some, as just mentioned, may be associated with less-massive embedded YSOs.

3. Maser disks vs. maser outflows

Some maser species, class II methanol in particular, have been suggested to be associated with circumstellar disks around massive young protostars. Norris *et al.* (1993) discovered that 45% of methanol masers groups tend to be distributed in the sky in linear structures. Occasionally these linearly distributed masers have apparent velocity gradients along the maser distributions that suggest rotation. Observations by Walsh *et al.* (1998) confirmed that methanol masers are linearly distributed in approximately half the cases, but that velocity gradients were not a general feature.

However, in the article by Norris *et al.* (1993) it was suggested that these linearly distributed methanol masers exist in, and delineate, edge-on circumstellar disks around massive stars. In Brazil I presented results from my low resolution (1.0–2.0") mid-IR imaging survey of star forming regions with methanol maser emission, which contained

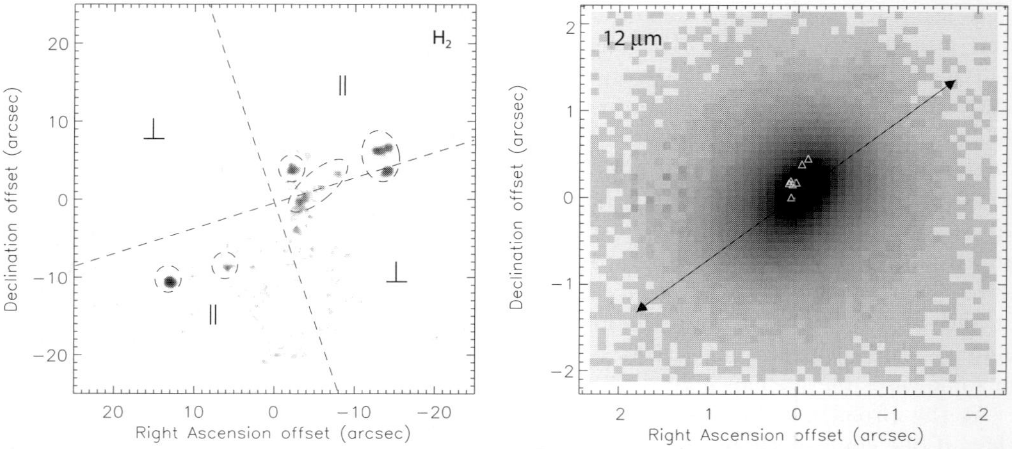

Figure 2. (Left) The H_2 line image from De Buizer (2003) of the ∼50" field of G318.95-0.20 centered on the methanol maser location. The dashed straight lines intersect at the maser location and show the parts of the field perpendicular to (⊥) and parallel to (||) the methanol maser distribution angle. Dashed ellipses show the location of real H_2 emission (other faint signatures are due to poor subtraction of continuum emission). The H_2 emission is distributed in the quadrants parallel to the linear methanol maser distribution angle. **(Right)** The 12 μm image of the ∼5" field of G318.95-0.20. The methanol masers of Norris *et al.* (1993) are shown as white triangles. The mid-IR emission is elongated at the same angle as the methanol maser distribution, however this is the direction of the outflow seen in H_2. The arrow shows the direction of the SiO outflow (De Buizer *et al.* in prep). The similarity of the methanol maser distribution angle and mid-IR emission elongation angle with the outflow angle suggest that both the maser and the mid-IR emission are outflow related, and not directly tracing a circumstellar disk.

10 sites of linearly distributed masers. Three of these maser locations were coincident with young stellar objects with extended mid-IR emission elongated at the same position angle as their methanol masers distributions (De Buizer *et al.* 2000). These sources were suggested to be dusty circumstellar disks, thereby apparently adding credence to the disk hypothesis for linearly distributed masers. However, one of these sources was later observed at high angular resolution (<0.5") with Keck and found to be three individual mid-IR sources arranged in a linear fashion, and not a disk (De Buizer *et al.* 2002b).

To further test the maser/disk hypothesis, several sites of linearly distributed methanol masers were observed to search for outflows. According to the standard model of accretion, during the phase of stellar formation where the star is being fed from an accretion disk, it is also undergoing mass loss through a bipolar outflow. The bipolar outflows are perpendicular to the plane of the accretion disk, and along the axis of rotation. Wide-field images of the sites of linearly distributed methanol masers were obtained using the 2.12 μm H_2 (ν=1-0) S(1) line as the outflow diagnostic (De Buizer 2003). Surprisingly, in 12 of 14 fields where H_2 emission was detected the emission was not aligned perpendicular to the maser distributions as expected, but instead was parallel (Figure 2). This seemed to suggests that the methanol masers delineate outflows rather than circumstellar accretion disks.

The interpretation of the results from De Buizer (2003) remained ambiguous because 2.12 μm H_2 line emission can be excited both by outflow shocks and by UV excitation. The overall morphology of the H_2 emission seen in the fields imaged by De Buizer (2003) does not resemble the bipolar outflows seen around young, low-mass stars. It was cautioned that perhaps some of the H_2 emission was perhaps due to excitation from UV emission by other massive stars in the star forming region. Without additional evidence, it could not be said with certainty which mechanism is stimulating the H_2 emission near these high-mass protostars, nor definitively link the outflows to the methanol masers. However, 86% of the sources showed H_2 emission parallel to their maser distribution angles, and the probability of this occuring by chance is low. Nonetheless, there was sufficient doubt to justify observing these sources in an independent outflow indicator.

SiO is a good shock tracer because its abundance is enhanced by factors of up to 10^6 behind strong shocks (i.e., Avery & Chiao 1996). Using the JCMT to try to detect the presence of outflows in the SiO (6-5) transition, nine sources from the H_2 survey of De Buizer (2003) were observed and SiO emission was found in seven of those fields (Feldman *et al.* 2005). Many of the stronger detections have line profiles with a relatively narrow core on top of wide wings which are characteristic of outflows. To follow this up, five of these JCMT sources were recently observed with the Australia Telescope Compact Array (ATCA) to map out the outflows in the SiO (2-1) transition. Preliminary results from these observations show that the SiO emission is in all cases distributed at an angle close to that of the maser distribution angles and H_2 emission position angles for each source. In four of the five cases these SiO maps show clear signs of a single outflow with red- and blue-shifted lobes. In the fifth case the velocity structure is more complex, indicating perhaps multiple outflows are present, however the overall structure of the SiO is parallel to the maser position angle (De Buizer *et al.* in prep). These SiO observations clearly indicate that the H_2 emission observed in Buizer (2003) is indeed outflow related. These observations also lend convincing support to the idea that the methanol masers in linear distributions are directly associated in some way with the outflow from their parent stars.

Recent high angular resolution observations of individual young massive stellar objects with masers (De Buizer 2006, De Buizer 2007, and as yet unpublished data) have revealed that many sources that are elongated in their mid-IR emission are not disks, but instead are elongated in the direction of outflow. The best example of this is G35.20-0.74 (De Buizer 2006), which shows an outflow cavity in mid-IR continuum emission unmistakeably similar to the outflow jet seen at various other wavelengths. It is believed that sources such as these are too embedded to directly detect their accretion disks even at mid-IR wavelengths. However if the outflows clear away enough material in the natal cloud surrounding a massive young stellar object, we are then able to see these cavities and the mid-IR emission of the warm cavity walls. More developed bipolar cavities should exist where the material above and below the accretion disk is well-cleared as the opening angles of the outflow cavities widen. In this case the definition of what is the surface of a outflow cavity and what is the surface of a flared accretion disk may become blurred, and the source could appear as a mid-IR silhouette disk. These disks would be similar to the near-IR silhouette disks seen in Orion (i.e., McCaughrean *et al.* 1998). However, the accretion disks would still be so optically thick at their mid-planes that one could not detect the direct emission from the disk itself. Detections of such sources may have already been made; possible examples of such sources are the massive young stellar objects in M17 (Chini *et al.* 2004) and IRAS 20126+4104 (De Buizer 2007).

In the case of G35.20-0.74 the OH, water, and methanol masers appear to delineate the edges of these outflow cavity walls as seen in the mid-IR. Therefore, it is a possibility that the linear arrangements of water and methanol masers associated with other sources are tracing cavity walls where there are oblique shocks for the collisional pumping of water masers and a sufficient mid-IR thermal bath of photons to radiatively pump the methanol masers.

Observations in the mid-IR of massive young stellar objects that only show signs of emission from their outflow cavities may also help interpret the mid-IR emission from more embedded HMPOs. Some HMPOs are not mid-IR bright, while others are. In many cases this may simply be a temperature issue, i.e. that some HMPOs are too cold to be seen at mid-IR wavelengths. However, HMPOs that are detectable in the mid-IR may just be sources where the outflows from the central stars are more or less pointed toward us. In this way, we are seeing through the cavity and closer to the central heating source since the outflow is clearing out material along our line of sight. Therefore the mid-IR emission observed to be coming from HMPOs like G29.96-0.02 may be tracing beamed emission in outflow cavities and not reprocessed emission coming from the entire core, as is seen at longer wavelengths. If this is true, SED models of such sources would be influenced by this and could lead to poor fits unless accounted for.

4. Conclusions

A lot has been learned about the infrared environments of masers since the 2001 Brazil maser meeting. The IR regime has been shown to complement radio continuum observations and offer new insights into maser environments.

Methanol, hydroxyl, and water masers appear to be most closely associated with regions of hot to warm (300-30 K) thermal dust emission traced by mid-IR and sub-mm emission. As such, studies of the IR environments of masers will continue to offer much insight into the relationship between masers and the massive star formation process, *especially at high (<0.5") angular resolution* where one can hope to separate IR emission from disks, envelopes, outflows, and other nearby stellar sources.

It has been found that water and methanol masers offset from UC HII regions trace mid-IR sources that are in some cases HMPOs. Therefore masers that appear to be isolated from traditional massive star formation tracers such as radio continuum emission may be great locations to find more HMPOs. If enough observations are made over the full SEDs of HMPOs, physical properties of these sources can be derived from accretion models. However, such SED modeling alone can only place constraints and limits on most of these physical parameters.

Evidence is building that shows that linear distributions of methanol masers are not generally tracing disks around young massive stars. These maser distributions were found to be dominantly at the same angle as extended near-IR H_2 emission, which is a potential outflow indicator. Follow-up observations in the mm using SiO as an outflow tracer appear to confirm the notion that outflows are dominantly at the same position angles as the linear methanol maser distributions of massive young stellar objects. This further implies a *general, direct physical relationship between the methanol masers and outflows*, rather than disks, however these observations do not rule out the possibility that masers may trace circumstellar disks or other phenomena in some cases.

Finally, high-resolution imaging is showing circumstellar mid-IR continuum emission commonly comes from outflows, not disks. This mid-IR emission is often found to be co-spatial with masers, and therefore further suggests that there is a general tendency for masers, especially water and methanol, to be often directly associated with outflows.

References

Avery, L. W., & Chiao, M. 1996, *ApJ* 463, 642

Beuther, H., Walsh, A., Schilke, P., Sridharan, T. K., Menten, K. M., & Wyrowski, F. 2002, *A&A* 390, 289

Cesaroni, R., Churchwell, E., Hofner, P., Walmsley, C. M., & Kurtz, S. 1994, *A&A* 288, 903

Chini, R., Hoffmeister, V., Kimeswenger, S., Nielbock, M., Nürnberger, D., Schmidtobreick, L., & Sterzik, M. 2004, *Nature* 429, 155

De Buizer, J. M., Pi na, R. K., & Telesco, C. M. 2000, *ApJ* (Supplement) 130, 437

De Buizer, J. M. 2002a, *Cosmic Masers: From Proto-Stars to Black Holes*, IAU Symposium 206 (San Francisco: ASP), vol. 206, p. 18

De Buizer, J. M., Watson, A. M., Radomski, J. T., Pi na, R. K., & Telesco, C. M. 2002a, *ApJL* 564, 101

De Buizer, J. M., Walsh, A. J., Pi na, R. K., Phillips, C. J., & Telesco, C. M. 2002b, *ApJ* 564, 327

De Buizer, J. M. 2003, *MNRAS* 341, 277

De Buizer, J. M., Radomski, J. T., Telesco, C. M., & Pi na, R. K. 2003, *ApJ* 598, 1127

De Buizer, J. M., Radomski, J. T., Telesco, C. M., & Pi na, R. K. 2005, *ApJ* (Supplement) 156, 179

De Buizer, J. M., Osorio, M., & Calvet, N. 2005, *ApJ* 635, 452

De Buizer, J. M. 2006, *ApJL* 642, 57

De Buizer, J. M. 2007, *ApJL* 654, 147

De Buizer, J. M., Redman, R. O., Feldman, P. A., Longmore, S., & Caswell, J. 2007, *in preparation*

Feldman, P. A., Redman, R. O., De Buizer, J. M., di Francesco, J., & Carey, S. J. 2005, *Astrochemistry: Recent Successes and Current Challenges*, IAU Symposium 231 (Poster Session), p. 111

Hofner, P., & Churchwell, E. 1996, *A&A* (Supplement) 120, 283

McCaughrean, M. J., *et al.* 1998, *ApJL* 492, 157

Maxia, C., Testi, L., Cesaroni, R., & Walmsley, C. M. 2001, *A&A* 371, 287

Norris, R. P., Whiteoak, J. B., Caswell, J. L., Wieringa, M. H., & Gough, R. G. 1993, *ApJ* 412, 222

Olmi, L., Cesaroni, R., Hofner, P., Kurtz, S., Churchwell, E., & Walmsley, C. M. 2003, *A&A* 407, 225

Osorio, M., Lizano, S., & D'Alessio, P. 1999, *ApJ* 525, 808

Robitaille, T. P., Whitney, B. A., Indebetouw, R., & Wood, K. 2007, *ApJ* (Supplement) 169, 328

Su, Y.-N., *et al.* 2004, *ApJL* 616, 39

Walsh, A. J., Burton, M. G., Hyland, A. R., & Robinson, G. 1998, *MNRAS* 301, 640

Walsh, A. J., Bertoldi, F., Burton, M. G., & Nikola, T. 2001, *MNRAS* 326, 36

Walsh, A. J., Macdonald, G. H., Alvey, N. D. S., Burton, M. G., & Lee, J.-K. 2003, *A&A* 410, 597

Astrophysical Masers and their Environments
Proceedings IAU Symposium No. 242, 2007
J.M. Chapman & W.A. Baan, eds.

doi:10.1017/S1743921307012653

A review of H_2CO 6 cm masers in the Galaxy

E. Araya[1,2], **P. Hofner**[1,2] **and W. M. Goss**[2]

[1]National Radio Astronomy Observatory, P.O. Box 0, Socorro, NM 87801, USA
[2]New Mexico Institute of Mining and Technology, Physics Department, 801 Leroy Place, Socorro, NM 87801, USA
email: earaya@nrao.edu

Abstract. We present a review of the field of formaldehyde (H_2CO) 6 cm masers in the Galaxy. Previous to our ongoing work, H_2CO 6 cm masers had been detected in the Galaxy only toward three regions: NGC 7538 IRS1, Sgr B2, and G29.96–0.02. Current efforts by our group using the Very Large Array, Arecibo, and the Green Bank Telescope have resulted in the detection of four new H_2CO 6 cm maser regions. We discuss the characteristics of the known H_2CO masers and the association of H_2CO 6 cm masers with very young regions of massive star formation. We also review the current ideas on the pumping mechanism for H_2CO 6 cm masers.

Keywords. masers, stars: formation, ISM: molecules, HII regions, radio lines: ISM

1. Introduction

Formaldehyde (H_2CO) was the first organic polyatomic molecule discovered in the interstellar medium. H_2CO is an asymmetric top molecule, however the asymmetry is small; the moment of inertia for rotation of the molecule about the 'c' axis is slightly greater than the moment of inertia about the 'b' axis (Figure 1a). In the case of ortho-H_2CO (i.e., when the nuclear spins of the hydrogen atoms are parallel), the small asymmetry causes splitting of the rotational states into closely spaced energy levels known as K-doublets (Figure 1b). Electric dipole transitions between low energy K-doublets ($\Delta K_c = \pm 1$, $\Delta J = 0$, Q-branch transitions, e.g., Townes & Schawlow 1975) result in radio-frequency lines.

The first detection of H_2CO was reported by Snyder *et al.* (1969). They found H_2CO absorption in the 6 cm line ($J_{K_aK_c} = 1_{11} - 1_{10}$; $\nu_o = 4829.6596$ MHz for the F=2–2 hyperfine component, Tucker *et al.* 1971), i.e., the K-doublet transition from the lowest ortho-H_2CO energy levels (Figure 1b). Soon after the first detection of H_2CO, Palmer *et al.* (1969) discovered H_2CO 6 cm absorption against the 2.7 K Cosmic Microwave Background (CMB), implying an excitation temperature $T_{ex} < 2.7$ K for the H_2CO 6 cm K-doublet in Galactic dark clouds. The detection of H_2CO absorption against the CMB (the so called *anomalous* absorption of H_2CO) required an anti-inversion (cooling) mechanism that was promptly recognized to be caused by H_2 collisions (Townes & Cheung 1969, Garrison *et al.* 1975, Evans *et al.* 1975a, Green *et al.* 1978).

Almost four decades after its discovery, the H_2CO 6 cm line has been detected toward hundreds of regions in the Galaxy. H_2CO has been observed in *absorption* against the CMB, Galactic and extragalactic radio continuum sources (e.g., Rodríguez *et al.* 2006, Young *et al.* 2004, Sewiło *et al.* 2004b, Watson *et al.* 2003, Araya *et al.* 2002, Downes *et al.* 1980). In sharp contrast to the ubiquitous H_2CO 6 cm absorption line, H_2CO 6 cm *emission* is an extremely rare phenomenon: H_2CO 6 cm emission has been confirmed as megamaser emission only toward four extragalactic objects (Araya *et al.* 2004a, Baan private communication), found as thermal emission only toward the Orion BN/KL region

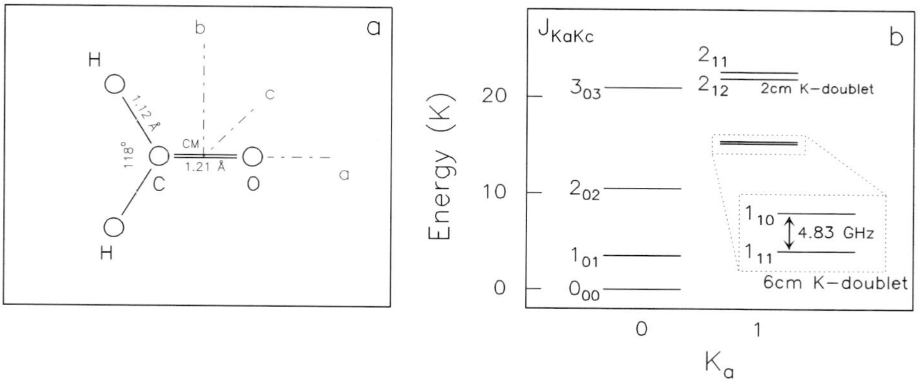

Figure 1. *Left:* Geometry of the H_2CO molecule (Townes & Schawlow 1975; for aesthetic reasons we show a 'left-handed' a-b-c coordinate axis). *Right:* H_2CO energy level diagram of states with $E(J_{K_a K_c}) < 30$ K (Jaruschewski *et al.* 1986).

(Zuckerman *et al.* 1975, see also Araya *et al.* 2006b)†, and only seven Galactic maser regions have been reported (Forster *et al.* 1980, Whiteoak & Gardner 1983, Pratap *et al.* 1994, Araya *et al.* 2005, 2006a, 2007 *in prep.*). In this article we review the field of H_2CO 6 cm masers: the characteristics of the known masers, the astrophysical environments where the masers are found, and the current ideas on the excitation of H_2CO 6 cm masers.

2. H_2CO 6 cm maser surveys

The first H_2CO maser region detected was NGC 7538. Using the Effelsberg 100 m telescope, Downes & Wilson (1974) detected a double peak emission profile superimposed with an absorption feature (see Figure 2 for a recent spectrum of the maser). Aperture synthesis observations by Forster *et al.* (1980) and Rots *et al.* (1981) demonstrated the maser nature of the H_2CO 6 cm emission. Approximately 10 years after the detection of H_2CO emission in NGC 7538, Whiteoak & Gardner (1983) using the VLA detected maser emission from five locations in Sgr B2. Mehringer *et al.* (1994) conducted further VLA observations ($\theta_{syn} \sim 1''$) of the region and detected four more H_2CO maser sites, resulting in a total of nine H_2CO 6 cm maser sites in Sgr B2.

Since the detections by Downes & Wilson (1974) and Whiteoak & Gardner (1983), there have been six surveys specifically focused on the search for H_2CO masers (see Table 1). Given the ubiquitous H_2CO 6 cm absorption in molecular clouds (e.g., Watson *et al.* 2003) and the weak intensity of the known H_2CO 6 cm masers (see §3), surveys for H_2CO maser emission have been conducted using large single dish radio telescopes and interferometers to detect weak lines ($\sim$ 100 mJy) and to avoid confusion due to H_2CO absorption.

The survey by Forster *et al.* (1985) focused on OH maser sources, including not only massive star forming regions but also OH maser stars. The survey yielded no new detections. Pratap *et al.* (1994) and Mehringer *et al.* (1995) conducted observations of active regions of massive star formation known to harbor ultra-compact H II regions as well as

† H_2CO 6 cm thermal emission was also reported toward comets Halley and Machholz (1988j) (Snyder *et al.* 1989, 1990; see however Bockelée-Morvan & Crovisier 1992). Emission of the 2 cm K-doublet is also rare (Martín-Pintado *et al.* 1985, Johnston *et al.* 1984, Loren *et al.* 1983, Wilson *et al.* 1982, Evans *et al.* 1975b), and maser emission from the 2 cm transition has not been observed.

Table 1. H_2CO 6 cm Galactic masers

Reference	Sample	Telescope	Detections	Selection Criteria
Downes & Wilson (1974)	1	Effelsberg	NGC 7538*	NGC 7538
Whiteoak & Gardner (1983)	1	VLA	Sgr B2*	Sgr B2
Forster *et al.* (1985)	19	WSRT	—	OH Maser Sources
Pratap *et al.* (1994)	7	VLA	G29.96−0.02*	UCHII Regions
Mehringer *et al.* (1995)	22	VLA	—	MSFR
Araya *et al.* (2004b)	15	Arecibo	IRAS 18566+0408	Weak Cont. MSFR
Araya *et al.* (2007a)	58	GBT/VLA	G23.71−0.20*	MSFR, H_2CO Spectra
Araya *et al.* (2007 *in prep.*)	14	VLA	G23.01−0.41 & G25.83−0.18	MSFR, H_2CO Spectra

* Sources that have been observed with MERLIN and/or the VLBA (see §3).

maser emission from a variety of molecules. Out of 29 sources observed with the VLA in these surveys, only G29.96−0.02 was found to harbor H_2CO maser emission.

Recently, Araya and collaborators conducted three surveys exploring different search strategies: 1. they observed regions of weak radio continuum to reduce confusion due to H_2CO absorption and focused on massive star forming regions thought to be in an evolutionary stage prior to the ultra-compact H II phase (Arecibo and GBT; Araya *et al.* 2004b, 2007a), 2. they observed massive star forming regions independently of the radio continuum to search for strong and potentially variable masers (GBT; Araya *et al.* 2007a), and 3. they conducted VLA observations of sources that had been previously observed with the GBT or Arecibo and that showed complex absorption line profiles consistent with H_2CO emission blended with absorption (Araya *et al.* 2007a, 2007 *in prep.*). The three surveys resulted in detection of four new maser regions: IRAS 18566+0408 (Araya *et al.* 2004b, 2005), G23.71−0.20 (Araya *et al.* 2006a, 2007a), G23.01−0.41 and G25.83−0.18 (Araya *et al.* 2007 *in prep.*).

3. Physical properties

H_2CO masers are weaker in comparison with most OH, H_2O, and CH_3OH masers; the flux density range of the known H_2CO masers is between 10 mJy (for the maser in IRAS 18566+0408; see the poster contribution by Araya *et al.* in these proceedings) and ~2 Jy (for the brightest masers in Sgr B2 and NGC 7538; Hoffman *et al.* 2007, Araya *et al.* 2007a), while most masers have flux densities of the order of ~ 100 mJy. Two maser regions have been observed at ~ 50 mas angular resolution with MERLIN: NGC 7538 (Hoffman *et al.* 2003), and G23.71−0.20 (Araya *et al.* 2007 *in prep*). MERLIN observations recover most ($\gtrsim$ 70%) of the flux density detected at lower angular resolutions, and the masers are unresolved or barely resolved. Araya *et al.* (2007 *in prep.*) find a brightness temperature $\gtrsim 10^5$ K for the maser in G23.71−0.20.

Three sources have been observed at ~ 10 mas angular resolution with the VLBA: NGC 7538, G29.96−0.02, and Sgr B2. Hoffman *et al.* (2003) report brightness temperatures between 10^6 and 10^8 K for the masers in G29.96−0.02 and NGC 7538. Two of the nine maser components in Sgr B2 were observed with the VLBA by Hoffman *et al.* (2007). They also found brightness temperatures in the 10^8 K range. In general, only a fraction of the flux density is recovered with the VLBA, and as in the case of other astrophysical masers, the lines are narrower in the VLBA observations compared with VLA or single dish observations. Hoffman *et al.* (2007) discuss these results in the context of a core-halo model, where the maser brightness distribution is the result of the superposition of two Gaussian components, one compact (~ 10 mas) saturated component that is

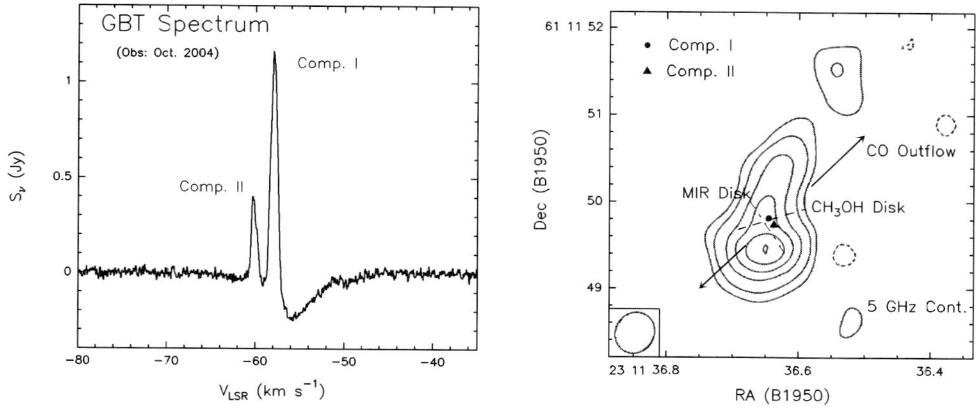

Figure 2. *Left:* H_2CO 6 cm spectrum of NGC 7538 IRS1 (Araya *et al.* 2007a). Two maser components are blended with an H_2CO absorption line. *Right:* Location of the two H_2CO maser components (dot and triangle) superimposed on the 6 cm radio continuum in NGC 7538 IRS1 (Rots *et al.* 1981). Re-reduction of the Rots *et al.* (1981) data using a more accurate position of the phase calibrator results in H_2CO maser positions that are consistent with the VLBA and MERLIN values reported by Hoffman *et al.* (2003). The direction of the CO (2–1) outflow is indicated by arrows (Davis *et al.* 1998), and two possible orientations of a circumstellar disk in the region are shown with dot-dashed (MIR Disk, De Buizer & Minier 2005) and dashed (CH_3OH Disk, Pestalozzi *et al.* 2004) lines.

detected with the VLBA ($T_b \sim 10^8$ K), and one extended and unsaturated halo that is resolved out by the VLBA observations ($T_b \sim 10^5$ K). Based on the VLBA and MERLIN results, the projected physical size of the masers is between 30 and $\sim$ 200 AU, while the maser gains range between -6 and -12; the emission is unpolarized within the current sensitivity limits (Hoffman *et al.* 2003, 2007).

4. Line profiles and velocity gradients

With the exception of Sgr B2 where nine H_2CO maser spots have been found (some of them showing multiple-peaked and broad line profiles, e.g., Mehringer *et al.* 1994), the H_2CO maser line profiles are relatively simple. A peculiar characteristic of the known H_2CO maser regions is that double peaked profiles are very common; double peaked profiles have been detected in NGC 7538 (e.g., Figure 2), G29.96$-$0.02 (Pratap *et al.* 1994, Hoffman *et al.* 2003), IRAS 18566+0408 (Araya *et al.* 2007c), G25.83$-$0.18 (Figure 3), and possibly toward G23.01$-$0.41 (Araya *et al.* 2007 *in prep.*). In all of these cases the velocity separation of the maser components is less than 3 km s^{-1}, and the components are (in most cases) spatially coincident in VLA observations. The double peaked profiles are unlikely caused by the hyperfine structure of the 6 cm H_2CO transition. Recent high spectral resolution (0.1 km s^{-1} channel width) VLA observations of the masers in G23.01$-$0.41 and G25.83$-$0.18 (Araya *et al.* 2007 *in prep.*) show very narrow maser components (FWHM$\sim$0.3 km s^{-1}), possibly due to the line narrowing effect of unsaturated masers.

In the case of the H_2CO masers in NGC 7538, the components are oriented in a NE–SW direction, with a projected separation of 79 mas (240 AU). VLBA observations of the red shifted component show a 1900 km s^{-1}pc^{-1} velocity gradient also in a NE–SW orientation (see figures 4 and 5 of Hoffman *et al.* 2003).

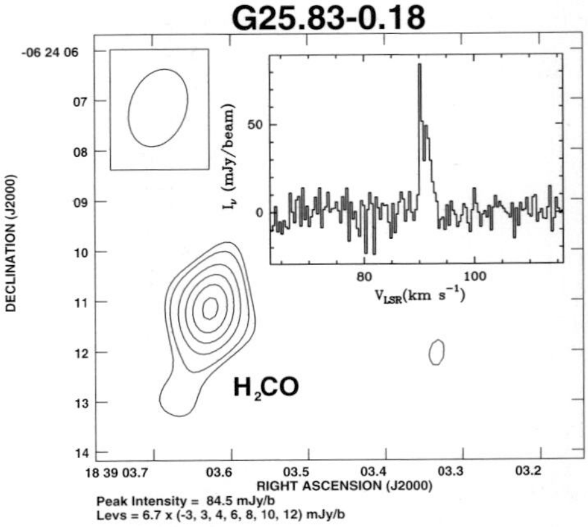

Figure 3. H_2CO 6 cm maser emission in G25.83−0.18 detected with the VLA by Araya *et al.* (2007 *in prep.*). Excluding the masers in Sgr B2, five out of six sources show double peak profiles.

5. Variability

Prior to the recent work by Araya *et al.* (2007c), variability of H_2CO masers had been observed only in some of the Sgr B2 masers and in the NGC 7538 masers; only long time-scale variability (> 1 yr) had been reported (e.g., Mehringer *et al.* 1994, Hoffman *et al.* 2003). In the case of the variability of the H_2CO masers in NGC 7538 and based on the similar variability-rate of the two maser components (Figure 4 *left panel*), Araya *et al.* (2007a) suggested that the variability of the masers may be caused by a perturbation that took ∼ 14 yr† to reach Comp. II after having reached Comp. I. If that were the case, an increase in the rate-of-change of the intensity of Comp. II would be expected after the year 2009. The variability of the two maser components could be related to the precessing jet reported by Kraus *et al.* (2006).

Araya *et al.* (2007c) have recently found a new type of H_2CO maser variability, namely, short term flares. Using Arecibo, VLA, and GBT data, Araya *et al.* (2007c) reported occurrence of an outburst of the H_2CO 6 cm maser in IRAS 18566+0408; the maser flare lasted for less than three months and decayed to the pre-flare intensity within a month. The H_2CO maser in IRAS 18566+0408 has a double peak profile. Both components varied by approximately the same factor and in the same time period; no change in the line widths and peak velocities was detected. Araya *et al.* (2007c) discussed the implications of the flare on the possible excitation mechanism of the maser, and concluded that if the flare were due to a maser gain change, then (independently of the saturation state of the maser) the pumping mechanism is likely radiative; whereas if the maser is unsaturated, then a change in the background 6 cm radio continuum might have been amplified by the maser gas (independently on the maser pumping mechanism).

A monitoring program of the maser with Arecibo has recently resulted in the detection of a second H_2CO maser burst in IRAS 18566+0408 (Figure 4, *right panel*); showing that the flares are recurrent in this source. Araya and collaborators are also monitoring the CH_3OH 6.7 GHz masers in the region, and found that one of the CH_3OH maser components showed the same outburst as the H_2CO 6 cm maser (Figure 4, *right panel*). The CH_3OH maser peak that showed the flare does not correspond in velocity to the H_2CO maser, hence the masers originate in different regions. It is possible that both masers are unsaturated and that a change in the background radio continuum was amplified by the

† Curiously, the period of the long-term variability of H_2O masers in the region reported by Lekht *et al.* (2004) is ∼ 13 yr.

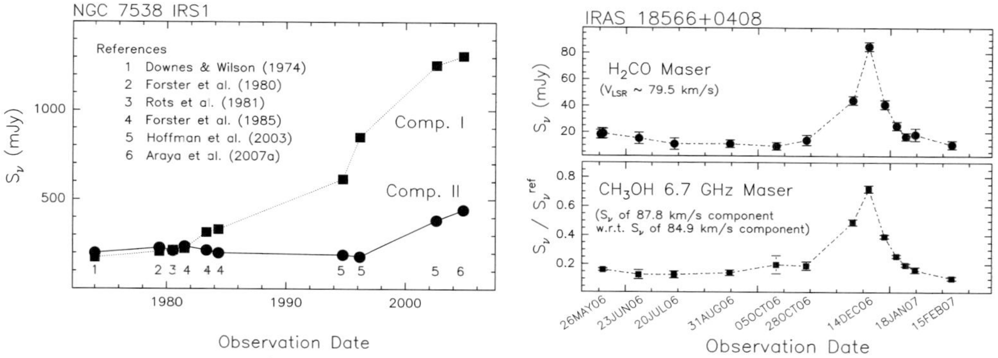

Figure 4. *Left:* Long term variability of the H_2CO masers in NGC 7538. The similar intensity rate-of-change of the two maser components after the onset of the variability lead Araya *et al.* (2007a) to propose that the variability of both components may have a common origin. *Right:* Arecibo light-curve of the second short-term H_2CO maser flare detected (Araya *et al.* 2007 *in prep*). Araya and collaborators are also monitoring with Arecibo the CH_3OH 6.7 GHz masers; one of the CH_3OH maser components showed the same outburst as the H_2CO 6 cm maser (see the poster contribution by Araya and collaborators in these proceedings).

CH_3OH and H_2CO masers (see poster contribution by Araya and collaborators in these proceedings).

Variability of some of the H_2CO masers in Sgr B2 has also been found (Mehringer *et al.* 1994, Hoffman *et al.* 2007); however the available data are insufficient to establish whether the masers show long term variability or maser flares as in IRAS 18566+0408.

6. Astrophysical environments: H_2CO masers pinpointing disk candidates around young massive stars

Motivated by the detection of H_2CO maser emission toward NGC 7538 and Sgr B2 (both massive star forming regions) the subsequent surveys for H_2CO maser emission have been conducted mainly toward regions of massive star formation (Table 1). However, H_2CO 6 cm observations have also been carried out toward a number of non-massive star forming regions and no new maser has been reported (e.g., Araya *et al.* 2006b, 2003; Rodríguez *et al.* 2006; Young *et al.* 2004). Thus, H_2CO 6 cm masers appear to be exclusively associated with massive star formation.

Except for some of the H_2CO masers in Sgr B2, H_2CO masers are mostly found along line-of-sights devoid of strong compact radio continuum (though continuum regions may be found nearby, e.g., Pratap *et al.* 1994); they are located close to CLASS II CH_3OH and H_2O masers (in many cases coincident within a synthesized beam), deeply embedded infrared sources and/or other evidence of massive star formation such as hyper-compact H II regions and hot molecular cores (e.g., Araya *et al.* 2007b, 2006a, 2005; Hoffman *et al.* 2007, 2003; Pratap *et al.* 1994). Thus, H_2CO masers appear to trace young massive stellar objects before the onset of a bright ultra-compact HII region. Moreover, in the case of three of the H_2CO maser sources that have been studied in detail, there is some evidence for an association between H_2CO maser emission and circumstellar disks:

– **G29.96−0.02** is a massive star forming region that harbors an ultra-compact HII region and a hot molecular core (e.g., Cesaroni *et al.* 1994). The H_2CO maser is coincident with the hot molecular core (Pratap *et al.* 1994). Hot molecular cores are believed to be an evolutionary phase prior to the formation of an ultra-compact H II region (e.g.,

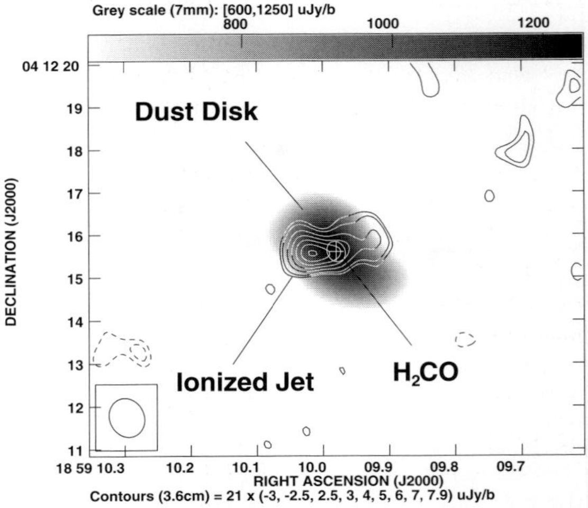

Figure 5. Location of the H_2CO 6 cm maser in IRAS 18566+0408 with respect to 7 mm (gray scale) and 3.6 cm (contours) continuum detected with the VLA. Araya *et al.* (2007b) concluded that the 3.6 cm emission is tracing an ionized jet whereas the 7 mm emission is dominated by dust emission from a possible massive circumstellar disk (torus). The H_2CO maser is coincident with the massive disk candidate.

Churchwell 2002), thus the H_2CO maser pinpoints a very young region of massive star formation. Olmi *et al.* (2003) report evidence of infall and a massive rotating disk in the hot molecular core. Thus, the H_2CO 6 cm emission may be associated with a disk around a massive young stellar object. However, the assumption of a single massive disk might be over-simplistic given the complex sub-mm morphology found by Beuther *et al.* (2007).

– **NGC 7538 IRS1** is a massive star forming region which harbors a hyper-compact H II region (e.g., Sewiło *et al.* 2004a), and a CO outflow centered at the NGC 7538 IRS1 position (Figure 2, right panel). NGC 7538 IRS1 is one of the few sources where a circumstellar disk around a massive (proto)star has been reported. However, the orientation of the disk is controversial. Based on CH_3OH maser data by Minier *et al.* (1998, 2000, 2001), Pestalozzi *et al.* (2004) report a possible Keplerian disk oriented $\sim$ SE–NW. However, based on mid-IR observations, De Buizer & Minier (2005) considered that the CH_3OH masers are tracing the outflow and that the disk is oriented in a NE–SW direction (i.e., perpendicular to the CO outflow, see Figure 2, right panel). As mentioned in §4, the two H_2CO maser spots are oriented NE–SW, and the velocity gradient of Comp. I is also in the NE–SW direction, i.e., perpendicular to the CO outflow and parallel to the MIR disk. The H_2CO 6 cm masers appear to trace material very close to (within 1000 AU) or directly associated with a circumstellar disk.

– **IRAS 18566+0408** was classified by Zhang (2005) as a massive circumstellar disk candidate. Based on high sensitivity and angular resolution 6, 3.6, 1.3, and 0.7 cm VLA continuum observations, Araya *et al.* (2007b) recently found supporting evidence for the presence of a massive circumstellar disk in IRAS 18566+0408 (see Figure 5). The massive disk (torus) is traced by 7 mm dust emission and has an elongation almost perpendicular to an ionized jet traced by cm radio continuum. The H_2CO maser is coincident with the center of the massive disk candidate.

7. The pumping mechanism of H_2CO 6 cm masers

Although the number of H_2CO 6 cm maser sources is still small, in the past few years significant progress in the characterization of H_2CO masers and their environments has been made; however, a theoretical understanding of the excitation mechanism of H_2CO masers is still lacking. Even before the detection of the first H_2CO maser, several authors mentioned and/or discussed possible excitation mechanisms that would result in maser emission of the 6 cm line; including collisional excitation with H_2 molecules and electrons (Thaddeus 1972; see also Fig. 14 of Evans *et al.* 1975a), and infrared pumping (Litvak 1970). However, only Boland & de Jong (1981) developed a specific model to explain one of the known H_2CO masers (NGC 7538). This model is based on inversion via background radio continuum radiation; however, the model appears to be incapable of explaining most of the known H_2CO masers (Araya *et al.* 2007b, Hoffman *et al.* 2007, 2003, Pratap *et al.* 1994, Mehringer *et al.* 1994; see however Pratap *et al.* 1992).

Besides inversion by radio continuum, other proposed excitation mechanisms appear to be possible: 1. Hoffman *et al.* (2007, 2003) and Araya *et al.* (2005) find indications that the masers could be collisionally excited in shocked regions (see also Martín-Pintado *et al.* 1999), 2. some H_2CO masers are found close to deeply embedded infrared objects and thus infrared pumping could be possible (Araya *et al.* 2006a), and 3. H_2CO masers are located close to very young massive stellar objects where a high ionization fraction is expected, and thus electron collision may play a role in the pumping.

Araya *et al.* (2007 *in prep.*) explore excitation of H_2CO masers via H_2 and electron collisions and have found that collision with electrons can indeed produce an inversion of the 6 cm K-doublet (see also Thaddeus 1972). However, preliminary results by Araya *et al.* (2007 *in prep.*) appear to require long path lengths ($\sim$ parsec scales) to reproduce the brightness temperature of the known H_2CO masers. Parsec-scale path lengths of coherent velocity and homogeneous physical conditions in massive star forming regions are unlikely. However pumping of H_2CO by electron collisions appears to be a promising mechanism to explain extragalactic megamasers (Araya *et al.* 2007 *in prep.*, see also Araya *et al.* 2004a; Baan & Haschick 1995). Araya *et al.* (2007 *in prep.*) also find that including radiation trapping, the 6 cm K-doublet may be inverted at a molecular density of $\sim 10^6$ cm^{-3} (i.e., in the transition between anomalous H_2CO absorption and thermalization). However, the model depends on accurate H_2(ortho/para) – H_2CO collision rates which are not available at present (e.g., Green 1991; see also Hoffman *et al.* 2003).

8. Why are H_2CO masers so rare?

In spite of a number of surveys specifically focused on the search for H_2CO masers and hundreds of sources for which H_2CO 6 cm absorption studies have been conducted (e.g., Table 1, Araya *et al.* 2002, Watson *et al.* 2003, Sewiło *et al.* 2004), H_2CO maser emission has been detected only toward seven regions (and in a total of 15 maser spots at $1''$ resolution). Thus, H_2CO maser emission is indeed a rare phenomenon. Reformulating the ideas presented by Mehringer *et al.* (1995), Pratap *et al.* (1992), and Forster *et al.* (1985), H_2CO masers may be uncommon because: 1. they are weak in comparison with other astrophysical masers and/or are highly beamed: the brightest known H_2CO maser is just $\sim$ 2 Jy; 2. they occur at LSR velocities close to the systemic velocity of the star forming regions and thus the maser emission is highly attenuated by large optical depths of the H_2CO absorption line: H_2CO masers are typically found within $\sim 5\,\mathrm{km\,s^{-1}}$ from the systemic cloud velocity as traced by H_2CO absorption (e.g., Figure 2, Araya *et al.* 2004b) and are not found at high velocities, indicating that H_2CO masers are not associated with high velocity outflows; and 3. the physical conditions needed for the excitation of the masers are very specific, and thus, short-lived in massive star forming region environments. For example, the H_2CO masers (and H II regions) in Sgr B2 are distributed in a N–S direction suggesting that star formation in Sgr B2 was triggered by a cloud collision event (see Sato *et al.* 2000) and thus the masers may be tracing an isochrone of the physical conditions during the massive star formation process (see Gardner *et al.* 1986).

9. Summary

Despite the ubiquitous presence of H_2CO 6 cm absorption, there are only seven known H_2CO maser regions in the Galaxy. Recent VLBA and MERLIN observations toward four sources show brightness temperatures between $10^5 - 10^9$ K, and physical sizes between 30–200 AU. At least two masers show long-term variability (>1 year), and one maser source was recently shown to exhibit recurrent short-term (< 3 months) bursts. All known H_2CO 6 cm masers are found in massive star forming regions, and in the case of the three sources that have been studied in detail (NGC 7538 IRS1, G29.96−0.02, and IRAS 18566+0408; excluding Sgr B2) the H_2CO masers pinpoint the location of candidate disks around massive protostars. However, the H_2CO maser mechanism has to be clarified before the masers can be used as an astrophysical probe.

Acknowledgements

The research on H_2CO 6 cm masers by the author (E.A.) has greatly benefited from collaboration with H. Linz, S. Kurtz, W. Baan, M. Sewiło, E. Churchwell, L. Olmi, C. Watson, I. Hoffman, L. F. Rodríguez, G. Garay, & P. Palmer. E. A. is supported by a NRAO predoctoral fellowship. We thank H. Linz and I. Hoffman for comments that improved the manuscript, and H. Beuther for sharing information about the G29.96−0.02 small scale structure based on SMA sub-mm data prior to publication. E. A. thanks the IAU for a travel grant to attend this symposium.

References

Araya, E., Baan, W., & Hofner, P. 2004a, *ApJS* 154, 541
Araya, E., Hofner, P., Churchwell, E., & Kurtz, S. 2002, *ApJS* 138, 63
Araya, E., *et al.* 2007b, *ApJ* submitted
Araya, E., Hofner, P., Goldsmith, P., Slysh, S., & Takano, S. 2003, *ApJ* 596, 556
Araya, E., Hofner, P., Goss, W. M., Kurtz, S., Linz, H., & Olmi, L. 2006a, *ApJ* (Letters) 643, L33
Araya, E., Hofner, P., Goss, W. M., Linz, H., Kurtz, S., & Olmi, L. 2007a, *ApJS*, 170, 152
Araya, E., Hofner, P., Kurtz, S., Linz, H., Olmi, L., Sewiło, M., Watson, C., & Churchwell, E. 2005, *ApJ* 618, 339
Araya, E., Hofner, P., Linz, H., Sewiło, M., Watson, C., Churchwell, E., Olmi, L., & Kurtz, S. 2004b, *ApJS* 154, 579
Araya, E., Hofner, P., Olmi, L., Kurtz, S., & Linz, H. 2006b, *AJ* 132, 1851
Araya, E., Hofner, P., Sewiło, M., Linz, H., Kurtz, S., Olmi, L., Watson, C., & Churchwell, E. 2007c, *ApJ* (Letters) 654, L95
Baan, W. A., & Haschick, A. D. 1995, *ApJ* 454, 745
Beuther, H., Zhang, Q., Bergin, E. A., *et al.* 2007, *A&A*, accepted (arXiv:0704.0518)
Bockelée-Morvan, D., & Crovisier, J. 1992, *A&A* 264, 282
Boland, W., & de Jong, T. 1981, *A&A* 98, 149
Cesaroni, R., Churchwell, E., Hofner, P., Walmsley, C. M., & Kurtz, S. 1994, *A&A* 288, 903
Churchwell, E. 2002, *ARA&A* 40, 27
Davis, C. J., Moriarty-Schieven, G., Eislöffel, J., Hoare, M. G., & Ray, T. P. 1998, *AJ* 115, 1118
De Buizer, J. M., & Minier, V. 2005, *ApJ* (Letters) 628, L151
Downes, D., & Wilson, T. L. 1974, *ApJ* (Letters) 191, L77

Downes, D., Wilson, T. L., Bieging, J., & Wink, J. 1980, *A&AS* 40, 379

Evans, N. J. II, Zuckerman, B., Morris, G., & Sato, T. 1975a, *ApJ* 196, 443

Evans, N. J. II, Zuckerman, B., Sato, T., & Morris, G. 1975b, *ApJ* 199, 383

Forster, J. R., Goss, W. M., Gardner, F. F., & Stewart, R. T. 1985, *MNRAS* 216, 35

Forster, J. R., Goss, W. M., Wilson, T. L., Downes, D., & Dickel, H. R. 1980, *A&A* (Letters) 84, L1

Gardner, F. F., Whiteoak, J. B., & Forster, J. R. 1986, *MNRAS* 218, 385

Garrison, B. J., Lester, W. A. Jr., Miller, W. H., & Green, S. 1975, *ApJ* (Letters) 200, L175

Green, S. 1991, *ApJS* 76, 979

Green, S., Garrison, B. J., Lester, W. A. Jr., & Miller, W. H. 1978, *ApJS* 37, 321

Hoffman, I. M., Goss, W. M., & Palmer, P. 2007, *ApJ* 654, 971

Hoffman, I. M., Goss, W. M., Palmer, P., & Richards, A. M. S. 2003, *ApJ* 598, 1061

Jaruschewski, S., Chandra, S., Varshalovich, D. A., & Kegel, W. H. 1986, *A&AS* 63, 307

Johnston, K. J., Henkel, C., & Wilson, T. L. 1984, *ApJ* (Letters) 285, L85

Kraus, S., Balega, Y., Elitzur, M., Hofmann, K.-H., Preibisch, Th., Rosen, A., Schertl, D., Weigelt, G., & Young, E. T. 2006, *A&A* 455, 521

Lekht, E. E., Munitsyn, V. A., & Tolmachev, A. M. 2004, *Astron. Rep.* 48, 200

Litvak, M. M. 1970, *ApJ* (Letters) 160, L133

Loren, R. B., Sandqvist, A., & Wootten, A. 1983, *ApJ* 270, 620

Martín-Pintado, J., Gaume, R. A., Rodríguez-Fernández, N., de Vicente, P., & Wilson, T. L. 1999, *ApJ* 519, 667

Martín-Pintado, J., Wilson, T. L., Gardner, F. F., & Henkel, C. 1985, *A&A* 142, 131

Mehringer, D. M., Goss, W. M., & Palmer, P. 1994, *ApJ* 434, 237

Mehringer, D. M., Goss, W. M., & Palmer, P. 1995, *ApJ* 452, 304

Minier, V., Booth, R., & Conway, J. 1998, *A&A* (Letters) 336, L5

Minier, V., Booth, R., & Conway, J. 2000, *A&A* 362, 1093

Minier, V., Conway, J., & Booth, R. 2001, *A&A* 369, 278

Olmi, L., Cesaroni, R., Hofner, P., Kurtz, S., Churchwell, E., & Walmsley, C. M. 2003, *A&A* 407, 225

Palmer, P., Zuckerman, B., Buhl, D., & Snyder, L. E. 1969, *ApJ* (Letters) 156, L147

Pestalozzi, M. R., Elitzur, M., Conway, J. E., & Booth, R. S. 2004, *ApJ* (Letters) 603, L113

Pratap, P., Menten, K. M., & Snyder, L. E. 1994, *ApJ* (Letters) 430, L129

Pratap, P., Snyder, L. E., & Batrla, W. 1992, *ApJ* 387, 241

Rodríguez, M. I., Allen, R. J., Loinard, L., & Wiklind, T. 2006, *ApJ* 652, 1230

Rots, A. H., Dickel, H. R., Forster, J. R., & Goss, W. M. 1981, *ApJ* (Letters) 245, L15

Sato, F., Hasegawa, T., Whiteoak, J. B., & Miyawaki, R. 2000, *ApJ* 535, 857

Sewiło, M., Churchwell, E., Kurtz, S., Goss, W. M., & Hofner, P. 2004a, *ApJ* 605, 285

Sewiło, M., Watson, C., Araya, E., Churchwell, E., Hofner, P., & Kurtz, S. 2004b, *ApJS* 154, 553

Snyder, L. E., Buhl, D., Zuckerman, B., & Palmer, P. 1969, *Phys. Rev. Lett.* 22, 679

Snyder, L. E., Palmer, P., & de Pater, I. 1990, *ICARUS* 86, 289

Snyder, L. E., Palmer, P., & de Pater, I. 1989, *AJ* 97, 246

Thaddeus, P. 1972, *ApJ* 173, 317

Townes, C. H., & Cheung, C. 1969, *ApJ* (Letters) 157, L103

Townes, C. H., & Schawlow, A. L. 1975, *Microwave Spectroscopy* (New York: Dover)

Tucker, K. D., Tomasevich, G. R., & Thaddeus, P. 1971, *ApJ* 169, 429

Watson, C., Araya, E., Sewiło, M., Churchwell, E., Hofner, P., & Kurtz, S. 2003, *ApJ* 587, 714

Whiteoak, J. B., & Gardner, F. F. 1983, *MNRAS* 205, 27

Wilson, T. L., Martín-Pintado, J., Gardner, F. F., & Henkel, C. 1982, *A&A* (Letters) 107, L10

Young, K. E., Lee, J. E., Evans, N. J., II, Goldsmith, P. F., & Doty, S. D. 2004, *ApJ* 614, 252

Zhang, Q. 2005, in: R. Cesaroni, M. Felli, E. Churchwell, & M. Walmsley (eds.), *Massive Star Birth: A Crossroads of Astrophysics* (Cambridge: Cambridge Univ. Press), p. 135

Zuckerman, B., Palmer, P., & Rickard, L. J. 1975, *ApJ* 197, 571

Astrophysical Masers and their Environments
Proceedings IAU Symposium No. 242, 2007
J.M. Chapman & W.A. Baan, eds.

doi:10.1017/S1743921307012665

Profiling young massive stars

Tracey Hill[1,2]†, M. G. Burton[2], M. R. Cunningham[2] and V. Minier[3]

[1] Leiden Observatory, Leiden University, PO BOX 9513, 2300RA, Leiden, the Netherlands.
email: thill@strw.leidenuniv.nl

[2] School of Physics, University of New South Wales, Kensington, NSW 2052, Australia

[3]Service d'Astrophysique, DAPNIA/DSM/CEA Saclay, 91191 Gif-sur-Yvette, France.

Abstract. We present the results of spectral energy distribution analysis for 162 of the 405 sources reported in the SIMBA survey of Hill *et al.* (2005). The fits reveal source specific parameters including: the luminosity, mass, temperature, H_2 number density, the surface density and the luminosity-to-mass ratio. Each of these parameters are examined with respect to the four classes of source present in the sample. Obvious luminosity and temperature distinctions exist between the mm-only cores and those cores with methanol maser and/or radio continuum emission, with the former cooler and less luminous than the latter. The evidence suggests that the mm-only cores are a precursor to the methanol maser in the formation of massive stars. The mm-only cores comprise two distinct populations distinguished by temperature. Analysis and conclusions about the nature of the cool-mm and warm-mm cores comprising the mm-only population are drawn.

Keywords. masers, stars: formation, stars: fundamental parameters, stars: early-type, radio continuum: stars, methods: data analysis

1. Introduction

Despite the wealth of attention that massive stars have received of late, their formation scenario is still largely unclear. This may be attributed to the fact that massive stars form in clustered mode, deeply embedded in their natal molecular envelope where their development is optically obscured prior to main sequence evolution. This, coupled with the rapid timescales over which they evolve, as well as the fact that high-mass stellar candidates are typically at further distances than their lower mass counterparts, hinders the study of these objects.

Massive star formation regions are often found coincident with maser sources – in particular the methanol variant (Pestalozzi *et al.* 2005), UC HII regions (Thompson *et al.* 2006), *IRAS* colour selected sources (Wood & Churchwell 1989a) and *MSX* colour selected sources (Lumsden *et al.* 2002). More specifically, methanol masers and UC HII regions are thought to feature prominently in the earliest stages of massive star formation (cf. Batrla *et al.* 1987, Caswell *et al.* 1995, Walsh *et al.* 1997, Minier *et al.* 2001, Beuther *et al.* 2002, Faúndez *et al.* 2004, Thompson *et al.* 2006).

Methanol masers have been found associated with strong radio continuum emission (i.e. HII regions), *IRAS* far-infrared colour selected sources (e.g. Wood & Churchwell 1989b), H_2O and OH masers (Caswell *et al.* 1995) as well as millimetre and submillimetre continuum emission (Hill *et al.* 2005, Walsh *et al.* 2003 respectively).

† Present address: Leiden Observatory, Leiden University, PO BOX 9513, 2300RA Leiden, the Netherlands.

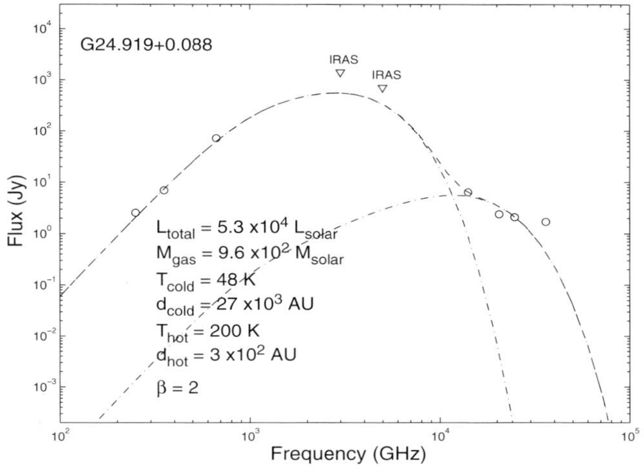

Figure 1. Spectral energy distribution of the mm-only core G24.919+0.088. *IRAS* 60 and 100μm upper limits are indicated on the diagram. Source specific parameters are as listed.

Early work (e.g. Walsh *et al.* 1998) suggested that the methanol maser was indicative of the earliest stages of massive star formation prior to the onset of radio continuum emission. More recent work has focused on finding a precursor to the methanol maser, which would mark the *very* earliest stages of massive star formation. The hot molecular core (HMC) is one such object proposed to satisfy this criterion (cf. Olmi *et al.* 1996, Osorio *et al.* 1999, Minier *et al.* 2005).

Hill *et al.* (2005), herefater, Paper I undertook a millimetre continuum emission survey toward regions displaying evidence of massive star formation in search of cold cores that would mark the earliest stages of their formation. This survey revealed evidence of star formation clearly offset from, and devoid of, both the methanol maser and radio continuum sources targeted. Follow-up submillimetre observations of these 'mm-only' cores Hill *et al.* (2006), hereafter Paper II revealed each of them to be associated with submillimetre continuum emission and hence cold deeply embedded objects.

2. Spectral energy distribution fitting

Combining the SIMBA and SCUBA data from Papers I and II, together with existing submillimetre SCUBA data (Pierce-Price *et al.* 2000, Walsh *et al.* 2003, Thompson *et al.* 2006), as well as archival *MSX* data where applicable and in some instances *IRAS*, we have drawn spectral energy distribution (SED) diagrams for the sources in our sample. The SED is drawn using a two-component Levenberg-Marquardt least squares fit according to equation 2.1. This relation models emission from a hot core embedded in a larger cold or warm envelope (cf. Minier *et al.* 2005).

$$F_\nu = [\Omega_h B_\nu(T_{hot})] + \Omega_c B_\nu(T_{cold})\, \epsilon_\nu \tag{2.1}$$

where F_ν is the flux density of the source, Ω_h and Ω_c are the source solid angles for the hot and the cold component of the source, respectively, B_ν is the Planck function for a temperature of T_{hot} and T_{cold}, and ϵ_ν is the emissivity for the cold component, which is equivalent to $(1 - e^{-\tau_\nu})$. For optically thin regions, ϵ_ν approximates to τ_ν where τ_ν is given by $\tau_0(\nu/\nu_0)^\beta$. The dust grain emissivity index (β) is assumed to be 2 for the sample, as per Paper II.

SED fits were applied to 162 of the 405 sources in the SIMBA sample (see Fig. 1).

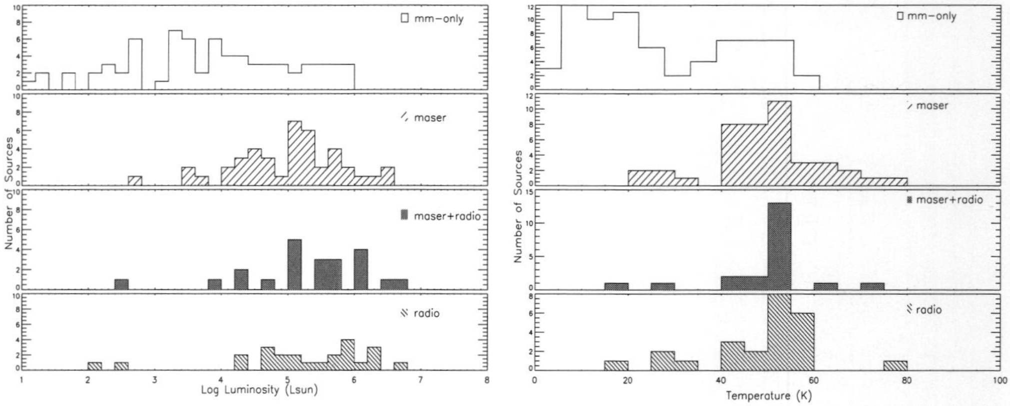

Figure 2. Histogram distribution of the luminosity (*left*) and temperature (*right*) for the four classes of source in the SIMBA sample.

3. Results and analysis

The SED fit provided three main parameters: the dust temperature, luminosity and the mass. These were then used to determine the hydrogen number density (n_{H_2}), the surface density (Σ) and the luminosity-to-mass ratio (L/M). Together with the radius and distance from our earlier work, eight parameters are known for each source.

We have compared each of these eight parameters with respect to the four classes of source found in the sample (see Paper I) through histogram and cumulative distribution plots (see figures 2 and 3) as well as Kolmogorov-Smirnov (KS) tests.

Analysis revealed (Hill 2006) the mm-only sample to be comparable in terms of mass and radii to sources harbouring a methanol maser and/or radio continuum source. Intriguingly, the mm-only sample are also cooler and less luminous (see Fig. 2 and 3) with smaller luminosity-to-mass ratios than these sources. As well as corroborating these results, the KS tests for the temperature, luminosity and L/M indicate that the mm-only cores are distinctly different from the methanol maser and/or radio continuum associated cores for these parameters. At least 45 per cent of the mm-only cores are also without mid-infrared *MSX* emission.

These results are consistent with cooler, younger examples of massive star formation, suggesting that the mm-only core is indicative of the very earliest stages of massive star formation, prior to the onset of methanol maser emission.

4. A bimodal distribution for the mm-only core?

The mm-only sample was identified as comprising two populations which are distinguished by temperature (Fig. 2). For the following analysis, the mm-only sample was segregated into cool-mm and warm-mm components, creating five classes of sources in total.

According to the cumulative distribution plots (cf. Fig. 4) the cool-mm sources are less luminous with smaller L/M and radii, and higher densities (n_{H_2} and Σ) than the warm-mm sources as well as those sources with a methanol maser and/or radio continuum source. The KS tests indicate that the cool-mm sources are not drawn from the same population as the warm-mm sources, nor the sources with radio continuum and/or methanol maser emission for the luminosity, luminosity-to-mass ratio, H_2 number density and the surface density. However no such distinction could be discerned for the mass.

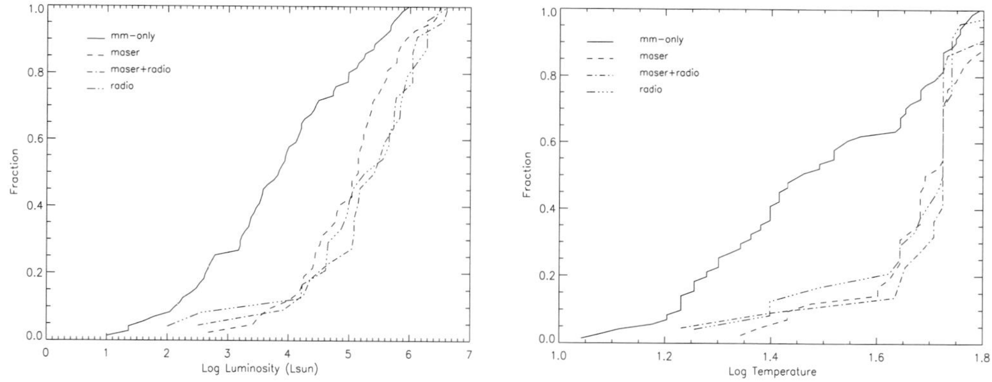

Figure 3. Cumulative distribution plots of the luminosity (*left*) and temperature *right*) for the four classes of source in the SIMBA sample.

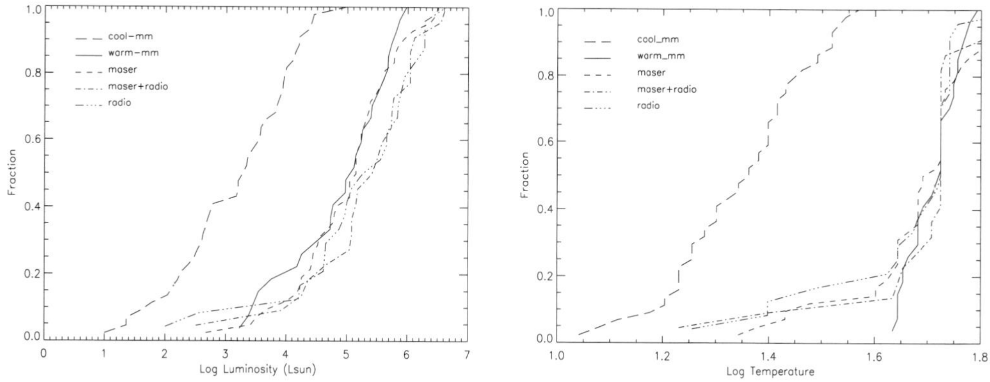

Figure 4. Cumulative distribution plots of the luminosity (*left*) and temperature (*right*) highlighting the two populations of mm-only core (cool-mm and warm-mm).

The warm-mm sources on the other hand display little distinction from the methanol maser and/or radio continuum sources for all of the parameters tested. These results led us to conclude that the cool-mm and warm-mm sources are two distinct populations of the mm-only core. The cool-mm sources are the least luminous sources in the sample, with lower luminosity-to-mass ratios as well as higher H_2 and surface densities. However they are of comparable mass and radii to the warm-mm and maser sources.

We propose that the warm-mm sources are indicative of the earliest stages of massive star formation prior to the onset of methanol maser emission, whilst the cool-mm sources are examples of starless cores as described by (Vázquez-Semadeni *et al.* 2005). High resolution observations of these cores, to search for embedded sources, are necessary in order to ascertain which of them are forming stars, and which will remain starless, and so test this hypothesis.

5. Future work

A caveat of this work is that the separation of the mm-only core into two distinct populations was concluded from a simple greybody fit to the SED, which enabled a temperature estimation in addition to the mass and luminosity of the dust cores. This distinction may be in error, a result of the limited number of data points available to fit the SEDs for these sources.

Ammonia (NH_3) is an excellent tracer of dense molecular gas. It can be used to probe the physical conditions of the molecular cloud in which cores are forming and can be used to determine physical conditions of the gas, such as temperature, density and optical depth (cf. Longmore *et al.* 2006). Intriguingly, Longmore *et al.* (these proceedings) report observations where the youngest star formation regions in their sample are devoid of methanol maser emission, traced solely by their NH_3 emission.

We anticipate a direct correlation between future NH_3 observations of our sample, with the millimetre continuum emission detected by SIMBA, as per the results of Pillai *et al.* (2006)). Ammonia will therefore provide an independent determination of the temperature of our mm-only cores, with insights into their role in the formation of massive stars.

Acknowledgements

The authors would like to thank S. Lumsden for his *MSX* script, as well as C. Purcell and S. Longmore for assistance with converting the *MSX* and *IRAS* images. T. H. was supported by the UNSW Faculty Research Grant Program (FRGP) during her thesis.

References

Batrla, W., Matthews, H. E., Menten, K. M., & Walmsley, C. M. 1987, Nature, 326, 49
Beuther, H., Schilke, P., Menten, K. M., Motte, F., Sridharan, T. K., & Wyrowski, F. 2002, ApJ, 566, 945
Caswell, J. L., Vaile, R. A., Ellingsen, S. P., Whiteoak, J. B., & Norris, R. P. 1995, MNRAS, 272, 96
Faúndez, S., Bronfman, L., Garay, G., Chini, R., Nyman, L.-Å., & May, J. 2004, A&A, 426, 97
Hill, T., Burton, M. G., Minier, V., Thompson, M. A., Walsh, A. J., Hunt-Cunningham, M., & Garay, G. 2005, MNRAS, 363, 405
Hill, T., Thompson, M. A., Burton, M. G., Walsh, A. J., Minier, V., Cunningham, M. R., & Pierce-Price, D. 2006, MNRAS, 368, 1223
Hill, T. 2006, PhD thesis, University of New South Wales
Minier, V., Burton, M. G., Hill, T., Pestalozzi, M. R., Purcell, C. R., Garay, G., Walsh, A. J., & Longmore, S. 2005, A&A, 429, 945
Longmore, S.N, Burton, M.G, Barnes, P.J, Wong, T, Purcell, C.R, & Ott, J. 2006, MNRAS, accepted
Lumsden, S. L., Hoare, M. G., Oudmaijer, R. D., & Richards, D. 2002, MNRAS, 336, 621
Minier, V., Conway, J. E., & Booth, R. S. 2001, A&A, 369, 278
Olmi, L., Cesaroni, R., Neri, R., & Walmsley, C. M. 1996, A&A, 315, 565
Osorio, M., Lizano, S., & D'Alessio, P. 1999, ApJ, 525, 808
Pestalozzi, M. R., Minier, V., & Booth, R. S. 2005, A&A, 432, 737
Pillai, T., Wyrowski, F., Carey, S. J., & Menten, K. M. 2006, A&A, 450, 569
Pierce-Price, D., *et al.* 2000, ApJL, 545, L121
Thompson, M. A., Hatchell, J., Walsh, A. J., MacDonald, G. H., & Millar, T. J. 2006, A&A, 453, 1003
Walsh, A. J., Hyland, A. R., Robinson, G., & Burton, M. G. 1997, MNRAS, 291, 261
Walsh, A. J., Burton, M. G., Hyland, A. R., & Robinson, G. 1998, MNRAS, 301, 640
Walsh, A. J., Macdonald, G. H., Alvey, N. D. S., Burton, M. G., & Lee, J.-K. 2003, A&A, 410, 597
Wood, D. O. S. & Churchwell, E. 1989, ApJSS, 69, 831
Wood, D. O. S., & Churchwell, E. 1989, ApJ, 340, 265
Vázquez-Semadeni, E., Kim, J., Shadmehri, M., & Ballesteros-Paredes, J. 2005, ApJ, 618, 344

Astrophysical Masers and their Environments
Proceedings IAU Symposium No. 242, 2007
J.M. Chapman & W.A. Baan, eds.

doi:10.1017/S1743921307012677

The molecular environment of massive star forming cores associated with Class II methanol maser emission

S. N. Longmore[1,2]†, M. G. Burton[1], P. J. Barnes[3], T. Wong[1,2,5], C. R. Purcell[1,4], and J. Ott[2,6]

[1]School of Physics, University of New South Wales, Kensington, NSW 2052, Sydney, Australia
[2]Australia Telescope National Facility, CSIRO, PO Box 76, Epping, NSW 1710, Australia
[3]School of Physics A28, University of Sydney, NSW 2006, Australia
[4]University of Manchester, Jodrell Bank Observatory, Macclesfield, Cheshire SK11 9DL, UK
[5]Department of Astronomy, University of Illinois, Urbana IL 61801, USA
[6]National Radio Astronomy Observatory, 520 Edgemont Road, Charlottesville, VA 22903, USA

Abstract. Methanol maser emission has proven to be an excellent signpost of regions undergoing massive star formation (MSF). To investigate their role as an evolutionary tracer, we have recently completed a large observing program with the ATCA to derive the dynamical and physical properties of molecular/ionised gas towards a sample of MSF regions traced by 6.7 GHz methanol maser emission. We find that the molecular gas in many of these regions breaks up into multiple sub-clumps which we separate into groups based on their association with/without methanol maser and cm continuum emission. The temperature and dynamic state of the molecular gas is markedly different between the groups. Based on these differences, we attempt to assess the evolutionary state of the cores in the groups and thus investigate the role of class II methanol masers as a tracer of MSF.

Keywords. stars: formation, ISM: evolution, ISM: molecules, line: profiles, masers, molecular data, stars: early-type, radio continuum: stars

1. Introduction

In terms of luminosity, energetics and chemical enrichment, massive stars exert a disproportionate influence compared to their number on Galactic evolution. However, the collective effects of their rarity, short formation timescales and heavy obscuration due to dust, make it difficult to find large samples of massive young sources at well constrained evolutionary stages needed to develop an understanding of their formation mechanism. The $5_1 \rightarrow 6_0$ A$^+$ class II methanol (CH_3OH) maser transition at 6.7 GHz is one of the most readily observable signposts of MSF (Menten 1991).

The specific conditions required for the masers to exist makes them a powerful probe of the region's evolutionary stage. While masers probe spatial scales much smaller than their natal cores, the numerous feedback processes from newly formed stars [e.g. turbulence injection from jets/outflows, ionisation (stars $M>8M_\odot$) and heating from radiation etc.] must significantly alter the physical conditions of the surrounding region. We have completed an observing program with the Australia Telescope Compact Array (ATCA) to derive properties of molecular and ionised gas towards MSF regions traced by 6.7 GHz methanol maser emission (Longmore *et al.* 2007). In this contribution, we use these

† E-mail:snl@phys.unsw.edu.au

results to investigate the use of class II methanol masers as a diagnostic of the evolutionary stage of MSF.

2. Observations

Observations of NH_3(1,1), (2,2), (4,4) & (5,5) and 24 GHz continuum were carried out using the ATCA towards 21 MSF regions traced by 6.7 GHz methanol maser emission (selected from a similar sample to Hill *et al.* (see these proceedings)). The H168 [NH_3(1,1)/(2,2)] and H214 [NH_3(4,4)/(5,5)] antenna configurations with both East-West and North-South baselines, were used to allow for snapshot imaging. Primary and characteristic synthesised beam sizes were $\sim 2.'2$ and $\sim 8 - 11''$ respectively. Each source was observed for 4×15 minute cuts in each transition separated over 8 hours to ensure the best possible sampling of the *uv*-plane. The data were reduced using the MIRIAD package (see Sault *et al.* 1995). Characteristic spectra were extracted at every transition for each core at the position of the peak NH_3(1,1) emission, baseline subtracted and fit using the *gauss* and *nh3(1,1)* methods in CLASS (see http://www.iram.fr/IRAMFR/GILDAS/). Continuum source fluxes and angular sizes were calculated in both the image domain and directly from the *uv* data.

3. Core separation

NH_3 detections within each region were separated into individual cores if offset by more than a synthesised beam width spatially, or more than the FWHM in velocity if at the same sky position. The same criteria were used to determine whether the NH_3, continuum and methanol maser emission in each of the regions were associated. In all but three cases, these criteria were sufficient to both unambiguously separate cores and determine their association with continuum and maser emission. We find 41 NH_3(1,1) cores (of which three are in absorption and two are separated in velocity) and 14 24 GHz continuum cores. Observationally the cores fall in to four groups: NH_3 only (Group 1); NH_3 + methanol maser (Group 2); NH_3 + methanol maser + 24 GHz continuum (Group 3); NH_3 + 24 GHz continuum (Group 4). The cores were distributed with 16, 16, 6 and 2 cores in Groups 1 to 4, respectively. Based on this grouping, most of the NH_3(1,1) cores are coincident with methanol maser emission (Groups 2 & 3), but there are a substantial fraction of NH_3 cores with neither 24 GHz continuum nor maser emission (Group 1).

Having separated the cores into these groups, we then considered observational biases and selection effects which may affect their distribution. The biggest potential hindrance was the difference in linear resolution and sensitivity caused by the factor of ~ 5 variation in distance to the regions. Despite this, the NH_3, continuum and methanol maser observations have the same sensitivity limit towards all the regions: therefore, the *relative* flux densities of these tracers in a given region are directly comparable. In addition, no correlation was found between a region's distance and the number of cores toward the region or their association with the different tracers. From this we conclude the distance variation does not affect the distribution of cores into separate groups. However, it should be remembered that any conclusions drawn about the cores are limited by the observational parameters used to define the groups.

4. Deriving physical properties

Properties of the molecular gas in each of the cores were derived from the NH_3 observations. The core size was calculated from the extent of the integrated NH_3(1,1) emission after deconvolving the synthesised beam response. The dynamical state of the molecular gas was derived from the line profiles of the high spectral resolution (0.197 $\mathrm{kms^{-1}}$) NH_3(1,1) observations after deconvolving the instrumental response. Preliminary gas kinetic temperatures were calculated by fitting the measured column densities of the multiple NH_3 transitions to the LVG models described in Ott *et al.* (2005). Finally, properties of the ionised gas were derived from the 24 GHz continuum emission following Mezger & Henderson (1967), assuming it was spherically symmetric and optically thin at an electron temperature of 10^4 K.

5. Results

In general the core properties are comparable to those derived from similar surveys towards young MSF regions. Below we outline differences, in particular between the cores in the different groups described in §3.

5.1. *Molecular Gas*

The measured NH_3(1,1) linewidth varies significantly between the groups, increasing from 1.43, 2.43, 3.00 $\mathrm{kms^{-1}}$ for Groups 1 to 3 respectively. This shows the NH_3-only cores are more quiescent than those with methanol maser emission.

The NH_3(1,1) spectra of some cores deviate significantly from the predicted symmetric inner and outer satellite brightness temperature ratios. These line profile asymmetries are often seen toward star forming cores and are understood to be caused by selective radiative trapping due to multiple NH_3(1,1) sub-clumps within the beam (see Stuzki & Winnewisser (1985) and references therein). The NH_3-only cores (Group 1) have by far the strongest asymmetries.

NH_3(4,4) emission is detected toward the peak of 13 NH_3(1,1) cores and 11 of these also have coincident NH_3(5,5) emission. The higher spatial resolution of the NH_3(4,4) and (5,5) images compared to the NH_3(1,1) observations ($8''$ vs $11''$) provides a stronger constraint to the criteria outlined in §3 as to whether this emission is associated with either methanol or continuum emission. In every case, the NH_3(4,4) and (5,5) emission is unresolved, within a synthesised beam width of the methanol maser emission spatially and within the FWHM in velocity. This shows the methanol masers form at the warmest parts of the core. Significantly, no NH_3(4,4) or (5,5) emission is detected toward NH_3-only sources.

As shown in Figure 1, cores with NH_3 and methanol maser emission (Groups 2 and 3) are generally significantly warmer than those with only NH_3 emission (Group 1). However, there are also a small number of cores with methanol maser emission that have very cool temperatures and quiescent gas, similar to the NH_3-only cores. Modelling shows pumping of 6.7 GHz methanol masers requires local temperatures sufficient to evaporate methanol from the dust grains ($T \gtrsim 90$K) and a highly luminous source of IR photons (Cragg *et al.* 2005) i.e. an internal powering source. It is therefore plausible that the cold, quiescent sources with methanol maser emission are cores in which the feedback from the powering sources have not had time to significantly alter the larger scale properties of the gas in the cores.

5.2. *Ionised Gas*

Of the 14 continuum cores detected at 24 GHz, 10 are within two synthesised beams of the 6.7GHz methanol maser emission. This is contrary to the results of Walsh *et al.* (1998), who found the masers generally unassociated with 8 GHz continuum emission. However, six of the 24 GHz continuum sources found at the site of the methanol maser

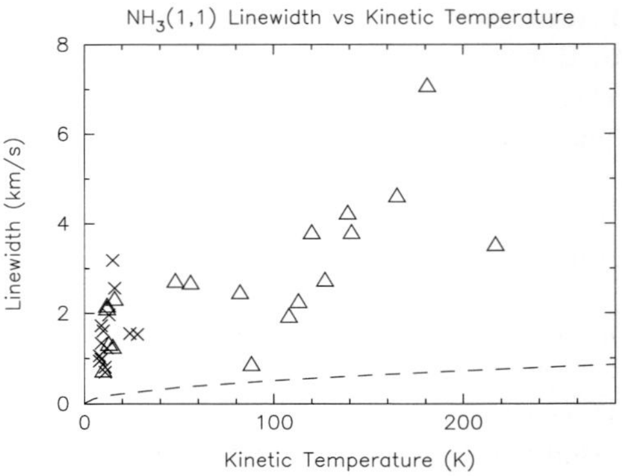

Figure 1. NH_3(1,1) linewidth vs gas kinetic temperature. Cores with NH_3 only (Group 1) are shown as crosses while those with NH_3 and methanol maser emission (Groups 2 and 3) are shown as triangles. The dashed line shows the expected linewidth due to purely thermal motions.

emission have no 8 GHz counterparts. A possible explanation for this may be that the continuum emission is optically thick rather than optically thin between 8 and 24 GHz and hence has a flux density proportional to ν^2 rather than $\nu^{-0.1}$. The seemingly low emission measures derived for the 24 GHz continuum are unreliable due to the large beam size of the observations. Alternatively, the 24 GHz continuum sources may have been too extended and resolved-out by the larger array configuration used at 8 GHz by Walsh *et al.* (1998). Further high spatial resolution observations at $\nu \geqslant 24$ GHz are required to derive reliable emission measures and spectral indexes to unambiguously differentiate between the two explanations.

6. Towards an evolutionary sequence

From the previous analysis, the core properties are seen to vary depending on their association with methanol maser and continuum emission. Making the reasonable assumption that cores heat up and becomes less quiescent with age, we now investigate what these physical conditions tell us about their evolutionary state.

As the core sizes are similar, the measured linewidths can be reliably used to indicate how quiescent the gas is, without worrying about its dependence on the core size (Larson 1981). It then becomes obvious that the isolated NH_3 cores (Group 1) contain the most quiescent gas. However, from the linewidths alone it is not clear if these cores will eventually form stars or if they are a transitory phenomenon. The fact that a large number of these Group 1 cores contain many dense sub-clumps (as evidenced by the NH_3(1,1) asymmetries) suggests the former is likely for at least these cores. The linewidth of sources with methanol maser emission (Groups 2 and 3) are significantly larger, and hence contain less quiescent gas, than those of Group 1. The larger linewidths combined with generally higher temperatures, suggests cores in Groups 2 and 3 are more evolved than in Group 1.

The detection of continuum emission suggests a massive star is already ionising the gas. With the current observations, the properties of the continuum sources are not well enough constrained to further separate their evolutionary stages. However, as all (with one possible exception) of the continuum sources only detected at 24GHz are associated with dense molecular gas and masers, this would suggest they are younger than those detected at both 8 + 24 GHz, despite their seemingly small emission measures. In this scenario, the cores with only 8 + 24GHz continuum and no NH_3 emission, may be sufficiently advanced for the UCHIIregion to have destroyed its natal molecular material.

From this evidence, the cores in the different groups do appear to be at different evolutionary stages, going from most quiescent to most evolved according to the group number.

7. Conclusions

What then, can we conclude about the role of methanol masers as a signpost of MSF? The observations show that 6.7 GHz methanol masers:

• are found at the warmest location within each core.

• generally highlight significantly warmer regions with less quiescent gas (i.e. more evolved sources) than those with only NH_3 emission.

• may also highlight regions in which the internal pumping source is sufficiently young that it has not yet detectably altered the large scale core properties.

Methanol masers therefore trace regions at stages shortly after a suitable powering source has formed right through to relatively evolved UCHII regions. While remaining a good general tracer of young MSF regions, the presence of a methanol maser does not single out any particular intermediate evolutionary stage.

Finally, these data confirm and strengthen the results of Hill *et al.* (these proceedings), that the youngest MSF regions appear to be molecular cores with no detectable methanol maser emission.

8. Acknowledgements

SNL is supported by a scholarship from the School of Physics at UNSW. We thank Andrew Walsh for comments on the manuscript. We also thank the Australian Research Council for funding support. The Australia Telescope is funded by the Commonwealth of Australia for operation as a National Facility managed by CSIRO.

References

Cragg D. M., Sobolev A. M., & Godfrey P. D. 2005, *MNRAS*, 360, 533

Larson R. B., 1981, *MNRAS*, 194, 809

Longmore S. N., Burton M. G., Barnes P. J., Wong T., Purcell C. R., & Ott J., 2007, *MNRAS*, in press.

Menten K. M. 1991, *ApJ*, 380, 75

Mezger, P. G. & Henderson, A. P. 1967, *ApJ* 147, 471

Ott J., Weiss A., Henkel C., & Walter F., 2005, *ApJ*, 629, 767

Sault, R. J., Teuben, P. J., & Wright, M. C. H. 1995, in: R.A. Shaw, H.E. Payne & J. J. E. Hayes (eds.), *Astronomical Data Analysis Software and Systems IV*, ASP Conf. Ser. 77, p. 433

Stutzki J. & Winnewisser G. 1985, *A&A*, 144, 13

Walsh A. J., Burton M. G., Hyland A. R., & Robinson G. 1998, *MNRAS*, 301, 640

Astrophysical Masers and their Environments
Proceedings IAU Symposium No. 242, 2007
J.M Chapman & W.A. Baan, eds.

doi:10.1017/S1743921307012689

A documentary of high-mass star formation: Probing the dynamical evolution of Orion Source I on 10–100 AU scales using SiO masers

L. D. Matthews[1], **C. Goddi**[1], **L. J. Greenhill**[1], **C. J. Chandler**[2], **M. J. Reid**[1] **and E. M. L. Humphreys**[1]

[1]Harvard-Smithsonian Center for Astrophysics, Cambridge, MA, USA
[2]National Radio Astronomy Observatory, Socorro, NM, USA

Abstract. A comprehensive picture of high-mass star formation has remained elusive, in part because examples of high-mass young stellar objects (YSOs) tend to be relatively distant, deeply embedded, and confused with other emission sources. These factors have impeded dynamical investigations within tens of AU of high-mass YSOs—scales that are critical for probing the interfaces where outflows from accretion disks are launched and collimated. Using observations of SiO masers obtained with the Very Large Array (VLA) and the Very Long Baseline Array (VLBA), the KaLYPSO project is overcoming these limitations by mapping the structure and dynamical/temporal evolution of the material 10-1000 AU from the nearest high-mass YSO: Radio Source I in the Orion BN/KL region. Our data include ~40 epochs of VLBA observations over a several-year period, allowing us to track the proper motions of individual SiO maser spots and to monitor changes in the physical conditions of the emitting material with time. Ultimately these data will provide 3-D maps of the outflow structure over approximately 30% of the outflow crossing time. Here we summarize recent results from the KaLYPSO project, including evidence that high-mass star formation occurs via disk-mediated accretion.

Keywords. masers, stars: pre-main sequence, techniques: high angular resolution, circumstellar matter, stars: formation, stars: winds, outflows, stars: individual (Source I)

---

1. The KaLYPSO Project: Introduction and motivation

Increasing observational evidence supports a picture in which high-mass stars ($M \gtrsim 10 M_{\odot}$) form through disk-mediated accretion, in a manner analogous to a scaled-up version of low-mass star formation (e.g., Zhang *et al.* 1998; Cesaroni 2002; Beuther *et al.* 2002; Patel *et al.* 2005; Beltrán *et al.* 2006). However, the details of this process remain poorly understood. For example, rapid formation timescales ($\ll 10^6$ years) seem to necessitate extremely high mass accretion rates. Meanwhile, strong radiation pressure from a high-mass YSO is expected to impede accretion, and the lack of a strong stellar dynamo in high-mass protostars necessitates an alternative mechanism for shedding large amounts of angular momentum. Although a number of theoretical studies have attempted to address these issues (e.g., McKee *et al.* 2002; Bonnell *et al.* 2005, Krumholtz *et al.* 2005), observations have been unable to readily distinguish between various models. Open questions include: What are the sizes and structures of accretion disks around high-mass YSOs? What is the driving mechanism of their outflows? What are the physical properties (density, temperature) of the disk-outflow interface? What is the role of magnetic fields in high mass star formation?

One of the chief obstacles for constraining models of high-mass star formation is that examples of high-mass YSOs tend to be relatively distant ($d \gtrsim 1$ kpc), deeply embedded, and confused with other emission sources. Moreover, since high-mass stars evolve rapidly, by the time an unobstructed view of the young star emerges, the disk and outflow structures may have been destroyed. Consequently, observations to date have been unable to probe the 10-100 AU spatial scales over which outflows from the accretion disks are expected to be launched and collimated. The KaLYPSO (Kleinmann-Low Young Proto-Stellar Object) project (http://www.cfa.harvard.edu/kalypso/) aims to overcome these limitations and provide the most detailed picture yet of a high-mass star in formation.

2. Our target: Radio Source I in the Orion BN/KL region

The nearest region of high-mass star formation is the Kleinmann-Low (KL) nebula in Orion, at a distance of ~450 pc. Within this nebula is located the radio continuum-emitting "Source I", a luminous, highly embedded YSO with a disk-like morphology (Reid *et al.* 2007; Figure 1).

3. Goals of the KaLYPSO project

The KaLYPSO team has launched a multi-faceted study of Source I with unprecedented angular and temporal resolution. Our goals include:

- Chart the time-varying distribution of ~1000 SiO maser spots within 10–100 AU of Source I
- Measure proper motions of ~100 individual SiO masers with precision <1 km s^{-1}.
- Produce geometric and dynamical models of the molecular gas surrounding Source I
- Compare distributions of different maser transitions surrounding Source I to probe local density and temperature conditions
- Constrain the physical processes (e.g., shocks, collisions, fragmentation, magnetic fields) that contribute to the process of forming a high-mass star
- Produce a movie documenting the 3-D evolution of the outflows surrounding a massive protostar over ~30% of the outflow crossing time

4. The Observations of SiO Masers toward Source I

Observations of masers using Very Long Baseline Interferometry offer a powerful means to study massive star formation. Masers are not affected by extinction, furnish information on the physical conditions of the emitting gas, and can supply kinematic information on the material surrounding YSOs at extremely high angular resolutions.

Using the VLBA of the National Radio Astronomy Observatory, we have obtained ~40 epochs of observations of the SiO $v = 1$ and v=2 masers toward Source I at one-month intervals over a multi-year period. The two transitions were observed simultaneously with a spectral resolution of 0.2 km s^{-1}. Our resulting images have an angular resolution of ~0.2 mas—corresponding to ~0.1 AU at the distance of Orion.

5. Preliminary results

5.1. *The large-scale picture*

Figure 1 shows a large-scale (~1000 AU) bipolar structure surrounding Radio Source I, as previously observed in thermal and maser emission from the SiO ground state ($v = 0$) with $\sim 0.5''$ resolution (Wright *et al.* 1995). Spectroscopy has confirmed similar velocity

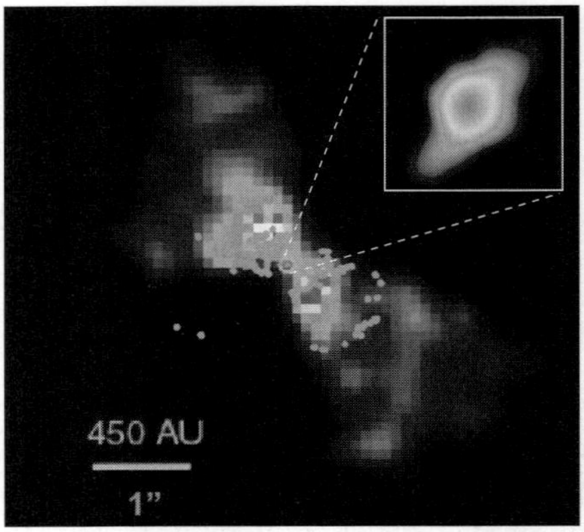

Figure 1. The large-scale (~1000 AU) bipolar structure surrounding Radio Source I, as observed in thermal and maser emission from the SiO ground state ($v = 0$) with $\sim 0.5''$ resolution by Wright *et al.* 1995. Locations of H_2O masers mapped with the VLA by Greenhill *et al.* 1998 are overplotted as pink spots (for color see the electronic version). The inset shows a λ7mm continuum map of Radio Source I from Reid *et al.* 2007.

spreads in both the north and south lobes, consistent with an outflow along the plane of the sky. H_2O masers also lie along the apparent outflow.

The Figure 1 inset shows a λ7mm continuum map of Radio Source I from Reid *et al.* (2007). The 7-mm emission appears to trace an edge-on, ionized disk with a diameter of ~100 AU, centered on the larger-scale outflow. The SiO $v = 1$ and $v = 2$ observations described here are providing key new information linking the observations shown in Figure 1, and further solidify the picture of Source I as a high-mass star in the process of formation via disk-mediated accretion and outflow.

5.2. *The Distribution and Kinematics of the SiO Masers around Source I*

Figure 2 shows a velocity field derived from one epoch of our combined VLBA SiO $v = 1$ and $v = 2$ observations toward Source I. The maser emission lies within an "X"-shaped distribution centered on Source I (whose position is indicated by an asterisk symbol). The SiO $v = 2$ emission tends to lie closer to Source I than the $v = 1$ emission, although there is considerable overlap. Emission is observed over the velocity range $-15 \lesssim V_{\rm LSR} \lesssim 30$ km s^{-1} (and is color-coded by Doppler shift in the electronic version of this article). Emission to the north/west is predominantly redshifted, while that to the east/south is predominantly blue-shifted, suggesting rotational motion. Consistent with this, material in the "bridge" between the south and west lobes lies near the systemic velocity. A similar but fainter bridge between the north and southeast lobes is also seen in some epochs. Proper motions of individual maser features (~10 km s^{-1}) are observed between consecutive epochs; their bulk directions are indicated by arrows. A more detailed analysis of these proper motions is forthcoming.

6. An emerging model for source I

The morphology and kinematics of the emission surrounding Source I provide strong evidence of both accretion and outflow. Figure 3 shows a schematic model for the source, based on a synthesis of the observations in the previous two figures (see the electronic version for a full-color rendition). At $r < 150$ AU (dashed circle) material is being driven into a wide-angle bipolar outflow. The edges of the "funnel" (representing the

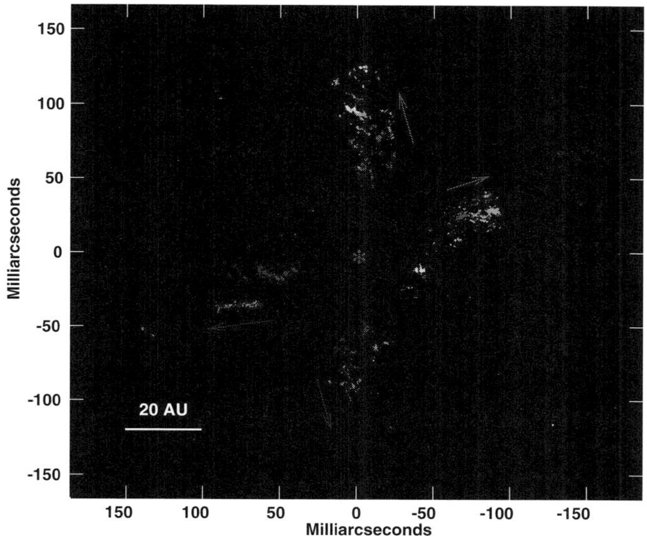

Figure 2. A velocity field derived from the combined SiO $v = 1$ and $v = 2$ emission toward Source I during one observational epoch. Emission is color-coded by Doppler shift (electronic version only) and spans the velocity range $-15 \lesssim V_{\rm LSR} \lesssim 30$ km s^{-1}. An asterisk indicates the position of Source I as determined from radio continuum observations. Directions of the bulk proper motions of individual maser spots observed over the course of multiple observing epochs are indicated by arrows.

loci of the Doppler-shifted SiO $v = 1$ & $v = 2$ emission; see Figure 2) may be infalling, rotating molecular material swept up by a wind/outflow (Cunningham *et al.* 2005), or may demarcate the edge-brightened portion of the wind from a highly inclined ($i \sim 80°$), flared, rotating accretion disk (traced at smaller radii by 7-mm continuum emission; see Figure 1). The canting of the individual SiO emission arms (see Figure 2) and the existence of a "bridge" of gas between the two sides of the funnel with a velocity gradient across it (see Figure 2) now seem to exclude a model of the masers as a biconical outflow along the northwest/southeast direction (cf. Greenhill *et al.* 1998; Doeleman *et al.* 1999).

At $r > 150$ AU (outside the dashed circle), cooler gas traced by SiO $v = 0$ maser+ thermal emission (represented by the shaded areas; cf. Figure 1) and H_2O maser emission (depicted as white spots) delineate a more extended bipolar outflow along a direction close to the plane of the sky.

7. Summary

Using the VLBA, our KaLYPSO team has obtained ~40 epochs of monthly observations of the SiO $v = 1$ and $v = 2$ masers toward Radio Source I in the Orion BN/KL region. Source I is believed to be the nearest example of a high-mass YSO. These observations are the first to probe the kinematics of the molecular gas within 10–100 AU of such an object. Our analysis of these data, together with complementary observations of the adjacent 7-mm radio continuum, H_2O masers, and SiO $v = 0$ masers/thermal emission, provide compelling evidence that disk-mediated accretion and outflow are fundamental to the process of high-mass star formation. Further data reduction and analysis are ongoing. Updates will be posted at http://www.cfa.harvard.edu/kalypso.

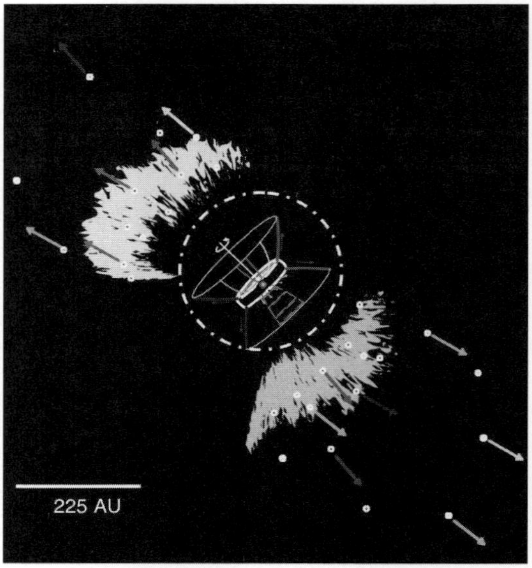

Figure 3. Schematic showing a working model for Source I. Details are described in the Text.

Acknowledgements

The Very Long Baseline Array (VLBA) of the National Radio Astronomy Observatory (NRAO) is a facility of the National Science Foundation, operated under cooperative agreement by Associated Universities, Inc. This work is support by grant 0507478 from the National Science Foundation.

References

Beltrán, M. T., Cesaroni, R., Codella, C., Testi, L., Furuya, R. S., & Olmi, L. 2006, *Nature*, 443, 427

Beuther, H., Kerp, J., Preibisch, T., Stanke, T., & Schilke, P. 2002, *A&A*, 395, 169

Bonnell, I. A. & Bate, M. R. 2005, *MNRAS*, 362

Cesaroni, R. 2002, Highlights of Astronomy, Vol. 12, ed. H. Rickman, (ASP:San Francisco), p. 156

Cunningham, A., Frank, A., & Hartmann, L. 2005, *ApJ*, 631, 1010

Doeleman, S. S., Lonsdale, C. J., & Pelkey, S. 1999, *ApJ*, 510, L55

Greenhill, L. J., Gwinn, C. R., Schwartz, C., Moran, J. M., & Diamond, P. J. 1998, *Nature*, 396, 650

Krumholz, M. R., McKee, C. F., & Klein, R. I. 2005, *Nature*, 438, 332

McKee, C. F. & Tan, J. C. 2002, *Nature*, 416, 59

Patel, N. A. *et al.* 2005, *Nature*, 437, 109

Reid, M. J., Menten, K. M., Greenhill, L. J., & Chandler, C. J. 2007, *ApJ*, submitted

Wright, M. C. H., Plambeck, R. L., Mundy, L. G., & Looney, L. W¿ 1995, *ApJ*, 455, L185

Zhang, Q., Hunter, T. R., & Sridharan, T. K. 1998, *ApJ*, 505, L151

Astrophysical Masers and their Environments
Proceedings IAU Symposium No. 242, 2007
J. M. Chapman & W. A. Baan, eds.

doi:10.1017/S1743921307012690

Massive star-formation in G24.78+0.08 studied by means of maser VLBI and thermal interferometric observations

L. Moscadelli[1], C. Goddi[2], R. Cesaroni[1] and M. T. Beltràn[3]

[1]INAF, Osservatorio Astrofisico di Arcetri, Largo E. Fermi 5, 50125 Firenze, Italy
email: mosca@arcetri.astro.it

[2]Harvard-Smithsonian Center for Astrophysics, 60 Garden Street, Cambridge, MA 02138, USA

[3]Departament d'Astronomia i Meteorologia, Universitat de Barcelona, Av. Diagonal 647, 08028, Barcelona, Catalunya, Spain

Abstract. This work presents the results of VLBI observations of 6.7 GHz methanol and 22.2 GHz water masers towards the mm core A in the massive star-forming region G24.78+0.08. Comparing the maser with previous millimeter interferometric and recent continuum VLA observations, the physical properties and the gas kinematics of the G24 A core on linear scales from ~100 AU to ~0.1 pc are determined.

Keywords. Masers, stars: formation, HII regions, techniques: interferometric

1. Introduction

The star forming region G24.78+0.08 is located at a distance of 7.7 kpc and has a large bolometric luminosity ($\leqslant 7\times10^4\ L_\odot$). Furuya *et al.* (2002) first detected three molecular cores (in continuum emission at 2 mm) and two HII regions (one of these embedded in one of the cores) spread over ~ 0.4 pc. The region contains at least four distinct centers of star formation, named A, B, C, and D, which Furuya *et al.* (2002) speculate to be in different evolutionary states. Both the molecular core A and the source B have associated HII regions (Codella *et al.* 1997) and, hence, are already ZAMS stars, whereas the other two appear to be less evolved. Cores A and C power bipolar molecular outflows (Furuya *et al.* 2002).

1.4 mm Plateau de Bure Interferometer (PdBI) one-arcsecond angular resolution observations by Beltrán *et al.* (2004) have resolved the structure of the G24 A core into two distinct subcores: G24 A1, towards the southeast (SE), and G24 A2, towards the northwest (NW). G24 A1 is associated with the most compact HII region, while G24 A2 shows strong pointlike emission from 3.6 to 8 μm (GLIMPSE) but weak radio 22 GHz continuum emission. Both G24 A1 and A2 are hot molecular cores (HMCs) (Beltrán *et al.* 2004), with a V_{LSR} gradient along a northeast-southwest (NE-SW) direction, approximately perpendicular to the axis of the ^{12}CO molecular outflow. This fact suggests that the two cores might be rotating around the outflow axis (Beltrán *et al.* 2004). Towards G24 A1 and A2 several species of molecular masers have been detected: 6.7 GHz CH_3OH, observed with the Australian Compact Array (ATCA) by Walsh *et al.* (1998), and 22.2 GHz H_2O and 1.6 GHz OH masers observed with the VLA by Forster & Caswell (1989).

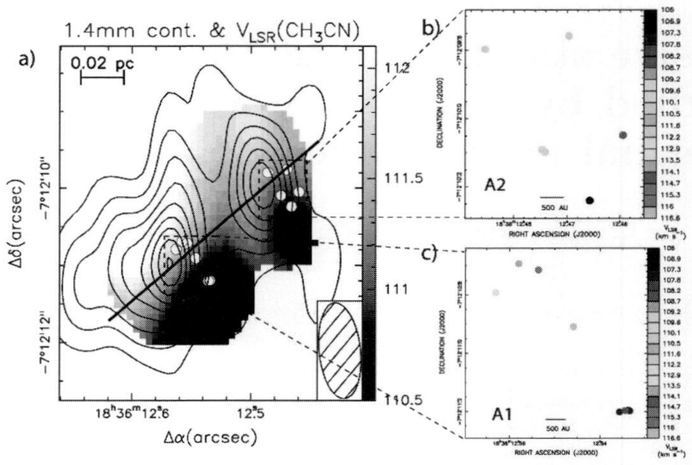

Figure 1. **a)** Map of the CH_3CN (12 → 11) line peak velocity towards G24 A, with the contour levels in kilometers per seconds indicated in the wedge to the right of the panel (Beltrán *et al.* 2004); overlaid on the CH_3CN velocity map, is shown the contour map of the 1.4 mm continuum emission (*solid lines*) and the positions of the CH_3OH 6.7 GHz masers (*filled spheres*); the straight black line represents the axis of the molecular outflow observed with PdBI in $^{12}CO(1–0)$ by Furuya *et al.* (2002). **b)** and **c)** Positions and line-of-sight velocities of the 6.7 GHz maser features observed towards the subcore A2 and A1, respectively, with the colors (see online version) denoting different line-of-sight velocities, according to the scale on the right-hand side of the panels.

2. Observational results

2.1. *Class II CH_3OH masers*

Figure 1 shows a map of the CH_3CN (12–11) line line-of-sight velocity overlaid on the contour map of the 1.4 mm continuum emission towards G24 A1 and G24 A2 (Beltrán *et al.* 2004). The absolute positions (accurate to within 50 mas) of the 6.7 GHz methanol masers observed with the EVN are plotted as filled black circles (Fig. 1, panel a). The 6.7 GHz maser emission occurs from two distinct groups of features, each associated with one of the mm subcores (A1 or A2). Both groups of maser features are elongated along a direction that agrees well with the axis of the velocity gradient measured in the CH_3CN 1.4 mm line.

Panels c and b of Fig. 1 show the line-of-sight velocities of the 6.7 GHz maser features, associated respectively with the A1 and A2 subcores. Across each maser group, the variation of velocities agrees with the velocity gradient observed using the CH_3CN (12–11) line in the harbouring subcore, with redshifted velocities to the NE and blueshifted velocities to the SW. The methanol maser features are distributed over regions ~three times smaller (0.023 and 0.02 pc) and cover an observed velocity range ~three times larger (9.1 and 5.7 km s^{-1}) than the emission of the CH_3CN thermal line in both subcores.

2.2. *H_2O masers*

By observing G24.78+0.08 at 22.2 GHz with the VLBA in phase-reference mode at four different epochs, absolute positions and proper motions of the water maser features have been derived. Figure 2 shows the water maser distribution in the subcore A1, which contains the strongest water maser features of the entire G24.78+0.08 region, with

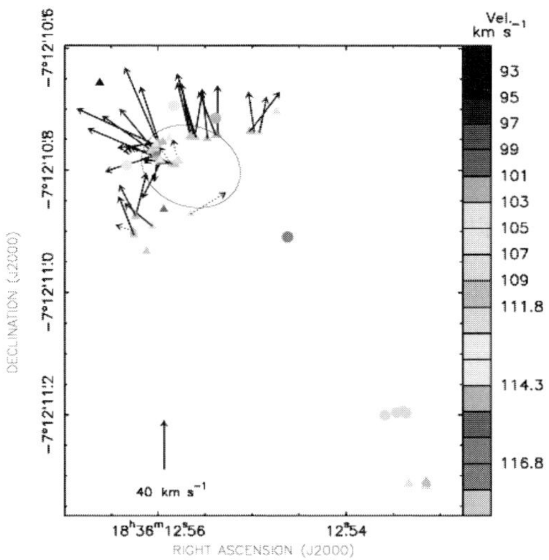

Figure 2. Comparison of the distribution of the VLBA 22.2 GHz water (*triangles*) and EVN 6.7 GHz methanol (*dots*) maser features in the mm subcore G24 A1. The colors (see online version) denote different velocities, according to the scale on the right-hand side of the panel. Black arrows show the measured absolute proper motions of water maser features, using dotted arrows to denote the most uncertain measurements; the amplitude scale for proper motions is reported at the bottom of the panel. The ellipse shows the position and the deconvolved FWHM size of the G24 A1 continuum source observed recently in K-band with the VLA by Beltran *et al.* (in preparation).

intensities ranging from ≈ 0.05 to ≈ 100 Jy beam^{-1}. Maser emission arises from two distinct regions found in the NE and SW corner of the plotted area. The strongest features of the NE maser cluster (called "A1N") are distributed along an EW line with a length of ≈ 100 mas, and along an arc-shaped structure of diameter ≈ 50 mas. With a few exceptions, most of the maser features have velocities within ± 5 km s^{-1} of the LSR systemic velocity (111 km s^{-1}) of the region. The measured absolute proper motions, with amplitudes in the range 20–60 km s^{-1}, have directions roughly perpendicular to the linear and arcuate spatial distributions of the maser features.

3. Rotation and expansion in the mm core G24 A

We now compare the VLBI CH_3OH and H_2O maser observations towards the core G24 A with previous millimeter PdBI and recent VLA observations, and discuss a model for explaining the gas kinematics from small (100 AU) to large (0.1 pc) scale.

3.1. *Evidence for rotation*

Figure 1 shows that, in both cores G24 A1 and A2, the velocity distribution of the methanol 6.7 GHz maser spots agrees well with the velocity gradient measured in the thermal CH_3CN (12–11) line. If, as suggested by Beltrán *et al.* (2004), the CH_3CN velocity gradient is due to rotation, the three times larger spread of velocities of the methanol masers across a three times smaller region, might be explained in terms of conservation of angular momentum from large to small scales inside the cores. Assuming that inside the region (of diameter ≈ 0.02 pc) traced by the 6.7 GHz maser emission in both cores, the gas is in rotational equilibrium, a dynamical mass of 55 and 19 $M_\odot$

for G24 A1 and G24 A2, is derived respectively. From the free-free continuum emission of the ultra-compact HII (UCHII) region in G24 A1, Codella *et al.* (1997) estimate a spectral type of O9.5 for the exciting star, corresponding to a stellar mass of $\approx 20\ M_\odot$. This must be added to the mass of the gas within 0.02 pc from the YSO, which is $\approx$ 30 $M_\odot$ for a mean density of 10^8 cm^{-3}. The total mass (star plus gas) is consistent with the mass of 55 $M_\odot$ computed from the methanol maser velocity range in G24 A1.

3.2. *Evidence for expansion*

The water maser proper motions measured in the A1N cluster are oriented approximately perpendicular to the geometrical distribution of the maser features, which are mainly concentrated in two structures, a line oriented along the EW direction and an arc (see Fig. 2). The separation from the LSR systemic velocity of the line-of-sight velocities of most maser features of the A1N cluster is quite small ($\leqslant 5$ km s^{-1}) compared with the sky-projected velocity components (20–60 km s^{-1}), implying that the velocity vectors of the maser features are close to the plane of the sky. The observed velocity distribution suggests that the A1N water masers are tracing gas *expanding* towards N-NE.

The ellipse drawn in Fig. 2 denotes the position and the deconvolved FWHM size of the continuum source in the G24 A1 core as recently determined using the VLA in K and Q-band by Beltran *et al.* (in preparation). The position of the continuum peak at 22.2 GHz is determined with an accuracy of $\approx$ 20 mas. With an angular resolution of $\approx 0\overset{''}{.}05$, the continuum source is partially resolved with a deconvolved radius (defined as the geometric mean of the HWHM major and minor axes) of $0\overset{''}{.}07$ (540 AU at a distance of 7.7 kpc). These recent VLA observations confirm that the most likely interpretation for the continuum source is in terms of an UCHII ($R_{HII} \leqslant 0.01$ pc), that is optically thick at frequencies below 22.2 GHz, becoming optically thin between 22.2 GHz and 44 GHz. The observed continuum flux is consistent with previous observations, indicating a ZAMS spectral type of O9.5 for the central ionizing star.

Figure 2 shows that the UCHII region falls close to the A1N maser cluster. Considering that the estimated size of the HII region is likely an upper limit, most of the water maser features appear to be distributed along the border of the ionized gas. Furthermore, the elongation of the continuum source (P.A. = 66°) agrees well with the average orientation of the water maser proper motions. Thus, the position and the geometry of the continuum source suggests that the water maser motions are driven by the expansion of the ionized gas. Freely expanding ionized gas should move at a maximum velocity comparable to the sound speed in the ionized gas, $\approx$10 km s^{-1} assuming a kinetic temperature of 10^4 K. The observed velocities of the water masers are significantly higher than 10 km s^{-1}, indicating that the ionized gas is *not* freely expanding, but that the expansion must be driven by some additional mechanism than thermal pressure.

Stellar winds are commonly observed towards early-type OB main-sequence stars and observational evidence is being accumulated that such winds play an important role during the early expansion phase of an UCHII region (Hoare *et al.* 2007). A detailed study by Lamers & Leitherer (1993) indicates wind velocities V_w = 1500 – 2500 km s^{-1} and mass loss rates $\dot{M}_w$ = 1–5 $\times$ 10^{-6} $M_\odot$ yr^{-1}, as appropriate for an O9.5 star. Such a powerful stellar wind might be already emitted during the early ZAMS evolution of a massive star. We have considered the case that the expansion of the water maser shell in G24 A1 is driven by a strong wind. Our VLBA water maser and VLA continuum radio observations allow us to set a value for the shell radius and velocity with $R_0 \approx 500$ AU, and $V_0 \approx 40$ km s^{-1} respectively. To model the dynamics of the wind-driven shell, we have used equations given in Castor *et al.* (1975) and Shull (1980). A detailed report of our calculations is given in Moscadelli *et al.* 2007 (in preparation). We find that, over the explored parameter space for the stellar wind properties, and fixing the wind mechanical luminosity, gives a unique value of ambient density for which the wind-driven shell can attain the observed values of radius and velocity. For the likely values of wind luminosities for a O9.5 star, the wind-driven shell models require the hydrogen particle ambient density to lie in the range 6×10^6 cm^{-3} $\leqslant n_{\rm H} \leqslant 2.5 \times 10^7$ cm^{-3}. The shell age, t_0, is predicted to be in the narrow interval $t_0 = 30$–35 yr.

3.3. *Rotation or expansion in the G24 A1 core?*

On the basis of the discussed observational evidences, we propose two alternative interpretations to explain the gas kinematics in the G24 A1 core on scales from 100 AU to 0.1 pc.

The expansion detected in the 22.2 GHz water masers on scales of ≈ 100 AU (see Fig. 2) might be the dominant motion across the whole molecular core. Thus, the regular variation of line-of-sight velocity observed in the thermal CH_3CN (12–11) line as well as in the CH_3OH 6.7 GHz maser emission (see Fig. 1)on linear scales of 0.05 pc, might be due to expansion rather than to rotation, as proposed by Beltrán *et al.* (2004). In particular, the regular velocity gradient can be accounted for only by a *collimated* flow. In this interpretation, the 22.2 GHz water masers would mark the inner portion of a molecular outflow, closer to the YSO and not yet collimated, while the CH_3OH 6.7 GHz maser and the CH_3CN thermal emissions would trace the same outflow at larger scale, where collimation is already effective.

Interpreting the motion of the gas in the G24 A1 core in terms of a collimated outflow oriented along the NE–SW direction, apparently contrasts with the PdBI ^{12}CO(1–0) observations by (Furuya *et al.* 2002), indicating the presence of a bipolar outflow on scale $\geqslant 0.1$ pc oriented SE–NW. We propose an alternative scenario, where expansion and rotation can coexist. Figure 2 shows that the detected CH_3OH 6.7 GHz maser spots are on average more distant from the center of the continuum source than the water maser spots and located beyond the border of the UCHII region. We speculate that methanol masers are tracing the pre-shock ambient gas radiatively heated by the YSO radiation. The value of pre-shock ambient gas density indicated by the wind-driven shell model fits well with the observations of both H_2O and CH_3OH masers around the G24 A1 continuum source, since a pre-shock gas density $n_{\rm H} \sim 10^7$ cm^{-3} is predicted in the excitation models of both water and methanol masers. The kinematics of methanol masers is expected to differ from that of water masers as the former are located near the G24 A1 core and are unaffected by expansion in the HII region. The water masers should mark the border of the dynamical interaction between the YSO and the surrounding ambient medium: gas moving beyond the water maser shell is not yet perturbed by the YSO evolution (and might be still rotating and slightly contracting), whereas gas inside the maser shell is rapidly expanding due to the YSO wind pressure.

References

Beltrán, M. T., Cesaroni, R., Neri, R., *et al.* 2004, *ApJL*, 601, L187

Castor, J., McCray, R., & Weaver, R. 1975, *ApJL*, 200, L107

Codella, C., Testi, L., & Cesaroni, R. 1997, *A&A*, 325, 282

Forster, J. R. & Caswell, J. L. 1989, *A&A*, 213, 339

Furuya, R. S., Cesaroni, R., Codella, C., *et al.* 2002, *A&A*, 390, L1

Hoare, M. G., Kurtz, S. E., Lizano, S., Keto, E., & Hofner, P. 2007, in Protostars and Planets V, ed. B. Reipurth, D. Jewitt, & K. Keil, 181–196

Lamers, H. J. G. L. M. & Leitherer, C. 1993, *ApJ*, 412, 771

Shull, J. M. 1980, *ApJ*, 238, 860

Walsh, A. J., Burton, M. G., Hyland, A. R., & Robinson, G. 1998, *MNRAS*, 301, 640

Astrophysical Masers and their Environments
Proceedings IAU Symposium No. 242, 2007
J.M. Chapman & W.A. Baan, eds.

doi:10.1017/S1743921307012707

A variability study of the H_2CO 6 cm maser in IRAS 18566+0408

E. Araya[1,2], P. Hofner[1,2], S. Kurtz[3], H. Linz[4], M. Sewilo[5], L. Olmi[6,7] and C. Watson[8]

[1]New Mexico Tech, Socorro, NM, USA. [2]NRAO, Socorro, NM, USA. [3]Centro de Radioastronomía y Astrofísica, UNAM, México. [4]MPIA, Heidelberg, Germany. [5] Space Telescope Science Institute, Baltimore, USA. [6]Istituto di Radioastronomia, INAF, Florence, Italy. [7]University of Puerto Rico at Rio Piedras, Puerto Rico. [8]Manchester College, USA.
email: earaya@nrao.edu

Abstract. We report the first detection of a short-duration (< 3 months) outburst of an H_2CO 6 cm maser based on multi-epoch observations of IRAS 18566+0408 obtained with Arecibo, the Green Bank Telescope, and the Very Large Array. The H_2CO maser was observed nine times between 2002 and 2005. In May 2006 we began a two-year program of monthly monitoring with the Arecibo Telescope. The H_2CO maser in IRAS18566+0408 is coincident with a young massive stellar object and the line profile of the maser suggests that the H_2CO emission originates from two different regions, perhaps associated with the kinematics of a circumstellar disk.

Keywords. masers, stars: formation, ISM: molecules, radio lines: ISM

1. Introduction

IRAS 18566+0408 is a massive star forming region located at a distance of 6.7 kpc (Araya *et al.* 2004). IRAS 18566+0408 was classified by Zhang (2005) as a massive circumstellar disk candidate. Araya *et al.* (2007a) report supporting evidence for the presence of a massive disk (torus) based on VLA continuum observations. H_2CO 6 cm emission in the region was first detected by Araya *et al.* (2004) with Arecibo, and shown to be a maser via VLA observations (Araya *et al.* 2005). Here we present results from a variability study of the H_2CO 6 cm maser in IRAS 18566+0408 based on Arecibo, GBT, and VLA data.

2. First detection of an H_2CO maser flare

Using the GBT, we detected in November 2002 an outburst of the H_2CO 6 cm maser in IRAS 18566+0408. The maser decayed to its pre-flare flux density value within a month based on Arecibo observations conducted one month after the GBT observations. This is the first detection of a short-duration (less than three months) outburst of an H_2CO 6 cm maser. The H_2CO maser has a double peaked Gaussian profile. Both components showed the maser flare and the relative intensity between the two components did not substantially change during the flare. An extensive discussion of this detection is presented by Araya *et al.* (2007b).

3. Arecibo monitoring of the H_2CO and CH_3OH 6.7 GHz masers

We are monitoring the H_2CO 6 cm maser in IRAS 18566+0408 with Arecibo. The monitoring program began in May 2006 and is expected to continue up to May 2008. We report the detection of a second H_2CO maser flare in IRAS 18566+0408 (Figure 1, upper panel). The superb sensitivity of Arecibo allows us to also monitor the CH_3OH 6.7 GHz masers with just one minute on-source integration per observing run (Figure 1,

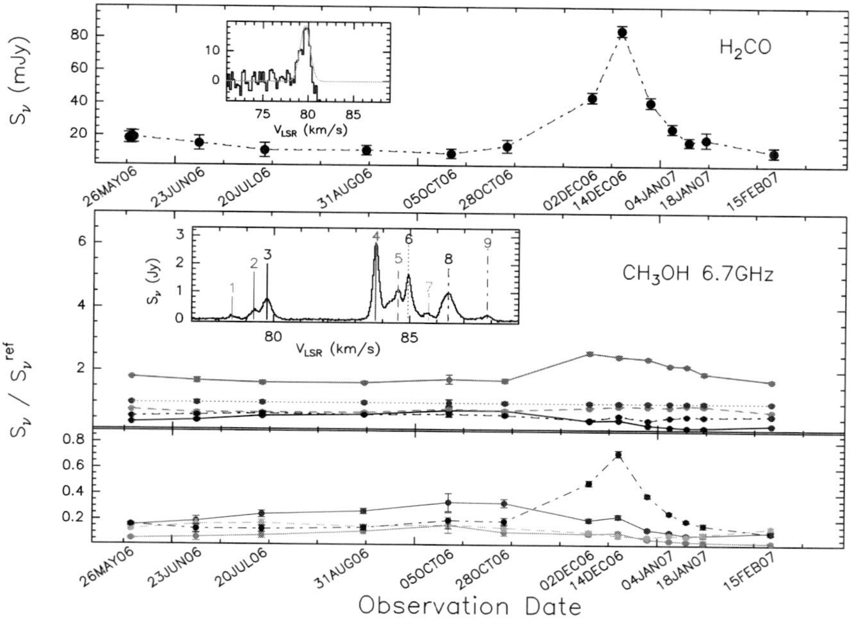

Figure 1. *Upper Panel:* Second detection of an H_2CO 6 cm maser flare. The spectrum of the maser together with a Gaussian fit is shown in the inset. *Lower Panel:* Light-curve of the CH_3OH 6.7 GHz masers in IRAS 18566+0408. The CH_3OH spectrum from one of the observing epochs is shown in the inset. We show the flux density of the different maser components normalized to the flux density of maser 6. For clarity, two normalized flux densities scales are displayed, showing the bright and dim masers separately. (See on-line edition for a color version.)

lower panel). The CH_3OH maser 9 showed the same outburst as the H_2CO maser, however, the LSR velocity of the two masers is substantially different, i.e., the masers arise from different volumes. If the CH_3OH maser 9 and the H_2CO maser are unsaturated and collinear, then the flares could have been caused by amplification of variable background 6 cm radio continuum from the disk/jet system in IRAS 18566+0408. Future high angular resolution observations of the CH_3OH masers are needed to verify the position coincidence between the H_2CO maser and the CH_3OH maser 9.

Acknowledgements

E. A. is supported by a NRAO predoctoral fellowship. The Arecibo Observatory is part of NAIC, operated by Cornell University under contract with the National Science Foundation (NSF). The 100m Green Bank Telescope is operated by NRAO, a facility of the NSF operated under cooperative agreement by Associated Universities Inc. The authors thank H. Hernández for the flexibility in scheduling of the observations. E. A. thanks the IAU for a travel grant to attend this symposium.

References

Araya, E., Hofner, P., Kurtz, S., Linz, H., Olmi, L., Sewiło, M., *et al.* 2005, *ApJ* 618, 339
Araya, E., Hofner, P., Linz, H., Sewiło, M., Watson, C., *et al.* 2004, *ApJS* 154, 579
Araya, E., Hofner, P., Sewiło, M., Goss, M.W., Linz, H., S. Kurtz, *et al.* 2007a, *ApJ* submitted
Araya, E., Hofner, P., Sewiło, M., Linz, H., Kurtz, S. , *et al.* 2007b, *ApJ* (Letters) 654, L95
Zhang, Q. 2005, in: R. Cesaroni, M. Felli, E. Churchwell, & M. Walmsley (eds.), *Massive Star Birth: A Crossroads of Astrophysics* (Cambridge: Cambridge Univ. Press), p. 135

Astrophysical Masers and their Environments
Proceedings IAU Symposium No. 242, 2007
J.M. Chapman & W.A. Baan, eds.

doi:10.1017/S1743921307012719

The variability of 6.7 and 12.2 GHz methanol masers

Leszek P. Błaszkiewicz[1], **A. Mikuć**[1] **and A. J. Kus**[2]

[1]Dept. of Math. and IT Sciences, University of Warmia and Mazury, Olsztyn, Poland
[2]Centre for Astronomy, Nicolaus Copernicus University, Torun, Poland

Abstract. Eight strong methanol masers were monitored using the 32-m Torun radio telescope at 6.7 and 12.2 GHz. The observations were taken in the period 2003 December to 2005 July at 1–2 week intervals. We have noted strong variability of single features but did not notice spectacular spectral shape changes. From our results we designated two groups of variability for corresponding features: Those where the 12.2 GHz flux density variations follow those at 6.7 GHz, and those where the 12.2 GHz flux density variations are opposite to 6.7 GHz.

Keywords. masers, radio lines: ISM, ISM: molecules, techniques: spectroscopic

---

1. Introduction

Methanol maser sources, associated with star foming regions (eg. Minier *et al.* 2003) often show emission structure which can be explained as rotating discs (Norris *et al.* 1998) or outflows (Minier *et al.* 2002). Changes in the dynamical environments can lead to changes in the maser regions which manifest as flux density changes. Methanol masers have previously been monitored by eg. Moscadelli *et al.* 1996, and Gaylard *et al.* 2004. Our monitoring programme follows the 12.2 GHz survey of Blaszkiewicz & Kus (2004a, 2004b). We have put particular stress on simultaneous observations of methanol masers at 6.7 and 12.2 GHz.

2. Observations

The selected sources were observed from December 2003 until August 2005 at intervals of 1–2 weeks for a period of 20 months with a gap between May and November 2004 due to work on the receivers. We monitored the sources: g25.71+0.04, g25.83-0.18, 18456-0129, 18470-0050, 18517+0437, 18556+0136, 18592+0108, CepA with W3(OH) as the calibrating source. Altogether there were more than 60 observing sessions.

The front-end was a modified 12.2 GHz TV Sat receiver (SMV XL 800A) maintained by X Line Comp. with LHC polarisation only. The T_{sys} was about 130K. For the 6.7 GHz observations we used a dual-channel (LHC and RHC) cooled receiving system with a T_{sys} of about 70K.

As the back-end we used a 2-level, 4 × 4096 channel digital autocorrelation spectrograph. The resolution for a bandwidth of 4 MHz was 22 m/s (12.2 GHz) and 44 m/s (6.7 GHz). Velocities were determined to about 0.2 km/s

3. Results

We have monitored many single features from the spectra obtained. In this paper we show three different sources (Fig. 1) where time series have been obtained for single features present at the same velocity for both frequencies. In general we find two types

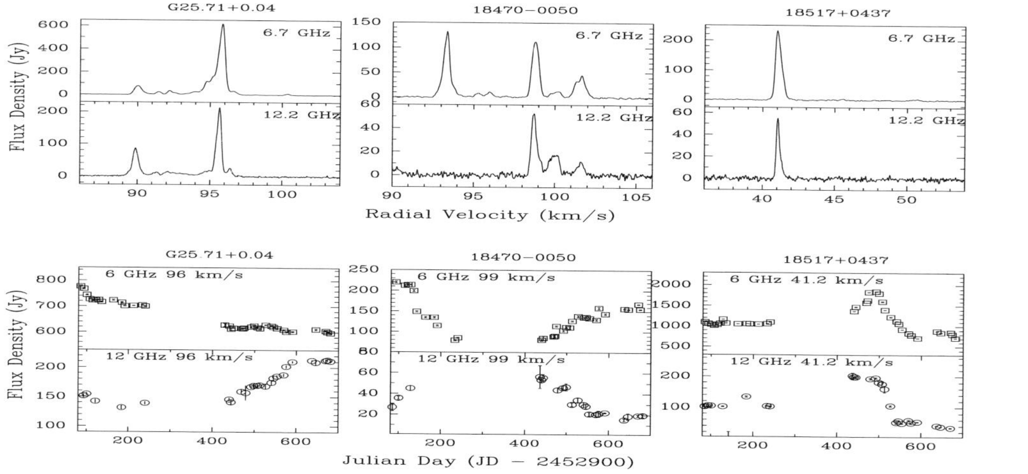

Figure 1. Three examples of variability of single features. The upper panels show spectral profiles at 6.7 GHz (top) and 12.2 GHz (bottom). The lower panels show time series of the flux density variations for selected features at the two frequencies.

of variability corresponding to whether the flux density changes are similar or opposite at the two frequencies.

4. Conclusions

We have observed eight strong methanol sources plus W3(OH) source at two methanol maser transitions: $2_0 \rightarrow 3_{-1}$E (12.178595 GHz) and $5_1 \rightarrow 6_0$ A^+ (6.668518 GHz). All sources were selected to be variable from a literature study.

The flux measurements were verified from observations of W3OH which is assumed to have very little variability at 6.7 and 12.2 GHz. We estimate that our flux densities are accurate to better than 10% at 6.7 GHz and 15% at 12.2 GHz.

The spectra of monitored sources (expect G25.83 0.18) were generally of reasonably high signal-to-noise with a one-sigma level below 4 Jy at 12.2 GHz and below 2 Jy at 6.7 GHz after 15 minutes of observations.

All of the observed sources revealed strong flux density variations. However, we did not observe a significant variability in the overall profile shapes for the monitored sources. In particular we did not detect new features or observe the disappearance of existing ones.

We have found two types of variability for corresponding pairs of spectral features. For one group the flux density variations increase and decrease in step, in the other group they are anti-correlated.

Acknowledgements

The author thanks J. Chapman for editorial assistance with this paper, the IAU for a grant and the UWM for financial support.

References

Blaszkiewicz L. & Kus A. J. 2004, *A&A* 413, 233
Blaszkiewicz L. & Kus A. J. 2004, BaltA 13, 474
Goedhart S., Gaylard M. J. & van der Walt D. J. 2004 *MNRAS* 355, 553
Minier V., Booth R. S. & Conway J. E. 2003, *A&A* 403, 1093
Moscadeli L. & Catarzi M. 1996 *A&AS* 116, 211
Norris R. P. *et al.* 1998 *ApJ* 508, 275

Astrophysical Masers and their Environments
Proceedings IAU Symposium No. 242, 2007
J.M. Chapman & W.A. Baan, eds.

doi:10.1017/S1743921307012720

Water masers within the G 333.2–0.6 giant molecular cloud

S. L. Breen,[1] S. P. Ellingsen,[1] M. Johnston-Hollitt,[1] S. Wotherspoon,[1] I. Bains,[2,3] M. G. Burton,[2] M. Cunningham[2] N. Lo,[2] C. E. Senkbeil[1] and T. Wong[2,4]

[1]School of Mathematics and Physics, University of Tasmania, Private Bag 37, Hobart, Tasmania 7001, Australia

[2]School of Physics, University of New South Wales, Sydney, NSW 2052, Australia

[3]Centre for Astrophysics and Supercomputing, Swinburne University of Technology, PO Box 218, Hawthorn, VIC 3122, Australia

[4]Australia telescope National Facility, CSIRO, PO Box 76, Epping, NSW 1710, Australia

Abstract. We report the results of a blind search for 22 GHz water masers in two regions, covering approximately half a square degree, within the G 333.2–0.6 giant molecular cloud. The complete search of the two regions was carried out with the 26 m Mount Pleasant radio telescope and resulted in the detection of nine water masers, five of which are new detections. Australia Telescope Compact Array (ATCA) observations of these detections have allowed us to obtain positions with arcsecond accuracy, allowing meaningful comparison with infrared and molecular data for the region. We find that for the regions surveyed there are more water masers than either 6.7 GHz methanol, or main-line OH masers. The water masers are concentrated towards the central axis of the star formation region, in contrast to the 6.7 GHz methanol masers which tend to be located near the periphery. The colours of the GLIMPSE point sources associated with the water masers are slightly less red than those associated with methanol masers. Statistical investigation of the properties of the ^{13}CO and 1.2 mm dust clumps with and without associated water masers shows that the water masers are associated with the more massive, denser and brighter ^{13}CO and 1.2 mm dust clumps. We present statistical models that can predict those ^{13}CO and 1.2 mm dust clumps likely to have associated water masers.

Keywords. Masers, ISM: molecules, radio lines: general, stars: formation.

---

1. Introduction

The G 333.2–0.6 giant molecular cloud (GMC) is located at a distance of 3.6 kpc (Lockman 1979) and extends approximately 1.2 x 0.6 degrees on the sky. The GMC has been the focus of numerous observations, including 1.2 mm dust continuum observations by Mookerjea *et al.* (2004) who identified 95 dust clumps within the region. Observations of ^{13}CO by Bains *et al.* (2006) identified 61 ^{13}CO clumps within the GMC and showed that the emission takes the form of a string of knots with the clumps arranged along an axis aligned NW to SE. Complete surveys of the region have been carried out by Ellingsen *et al.* (1996) for 6.7 GHz methanol masers and Caswell *et al.* (1980) for 1665 and 1667 MHz OH masers. These surveys resulted in the detection of four methanol and three OH masers within the regions surveyed for water masers. Here we present the results of an untargeted survey within two distinct regions of the G 333.2–0.6 GMC which has resulted in the detection of nine water masers, five of which are new detections. The two regions surveyed cover over half a square degree and encompass much of the high density gas and dust regions of the GMC. For greater detail of the observations and analysis see Breen *et al.* (2007).

2. Discussion and conclusions

2.1. *Association with other maser species and infrared sources*

We find that within the regions surveyed there are more water than either methanol or OH masers and that there are relatively few associations between the different maser species. Comparison between the relative locations of the water masers and the 6.7 GHz methanol masers shows that the water masers lie along the main axis of star formation within the GMC while the methanol masers tend to be located near the periphery. Four of the five newly discovered water masers have an associated GLIMPSE point source and the four previously detected water masers are all projected against regions of bright mid-infrared emission. Comparison of the [3.6]–[4.5] colours of the water maser associated GLIMPSE point sources with those associated with methanol masers (Ellingsen 2006) shows that the sources associated with the water masers are clustered toward the less red end of the range observed in the methanol associated sources. This coupled with the relative positions of the water and methanol masers within the GMC lends support to the idea that methanol masers trace an earlier evolutionary phase than water masers.

2.2. *^{13}CO and 1.2 mm dust clump analysis*

We have investigated the properties of the ^{13}CO (Bains *et al.* 2006) and the dust clumps (Mookerjea *et al.* 2004) that are associated with water maser emission by fitting a Binomial generalised linear model to the maser presence/absence data using the ^{13}CO and dust clump properties as predictors. This analysis has shown that in general water masers are associated with bigger, brighter, more massive and denser ^{13}CO and dust clumps. Stepwise model selection based on the Akaike Information Criteria was used to select the most parsimonious models with the greatest predictive properties for both the ^{13}CO and the dust clumps. The estimated regression relations for the ^{13}CO and dust clumps are

$$\log\frac{p_i}{1-p_i} = -21.2018 + 1.3037x_{integrated} + 8.0589x_{radius} - 1.2290x_{mass},$$

and

$$\log\frac{p_i}{1-p_i} = -11.477 + 9.174x_{radius},$$

respectively, where p_i is the probability of maser presence in the ith clump and $x_{integrated}$, x_{radius} and x_{mass} represent clump properties integrated flux density of the clump peak (10 K km s^{-1}), mass (10^3 M$_\odot$) and radius (pc). These models have low misclassification rates and allow the probability of maser presence within a given clump to be calculated, enabling efficient targeted searches to be carried out where suitable data is available.

References

Bains, I., Wong, T., Cunningham, M., Sparks, P., Brisbin, D., Calisse, P., Dempsey, J. T., Deragopian, G., Ellingsen, S., Fulton, B., Herpin, B., Jones, P., Kouba, Y., Kramer, C., Ladd, E. F., Longmore, S. N., McEvoy, J., Maller, M., Minier, V., Mookerjea, B., Phillips, C., Purcell, C. R., Walsh, A., Voronkov, M. A., Burton, M. G. 2006, *MNRAS* 367, 1609

Breen, S. L., Ellingsen, S. P, Johnston-Hollitt, M., Wotherspoon, S., Bains, I., Burton, M. G., Cunningham, M., Lo, N., Senkbeil, C. E., Wong, T. 2007 *MNRAS* in press

Caswell, J. L., Haynes, R. F., Goss, W.M 1980, *Aust. J. Phys.*, 33, 639

Ellingsen, S. P., von Bibra, M. L., McCulloch, P. M., Norris, R. P., Deshpande, A. A., Phillips, C. J. 1996, *MNRAS* 280, 378

Ellingsen, S. P. 2006, *ApJ* 638, 241

Lockman, F. J. 1979, *ApJ* 232, 761

Mookerjea, B., Kramer, C., Nielbock, M., Nyman, L. 2004, *A & A* 426, 119

Astrophysical Masers and their Environments
Proceedings IAU Symposium No. 242, 2007
J.M. Chapman & W.A. Baan, eds.

doi:10.1017/S1743921307012732

A sensitive survey for water masers towards Bok globules

I. de Gregorio-Monsalvo[1], J. F. Gómez[2], O. Suárez[3], T. B. H. Kuiper[4], G. Anglada[2], N. A. Patel[5] and J. M. Torrelles[6]

[1]European Southern Observatory, Alonso de Córdova 3107, Vitacura, Santiago, Chile , email: idegrego@eso.org.

[2]Instituto de Astrofísica de Andalucía (CSIC), Apartado 3004, E-18080 Granada, Spain.

[3]Laboratorio de Astrofísica Espacial y Física Fundamental (INTA), Apartado 50727, E-28080 Madrid, Spain.

[4]Jet Propulsion Laboratory, California Institute of Technology, 4800 Oak Grove Drive, Pasadena, CA 91109, USA.

[5]Harvard-Smithsonian Center for Astrophysics, 60 Garden Street, Cambridge, MA 02138, USA.

[6]Instituto de Ciencias del Espacio (CSIC) and Institut d'Estudis Espacials de Catalunya, Facultat de Física, Planta 7, Universitat de Barcelona, Av. Diagonal 647, E-08028 Barcelona, Spain

Abstract. In this work we report the most sensitive water maser survey towards Bok globules to date, using NASA's 70 m antenna in Robledo de Chavela (Spain). We observed 207 positions within the Clemens & Barvainis catalog that show indications of possible star formation or with a high probability of harboring a young stellar object. With this survey we have increased the number of Bok globules known to present water maser emission from three to nine. We have complemented these results with interferometric high-angular resolution observations towards some of our detections.

Keywords. ISM: globules, ISM:molecules, masers, stars:formation, surveys.

1. Motivation

Bok globules, which are the smallest and more isolated molecular clouds in our galaxy, are important laboratories for the study of star formation processes and protostar evolution with a lower chance of contamination from multiple generations of young stellar objects (YSOs) within the same region. They may span a wide range of evolutionary stages, and low- and intermediate-mass star formation can take place inside these clouds (Bok & Reilly 1947, Yun & Clemens 1992).

Water masers are powerful observational tools for characterizing the age of low-mass protostars (most of them are associated with Class 0 protostars) and the physical conditions of the gas surrounding YSOs (Moscadelli *et al.* 2006). In addition, these masers tend to be located very close (a few hundred AUs) to their powering sources in low-mass YSOs (Claussen *et al.* 1998) and this makes them well-suited to locating the central exciting sources of mass-loss activity in low-mass star forming regions.

Since Bok globules are associated with YSOs in different evolutionary stages and there are several evolutionary aspects related to water maser emission, we have searched for water masers at 22 GHz towards these clouds using both single-dish and interferometric techniques.

2. Single-dish survey for water masers towards Bok globules

We have used the NASA's 70 m antenna in Robledo de Chavela, near Madrid, Spain, to search for water maser emission at 22 GHz towards 207 positions that include radio continuum sources, centers of molecular outflows, peaks of high-density molecular tracers, and IRAS sources, i.e., targets that may indicate the presence of star formation activity.

As a result, we obtained seven detections (six of them are new) towards CB 3, CB 34, CB 54, CB 65, CB 101, CB 199, and CB 232, increasing from three known sources (CB 3, CB 39, and CB 205) to nine the number of known Bok globules in the Clemens & Barvainis (1988) catalog associated with this kind of emission (Gómez *et al.* 2006).

3. High-angular resolution observations

As a complement to our single-dish survey we have performed interferometric observations using the Very Large Array (VLA) towards some of our water maser detections and processed archive data in the case of CB 3. We have used these interferometric observations to study the physical conditions and kinematics of the innermost areas around the central YSOs, as well as to determine accurate positions for the masers and additional information about their powering sources.

We have detected water maser emission with the VLA in CB 3, CB 54, CB 101, and CB 232. All of these masers are associated with regions that show bipolar molecular outflows and young multiple stellar systems in different stages of evolution, with the exception of CB 101 which we propose to be associated with an evolved object, possibly a Mira star (see de Gregorio-Monsalvo *et al.* 2006).

Thanks to the accurate positions provided by the interferometric observations, we obtained information about the powering sources in CB 3 (where masers are located close to the Class 0 object CB 3-mm), CB 54 (where maser emission is associated to a near-IR elongated feature named CB 54 YC1-SW), and CB 232 (with the water maser emission located at the position of the near-IR source CB 232 YC1-I). We have confirmed that CB 54 YC1-SW is the powering source of the maser emission and possibly of the mass loss phenomenon in CB 54, with recent deep VLA-BnA observations that show a faint radio continuum source at that position (de Gregorio-Monsalvo *et al.* in preparation).

Acknowledgements

GA, JFG, and JMT acknowledge partial financial support from grant AYA2005-08523-C03 of the Spanish MEC (co-funded with FEDER funds). GA, IdG, JFG, and JMT acknowledge support from Junta de Andalucía grant FQM-1747. GA and JFG are also partially supported by Junta de Andalucía grant TIC-126.

References

Bok, B. J., & Reilly, E. F. 1947, ApJ, 105, 255

Claussen, M. J., Marvel, K. B., Wootten, A., & Wilking, B. A. 1998, ApJ, 507, L79

Clemens, D. P. & Barvainis, R. 1988, ApJS, 68, 257

de Gregorio-Monsalvo, I., Gómez, J. F., Suárez, O., Kuiper, T. B. H., Anglada, G., Patel, N. A., & Torrelles, J. M. 2006, AJ, 132, 2584

Gómez, J. F., de Gregorio-Monsalvo, I., Suárez, O, & Kuiper,T. B. H. 2006, AJ, 132, 1322

Moscadelli, L., Testi, L., Furuya, R. S., Goddi, C., Claussen, M., Kitamura, Y., & Wootten, A. 2006, A&A, 446, 985

Yun, J. L., & Clemens, D. P. 1992, ApJ, 385, L21

Astrophysical Masers and their Environments
Proceedings IAU Symposium No. 242, 2007
J.M. Chapman & W.A. Baan, eds.

doi:10.1017/S1743921307012744

Japanese VLBI network observations of 6.7-GHz methanol masers I. Array

Akihiro Doi[1]†, Kenta Fujisawa[1] Mareki Honma[2] Koichiro Sugiyama[1] Yasuhiro Murata[3] Nanako Mochizuki[3] and Yasuko Isono[1]

[1]Faculty of Science, Yamaguchi University,
1677-1 Yoshida, Yamaguchi, Yamaguchi 753-8512, Japan

[2]National Astronomical Observatory of Japan, 2-21-1 Osawa, Mitaka, Tokyo 181-8588, Japan

[3]The Institute of Space and Astronautical Science, Japan Aerospace Exploration Agency,
3-1-1 Yoshinodai, Sagamihara, Kanagawa 229-8510, Japan.

Abstract. The Japanese VLBI network (JVN) has begun observations of 6.7-GHz methanol masers associated with massive star-forming regions. The JVN is a newly-established VLBI array with baselines ranging from 50 to 2560 km spread across the Japanese islands. Three observing bands of 6.7, 8.4, and 22 GHz are now available. The array consists of ten antennas: VERA Mizusawa 20 m, VERA Ishigaki 20 m, VERA Iriki 20 m, Usuda 64 m, Yamaguchi 32 m, Tomakomai 11 m, Tsukuba 32 m, Kashima 34 m, VERA Ogasawara 20 m, and Gifu 11 m, the first five of which have 6.7-GHz receiving systems. In summer 2005, we obtained the first fringes at 6.7 GHz, and VLBI images of 12 methanol maser sites including seven that had not previously been imaged with VLBI at this band. In 2006 summer, we obtained phase-reference observations toward several methanol maser sites.

Keywords. masers, instrumentation: interferometers, radio lines: ISM

---

1. The Japanese VLBI network

The Japanese VLBI network (JVN) obtained the first fringes at 8.4 GHz in November 2004, at 22 GHz in May 2005, and at 6.7 GHz in September 2005. The array consists of ten antennas: VERA Mizusawa 20 m, VERA Ishigaki 20 m, VERA Iriki 20 m, VERA Ogasawara 20 m (Kobayashi et al. 2003), Usuda 64 m, Yamaguchi 32 m, Tomakomai 11 m, Tsukuba 32 m, Kashima 34 m, and Gifu 11 m, with baselines ranging from 50 to 2560 km. The VSOP-terminal system is currently used as a digital back-end; digitized data in 2-bit quantization are recorded onto magnetic tapes at a data rate of 128 Mbps. Correlation processing is performed with the Mitaka FX correlator (Shibata et al. 1998) at the National Astronomical Observatory of Japan.

2. The 6.7 GHz array

The receiving systems at 6.7 GHz are currently installed in five anntenas: Mizusawa, Iriki, Ishigaki, Usuda, and Yamaguchi, with baselines of 293–2270 km, resulting in synthesized beam sizes of ~ 5 milli-arcsecond. The individual baseline sensitivities are 2.4–28 Jy beam^{-1} for a 180-sec integration and 2.0-kHz bandwidth, and the resulting image sensitivity is $\sim$0.4 Jy beam^{-1} for a 1-hour integration.

Currently, the 6.7 GHz JVN array is in an experimental phase and: (1) the receivers on Yamaguchi 32 m and VERA 20 m antennas are not cooled, (2) there is only a single

† Present address: The Institute of Space and Astronautical Science, Japan Aerospace Exploration Agency, 3-1-1 Yoshinodai, Sagamihara, Kanagawa 229-8510, Japan.

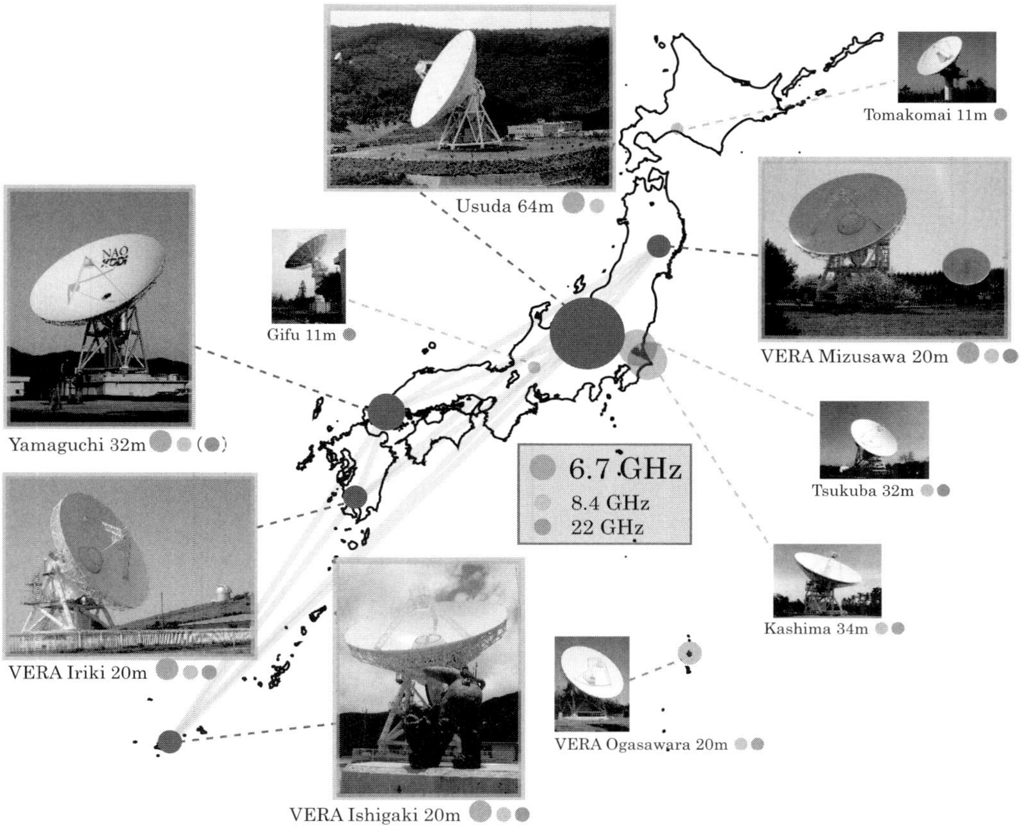

Figure 1. Japanese VLBI network and its 6.7 GHz array, which currently consists of Usuda 64 m, Yamaguchi 32 m, VERA Mizusawa 20 m, VERA Ishigaki 20 m, and VERA Iriki 20 m telescopes.

polarization receiving system on each antenna, and (3) the polarization is mismatched between antennas, with linear polarization for the VERA Mizusawa 20m and VERA Ishigaki 20m antennas and circular polarization for the other antennas. The corresponding loss of correlated amplitude is corrected in a data reduction process. Because of these factors, *at present*, observations at 6.7 GHz are not open to general users.

Acknowledgements

The JVN project is led by the National Astronomical Observatory of Japan (NAOJ) that is a branch of the National Institutes of Natural Sciences (NINS), Hokkaido University, Gifu University, Yamaguchi University, and Kagoshima University, in cooperation with Geographical Survey Institute (GSI), the Japan Aerospace Exploration Agency (JAXA), and the National Institute of Information and Communications Technology (NICT).

References

Kobayashi, H., et al. 2003, *ASP Conf. Ser.* 306, p. 367
Shibata, K. M., Kameno, S., Inoue, M., & Kobayashi, H. 1998, *ASP Conf. Ser.* 144, p. 413

Astrophysical Masers and their Environments
Proceedings IAU Symposium No. 242, 2007
J.M. Chapman & W.A. Baan, eds.

doi:10.1017/S1743921307012756

Tick tock – the 12.2 GHz methanol masers in G9.62+0.20

Michael J. Gaylard[1] **and Sharmila Goedhart**[1]

[1]Hartebeesthoek Radio Astronomy Observatory, PO Box 443, Krugersdorp 1740, South Africa
email: mike@hartrao.ac.za

Abstract. The bright interstellar methanol masers at 12.2 GHz and 6.7 GHz were discovered in 1987 and 1991 respectively. It was soon established that many were quite variable. Goedhart, Gaylard & van der Walt (2003) reported that one source, G9.62+0.20E, exhibited flares at 12.2 and 6.7 GHz that appeared to be periodic, repeating every 246 days. Since then, monitoring of this and other possibly periodic sources has continued with the 26-m Hartebeesthoek telescope. We discuss here the full 12.2 GHz time series data of G9.62+0.20 through 2006. The data quality has been much improved by telescope upgrades. Flares in the main maser peak continue, the repetition rate remains close to that originally determined.

Keywords. radio lines: ISM, stars: formation

1. Introduction

The discovery of apparently regular brightness variations in the 6.7 GHz methanol masers in G9.62+0.20 by Goedhart, Gaylard & van der Walt (2003), and in several other sources by Goedhart, Gaylard & van der Walt (2004), came as a surprise.

Monitoring of G9.62+0.20 at 12.2 GHz showed that the relative amplitude of the variations was much greater than at 6.7 GHz. The regularity of the flaring contrasted with the repetitive but aperiodic flaring of G351.78-0.54 at 6.7 GHz reported by Macleod & Gaylard (1996) and Goedhart, Gaylard & van der Walt (2004). A period of 246 days gave the best fit to the data available to Goedhart *et al.* (2003) and permitted the mapping of the source during a flare by Goedhart, Minier, Gaylard *et al.* (2005) at 12.2 GHz using the VLBA. This showed that existing masers spots brightened during a flare, but no changes in morphology occurred. Monitoring has continued at both 6.7 and 12.2 GHz, and we report on the latter here.

2. Observations

During this period of monitoring with the 26-m Hartebeesthoek telescope, the main reflecting surface was replaced with more accurate solid panels. The aperture efficiency at 12 GHz fell from 0.20 to 0.10 before rising to 0.45 when the new panels were aligned. In addition, the 12 GHz receiver was converted from single to dual polarization, the 1x256-channel spectrometer was replaced with a new 2x1024 channel spectrometer and the telescope control system was converted from minicomputers to PC's running Linux.

Calibration during these major changes was based on contemporaneous monitoring of Virgo A (which is partly resolved), 3C123, 3C218 and the maser source G351.42+0.64, which was previously found to show little variation. The minimal variation seen in weaker peaks of G9.62+0.20 provided an internal consistency check.

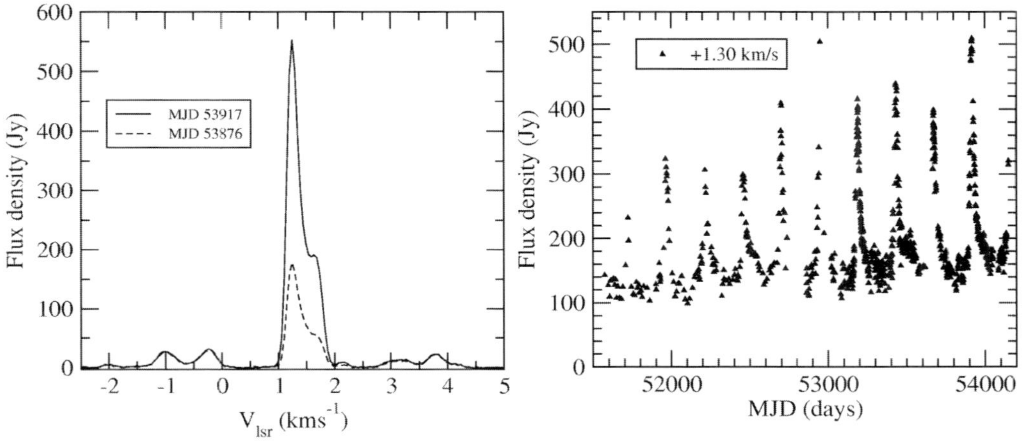

Figure 1. Spectra of G9.62+0.20 at the minimum and maximum of a recent flare, and the time series of the brightest maser peak.

3. Results

Typical spectra obtained with the upgraded 26-m telescope are shown in figure 1. It can be seen that only the maser peaks at velocities between 0.5 and 2.5 kms^{-1} flare strongly at 12.2 GHz. The program Period04 of Lenz & Breger (2005) was used to obtain the periodogram of the strongest peak using the time series shown in figure 1. The full data set gave a period of 243.0 days, the first half 242.2 days and the second half 242.5 days. Epoch-folding was used as an alternative period tester, the L-statistic of Davies (1990) being employed. The results are consistent with a period of 243 to 244 days. Plots of the folded time series showed that a shorter period, such as 242 days, provides the best alignment of the rising edges of the flares, but a longer period, such as 244 days, best aligns the falling edges, owing to differences in flare duration.

4. Discussion

Despite almost every piece of hardware and software involved in the monitoring having been replaced, the large amplitude flares remain. Within present experimental uncertainty they appear to have an unchanging period. The extreme regularity of the phenomenon over seven years argues for a strongly deterministic cause. The length of the interval between flares suggest orbital motion of a secondary object around the primary forming star could be the origin of the regular variations. However this possibility needs to be evaluated in the context of the different types of light curves seen in other methanol masers showing similar regularity.

References

Davies, S. R. 1990, *MNRAS* 244, 93
Goedhart S., Gaylard M. J. & van der Walt D. J. 2003, *MNRAS* 339, L33
Goedhart S., Gaylard M. J. & van der Walt D. J. 2004, *MNRAS* 355, 553
Goedhart S., Minier V., Gaylard M. J. & van der Walt D. J. 2005, *MNRAS* 356, 839
Lenz P., & Breger M. 2005, *Commun. Asteroseismology* 146, 53
MacLeod G. C. & Gaylard M. J. 1996, *MNRAS* 280, 868

Astrophysical Masers and their Environments
Proceedings IAU Symposium No. 242, 2007
J.M. Chapman & W.A. Baan, eds.

doi:10.1017/S1743921307012768

VLBI observations of H_2O and CH_3OH masers in two high-mass YSOs

C. Goddi[1], **L. Moscadelli**[2], **A. Sanna**[3], **R. Cesaroni**[2] **and V. Minier**[4]

[1]Harvard-Smithsonian Center for Astrophysics, 60 Garden St. Cambridge, MA 02138, USA

[2]INAF, Osservatorio Astrofisico di Arcetri, Largo E. Fermi 5, 50125 Firenze, Italy

[3]Dipartimento di Fisica, Universitá degli Studi di Cagliari, S.P. Monserrato-Sestu Km 0.7, 09042 Cagliari, Italy

[4]Service d'Astrophysique, DAPNIA/DSM/CEA Saclay, 91191 Gif-sur-Yvette, France

Abstract. We have conducted phase-reference VLBI observations of H_2O and CH_3OH masers toward two high-mass star forming regions, Sh 2-255 IR and AFGL 5142. In Sh 2-255 infrared water masers are aligned along a direction close to the orientation of a large scale H_2 jet, tracing possibly shocked material in a precessing jet, or, alternatively, the disk-wind emerging from the disk atmosphere. In AFGL 5142 water masers trace expansion at the base of a protostellar jet, whilst methanol masers are more probably tracing infalling than outflowing gas. Our results suggest that water and methanol masers trace different kinematic structures in the circumstellar gas.

1. Introduction

Most previous high-angular ($\leqslant 0.''1$) resolution studies of molecular masers in high-mass star forming regions (SFRs) have concentrated mainly on either H_2O or CH_3OH masers. While Very Long Baseline Interferometry (VLBI) multi-epoch observations have clarified that water masers originate from shocks associated with protostellar jets (e.g. Goddi *et al.* 2005), different environments have been proposed in several sources to explain the origin of CH_3OH masers (Keplerian disks, jets, hot molecular cores, and HII regions). Comparing the spatio-kinematic distribution of both maser species in the same YSO at high-angular resolution better constrains the interpretation of the kinematic structures traced by the masers. So far only a few studies have been performed in both maser types with sufficient angular resolution ($<0.''1$) to investigate the nature of the CH_3OH maser birthplace in SFRs and the association between the H_2O and CH_3OH maser emission in the same YSO. Here, we present the results of a multi-epoch VLBI campaign conducted on two high-mass SFRs, Sh 2-255 IR and AFGL 5142, where both the 6.7 GHz methanol and 22.2 GHz water masers are observed.

2. Observational results

We performed Very Long Baseline Array (VLBA) multi-epoch observations of the 22.2 GHz water masers and European VLBI Network (EVN) single-epoch observations of the 6.7 GHz methanol masers toward two high-mass YSOs: Sh 2-255 IR and AFGL 5142.

In Sh 2-255 IR, water maser emission consists of three main clusters of features aligned along a direction close to the orientation of a H_2 jet (Fig. 1, left panel) observed on angular scales of 1–10 arcsec (Howard *et al.* 1997). This might suggest that the water maser features are associated with the inner part of the jet. However, all the measured proper motions form large angles with the outflow axis. The observed spatial and velocity distribution of water masers might be explained considering either jet precession or rotation

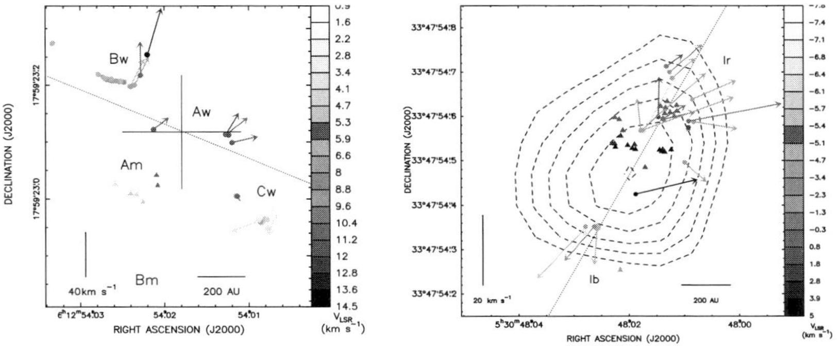

Figure 1. Positions and velocities of VLBA H_2O (*filled circles*) and EVN CH_3OH (*filled triangles*) masers, for Sh 2-255 IR (*left panel*) and AFGL 5142 (*right panel*), respectively. Different grayscale tones denote the maser line-of-sight velocities and the *arrows* indicate the measured *absolute* proper motions of the H_2O masers. The dotted lines indicate the axes of the H_2 jet (Howard *et al.* 1997) and the CO outflow (Zhang *et al.* 2007) observed in Sh 2-255 IR and AFGL 5142, respectively. The cross indicates the position uncertainty of the VLA 15 GHz continuum emission peak detected in Sh 2-255 IR (Rengarajan *et al.* 1996), whereas the *contour* map shows the 22 GHz VLA continuum emission in AFGL 5142 (Goddi & Moscadelli 2006).

around the jet axis. The data in hand do not allow us to discriminate unambiguously between the two scenarios. The 6.7 GHz methanol masers were observed with the EVN by Minier *et al.* (2000). The present data are too scarce even to attempt a qualitative interpretation of their birthplace. New multi-epoch EVN observations of the 6.7 GHz methanol masers are scheduled in order to determine their absolute positions and proper motions.

In AFGL 5142, the water masers show a bipolar spatial and line-of-sight (LOS) velocity distribution; their proper motions clearly indicate that H_2O masers trace expansion from the YSO along a NW-SE direction (Fig. 1, right panel). Recent SMA observations (Zhang *et al.* 2007) revealed a CO outflow oriented along a similar direction, suggesting that water masers might trace the innermost portion of the CO outflow. The methanol masers, which on the plane of the sky lie closer to the flow center than H_2O masers, show higher LOS velocities, in disagreement with a Hubble flow (as observed in the CO emission). Since the 6.7 GHz masers are seen in projection against the HII region, which is optically thick at 6.7 GHz (Goddi & Moscadelli 2006), the 6.7 GHz masers must be located in the foreground of the HII region. The fact that most of the spots present strongly red-shifted l.o.s. velocities suggests that they are tracing gas infalling toward the YSO.

3. Conclusions

Our VLBI data reveal a *true* association of both H_2O 22.2 GHz and CH_3OH 6.7 GHz masers with the same massive YSO in AFGL 5142. Notwithstanding that, the two species appear to trace different kinematic structures, in particular expansion at the base of a molecular outflow and infall in a molecular envelope, respectively.

References

Goddi, C., Moscadelli, L., Alef, W., Tarchi, A., & Pani, M. 2005, *A&A* 432, 161
Goddi, C., & Moscadelli, L. 2006, *A&A* 447, 577
Howard, E. M., Pipher, J. L., & Forrest, W. J. 1997, *ApJ* 481, 327
Minier, V., Booth, R. S., & Conway, J. E. 2000, *A&A* 362, 1093
Rengarajan, T. N., & Ho, P. T. P. 1996, *ApJ* 465, 363
Zhang, Q., *et al.* 2006, *astroph-12027*

Astrophysical Masers and their Environments
Proceedings IAU Symposium No. 242, 2007
W.A. Baan & J.M. Chapman, eds.

doi:10.1017/S174392130701277X

Discovery of polarized 6.7-GHz methanol masers in DR21/W75

L. Harvey-Smith[1], R. Soria-Ruiz[1], A. Duarte-Cabral[1,2] and R. J. Cohen[3]

[1]Joint Institute for VLBI in Europe, Postbus 2, 7990 AA, The Netherlands.

[2]University of Porto, Rua Campo Alegre 687, 4169-007 Porto, Portugal.

[3]Jodrell Bank Observatory, Macclesfield, Cheshire, SK11 9DL, UK. (Deceased)

Abstract. We present the first images of 6.7 GHz methanol masers in the DR21 star-forming complex. We have discovered two sites of Class II methanol maser emission; in DR21(OH) and DR21(OH)N. The emission comprises clusters of linearly extended masers which have velocity gradients along their length. There are four maser spots in DR21(OH)N, some of which show a small fraction of linear polarization. The twelve masers in DR21(OH) lie in a linear arrangement stretching approximately 0.7 arcseconds and show no significant linear polarization. We were not able to detect any circular polarization in the masers.

Keywords. masers, (ISM:) HII regions, ISM: molecules, radio lines: ISM, stars: formation

Dedication

This work was conceived by the late Jim Cohen. The authors acknowledge the leading role he had in the early stages of this project. His contribution to maser physics and his friendship will be greatly missed.

1. Introduction and observations

Despite playing a vital and dominant role in feedback processes in the interstellar medium, massive stars are still not understood in terms of their formative accretion processes. Methanol masers at 6.7 GHz are bright and widespread tracers of massive star-formation. High-resolution study of methanol masers, including polarization measurements, is the best way to characterise these environments.

The DR21 star-forming complex was observed using the Multi-Element Radio-Linked Interferometry Network (MERLIN) at 6.7 GHz in January 2005. The maximum baseline of MERLIN was 217 km, giving an angular resolution of approximately 50 mas at 6.7 GHz. Phase-referencing was used to ascertain positions to an absolute accuracy of 15 mas.

2. Results

We have found a cluster of four methanol masers in DR21(OH)N (also known as W75) with radial velocities ranging from +3.2 to +4.6 km s^{-1}. They lie very close both spatially and in velocity to the OH 4765-MHz masers reported in the region by Harvey-Smith & Cohen (2005). Assuming that we are seeing elliptical polarization of the σ components, the magnetic field runs *perpendicular* to the linear polarization vectors (shown in Figure 1, left). The resulting magnetic field direction is then broadly consistent with measurements

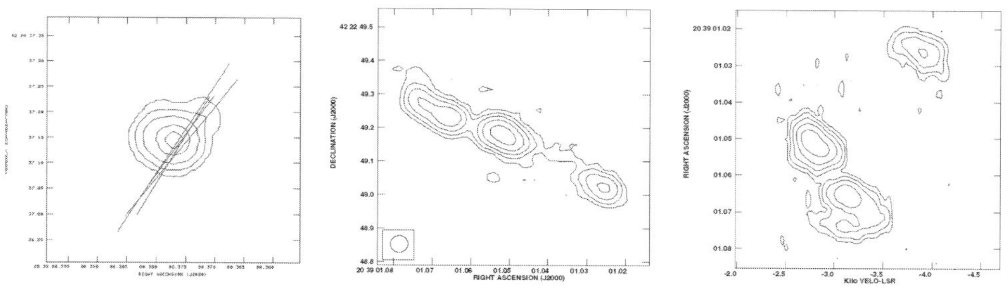

Figure 1. Left: Image of the 6.7 GHz methanol maser emission in DR21(OH)N, integrated between +3.0 and +5.0 km s^{-1}. We found four individual maser spots (corresponding to four distinct spectral peaks), but these were very closely spaced. The linear polarization vectors are shown. Middle: The 6.7 GHz methanol masers in DR21(OH), integrated between -2.5 and -4.0 km s^{-1}. A total of twelve individual spectral peaks make up this filamentary maser. Right: The R.A. – radial velocity plot of DR21(OH), showing that the methanol maser filament is composed of two distinct parts, separated in velocity. The longer section stretches some 780 AU and has a velocity gradient of 2.5 km s^{-1} arcsec^{-1}.

of the large-scale magnetic field direction perpendicular to the DR21 molecular ridge by Curran *et al.* (2006).

We have discovered a line of 6.7 GHz methanol masers in DR21(OH) (also known as W75S) close to the continuum source and the 1.7 GHz OH masers (Fish *et al.* 2005). Figure 1 (middle) shows the total intensity image, integrated over the full velocity range of the maser emission. Figure 1 (right) shows the velocity profile of the filament. Although the methanol masers in this region have no measurable polarization, the magnetic field direction has been extrapolated by measurements of CO and dust polarization (Lai *et al.* 2003, Curran *et al.* 2006), suggesting that the magnetic field runs along the major axis of the methanol maser filament, as is the case in W3(OH) (Vlemmings, Harvey-Smith & Cohen 2006).

3. Summary

We have detected a 1200 AU filament of unpolarized 6.7 GHz methanol masers in DR21(OH) using MERLIN. In DR21(OH)N we detected a small group of linearly polarized methanol masers, where the degree of linear polarization was 4.5%. Linear polarization measurements of other maser transitions in this region will enable us to make a physical interpretation of the maser polarization in terms of the direction of the magnetic field and its influence on the gas structures.

References

Curran R. L. *et al.* *Proceedings of Protostars and Planets V*, B. Reipurth, D. Jewitt, and K. Keil (Eds.), University of Arizona Press, Tucson, 2006, 8160
Fish V. L., Reid M. J., Argon A. L. & Zheng X-W. 2005, *ApJ*, 160, 220
Harvey-Smith L. & Cohen R. J. 2005, *MNRAS*, 356, 637
Harvey-Smith L. & Cohen R. J. 2006, *MNRAS*, 371, 1550
Lai S-P., Girart J. M. & Crutcher R. M. 2003, *ApJ*, 598, 392
Vlemmings W. H. T., Harvey-Smith L. & Cohen R. J. 2006, *MNRAS*, 371, L26

Astrophysical Masers and their Environments
Proceedings IAU Symposium No. 242, 2007
W.A. Baan & J.M. Chapman, eds.

doi:10.1017/S1743921307012781

Polarization of 6.0-GHz OH masers in W3(OH)

L. Harvey-Smith[1], R. Soria-Ruiz[1], A. M. S. Richards[2] and R. J. Cohen[3]

[1]Joint Institute for VLBI in Europe, Postbus 2, 7990 AA, The Netherlands.
[2]Jodrell Bank Observatory, Macclesfield, Cheshire, SK11 9DL, UK.
[3]Jodrell Bank Observatory, Macclesfield, Cheshire, SK11 9DL, UK (Deceased)

Abstract. We present preliminary results of a MERLIN polarization study of 6.0-GHz OH masers in the W3(OH) region of massive star-formation. We have detected a small but significant amount of linear polarization in the 6031-MHz masers, as well as a much larger fraction of circular polarization. We have found many Zeeman pairs, the analysis of which (along with the results at 6035-MHz) will be presented elsewhere. MERLIN is well-suited to detecting all the single-dish maser flux in these regions, owing to its range of short baselines (tens of kilometers) and resulting sensitivity to extended emission. We have found core-halo structure in the 6031-MHz OH maser emission in the region of the highly extended 4.7-GHz OH and 6.7-GHz methanol maser filament found by Harvey-Smith & Cohen (2005, 2006).

1. Introduction

The W3(OH) star-forming region consists of an ultra-compact HII region containing one or more massive OB stars, surrounded by molecular gas. Projected on the nearside western face of the ultra-compact HII region are masers of OH and methanol from every transition known to occur in the interstellar medium. High-resolution polarization studies of these masers will allow us to make independent measurements of the radial velocities and magnetic field strength and direction in this young massive star-forming region.

2. Observations

W3(OH) was observed at 6031- and 6035-MHz in January 2005 using the Multi-Element Radio-Linked Interferometry Network (MERLIN). The maximum baseline was 217 km, giving an angular resolution of approximately 50 mas. Phase-referencing was used on a nearby compact source in order to ascertain maser spot positions to an absolute accuracy of 15 mas.

3. Linearly polarized masers in W3(OH)

We have produced the first images of linearly polarized 6031-MHz masers in W3(OH). Linear polarization was detected in the broadline and southern regions. Sample channel maps are shown in Figure 1. There was also a high degree of circular polarization and several Zeeman pairs were found. Full Zeeman analysis will be presented elsewhere (Harvey-Smith *et al.* (2007) *in prep.*) and compared with the results of Moran *et al.* (1978), Desmurs *et al.* (1998) and Etoka, Cohen & Gray (2005).

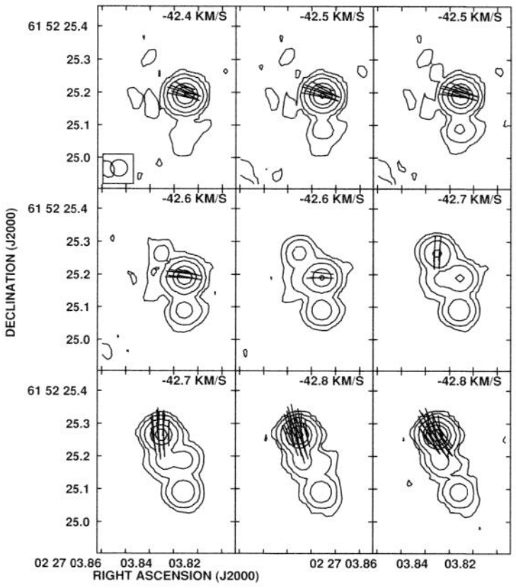

Figure 1. Sample channel maps in the broadline region of W3(OH) showing (contours) the OH flux density at 6031 MHz and (vectors) the linear polarization of the masers. The orientation of the linear polarization vectors varies significantly within this very compact region, as reported previously by Wright Gray & Diamond (2004) in ground-state OH, and Vlemmings, Harvey-Smith & Cohen (2006) in methanol masers.

4. Core-halo or extended maser structure

We have found that the excited OH 6031-MHz masers show extended structure in the form of halos and ordered structures extending far beyond the size of the restoring beam. High-resolution VLBI images of excited OH masers often miss a significant percentage of the flux observed with a single dish (e.g. Desmurs *et al.* 1998). In W3(OH), this missing flux has previously been found using MERLIN in OH at 4765-MHz (Harvey-Smith & Cohen 2005) and methanol at 6668-MHz (Harvey-Smith & Cohen 2006). When more examples of this phenomemon are known, it will be useful to compare co-spatial extended maser emission at different frequencies in order to assist parameter constraints on maser pumping models.

5. Summary

We have carried out a full linear and circular polarization study of excited OH 6.0-GHz masers in W3(OH), the full results of which will be published elsewhere. By comparing the linear polarization vectors with those of 6.7-GHz methanol masers, we remove the ambiguity in the magnetic field direction relative to the polarization vectors without encountering problems relating to (strongly frequency-dependent) Faraday rotation. The OH masers show extended structure beyond the compact VLBI-resolved 'spots'. Future work will involve comparing the relative positions of extended OH and methanol maser 'halo' emission at different frequencies between 1.7 and 6.7-GHz.

References

Desmurs J. F., Baudry A., Wilson T. L., Cohen, R. J., & Tofani G. 1998, *A&A*, 334, 1085
Etoka S., Cohen R. J. & Gray M. D. 2005, *MNRAS*, 360, 1162
Harvey-Smith L. & Cohen R. J. 2005, *MNRAS*, 356, 637
Harvey-Smith L. & Cohen R. J. 2006, *MNRAS*, 371, 1550
Moran J. M., Reid M. J., Lada C. J., Yen J. L., Johnston K. J., & Spencer J. H. 1978, *ApJ*, 224, L67
Vlemmings W. H. T., Harvey-Smith L. & Cohen R. J. 2006, *MNRAS*, 371, L26
Wright M. M., Gray M. D. & Diamond P. J. 2004, *MNRAS*, 350, 1253

Astrophysical Masers and their Environments
Proceedings IAU Symposium No. 242, 2007
J.M. Chapman & W.A. Baan, eds.

doi:10.1017/S1743921307012793

Astrometry of H_2O maser sources in nearby molecular clouds with VERA

Tomoya Hirota[1,2], Takeshi Bushimata[1], Yoon Kyung Choi[1,3], Mareki Honma[1,2], Hiroshi Imai[4], Kenzaburo Iwadate[5], Takaaki Jike[5], Seiji Kameno[4], Osamu Kameya[2,5], Ryuichi Kamohara[1], Yukitoshi Kan-ya[6], Noriyuki Kawaguchi[1,2], Masachika Kijima[2], Mi Kyoung Kim[1,3], Hideyuki Kobayashi[1,3,5], Seisuke Kuji[5], Tomoharu Kurayama[1], Seiji Manabe[2,5], Kenta Maruyama[4], Makoto Matsui[4], Naoko Matsumoto[4], Takeshi Miyaji[1], Takumi Nagayama[4], Akiharu Nakagawa[4], Kayoko Nakamura[4], Chung Sik Oh[1,3], Toshihiro Omodaka[4], Tomoaki Oyama[1], Satoshi Sakai[5], Tetsuo Sasao[7,8], Katsuhisa Sato[5], Mayumi Sato[1,3], Katsunori M. Shibata[1,2], Motonobu Shintani[4], Yoshiaki Tamura[2,5], Miyuki Tsushima[4] and Kazuyoshi Yamashita[2]

[1] National Astronomical Observatory of Japan, Mitaka, Tokyo 181-8588, Japan

[2] Graduate University for Advanced Studies, Mitaka, Tokyo 181-8588, Japan

[3] The University of Tokyo, Bunkyo-ku, Tokyo 113-0033, Japan

[4] Kagoshima University, Kagoshima 890-0065, Japan

[5] National Astronomical Observatory of Japan, Oshu-shi, Iwate 023-0861, Japan

[6] Yonsei University, Seoul 120-749, Republic of Korea

[7] Ajou University, Suwon 443-749, Republic of Korea

[8] Korean VLBI Network, Seoul 120-749, Republic of Korea

email: tomoya.hirota@nao.ac.jp

Abstract. We have carried out multi-epoch VLBI observations of the H_2O maser sources associated with young stellar objects (YSOs) in nearby molecular clouds with VERA (VLBI Exploration of Radio Astrometry), which is a newly constructed VLBI network in Japan (Kobayashi *et al.* 2003). The main goal of our study is to measure the absolute proper motions and distances to nearby molecular clouds within 1 kpc from the Sun, to reveal their 3-dimensional structures and dynamical properties. Using the VERA dual-beam receiving system (Honma *et al.* 2003), we have carried out phase-referencing VLBI observations and measured annual parallaxes and absolute proper motions of the H_2O maser features with respect to the extragalactic radio sources. We have successfully detected the annual parallax of one of the H_2O maser features in Orion KL to be 2.29±0.10 mas, corresponding to the distance of 437±19 pc from the Sun (Hirota *et al.* 2007). In addition, the annual parallax of SVS13 in NGC 1333 is also determined to be 4.10±0.17 mas, corresponding to the distance of 244±10 pc from the Sun, although the life time of the maser features are only 6 months. The absolute proper motions of the H_2O maser features associated with Orion KL and NGC 1333 are derived, possibly indicating the outflow motions from the YSOs as well as the systemic motions of the powering sources.

Keywords. Astrometry, ISM: individual (Orion KL, NGC 1333), masers (H_2O)

References

Hirota, T. *et al.* 2007, submitted to *PASJ*

Honma, M. *et al.* 2003, *PASJ*, 55, L57

Kobayashi, H. *et al.* 2003, in: Y. C. Minh (eds.), *New technologies in VLBI*, ASP Conf. Ser. (San Francisco: ASP), vol. 306, p. 367

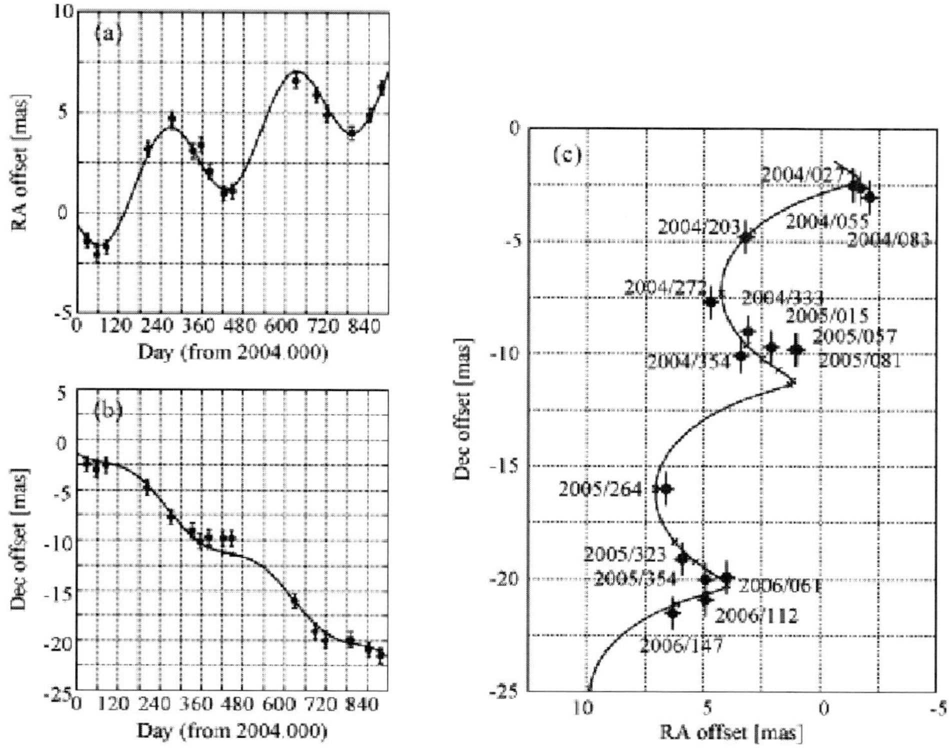

Figure 1. Results of the position measurements of the maser feature in Orion KL. (a) The movement of the maser feature in right ascension as a function of time. (b) The same as (a) in declination. (c) The movement of the maser feature on the sky. Solid lines represent the best fit model with the annual parallax (2.29±0.10 mas) and linear proper motion for the maser feature. Filled circles with error bars represent the observed positions of the maser features.

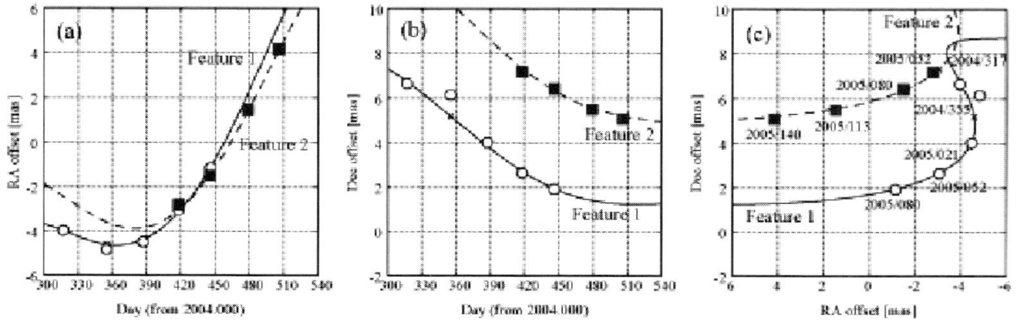

Figure 2. Results of the position measurements of the maser feature in NGC 1333 SVS13. (a) The movement of the maser feature in right ascension as a function of time. (b) The same as (a) in declination. (c) The movement of the maser feature on the sky. Solid and dashed lines represent the best fit model with the annual parallax (4.10±0.17 mas) and linear proper motion for the maser features 1 and 2, respectively. Open circles and filled squares represent the observed position of the maser features 1 and 2, respectively.

Astrophysical Masers and their Environments
Proceedings IAU Symposium No. 242, 2007
J.M Chapman & W.A. Baan, eds.

doi:10.1017/S174392130701280X

The 44 GHz methanol maser line in massive star forming regions

P. Hofner[1,2], E. Jordan[1], E. Araya[1,2] and S. Kurtz[3]

[1]New Mexico Tech, NM, USA
[2]National Radio Astronomy Observatory, NM, USA
[3]Centro de Radioastronomía y Astrofísica, UNAM, Morelia, México
email: phofner@nrao.edu

Abstract. We present the results of recent surveys in the 44 GHz methanol maser line toward regions of massive star formation using the Haystack 37 m telescope and the VLA. We discuss a possible shock origin of this maser line and present evidence for variability from multi-epoch observations of selected sources.

Keywords. masers, surveys, ISM:molecules, radio lines:ISM, stars: formation

1. Introduction

The CH_3OH $7_0 - 6_1$ A^+ line at a rest frequency of 44.069430 GHz is wide spread in regions of massive star formation (e.g. Haschick *et al.* 1990, ApJ 354, 556), however prior to our work (e.g. Kurtz *et al.* 2004, ApJS 155, 149), interferometric observations of this line were scarce. We are conducting single dish and VLA surveys in this line with the goal of establishing this maser as a probe for massive star formation research.

2. A catalogue of 44 GHz methanol masers in massive star forming regions

We used the VLA to survey 44 massive star-forming regions; 37 fields showed maser emission. Thirty-one sources were also observed in the 23 GHz $9_2 - 10_1$ A^+ methanol line, only two fields showed maser emission in this line. Although the 44 GHz line is a class I methanol maser, we find a large number of sources in relatively close association with H II regions and water masers: In a subsample of 22 fields which contain both 44 GHz maser emission and compact HII regions, the median projected distance between masers and ionized gas is 0.2 pc.

Several sources, most notably IRAS 20126 + 4104, G31.41 + 0.31, G34.26 + 0.15, and DR21(OH), show strong evidence for a correlation between 44 GHz masers and shocked molecular gas, supporting the interpretation that molecular outflows give rise to class I maser emission (Figure 1).

We provide maser positions with arcsecond accuracy that not only locate the masers with respect to other star formation phenomena, but also provide, for the stronger masers, phase referencing and pointing sources that can be used for observations in the 7 mm band of massive star forming regions.

3. Haystack observations of 44 GHz methanol masers towards massive protostars

Based on the results presented above we are continuing our 44 GHz CH_3OH masers survey. The sources observed in Kurtz *et al.* (2004) mostly contained UC HII regions, and

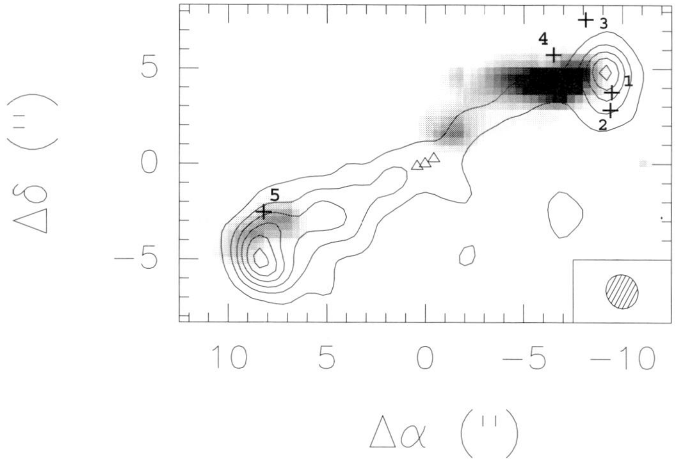

Figure 1. IRAS 20126+4104. Contours show SiO (2-1) emission, gray-scale shows H_2 emission, triangles indicate water masers while numbered crosses are 44 GHz methanol masers. The methanol masers appear to trace the edge of the shocked gas seen in H_2 (see Kurtz *et al.* 2004).

we are now surveying an earlier phase of massive star forming regions, containing high luminosity sources without detectable radio continuum. Our objective is to investigate whether the properties of the 44 GHz maser line will change with evolutionary phase. Since this maser line is a good tracer of molecular flows (see above) we expect to learn about the frequency of occurence and properties of molecular flows in a phase of massive star formation prior to that of UCHII regions. We will contrast the characteristics of the massive protostellar sample with the results from the UC HII region sample presented above.

To prepare for interferometric observations we have observed a sample of massive protostellar candidates with the 37m Haystack telescope (beamwidth at 44 GHz is 46″). We detect strong emission in 12 sources, and weak emission in many others. The typical rms noise of the Haystack spectra is 1 Jy. As a few deep integrations showed, most of the weak candidates are likely to be real detections when observed at lower rms noise levels. Thus, a very high detection rate is expected with the VLA given that the theoretical VLA rms noise in 10 min on-source observations is 0.025Jy/b. The VLA observations of the full sample are scheduled for spring 2007.

The Haystack data also allows us to investigate whether the CH_3OH 44 GHz transition is variable. Preliminary analysis indicates that this is the case for most sources.

Acknowledgements

This research was partially supported by NSF grant AST 0454665. E. A. is supported by the NRAO Pre-doctoral program. Student use of the MIT Haystack Observatory 37m radio telescope is made possible by a grant from the NSF.

References

Haschick, A. D., Menten, K., & Baan, W. A. 1990, *ApJ* 354, 556

Kurtz, S., Hofner, P. & Vargas-Alvarez, C. 2004, *ApJS* 155, 149

Astrophysical Masers and their Environments
Proceedings IAU Symposium No. 242, 2007
J.M. Chapman & W.A. Baan, eds

doi:10.1017/S1743921307012811

LBA observations of OH masers in the star-forming region OH 330.953–0.182

B. Hutawarakorn Kramer[1], **J. L. Caswell**[2], **A. Sukom**[3] **and J. E. Reynolds**[2]

[1]National Astronomical Research Institute of Thailand, Physics Building, Chiang Mai University 50200, Thailand
email: busaba@narit.or.th

[2]Australia Telescope National Facility, CSIRO, PO Box 76, Epping, NSW 2121, Australia
email: james.caswell@csiro.au; john.reynolds@csiro.au

[3]Department of Physics, Chiang Mai University 50200, Thailand

Abstract. OH masers are sensitive probes of the kinematics, physical conditions, and magnetic fields in star-forming regions. The maser site OH 330.953−0.182 has been studied using the Long Baseline Array of the Australia Telescope National Facility. Simultaneous observations of the 1665- and 1667-MHz hydroxyl ground-state transitions yield a series of maps at velocity spacing 0.09 km s^{-1}, in both right- and left-hand circular polarization, with tenth-arcsec spatial resolution. Several clusters of maser spots have been detected within a five-arcsec region. Eight Zeeman pairs were found, and in one case, at 1665 MHz, there is a nearby 1667-MHz pair indicating a similar value of magnetic field and velocity. Over the whole site, all magnetic field estimates are toward us (negative), and range from −3.7 to −5.8 mG. We also compared the morphology and kinematics of the 1665- and 1667-MHz maser spots with those from the excited state of OH at 6035 MHz and from methanol at 6668 MHz.

Keywords. OH masers, star-forming region, OH 330.953-0.182, magnetic field

1. Observations and data reduction

The present observations were made using the Australian Long Baseline Array (LBA) telescopes at Parkes, Mopra, and Narrabri on 16–17 August 2000. The signals from the telescopes forming the Compact Array at Narrabri were combined together to form a single telescope with an equivalent size of diameter 46 m. The three stations yield baselines of approximately 119 km, 203 km and 321 km in a predominantly north-south direction. The S2 tape system was used to record both right- and left-hand circular polarizations simultaneously with a band pass centred at 1667 MHz and limited to 4-MHz bandwidth by digital filters with an excellent flat response in both amplitude and phase. The target source, OH 330.953−0.182, was observed for 25-min periods, interleaved with another target source, and with 8-min periods on a phase and secondary amplitude calibrator. The data reduction, mapping and image analysis were done by using AIPS – see Caswell & Reynolds (2001) and Hutawarakorn & Cohen (1999).

2. Results and discussion

The positions of the 1665- and 1667-MHz OH masers are shown in Figure 1, revealing three clusters of maser spots. The north-west and north-east clusters correspond respectively to 330.953–0.182 and 330.954–0,182 (Caswell 1998). The 1665-MHz southern cluster, newly discovered in these observations, is at the same position as the 6035-MHz

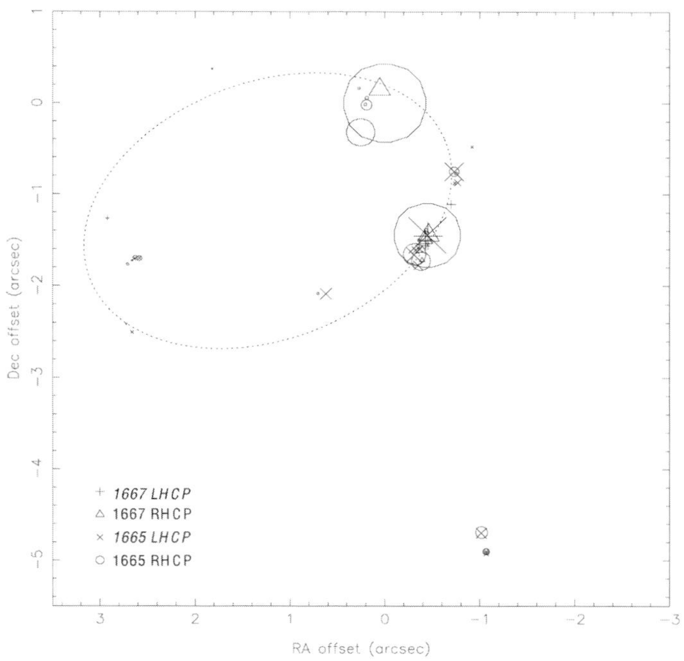

Figure 1. Intensities and relative positions of 1665- and 1667-MHz masers. The size of each symbol is proportional to the peak flux density.

OH maser (Caswell 2001), which in turn coincides with the 6668-MHz methanol maser (Caswell 2001).

The brightest 1665- and 1667-MHz features, in both LHCP and RHCP, are clustered in the north-west region (330.953-0.182). If we focus on the northern clusters, the position distribution can be fitted by a ring shape. At an assumed distance of $\sim$ 8.5 kpc, the diameter of the ring is $\sim$ 1700 au. The velocity distribution of the masers shows a gradient along the RA axis. This might be interpreted as a rotating disk, with central radial velocity $\sim -90\,\mathrm{km\,s^{-1}}$. Similar results have been found in recent studies of OH masers in massive star-forming regions associated with disks and outflows (Nammahachak, Asanok, Hutawarakorn Kramer *et al.* 2006 and references therein).

Eight Zeeman pairs were found, and in one case at 1665 MHz, there is a nearby 1667-MHz pair indicating a similar value of magnetic field and velocity. Over the whole site, all magnetic field estimates are toward us (negative), and range from -3.7 to -5.8 mG. A magnetic field of this strength is sufficient for the field to be dynamically important in the star-forming regions.

References

Caswell, J. L. 1998, *MNRAS* 297, 215

Caswell, J. L. 2001, *MNRAS* 326, 805

Caswell, J. L. & Reynolds, J. E. 2001, *MNRAS* 325, 1346

Hutawarakorn, B. & Cohen, R. J. 1999 *MNRAS* 303, 845

Nammahachak, S., Asanok, K., Hutawarakorn Kramer, B., Cohen, R. J., Muanwong, O., Gasiprong, N. 2006, *MNRAS* 371, 619

Astrophysical Masers and their Environments
Proceedings IAU Symposium No. 242, 2007
J.M. Chapman & W.A. Baan, eds.

doi:10.1017/S1743921307012823

Low-mass star formation in bright rimmed clouds

Migenes, V.[1], Trinidad, M. A.[1], Valdettaro, R.[2], Palla, F.[2] and Brand, J.[3]

[1]Department of Astronomy, University of Guanajuato, P.O. Box 144, Guanajuato, GTO. Mexico 36000
email: vmigenes@astro.ugto.mx, trinidad@astro.ugto.mx

[2]Osservatorio Astrofisico di Arcetri, Largo Enrico Fermi 5, Firenze, Italy 50125
email: rv@arcetri.astro.it, email: palla@arcetri.astro.it

[3]Istituto di Radioastronomia, Via Gobetti 101, Bologna, Italy I-40129
email: brand@ira.cnr.it

Abstract. Bright Rimmed Clouds (BRCs) are clouds that have been compressed by an external ionization-shock front which focuses the neutral gas into compact globules. The boundary layer between the neutral gas and the gas ionized by the incident photons is often called "bright rim" but the clumps are sometimes classified also as speck globules or cometary globules depending on their appearance. Small globules with bright rims have been considered to be potential sites of star formation and have been studied in several individual regions. We present the first high resolution VLA observations of 20 of these BRCs, but only three detections were obtained. The low detection rate seems to support the idea that BRCs produce mostly low-luminosity objects, for which maser emission is weak and episodic, and that the embedded sources are in a more advanced evolutionary phase than class 0 objects.

Keywords. ISM: clouds, star: formation, masers

1. Observations

The observations were made with the VLA of the National Radio Astronomy Observatory (NRAO) in its B configuration during four runs (approx 2 hours long each) on 2005 February 20, 24, 25 and March 30. We observed water maser emission, in LCP mode, during 15 minutes on each target source. We used a bandwidth of 3.125 MHz with 127 channels, giving a velocity resolution of 0.33 km s^{-1}, and a velocity range of $\sim$40 km s^{-1}. The data were reduced with standard techniques using the NRAO AIPS software package.

Though all 44 northern BRCs, from the list of Sugitani et al (1991), have been observed by Valdettaro *et al.* (2005) only twenty were observed with the VLA. Only three were detected, all towards the HII region IC1396, also known as S131, in the Cep OB2 association at a distance of 750 pc. IC 1396 N is an extended HII region ($\sim$3deg in diameter) excited by the O6 star HD 206267 and a cluster of B stars. This region is associated with OH and H_2O maser emission (Slysh *et al.* 1999 and Migenes *et al.* 1999) in addition to other molecular species.

2. Discussion: The detected sources

The IRAS source associated with this globule, IRAS 21346+5714 (BRC36 from Sugitani's catalog), has a luminosity of 110 $L_\odot$ and FIR-colors typical of protostellar candidates or precursors to UCHII regions. It is located in the tail of the globule, rather than

Table 1. Detected Sources

Name	*RA(J2000)* (*hms*)	*DEC(J2000)* (*d' "*)	*Flux* (*Jy/b*)	V_{LSR} (km/s)
IRAS 21346+5714	21 36 07.881	+57 26 43.43	32.32	−10.3
IRAS 21388+5622	21 40 28.107	+56 36 02.16	0.28	−3.0
IRAS 21445+5712	21 46 07.140	+57 26 30.85	0.17	−6.1

in its head, close to the bright rim as in most BRCs. Single-dish monitoring observations (Valdettaro *et al.* 2005) show very variable H_2O maser emission, which is common in low-luminosity sources. The VLA observations reveal a single spectral component at about −10.4 km s^{-1}. The map shows a spot with extended E-W structure, about 2 km s^{-1} wide, suggesting that higher resolution may show more than one component.

The globule associated with IRAS 21388+5622 (BRC37 from Sugitani's catalog) has a cometary rim morphology. The embedded IRAS source has a luminosity of 110 $L_\odot$ (Sugitani *et al.* 1991), classified as a Class 0 like object by Sugitani *et el.* (2000) on the basis of the ratio L_{bol}/L_{sub}. The IRAS source is also associated with the densest core of the whole globule. Duvert *et al.* (1990) detected a bipolar outflow with a 20" separation between the red and blue lobes and a N-S orientation. The VLA spectrum exhibits 2 features that the map reveals as 3 separate spots 1.0 km s^{-1} wide.

The source IRAS 21445+5712 (BRC39 from Sugitani's catalog) is an embedded IRAS source with a luminosity of 116 $L_\odot$. The globule has a tightly curved rim morphology and located near the center of the BRC (Schwartz *et al.* 1991). The VLA position of the H_2O maser is quite close to that of the IRAS source, indicating that the latter is the exciting source. Froebrich *et al.* (2005) found in the optical field a new giant flow emerging from the embedded IRAS source. These observations show a single-feature spectrum and the map presents a single spot 1.6 km s^{-1} wide.

3. Conclusions

The low detection rate (3/20) for water vapor maser emission in BRCs, is not completely due to the sensitivity of the observations, but it could also be due to lower FIR flux from the YSO than what is detected by IRAS. The warm dust within and close to the bright rim strongly contributes to its flux as measured by IRAS. The much lower intrinsic FIR flux of the YSO would suggest that BRCs are sites of low-mass-star formation. BRCs seem to produce mostly low-luminosity objects, which are presently in a more advance evolutionary phase than class 0 sources, for which the frequency of occurrence of maser emission is low and highly episodic. The high resolution positions of the water vapor maser emission will help us identify their exciting source.

References

Duvert, G. *et al.* 1990, *A&A* 233, 190
Froebrich, D. *et al.* 2005, *A&A* 432, 575
Migenes V., *et al.* 1999, *ApJS* 123, 487
Slysh, V. I. *et al.* 1999, *ApJ* 526, 238
Sugitani, K., Fukui, Y. & Ogura, K., 1991, *ApJS* 77, 59
Sugitani, K. *et al.* 2000, *AJ* 119, 323
Valdettaro, R. *et al.* 2005, *A&A* 443, 535
Schwartz, R. D., Wilking, B. A., & Giulbudagian, A. L., 1991, *ApJ* 370, 262

Astrophysical Masers and their Environments
Proceedings IAU Symposium No. 242, 2007
J.M. Chapman & W.A. Baan, eds.

doi:10.1017/S1743921307012835

An OH maser flare with a strong magnetic field in W75N

V. I. Slysh[1] and V. Migenes[2]

[1]Astro Space Center, Lebedev Physical Institute, 84/32 Profsoyuznaya, Moscow, 117997
email: vslsysh@asc.rssi.ru

[2]Department of Astronomy, University of Guanajuato, P.O. Box 144, Guanajuato, GTO. Mexico 36000,
email: vmigenes@astro.ugto.mx

Abstract. A flare of OH maser emission was discovered in W75N in 2000. Its location was determined with the VLBA to be within 110 AU from one of the ultra-compact HII regions, VLA2. The flare consisted of several maser spots. Four of the spots were found to form Zeeman pairs, all of them with a magnetic field strength of about 40 mG. This is the highest ever magnetic field strength found in OH masers, an order of magnitude higher than in typical OH masers. We discuss the possible source for the enhanced magnetic field and its relation to the flare event.

Keywords. ISM:magnetic fields, stars: flare, maser

1. Observations

The new observations of the OH maser emission in W75N were conducted on 2001 January 01 with the VLBA in a snapshot mode of 6-min duration. The velocity resolution was 0.176 km s^{-1} covering 45 km s^{-1} in each of the OH main lines at 1665 MHz and 1667 MHz. In addition, we reduced and analyzed archived observations from the EVN and VLBA for 2000 September 27 and 2000 November 22 and 2001 January 6, respectively. Both data sets had the same velocity resolution though different velocity coverage.

All three sets of data have been obtained during the maximum phase of the OH maser flare. The data were reduced in the standard way using NRAO AIPS package. Most of the maser spots were unresolved by the synthesized beam. The absolute position given in Table 1 was measured through fringe rates using AIPS task FRMAP.

2. Discussion: Flare and magnetic field

Two new strong spectral features have appeared since the 1998 observations, at the low-velocity side of the spectrum. On the other hand, two relatively strong spectral features from the 1998 spectrum became a factor of three weaker in the 2000–2001 spectra. It is also evident from the spectra that the new "flare" features were rapidly evolving in about a three months time interval, between 2000 September 27 (EVN) and 2001 January 1 (VLBA). Four months later, on 2001 April 12, these features had become considerably weaker as observed with the Bear Lake 64-m single dish telescope (Alakoz *et al.* 2005). In the same time interval the rest of the spectral features remained unchanged. All constant features are connected with the ultra-compact HII region VLA1 while the variable features are connected with VLA2.

Compared to the 1998 map (Slysh *et al.* 2002) several additional spots have been detected, partly due to a higher sensitivity of the new observations. The other features

Table 1. OH Zeeman pairs in W75N

Zeeman Pairs	ΔRA RCP	ΔDEC RCP	Velocity (km/s)	ΔRA LCP	ΔDEC LCP	Velocity (km/s)	B-field Strength (mG)	Separation (mas)	Associate HII Region
Z1[a]	0.0	0.0	12.45	0.16	0.03	9.28	5.4	0.2	VLA1
Z2	205.42	666.92	8.07	206.91	668.90	5.43	7.5[b]	2.5	VLA1
Z3	488.80	1345.93	7.24	489.13	1345.29	5.13	3.6	0.7	VLA1
Z4	765.77	-178.12	24.40	766.15	-177.75	-0.48	42.2	1.6	VLA2
Z5	768.89	-135.24	19.80	767.59	-132.00	4.76	42.5[c]	3.5	VLA2
Z6	770.30	-122.33	29.27	770.89	-125.00	4.71	41.6[d]	2.7	VLA2
Z7	770.41	-129.98	21.17	772.36	-128.62	8.42	36.3[b]	2.4	VLA2

[a] Z1 is spot A: with a measured absolute position of RA = 20h38m36s.416; DEC = 4237'34".42 (0".01) (J2000). [b] 1667 MHz. [c] 1667 MHz (Fish *et al.* 2005, table 15). This Zeeman pair was probably overlooked by the authors, or dismissed as showing too large a velocity separation. [d] From EVN data.

are really new as they are related to the flare which took place between 1998 and 2000. Also, more accurate absolute positions of OH spots were obtained and are given in Table 1, as well as the positions of the 1667 MHz spots relative to the 1665 MHz spots.

H_2O-masers have also been found near the OH masers in W75N, located in two clusters around VLA1 and VLA2. Torrelles *et al.* (2003) have found a shell of water masers around the ultra-compact HII region VLA2 with a radius of 160 AU. The shell is expanding with a velocity of 28 km s^{-1}, perhaps episodically as in a recurrent outflow. The high magnetic field OH maser spots Z4-Z7 are located very close to VLA2, at a distance of 55 mas (±40 mas), or at the projected distance of 110 AU (±80 AU). Therefore, the OH masers may well be located in the same shell as the water masers. The magnetic field in water masers associated with star-forming regions is typically around 100 mG, which is about the same order of magnitude as in the OH maser flare reported here.

The appearance of new strong maser features and the simultaneous dimming of nearby features can be interpreted as originating from the passage of a magnetohydrodynamic (MHD) shock (Alakoz *et al.* 2005). The shock was probably generated by the exciting star of VLA2 and propagated in the gas of the stellar wind.

3. Conclusions

A very strong magnetic field of 40 mG has been detected in several OH masers spots which have appeared during a flare of OH maser emission in 2000, within 110 AU from the ultra-compact HII region. The magnetic field probably originates in the exciting star where its intensity is about 500 G, or from the compression of interstellar gas by MHD shock, or in icy planetary bodies serving as nuclei for the maser spot emission. More frequent high angular resolution observations of future flares may help to distinguish between these models.

References

Alakoz A. V., Slysh V. I., Popov M. V. & Val'tts I. E., 2005, *Astron. Letters* 31, 375
Fish V. L., Reid M. J., Argon A. L. & Zheng X.-W., 2005, *ApJS* 160, 220
Slysh V. I., Migenes V., Val'tts I. E., Lyubchenko S. Yu., Horiuchi S., Altunin V. I., Fomalont E. B. & Inoue M., 2002, *ApJ* 564, 317
Torrelles J. M. *et al.*, 2003, *ApJL* 598, L115

Astrophysical Masers and their Environments
Proceedings IAU Symposium No. 242, 2007
J.M. Chapman & W.A. Baan, eds.

doi:10.1017/S1743921307012847

A precise distance to IRAS 00420+5530 via H_2O maser parallax with the VLBA

G. A. Moellenbrock, M. J. Claussen and W. M. Goss

National Radio Astronomy Observatory, 1003 Lopezville Road, Socorro, NM 87801, USA
email: gmoellen@nrao.edu, mclausse@nrao.edu, mgoss@nrao.edu

Abstract. We have used the VLBA to measure the annual parallax of the H_2O masers in the star-forming region IRAS 00420+5530 ($l = 122.0°, b = -7.1°$). This measurement yields a direct distance estimate of 2.17 ± 0.05 kpc, which disagrees substantially with most other published estimates (a range from 1.7 to 7.7 kpc). This distance is consistent with recent parallax-based distances reported for W3(OH) (Xu *et al.* 2006, Hachisuka *et al.* 2006), lending additional support to the conclusion that the Perseus Spiral Arm of the Milky Way is a factor of at least two closer than kinematic distance estimates imply at galactic longitude ~130 degrees.

1. Introduction

Very Long Baseline Interferometry extends the range of the direct distance measuring technique of annual parallax from ~100 pc optically, to ~10 kpc with radio telescopes such as the VLBA of the National Radio Astronomy Observatory†. Compact radio-emitting objects such as pulsars and masers are thus choice beacons for distance measurements across the Galaxy (e.g., van Langevelde *et al.* 2000, Brisken *et al.* 2002, Chatterjee *et al.* 2004). Such distance measurements will enable enormous improvements in our understanding of Galactic structure and the physics of individual objects.

IRAS 00420+5530 is a star formation region with a high-velocity outflow (Zhang *et al.* 2005), 3.6 cm and 3 mm continuum emission (Molinari *et al.* 2002), and H_2O masers (e.g., Brand *et al.* 1994). Distances of 4.3 to 7.7 kpc (usually kinematic) have been reported and used in these and other recent papers (e.g., Kumar *et al.* 2006, Brand *et al.* 2001), in substantial disagreement with an early photometric distance of 1.7 kpc (Neckel & Staude 1984). Clearly, a better distance estimate is desirable for this object.

2. VLBA parallax observations of IRAS 00420+5530

To resolve this distance uncertainty, we have conducted 12 four-hour VLBA observations (2005 Nov–2006 Sep) of the 22 GHz H_2O masers in IRAS 00420+5530. Two point-like radio-loud background quasars were observed alternately with the target: J0042+5708 (~190 mJy) for phase-referencing calibration and position registration and J0047+5657 (~85 mJy) to check astrometric precision. Additionally, seven strong (>1 Jy) calibrators distributed in zenith angle were observed to enable correction for errors in the VLBA correlator's coarse macroscopic tropospheric delay model. The resulting images of the IRAS 00420+5530 reveal the Earth-relative motion of the maser sources on the plane of the sky. The corresponding images of J0047+5708 show that it is stationary to within ~40 microarcseconds, indicating a per-epoch phase-referencing astrometric precision of this magnitude.

† The National Radio Astronomy Observatory is a facility of the National Science Foundation operated under cooperative agreement by Associated Universities, Inc.

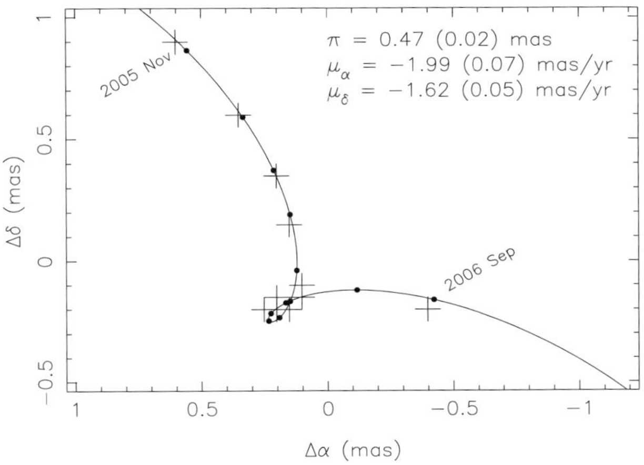

Figure 1. The motion of a maser spot in IRAS 00420+5530 at V_{lsr} = −44.1 km/s. Crosses: observations, Dots and curve: proper motion and parallax fit. Ten such fits were combined to form the mean parallax estimate described in the text.

We have fitted the sequence of peak positions for four maser spot regions (in 19 channels, ΔV_{lsr} = 0.21km/s/channel) in IRAS 00420+5530 for linear proper motion and annual parallax. An example of one such fit (for the V_{lsr} = −44.1 km/s) is shown in Figure 1. The accuracy of these fits is limited not by the phase-referencing accuracy, but rather by systematic spatial and velocity variability in the maser spots, which make it difficult to identify the same *physical* source components from epoch to epoch to better than ∼50 microarcseconds, in the best cases. Two of the maser spot groups were excluded on this basis. The parallax fits for the remaining two groups (10 channels) are statistically indistinguishable (though their proper motions differ significantly), and have been combined to yield a net parallax estimate of $\pi = 0.46 \pm 0.01$ milliarcseconds.

3. The distance to IRAS 00420+5530

Our parallax yields a distance for IRAS 00420+5530 of $d = 2.17 \pm 0.05$ kpc. This 3% distance disagrees by at least a factor of two compared to kinematic estimates. It is consistent with similarly-measured distances reported for W3(OH) (Xu *et al.* 2006, Hachisuka *et al.* 2006), supporting the conclusion that the Perseus Spiral Arm of the Milky Way is a factor of at least two closer than kinematic distance estimates imply at galactic longitude ∼130°.

References

Brand, J., *et al.* 1994, *A&AS* 103, 541
Brand, J., *et al.* 2001, *A&A* 370, 230
Brisken, W., *et al.* 2002, *ApJ* 571, 906
Chatterjee S., *et al.* 2004, *ApJ* 604, 339
Kumar, M. S. N., *et al.* 2006, *A&A* 449, 1033
Hachisuka, K., *et al.* 2006, *ApJ* 645, 337
Molinari, S., *et al.* 2002, *ApJ* 570, 758
Neckel, T. & Staude, H. J. 1984, *A&A* 131, 200
van Langevelde, H. J., *et al.* 2000, *A&A* 357, 945
Xu, Y. *et al.* 2006, *Science* 311, 54
Zhang, Q. *et al.* 2005, *ApJ* 625, 864

Astrophysical Masers and their Environments
Proceedings IAU Symposium No. 242, 2007
J.M. Chapman & W.A. Baan, eds.

doi:10.1017/S1743921307012859

Absolute proper motions of water masers in NGC 281 measured with VERA

Mayumi Sato,[1,2] Tomoya Hirota,[1] Mareki Honma,[1] Hideyuki Kobayashi[1] and the VERA Project Team[1]

[1]VERA Office, National Astronomical Observatory, 2-21-1 Osawa, Tokyo 181-8588 JAPAN
email: mayumi.sato@nao.ac.jp

[2]Department of Astronomy, The University of Tokyo, Hongo 7-3-1, Tokyo 113-0033, JAPAN

Abstract. We report on absolute proper-motion measurements of H_2O maser features in the NGC 281 West molecular cloud, located ~320 pc above the Galactic plane and associated with an HI loop extending from the Galactic plane. We conducted six-epoch phase-referencing observations of the maser source with VERA (VLBI Exploration of Radio Astrometry) over six months since May 2006. The H_2O maser features are found to be systematically moving toward the southwest and further away from the Galactic plane with a vertical velocity of ~20–30 km s $^{-1}$ at its estimated distance of 2.2–3.5 kpc. Our new results provide the most direct evidence that the gas in the NGC 281 region was blown out from the Galactic plane, most likely in a superbubble driven by multiple or sequential supernova explosions in the Galactic plane.

Keywords. astrometry, Galaxy: kinematics and dynamics, ISM: bubbles, masers

---

Introduction and results

NGC 281 (α_{2000} =00^h52^m, δ_{2000} = +56°34' or l =123°.07, b = −6.°31) is an HII region situated z ~320 pc above the midplane of the Perseus arm of the Galaxy at a heliocentric distance of 2.9 kpc (Guetter & Turner 1997), surrounded by a giant molecular cloud complex. Megeath *et al.* (2002) identified this region on a large-scale HI loop extending from the Galactic plane, and found a 270-pc ring of the clouds of the region, expanding at a velocity of 22 km s^{-1} parallel to the plane. They suggested that the clouds have been formed in a fragmenting superbubble shell. In order to investigate the systemic motion of the region, we have conducted six-epoch phase-referencing observations of an H_2O maser source in the NGC 281 West cloud with VERA over six months since May 2006.

Figure 1 shows the results of relative and absolute proper-motion measurements of the H_2O maser features in NGC 281 West. We detected 10 maser features persistent for more than two epochs, with radial velocities $V_{\rm LSR}$ of −39 to −22 km s^{-1}. These include eight features in C3 and 2 features in C1, corresponding to the VLA observations by Tofani *et al.* (1995). We first examined the internal motions of the maser features in C3, relative to the brightest maser, feature 4 with $V_{\rm LSR}$ = −31.8 km s^{-1}, and they showed expanding motions with velocities from 10 to 20 km s^{-1} (1 mas yr^{-1} corresponds to 13.7 km s^{-1} at a distance of 2.9 kpc). We then measured the absolute proper motions of the maser features 4 (in C3), 9 and 10 (in C1), assuming the distance to NGC 281 to be 2.9 kpc and subtracting the expected parallax from the observed motions. Adopting a different distance of 2.2−3.5 kpc instead of 2.9 kpc in the parallax subtraction yields a difference of 10−15% in the resultant proper motions. As shown in Figure 1, the mean absolute proper motions of the features in C3 and C1 show a systematic motion toward the southwest. This direction, southwestward, agrees well with that of the proper motion of HD 5005 (the ionizing source of NGC 281), measured with *Hipparcos* (Perryman *et al.* 1997). Figure 2 shows a plot of the mean motion of the maser features relative to the

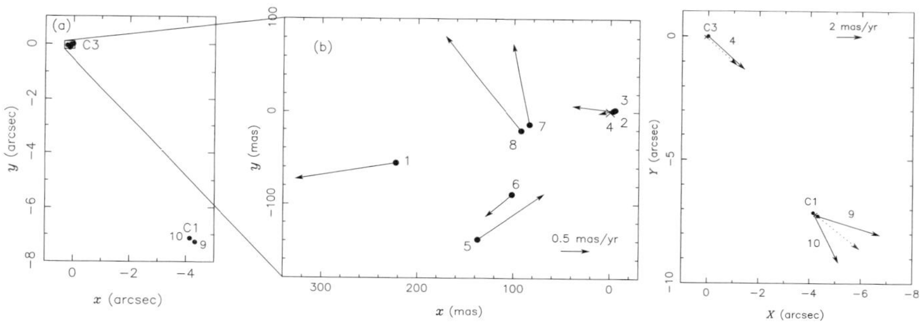

Figure 1. *(left)(a)* The maser feature distribution; *(b)* the internal motions of features in C3 relative to feature 4 (denoted by X). *(right)* A map of the absolute proper motions of features 4, 9 and 10 (solid arrows) and the mean proper motions in C3 and C1 (dotted arrows). The map origin is set at RA(J2000.0)=$00^h52^m24.^s70086$ and DEC(J2000.0)= $+56°33'$ $50.''5270$. From Sato *et al.* (2007), ©2007 Astronomical Society of Japan.

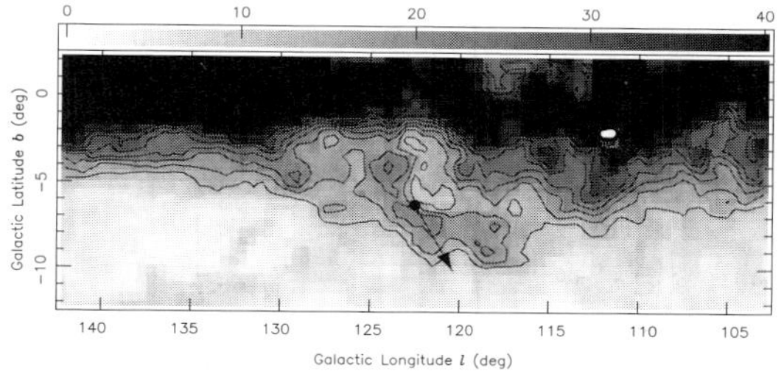

Figure 2. A plot of the mean motion of H_2O maser features in NGC 281 West relative to the simulated Galactic rotation, superimposed on gray-scale and contour maps of the HI data from Hartmann & Burton (1997), integrated over $V_{\rm LSR} = -65$ to -25 km s^{-1}. From Sato *et al.* (2007), ©2007 Astronomical Society of Japan.

simulated Galactic rotation, obtained by subtracting the expected apparent motion of NGC 281 due to the Galactic rotation (assumed to be a flat rotation) and the solar motion from the detected proper motion. This mean motion is considered to trace the systemic motion of the region away from the Galactic plane at a transverse velocity of 20-30 km s^{-1} perpendicular to the plane, which is comparable to the ring expansion velocity of 22 km s^{-1} parallel to the plane (Megeath *et al.* 2002), thus yielding a comparable kinetic energy ($\sim 4-8\times10^{51}$ ergs) of the clouds for motions perpendicular to the plane as well as parallel. Our measurements provide the most direct evidence that the gas in the NGC 281 region was blown out from the Galactic plane, most likely in a superbubble driven by multiple or sequential supernova explosions in the Galactic plane.

This work is based on Sato *et al.* (2007) and reproduced with authorization from Astronomical Society of Japan. See Sato *et al.* (2007) for more details on the measurements and discussions.

References

Guetter, H. H., & Turner, D. G. 1997, *AJ* 113, 2116

Hartmann, D., & Burton, W. B. 1997, *Atlas of Galactic Neutral Hydrogen* (Cambridge: Cambridge Univ. Press)

Megeath, S. T., Biller, B., Dame, T. M., Leass, E., Whitaker, R. S., & Wilson, T. L. 2002, in: P.A. Crowther (ed.), *ASP Conf. Ser. 267, Hot Star Workshop III: The Earliest Stages of Massive Star Birth* (San Francisco: ASP), p. 257

Perryman, M. A. C., *et al.* 1997, *A&A* (Letters) 323, L49

Sato, M., Hirota, T., Honma, M., Kobayashi, H. *et al.* 2007, *PASJ* 59, No. 4, in press

Tofani, G., Felli, M., Taylor, G. B., & Hunter, T. R. 1995, *A&AS* 112, 299

Astrophysical Masers and their Environments
Proceedings IAU Symposium No. 242, 2007
J.M. Chapman & W.A. Baan, eds.

doi:10.1017/S1743921307012860

Groundstate OH masers associated with Herbig-Haro objects

A. de Witt and D. P. Smits

Department of Mathematical Sciences, University of South Africa, Pretoria, 0003, South Africa

email: aletdewitt@webmail.co.za, dps@astro.unisa.ac.za

Abstract. Jets from newly-formed stars plow into the surrounding ISM, creating nebulous regions known as Herbig-Haro (HH) objects. Signatures of C-type shocks have been found in many HH objects. Models of shock-excited 1720-MHz OH masers associated with SNRs suggest that the masers could be signposts of C-type shocks. We have conducted a 1720 MHz survey towards a number of HH objects to see if any shock-excited masers occur in association with these objects. Some results of our search are reported. When present, the 1720-MHz OH line can occur as emission, absorption or as P Cygni profiles. OH mainline emission is found in many of our sources. We have identified several potential 1720-, 1665- and 1667-MHz OH masers.

Keywords. ISM: Herbig-Haro objects, ISM: jets and outflows, masers, line: profiles

1. Introduction

Herbig-Haro (HH) objects are shock-excited nebulae identified from characteristic line ratios in their optical spectra. They are associated with outflows, usually in the form of jets, often (but not exclusively) from protostars with $M_* \leqslant 4M_\odot$. When these jets impact the interstellar medium (ISM), bow-shaped structures are formed which show signatures of J-type shocks at the apex and C-type shocks on the wings. For example, modelling of HH 7 by Smith, Khanzadyan & Davis (2003) successfully explained their near infrared observations of 2.12 μm H_2 line emission in terms of a C-type shock on the wings of the bow structure.

About 10% of known Galactic supernova remnants (SNRs) have been found to support 1720-MHz OH masers at their shock interface. Lockett, Gauthier & Elitzur (1999) modelled these regions and showed that under a restrictive set of physical conditions, 1720-MHz OH masers can be collisionally excited in a C-type shock formed when a SNR interacts with a dense molecular cloud within the ISM. Observationally the masers have been found to occur along a ridge where the 2.12 μm H_2 line emission peaks. Based on the tight constraints under which the masers form, Lockett *et al.* (1999) suggested that 1720 MHz OH masers could be an important signpost of C-type shocks. This report presents some results of a survey carried out on the 100 m Effelsberg and Greenbank telescopes to search for 1720-MHz OH masers towards HH objects.

2. Observations

2.1. *Source selection*

Based on SNR statistics, only 10 % of SNRs searched have 1720-MHz OH masers. If the same statistics apply to HH objects, a large number of sources need to be surveyed to find at least a few maser detections. Furthermore, shock-excited masers are not as intense as those in H II regions and therefore a survey with high sensitivity is required. The most efficient way to search a large number of sources for masers is to use single-dish telescopes with large collecting surface areas. Sources with positive detections can be re-observed with interferometers to confirm the presence of masers.

A sample of HH objects were selected from Reipurth's catalogue (Reipurth, 1999) to search for 1720-MHz OH emission. The nearest HH objects were our primary targets, as they should have the strongest intensity. Sources for which a case has been made in the literature that C-type shocks are present were also included in our list. A total of 231 HH objects have been observed with either the Greenbank or Effelsberg telescopes. Sometimes more than one object fell within the $7'$ beam of the telescope so that the number of objects observed exceeds the number of pointings.

2.2. *Greenbank 100m Telescope*

Observations using the Greenbank 100m telescope were performed on 04 and 07 February 2005. We simultaneously observed the 1720-MHz OH satellite line and the OH ground-state mainlines at 1665 and 1667 MHz in both LCP and RCP towards 45 positions. With a bandwidth of 12.5 MHz this gave a velocity coverage of $\pm 1090\,\mathrm{km\,s^{-1}}$ at a resolution of $\pm 0.13\,\mathrm{km\,s^{-1}}$. The total integration time per source was 15 mins per pointing, giving a typical noise level of $\sim$ 30 mJy.

2.3. *Effelsberg 100m Telescope*

Observations using the Effelsberg 100m telescope were performed on 28 February, 02 March, 05 May and 20 May 2005. Only the 1720-MHz OH line was observed in both LCP and RCP towards 74 positions. With a bandwidth of 1.56 MHz, the 1024 channel correlator gave a velocity coverage of $\pm 84\,\mathrm{km\,s^{-1}}$ at a resolution of $\pm 0.53\,\mathrm{km\,s^{-1}}$ after folding the spectra. Integration times varied between 10 and 15 mins per pointing, giving a typical noise level of $\sim$ 40 mJy.

3. Results

For the Greenbank data at 1720 MHz there were 18 non-detections, 16 pointings show lines in emission, 4 have absorption and 7 have P Cyg-type profiles. Two of the P Cyg profiles have blue-shifted absorption and red-shifted emission indicating infall rather than outflow. Line widths of $\Delta v \sim 0.5$–$3.5\,\mathrm{km\,s^{-1}}$ are consistent with those of the 1720-MHz OH masers in SNRs. These are broader than would be expected for a single maser spot but could be multiple lines unresolved in velocity.

A surprising result is that the Greenbank data show 1665- and 1667-MHz OH emission lines at all but 4 of the pointings, 14 of which have no corresponding 1720 MHz lines. None of the sources show unambiguous 1720-MHz OH maser emission, i.e. 1720 MHz in emission and 1667 MHz in absorption.

For the OH mainlines, optically thin thermal emission has a flux density ratio $F(1665) : F(1667) = 5 : 9$, while for optically thick conditions the ratio is $1 : 1$. If observed flux densities have a ratio $F(1667) : F(1665)$ outside this range, this is suggestive of non-thermal emission such as a maser. A number of our sources have non-thermal line ratios for the mainlines. All of the sources with P Cyg profiles at 1720 MHz have non-thermal mainline ratios less than 1, indicating that $F(1665) > F(1667)$.

In Fig. 1 some examples of the Greenbank data are presented. HH33 is an example of a back-to-front P Cyg profile. The 1720 MHz absorption peak corresponds in velocity to that of one peak in the 1665 and 1667 MHz spectra, while the 1720 MHz emission peak is offset from that of the mainlines by $\sim 1\,\mathrm{km\,s^{-1}}$. HH 167 and HH 279 have no detectable features at 1720 MHz, but have $F(1667) : F(1665)$ of 2.40 and 2.80 respectively. HH 200

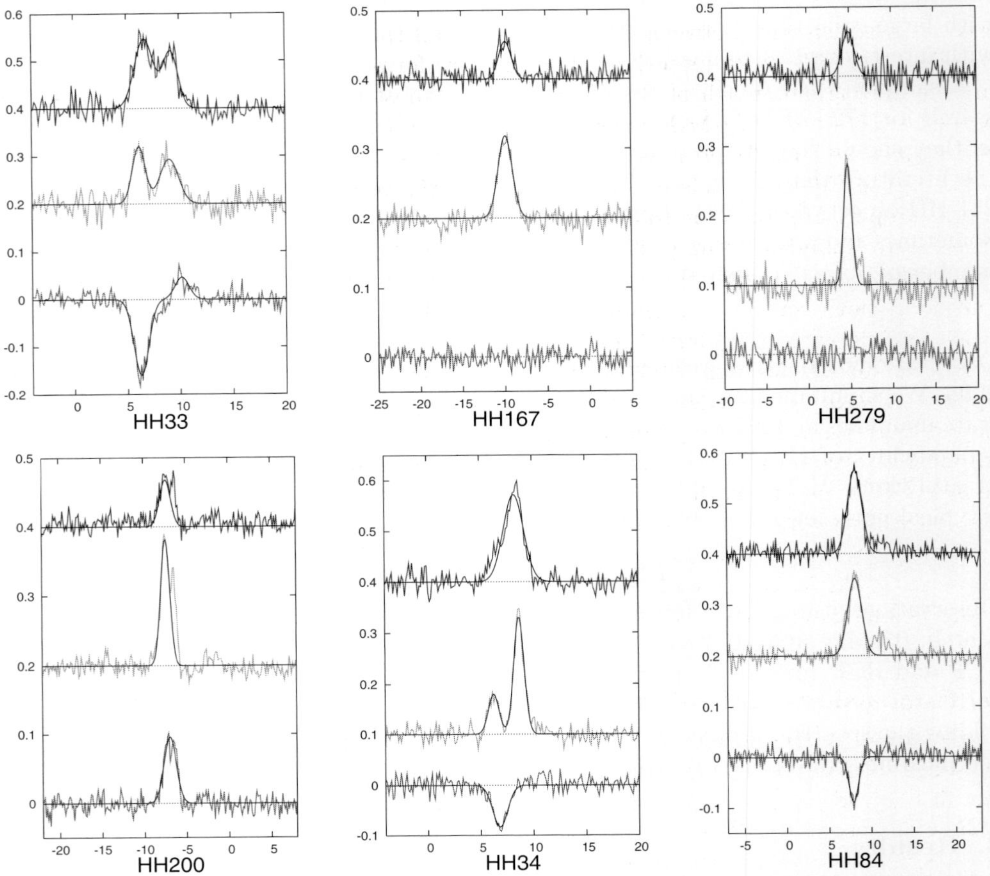

Figure 1. Some example spectra from Greenbank, plotting flux density in Jy against velocity in $\mathrm{km\,s^{-1}}$. In each panel from top bottom the 1665-, 1667- and 1720-MHz data are plotted together with a fitted Gaussian.

appears to have two peaks, both of which are narrow, but they cannot be resolved when fitting Gaussians. In this object, 1720 MHz appears in emission and the mainlines have a flux ratio $F(1667) : F(1665) = 2.60$. HH 34 and HH 84 are examples in which the 1720 MHz is in absorption while the mainlines are in emission. In HH 34 there could be structure in the blue wing of the 1665 MHz line which we cannot resolve; the fitted Gaussians for the 1665 MHz line and the red peak in the 1667 MHz spectrum have a thermal line ratio. The velocity of the 1720 MHz absorption dip is offset from the blue peak in 1667 MHz by $v \sim 0.6\,\mathrm{km\,s^{-1}}$. HH 84 has mainlines with $F(1667) : F(1665) = 0.88$. The fitted Gaussians all peak at slightly different velocities.

The 1720 MHz data from Effelsberg have 48 non-detections, 10 sources have emission, 11 have absorption and 5 have P Cyg-type profiles. 1 out of 5 sources with P Cyg profiles indicate infall rather than outflow. Gaussian fits to the line profiles show similar properties to those from the Greenbank data.

4. Discussion

With flux densities between 0.03 and 0.4 Jy, all the lines are weak, particularly compared to the OH maser intensities found in star-forming regions. Weak OH emission is present throughout much of the Galaxy, but in most cases it is from extended sources with the 1665- and 1667-MHz lines having fluxes indicative of thermal emission. Many of the sources we have found have mainline flux ratios indicative of non-thermal emission.

Using the Effelsberg telescope in January 1986, Wouterloot, Brand & Fiegel (1993) did not report any 1665- or 1667-MHz OH emission to a limit of < 50 mJy from HH288, whereas we found a 60 mJy line at 1665-MHz with $F(1665) > F(1667)$. According to the Lockett *et al.* (1999) theory, shock-excited 1720-MHz OH masers are always weak and hence never saturate. Variability on time scales of weeks to months is quite possible when dealing with unsaturated masers.

Argon, Reid & Menten (2003) reported weak OH masers in the W3 OH region that appear to form at the edges of the outflow of a protostellar source, the Turner-Welch object, which also drives H_2O masers. They suggest that protostellar outflows may be an important mechanism for the production of weak OH masers when no ionizing stars are present. They point out that it is possible, due to bandwidth and sensitivity limitations, that most surveys could have missed weak maser features. Some of the sources mapped by Forster & Caswell (2000) containing both OH and H_2O masers but no continuum emission could be associated with protostellar outflows. The emission we have found could belong to this new class of maser.

5. Conclusions and future work

The maser model developed for SNRs indicate that these masers are produced in the presence of a C-type shock. The shocks in HH objects have a wide range of physical conditions and hence it is possible that some of them could support shock-excited 1720-MHz OH masers. In SNRs soft x-rays are required to dissociate H_2O behind the C-shock to produce a sufficient OH column density to support 1720-MHz masers. The intensity and spectrum of x-rays associated with HH objects still needs to be investigated. Most observed SNR masers lie along the edge of the shock where the maser path length is maximized. Because the jets in HH objects do not necessarily lie in the plane of the sky, detection rates could be limited.

In our OH survey towards HH objects we have not found any unambiguous maser detections but we have identified several potential 1720-MHz and mainline OH masers. Follow-up observations of these sources are required to confirm the presence of masers. If any OH masers are found to be associated with outflows, they should provide useful constraints on the physical conditions in the regions where they occur, and will be used in hydrodynamic models to try and understand what physical and chemical processes are shaping these regions.

References

Argon, A. L., Reid, M. J. & Menten, K. M. 2003, *ApJ* 593, 925

Foster, J. R. & Caswell, J. L. 2000, *ApJ* 530, 371

Lockett, P., Gauthier, E. & Elitzur, M. 1999, *ApJ*, 511, 235

Reipurth, B. 1999 A General Catalogue of Herbig Haro Objects 2nd Edition, http://casa.colorado.edu/hhcat

Smith, M. D., Khanzadyan, T. & Davis, C. J. 2003, *ApJ* 339, 524

Wouterloot, J. G. A., Brand, J. & Fiegel, K. 1993, *A&AS* 98, 589

Astrophysical Masers and their Environments
Proceedings IAU Symposium No. 242, 2007
J.M. Chapman & W.A. Baan, eds.

doi:10.1017/S1743921307012872

Japanese VLBI network observations of 6.7-GHz methanol masers II. Results

Koichiro Sugiyama[1], Kenta Fujisawa[1], Mareki Honma[2], Akihiro Doi[3], Nanako Mochizuki[3], Yasuhiro Murata[3] and Yasuko Isono[1]

[1]Faculty of Science, Yamaguchi University,
1677-1 Yoshida, Yamaguchi, Yamaguchi 753-8512

[2]National Astronomical Observatory of Japan, 2-21-1 Osawa, Mitaka, Tokyo 181-8588

[3]The Institute of Space and Astronautical Science, Japan Aerospace Exploration Agency,
3-1-1 Yoshinodai, Sagamihara, Kanagawa 229-8510

Abstract. We are using a Japanese VLBI network (JVN) for VLBI observations of 6.7-GHz methanol masers associated with massive star-forming regions. Here we present results for Cepheus A (Cep A) from observations taken on September 9, 2006. The distribution of the maser spots indicates either a spherical bubble or a disk having an inclination of several tens degree. We construct a disk model with an inclination of $\sim70°$ and a radius of ~700 AU. From a luminosity of $1.7 \times 10^4\ L_\odot$ for a source observed in this region, the excitation of the maser is probably radiative.

Keywords. masers, stars: formation, HII regions, ISM: individual (Cepheus A)

1. Introduction

Methanol masers at 6.7 GHz provide a useful probe for investigating star formation associated with massive young stellar objects (YSOs). The masers often show linear structures and linear velocity gradients, and these are often interpreted as evidence for edge-on disks (e.g., Pestalozzi *et al.* 2004). However, other morphologies could also explain the maser distributions.

There are 519 known sites of 6.7 GHz methanol maser listed (Pestalozzi *et al.* 2005), while the number of sources observed with VLBI at milli-arcsecond (mas) resolution is, currently, less than 30. We have begun VLBI observations of maser sources at 6.7 GHz with a Japanese VLBI network (JVN; Doi *et al.* 2006) with five antennas, Yamaguchi 32 m, Usuda 64 m, VERA-Mizusawa, Ishigaki and Iriki 20 m.

In this paper, we describe results from observations of Cepheus A (Cep A). The observations were taken on September 9, 2006. Cep A has a CO condensation at a distance of 730 pc (Johnson 1957) and is known to have some massive star forming regions. One of these is an UC HII region, Cep A-HW 2, which is the site of the methanol maser emission. It is also known that a biconical ionized jet exists in Cep A-HW 2 from radio continuum observations (Torrelles *et al.* 1996).

2. Results and discussion

The distribution of maser spots of Cep A (Figure 1) shows an incomplete arc which could be part of a spherical bubble, or a disk with an inclination of several tens degree. We assume here that the distribution of spots traces a tilted disk because of the nearly perpendicular direction to the radio jet. We also assume that the maser spots are excited in regions of the same radiation temperature, i.e. that they are at the same distance from

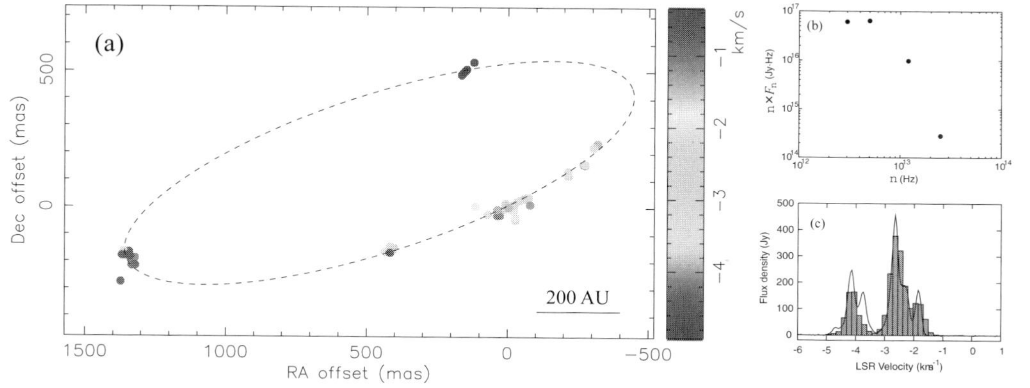

Figure 1. Cep A: (a) distribution of methanol maser at 6.7 GHz. (b) IRAS flux at each wavelength (12, 25, 60, 100 μm) versus frequency for IRAS 22543+6145. (c) The filled box shows the CLEANed component spectrum while the solid line shows the single-dish (total-power) spectrum.

an excitation source. According to this model, we have fitted the distribution with an ellipse and obtained a radius of R ~700 AU with an inclination of ~70°. The radiation temperature at this radius is estimated as follows. We first calculate the luminosity of the (suspected) excitation source of an IRAS point source (IRAS 22543+6145) located at Cep A-HW 2, by using formula (3) from Walsh *et al.* (1997). For a distance to Cep A of 730 pc, a luminosity of $L = 1.7\times10^4\ L_\odot$ is obtained. We can then derive the radiation temperature $T_{\rm d}$ with:

$$T_{\rm d} = (16\pi\sigma)^{-1/4} \cdot R^{-1/2} \cdot L^{1/4} \qquad (2.1)$$

where σ is the Stefan-Boltzmann constant. The temperature obtained, $T_{\rm d}$ ~120 K, agrees well with a theoretical value >100 K (Cragg *et al.* 2002). This is consistent with a single excitation source and the methanol maser spots in Cep A distributed within a disk of diameter 1400 AU.

Acknowledgements

The JVN is led by the National Astronomical Observatory of Japan (NAOJ), Hokkaido University, Gifu University, Yamaguchi University, and Kagoshima University, in co-operation with Geographical Survey Institute (GSI), the Japan Aerospace Exploration Agency (JAXA), and the National Institute of Information and Communications Technology (NICT).

References

Cragg, D. M., Sobolev, A. M., & Godfrey, P. D. 2002, *MNRAS*, 331, 521
Doi, A., *et al.* 2006, Proc. *8th European VLBI Network Symposium*, PoS(8thEVN) 071
Johnson, H. L. 1957, *ApJ*, 126, 121
Pestalozzi, M. R., Elitzur, M., Conway, J. E., & Booth, R. S. 2004, *ApJ*, 603, L113
Pestalozzi, M. R., Minier, V., & Booth, R. S. 2005, *A&A*, 432, 737
Torrelles, J. M., Gomez, J. F., Rodriguez, L. F., Curiel, S., Ho, P. T. P., & Garay, G. 1996, *ApJ*, 457, L107
Walsh, A. J., Hyland, A. R., Robinson, G., & Burton, M. G. 1997, *MNRAS*, 291, 261

Astrophysical Masers and their Environments
Proceedings IAU Symposium No. 242, 2007
J.M. Chapman & W.A. Baan, eds.

doi:10.1017/S1743921307012884

Where methanol masers spring

Karl J. E. Torstensson[1,2], Huib Jan van Langevelde[2,1] and Stephen Bourke[3]

[1]Leiden Observatory
P.O. Box 9513, 2300 RA Leiden, The Netherlands
email: kalle@strw.leidenuniv.nl

[2]Joint Institute for VLBI in Europe
Postbus 2, 7990 AA Dwingeloo, The Netherlands
email: langevelde@jive.nl

[3]National University of Ireland, Galway
Department of Information Technology, National University of Ireland, Galway, Ireland
email: stephen@it.nuigalway.ie

Abstract. We are carrying out a program to observe the 6.7 GHz methanol maser emission in high-mass star-forming regions using large FOV (~2'), astrometric, VLBI data. Here we report on the first results of the inner few arc seconds in Cepheus A East. We find a maser filament extending over ~1.7" (1200 AU), straddling the waist of Cep A HW2. The region in which the CH_3OH masers are found contains several YSO's and it is not clear whether the CH_3OH masers are associated with several different objects or rather the larger scale surrounding environment.

Keywords. masers, stars: formation, stars: individual (Cepheus A)

1. Introduction

At a distance of 725 pc, Cepheus A East is a nearby complex high-mass star-forming region. At large scale several outflows are seen in many molecular species (e.g. Codella *et al.* 2005). In the centimetre regime the continuum emission is dominated by several compact sources and the bipolar jets associated with Cep A HW2. It is not yet clear how many of the compact sources are associated with individual YSO's. Also, in the region close to HW2 several YSO's have been reported in a number of tracers (Martín-Pintado *et al.* 2005, Curiel *et al.* 2002, Patel *et al.* 2005, Brogan *et al.* 2007). In this poster we report on recent results of the 6.7 GHz methanol masers in the region of Cep A HW2.

2. Observations

The observations were carried out in November 2004 with 8 telescopes of the EVN. Cep A was observed for ~2h including calibrators. The observations were split into two 1h blocks, separated by a few hours to improve the uv-coverage. The experiment was setup at a centre frequency of 6668.5142 MHz (rest frequency) with 2 MHz bandwidth and 1024 channels, resulting in a velocity resolution of ~0.1 km s^{-1}. Both RCP and LCP were recorded. The data was correlated at JIVE with an integration time of 0.25 s.

All calibration and imaging was done in AIPS. Amplitude and bandpass calibration was done on 3C345 and DA193, observed at the beginning and end of each block. For phase calibration and initial self-calibration J2302+6405 was observed every three minutes. The brightest maser feature was then used to determine the absolute position and as a source for further self-calibration, resulting in an rms of ~3.5 mJy $beam^{-1}$ in the emission free channels.

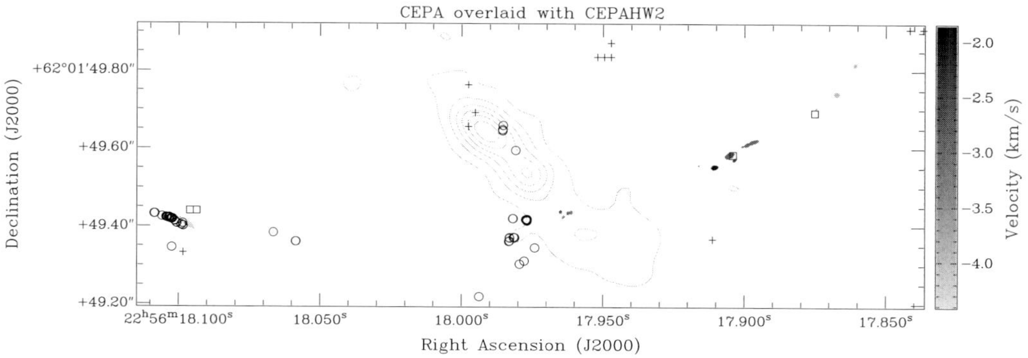

Figure 1. The 6.7 GHz CH_3OH masers (grey scale) overlaid on K band continuum (contours) (Torrelles *et al.* 1998). Also shown are the 12.2 GHz CH_3OH masers (boxes) (Minier *et al.* 2001), water masers (circles) (Vlemmings *et al.* 2006) and OH masers (pluses) (Bartkiewicz *et al.* 2005).

3. Discussion

In Cep A we find the CH_3OH maser emission to be arising from an extended ∼1.7" (∼1200 AU) narrow filamentary structure, which seems to suggest a single underlying structure. Across the whole structure there is a small velocity gradient but without obvious signs of rotation. The maser positions to the west correspond well to the positions of the 12.2 GHz CH_3OH masers measured by Minier *et al.* (2001). To the east the positional agreement is poorer, this could simply be an effect of observations at different epochs. Overall, the velocity of the 6.7 GHz masers agrees well with that of the 12.2 GHz masers.

In all regions except one the CH_3OH masers are spatially offset from other maser species in the region (water and OH). Interestingly, the methanol masers to the east are spatially coincident with the water masers observed by Vlemmings *et al.* (2006), although they show a velocity offset of ~ 10 km s^{-1}. The weak maser emission arising from the very waist of HW2 is very interesting as it may point toward the very object responsible for the outflows/jets. It is however not coincident with the 7 mm continuum object identified by Curiel *et al.* (2006), but a position further to the SW in the jet.

In fact, none of the proposed YSO's in this region (Martín-Pintado *et al.* 2005, Curiel *et al.* 2002, Patel *et al.* 2005, Brogan *et al.* 2007) show a direct, small scale, association with any CH_3OH maser. If the masers are arising in a single structure it seems difficult to attribute their origin to a single object. It is remarkable that a structure encompassing several young stars has such a small velocity gradient.

References

Bartkiewicz, A., Szymczak, M., Cohen, R. J., *et al.* 2005, *MNRAS* 361, 623

Brogan, C. L., Chandler, C. J., Hunter, T. R., *et al.* 2007, *arXiv:astro-ph/0703626v1*

Codella, C., Bachiller, R., Benedettini, M., *et al.* 2005, *MNRAS* 361, 244

Curiel, S., Trinidad, M. A., Cantó, J., *et al.* 2002, *ApJL* 564, L35

Curiel, S., Ho, P. T. P., Patel, N. A., *et al.* 2006, *ApJ* 638, 878

Martín-Pintado, J., Jiménez-Serra, I., Rodríguez-Franco, A., *et al.* 2005, *ApJL* 628, L61

Minier, V., Conway, J. E. & Booth, R. S. 2001, *A&A* 369, 278

Patel, N. A., Curiel, S., Sridharan, T. K., *et al.* 2005, *Nature* 437, 109

Torrelles, J. M., Gómez, J. F., Garay, G., *et al.* 1998, *ApJ* 509, 262

Vlemmings, W. H. T., Diamond, P. J., van Langevelde, H. J., *et al.* 2006, *A&A* 448, 597

Astrophysical Masers and their Environments
Proceedings IAU Symposium No. 242, 2007
J.M. Chapman & W.A. Baan, eds.

doi:10.1017/S1743921307012896

Water maser and radio continuum emission towards IRAS 23139+5939

M. A. Trinidad[1], S. Curiel[2], J. M. Torrelles[3], L. F. Rodríguez[4], V. Migenes[1] and N. Patel[5]

[1]Depto. Astronomia, Univ. de Guanajuato, Apdo. Postal 144, Guanajuato, Gto. Mexico
email: trinidad@astro.ugto.mx

[2]Instituto de Astronomia, UNAM, Apdo. Postal 70-264, D.F. 04510, Mexico

[3]Instituto de Ciencias del Espacio and Institut d'Estudis Espacials de Catalunya, Facultat de Fisica, Planta 7a, Universitat de Barcelona, Av. Diagonal 647, 08028 Barcelona, Spain

[4]CRyA, UNAM, Apdo. Postal 3-72 (Xangari) 58089 Morelia, Michoacan, Mexico

[5]Harvard-Smithsonian Center for Astrophysics, 60 Garden Street, Cambridge, MA 02138, USA

Abstract. We present simultaneous observations of continuum (3.5 and 1.3 cm) and water maser line emission (1.3 cm) carried out with the VLA-A toward the high-mass object IRAS 23139+5939. We detected two radio continuum sources at 3.5 cm separated by $0''.5$ ($\sim$ 2400 AU), I23139 and I23139S. Based on the observed continuum flux density and the spectral index, we suggest that I23139 is a thermal radio jet associated with a high-mass YSO. On the other hand, based on the spatio-kinematical distribution of the water masers, together with the continuum emission information, we speculate that I23139S is also a jet source powering some of the masers detected in the region.

Keywords. ISM: jets and outflows, masers

1. Introduction

There is a deficit in the detection of thermal radio jets in high-mass YSOs with just a few clear cases (Cepheus A-HW2: Rodriguez *et al.* 1994, Curiel *et al.* 2006) and therefore it is unclear whether molecular outflows associated with high-mass objects also include at their base highly collimated thermal radio jets as low-mass stars do. In order to address this issue, we study the source IRAS 23139+5939, which is a high-mass star-forming region located near S157 at a distance of 4.8 kpc and has a luminosity of $\sim 2 \times 10^4$ $L_\odot$ (Sridharan *et al.* 2002).

2. Observations and results

The observations were made with the VLA of the National Radio Astronomy Observatory (NRAO)† in its A configuration. We observed simultaneously 1.3 cm continuum and water maser emission. The bandwidth for the line observations was centered at the frequency of the H_2O $6_{16} \rightarrow 5_{23}$ maser line with $V_{LSR} = -52.7$ km s^{-1}, covering from -73.1 to -32.3 km s^{-1} in velocity. The maser line and continuum data were reduced with standard techniques using the NRAO AIPS software package.

A radio continuum source, I23139, is only detected at 1.3 cm in the region, while two continuum sources, I23139 and I23139S, are detected at 3.5 cm (see Figure 1). The two

† NRAO is a facility of the National Science Foundation operated under cooperative agreement by Associated Universities Inc.

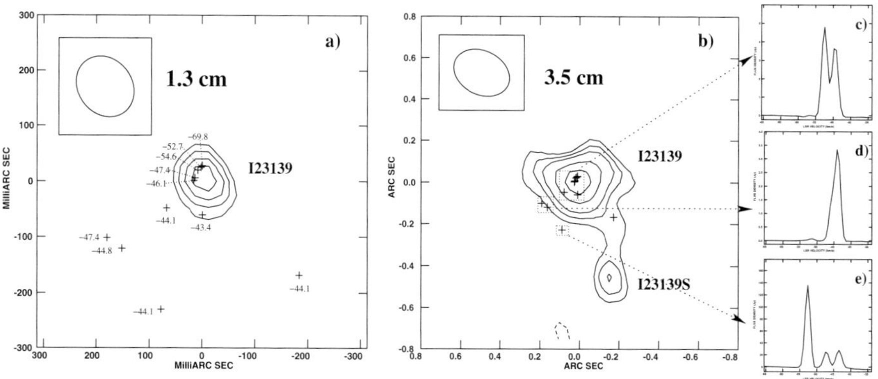

Figure 1. a) VLA 1.3 cm contour map toward IRAS 23139+5939. Contours are -4, -3, 3, 4, 5, and 6 times 140 μJy beam^{-1}, the rms noise of the map. The beam size is $0\farcs12 \times 0\farcs10$. The crosses show the positions of the water masers observed in the region. b) VLA contour map of the continuum emission at 3.5 cm toward IRAS 23139+5939. Contours are -4, -3, 3, 4, 5, 7, and 9 times 43 μJy beam^{-1}, the rms noise of the map. The beam size is $0\farcs30 \times 0\farcs19$. In both panels the angular offset is measured with respect to the I23139 peak position at 1.3cm. In order to be able to coincide both 1.3 and 3.5 cm peaks of I23139, we have applied an offset of 0.03" (northwest) to the position of I23139 at 3.5 cm, which is within the positional errors. c), d) and e) Spectra of the water masers enclosed in the box marked in panel b).

sources are separated by $0\farcs5$ ($\sim$ 2400 AU), with I23139S a new detection. No previous spectral index information for I23139 has been available. From the flux densities at 1.3 and 3.5 cm, we estimate a spectral index ($S_\nu \propto \nu^\alpha$) of 0.64 ± 0.36 for the source I23139, which is consistent with free-free emission from a thermal radio jet (Reynolds 1986).

We also find water maser emission toward IRAS 23139+5939, which is mainly distributed in two groups. One group, with seven masers, is spatially associated with I23139, while the other, with three masers, may be associated with I23139S. The precision of the relative positions between the radio continuum sources and the water masers is of the order of the 10 milliarcseconds. The water masers associated with I23139 trace an arc-like structure, while those associated with I23139S trace a linear structure. From the spatial and velocity distribution of the water masers, both groups of masers probably trace expanding motions. This appears to be consistent with Goddi *et al.* (2005), who measured proper motions of two water masers toward this region and found that the proper motions could be tracing the inner part of a conical outflow. In addition, I23139 is the best candidate for the driving source of the extended molecular outflow observed in the region. On the other hand, the second maser group seems to be pointing towards the source I23139S, which is located $0\farcs35$ to the southwest from the masers. This allows us to speculate that these masers could be associated with, and are pumped by, a jet of I23139S, although we do not rule out I23139, or even another undetected embedded source, as the pumping source.

References

Curiel, S., Ho, P. T. P., Patel, N. A., Torrelles, J. M., Rodríguez, L. F., *et al.* 2006, ApJ, 638, 878

Goddi, C., Moscadelli, L., Alef, W., Tarchi, A., Brand, J., & Pani, M. 2005, A&A, 432, 161

Reynolds, S. P. 1986, ApJ, 304, 713

Rodríguez, L. F., Garay G., Curiel, S. Ramírez, S., Torrelles J. M., *et al.* 1994, ApJ, 430, L65

Sridharan, T. K., Beuther, H., Schilke, P., Menten, K. M., & Wyrowski, F. 2002, ApJ, 566, 931

Astrophysical Masers and their Environments
Proceedings IAU Symposium No. 242, 2007
J.M. Chapman & W.A. Baan, eds.

doi:10.1017/S1743921307012902

The Australia Telescope campaign to study southern class I methanol masers

M. A.Voronkov[1], K. J. Brooks[1], A. M. Sobolev[2], S. P. Ellingsen[3], A. B. Ostrovskii[2] and J. L. Caswell[1]

[1]Australia Telescope National Facility, PO Box 76, Epping, NSW 1710
email: Maxim.Voronkov@csiro.au

[2]Ural State University, Lenin ave. 51, 620083 Ekaterinburg, Russia

[3]University of Tasmania, GPO Box 252-37, Hobart, Tasmania 7000, Australia

Abstract. The Australia Telescope Compact Array (ATCA) and the Mopra facility have been used to search for new southern class I methanol masers at 9.9, 25 (J = 5) and 104 GHz, which are thought to trace more energetic conditions in the interface regions of molecular outflows, than the widespread class I masers at 44 and 95 GHz. One source shows a clear outflow association.

Keywords. Methanol masers, outflows

A number of surveys have been carried out to investigate various class I methanol masers and to search for an association with the other constituents of a typical massive star-forming region. The ATCA 9.9 GHz and Mopra 104 GHz surveys indicate that these masers are rare (three detections out of 40 targets and two detections out of 69 targets, respectively) confirming the results of the previous searches (Slysh *et al.* 1993; Voronkov *et al.* 2005). In contrast, there were 67 detections at 25 GHz (J = 5) out of 102 targets observed (ATCA observations, but no accurate positions were measured for the majority of the sources), although these masers were also previously believed to be rare following a survey of Menten *et al.* (1986). Unlike the spectra of the 44 GHz masers, the spectra of 9.9, 25 and 104 GHz masers appeared to be quite simple (small number of components) and no significant correlation between the flux densities has been found. This is in line with the theoretical prediction that these masers require more energetic conditions (higher temperatures and densities) to form (Sobolev *et al.* 2005). Only one source, G343.12-0.06 (IRAS 16547-4247), was active in all observed transitions. Its detailed observations showed a remarkable association with a jet-driven molecular outflow (see Voronkov *et al.* 2006 for details). The spectra of detections are given in Fig. 1. The 25 GHz spectra are given only for the sources with measured accurate positions. A table with the absolute positions of maser spots is available in the poster which is available from the M. Voronkov, or can be downloaded from the maser meeting website.

References

Menten, K. M., Walmsley, C. M., Henkel, C., & Wilson, T. L. 1986, *A&A* 157, 318

Slysh, V. I., Kalenskii, S. V., & Val'tts, I. E. 1993, *ApJ* 413, L133

Sobolev, A. M., Ostrovskii, A. B., Kirsanova, M. S., Shelemei, O. V., Voronkov, M. A., & Malyshev, A. V. 2005, in: R. Cesaroni, E. Churchwell, Walmsley M., *Massive star birth: A crossroads of Astrophysics*, Proc. IAU Symposium No. 227 (Cambridge: Cambridge University Press), p. 174

Voronkov, M. A., Sobolev, A. M., Ellingsen, S. P., Ostrovskii, A. B., & Alakoz A. V. 2005, *Ap&SS* 295, 217

Voronkov, M. A., Brooks, K. J., Sobolev, A. M., Ellingsen, S. P., Ostrovskii, A. B., & Caswell, J. L. 2006, *MNRAS* 373, 411

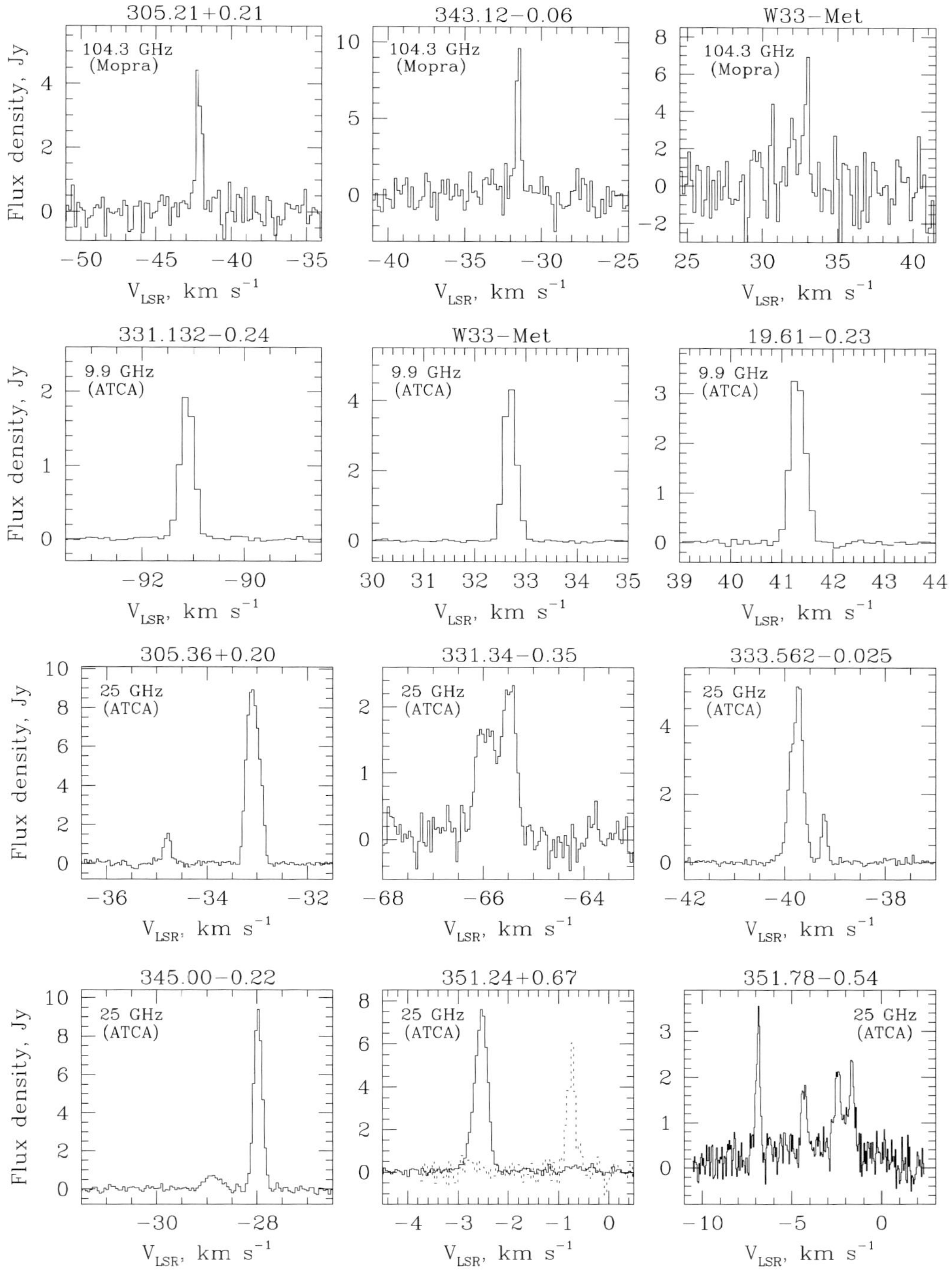

Figure 1. The spectra of the 104 GHz (top row), 9.9 GHz (second row) and 25 GHz (two bottom rows) maser detections. The 25 GHz spectra are given only for the sources with measured accurate positions. The dotted (blue) spectrum for G351.24+0.67 corresponds to an offset component (see the poster referred to in the text).

Astrophysical Masers and their Environments
Proceedings IAU Symposium No. 242, 2007
J.M. Chapman & W.A. Baan, eds.

doi:10.1017/S1743921307012914

Imaging of selected sources from the Methanol Multibeam Survey

D. Wong-McSweeney[1], G. A. Fuller, S. Etoka

School of Physics and Astronomy, University of Manchester, Sackville Street Building, Sackville Street, PO Box 88, Manchester, M60 1QD
[1]email: d.wong-mcsweeney@postgrad.manchester.ac.uk

Abstract. High angular resolution observations are essential for understanding the nature of maser emission and the sources which excite it. Here we present preliminary results from MERLIN observations of three methanol masers from the Toruń survey. These MERLIN observations are being analysed as part of the interferometric component of the Methanol Multibeam (MMB) Survey which is surveying the Galactic plane at $|b| \leqslant 2°$ for 6.67 GHz methanol maser sources.

Keywords. masers, surveys, stars: formation, ISM: individual (G20.24+0.07, G22.45-0.17, G23.01-0.41) galaxies: individual (Milky Way)

1. Introduction

The process of high mass star formation is poorly understood relative to that for low mass stars. The 6.67 GHz methanol maser is known to be associated exclusively with regions of massive star formation and therefore searching for this maser will aid in the location and further study of these regions. The Galactic Methanol Multibeam Survey (MMB, Green *et al.* 2007) aims to find all (to a noise level of ~0.2 Jy) of the 6.67 GHz methanol masers in the Galactic plane in the range $|b| \leqslant 2°$. In parallel with single dish observations made using the Parkes telescope and the Lovell telescope at Jodrell Bank, interferometer observations using the ATCA and MERLIN are being used to determine source positions more accurately. An earlier methanol maser survey (Szymczak *et al.* 2002), using the 32 m dish at Toruń in Poland, searched the region $20° \leqslant l \leqslant 40°$, $|b| \leqslant 0°.52$ of the Galactic plane, an area that is being covered by the MMB survey. Of the 100 sources found in the Toruń survey, 60 needed interferometric follow up,of which 44 have now been observed. The preliminary results of the analysis of three sources, G20.24+0.07, G22.45 − 0.17, and G23.01 − 0.41 from both archive and new MERLIN observations, are presented here. The high angular resolution of these observations (~0.04″) allows the reliable association of the masers with near and mid IR sources. The aim of this project is to determine the properties of these sources and the relationship between different maser species.

2. Sources and results

All three sources show complex velocity structure in their methanol spectra. G22.45-0.17, with the most complex velocity structure, has eight velocity components whose peaks are separated by ~2 km s^{-1}, and spread over a range of ~18 km s^{-1}. These features correspond to four spatial components (Figure 1). G23.01 − 0.41, conversely, has a more simple velocity structure and a more complex spatial structure. G20.24+0.07 has been resolved from a single source in the Toruń survey, into two sources separated by ~8.6″ in the north-east south-west direction, and ~1 km s^{-1} in velocity.

The sources are at near/far distances of 6/~12, 5/12.8 and 2.7/15.8 kpc for G20.24 +0.07, G22.45-0.17, and G23.01-0.41 respectively (Caswell & Haynes 1983).

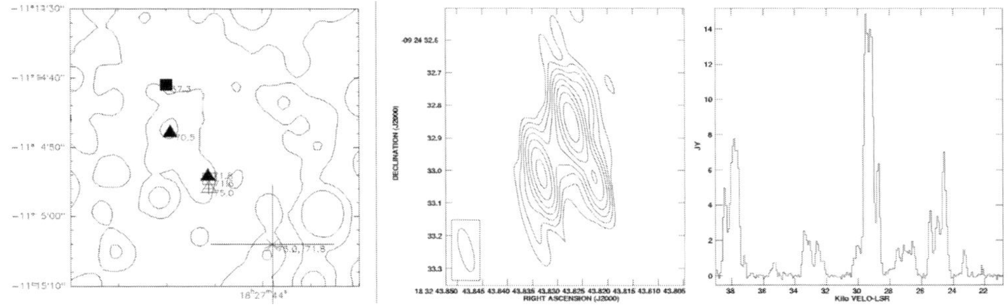

Figure 1. Left: G20.24+0.07 region, RA/Dec in J2000, Grey and contours - 2MASS K-band 2.2 μm. Masers: ▲ MERLIN 6.67 GHz CH_3OH; ■ 22 GHz H_2O; △ 1720 MHz OH; □ 6030/6035 MHz OH; 3 line cross 4660/4750/4765 MHz OH; Diagonal Cross 12 GHz CH_3OH; The size of the '+'s indicate the position accuracy. Velocities, in km s^{-1}, are marked where known. Refs: Caswell *et al.* 1995, Caswell *et al.* 2000, Caswell 2003, Smitts 2003, Szymczak & Gérard 2004, Testi *et al.* 1998. **Middle:** G22.45-0.17 MERLIN 6.67 GHz integrated intensity map with beam shape in the lower left. **Right:** G22.45-0.17 MERLIN spectrum with velocity in km s^{-1}.

Both G22.45-0.17 and G23.01-0.41 are seen in the J=2–1 transition of SiO, which is suggestive of shocked material (Harju *et al.* 1998) as SiO is usually found on grains rather than as a gas. Of the three sources only G20.24+0.07 is coincident with K-band, 2.2μm 2MASS emission, and it is also found to have emission from $C^{32}S$ (J=2–1) which is a tracer of dense gas. The $C^{32}S$ is observed at a velocity of 71.9 km s^{-1} which is comparable to the other maser species in the region with the exception of the 22GHz water maser (Szymczak *et al.* 2005) suggesting that this traces different material, for example an outflow. Only G20.24+0.07 is associated with an IRAS point source (IRAS 18249-1116), which has a luminosity of $2.8 \times 10^4 L_\odot$ (Larionov *et al.* 1999).

Acknowledgements

Thanks to MMB collaborators: Jim Cohen, Phil Diamond, Malcolm Gray, Jimi Green, Lyshia Quinn, Manchester; Antonio Chrysostomou, Michele Pestalozzi, Mark Thompson, Hertfordshire; Melvin Hoare, Leeds; Mike Masheder, Bristol; Derek Ward-Thompson, James Cox, Cardiff; Jeremy Yates, University College, London; James Caswell, Kate Brooks, Naomi McClure-Griffiths, Maxim Voronkov, Chris Phillips, ATNF, CSIRO; Michael Burton, New South Wales; Andrew Walsh, James Cook; Simon Ellingsen, Tasmania.

References

J. L. Caswell & R. F. Haynes 1983, *Aus.J.Phy* 36, 417
J. L. Caswell, R. A. Vaile, S. P. Ellingsen & R. P. Norris 1995, *MNRAS* 274, 1126
J. L.Caswell, Jiynue Yi, R. S. Booth, & D. M.Cragg 2000, *MNRAS* 313, 599
J. L. Caswell 2003, *MNRAS* 341, 551
J. Green, R. J. Cohen, J. L. Caswell, G. A. Fuller *et al.* 2007, *This volume*
J. Harju, K. Lehtinen, R. S. Booth, & I. Zinchenko 1998, *A&A* 132, 211
G. M. Larionov, I. E. Val'tts, A. Winnberg, L. E. B. Johansson, *et al.* 1999, *A&A* 139, 257
D. P. Smitts 2003, *MNRAS* 339, 1
M.Szymczak, A. J. Kus, G. Hrynek, A. Kepa & E. Pazderski 2002, *A&A* 392, 277
M. Szymczak & E. Gérard 2004, *A&A* 414, 235
M. Szymczak, T. Pillai, & K. M. Menten 2005, *A&A* 434, 613
L. Testi, M. Felli, P. Persi, & M. Roth 1998, *A&A* 129, 495

Astrophysical Masers and their Environments
Proceedings IAU Symposium No. 242, 2007
J.M. Chapman, & W.A. Baan, eds.

doi:10.1017/S1743921307012926

Spectral and VLBI-structure monitoring of an OH maser flare in W75N

A. V. Alakoz[1], V. I. Slysh[1] and V. Migenes[2]

[1]Astro Space Center, Lebedev Physical Institute, Profsouyzanya 84/32, Moscow, Russia
email: rett@asc.rssi.ru; vslysh@asc.rssi.ru

[2]Departamento de Astronomia, Universidad de Guanajuato, Apdo. Postal 144, Guanajuato, CP 36000, GTO Mexico
email: vmigenes@astro.ugto.mx

Abstract. We present VLBI and single dish observations of a strong (1000 Jy) OH maser flare at 1665 MHz in the star-forming region W75N. This new flare appeared in 2003, after a 100 Jy flare in 2000. Three major spectral features N_1, N_2, and N_3 were present at all epochs, with N_1 declining and N_2 increasing in flux density during observations taken in 2003–2006. All three features have strong linear polarization approaching 90 %. VLBI maps show that along with the strong features N_{1-3}, many weaker features have appeared within 100 mas of the ultra compact HII region VLA 2. The proper motion of the flare feature N_1 was measured: $\Delta\alpha = 1.36 \pm 0.3$; $\Delta\delta = -0.3 \pm 0.3$ mas in 1.75 years, corresponding to a tangential velocity of about 8 km s^{-1} relative to the rest of the features, and is comparable to the radial velocity range in the VLA 2 disk.

Keywords. ISM: individual (W75 N) – masers – stars: formation – radio lines: ISM – ISM: molecules – ISM: kinematics and dynamics

1. Introduction

OH maser spots in W75N are grouped along two almost perpendicular straight-line segments, with the HII regions located approximately in the middle of the lines. The overall picture resembles two edge-on discs, which contain OH masers and are centered on VLA1 and VLA2. VLBA observations in November 2000 and January 2001 have shown that two new spectral features appeared with a flux density 145 and 35 Jy. One can suppose that the flare of these features (called P_1 and P_2) was a precursor of a stronger flare which was first observed 2.5 years later, on October 24, 2003 with the Nançay radio telescope (Alakoz *et al.* 2005). This time very strong features N_1 and N_2, with a flux density for N_1 up to 1,000 Jy, have emerged in the spectrum of 1665 MHz line. This is possibly a record flux in the history of observations of OH masers. In early 2004 we started a spectral monitoring campaign with the 64-m Kalyazin radio telescope. Also, we have used new EVN 1665 MHz multi-epoch observations in combination with EVN and VLBA archive data (Slysh & Migenes 2006, Fish *et al.* 2005, Fish & Reid 2007) to measure absolute positions and proper motions of flaring OH maser components. In this paper we present a short summary of the monitoring results.

2. Results

Fig. 1 shows 1665 MHz RCP spectra of W75N for the period from July 2004 to July 2006. Strong variability, by about two order of magnitudes was observed in the flare features N_1 – N_4, without significant changes in the other features. Mapping of the maser has revealed that N_1 – N_4 are new features, not coinciding in position and radial

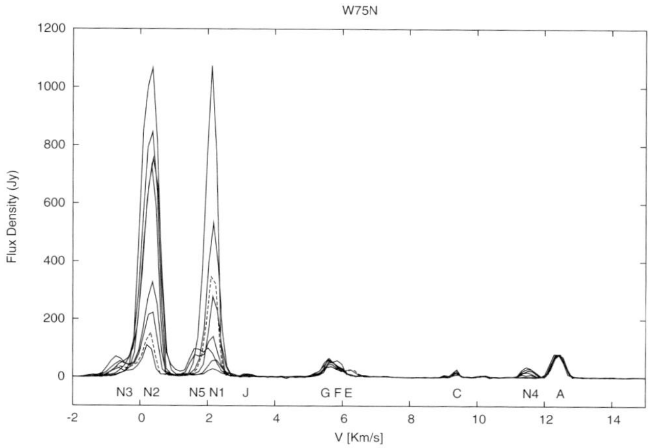

Figure 1. Overlapped spectra from eight observing sessions in Kalyazin (Jul. 2004 – Jul. 2006). 1665 MHz maser emission in RCP.

Table 1. Proper motion of OH maser spots in W75N from September 2004 to June 2006

Maser Spot	N1	N2	N3
ΔR.A., (mas)	1.36	−0.41	−0.8
ΔDec., (mas)	−0.31	1.29	3.6

velocity with P1 and P2, which have already faded and are now not detectable. The proper motion measurements exclude a possibility that they are the same features, with large tangential velocity. In Table 1 the variation of the relative distance between the strongest features in two years, from 2004 to 2006, is given. The relative displacement of the maser spots is very small (Table 1), with a maximum value of 1.4 ± 0.3 milliseconds corresponding to a transverse velocity 7.8 ± 1.7 km s^{-1}. For N_3 the errors are larger, about 2 mas because of the extended image of this feature., and the values given in Table 1 must be considered as upper limits. Therefore, the maser spots are stationary or slowly moving objects. Such a velocity is consistent with a Keplerian velocity in the circumstellar disc, but is too small for a shock or stellar wind. The OH maser flare can be due to a MHD-shock travelling from the young star and energizing maser emission in clumps encountered on its way out.

Acknowledgements

We would like to acknowledge the staff of EVN, NRAO, Kalyazin and Nançay Observatories who made these observations possible. This research was conducted under partial support from the Russian Foundation for Basic Research (project 04-02-17057).

References

Alakoz, A. V., Slysh, V. I., Popov, M. V., & Val'tts, I. E. 2005, *Astron. Lett.*, 31, 375
Fish, V. L., & Reid, M. J. 2007, *ApJ*, 656, 943
Fish, V.L., Reid, M.J., Argon, A.L., & Zheng, X.-W. 2005, *ApJS*, 160, 220
Slysh, V. I., & Migenes, V. 2006, *MNRAS*, 369, 1497

Astrophysical Masers and their Environments
Proceedings IAU Symposium No. 242, 2007
J.M. Chapman & W.A. Baan, eds.

doi:10.1017/S1743921307012938

A MERLIN study of 6 GHz excited OH & 6.7 GHz methanol masers in ON1

James A. Green†,
A. M. S. Richards, H. Flood, W. H. T. Vlemmings and R. J. Cohen

University of Manchester, Jodrell Bank Observatory, Macclesfield, Cheshire, SK11 9DL, UK

Abstract. MERLIN observations of 6.668-GHz Methanol and 6.035-GHz OH emission from the known massive star-formation region ON1 are presented. Maser components are found to lie at the southern edge of the UCHII with consistent polarization angles across the strongest features. Zeeman splitting of OH shows magnetic field strengths between +0.4 to −5.3 mG and from cross-correlation a tentative methanol magnetic field of −18mG is detected.

Keywords. Star-Formation, Masers, Magnetic Fields, Polarization, ISM: individual (ON1)

1. Introduction

ON1 is an ultra-compact HII region (UCHII) associated with the IRAS source 20081+3122 and is believed to contain a central binary massive proto-star of type B0.3 surrounded by a young stellar cluster. Its near kinematic distance is 1.8 kpc. Recent MERLIN studies of Nammahachak *et al.* (2006) identified 40 groundstate OH maser features with a linear distribution at the southern edge of the UCHII. No previous high resolution 6.668-GHz methanol emission observations exist for ON1, but Szymczak, Hrynek & Kus (2000) observed the associated IRAS source with the 32m Torun dish and found emission with a peak flux of 109 Jy.

MERLIN observations at 6.031, 6.035 and 6.668 GHz were taken in January 2005. Observations were taken in dual polarization with a 0.5 MHz bandwidth over 512 frequency channels, giving velocity widths of ~0.05 km s^{-1}.

2. Results

10 methanol components were observed with velocities between 14.2 km s^{-1} and 15.6 km s^{-1}. 1 OH RHC and 1 LHC component were found at 6.031 GHz with velocities of 13.87 km s^{-1} and 14.19 km s^{-1} respectively. 10 OH LHC and 10 RHC components were found at 6.035 GHz with velocities between 13.52 km s^{-1} and 15.50 km s^{-1}. In concurrence with the groundstate OH, the strongest components lie at the southern edge of the UCHII, perpendicular to the $H^{13}CO^+$ outflow identified by Kumar, Tafalla & Bachiller (2004). The methanol components cover a range of about 0.6″, whilst the OH show a wider spread across about 1.3″, with one component, perpendicularly offset, by about 0.5″. The RHC and LHC Zeeman pair at 6.031 GHz coincides with a 6.035-GHz Zeeman pair to <5.5 mas spatially and <0.049 km s^{-1} in velocity, implying co-propagation. The closest coincidence between methanol and OH is one pair to about 25 mas, with a velocity coincidence of <0.18 km s^{-1}.

† Email: greenjimi@gmail.com

2.1. *Polarization*

The excited OH at 6.035 GHz showed linear polarization varying between 0.9% and 18.5%, whilst the 6.031-GHz frequency had a value of 1.2%. The methanol in comparison varied between 0.2% and 3.6%, with a mean value of 1.2%. The level of methanol polarization is consistent with that found in W3(OH) by Vlemmings, Harvey-Smith & Cohen (2006) and that found in NGC6334F by Ellingsen (2002), where both had the majority of sources displaying less than 5% linear polarization.

The polarization angles of the strongest linearly polarized OH components are consistent across the region, aligned to within 25° of one another. The polarization angle is also consistent with the line of maser distribution (which is itself perpendicular to the known $H^{13}CO^{+}$ outflow).

2.2. *Magnetic fields*

OH Zeeman pairs were confirmed by close spatial association of opposite circular components, all to within 6 mas. Altogether seven pairs were found, six at 6.035 GHz and one at 6.031 GHz. Splitting factors of 0.0564 $\mathrm{kms^{-1}\,mG^{-1}}$ and 0.0790 $\mathrm{km\,s^{-1}\,mG^{-1}}$ were assumed respectively Fish *et al.* (2006). This gave a magnetic field strength ranging from +0.4 to −5.3 mG for the pairs at 6.035 Ghz and −3.94 mG for the single pair at 6.031 GHz. All the magnetic fields are directed towards us, with the exception of one, and they concur in magnitude with those found previously by Desmurs & Baudry (1998), which varied between −3.6 mG and −6.3 mG. They also compare favourably with the field strengths found for the ground state OH by Nammahachak *et al.* (2006), which varied between −0.4 and −4.6 mG.

For methanol we employed the cross-correlation method of Modjaz *et al.* (2005), modified as per Vlemmings, Harvey-Smith & Cohen (2006). Compared with OH, methanol has a much lower Zeeman splitting factor of 0.0493 $\mathrm{km\,s^{-1}\;G^{-1}}$. Of the two brightest methanol features, the weaker of the two demonstrated a field strength of −18 mG. Without the dynamic range effect this would be a clear detection, however because of it we find the result is at exactly 3σ. As such, the result can be taken as an upper limit, but could represent the first tentative detection of Zeeman splitting in 6.7 GHz methanol.

Acknowledgements

J. A. G. is supported by a PPARC studentship. The authors would like to dedicate this poster paper to the memory of R. J. Cohen.

References

Baudry, A. & Desmurs, J. F. 2002, *A&A* 394, 107
Baudry, A. & Diamond, P. J. 1991, *A&A* 247, 551
Desmurs, J. F. & Baudry, A. 1998, *A&A* 340, 521
Ellingsen, S. P. 2002 in: V. Mineese & M. Reid (eds.), *Cosmic Masers: From Proto-Stars to Black Holes* (IAU Symposium 206), p. 151
Fish, V. L., Reid, M. J., Menten, K. M., Pillai, T. 2006, *A&A* 458, 485
Kumar, M. S. N., Tafalla, M., & Bachiller, R. 2004, *A&A* 426, 195
Kurtz, S., Hofner, P., & Alvarez, C. V. 2004, *ApJSS* 155, 149
Modjaz, M., Moran, J. M., Kondratko, P. T., & Greenhill, L. J. 2005, *ApJ* 626, 104
Nammahachak, S., Asanok, K., Hutawarakorn Kramer, B., Cohen, R. J., Muanwong, O., Gasiprong, N. 2006, *MNRAS* 371, 1550
Szymczak, M., Hrynek, G., & Kus, A. J. 2000, *A&AS* 143, 269
Vlemmings, W. H. T., Harvey-Smith, L., & Cohen, R. J. 2006, *MNRAS* (Letters) 371, L26

Astrophysical Masers and their Environments
Proceedings IAU Symposium No. 242, 2007
J.M. Chapman & W.A. Baan, eds.

doi:10.1017/S174392130701294X

G23.657−0.127, what can we learn from a perfect methanol maser source?

Anna Bartkiewicz[1], **Huib Jan van Langevelde**[2,3], **Marian Szymczak**[1] **and Andreas Brunthaler**[4]

[1]Centre for Astronomy, Nicolaus Copernicus University, Gagarina 11, Torun, Poland
email: annan@astro.uni.torun.pl

[2]Joint Institute for VLBI in Europe, Postbus 2, 7990 AA Dwingeloo, email: langevelde@jive.nl,
[3] Sterrewacht Leiden, Postbus 9513, 2300 RA Leiden, The Netherlands

[4] Max Planck Institute fur Radioastronomie, Auf dem Hugel 69, 53121 Bonn, Germany
email: brunthal@mpifr-bonn.mpg.de

Abstract. In the course of following up compact methanol masers at 6.7 GHz which were found in the Torun blind survey (Szymczak *et al.* 2002), we discovered a ring structure in the source G23.657−00.127. This source provides interesting insights into whether methanol masers arise in rotating disks around massive stars, in their outflows, or behind shocks. By monitoring the 12.2 GHz masers, which fortunately follow the same structure, we hope to resolve the kinematics of the ring. Moreover, the symmetry of the source points to the existence of a central source. Early results on the nature of this source indicate the existence of a hyper–compact H II region.

Keywords. methanol maser, stars: formation, instrumentation: high angular resolution

---

Methanol masers at 6.7 GHz are a valuable tool in modern studies of *how massive stars are born* as they enable us to investigate the surrounding regions around protostars on scales of 100–1000 AU. Recent VLBI observations at 6.7 GHz with milliarcsecond (mas) resolutions revealed a new class of spherically symmetric methanol maser sources (Bartkiewicz *et al.* 2005). The maser ring in G23.657−00.127 has become a unique laboratory for detailed research into how and where methanol masers arise in an isolated massive star forming region. The morphology and the dynamics of the maser spots at 6.7 GHz have not allowed us to establish the origin of the ring-like structure yet. To do this we have begun a wide range of follow–up observations including detection of 12.2 GHz methanol maser line, proper motion studies at both 6.7 and 12.2 GHz, and searching for the continuum counterpart of the central source.

Using the VLBA we have recently imaged the 12.2 GHz methanol masers. These show good correlations in space and velocity with the 6.7 GHz ring (Fig. 1). The brightness of the co-propagating masers implies a cold, $T_{kin} \sim 30$ K, but dense, $n_H \sim 10^7$ cm^{-3}, environment (Sobolev *et al.* 1997). Currently, we are carrying out multi-epoch observations, spread over a year, at 12.2 GHz. These should allow us to determine the nature of the ring and to distinguish between rotation and expansion. We also expect to estimate an exact distance to the source from the simultaneous measurement of its parallax.

The spherically symmetric structure of the masers clearly points to the existence of a central source. Our VLA observations at 22 and 43 GHz towards the centre of the ring show weak emission. This could originate from a hyper-compact H II region, or from dust continuum. These new detections of counterparts at sub-cm wavelengths fit well with the spectral energy distribution of the infrared counterparts (1.25–100 μm), which we presented earlier (Fig. 2). A two black-body model by Walsh *et al.* (1999) fitted to

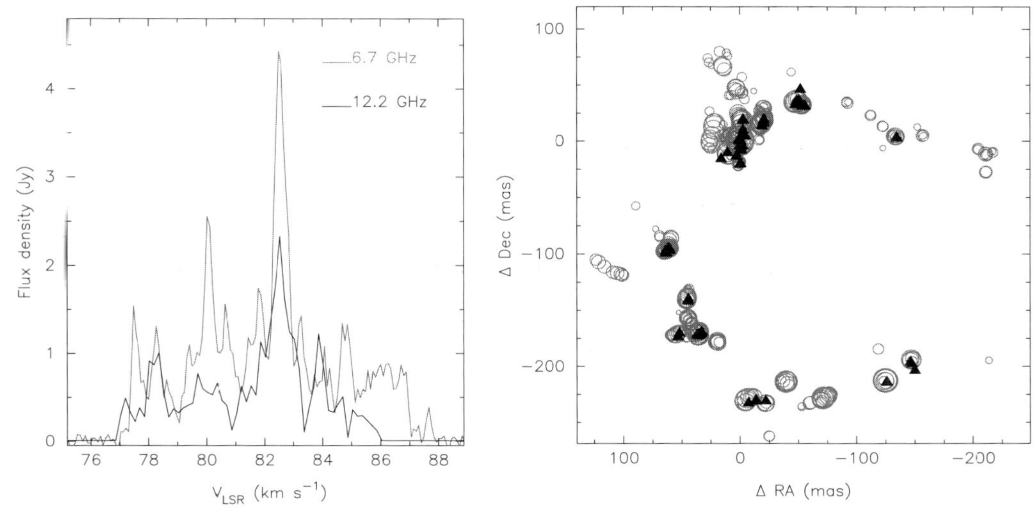

Figure 1. Spectra and distributions of 6.7 (circles) and 12.2 GHz (triangles) methanol masers seen towards G23.657−00.127.

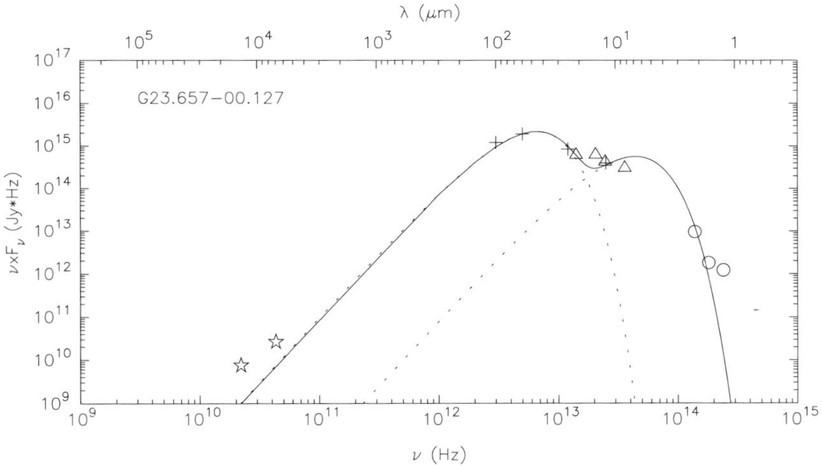

Figure 2. Spectral energy distribution towards the ring. Stars trace the tentative radio emission detected using VLA, while 2MASS, MSX6C and IRAS objects are marked by circles, triangles and crosses, respectively. Fitted lines represent a model of two black-body components (Walsh *et al.* 1999).

the data implies temperatures of 80 K and 540 K for the cold and hot dust, respectively (Bartkiewicz *et al.* 2005).

We hope to complete the observations soon and to be able to establish what is going on in G23.657−00.127.

References

Bartkiewicz, A., Szymczak, M. & van Langevelde, H. J. 2005, *A&A* 442, L61
Sobolev, A. M., Cragg, D. M. & Godfrey, P. D. 1997, *MNRAS* 288, L39
Szymczak, M., Kus, A. J., Hrynek, G., Kepa, A. & Pazderski, E. 2002, *A&A* 392, 277
Walsh, A. J., Burton, M. G., Hyland, A. R. & Robinson, G. 1999, *MNRAS* 309, 905

Astrophysical Masers and their Environments
Proceedings IAU Symposium No. 242, 2007
J.M. Chapman & W.A. Baan, eds.

doi:10.1017/S1743921307012951

Towards constraining the environments of methanol masers

M. Szymczak[1], **A. Bartkiewicz**[1] **and A. M. S. Richards**[2]

[1]Toruń Centre for Astronomy, Nicolaus Copernicus University, Poland
email: msz@astro.uni.torun.pl, annan@astro.uni.torun.pl

[2]Jodrell Bank Observatory, University of Manchester, UK
email: amsr@jb.man.ac.uk

Abstract. Observations of 10 transitions of HCO^+, CO and CS isotopomers were taken with the IRAM 30 m telescope for a sample of 28 methanol maser sources detected in the Torun unbiased survey of the galactic plane. About 64% of the sources show line wings in one or more transitions, indicating the presence of molecular outflows. Two sources show evidence for infall motion. Comparison of the widths of line wings and methanol maser emission suggests that the 6.7 GHz maser line traces the environment of massive young stellar objects of various kinematic regimes. Calculations based on the CO and HCO^+ lines and the CS line intensity ratios refine the physical parameters. Specifically, a gas density of $< 10^7$ cm^{-3} is sufficient for strong maser emission and a high methanol fractional abundance ($> 5 \times 10^{-7}$) is required.

Keywords. masers, ISM: molecules, radio lines: ISM, stars: formation

---

Unbiased surveys of the 6.7 GHz methanol maser line appear to provide an efficient method of forming complete samples of candidates of on-going or recently formed massive stars (Ellingsen *et al.* 1996; Szymczak *et al.* 2002; Pandian *et al.* 2007). Homogeneous samples are important for understanding when and where the methanol masers arise. However, due to the non-linear nature of maser amplification it is quite difficult to extract information about physical conditions in the maser environments from the maser data alone. In this report we present the main results of our observations of candidate massive stars in thermal molecular lines in order to constrain the range of environments probed by methanol masers.

A sample of 28 sources were selected from Szymczak *et al.* (2002); 25 of them were undetected prior to the Torun survey. Therefore, our selection excludes previously known objects associated with OH masers or with IRAS-selected bright ultra-compact HII regions. The astrometric positions of the targets were measured with the Mark II – Cambridge baseline of MERLIN (Niezurawska *et al.* 2005). The absolute positions of the objects were determined with an accuracy of $0.18''$ and $0.53''$ in right ascension and declination, respectively. The sample were observed in ten transitions of HCO^+, CO and CS isotopomers using three receivers at 3, 2 and 1.3 mm of the IRAM 30 m telescope. The half power beam widths ranged from $10{-}27''$ and the spectral resolutions were 0.10–0.16 km s^{-1}.

In 18 targets the wing emission was identified at least in one of the optically thick ($^{13}CO(2{-}1)$, $HCO^+(1{-}0)$, CS(2–1), CS(3–2)) and/or thin ($C^{18}O(2{-}1)$, $H^{13}CO^+(1{-}0)$) lines. The source G25.410+0.105 is the best example of sources which show evidence for the line wings in several transitions (Fig. 1). Because the line wings are considered as signatures of outflow motion we suggest that the maser emission may be associated with molecular outflows in about 64% of the sources in our sample. A second source, G26.598−0.024, is one of the two objects in the sample with an excess of blue-shifted

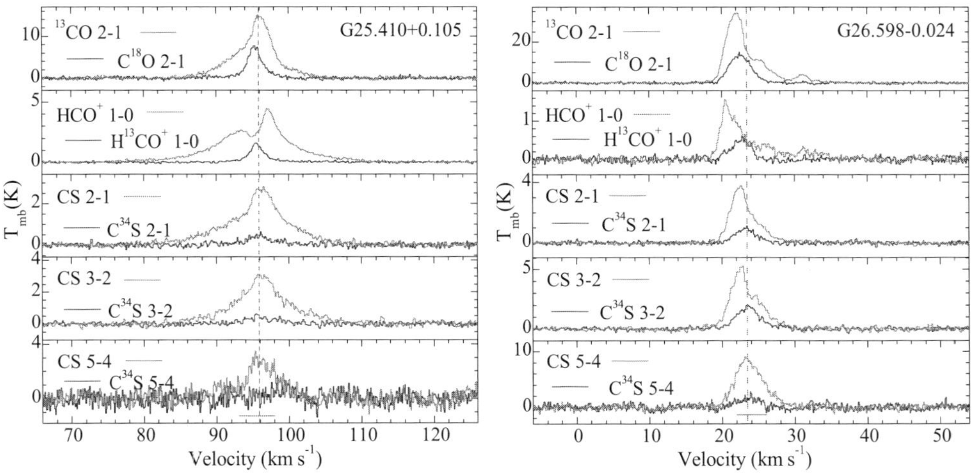

Figure 1. Molecular spectra of selected sources. The main beam brightness temperature (T_{mb}) is plotted against velocity (V_{LSR}) for one of the eighteen outflow candidates (left) and one of the two infall candidates (right). Source names, molecules and transitions are indicated. The dashed vertical line shows the systemic velocity (inferred from a Gaussian fit of one of the optically thin transitions) for each target. The thick bottom bar indicates the velocity range of the 6.7 GHz methanol maser emission (Szymczak *et al.* 2002) for each source.

emission in the optically thick lines (Fig. 1). Such a blue asymmetry of line profiles was postulated as evidence of inward motion of the gas and we suggest that the minority of the methanol maser sources may be infall candidates. The systemic velocity determined from the optically thin lines agrees within $\pm 3\,\mathrm{km\,s^{-1}}$ with the central velocity of the methanol maser emission for almost all the sources.

Using an escape-probability model we calculated the density and temperature of the gas required for the observed line intensity ratios of CS and $C^{34}S$. For the 16 sources for which all three CS transitions were detected, we found that the range of gas density is $10^5 - 10^7\,\mathrm{cm^{-3}}$ while the H_2 column density is between 10^{22} and $2 \times 10^{23}\,\mathrm{cm^{-1}}$ and the kinetic temperature varies from 30 to 100 K. The H_2 column density range is consistent with that inferred from our data for two pairs of molecular lines $HCO^+(1-0)$, $H^{13}CO^+(1-0)$ and $^{13}CO(2-1)$, $C^{18}O(2-1)$ using the method outlined in Purcell *et al.* (2006) and references therein. This provides a significant refinement to the input parameters of methanol maser models.

Acknowledgements

The work has been supported by the Polish MNiI grant 1P03D02729.

References

Ellingsen, S. P., von Bibra, M. L., McCulloch, P. M. *et al.* 1996, *MNRAS* 280, 378
Niezurawska, A., Szymczak, M., Richards, A. M. S., & Cohen, R. J. 2005, *BaltA* 14, 429
Pandian, J. D., Goldsmith, P. F., & Deshpande, A. A. 2007, *ApJ* 656, 255
Purcell, C.R., Balasubramanyam, R., Burton, M.G. *et al.* 2006 *MNRAS* 367, 553
Szymczak, M., Kus, A. J., Hrynek, G., Kepa, A., & Pazderski, E. 2002, *A&A* 392, 277

Astrophysical Masers and their Environments
Proceedings IAU Symposium No. 242, 2007
J.M. Chapman & W.A. Baan, eds.

doi:10.1017/S1743921307012963

Recent southern maser surveys

J. L. Caswell[1]

[1] Australia Telescope National Facility

Abstract. This discussion of recent maser surveys in the southern sky will allow us to see the context of the new methanol multibeam survey, which is currently in progress and described elsewhere at this symposium. My emphasis will be on masers that arise in star-forming regions in the disc of our Galaxy. I discuss outcomes from existing surveys, and comparison of different maser species, with special attention to the recognition of blue-shifted outflows.

Keywords. masers, surveys, stars: formation, ISM: molecules, radio lines: ISM

1. Introduction

A listing of 'Southern Galactic Plane surveys of OH and H_2O Masers' (Caswell & Haynes 1983) was given at a conference on 'Surveys of the Southern Galaxy' – paradoxically held at a rather northerly site – in the Netherlands! This meeting, in Australia, gives me the opportunity to redress the imbalance. Notably, 1983 was before the discovery of the widespread methanol masers at 6668 MHz. And it was before the completion of the Australia Telescope Compact Array, and thus positional accuracy of most known southern masers was poorer than 10 arcsec. Today I plan to bring the record up to date as to what is now available in 2007.

1.1. *The need for surveys*

Masers of OH, methanol and water play three special roles in studying the birth of massive stars.

1. They are one of the earliest and most prominent indicators of the recent birth of a massive star, and they are a principal means of discovering such sites.

2. Each maser site is powered by a single embryonic massive star, but comprises many high brightness maser spots of small diameter and narrow velocity width, rendering them excellent kinematic probes.

3. Conditions necessary for maser excitation yield information on the chemical and physical conditions pertaining to the site at an early stage before the fully developed star has blown the evidence away.

I will cover three main topics:

firstly, I will present a brief summary of the more extensive surveys of the recent past, and their particular strengths; secondly, I will draw attention to some direct outcomes from the surveys, and additional outcomes that rely on comparisons of surveys; and finally I will discuss some recent surveys that have been conducted, but not yet published, and some preliminary results from them.

2. Recent surveys of methanol, OH and water

2.1. *Methanol maser surveys*

For methanol, the Class II maser transition at 6668.5192 has received the most observing effort, and the compilation of survey data by Pestalozzi, Minier, Booth (2005) is an excellent starting point, in view of its extensive references. For that reason, I will not attempt to be exhaustive in my listing of references. However, it is important to stress that the original source material needs to be consulted, since it is very heterogeneous in its sensitivity, positional accuracy and general reliability; these limitations cannot be fully conveyed in any compilation and, furthermore, they lead to the occasional double listing of sources due to the apparently different positions in catalogues with poor positional accuracy.

I will distinguish between targeted surveys and 'blind' surveys. Menten's (1991) initial list of discoveries were the result of searches towards a variey of targets loosely regarded as tracers of star formation regions. The series of Hartebeesthoek surveys (starting with MacLeod, Gaylard, Nicholson (1992) and continuing to MacLeod, van der Walt, North *et al.* (1998)) primarily chose IRAS targets in the southern sky, and its detections usefully included some at Galactic latitudes too large to have been detected in typical blind surveys that are normally confined close to the Galactic plane; interestingly, it was also valuable in revealing the rather limited value of IRAS as a maser site predictor. From Torun, there was a similar search of IRAS sources in the northern sky (Szymczak, Hyrnek, Kus 2000). From Parkes there was a search of several hundred OH maser targets, with high success rate further demonstating the clear association of the two maser species, and providing the first extensive atlas of spectra with high sensitivity (Caswell, Vaile, Ellingsen *et al.* 1995a).

'Blind' surveys were conducted at Hobart with a 25m dish looking in the southern sky from Galactic longitude 325 to 335 degrees (Ellingsen, von Bibra, McCulloch, *et al.* 1996b). This was later complemented by the Torun survey in the northern sky (Szymczak, Kus, Hrynek *et al.* 2002). More recently, Arecibo has been used in the small northern region accessible to it (Pandian, Goldsmith, Deshpande 2007), with the large collecting area providing a rapid surveying capability, combined with good sensitivity and fairly good positions.

In addition to the single dish surveys, we have an exploratory 'blind' survey from longitude 330 to 340 degrees made with the ATCA (Caswell 1996), followed by a similar survey of the Galactic Centre. ATCA follow-up to provide precise (sub-arcsecond) positional accuracy for masers discovered in single dish surveys has been made by Caswell, Vaile, Forster (1995b); Ellingsen, Norris, McCulloch (1996a); Caswell (1997); Phillips, Norris, Ellingsen, McCulloch (1998); Walsh, Burton, Hyland, Robinson (1998), and others.

2.2. *OH maser surveys*

In the southern sky, a series of Parkes surveys (culminating with Caswell & Haynes 1987) made an extensive unbiassed 'blind' survey of 1665 and 1667-MHz emission along the Galactic plane, and were followed by additional surveying, plus precise positioning of known masers, using the ATCA (Caswell 1998). The rarer OH transitions at 6035 and 6030 MHz were surveyed at the locations of the main line (Caswell 2003), and likewise the 1720-MHz transition (Caswell 2004a). At 1612 MHz, a survey by Sevenster, Chapman, Habing *et al.* (1997a), Sevenster *et al.* (1997b), although optimised for late-type stars, is a valuable resource for recognising 1612-MHz emission associated with SFR main line emission, and emphasises that in SFR regions the masers at this transition are quite rare, with comparable occurrence to the 1720-MHz transition (see also Caswell 1999).

2.3. *Water maser surveys*

'Blind' surveys for water masers have been extremely rare - essentially confined to a small pilot study done more than 20 years ago (Matthews, Olnon, Winnberg, Baud 1985), and a very recent survey of a small southern region (Breen *et al.* 2007), as shown in Shari Breen's poster. And yet large numbers of water masers are known, comparable to that of all of the other maser species. The water maser detections have accumulated over the past 30 years as a result of many very heterogeneous targeted surveys (e.g. Batchelor, Caswell, Goss *et al.* 1980). As noted later, there are new opportunities for investing effort into significant new unbiassed 'blind' surveys, and these promise to be very productive.

3. Outcomes of the maser surveys

3.1. *Direct outcomes*

Outcomes from even quite small targeted surveys can be very useful, but some classes of investigation require extensive 'blind' surveys.

1. From the only 'blind' survey covering an extensive portion of the Galactic plane, the series of OH surveys has yielded a Galactic distribution (Caswell & Haynes 1987).

2. A luminosity function has also been derived for the OH masers, and is useful in estimating the productivity expected from future, deeper and more extensive surveys.

3. Estimations of a water maser luminosity function have been attempted, but mostly demonstrate the inadequate sensitivity and poor completeness of the input data.

4. Surveys of OH at 1720, 6035 and 6030 MHz have revealed the enormous potential of these transitions to map magnetic fields by utilizing the Zeeman effect; the effect is usually much easier to interpret in the case of these transitions, for which the splitting is inherently smaller, and propagation effects are far less prone to destroy one of the Zeeman components, than is the case for 1665- and 1667-MHz masers.

5. Variability over many timescales is common amongst the majority of the maser species, and interpreted as a consequence of a variety of intrinsic and extrinsic effects. Some of the most striking variability occurs amongst the methanol masers, with remarkable outcomes as shown by Sharmila Goedhart at this meeting.

6. The typical maser velocity spread is another important parameter. To be valid, a study of this property is best confined to data from observations with good spatial resolution, so as to distinguish blends of nearby sources, and prevent the erroneous listing of wide velocity spreads that are merely blends.

7. Targeted surveys can provide a useful adjunct to pure 'blind' surveying. When targeted searches are conducted with a large beam, and are subsequently followed up with derivation of precise positions, masers that are offset from the targets are found, and these offset sources are rightly regarded as the outcome of an element of blind surveying.

8. A special case of the above occurs in the case of searches for Class I methanol masers at the location of known Class II methanol masers. Despite the known exclusivity of the two varieties of maser, such searches have been remarkably productive, with Class I masers generally offset from those of the Class II targets spatially (in most cases), and slightly offset in velocity. This conundrum of the relationship between Class I and Class II methanol masers is explored at this meeting by Simon Ellingsen.

3.2. *Discoveries from comparisons of surveys*

After excluding the spuriously large apparent velocity spreads (arising from spatially nearby but distinct masers), there remain some OH and some methanol masers (and many water masers) with large velocity spreads. What then is the systemic velocity?

Among the methanol masers, there are relatively few with large widths (exceeding $12\,\mathrm{km\,s^{-1}}$) and most of these arise in sources where there is very strong emission near the median velocity and, extending from this, there are wings that are weak and roughly symmetrical. The systemic velocity then seems most likely to be at the median velocity. Overall, methanol seems to be an excellent indicator of systemic velocity.

In contrast to the methanol, there are some striking examples of OH masers with wide velocity spreads where the emission is concentrated at the velocity extremes. Where, in these cases, is the systemic velocity? Is it in the middle where there is no emission (as in the late-type stars where the front and back of an expanding shell dominate the emission)? Or is it at one extreme or the other; and if so, which extreme? And why?

To understand these OH masers, we can compare them with their methanol counterparts, with interesting results. Assuming that the methanol defines the systemic velocity, we find a number of OH masers showing outflows, with the most prominent ones being blue-shifted outflows. An example is $337.405 - 0.402$, first discovered by Caswell, Haynes & Goss (1980), where features are clustered around velocity $-38\,\mathrm{km\,s^{-1}}$ (now known to be the same velocity as the methanol, and thus regarded as the systemic velocity) and a group separated in velocity (but not spatially) extending to at least $-56\,\mathrm{km\,s^{-1}}$. We interpret this as a blue-shifted outflow, and it is detected at 1667 MHz as well as 1665 MHz. Water maser emission is present but is confined to the systemic velocity range.

4. Surveys in progress but not yet published

With Chris Phillips, I have been conducting a water maser study at the locations of the southern SFR OH maser sites, using the ATCA. The use of the ATCA rather than a single dish has allowed immediate measurement of precise positions, and confirms that 170 of the 200 OH masers searched have closely associated water masers; additional water masers found within the fields of view are quite common.

A 'blind' water maser search has been made, again with the ATCA, over two small (each 0.5 square degrees) regions of sky, chosen to straddle the Galactic plane, but to have only a few previously known masers. Preliminary results reveal more than 20 masers in the regions. Although these detections are not necessarily SFR masers, our initial expectation is that they are likely to be so, since, in a region close to the Galactic plane, SFR masers are likely to dominate in such a flux-limited sample.

Plans are in place to conduct a much more extensive water maser survey with the ATCA when the new correlator (the CABB) becomes operational in 2008. This will allow simultaneous coverage of a wide velocity range with no need to sacrifice spectral resolution, and continuum observations, recombination lines, and other spectral lines to be decided after preliminary tests.

Extensive water maser surveys are rightly regarded as the last frontier. They are difficult to conduct on account of the small beams at high frequency, and the discoveries are especially difficult to interpret without the benefit of pre-existing surveys of OH and methanol, since there may be significant numbers that are not of the SFR variety.

Interesting discoveries from the new water maser surveys so far include several that are DOMINATED by emission from a blue-shifted outflow. It will be interesting to monitor these to see whether any are of the exotic variety that show acceleration of the blue shifted outflow, as found for 291.284 − 0.716 and 291.270 − 0.719 (Caswell 2004b).

I conclude by mentioning one more 'survey' of OH masers - in this case, not searching for new masers, but conducting full spectropolarimetry on the full sample of known southern masers. This has been done with Chris Phillips using the Parkes telescope, with processing from the new spectral line processing package ASAP as described in a poster at this meeting by Chris Phillips & Malte Marquarding. The resulting spectra from this survey reveal that linear polarization can be very prominent, despite the use of a single dish telescope (for which we might expect a tendency for polarization to be washed out when features are not spatially resolved). The atlas of these OH spectra therefore reveals which ones have especially interesting polarization, and provides an excellent way of determining priorities for observing at long baselines with the Australian Long Baseline Array (see Caswell & Reynolds 2001, and poster papers at this meeting from Busaba Hutawarakorn Kramer and Indra Bains).

Acknowledgements

I thank my collaborators in the methanol multibeam (MMB) survey (co-authors of the paper by James Green in this symposium) for enthusiastic discussion of the role of maser surveys. I especially value the memory of time spent with Jim Cohen at Jodrell Bank, at Rose Cottage, and at Epping, Narrabri and Parkes, as we planned and began the MMB survey which, when completed, will be a lasting tribute to Jim.

References

Batchelor, R. A., Caswell, J. L., Goss, W. M., Haynes, R. F., Knowles, S. H., & Wellington, K. J. 1980, *AustJPhys* 33, 139

Breen, S. L., Ellingsen, S. P., Johnston-Hollitt, M., Wotherspoon, S., Bains, I., Burton, M. G., Cunningham, M., Lo, N., Senkbeil, C. E., & Wong, T. 2007, *MNRAS* 377, 491

Caswell, J. L. 1996, *MNRAS* 279, 79

Caswell, J. L. 1997, *MNRAS* 289, 203

Caswell, J. L. 1998, *MNRAS* 297, 215

Caswell, J. L. 1999, *MNRAS* 308, 683

Caswell, J. L. 2003, *MNRAS* 341, 551

Caswell, J. L. 2004a, *MNRAS* 349, 99

Caswell, J. L. 2004b, *MNRAS* 351, 279

Caswell, J. L., Haynes, R. F., & Goss, W. M., 1980, *AustJPhys* 33, 639

Caswell, J. L., Haynes, R. F., 1983, in: W. B. Burton & F. P. Israel (eds.), *Surveys of the Southern Galaxy*, Proceedings of a workshop, Leiden, Netherlands August 4-6 1982 (Dordrecht: D.Reidel Publishing) pp. 25–30

Caswell, J. L., Haynes, R. F., 1987, *AustJPhys* 40, 215

Caswell, J. L., Reynolds, J. E, 2001, *MNRAS* 325, 1346

Caswell, J. L., Vaile, R. A., Ellingsen, S. P., Whiteoak, J. B., & Norris, R. P. 1995a, *MNRAS* 272, 96

Caswell, J. L., Vaile, R. A., & Forster, J. R. 1995b, *MNRAS* 277, 210

Ellingsen, S. P., Norris, R. P., & McCulloch, P. M. 1996a *MNRAS* 279,101

Ellingsen, S. P., von Bibra, M. L., McCulloch, P. M., Norris, R. P., Deshpande, A. A., & Phillips, C. J. 1996b *MNRAS* 280, 378

MacLeod, G. C., Gaylard, M. J., & Nicholson, G. D. 1992, *MNRAS* 254, 1P

MacLeod, G. C., van der Walt, D. J., North, A. J., Gaylard, M. J., Galt, J.A., & Moriarty-Schieven, G. H. 1998, *AJ* 116, 2936

Matthews, H. E., Olnon, F. M., Winnberg, A., & Baud, B. 1985, *A&A* 149, 227

Menten, K. M. 1991, *ApJ* 380, L75

Pandian, J. D., Goldsmith, P. F., & Deshpande, A. A. 2007, *ApJ* 656, 255

Pestalozzi, M. R., Minier, V., & Booth, R. S. 2005, *A&A* 432, 737
Phillips, C. J., Norris, R. P., Ellingsen, S. P., & McCulloch, P. M. 1998 *MNRAS* 300, 1131
Sevenster, M. N., Chapman, J. M., Habing, H. J., Killeen, N. E. B., & Lindqvist, M. 1997a, *A&AS* 122, 79
Sevenster, M. N., Chapman,J. M., Habing, H. J., Killeen, N. E. B., & Lindqvist, M. 1997b, *A&AS* 124, 509
Szymczak, M., Hyrnek, G., Kus, A. J. 2000, *A&AS* 143, 269
Szymczak, M., Kus, A. J., Hrynek, G., Kepa, A., & Pazderski, E. 2002, *A&A* 392, 277
Walsh, A. J, Burton, M. G., Hyland, A. R., & Robinson, G. 1998, *MNRAS* 301, 640

Astrophysical Masers and their Environments
Proceedings IAU Symposium No. 242, 2007
J.M. Chapman & W.A. Baan, eds.

doi:10.1017/S1743921307012975

SiO maser survey of evolved stars in the Galaxy: various environments of maser sources

Shuji Deguchi[1,2]

[1]Nobeyama Radio Observatory, National Astronomical Observatory, Minamimaki, Minamisaku, Nagano, 384-1305, Japan
email: deguchi@nro.nao.ac.jp

[2]Graduate University for Advanced Studies, National Astronomical Observatory, Minamimaki, Minamisaku, Nagano, 384-1305, Japan

Abstract. Using the SiO $J = 1$–0 $v = 1$ and $v = 2$ lines near 43 GHz, we have detected about 2000 of 3600 sources observed with the Nobeyama 45-m radio telescope. The sources were chosen from IRAS/MSX/2MASS catalogs using color-selection criteria to pick up mass-losing oxygen-rich AGB stars and some post-AGB objects. A number of interesting sources were also found: supergiants in a massive star cluster, a nova with light echo (V838 Mon), AGB stars in globular clusters, and AGB candidates associated with dwarf galaxies. With the exception of the stars in the massive open star cluster, these 'unusual' objects are associated with metal poor environments where mass losing oxygen-rich AGB stars are unexpected. It is inferred that these objects were created by stellar merging which can occur in dense star clusters.

Keywords. Masers, SiO, Evolved stars, Mergers

1. Introduction

SiO maser sources are useful probes of Galactic structure and stellar evolution. Using the Nobeyama 45-m telescope, we have so far observed about 3600 objects and have obtained about 2000 detections of the SiO $J = 1$–0 $v = 1$ or $v = 2$ lines at 43.122 and 42.821 GHz respectively. For a summary of disk objects, see Deguchi *et al.* (2004b). Figure 1 shows a sky distribution of the observed and detected objects in Galactic coordinates. A large concentration of detected sources can be seen at the center, which is the Galactic bulge, while a feature extended to the left is the Galactic disk ($l \lesssim 0°$). The almost vacant part on the right ($-270° \gtrsim l \gtrsim 0°$) is the region of southern sky that we cannot access from Nobeyama. However, we added a few sources observed using SEST and Mopra telescopes (Deguchi *et al.* 1990, 2001), so this is not totally empty. Our surveys are mostly targeted surveys based on infrared catalogs, namely the IRAS catalog. However, the IRAS catalog is not complete in the Galactic center region. An unbiased ('blind') survey of the inner bar region of the Galactic bulge (Fujii *et al.* 2006) was made recently based on the MSX and 2MASS catalog of infrared sources (Egan *et al.* 2003), resulting in ~ 160 detections. Another unbiased survey of the Galactic center area of $24' \times 24'$ was made in 2001–2003, resulting in ~ 200 SiO detections (Deguchi *et al.* 2004a). Our surveys in the Galactic center and disk are summarized in Table 1. The survey depths and color selection criteria are not uniform among these surveys. From our experience, we have found that about 1/3 of SiO detections have corresponding OH 1612 MHz objects while the other 2/3 of SiO sources have no OH detections (Deguchi *et al.* 2004a; Sevenster

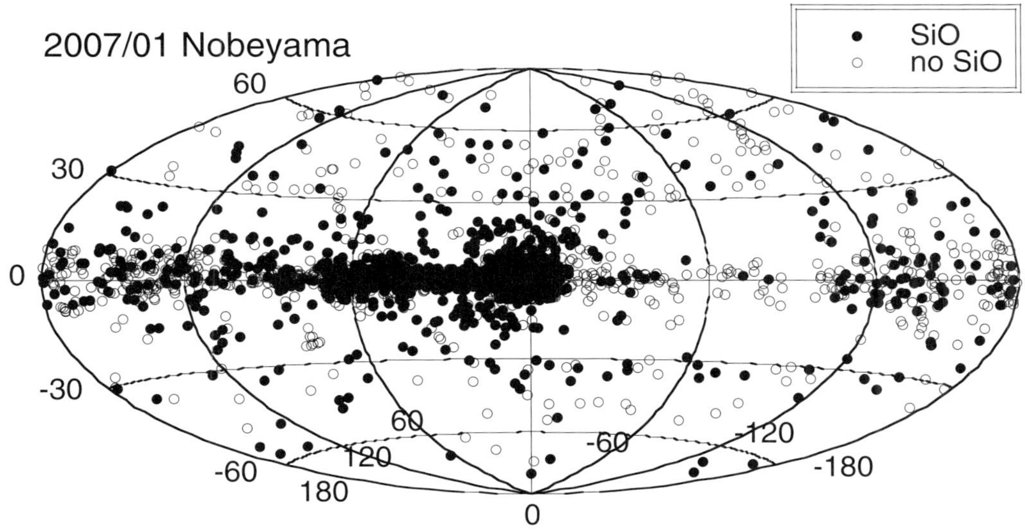

Figure 1. Distribution of observed objects in the Galactic coordinates in the Hammer-Aitoff projection. The filled and open circles indicate SiO detections and non-detections.

Table 1. Recent SiO maser surveys in the Galactic center and disk

name	area	survey mode	detections	paper
N. Gal. Cap	$b > 30°$	targeted	24	Ita *et al.* (2001)
G.C.	$7' \times 13'$	unbiased	9	Miyazaki *et al.* (2001)
G.C.	$200'' \times 100''$	unbiased	15	Deguchi *et al.* (2002)
disk	$40 < l < 70°$	targeted	134	Nakashima *et al.* (2003)
LAV G.C.	$24' \times 24'$	targeted	196	Deguchi *et al.* (2004a)
Inner disk	$-10 < l < 40°$	targeted	254	Deguchi *et al.* (2004b)
Inner bar	$7° \times 2°$	targeted	163	Fujii *et al.* (2006)
Off-plane	$\|b\| > 3°$	targeted	119	Deguchi *et al.* (2007a)

et al. 1997). This is the main reason that we do extensive SiO maser surveys in various parts of the Galaxy.

2. Observations and color selection criteria of objects

The 45-m millimeter-wave telescope at Nobeyama operates at frequencies between 22 and 115 GHz. The beam size of the telescope is about $40''$ at 43 GHz. We use a HEMT receiver at 43 GHz with a bandwidth of 2 GHz, and system temperature of about 200 K. We can observe simultaneously four lines of SiO $J = 1$–0 from $v = 0$ to $v = 3$, and the ground rotational line of ^{29}SiO. However, for distant objects, we usually detect only the two strongest lines in the first and second vibrationally excited lines. Therefore, this paper discuss only the SiO $J = 1$–0 $v = 1$ and $v = 2$ lines. Our recent observations are summarized in Table 1.

2.1. *Color selection criteria from MSX catalog and 2MASS*

Van der Veen & Habing (1988) classified IRAS point sources by the IRAS two-color diagram; mass-losing AGB stars fall into type IIIa and IIIb regions, and post-AGB stars fall into type VIb for which SiO masers are unlikely. After 2000, we shifted to using the

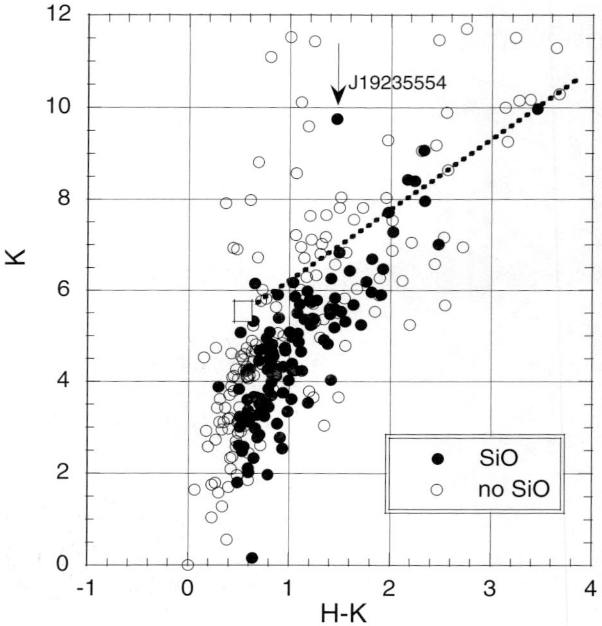

Figure 2. Plot of 2MASS $H-K$ against K for an off-Galactic-plane sample Deguchi *et al.* (2007). Filled and unfilled circles indicate SiO detections and no detections. The square indicates M5III without reddening at a distance of 8 kpc. Interstellar and circumstellar extinctions move the position of the star along the dotted line.

MSX and 2MASS catalogs. The MSX color criteria are similar to the IRAS selection criteria [$-0.4 < log(F_E/F_C) < 0.4$, where F_C and F_E are MSX 12 and 21 μm flux densities, respectively]. In addition, we check the 2MASS images and find a near-infrared counterpart for the selected MSX source and require the counterpart to have a color index $H-K > 0.8$. Figure 2 shows the diagram of the NIR color index $H-K$ against K magnitude. A typical M5 giant at a distance of 8 kpc occupies the position indicated by the square, and both interstellar and circumstellar reddening move the position of the M5 star along the line with inclination of about 1.5. Therefore, if we correct the interstellar and circumstellar extinction by a formula, $m_K = K-[A_K/E(H-K)]\times(H-K-[H-K]_0)$ (Nishiyama *et al.* 2006; $[A_K/E(H-K)] \sim 1.5$ and $[H-K]_0 \sim 0.5$), we can use the corrected K magnitude (m_K) as a distance measure (Fujii *et al.* 2006). If an object is located lower than this line corresponding to a constant distance, it is a nearby foreground object; if it is located higher than that line, it is very distant object or non-AGB star, probably a young stellar object. In this way we can select oxygen-rich mass-losing AGB stars quite effectively. With these selection criteria, we get an average SiO detection rate of about 60 %.

3. Various environments of maser sources

Here we will describe individually interesting sources in various environments in the Galaxy, that were found in the SiO surveys.

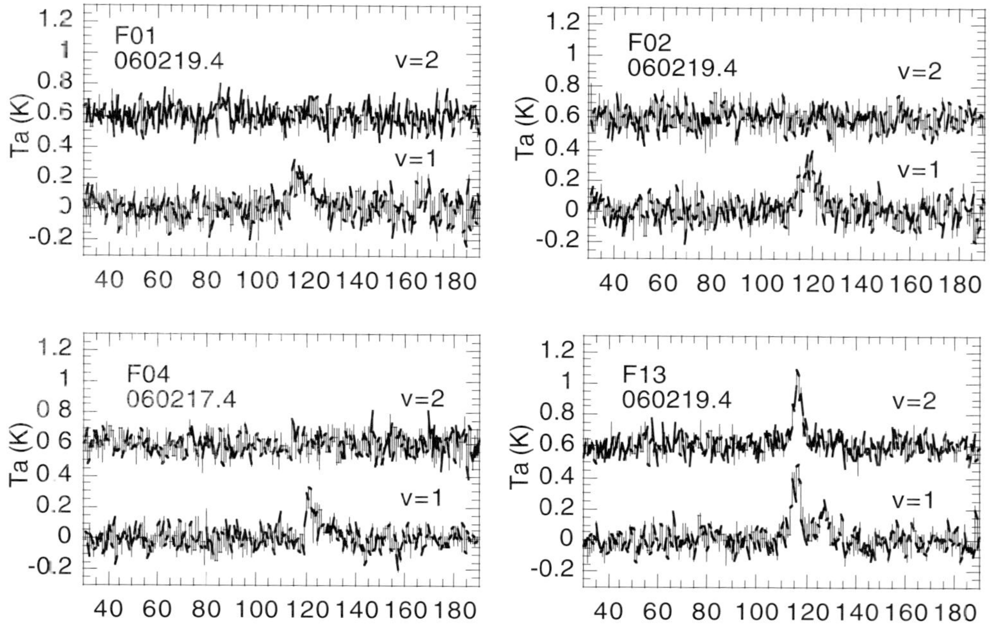

Figure 3. SiO $J = 1\text{–}0$ $v = 1$ and 2 spectra toward M supergiants in the Scutum young massive star cluster.

3.1. *Scutum young massive star cluster*

The young open star clusters often contain several M-supergiants. Measuring equivalent widths of the CO first-overtone bands at 2.3 μm, Figer *et al.* (2006) identified 14 M-supergiants in the open cluster called # 122 in Bica's infrared cluster catalog (Bica *et al.* 2003). Assuming the Saltpeter's initial mass function, they found that, if the total mass of the cluster is about 3×10^4 $M_\odot$ and the age approximately 10^7 years, such a young massive star cluster can contain as many as 14 M-supergiants. Some of these supergiants are IRAS or MSX objects. So we observed all of these M-supergiants with our telescope, and in addition a few more objects nearby. We detected SiO masers in five of these, but one, we call X18, turned out to be non-member from its radial velocity (Nakashima & Deguchi 2006). The radial velocities of the other four objects are about 120 km s^{-1} (see Fig. 2). We also detected the 22.235 GHz water maser line in one object.

What can we obtain from these observations? First, the radial velocity of the star cluster is 118.1 ± 2.8 km s^{-1}. Previously one OH 1612 MHz maser was detected at 102 km s^{-1} (Blommaert *et al.* 1994), but this was emission from a single OH peak so that the true radial velocity of the star was uncertain. The radial velocity we obtained gives the kinematic distance of 6.5 kpc, suggesting that this cluster is located at a tip of the Scutum-Crux spiral arm beyond the bulge bar. Second, we calculated the velocity dispersion from this limited data. Assuming an intrinsic velocity dispersion of 2 km s^{-1} in the SiO measurements, the true velocity dispersion of this cluster is inferred to be 2 – 2.5 km s^{-1} (Jiang *et al.* 1996). From this dispersion and a radius of 1 arc minute of the cluster size, we obtain a virial mass of about 10^4 $M_\odot$ for the cluster showing that it has a large mass. SiO observation are clearly useful for obtaining the physical parameters of massive stars in star clusters.

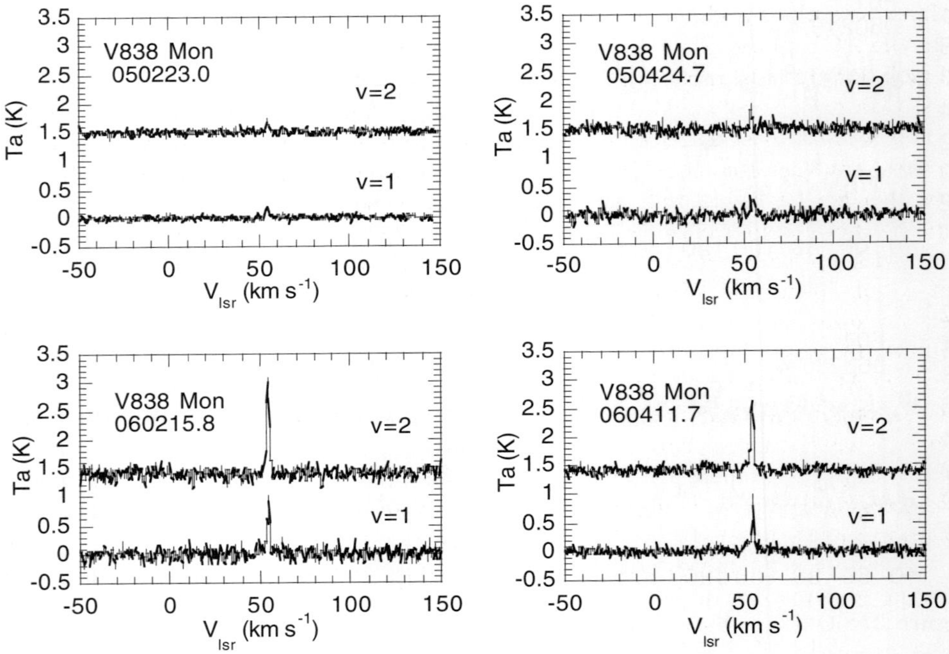

Figure 4. Intensity variation of the SiO $J = 1$–0 $v = 1$ and 2 toward V838 Mon. Dates of observations are 2005 Feb. 22, 2005 April 24, 2006 Feb. 15, and 2006 April 11.

3.2. *V838 Mon*

The second example is V838 Monocerotis. This nova appeared in January 2002 and a light echo was observed a month and a half later. A movie showing the expansion of the light echoes was created from HST images by Bond *et al.* 2003. In this picture we can see a red central star. In fact, spectroscopic observations revealed that the central star left after the nova eruption was an M supergiant (Munari *et al.* 2002). We detected SiO masers in this star three years after the eruption (Deguchi *et al.* 2005), in February 2005. The SiO maser intensities became stronger by 50% two months later. A half year later, we found that SiO maser intensity was about five times stronger (see Fig. 3). However, we could not detect any other transitions of SiO, including the 86 GHz transition apart from the $J = 1$–0 $v = 1$ and 2 lines. Monitoring observations of SiO emission from this star have been made by Mark Claussen using VLA and VLBA. Soker & Tylenda (2003) proposed that the central star, an M-supergiant, was created by a merging binary with 7 and 0.5 $M_\odot$ main-sequence objects. This star turned out to be in a star cluster, because the surrounding blue stars are all B stars located nearly at the same distance of 6–7 kpc (Afsar & Bond 2007). Such an environment probably helps binary merging to occur easily (Lombardi *et al.* 2003). Such binary merging can create AGB-like stars (similar to a pre-main sequence star going down along the Hayashi track; Tylenda & Soker 2006). A very rough estimation of the birth rate and life time of this kind of phenomena indicates that a few percent of OH/IR stars originated from binary merging (Deguchi *et al.* 2005).

3.3. *SiO maser sources in globular clusters*

The third example is a globular cluster. Low-mass stars usually evolve into carbon-rich stars in the AGB phase (Iben 1981). Therefore, SiO masers were not expected to occur in old globular clusters. In fact, globular clusters are generally not associated with IRAS sources. One day, Noriyuki Matsunaga, a student of University of Tokyo, came to my place and wished to search for SiO masers in Miras in globular clusters. One of the candidates is in the famous bulge cluster Terzan 5. which is an MSX source with a flux density of a few Jy at 12 μm. We detected SiO masers in this star as well as three objects in other globular clusters (Matsunaga *et al.* 2005). The observed radial velocities coincide with optically measured average radial velocities of these globular clusters to within about 30 km s^{-1}. We found that the optical radial velocities of these globular clusters are quite inaccurate, because the average radial velocities of these clusters are determined by observing only a few cluster-member stars. A big question is how are oxygen-rich AGB stars created in such old clusters? They are similarly bright objects to the disk or bulge AGB stars in terms of K-magnitude and IRAS flux densities. Therefore, their masses can be more than 1 $M_{\odot}$, which should not be found in these old globulars. One explanation is a binary merger hypothesis. In globular clusters, there are a number of blue stragglers which are luminous and are considered to have masses of more than 1 $M_{\odot}$ (Piotto *et al.* 2004). These stars have been created by binary merging due to three-body encounters in globulars. After a while, they can evolve into AGB stars, or as in V838 Mon, merging can directly make M-giants.

3.4. *SiO maser sources in dwarf galaxies or their tidal tails*

We spent a considerable time searching for very distant SiO maser sources. So far the most distant SiO maser source ever detected is an OH/IR star in the Large Magellanic cloud (van Loon *et al.* 1996). One of the most nearby galaxies accessible from Northern hemisphere is the Sgr dwarf elliptical galaxy (Ibata *et al.* 1994), also called SagDEG. The distance to this galaxy is about 24 kpc from the Sun. Another one is the Canis Majoris dwarf galaxy (Martin *et al.* 2005), or it is sometimes called Monoceros over-density (Martin *et al.* 2004a). The center of SagDEG is located south of the Galactic bulge. The orbital plane of this galaxy is almost perpendicular to the Galactic plane (Ibata *et al.* 2001b), and a number of globular clusters, tidal debris, M-giants and carbon stars (Ibata *et al.* 2001ab), have been observed along the tidal stream (Majewski *et al.* 2003, 2004). We used the 45-m telescope to search for SiO masers in candidate stars along the tidal tails. Because of its distance, the candidate stars are expected to be quite faint in the near- and middle-infrared. A problem is that there are many contaminating foreground objects. Radial velocities can be used to judge the association with these dwarf galaxies. Figure 5 shows SiO maser spectra of our probable candidates for associations with SagDEG and CMa dwarf. The radial velocity of 110 km s^{-1} in the right-panel spectrum fits well with the optically measured average velocity of M giants of CMa dwarf (Martin *et al.* 2005). Also this is a faint object with the IRAS 12 micron flux density of 2 Jy and 2MASS K magnitude of 6.2. A luminosity distance of about 7 kpc coincides well with the estimated distance of CMa dwarf. The left panel shows the SagDEG candidate. This is a very faint object of 1 Jy at 12 μm, and has an estimated luminosity distance (from 2MASS $K = 9.7$) of about 20 kpc from the Sun, far beyond the Galactic bulge. The radial velocity of this object is slightly different with the velocity of SagDEG. However, SagDEG has several tidal streams with different velocities, so that, we think that it is associated with one of SagDEG tidal streams. A question here is why such oxygen-rich AGB stars are associated with these dwarf galaxies of old population. One explanation is that it is a binary merger product.

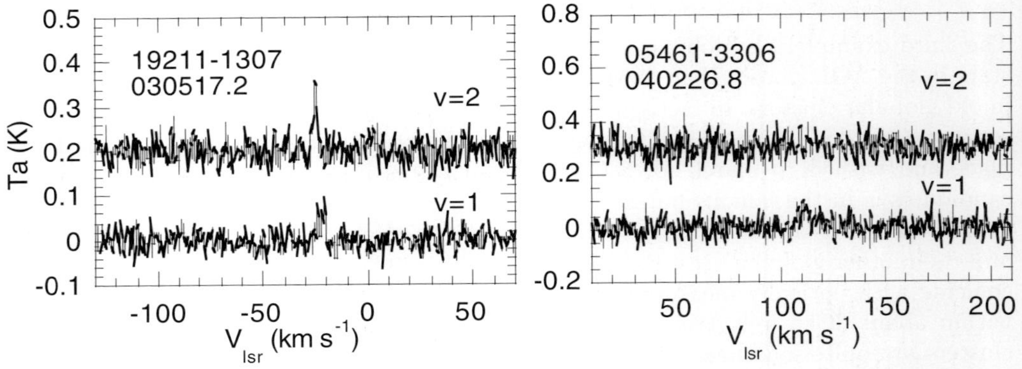

Figure 5. SiO $J = 1$–0 $v = 1$ and 2 spectra toward candidates associated with dwarf galaxies.

4. Conclusions

SiO masers turned out to be a powerful probe of Galactic structure as well as a probe of stellar evolution. V838 Monocerotis, which emits SiO masers, is an example of stellar mergers. SiO maser phenomena can be a tool to investigate massive open star clusters and globular clusters and the tidal tails of dwarf galaxies. The SiO maser surveys demonstrate that SiO maser sources can exist in metal-rich environments in the Galaxy as well as in the metal-poor environments of old populations such as globular clusters. For such old populations, they are possibly binary merger products or their progeny.

Acknowledgements

I thank the SiO maser survey team, Dr. T. Fujii, Y. Ita, H. Izumiura, O. Kameya, A. Miyazaki, N. Matsunaga, Y. Nakada, and J. Nakashima for help with long-term observations. The team also thanks Dr. R. Forster, S. Kwok, B. M. Lewis, A. Winnberg, and P. Wood for temporal cooperation.

References

Afsar, M. & Bond, H. E. 2007, *AJ*, 133, 387
Bica, E., Dutra, C. M., Soares, J., & Barbuy, B. 2003, *A&A*, 404, 223
Blommaert, J. A. D. L., van Langevelde, H. J., & Michiels, W. F. P. 1994, A&Ap, 287, 479
Bond H. E., *et al.* 2003, *Nature* 422, 405
Deguchi, S., Nakada, Y., & Sahai, R. 1990, *A&A*, 230, 339
Deguchi, S., Fujii, T., Miyoshi, M., & Nakashima, J. 2002, *PASJ*, 54, 61
Deguchi, S., Nakashima, J., & Balasubramanyam, R. 2001, *PASJ*, 53, 305
Deguchi, S., *et al.* 2004a, *PASJ*, 56, 261
Deguchi, S., *et al.* 2004b, *PASJ*, 56, 765
Deguchi, S., Matsunaga, N., & Fukushi, H. 2005, *PASJ*, 57, L25
Deguchi, S., *et al.* 2007, *PASJ*, 59, No. 3 (in press)
Egan M. P., Price S. D., Kraemer K. E., Mizuno D. R., Carey S. J., Wright C. O., Engelke C. W., Cohen M., & Gugliotti G. M. 2003, Air Force Research Laboratory Technical Report AFRL-VS-TR-2003-1589 available at *http://vizier.nao.ac.jp/viz-bin/VizieR?-source=V/114*
Figer, D. F., MacKenty, J., Robberto, M., Smith, K., Najarro, F, Kudritzki, R. P., & Herrero, A. 2006, *ApJ*, 643, 1166
Fujii, T., Deguchi, S., Ita, Y., Izumiura, H., Kameya, O., Miyazaki, A., & Nakada, Y. 2006, *PASJ*, 58, 529
Ibata, R. A., Gilmore, G., & Irwin, M.J. 1994, *Nature* 370, 194
Ibata, R., Irwin, M., Lewis, G., & Stolte, A. 2001a, *ApJ*, 547, L133

Ibata, R., Lewis, G., Irwin, M., Totten, E., & Quinn, T. 2001b, *ApJ*, 551, 294
Iben, I., Jr. 1981, ApJ, 246, 278
Ita, Y., Deguchi, S., Fujii, T., Kameya, O., Miyoshi, M., Nakada, Y., Nakashima, J., & Parthasarathy, M. 2001, *AAp*, 376, 112
Jiang, B. W., Deguchi, S., Yamamura, I., Nakada, Y., Cho, S. H., & Yamagata, T. 1996, *ApJs*, 106, 463
Lombardi, J. C., Thrall, A. P., Deneva, J. S., Fleming, S. W., Grabowski, P. E. 2003, *MNRAS*, 345, 762
Majewski, S. R., Skrutskie, M. F., Weinberg, M. D., & Ostheimer, J. C. 2003, *ApJ*, 599, 1082
Majewski , S. R. *et al.* 2004, *ApJ*, 128, 245
Martin, N. F., Ibata, R. A., Bellazzini, M., Irwin, M. J., Lewis, G. F., & Dehnen, W. 2004, *MNRAS*, 348, 12
Martin, N. F., Ibata, R. A., Conn, B.C., Lewis, G. F., Bellazzini, M., & Irwin, M. J. 2005, *MNRAS*, 362, 906
Matsunaga, N., Deguchi, S., Ita, Y., Tanabe, T., & Nakada, Y. 2005, *PASJ*, 57, L1
Miyazaki, A., Deguchi, S., Tsuboi, Ma., Kasuga, T., & Takano, S. 2001, *PASJ*, 53, 501
Munari, U., *et al.* 2002, *A&A*, 389, L51
Nakashima, J., & Deguchi, S. 2003, *PASJ*, 55, 203
Nakashima, J., & Deguchi, S. 2006, *ApJL*, 647, L139
Nishiyama, S., Nagata, T., Kusakabe, N. *et al.* 2006, *ApJ*, 638, 839
Piotto, G. *et al.* 2004, *ApJ*, 604, L109
Soker, N., & Tylenda, R. 2003, *ApJ*, 582, L105
Sevenster, M. N., Chapman, J. M., Habing, H. J., Killeen, N. E. B., & Lindqvist, M. 1997, *A&AS*, 122, 79
Tylenda, R. & Soker, N. 2006, *A&A*, 451, 223
van der Veen, W. E. C. J., & Habing, H. 1988, *A&A*, 194, 125
van Loon, J. T., Zijlstra, A. A., Bujarrabal, V., & Nyman, L.-Å. 1996, *A&A*, 306, L29

Discussion

GOSS: Pulsers have been found in Terzan 5. Do these pulsers influence the SiO maser sources?

DEGUCHI: More than 20 pulsers were found in this globular cluster. Among them, several millisecond pulsers were found and they are known as close-binary systems. Close binaries are considered easily merged in a dense stellar system. Stellar merging can make blue struggler, which finally evolved into mass-losing oxygen-rich AGB stars with masers.

RODRIGUEZ: The radial velocity of the extra object ($V_{lsr} = 84\,\mathrm{km\,s^{-1}}$) is not far from the average velocity of the Scutum star cluster. Is it still not a member of the cluster ?

DEGUCHI: No, a separation of $40\,\mathrm{km\,s^{-1}}$ is too large for this object to be a member of the cluster. The velocity dispersion of the cluster is just a few $\mathrm{km\,s^{-1}}$. It is just a foreground object.

MENTEN: Concerning V838 Mon, were any other molecules detected?

DEGUCHI: No, part from two vibrationally excited SiO J = 1-0 transitions, nothing has been detected towards this star. Mark Claussen has been monitoring the SiO maser intensities of these transitions. It is a pity that he could not come to this conference.

Astrophysical Masers and their Environments
Proceedings IAU Symposium No. 242, 2007
J.M. Chapman & W.A. Baan, eds.

doi:10.1017/S1743921307012987

The Arecibo Methanol Maser Galactic Plane Survey

Jagadheep D. Pandian[1], Paul. F. Goldsmith[2] and Avinash A. Deshpande[3]

[1]Max Planck Institute for Radio Astronomy, Auf dem Hügel 69, 53121 Bonn, Germany
e-mail: jpandian@mpifr-bonn.mpg.de

[2]Jet Propulsion Laboratory, Pasadena, CA 91109, USA

[3]Raman Research Institute, Bangalore 560080, India

Abstract. The Arecibo Methanol Maser Galactic Plane Survey (AMGPS) is a blind survey for 6.7 GHz methanol masers in a section of the Galactic plane visible from the Arecibo radio telescope. The survey for these signposts of massive star formation is complete at a flux density level of 0.27 Jy making it the most sensitive blind survey for methanol masers carried out to date, and resulted in the detection of 86 methanol masers, 48 of which are new discoveries. The properties of methanol masers discovered in the survey are consistent with their being associated with early phases of massive star formation. The data also show the tangent point of the Carina-Sagittarius spiral arm to be around a Galactic longitude of 49.6°.

Keywords. surveys, stars: formation, Galaxy: structure

1. Introduction

The 6.7 GHz methanol maser, discovered by Menten (1991), is the strongest of Class II methanol masers. Unlike OH and H_2O masers, the 6.7 GHz maser has been observed exclusively towards regions containing young, massive stars. A number of recent studies give circumstantial evidence for 6.7 GHz methanol masers being associated with early phases of massive star formation (e.g. Ellingsen 2006; Minier *et al.* 2005; Minier *et al.* 2001). This makes the 6.7 GHz maser a powerful tool to probe Galactic structure and massive star formation which is poorly understood.

However, many questions regarding 6.7 GHz maser emission are still unanswered. Which phases of massive star formation are associated with the maser emission? Where in relation to the massive young stellar objects (MYSOs) are the masers excited? Studies on a statistically complete sample derived from a blind survey are required to address this question. Furthermore, do faint methanol masers have different properties compared to the bright ones? Szymczak & Kus (2000) discovered that some IRAS sources associated with relatively weak methanol masers did not satisfy Wood & Churchwell (1989) (hereafter WC89) and Hughes & MacLeod (1989) (hereafter HM89) color criteria for ultracompact H II regions. Here, we present the results of a sensitive blind survey of a portion of the Galactic plane that is visible to the 305 m Arecibo radio telescope.

2. The Arecibo Methanol Maser Survey

The observations were made between June 2004 and March 2006 using the C-Band High receiver at the 305 m Arecibo radio telescope. The telescope has an rms pointing error of 7″ and an average full width half maximum (FWHM) beamwidth of 40″. The backend was configured to cover an LSR velocity range of $-70\,\mathrm{km\,s^{-1}}$ to $+110\,\mathrm{km\,s^{-1}}$

at a velocity resolution of $0.14\,\mathrm{km\,s^{-1}}$ after Hanning smoothing. The target region was covered by mosaicing 100 sub-regions, each sub-region being mapped in a rectangular grid (in right ascension and declination), the separation between grid points being $15''$ in right ascension and $20''$ in declination. The integration time per grid point was typically 0.5 sec, and the resulting noise in a spectral channel was $\sim$ 85 mJy after averaging the two orthogonal linear polarizations. Candidate sources were extracted from the data cubes through automated matched filter algorithms. Candidate sources were re-observed using one or two minute position switched observations at a much higher velocity resolution of $0.03\,\mathrm{km\,s^{-1}}$.

We detected a total of 86 sources, 48 of which are new detections. The region $35.2° \leqslant l \leqslant 53.7°$, $|b| \leqslant 0.41°$ was completely sampled, while a larger area of $34.7° \leqslant l \leqslant 54.3°$, $|b| \leqslant 0.54°$ was incompletely sampled. The peak flux density of the sources ranges from 0.11 Jy to over 750 Jy. Using simulations, we found that the probability of detection of a source in our source extraction algorithm was greater than 95% for signals that have signal to noise ratio higher than 3.0. Since the noise in the data cubes is no greater than 0.09 Jy, we conclude that our source catalog is complete at the level of 0.27 Jy. The source catalog is available in Pandian *et al.* (2007).

3. Distribution in the Galaxy

To consider the distribution of methanol masers in the Galaxy, we exclude sources that lie outside the region that is fully sampled by the survey, along with sources that have peak flux densities below our completeness limit of 0.27 Jy. This gives a sample of 72 methanol masers. Figure 1 shows the distribution of these methanol masers as a function of Galactic latitude. A Gaussian fit to the distribution has a mean at a latitude of $-0.09°$ and a FWHM of $0.49°$. The width of the distribution is comparable to that of embedded massive stars determined from the IRAS point source catalog (Wood & Churchwell 1989). The fit parameters to the distribution are very similar to those obtained by Pestalozzi *et al.* (2005) for the general catalog of methanol masers which contains over 500 sources. Over half of the sources in the general catalog were detected in targeted surveys towards tracers of massive star formation such as ultracompact HII regions and OH masers. The close agreement in the distribution of AMGPS sources with that of Pestalozzi *et al.* (2005) gives further evidence for the association of 6.7 GHz methanol masers with massive star formation.

4. Implications for Galactic Structure

Since massive stars form preferentially along spiral density waves which compress molecular gas to high densities, 6.7 GHz methanol masers are potential probes of the spiral structure of the Galaxy. As distances to masers in the AMGPS catalog are not known precisely, and most kinematic distances suffer from the distance ambiguity, we use an $l - v$ diagram to see what information methanol masers provide regarding the structure of our Galaxy. Two constructs of the $l - v$ diagram are shown in Figure 2. The observed radial velocities are assumed to have deviations from the values observed from pure circular rotation by up to $\pm 5\,\mathrm{km\,s^{-1}}$. The left and right panels of Figure 2 show the loci of various spiral arms in the models of Vallee (1995) and the NE2001 model of Cordes & Lazio (2002) respectively. The spiral arm loci are calculated using the rotation curve of Clemens (1985).

The clumping of sources around $l \sim 49.6°$ seems to arise from the tangent point of the Carina–Sagittarius spiral arm. The model of Vallee (1995) fits this feature reasonably

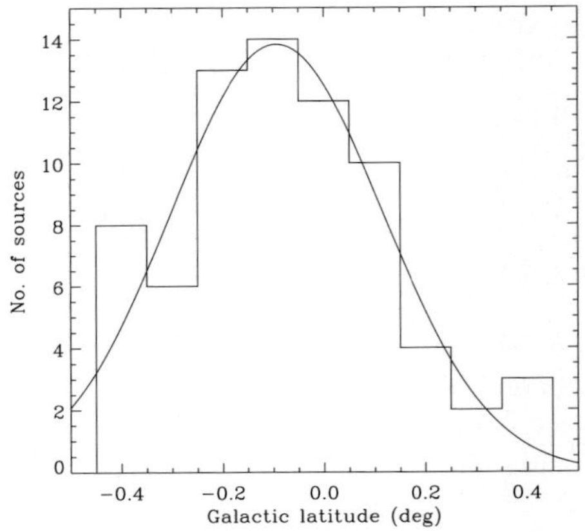

Figure 1. Distribution of methanol masers as a function of Galactic latitude. A Gaussian fit to the distribution is overlaid. The Gaussian fit has a mean at -0.09° and a FWHM of 0.49°.

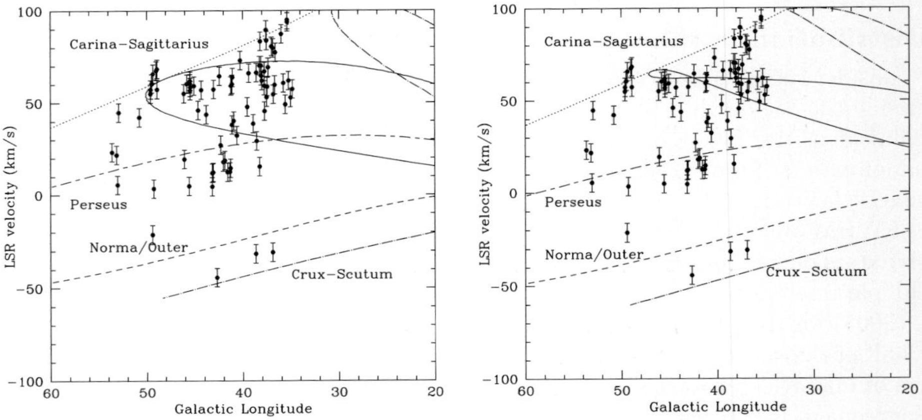

Figure 2. The methanol maser sample of AMGPS on an $l - v$ diagram. The left and right panels show the loci of spiral arms from the model of Vallee (1995) and the NE2001 model of Cordes & Lazio (2002) respectively. The dotted line shows the tangent point velocity.

well, while the tangent point in the model of Cordes & Lazio (2002) occurs at approximately 1° lower Galactic longitude which does not fit our data. Figure 2 also shows that it is quite challenging to extract information on Galactic spiral structure from 6.7 GHz methanol masers, since a significant fraction of sources do not lie near spiral arm loci.

5. The number of 6.7 GHz methanol maser sources in the Galaxy

The number of 6.7 GHz methanol masers in the Galaxy has been estimated by van der Walt (2005). The procedure followed is to model the distribution of young massive star forming regions as a function of Galactic longitude using a star formation law, initial mass function, and a spiral arm model of the Galaxy. This distribution is then scaled such that the number of maser sources in a chosen longitude bin coincides with the statistics from an existing survey. Since all surveys are flux limited, a minimum number of methanol masers can be derived by summing over all longitude bins. Thus, the minimum number of methanol masers was determined by van der Walt (2005) to be 845, by scaling the model distribution to the observations between Galactic longitudes of 330° and 340°.

Since our survey is an order of magnitude more sensitive more than that of existing surveys, our observations can be used to revise the minimum number of methanol masers in the Galaxy. The AMGPS detected 44 methanol masers brighter than 0.27 Jy between longitudes of 40° and 50°. Using this number as the normalization for the longitude distribution of van der Walt (2005), the number of methanol masers in the Galaxy, brighter than 0.27 Jy, is at least 1125. This number is a lower limit as the region $|b| > 0.41°$ is incompletely sampled by AMGPS.

Furthermore, van der Walt (2005) estimated the completeness of a survey as a function of flux density through Monte Carlo simulations. This was then used to derive the total number of methanol maser sources in the Galaxy as 1200 ± 84. Applying the same procedure to statistics from our survey results in the total number of maser sources to be 1197–1264, which is consistent with the analysis of van der Walt (2005).

6. Counterparts in IRAS, MSX and NVSS catalogs

As mentioned earlier, the rms pointing error of the telescope is 7″, and the statistical distribution of this error is a Gaussian to good approximation. The error circle associated with any methanol maser, defined at the 95% confidence-level, is then ∼ 18″. We thus search for counterparts within 23″ of the methanol maser (5″ being taken as the limit for association). We exclude sources in W49 and W51 from this analysis since the density of methanol masers in these regions is high enough that a given IRAS, MSX or NVSS source will lie within the search radius of more than one maser.

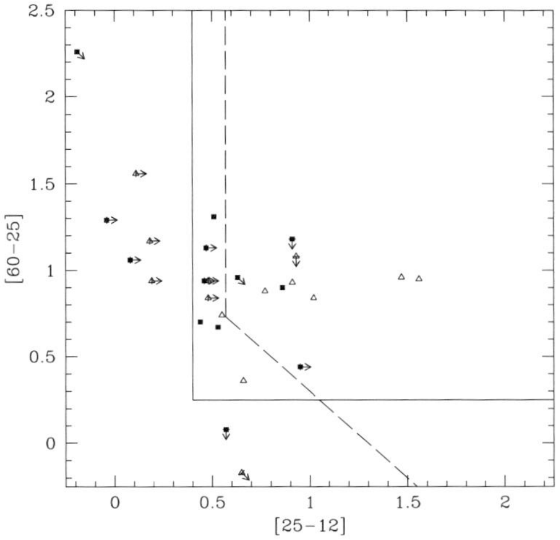

Figure 3. The color-color diagram for possible IRAS counterparts of 6.7 GHz methanol masers discovered in AMGPS. The x- and y-axes show the [25-12] and [60-25] colors defined as $\log_{10}(F_{25}/F_{12})$ and $\log_{10}(F_{60}/F_{25})$ respectively. The colors for previous detections are shown in open triangles, while those of new detections are shown in filled squares. Color limits pointing at 45° indicate upper limits in both [25–12] color and [60–25] colors. Sources to the right of the dashed lines satisfy WC89 colors for ultracompact HII regions, while sources to the right of the bold lines satisfy HM89 criteria for the same.

Carrying out this analysis, only 26 out of 76 sources have possible IRAS counterparts. This is more dramatic than results from the previous blind surveys of Ellingsen *et al.*

(1996) and Szymczak *et al.* (2002), who found IRAS–based surveys would miss about half of the methanol maser population. The color-color diagram of these IRAS sources is shown in Figure 3. The WC89 and HM89 criteria for embedded massive stars are shown in the figure. Only six sources satisfy WC89 criteria, while a larger number of sources satisfy HM89 criteria. This shows that targeted searches for methanol masers towards IRAS color selected sources will miss a significant population of masers in the Galaxy. One of the reasons for the poor correspondence between methanol masers and IRAS sources could be source confusion in crowded fields of the Galactic plane, which resulted in many sources being not included in the point source catalog. It is to be noted that Figure 3 does not show any segregation between the previous and new detections (new detections are typically weaker); this is contrary to the result of Szymczak & Kus (2000) who found that IRAS sources associated with weak methanol masers populated the upper left part of Fig. 3.

Only 41 out of 76 methanol masers have possible MSX point source counterparts. Visually inspecting MSX image fields, four sources are clearly associated with MSX dark clouds, while the association with dark clouds is more uncertain for an additional four sources. The lack of point source counterparts for many sources could be in part a sensitivity effect; many masers lie close to bright mid-infrared sources. Seven out of 76 sources have NVSS point sources within $23''$. This is most probably due to any HII regions associated with the methanol masers being too compact and optically thick to be detectable at 21 cm. Thus, the lack of NVSS counterparts to most of the methanol masers suggests that they are primarily associated with phases of massive star formation prior to an ultracompact HII region. A few methanol masers do not have any tracer of star formation detected to date within $1'$; the nature of these intriguing sources is not known at present.

7. Conclusions

The AMGPS has detected 48 new 6.7 GHz methanol masers between Galactic longitudes of 34.8° and 54°. The properties of the masers in the AMGPS catalog are consistent with them being associated with early phases of massive star formation. The total number of methanol maser sources in the Galaxy is estimated to be about 1200.

References

Clemens, D. P. 1985, *ApJ* 295, 422

Cordes, J. M., & Lazio, T. J. W. 2002, astro-ph/0207156

Ellingsen, S. P. 2006, *ApJ* 638, 241

Ellingsen, S. P., von Bibra, M. L., McCulloch, P. M., Norris, R. P., Deshpande, A. A., & Phillips, C. J. 1996, *MNRAS* 280, 378

Hughes, V. A., & MacLeod, G. C. 1989, *AJ* 97, 786

Menten, K. M. 1991, *ApJ* 380, L75

Minier, V., Burton, M. G., Hill, T., Pestalozzi, M. R., Purcell, C. R., Garay, G., Walsh, A. J., & Longmore, S. 2005, *A&A* 429, 945

Minier, V., Conway, J. E., & Booth, R. S. 2001, *A&A* 369, 278

Pandian, J. D., Goldsmith, P. F., & Deshpande, A. A. 2007, *ApJ* 656, 255

Pestalozzi, M. R., Minier, V., & Booth, R. S. 2005, *A&A* 432, 737

Szymczak, M., & Kus, A. J. 2000, *A&A* 360, 311

Szymczak, M., Kus, A. J., Hrynek, G., Krępa, A., & Pazderski, E. 2002, *A&A* 392, 277

Vallee, J. P. 1985, *ApJ* 454, 119

van der Walt, J. 2005, *MNRAS* 360, 153

Wood, D. O. S., & Churchwell, E. 1989, *ApJ* 340, 265

Astrophysical Masers and their Environments
Proceedings IAU Symposium No. 242, 2007
J.M. Chapman & W.A. Baan, eds.

doi:10.1017/S1743921307012999

Investigating high-mass star formation through maser surveys

S. P. Ellingsen[1], M. A. Voronkov[2], D. M. Cragg[3], A. M. Sobolev[4], S. L. Breen[1] and P. D. Godfrey[3]

[1]School of Mathematics and Physics, University of Tasmania, Private Bag 21, Hobart 7001, TAS, Australia
email: Simon.Ellingsen@utas.edu.au

[2]Australia Telescope National Facility CSIRO, PO Box 76, Epping 1710, NSW, Australia

[3]School of Chemistry, Building 23, Monash University, Victoria 3800, Australia

[4]Ural State University, Lenin ave. 51, 620083 Ekaterinburg, Russia

Abstract. Interstellar masers are unique probes of the environments in which they arise. In studies of high-mass star formation their primary function has been as signposts of these regions and they have been used as probes of the kinematics and physical conditions in only a few sources. With a few notable exceptions, we know relatively little about the evolutionary phase the different maser species trace, nor their location with respect to other star formation tracers. While detailed studies of a small number of maser regions can reveal much about them, other information can only be obtained through large, systematic searches. In particular, such surveys are vital in efforts to determine an evolutionary sequence for the common maser species, and there is growing evidence that methanol masers may trace an earlier phase than the other common maser species of OH and water.

Keywords. stars: formation, Masers, Surveys

1. Introduction

Interstellar masers are one of the best, if not *the* best signpost of young high-mass star formation regions. They are relatively common, intense and because they arise at centimetre wavelengths, are not affected by the high extinction that plagues observations in other wavelength ranges. The 6.7 GHz transition of methanol has been found to be a particularly useful signpost as it is both common and strong and appears to only be associated with high-mass star formation regions (Minier *et al.* 2003). The methanol multibeam (MMB) survey currently being undertaken with the Parkes and Jodrell Bank telescopes is using observations of the 6.7 GHz transition of methanol to trace the Galactic distribution of star formation and investigate the structure of our Galaxy (see the papers by Green *et al.* and Pestalozzi *et al.* in these proceedings).

In addition to being excellent signposts of star formation regions, masers have characteristics that make them very well suited to be utilised as probes of the regions. The maser emission is compact and intense, which means that it can be studied at milliarcsecond resolution using very long baseline interferometry (VLBI) techniques. This makes masers ideal probes of the kinematics of the gas in which they arise (e.g. Torrelles *et al.* 2005). They are potentially equally powerful probes of the physical conditions in these same regions (e.g. Cragg *et al.* 2001), however, the complexity of maser pumping schemes provides significant challenges for robust interpretation of the results. Although progress has been slow towards the overall goal of utilizing masers as tools to study star formation, the pace of advance has recently accelerated. The proliferation of complementary high-resolution observations of star formation regions at millimetre through mid-infrared wavelengths means that this trend is likely to continue.

Some of the questions relating to masers are best addressed through detailed studies of many transitions, molecules and wavelength bands towards a small number of sources. For example this is the best means of determining where the different maser species lie with respect to other objects within star formation regions. However, favourite sources which show strong emission from a wide range of maser transitions are atypical of high-mass star formation regions and so there are some questions for which a statistical approach is a better way to proceed. Uncommon phenomena must be associated with unusual sources, or brief evolutionary phases. What we really want to know are the common properties of masers in star formation regions, as they can give us an insight into the general characteristics of the process.

2. An Evolutionary Sequence for Masers

There are four species of interstellar maser which are common in high-mass star formation regions. Mainline OH masers at 1.6 GHz and water masers at 22 GHz have been studied since the 1960s, while the 1980s and 1990s saw the discovery of a number of strong transitions from two different types of methanol masers. The OH, water and class II methanol masers are found close to newly formed high-mass stars, as indicated by the presence of HII regions and bright infrared sources. The class I methanol masers are found towards high-mass star formation regions, but are usually offset from the high-mass protostars and other maser species. The presence of an interstellar maser within a star formation region signifies "special" physical conditions. They are special in the sense that masing is only predicted to occur for certain ranges of temperature, density, molecular abundance etc. Theoretical models find the most common maser species - mainline OH, 6.7 GHz methanol and 22 GHz water masers are inverted over a wide range of physical conditions (e.g. Cragg, Sobolev & Godfrey 2002).

At present there remains significant uncertainty about where within star formation regions the different maser species arise and the evolutionary phase they are associated with. However, as relatively easily observed signposts of star formation regions, it would be very desirable to be able to use the presence or absence of the different maser species as an evolutionary clock. Towards many star formation regions more than one (sometimes all) of the common maser species are observed, which implies that there is significant overlap between the evolutionary phases traced. The complexity of the goal of finding an evolutionary sequence for masers in high-mass star formation regions should not be underestimated, indeed De Buizer et al. (2005) suggest that the different maser species trace a variety of stellar phenomena and are hence associated with multiple evolutionary phases. It is only through statistical studies of large samples of maser sources that a definitive answer will be obtained.

There are a few facts that are commonly agreed for the common maser species - in the words of former US Secretary of Defense Donald Rumsfeld "Things we know we know" :

(*a*) Water masers usually trace outflows (e.g. Felli, Palagi & Tofani 1992).

(*b*) Water masers trace a generally earlier evolutionary phase than OH masers (Forster & Caswell 1989).

(*c*) Where OH and methanol masers are observed towards the same regions the clusters arise from the same general region (Caswell 1997), although there is no coincidence between the individual maser spots of the different species (Menten *et al.* 1992).

A number of independent lines of evidence suggest that in many cases the class II methanol masers are associated with a very early stage of the high-mass star formation process. In particular, only a small percentage are associated with ultra-compact HII regions (Phillips *et al.* 1998, Walsh *et al.* 1998), but all are associated with millimetre (Pestalozzi, Humphreys & Booth 2002) or submillimetre sources (Walsh *et al.* 2003) and many are embedded within infrared dark clouds (Ellingsen 2006). They are associated with sources which have a spectral energy distribution consistent with them being deeply embedded (Minier *et al.* 2005). This raises the obvious question, are some of the class II methanol masers associated with an earlier phase of high-mass star formation than water masers?

2.1. *Water and Class II methanol*

There have been a small number of comparative studies of water and class II methanol masers. Sridharan *et al.* (2002) searched for water and methanol masers towards an infrared-based sample of candidate high-mass star formation regions and found that the majority of sources which had associated masers showed emission from both species. While a search for water masers towards a sample of methanol masers by Szymczak, Pillai & Menten (2005) achieved a detection rate of approximately 50%. These studies suggest that water and methanol masers are commonly observed towards the same star formation regions, however, targeted searches inevitably suffer potential biases arising from the selection criteria. A recent complete survey for water maser emission towards the G333.6−0.2 giant molecular cloud (Breen *et al.* 2007) complements previous unbiased surveys of 6.7 GHz methanol (Ellingsen *et al.* 1996) and main-line OH masers (Caswell, Haynes & Goss 1980) in the same region. For the first time we have been able to compare in detail the distribution of these different maser species over an area of more than a square degree. Figure 3 of Breen *et al.* shows a striking dichotomy between the water and methanol masers in the region. All the water masers lie very close to the main axis of the cloud, as traced by the bright mid-infrared emission, while the methanol masers generally lie near the interface between the stronger and more diffuse emission.

That the water masers are generally projected against bright mid infrared emission along the main axis of the region suggests that they are associated with relatively evolved high-mass star formation. The presence of methanol masers offset from the brightest mid-infrared emission and their absence along the main axis suggests that they may be tracing a second epoch of star formation in the region, triggered by energetic winds and outflows from the mid infrared bright sources. The water maser search of Breen *et al.* was only sensitive to water masers with a peak flux density stronger than 5 Jy and searched a smaller fraction of the giant molecular cloud than the methanol and OH searches, so it is likely that there are some weak water maser sources in the region that have not yet been detected. It is not possible to draw general conclusions on the basis of observations of a single giant molecular cloud, however, the results strongly suggest that 6.7 GHz methanol masers trace an earlier evolutionary phase of high-mass star formation than do luminous water masers.

2.2. *Class I and class II methanol*

Class I methanol masers are the least well understood of the common maser species. They occur at the interface between outflows and the ambient molecular cloud environment (Plambeck & Menten 1990), offset from the other maser species. One of the main difficulties in studying class I methanol masers has been the lack of good criteria on which to base targeted searches. Ellingsen (2005) showed that approximately 40% of class II methanol masers have an associated class I methanol maser within 30 arcseconds. Recent ATCA observations have shown that in many cases the class I maser sources are not merely chance detections towards other objects within the larger star formation regions,

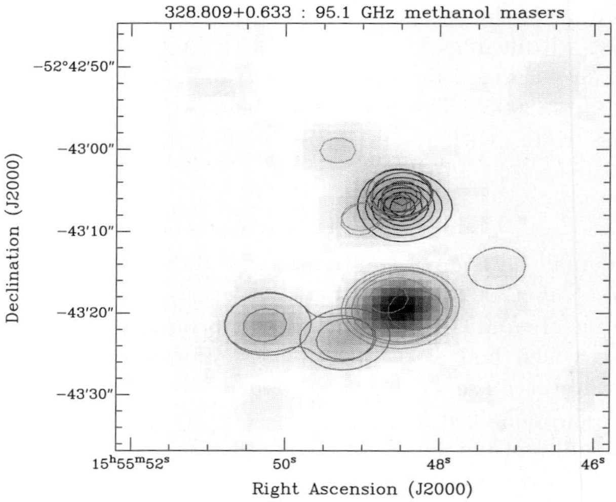

Figure 1. G328.81 + 0.63 shows strong (400 Jy) class II methanol maser emission projected against the ultra-compact HII region (thick black contours - centred at approximately $\alpha = 15:55:48.5$, $\delta = -52:43:07$). The greyscale image shows the integrated 95.1 GHz class I methanol maser emission and the thin grey contours are a renzogram representing the velocity of the emission.

but are intimately associated with the class II masers. Figure 1 shows the 95.1 GHz class I methanol masers towards the well known class II methanol maser source G328.81+0.63. The class I masers are spread over 8 or more clusters covering an extent of more than 30 arcseconds, the strongest emission being to the south of the HII region. In contrast the class II methanol masers lie within a single cluster projected against the north-east of the HII region. There is class I methanol maser emission from the same projected location as the class II masers and encompassing the same velocity range. Theoretically strong class I and class II methanol masers cannot coexist - the former are collisionally excited, while the latter are radiatively pumped. It may be that the class I maser are associated with an outflow directed nearly perpendicular to the line of sight from the source pumping the class II masers. Regardless of the detailed interpretation, these and other ATCA observations demonstrate that in many cases class I and class II methanol masers are associated with the same driving source, rather than a coincidental association.

The presence of molecular outflows is thought to be one of the earliest indications of the commencement of star formation and so we might naively expect class I masers to trace an even earlier evolutionary phase than either water or class II methanol masers. Ellingsen (2006) investigated the mid infrared colours (from the GLIMPSE survey) of a sample of class II methanol masers with and without associated class I methanol masers. Although the sample is relatively small, there is a strong tendency for the sources with an associated class I methanol maser to have redder colours. Redder colours are normally associated with more deeply embedded, and hence younger sources.

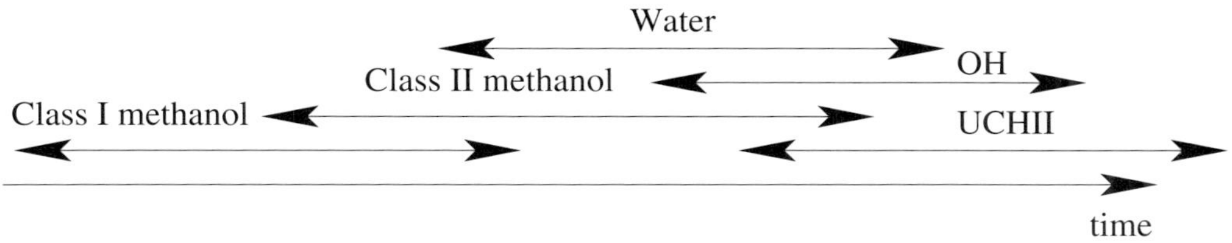

Figure 2. A "straw man" evolutionary sequence for masers in high-mass star formation regions.

3. Conclusions

Figure 2 outlines a "straw man" evolutionary sequence for masers in star formation regions that is consistent with the results outlined above and previously established facts. This sequence appears consistent with the general properties of masers in high-mass star formation regions, however, there are specific sources which are not consistent with this sequence. For example the Turner-Welsh object near W3(OH) has associated water and OH maser emission, but no methanol masers (Argon, Reid & Menten 2003), while IRAS16547-4247 has class I methanol, water and OH, but no class II methanol masers (Voronkov *et al.* 2006). These may be rare sources, but only rigorous statistical testing will show whether the proposed sequence usefully describes the majority of cases.

References

Argon, A. L., Reid, M. J., & Menten, K. M. 2003 *ApJ* 593, 925
Breen, S. L., Ellingsen, S. P., Johnston-Hollitt, M., Wotherspoon, S., Burton, M. G., Cunningham, M., Lo, N., Senkbeil, C. E., & Wong, T. 2007 *MNRAS* in press
Caswell, J. L. 1997 *MNRAS* 289, 203
Caswell, J. L., Haynes, R. F., & Goss, W. M. 1980 *Aust. J. Phys.* 33, 639
Cragg, D. M., Sobolev, A. M., Ellingsen, S. P., Caswell, J. L., Godfrey, P. D., Salii, S. V., & Dodson, R. G. 2001 *MNRAS* 323, 939
Cragg, D. M., Sobolev, A. M., & Godfrey, P. D. 2002 *MNRAS* 331, 521
De Buizer, J. M., Radomski, J. T., Telesco, C. M., & Piña, R. K. 2005 *ApJS* 156, 179
Ellingsen, S. P. 2005 *MNRAS* 359, 1498
Ellingsen, S. P. 2006, *ApJ* 638, 241
Ellingsen, S. P., von Bibra, M. L., McCulloch, P. M., Norris, R. P., Deshpande, A. A., & Phillips, C. J. 1996, *MNRAS* 280, 378
Felli, M., Palagi, F., & Tofani, G. 1992, *A&A* 255, 293
Forster, J. R., & Caswell, J. L. 1989, *A&A* 213, 339
Menten, K. M., Reid, M. J., Pratap, P., Moran, J. M., & Wilson, T. L. 1992, *ApJ* 401, L39
Minier, V., Burton, M.G., Hill, T., Pestalozzi, M.R., Purcell, C.R., Garay, G., Walsh, A.J., & Longmore, S. 2005 *A&A* 429, 945
Minier, V., Ellingsen, S. P., Norris, R. P., & Booth, R. S. 2003, *A&A* 403, 1095
Pestalozzi, M. R., Humphreys, E. M. L., & Booth, R. S. 2002 *A&A* 384, L15
Phillips, C. J., Norris, R. P., Ellingsen, S. P., & McCulloch, P. M. 1998, *MNRAS* 300, 1131
Plambeck, R. L., Menten, K. M. 1990 *ApJ* 364, 555
Sridharan, T. K., Beuther, H., Schilke, P., Menten, K. M., & Wyrowski, F. 2002, *ApJ* 566, 931
Szymczak, M., Pillai, T., & Menten, K. M. 2005, *A&A* 434, 613
Torrelles, J. M., Patel, N., Gómez, J.F., Anglada, G., & Uscanga, L. 2005, *Ap&SS* 295, 53
Voronkov, M. A., Brooks, K. J., Sobolev, A. M., Ellingsen, S. P., Ostrovskii, A. B., & Caswell, J. L. 2006 *MNRAS* 373, 411
Walsh, A. J., Burton, M. G., Hyland, A. R., & Robinson, G. 1998 *MNRAS* 301, 640
Walsh, A. J., Macdonald, G. H., Alvey, N. D. S., Burton, M. G., & Lee, J.-K. 2003 *A&A* 410, 597

Astrophysical Masers and their Environments
Proceedings IAU Symposium No. 242, 2007
J.M. Chapman & W.A. Baan, eds.

doi:10.1017/S1743921307013002

The Methanol Multibeam Survey

James A. Green[1], **R. J. Cohen**[1], **J. L. Caswell**[2], **G. A. Fuller**[3], **K. Brooks**[2], **M. G. Burton**[6], **A. Chrysostomou**[4], **P. J. Diamond**[1], **S. P. Ellingsen**[5], **M. D. Gray**[3], **M. G. Hoare**[7], **M. R. W. Masheder**[8], **N. McClure-Griffiths**[2], **M. Pestalozzi**[4], **C. Phillips**[2], **L. Quinn**[1], **M. A. Thompson**[4], **M. Voronkov**[2], **A. Walsh**[1]1, **D. Ward-Thompson**[9], **D. Wong-McSweeney**[3], **J. A. Yates**[10], **J. Cox**[9].

[1]Jodrell Bank Observatory, The University of Manchester;
[2]Australia Telescope National Facility; [3]The University of Manchester; [4]University of Hertfordshire; [5]University of Tasmania; [6]University of New South Wales; [7]University of Leeds; [8]Bristol University; [9]University of Cardiff; [10]University College, London. [11]James Cook University

Abstract. A new 7-beam methanol multibeam receiver is being used to survey the Galaxy for newly forming massive stars, that are pinpointed by strong methanol maser emission at 6.668 GHz. The receiver, jointly constructed by Jodrell Bank Observatory (JBO) and the Australia Telescope National Facility (ATNF), was successfully commissioned at Parkes in January 2006. The Parkes-Jodrell survey of the Milky Way for methanol masers is two orders of magnitude faster than previous systematic surveys using 30-m class dishes, and is the first systematic survey of the entire Galactic plane. The first 53 days of observations with the Parkes telescope have yielded 518 methanol sources, of which 218 are new discoveries. We present the survey methodology as well as preliminary results and analysis.

Keywords. stars: formation, Masers, Surveys, (galaxies:) Magellanic Clouds

1. Introduction

The 6668-MHz methanol maser has an intrinsic relationship with high-mass star-formation, having to date been observed exclusively in association with known high-mass star-formation tracers (Minier *et al.* 2003). The theory of massive ($>8\ M_{\odot}$) star-formation is not clearly defined, as the accretion processes of low-mass star-formation encounter conflicts between radiation pressure, thermal pressure and the accretion flow when scaled to high masses (Keto 2003, Yorke 2004). Within the context of massive star-formation, gaseous A^{+} type methanol is stimulated to produce a strong maser from the $5_1 \rightarrow 6_0$ transition at 6668.5192 MHz, first observed by Menten (1991). The 6.668-GHz methanol maser is therefore a useful tool to trace the Galactic distribution of massive star formation (and indeed the structure of the Galaxy itself).

The Methanol Multibeam (MMB) Survey is a Galactic plane survey for both 6.668-GHz methanol and 6.035-GHz excited OH, observing $0° < l < 360°$, $b \pm 2°$. The survey will take about 200 days split between the Parkes telescope in the southern hemisphere and the Lovell telescope in the northern. All detections will have high-resolution follow-up observations using the Australian Telescope Compact Array (ATCA) and the Multi-Element Radio Linked Interferometer Network (MERLIN). In addition to the Galactic plane survey, the MMB project will incorporate a survey of the Small and Large Magellanic Clouds and a pulsar survey "piggyback".

2. MMB Survey

2.1. *Multibeam Receiver*

The MMB survey is being conducted with a new 7-beam receiver jointly constructed by Jodrell Bank Observatory and the ATNF. This receiver comprises an hexagonal layout of 7 identical feeds capable of dual circular polarization over a large bandwidth. The low noise amplifiers are indium-phosphide High Electron Mobility Transistors (HEMTS), and along with 7 polarizers and 14 frequency downconverters, were constructed at JBO. The ATNF designed and constructed the feeds for the Parkes dish, along with the waveguide interface, provision of a 500-Hz switched noise diode calibration system, and the dewar assembly. The ATNF team also conducted all the receiver integration and rebuilds. The receiver was successfully commissioned at Parkes on the 21st of January 2006 and took first light on the 22nd January (Figure 1).

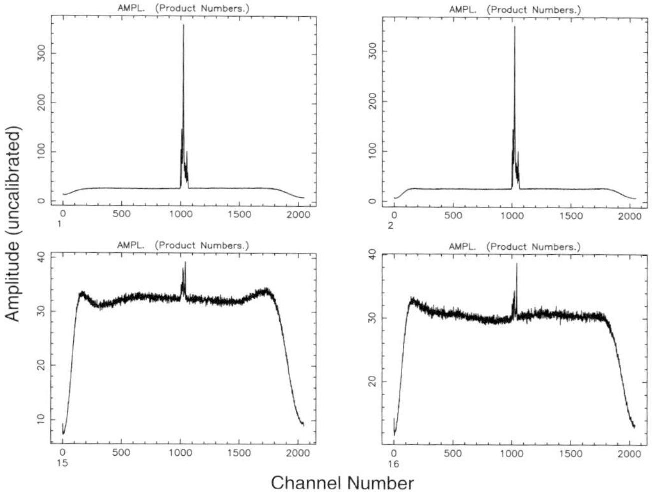

Figure 1. First light of the MMB survey, maser source G309.21+0.48. Top spectra are 6.668-GHz methanol in both right and left polarizations, bottom are 6.035-GHz OH in both polarizations.

2.2. *Survey Characteristics*

For the Parkes observations the correlator is configured for a bandwidth of 4 MHz, which corresponds to $180\,\mathrm{kms}^{-1}$ at 6.668 GHz, but $150\,\mathrm{kms}^{-1}$ is taken as fully usable. The centre velocities and range for observations have been determined from the Galactic CO emission of Dame, Hartmann & Thaddeus (2001); to fully cover the required velocities there is one setting for $20° < | l |$, two for $6° < | l | < 20°$, three for $2° < | l | < 6°$ and finally 4 for $| l | < 2°$. Doppler tracking is incorporated via the first 2 local oscillators. The correlator has 2048 frequency channels for each IF and each polarization, and the resultant 28*2048 spectra are dumped every 5 seconds. The 2048 channels across 4 MHz gives a velocity resolution of $0.0879\,\mathrm{kms}^{-1}$ at 6.668 GHz ($0.0971\,\mathrm{kms}^{-1}$ at 6.035 GHz). This is well within the established velocity width of methanol masers (0.2 - $0.3\,\mathrm{kms}^{-1}$). The system temperature is typically <20 K. The FWHM beam for methanol is 3.2 arcmin and for the excited OH it is 3.4 arcmin. The beams are spaced at 6.46 arcmins, so the total multibeam footprint is 15 arcmins wide. The beam pattern (Figure 2), is oriented at an optimal 19.1° to the scan direction (the plane of the Galaxy) which maintains equal beam spacings (Condon, Broderick & Seielstad 1989).

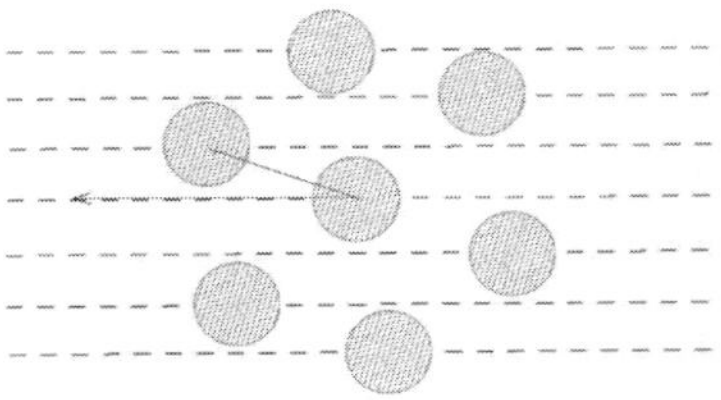

Figure 2. Multibeam Footprint, oriented at 19.1° to the Galactic plane.

Scans are conducted across 2° in longitude at a rate of 0.1° per min. The beams are then displaced by 1.23 arcmin in latitude before scanning back. There is then a larger displacement of 15 arcmins to the next scan pair, thus fully sampling a block of Galaxy of 2° in longitude by 4° in latitude requires 32 scans. One scan takes 20 mins, therefore a 2° by 4° block is completed in ~10 hours of observing time. Only one pass is needed as there is minimal RFI at 6.668 and 6.035 GHz at the Parkes site. The beam sensitivity varies by at worst ±3.5% across a 2° by 4° block.

For the purposes of calibrating the noise diode equivalent flux density values seen by each beam, and to measure pointing offsets, the continuum source 1934-638 is observed daily, using a SPOT schedule which scans each beam across the source in RA and Dec. Calibration stability is further checked by observing on most days a known maser source in MX mode (in which a source is tracked, and the pointing centre cycled through each of the beams). Commonly used sources are 232.621+0.996 for methanol, and 300.069+1.147 for both methanol and excited OH.

2.3. *Data Reduction*

The raw correlator data is processed through the package LIVEDATA, developed by Mark Calabretta, to facilitate processing of multibeam spectral line data (http://aips2.nrao.edu /docs/cookbook/cbvol3/node5.html). This removes a bandpass estimate and calibrates the flux density using a table of values derived from the SPOT observation data. LIVEDATA processing is done in realtime, automatically streaming the raw data from the correlator. The resultant Single Dish FITS files are then gridded using a top-hat smoothing kernel with a radius of 4.4 arcmin and a cutoff of 2.2 arcmin, to form Galactic longitude-latitude-velocity cubes with GRIDZILLA (also developed by Mark Calabretta). These data cubes are first examined by eye using the Kvis visualization tool (which is part of the Karma library) and then we run an automated detection algorithm developed by Maxim Voronkov. This gives us a typical rms noise of 0.17 Jy, and hence a 3.5σ detection limit of 0.6 Jy.

3. Results

In the first 12 months since commissioning the new receiver, observations have been made at Parkes over 53 days, completing 106 degrees of longitude. This represents 64% of the Parkes part of the survey, 32% of the total survey. The observations have yielded 518 methanol maser detections, of which 218 represent new sources (Figure 3). Most of our new detections are below 4 Jy, typically around 1 Jy peak intensity, with linewidth FWHMs of less than 0.2 $\mathrm{km s^{-1}}$. The detections include several far-side outer Galaxy masers.

4. Statistics

In terms of survey statistics Figure 4 shows a plot (using data from the first 5 observing sessions) of the Galactic distance as calculated using the rotation curve of Brand & Blitz

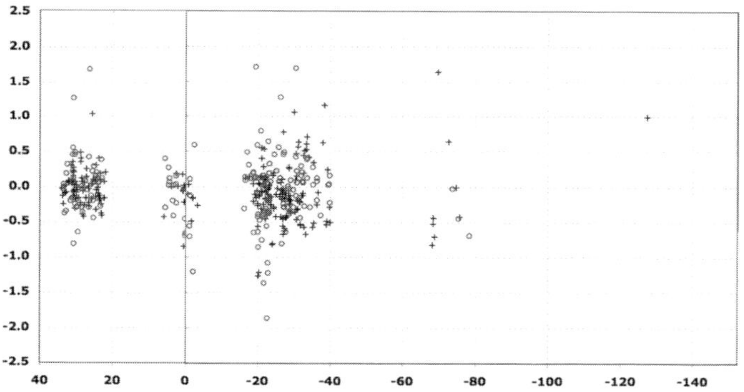

Figure 3. Longitude vs latitude plot for the Parkes survey region showing the MMB detections to date. Crosses represent previously known, circles new detections.

(1993) against number of sources, employing 1 kpc bins. This distribution is comparable to that found by Pestalozzi, Minier & Booth (2005) with the peak at the molecular ring distance. Figure 5 shows the flux density distribution, clearly highlighting the abundance of new detections at low flux densities. It is also possible to see that the peak in total detections at approximately 2 Jy, agrees with the observations of the Arecibo Galactic Plane Survey (Pandian, Goldsmith & Deshpande 2007), which has an rms noise level of ~70 mJy per spectral channel.

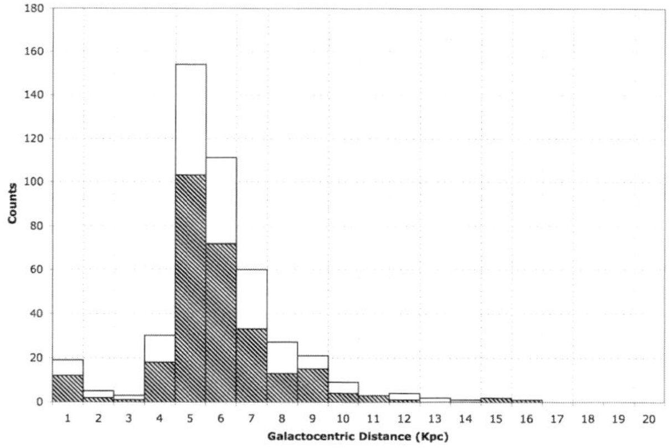

Figure 4. Galactocentric distance against number counts based on 1 kpc bins for the MMB survey. Shaded represents previously known detections from Pestalozzi, Minier & Booth (2005), white represents new detections from this survey.

5. High Resolution Follow-up

The Parkes detections can readily be positioned to an accuracy of the order of 30 arcsec, which is adequate for Galactic distribution studies, but it is not enough to determine the characteristics of massive star-formation, the association of the 6.668-GHz methanol maser with other species of masers, or the detailed structure of individual sources. The positions of the Parkes detections are then observed with MERLIN and the ATCA to refine the positions to the sub-arcsec accuracy achievable with these telescopes. So far, all 218 new detections have been observed and positioned using ATCA observations and 30 sources have been observed with MERLIN in the $20° < l < 30°$ region, positioning 25.

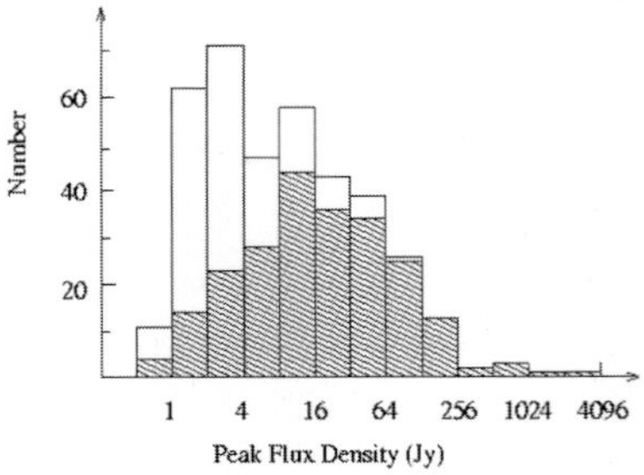

Figure 5. Peak flux density distribution for the known (shaded) and new (white) detections for the MMB survey.

6. Pulsar Survey "Piggyback"

In addition to the main MMB survey, we are also conducting "piggyback" observations on a Pulsar survey using our receiver. This is just at the methanol frequency, so as to allow us to use 4096 channels across an 8 MHz bandwidth, thus giving the same spectral resolution as the main survey, but across a wider velocity range. The pulsar survey covers a quite small area, primarily near the Galactic centre, but may extend out to $-45^\circ < l < 60^\circ$. The principal advantage of this "piggybacking" is to provide long integrations, which taking into account the higher background noise, will give an order of a magnitude improvement in sensitivity, and thus should provide a good indication of the completeness.

7. Conclusions

The Parkes observations of the MMB are over 60% complete and have yielded 518 detections, 218 of which are new and have been positioned to 0.1 arcsec with ATCA. The Parkes observations will be complete in 2008, at which point the receiver will move to Jodrell to complete the Galactic survey by 2010. The final MMB catalogue of 6.668-GHz methanol should thus be complete by 2011. With the complete catalogue the authors will also provide IR identifications and complementary survey associations (such as the CORNISH VLA continuum survey), distances to the detections, and an analysis of the Galactic distribution. The deep integration pulsar survey "piggyback" data will provide information on the completeness of the MMB survey. The MMB can be found online at http://www.jb.man.ac.uk/research/methanol/.

Acknowledgements

J. A. G. is supported by a PPARC studentship. The MMB project is dedicated to the memory of R.J.Cohen.

References

Brand, J., & Blitz L. 1993, *A&A* 275, 67
Condon, J. J., Broderick, J. J., & Seielstad, G. A. 1989, *ApJ* 97, 1064
Dame, T. M., Hartmann, D., & Thaddeus, P. 2001, *ApJ* 547, 792
Keto, E. 2003, *ApJ* 599, 1196
Menten, K. M. 1991, *ApJ* (Letters) 380, L75
Minier, V., Ellingsen, S. P., Norris, R. P., & Booth, R. S. 2003, *A&A* 403, 1095
Pestalozzi, M. R., Minier, V., & Booth, R. S. 2005, *A&A* 432, 737
Pandian, J. D., Goldsmith, P. F., & Deshpande, A. A. 2007, *ApJ* 656, 255
Yorke, H. W. 2004, in: R. Jayawardhana, M. G. Burton, & T. L. Bourke (ed.), *Star-Formation at High Angular Resolution* (IAU Symposium 221), p. 141

Astrophysical Masers and their Environments
Proceedings IAU Symposium No. 242, 2007
J.M. Chapman, & W.A. Baan, eds.

doi:10.1017/S1743921307013014

A 20-year H_2O maser monitoring program with the Medicina 32-m telescope

J. Brand[1], M. Felli[2], R. Cesaroni[2], C. Codella[3], G. Comoretto[2], S. Di Franco[4], F. Massi[2], L. Moscadelli[2], R. Nesti[2], L. Olmi[3], F. Palagi[3], F. Palla[2], D. Panella[2], and R. Valdettaro[2]

[1]INAF-Istituto di Radioastronomia, Bologna, Italy
email: brand@ira.inaf.it

[2]INAF-Osservatorio Astrofisico di Arcetri, Florence, Italy

[3]INAF-Istituto di Radioastronomia, Sez. di Firenze, Florence, Italy

[4]Dip. di Astronomia, Università degli Studi di Firenze, Florence, Italy

Abstract. The Arcetri/Bologna H_2O maser group has been monitoring the 1.3-cm water maser emission from a sample of 43 star-forming regions (SFRs) and 22 late-type stars for about 20 years at a sampling rate of 4-5 observations each year, using the 32-m Medicina Radio Telescope (HPBW 1$'$.9 at 22 GHz). For the late-type stars we observe representative samples of OH/IR-stars, Mira's, semi-regular variables, and supergiants. The SFR-sample spans a large interval in FIR luminosity of the associated Young Stellar Object (YSO), from 20 $L_\odot$ to 1.5 $\times 10^6$ $L_\odot$, and offers a unique data base for the study of the long-term (years) variability of the maser emission in regions of star formation.

This presentation concerns only the masers in SFRs. The information obtained from single-dish monitoring is complementary to what is extracted from higher-resolution (VLA and VLBI) observations, and can better explore the velocity domain and the long-term variability therein. We characterize the variability of the sources in various ways and we study how it depends on the luminosity and other properties of the associated YSO and its environment.

Keywords. masers, astronomical data bases: miscellaneous

1. Introduction

Twenty years ago in 1987, in the same month that this Symposium is held, we registered the first spectrum of many in our H_2O maser monitoring campaign. Since then, four or five times a year we have taken a spectrum of 43 masers in star-forming regions and in 22 circumstellar envelopes of late-type stars. The latter will not be discussed here - instead we refer to the poster by Engels, Winnberg, Brand, & Jimenez-Esteban in these proceedings.

After the presentation and analysis of a sub-set of our database of masers in SFRs (Valdettaro *et al.* 2002; Brand *et al.* 2003; Brand *et al.* 2005) we are now preparing to release our entire archive.

Interstellar water masers are highly variable, regarding both the velocities at which emission is detected and the intensity of that emission. The importance of the maser monitoring database lies above all in the opportunity it offers to study the *long-term variability*, i.e. on a scale of years. Valdettaro *et al.* (2002) presented a number of tools with which to visualize and quantify the maser data. One of the visual tools is the velocity-time-flux density diagram (hereafter: VtF-diagram), which allows a bird's eye view of the behaviour of the emission during the monitoring period (cf. Fig. 3). In view of the complex patterns

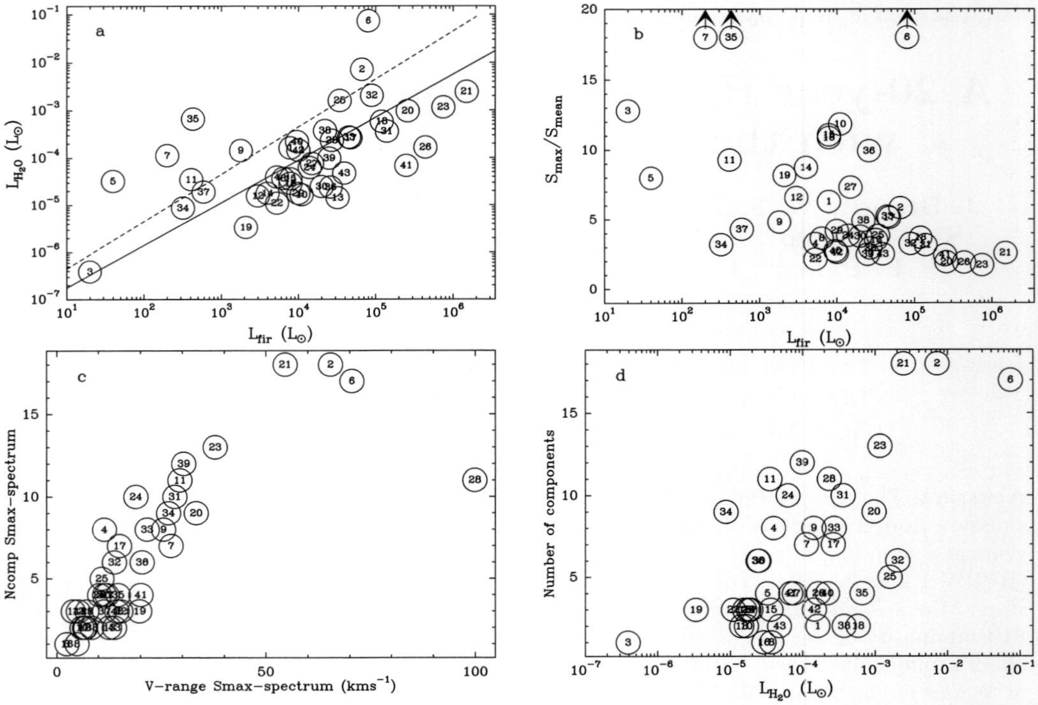

Figure 1. a. The luminosity of the maser, derived from the spectrum with the maximum integrated flux density during the monitoring, as a function of L_{fir}. The drawn line is a bisector least-squares fit to the data. The dashed line indicates the relation between the two luminosities found by Wouterloot *et al.* (1995). **b.** Variability index S_{max}/S_{mean} versus L_{fir}.. Three objects that have values outside the range shown are indicated by the symbols with the arrows. The deviation of nr. 6 (KLIRc2) is caused by a particularly strong burst (cf. Fig. 2b). **c.** The number of emission components ($> 5\sigma$) versus the range in velocity they cover, for the spectra with maximum integrated flux density of each source. **d.** As c, but showing the number of components as a function of the H_2O luminosity derived from those spectra.

that emerge from these diagrams which we derive for each source, the simple standard analysis of the variation of the integrated intensity, although still indicative of the overall maser emission, is clearly limited and may miss more physically interesting aspects of the activity within star-forming regions. For example: a) velocity drifts, either steady and continuous over the entire monitoring period or occasional, preferentially after bursts; b) bursts occuring at different velocities and their frequency of occurrence; c) correlated or anti-correlated intensity variations at different velocities, d) the fraction of time in which the maser is active, either in an absolute sense or in each individual velocity channel. Below we highlight these and other aspects, to illustrate how the database can be mined to yield useful physical information.

2. Results

2.1. *General properties*

The general properties of the maser emission, especially as a function of the luminosity L_{fir} of the associated IRAS (pumping) source, were extensively analyzed by Brand *et al.* (2003). Here we present four of the figures from that paper, updated for the whole sample. The properties and trends found for the sub-sample are confirmed: there is a correlation

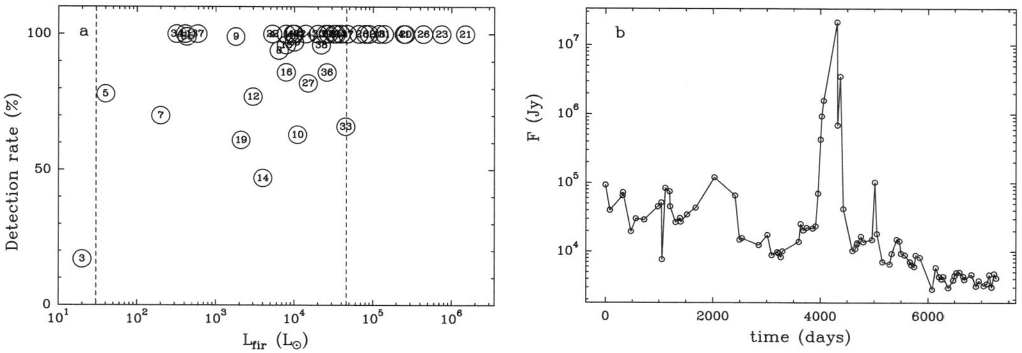

Figure 2. a. Detection rate (at levels $> 5\sigma$) of masers as function of $L_{\rm fir}$. The dashed lines separate $L_{\rm fir}$-ranges with different maser detection rates (see text). **b.** Monster-burst in KLIRc2 (1997-1999): flux density versus time.

between the maximum maser luminosity and $L_{\rm fir}$ [$\log L_{\rm H_2O} = (-7.66 \pm 0.47) - (0.90 \pm 0.12)\log L_{\rm fir}$], in good agreement with the upper limit found from single-epoch observations of many maser sources (Fig. 1a; cf. Wouterloot *et al.* 1995). The ratio between the maximum and the mean integrated flux density can be used as a variability index; its decrease with increasing $L_{\rm fir}$ (Fig. 1b) shows that more luminous sources are associated with more stable masers. The correlations found between the number of emission components and their velocity range (Fig. 1c) and the maser luminosity (Fig. 1d) indicate that higher maser power goes into more emission channels, spread over a larger range in velocity. Figure 2a shows the detection rate ($> 5\sigma$) of maser emission as a function of $L_{\rm fir}$. For masers associated with low-luminosity pumps ($L_{\rm fir} < 30$ L$_\odot$) the maser is undetected much of the time (det. rate 17%). Although there is only one object in this range, this luminosity limit is consistent with the 25 L$_\odot$ found by Claussen *et al.* (1996). For 30 L$_\odot < L_{\rm fir} < 4.6 \times 10^4$ L$_\odot$, the average detection rate is 90%, with a dispersion of 15%. This can be understood if for an individual source the masing conditions depend on the orientation of the outflow(s) and on the presence of high-density molecular gas rather than on $L_{\rm fir}$; only for $L_{\rm fir} > 4.6 \times 10^4$ L$_\odot$, the exciting source is the dominant factor and the maser is always detected.

2.2. *Bursts and velocity drifts*

Bursts of maser emission can easily be spotted in the VtF-diagram, and appear to be quite common. Often they also stand out in the maser light curve, as shown in Fig. 2b for KLIRc2. This burst began on Dec. 14, 1997 and reached a maximum flux density of 2.1×10^7 L$_\odot$ (an increase by a factor of almost 1300) after 400 days. After another 284 days the source was back to its normal level. Note that the general trend over the years is for the flux density to *decrease*. Figure 3 shows four examples of VtF-diagrams in which we have indicated possible velocity gradients of maser components. The indicated drifts (both acceleration and decelaration) are of the order of $0.3-3.5$ km s^{-1} yr^{-1}. In the case of IRAS20126+4104 , component (2) (Fig. 3b) the reality of the indicated drift could be verified by means of multi-epoch high spatial resolution observations (see Moscadelli *et al.* 2005): from 3-epoch VLBI observations absolute positions and proper motions of various maser spots could be determined. Extrapolating the proper motions backwards and forwards in time showed the spots' positions to be in good agreement with those actually measured in observations with VLBA (earlier) and MERLIN (later). Moscadelli *et al.* (2005) interpreted these observations in terms of maser spots being located on the walls of a conical outflow originating at the YSO where a stellar jet impacts with the

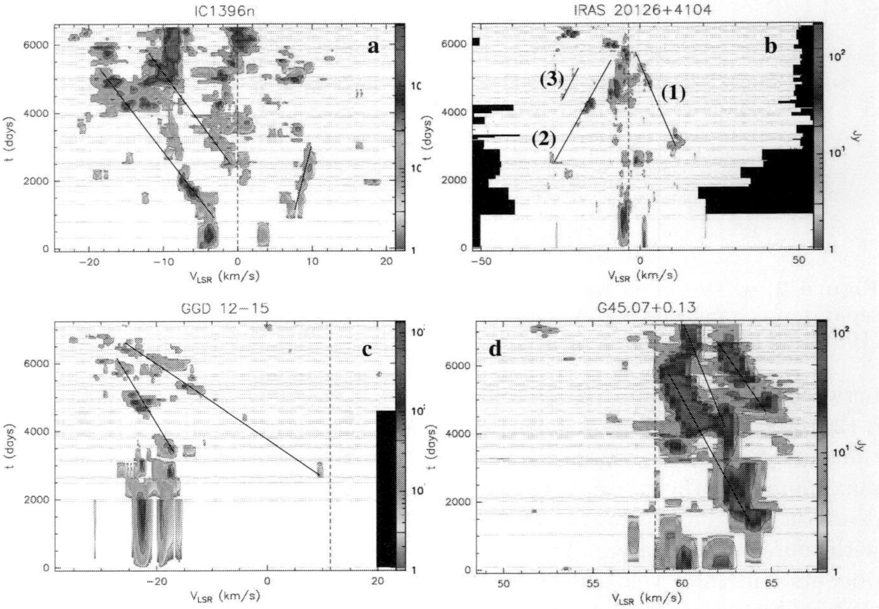

Figure 3. Velocity gradients in VtF-diagrams. Drawn lines indicate possible velocity drifts of maser components. **a.** IC1396N **b.** IRAS20126+4104 **c.** GGD12–15 **d.** G45.07+0.13. (See the online version for colour contours).

surrounding material. Subsequent deceleration leads to the temporal behaviour of the maser features as seen in the single-dish observations of Fig. 3b. We note, by the way, that a look at Fig. 3 already makes it evident that an instantaneous maser spectrum of a source will usually not be a "typical" representation of the maser emission of that source; in fact, a "typical" maser spectrum is unlikely to exist.

2.3. *Super-variability*

Apart from variations on relatively short timescales, longer-period "super-variability" has been found as well. Several examples are shown in Fig. 4, where the total maser output is shown as a function of time. This phenomenon, with periods between 4 and >16 years, is only brought to light in monitoring programs like this - in fact it has also been found in the Pushchino monitoring data (e.g., Lekht *et al.* 2001; Lekht *et al.* 2002). It may be caused by periodic variations in the wind of the Young Stellar Object.

3. Future Work

The 20-year database is presented in Felli *et al.* (2007); plots of all spectra and all derived information (VtF-diagrams, upper envelopes, light curves etc.) will be available on-line (http://www.arcetri.astro.it/science/Radio). This will be followed by a general analysis of the database, and by studies of individual objects. We think that a detailed knowledge of the immediate surroundings of the YSOs is essential in understanding the variability of the masers. This is illustrated for instance by the predominantly blue

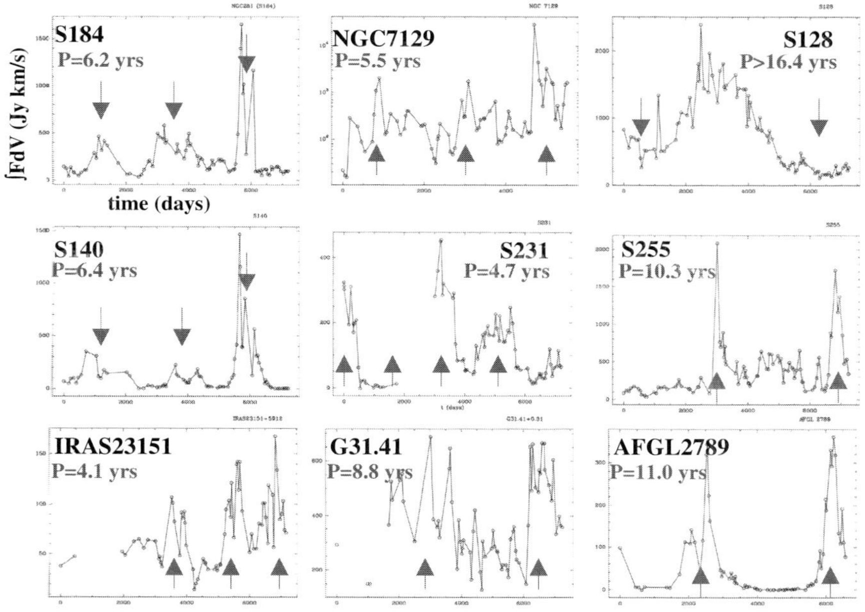

Figure 4. Long-term ("super") variability in maser sources. We show maser lightcurves ($\int FdV$ versus time) for nine objects. Approximate periods are indicated in each panel. The arrows indicate the location of the maxima (minima, for S128) from which the periods were derived.

tails sometimes seen in the detection rate histograms (see Brand *et al.* 2003), which we interpret as indicating a fortuitous alignment along the line-of-sight to the YSO of a highly collimated outflow, allowing one to see the amplification of the radio continuum emission of a compact HII region surrounding that YSO. Therefore we shall continue the monitoring, and carry out follow-up observations with interferometers.

References

Brand, J., Cesaroni, R., Comoretto, G., Felli, M., Palagi, F., Palla, F., & Valdettaro, R. 2003, *A&A* 407, 573

Brand, J., Cesaroni, R., Comoretto, G., Felli, M., Palagi, F., Palla, F., & Valdettaro, R. 2005, *Ap&Sp. Sci* 295, 133

Claussen, M. J., Wilking, B. A., Benson, P. J., *et al.* 1996, *ApJS*, 106, 111

Felli, M., Brand, J., Cesaroni, R., *et al.* 2007, *A&A, in preparation*

Lekht, E. E, Mendoza-Torres, J. E., & Berulis, I. I. 2002, *Astron. Rep.* 46, 57

Lekht, E. E, Paschenko, M. I., & Berulis, I. I. 2001, *Astron. Rep.* 45, 949

Moscadelli, L., Cesaroni, R., & Rioja M. J. 2005, *A&A* 438, 889

Valdettaro, R., Palla, F., Brand, J., Cesaroni, R., Comoretto, G., Felli, M., & Palagi, F. 2002, *A&A* 383, 244

Wouterloot, J. G. A, Fiegle K., Brand, J., & Winnewisser, G. 1995, *A&A* 301, 236 (Erratum: 1997, A&A 319, 360)

Astrophysical Masers and their Environments
Proceedings IAU Symposium No. 242, 2007
J.M. Chapman & W.A. Baan, eds.

doi:10.1017/S1743921307013026

The large-scale distribution and physical properties of massive star forming regions in the Milky Way

Jarken Esimbek[1], Zhou Jianjun[1] and Zheng Xingwu[2]

[1]National Astronomical Observatories/Urumqi Observatory, CAS, Urumqi 830011, China
email: Jarken@ms.xjb.ac.cn

[2]Department of Astronomy, Nanjing University, Nanjing 210093, China,
email: xwzheng@nju.edu.cn

Abstract. We are conducting a large survey with the Urumqi 25 m radio telescope to study the distribution and physical properties of massive star forming regions in the Milky Way. We will accomplish this by observing the $H_{110\alpha}$ hydrogen recombination line at 4.874 GHz and the H_2CO absorption line at 4830 GHz. These lines are associated with compact HII regions and dense molecular clouds. As a test of the 6 cm spectral line receiving system, we observed the two spectral lines toward massive star forming regions. We plan to study the large scale distribution and physical properties of massive star forming regions in the Milky way.

Keywords. ISM: HII regions,Galaxy: kinematics and dynamics,Galaxy: structure

---

1. The spectral system and observational tests

Formadehyde (H_2CO) was one of the first poly-atomic molecules to be detected in interstellar space and has since been found to be a common constituent of interstellar molecular clouds. H_2CO is associated with HII regions and is therefore a good tracer and probe of regions of star formation. HII regions can be expected to provide crucial evidence for any spiral structure in own Galaxy. This was recognized in the bold attempt by Georgelin & Georgelin (1976) to derive a grand spiral pattern for our Galaxy from studies of HII regions. Valuable additions to the northern hemisphere data were made by Downes *et al.* (1980) and to the southern hemisphere data were made by Caswell & Haynes(1987).

We will measure the line velocities, and hence derive kinematic distances. The kinematic distance ambiguity may partly be resolved by observing the two lines simultaneously in the same band. From the survey we aim to establish the Galactic distribution of massive star forming regions. The study of global star formation will provide important information for the environment of young stars and clusters. However, global star formation still remains poorly understood.

The 25-meter radio telescope of Urumqi Astronomical Station of NAOC, is located at Nanshan Mountain near the sourth suburb of Urumqi city, Xinjiang province. It is situated 2080 m above sea level, at a very dry site for short centimeter observations. The Urumqi Observatory 25 m radio telescope has a 4096-channel digital correlator as the back end. The correlator, used with a 80 MHz bandwidth, can simultaneously observe the $H_{110\alpha}$ hydrogen recombination line at 4.874157 GHz and the formaldehye (H_2CO) absorption line at 4.82966 GHz. In the C band, the 25 m telescope has a spatial resolution of about $10'$, and a system temperature of about 22 K. This is well suited to observing the two spectral lines simultaneously. We are using this system to survey the large-scale distribution and physical properties of massive star forming regions in the Milky Way.

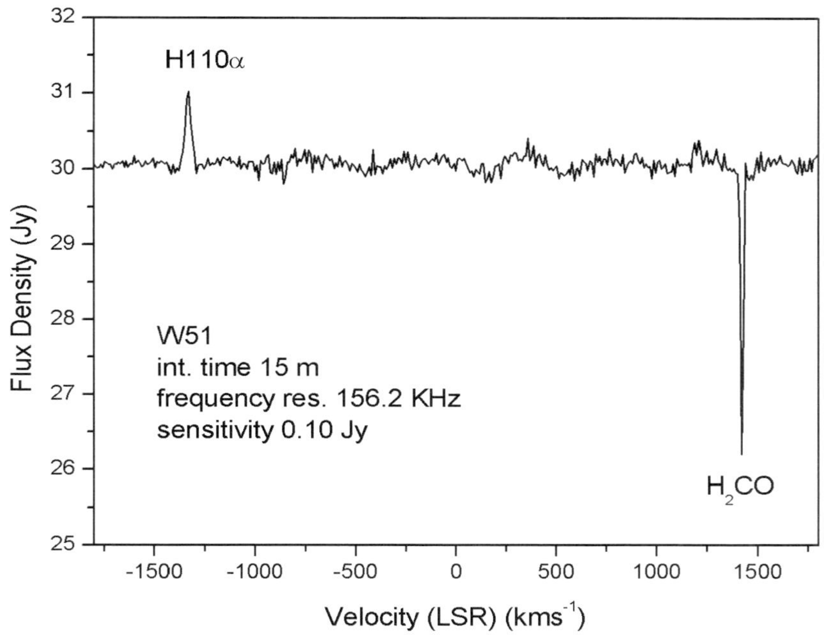

Figure 1. As a testing of 6 cm spectral line receiving system, we observed the two spectral lines toward W51 massive star forming regions, integration times is 15 m, the system temperature was about 22 K.

As a test of the 6 cm spectral line receiving system, we observed the two spectral lines toward massive star forming regions. An example is shown in Figure 1.

Acknowledgements

This work was financially supported by funds from the National Natural Science Foundation of China (Grant Nos. 10203003 and 10073004), and supported by the Program of the Light in China's Western Region (LCRW) under Grant No.RCPY200605.

References

Caswell, J. L. & Haynes, R. F 1987, *A&A* 171, 261C
Downes, D., Wilson, T. L., Bieging, J., & Wink, J. 1980, *A&AS* 40, 379D
Georgelin, Y. M. & Georgelin, Y. P. 1976, *A&A* 49, 57G

Astrophysical Masers and their Environments
Proceedings IAU Symposium No. 242, 2007
J.M. Chapman & W.A. Baan, eds.

doi:10.1017/S1743921307013038

Wide field imaging of Giant Molecular Clouds with the ATA

James R. Forster
UC Berkeley, Hat Creek Radio Observatory, 42233 Bidwell Rd, Hat Creek, CA, 96040, USA

Abstract. Galactic Giant Molecular Cloud (GMC) complexes are among the largest objects in the sky. In order to get the full picture of Galactic star formation a large field of view is important. The Allen Telescope Array (ATA) is the first of a new generation of large-N, small-D radio telescopes designed to provide high quality images on large angular scales. We present here the first wide field images of GMCs made with the ATA.

Keywords. masers, stars: formation, ISM: individual (W3), ISM: molecules

1. Introduction

The Allen Telescope Array is a joint project between the SETI Institute and the University of California at Berkeley. Among its features are a wideband dual-polarization feed which provides continuous spectral coverage from 0.5 to 11.3 GHz, four simultaneous spectral windows tunable anywhere in the band, and eight dual-polarization beamformers for simultaneous observation of multiple target stars with the SETI 1-Hz resolution backend. This paper shows some initial science observations made with the partially completed system.

2. Instrument and observations

The completed ATA will have 350 six-meter dishes, a total extent of 900 m and a minimum baseline of 9.8 m. The final ATA correlator will allow full Stokes observations using all antennas, or simultaneous observation at up to four frequencies using subarrays. The ATA is currently (March 2007) operating with 20 antennas and two interim 4-antenna, 50 MHz, 512-channel correlators.

The GMC observations were made in January 2007 using four antennas connected to one of the two correlators, each providing six baselines at a single polarization and sky frequency. Observations at sky frequencies of 1666 (OH), 4830 (H_2CO) and 6667 (CH_3OH) MHz were made for each GMC observed. Tsys is below 55K at frequencies up to 6667 MHz. The length of each observation was typically a few hours at each antenna configuration. A continuum image was formed from the average of line-free channels in a 12 MHz bandwidth, and data cubes were made for channels containing line emission. The continuum sensitivity was between 7 and 17 mJy/beam, and 80 to 190 mJy/beam in the channel maps.

3. Results

Figure 1 shows ATA observations at 1.6, 4.8 and 6.7 GHz toward the W3 massive star-forming region, centered on the compact HII region W3(OH). The left (1666 MHz) panel shows the field of view of the 6-meter ATA antennas at all three frequencies. Continuum images are shown in grayscale at each frequency, and spectra are shown for

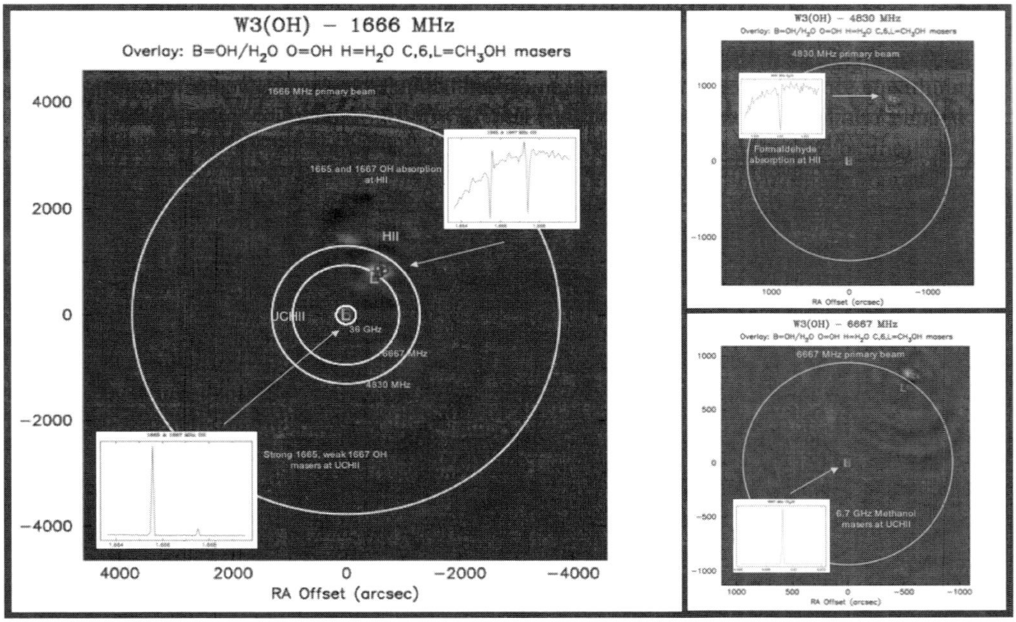

specific locations in the ATA fields. All panels show the FWHM beam at the frequency of observation as a white circle. Symbols representing known positions of OH, H_2O and CH_3OH masers are overlaid.

At the 2.0 kpc distance to W3 the 1.6 GHz FWHM of the ATA antennas (2.1 deg) is 73 pc across. This field contains two clusters of maser emission, one at the field center and another 17′ away near the edge of the 6.7 GHz primary beam. The cluster of masers at 17′ is associated with a continuum source detected at all three ATA frequencies, suggesting that it is an optically thin HII region. The OH spectrum taken at this location shows absorption by the 1665 and 1667 MHz hyperfine transitions of the OH radical, along with weak emission at slightly different velocities. Formaldehyde at 4.8 GHz is detected in absorption toward this HII region. The OH spectrum at the field center shows strong maser emission in the 1665 MHz line and weak emission at 1667 MHz. At 6.7 GHz a strong methanol line is also detected there, in agreement with previous observations.

4. Conclusions

Multi-frequency wide field imaging is a powerful way of assimilating complex astronomical process like the formation of massive stars and their association with masers, molecular clouds and HII regions. With sufficient sensitivity and dynamic range a full census of maser emission over a large field can be taken with a single pointing, and large scale surveys can be undertaken with much greater speed and efficiency than with single-dish telescopes or arrays with larger diameter antennas.

Acknowledgements

I would like to acknowledge the staff at HCRO, and the scientists and engineers at UC Berkeley and the SETI Institute for their part in designing and building the ATA. These people are too numerous to mention individually, but all have made important contributions to the success of the instrument.

Astrophysical Masers and their Environments
Proceedings IAU Symposium No. 242, 2007
J.M. Chapman & W.A. Baan, eds.

doi:10.1017/S174392130701304X

A search for OH 6 GHz maser emission towards southern supernova remnants

Korinne E. McDonnell, Alan E. Vaughan and Mark Wardle

Department of Physics, Macquarie University, Sydney, NSW 2109, Australia
email: {korinnem, alanv, wardle}@physics.mq.edu.au

Abstract. OH masers at 1720 MHz have proven to be excellent indicators of interactions between supernova remnants and molecular clouds. Recent calculations suggest that the 6049 MHz OH maser line is excited at higher column densities than for the 1720 MHz line. It is therefore a potentially valuable indicator of remnant-cloud interaction.

We present preliminary results of a survey using the Parkes Methanol Multibeam receiver for 6049 MHz and 6035/6030 MHz OH masers towards 36 supernova remnants and four fields in the Large and Small Magellanic Clouds. While no 6049 MHz masers have been found, three new sites of 6035 and 6030 MHz OH maser emission have been discovered in star-forming regions.

Keywords. supernova remnants, masers, stars: formation, radio lines: ISM

OH masers at 1720 MHz have proven to be an almost unambiguous indicator of the interaction between supernova remnants and molecular clouds (Frail *et al.* 1994). OH excitation calculations (Pavlakis & Kylafis 2000; Wardle, these proceedings) suggest that the 6049 MHz satellite line may be present at higher OH column densities where the 1720 MHz line is weak or absent.

A survey using the Parkes Methanol Multibeam receiver at 6049 and 6035/6030 MHz was conducted towards 36 supernova remnants and four fields in the Large and Small Magellanic Clouds. An 8 MHz bandwidth with 2048 channels was used, allowing a channel width of 0.2 km s^{-1}. The beam size is approximately 3.3 arcminutes and both circular polarisations were observed. Scans in right ascension and declination were conducted over the supernova remnants, with a total observing time in one direction of $\sim$100 minutes per square degree. The data was flux-calibrated, continuum subtracted and gridded using the programs Livedata and Gridzilla. The data cubes produced were searched for maser emission using the program Duchamp (Whiting 2007).

While preliminary analysis has not discovered any 6049 MHz maser emission, five maser sites at 6035 MHz have been identified (two containing 6030 MHz emission as well). Masers 6.86-0.09, 34.27-0.20 and 48.98-0.30 are new discoveries (see Fig. 1), while 336.941-0.156 and 337.705-0.053 are already known (Caswell & Vaile 1995; Caswell 2001).

As left and right-hand circular polarisations (LCP and RCP) were observed, Zeeman pairs can be recognised. A 1 mG magnetic field produces splittings equivalent to 0.079 km s^{-1} and 0.056 km s^{-1} in the 6030 and 6035 MHz transitions respectively. The magnetic fields calculated can be found in Table 1 and have an uncertainty of approximately 2 mG.

The 6035 MHz maser discovered at 48.98-0.30 is approximately coincident with two H II regions and is likely to be associated with one of them. It is expected that the other 6030 and 6035 MHz masers are associated with star forming regions.

Table 1. Masers at the 6035- and 6030- MHz OH transitions. The velocity and flux of the peaks are taken at the brightest peak, if multiple peaks are observed.

OH maser (l b) (° °)	RA (2000) (h m s)	Dec (2000) (° ′ ″)	6035 MHz Velocity Peak LCP (km s^{-1})	RCP	6035 MHz Peak flux LCP (Jy)	RCP	Magnetic Field (mG)	6030 MHz Peak (L and/or R) (Jy)	
336.941-0.156	16 35 55.20	-47 38 45.4	-65.6	-65.1	3.35	1.86	+8.9	1.03	0.46
337.705-0.053	16 38 29.67	-47 00 35.8	-53.6	-50.7	1.63	2.22	-	-	
6.86-0.09	18 00 48	-22 58 14	-2.37	-1.98	4.23	1.03	+7.0	-	
34.27-0.20	18 54 36	+01 05 54	54.3	54.5	4.60	2.33	+3.6	1.76 (L)	
48.98-0.30	19 22 27	+14 06 53	67.5	67.7	1.55	4.27	+3.6	-	

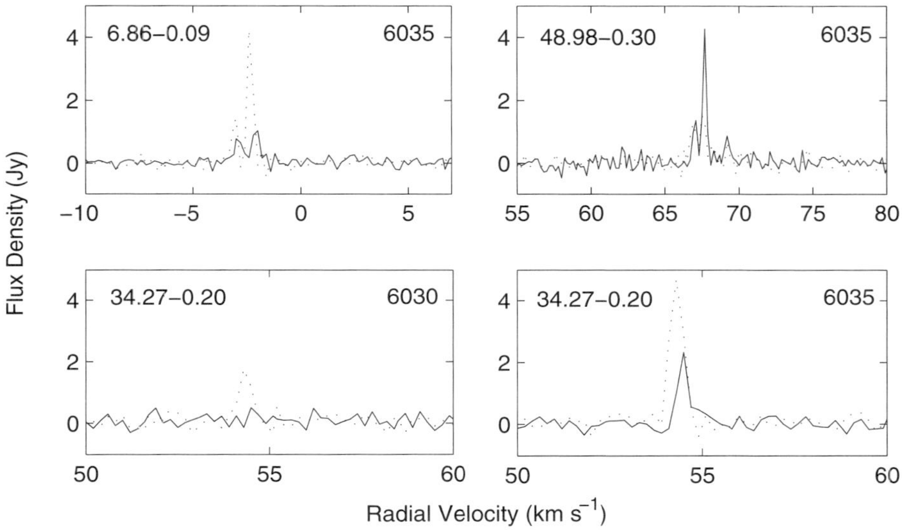

Figure 1. Spectra of new OH masers at the 6030- and 6035-MHz transitions. LCP and RCP are displayed as dashed and solid lines, respectively. The channel resolution is 0.2 km s^{-1} and the beamsize is approximately 3 arcminutes.

These results are preliminary and further analysis may yield weaker 6 GHz masers. In addition, data from ATCA observations of seventeen supernova remnants remain to be analysed.

Acknowledgements

We would like to thank Catherine Braiding for helping with the observing, Stacy Mader for observing assistance and help with data reduction, and James Caswell for assistance with maser identification.

References

Caswell, J. L. 2001, *MNRAS* 341, 551
Caswell, J. L. & Vaile, R. A. 1995, *MNRAS* 273, 328
Frail, D. A., Goss, W. M. & Slysh, V. I. 1994, *ApJ* 424, L111
Pavlakis, K. G. & Kylafis, N. D. 2000, *ApJ* 534, 770
Whiting, M. 2007, http://www.atnf.csiro.au/people/Matthew.Whiting/Duchamp/

Astrophysical Masers and their Environments
Proceedings IAU Symposium No. 242, 2007
J.M. Chapman & W.A. Baan, eds.

doi:10.1017/S1743921307013051

Millimetre wavelength methanol masers survey towards massive star forming regions

T. Umemoto[1], N. Mochizuki[2], K. M. Shibata[3], D.-G. Roh[4], and H.-S. Chung[4]

[1]Nobeyama Radio Observatory, Minamimaki, Minamisaku, Nagano 384-1305, Japan
email: umemoto@nro.nao.ac.jp

[2]Institute of Space and Astronautical Science, Sagamihara, Kanagawa, 229-8510, Japan

[3]National Astronomical Observatory, Mitaka, Tokyo 181-8588, Japan

[4]Taeduk Radio Astronomy Observatory, Hwaam-Dong, Yusong-Gu, Daejon 305-348, Korea

Abstract. We present the results of a mm wavelength methanol maser survey towards massive star forming regions. We have carried out Class II methanol maser observations at 86.6 GHz, 86.9 GHz and 107.0 GHz, simultaneously, using the Nobeyama 45 m telescope. We selected 108 6.7 GHz methanol maser sources with declinations above −25 degrees and fluxes above 20 Jy. The detection limit of maser observations was ∼3 Jy. Of the 93 sources surveyed so far, we detected methanol emission in 25 sources (27%) and "maser" emission in nine sources (10%), of which thre "maser" sources are new detections. The detection rate for maser emission is about half that of a survey of the southern sky (Caswell et al. 2000). There is a correlation between the maser flux of 107 GHz and 6.7 GHz/12 GHz emission, but no correlation with the "thermal" (non maser) emission. From results of other molecular line observations, we found that the sources with methanol emission show higher gas temperatures and twice the detection rate of SiO emission. This may suggest that dust evaporation and destruction by shock are responsible for the high abundance of methanol molecules, one of the required physical conditions for maser emission.

Keywords. masers, surveys, ISM: molecules, radio lines: ISM, stars: formation

---

1. Introduction

It is known that class II methanol maser emission is a good tracer for the earliest stages of massive star formation. Extensive surveys of methanol maser emission at 6.7 GHz have been carried out in the southern hemisphere (Val'tts *et al.* 1999; Caswell *et al.* 2000). Caswell *et al.* (2000) detected maser emission at 107 GHz in 22 of 84 sources (26%) and detected 157 GHz emission from only four sources (5%). However, no systematic surveys have been carried out in the northern hemisphere: Minier & Booth (2002) have surveyed a limited sample of 23 maser sources in the northern hemisphere. They detected maser emission at 107 GHz from six sources (26%) and at 86.6 GHz from two sources (9%). In this paper, we present the results of a systematic millimetre methanol maser survey in the northern hemisphere.

2. Observations

Methanol emission of 7_2-6_3 A^- at 86.6 GHz, 7_2-6_3 A^+ at 86.9 GHz and 3_1-4_0 A^+ at 107.0 GHz were simultaneously observed using the 45 m telescope of the Nobeyama Radio Observatory. In addition, we have made observations of SiO ($J=2$–1, $v=0$) and $H^{13}CO^+$ ($J=1$–0) lines simultaneously, to distinguish between maser and non-maser emssion by comparing the methanol data with thermal emission lines. We selected 108 sources of 6.7 GHz methanol maser sources with declinations > -25 degrees and fluxes > 20 Jy from the source list of Szymczak *et al.* (2000), Menten (1991) and others. We used two 3 mm SIS receivers. The backend was a bank of eight acousto-optical spectrometers with 37 kHz resolution (0.10 km s^{-1} at 107 GHz). The detection limit of our maser survey was $\sim$3 Jy, the same as for the southern survey by Caswell *et al.* (2000).

3. Results and discussion

So far we have surveyed 93 sources, with methanol emission detected in 25 sources (27%) and "maser" emission in nine sources (10%). Three "maser" sources, 18556+0136, 19110+1045 and 19216+1429, are new discoveries. Including previously known maser sources in the area we still need to survey, a total of 13 maser sources (14%) have been detected. Our detection rate of maser emission is about half that of the southern sky survey (Caswell *et al.* 2000). However, $H^{13}CO^+$ emission has been detected in only 48 sources (52%) in our sample, because of uncertainties in the 6.7 GHz maser positions and the smaller sizes of the emitting regions. If we consider only sources with detections of $H^{13}CO^+$ emission, then the detection rate for maser emission is 25%, the same as for the southern hemisphere sample.

We have compared the maser flux at 107 GHz and at 6.7 GHz/12 GHz including the southern survey data. The maser flux at 107 GHz is about (1–10)% of the flux at 6.7 GHz, and there is a correlation between the maser flux at 107 GHz and at 6.7 GHz and also at 12 GHz. However there is no correlation with the "thermal" (non maser) emission. We found that sources with methanol emission show higher gas temperature (determined from our NH_3 observations) and have twice the detection rate of SiO emission. In addition, sources with maser emission have wider thermal linewidths from SiO than from $H^{13}CO^+$. This may suggest that dust evaporation and destruction by shocks from outflows is responsible for the high abundance of methanol molecules, one of the required physical conditions of maser emission.

References

Caswell, J. L., Yi, J., Booth, R. S. & Cragg, D. M. 2000, *MNRAS*, 313, 599

Menten, K. M. 1991, *ApJ* (Letters) 380, L75

Minier, V. & Booth, R. S. 2002, *A&A*, 387,179

Szymczak, M., Hrynek, G. & Kuss, A. J. 2000, *A&AS*, 143, 269

Val'tts, I. E., Ellingsen, S. P., Slysh, V. I., kalenskii, S. V., Otrupcek, R. & Voronkov, M. A. 1999, *MNRAS*, 310, 1077

Astrophysical Masers and their Environments
Proceedings IAU Symposium No. 242, 2007
J.M. Chapman & W.A. Baan, eds.

doi:10.1017/S1743921307013063

Stellar masers, circumstellar envelopes and supernova remnants

Athol J. Kemball
University of Illinois at Urbana-Champaign, USA

Abstract. This paper reviews recent advances in the study or circumstellar masers and masers found toward supernova remnants. The review is organized by science focus area, including the astrophysics of extended stellar atmospheres, stellar mass-loss processes and outflows, late-type evolved stellar evolution, stellar maser excitation and chemistry, and the use of stellar masers as independent distance estimators. Masers toward supernova remnants are covered separately. Recent advances and open future questions in this field are explored.

Keywords. masers, stars:AGB and post-AGB, stars:atmospheres, supergiants, planetary nebulae, supernova remnants

1. Introduction

This review concerns circumstellar masers and the masers associated with supernova remnants. The material covered focuses primarily on results published since the last maser symposium, IAU 206 (Migenes & Reid 2002), but avoids those areas covered separately by other papers in these proceedings.

In considering circumstellar masers, it is appropriate to start by first examining the host stars and the basic properties of the most common masers found in their circumstellar shells. The late-type evolved stars that host circumstellar masers are either red supergiants or large-amplitude long-period variables (LALPV), such as thermally-pulsing asymptotic giant branch (TP-AGB) stars. These stars are astrophysically important for well-known, but still important, reasons (Habing 1996):

(*a*) Through their mass-loss they enrich the chemical and dust composition of the interstellar medium from which new stars form.

(*b*) They are high-luminosity tracers of faint and obscured stellar populations and their main sequence progenitors; intermediate mass stars, for example, achieve their highest brightness at the tip of the AGB.

(*c*) They highlight an important short-lived stage of stellar evolution in the transition from the AGB to planetary nebulae.

The key properties of the most common circumstellar maser transitions are summarized in Table 1. Numerical values in this table are approximate only; representative temperatures and densities are drawn from Reid (2002). In the most basic model of circumstellar masers, the maser species enumerated in Table 1 are assumed concentrically distributed about the central star, at radii broadly correlated with the transition excitation temperature.

Given the late-type evolved stars and the basic properties of circumstellar masers they host, it is appropriate to ask the broader question of what larger astrophysical questions can be explored by stellar maser studies. We can answer this in part by noting first that masers are unique probes of obscured circumstellar astrophysics due to both their high brightness temperature and compact spatial structure. As such, they give us unique insights into the extended stellar atmosphere, the region between the photosphere and

Table 1. Common masers in late-type, evolved stars

	SiO	H_2O	OH
Transition	$v = 0, 1, 2, 3, ..; \triangle J = 1$	$6_{16} - 5_{23}$	$^2\Pi_{3/2}, J = 3/2, \triangle F = \{0, 1\}$
Radius (AU)	$\sim$ 3-6	$\sim$ 100	$\sim$ 1000
n_F (cm^{-3})	5.10^{10}	10^8	10^7
Temperature (K)	1500	750	450
Pumping mechanism	Radiative or collisional	Collisional	Radiative

dust formation point. They also allow the study of the mass-loss process both near the star and in the outflow at larger radii. The masers allow an exploration of circumstellar excitation and pumping conditions as well as circumstellar chemistry. They also allow independent methods of stellar astrometry and distance measurement. We consider each of these scientific areas in more detail below.

In addition to the topics listed here, stellar masers provide insights into stellar magnetic fields, basic maser theory, surveys for galactic population and dynamical studies, and are also used in specialized galactic center dynamics studies. Though each is extremely important, these issues are not included further in this review as they are covered by separate papers in these proceedings. Masers toward supernova remnants are covered in Section 7 below.

2. Extended atmosphere

The near-circumstellar environment is shown schematically in Figure 1, taken from Reid (2002). This region is dominated by mass-loss, shocks from the centrally-pulsating star, and is expected to have a complex kinematic and dynamical structure (Elitzur 1992). High-resolution imaging of this region is essentially only provided by SiO maser observations; a capability unmatched by contemporary telescopes in other wave-bands.

A number of astrophysical sub-themes are of interest and importance in studies of the extended atmosphere. These can be broadly categorized as issues of kinematics, dynamics, and the physical conditions supporting individual maser transitions. In kinematics, there are open questions concerning localized flows, their connection to the central star and convective envelope, the nature and extent of any global asymmetry established at the base of the mass-loss process, and the matching of gas kinematics in this region to pulsation hydrodynamic models. The dynamical issues concern the nature of any shaping forces, and the magnitude and dynamical influence of local or global stellar magnetic fields. Physical conditions cover issues of excitation and chemistry.

Knowledge of the convective stellar envelope is limited primarily to numerical simulation studies (Freytag Steffen & Dorch 2002; Porter, Anderson & Woodward 1997). In the 3-D heat convection study of a red-giant atmosphere by Porter, Anderson & Woodward (1997), in which heat was supplied steadily to the central core to drive the convection, two unexpected behaviors arose, namely pulsation and large-scale convective flows. Complex convective structures and phenomena in the stellar envelope are likely to have a significant impact on the extended atmosphere.

The extended atmosphere can be studied uniquely using VLBI observations of SiO masers, which act as ultra-compact astrophysical probes of this region. Their importance as probes of this inner CSE was realized soon after their first detection in the 1970s but early imaging studies faced difficult technical challenges (Moran *et al.* 1979; Genzel *et al.* 1979), primarily due to the low sensitivity of millimeter-wavelength VLBI arrays. The advent of improved VLBI sensitivity enabled new observations (Colomer *et al.* 1992; Miyoshi *et al.* 1994); in particular, the commissioning of the VLBA†, a little over ten years ago, provided dramatically-improved imaging capabilities compared to the challenges faces by the earliest observations (Diamond *et al.* 1994; Greenhill *et al.* 1995).

A particularly important result in the last several years has been the conduct of simultaneous near-infrared (NIR) interferometry and radio interferometry imaging campaigns of stellar SiO maser sources using the VLBA both in conjunction with the VLTI‡ (Boboltz & Wittkowski 2005; Wittkowski *et al.* 2005) and with IOTA/FLUOR¶ (Cotton *et al.* 2004, 2006). These observations have allowed the unambiguous co-location of photospheric and SiO radii at $\frac{R_{SiO}}{R_*} \sim 1.5-3.0$, without having to rely on model-based photospheric diameter estimates. The monitoring campaign reported by Cotton *et al.* (2004) also hints at a possible relationship between the relative location of the 3.6 μm photosphere and a dependence of the SiO maser radius on the relative 3.6 to 2.2 μm photospheric radii. Those stars in this monitoring sample with higher 3.6 μm opacity in the molecular photosphere also appear to have large dust condensation radii.

At longer IR wavelengths, the ISI|| has recently reported an important direct measurement of radial pulsation in the diameter of Mira at 11 μm, showing a fluctuation in the continuum photosphere of +11% between stellar pulsation phase $\phi = -0.08$ to 0.15 (Weiner, Hale & Townes 2003), in good agreement with theoretical models for Mira-class pulsation.

Pulsation affects the extended atmosphere directly, driving periodic shocks into the near-circumstellar environment (NCSE) (Bertshinger & Chevalier 1985). The shock

† http://www.vlba.nrao.edu
‡ http://www.eso.org/projects/vlti
¶ http://tdc-www.harvard.edu/IOTA
|| http://isi.ssl.berkeley.edu/index.htm

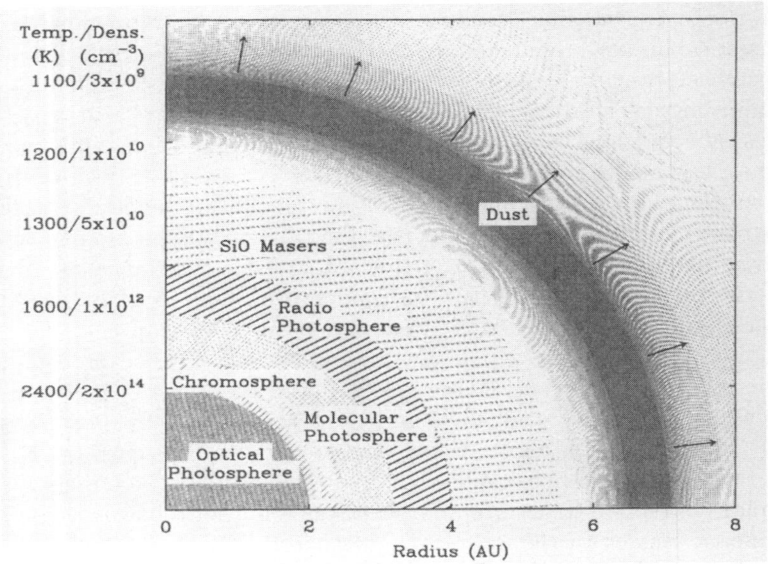

Figure 1. Schematic representation of the circumstellar envelope of a stellar maser from the inner envelope out to the dust formation zone; Figure 3 from Reid (2002)

emerges at pre-maximum, accelerating gas outwards, which subsequently falls back under the influence of gravity (Hinkle, Hall & Ridgway 1982; Hinkle, Lebzelter & Scharlach 1997; Alvarez *et al.* 2000). This produces a double-lined S-shaped velocity profile in photospheric infrared lines such as 1.6 μm CO $\triangle\nu = 3$ absorption (Hinkle, Hall & Ridgway 1982). The shocks levitate material above the the hydrostatic atmosphere and subsequent radiation pressure on dust couples to the gas and carries it further outwards. Shock formation in the NCSE is studied in a number of spherically-symmetric, piston-driven hydrodynamic models of the stellar atmosphere (Bowen 1988; Bessel, Scholz & Wood 1996; Humphreys *et al.* 2002). Shock propagation through the NCSE in these models leads to a saw-tooth velocity profile with radius. The models also accurately predict a net mass-loss of material to the outer circumstellar environment over time and a complex interaction between accelerating and decelerating gas from successive pulsation cycles.

In the time since the last maser conference in 2001, the number of synoptic imaging campaigns of stellar SiO masers has increased significantly and yielded new results. These imaging campaigns are enabled by the technical advances in VLBI imaging at this wavelength, as noted above. Cotton and collaborators are monitoring a small LALPV sample in a snapshot imaging campaign (Cotton *et al.* 2004, 2006). Synoptic imaging of the Mira variable TX Cam has also continued both at short-spacing in pulsation phase (Diamond & Kemball 2003; Gonidakis 2005), and over longer phase increments (Yi *et al.* 2005). Gonidakis (2005) has shown that the median SiO maser component lifetime in TX Cam is approximately 150-200 days, a substantial fraction of the 557 day stellar pulsation period. This permits detailed proper motion studies, as reported by Diamond & Kemball (2003); Gonidakis (2005). A global proper motion analysis shows outer components falling back from earlier pulsation cycles simultaneously with outflow during the current pulsation cycle. This is consistent with the saw-tooth velocity profile discussed above; however, there are significant local departures from a globally ordered flow in the measured proper motions in NSEW quadrants of the projected shell (Diamond & Kemball 2003). The mean velocities are broadly consistent with shock damping in the radio photosphere however, as deduced from upper limits to radio continuum stellar variability by Reid & Menten (1997).

An important astrophysical question in this area is the degree of agreement between measured mean global radial motions and the predictions of pulsation hydrodynamic models of the type discussed above. Recent reported mean global motions are summarized in Table 2 against the pulsation phase interval $\triangle\phi$, and whether the outflow was generally contracting, expanding, or neither. This table show that there are disagreements between individual stars in the same pulsation phase range (e.g. R Aqr, TX Cam, and VX Sgr), where both expansion and contraction are variously reported. However, note that this is comparing a known binary, a Mira variable, and a supergiant. Some authors (Cotton *et al.* 2004, 2006) have also reported no systematic motions with pulsation phase for their sample. However, it is clear that this picture is likely more complex. If we examine TX Cam over multiple pulsation periods we can argue for inter-cycle variability and possible interaction between two competing time-scales, the pulsation period and the free-fall gravitational time-scale, possibly causing a non-strictly repeating pattern of global motions. We also note that there are systematic measurement effects that need to be carefully addressed. The inner shell is not circular and robust estimators are needed to determine the mean inner radius (Diamond & Kemball 2003). For at least one pulsation phase interval of TX Cam, there is clear evidence for ballistic deceleration with physically reasonable parameter values (Diamond & Kemball 2003). It is not true for all pulsation phase cycles however.

Table 2. Measured pulsation kinematics

Reference	Sources	$\triangle\phi$	Projected velocity (kms^{-1})	Outflow or infall
Boboltz *et al.* (1997)	R Aqr	0.78-1.04	-4.2	Contraction
Diamond & Kemball (2003)	TX Cam	0.7-1.5	+7	Expansion
Cotton *et al.* (2004)	See below[1]	Various		No pattern
Yi *et al.* (2005)	TX Cam	0.60-1.05		Slowing expansion
Gonidakis (2005)	TX Cam	1.5-2.7		Both
Cotton *et al.* (2006)	See below[2]	Various		No pattern
Chen *et al.* (2006)	VX Sgr	0.75-0.80	-4	Contraction

1: R And, Mira, U Ori, R Leo, W Hya, S Crb, U Her, R Aqr, R Cas.
2: Mira, U Ori, R Aqr.

The issue of rotation in the extended atmosphere is also important from a kinematic and evolutionary perspective. Two detections of SiO maser shell rotation, including differential rotation, have been reported since the last maser meeting, namely for R Aqr (Hollis *et al.* 2001) and Mira (Cotton *et al.* 2006), both known binaries. This is, however, not common for other sources, as evident in the sector-averaged velocity plots reported by Cotton *et al.* (2006) for their LALPV sample. Substantial rotation is difficult to explain in evolved late-type stars that are not part of multiple star systems.

The expanded range of SiO imaging studies, such as those enumerated in Table 2, continue to reveal clear evidence for spatially-coherent arcs and filaments in the SiO maser region (Kemball & Diamond 1997). These local features have recently been argued by Yi *et al.* (2005) to be from spokes of SiO flows with line-of-sight velocities either slightly red- or blue-shifted with respect to the systemic stellar velocity. This kinematic structure maximizes the line-of-sight velocity coherence. The radial spokes are especially visible in several TX Cam image epochs reported by Yi *et al.* (2005), one of which is reproduced here as Figure 2.

3. Outflows

The water and hydroxyl masers probe intermediate and outer circumstellar radii (Chapman & Cohen 1986; Booth *et al.* 1981). These masers trace the mass-loss process beyond the dust formation point; in this region, the flow is driven by radiation pressure on dust. The astrophysical sub-themes that are relevant to this region of the outflow include the kinematic structure of the mass-loss, including the degree of asymmetry on local and global scales, and the connection of this kinematic pattern to the inner NCSE. For cases where a dominant kinematic pattern is clear, the science issue becomes that of identifying any shaping processes. This is also a particularly important area of the outflow in which to explore the evolutionary onset of more extreme asymmetry related to post-AGB evolution, including stellar jets.

Some of these science questions have been explored in recent MERLIN† observations of stellar water masers; Bains *et al.* (2003) report the first connected-element resolution of individual maser spots, and derive important parameters relating to the water maser region surrounding four LALPV stars: IK Tau, U Ori, RT Vir, and U Her. They report significant departures from spherical symmetry in tori that have an inner radius of 6-16 AU and an outer radius of 24-54 AU. Individual maser components are resolved out at 2-4 AU, and are found to be density-bounded ($\sim 10^{1-2}$ times ambient), with

† http://www.merlin.ac.uk

a volume filling factor $< 0.5\%$. In related MERLIN observations of the red-supergiant, VX Sgr, Murakawa *et al.* (2003) have performed a proper motion study of the water maser emission toward this source over a five year interval. A bi-conical outflow is used to describe the water maser kinematics. A similar over-density of maser components and a comparable extrapolated cloud size at the stellar surface of $\sim 0.07R_*$ is reported, relative to that given by Bains *et al.* (2003) for the sample of low-mass AGB stars. Both observations confirm an accelerating flow across the water maser region, with a logarithmic velocity gradient of $0.5 \leqslant \epsilon \leqslant 1$.

Concerning the role of water maser studies as a probe of early-onset post-AGB asymmetry, a particularly important result in recent years has been the discovery and exploration of a rare sub-class of AGB stars showing highly-collimated stellar jets (Imai *et al.* 2004; Vlemmings, Diamond & Imai 2006). The dynamical age of the jets is very young ($< 35 - 100$ years) and only a handful of such sources are known, consistent with this dynamical age. This rare class of "water fountain" sources is described in separate review papers in these proceedings by Imai and Vlemmings.

A particularly unusual and rare source studied by Babkovskaia *et al.* (2006); Szczerba *et al.* (2006) presents water masers in a warped disk around a companion to the SiC star V778 Cyg.

OH observations of the supergiant NML Cyg are also reported by Etoka & Diamond (2004). In this source, they present evidence for two shells, an older spherical shell, presumed from an earlier mass-loss phase, and a younger asymmetric flow.

4. Late-type stellar evolution

In this section, we consider the issue of post-AGB evolution, and how stellar masers can inform our understanding of this phase of stellar evolution. Planetary nebulae (PNe) show a rich diversity of point- and axi-symmetric structure (Balick & Frank 2002).

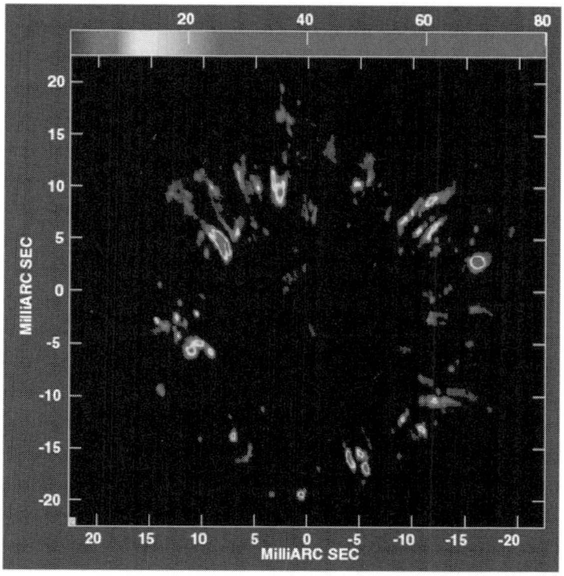

Figure 2. VLBI maps of $v = 1, J = 1 - 0$ SiO maser emission toward TX Cam, summed over all channels of emission. The bar shows the flux scale in Jy/beam (Figure 3 from Yi *et al.* (2005)).

The immediate post-AGB (PAGB) evolution is heavily shrouded and masers provide a powerful method to study this transition phase, complementary to imaging studies in other wave-bands, such as infrared.

The astrophysical sub-themes that are important in this area concern how to explain the prevalence of point- and axi-symmetry in PNe. Four major shaping mechanisms have been proposed including binarity, equatorial density enhancement, a globally-acting magnetic field and sculpting by highly-collimated stellar jets (Morris, Sahai & Claussen 2003). It is of significant interest to determine when the shaping mechanism is activated on the AGB.

An over-arching issue in this area is also how to identify and select post-AGB objects in samples of OH-IR stars, such as the complete 1612 MHz OH survey conducted by Sevenster *et al.* (1997a,b, 2001). PAGB identification in such surveys is considered by Sevenster (2002) based on clustering in color-color IR plots, using both IRAS and MSX data. Follow-up observations of candidate PAGB stars in the Sevenster survey are presented by Deacon, Chapman & Green (2004); their observations show a significantly higher incidence of asphericity in the candidate PAGB sample and further strengthen the IR color-color selection analysis by Sevenster (2002).

Kinematic models for OH outflow sources are presented by Zijlstra *et al.* (2001) using a two component model, consisting of a shell/torus (20-25 kms^{-1}) and higher-velocity linear outflow (10-80 kms^{-1}).

Masers are particularly rare in young PNe, and only a handful of such sources are known although they are very important. These sources are discussed by Gómez (these proceedings).

5. Excitation and chemistry

The question of maser excitation, pumping, and circumstellar chemistry cuts across both theory and observation. It is needed to understand physical conditions in the masering regions and most importantly to properly interpret observations. In addition, these issues are central to predicting new masering transitions that may be accessible using future telescopes.

Both collisional (Lockett & Elitzur 1991) and radiative (Deguchi & Iguchi 1976) mechanisms have been proposed for SiO maser pumping. There are observational discriminants between the two pumping mechanisms, primarily the relative spatial location of v=1 and v=2 masers, variability studies, and linear polarization morphology. There have been important recent observational results in this area, but conclusions drawn in the literature remain mixed. Based on the location of the v=1 and v=2 J=1-0 masers toward TX Cam, Yi *et al.* (2005) cite evidence for collisional pumping. However, Soria-Ruiz *et al.* (2004) draw the opposite conclusion from v=1 and v=2 J=1-0 imaging of the SiO masers toward IRC+10011 and χ Cyg. In recent variability studies, Pardo *et al.* (2004) reports finding a zero phase lag between IR and SiO flux density over an 11-year short time-spacing monitoring campaign, and McIntosh (2006) finds no time-lag between several vibrationally-excited J=1-0 transitions toward Mira. Both authors argue for radiative SiO pumping as a result. Asensio Ramos, Landi Degl'Innocenti & Trujillo Bueno (2005) consider the effect of anisotropic radiative pumping and the Hanle effect on the linear polarization morphology of SiO stellar masers.

Studies of the relative location of different rotational transitions in the same vibrational state, primarily J=2-1 relative to J=1-0, have shown perplexing results (Soria-Ruiz *et al.* 2004; Phillips *et al.* 2003). In the handful of sources imaged so far, the morphology is significantly different, contrary to basic theoretical expectations. Line overlaps may strongly influence the theoretical predictions about spatial location however, as discussed by Soria-Ruiz *et al.* (2004).

Another important recent result in SiO maser excitation has been the imaging of v=0, J=1-0 SiO maser emission towards a sample of late-type, evolved stars by Boboltz & Claussen (2004). A predominance of weak masers is confirmed and one source was found to have thermal emission. Although their spatial resolution is limited, they propose that v=0 is found at twice the radius of the v=1 emission.

Several new possible masers have been reported toward C-stars since our last meeting, including new HCN lines toward a sample of carbon stars (Bieging 2001), and a possible weak OH maser detection toward IRC+10216 (Ford *et al.* 2003).

Menten *et al.* (2006) also report the possible new detection of maser action in the H_2O $v_2 = 1, 6_{61} - 7_{52}$ line at 294 GHz toward VY CMa. In addition, Thorwirth *et al.* (2003) report weak maser emission in the J=9 l-type HCN transition toward CRL 618.

6. Independent distance estimates

Stellar masers can also be used to provide independent stellar distances. This technique, demonstrated by van Langevelde *et al.* (2000), is based on the assumption that the compact, blue-shifted maser emission is assumed coincident with the central star. The technique was applied to further AGB stars by Vlemmings *et al.* (2003).

OH stellar astrometry is likely limited to distances less than 1 kpc (Vlemmings *et al.* 2003); astrometry using H_2O and SiO masers can reach larger distances. Recently, Kurayama, Sasao & Kobayashi (2005) demonstrated stellar water maser astrometry in observations of UX Cyg. Vlemmings, van Langevelde & Diamond (2003) also showed positional coincidence between the brightest water maser feature toward U Her and the stellar Hipparcos position.

7. 1720 MHz OH masers toward supernova remnants

OH 1720 MHz masers, shock-excited toward supernova remnants (SNR), were first detected by Goss & Robinson (1968). Their phenomenology is now well-known and a sound theoretical basis has been established for their modeling and pumping (Green 2002).

They are an important probe of SNR shock interactions with molecular clouds, and are also found in the galactic center circumnuclear disk. They are considered in more detail in separate invited papers in these proceedings by Brogan, Yusef-Zadeh, and Hewitt.

8. The future

There remain important open astrophysical problems in the area of circumstellar and SNR masers, requiring advances in both observation and theory. An integrated model for the kinematics and dynamics of the mass-loss process from the convective stellar envelope, through the extended atmosphere to the outer circumstellar shell will require further detailed observations, and more complex numerical hydrodynamic models in order to explain the observed local and global asymmetries and to identify the underlying shaping forces. This is particularly important in so far as it concerns stellar magnetic fields of late-type, evolved stars. Recent detections of stellar jets in water maser emission also open new windows on post-AGB evolution. The theory of astrophysical masers continues to remain central to this field, and recent observations have posed new theoretical questions, particularly in the area of SiO excitation, that are vital to proper interpretation of observations. We are also on the cusp of major new observational advances in this field, driven by the commissioning of VERA†, ALMA‡, and the eVLA¶ and look forward to significant future results from these instruments.

Acknowledgements

The author would like to gratefully acknowledge the support of the National Science Foundation under grant NSF AST 05-07473.

References

Alvarez, R., Jorissen, A., Plez, B., Gillet, D., & Fokin, A. 2000, *A&A* 362, 655

Asensio Ramos, A., Landi Degl'Innocenti, E., & Trujillo Bueno, J. 2005, *ApJ* 625, 985

Babkovskaia, N., Poutanen, J., Richards, A. M. S., & Szczerba, R. 2006, *MNRAS* 370, 1921

Balick, B. & Frank, A. 2002, *ARAA* 40, 439

Bains, I., Cohen, R. J., Louridas, A., Richards, A. M. S., Rosa-González, D., & Yates, J. A. 2003, *MNRAS* 342, 8

Bertshinger, E. & Chevalier, R. A. 1985, *ApJ* 299, 167

Bessel, M. S., Scholz, M., & Wood, P. R. 1996, *A&A* 307, 481

Bieging, J. H. 2001, *ApJ* Letters 549, L125

Boboltz, D. A., Diamond, P. J., & Kemball, A. J. 1997, *ApJ* Letters 487, L147

Boboltz, D. A. & Claussen, M. J. 2004, *ApJ* 608, 480

Boboltz, D. A. & Wittkowski, M. 2005, *ApJ* 618, 953

Booth, R. S., Norris, R. P., Porter, N. D., Kus, A. J. 1981, *Nature* 290, 382

Bowen, G. H. 1988, *ApJ* 329, 299

Chapman, J. M. & Cohen, R. J. 1986, *MNRAS* 220, 513

Chen, X., Shen, Z.-Q., Imai, H., & Kamohara, R. 2006, *ApJ* 640, 982

Colomer, F., Graham, D. A., Krichbaum, T. P., Ronnang, B. O., de Vicente, P., Witzel, A., Barcia, A., Baudry, A., Booth, R. S., Gomez-Gonzalez, J., Alcolea, J., & Daigne, G. 1992, *A&A* Letters 254, L17

Cotton, W. D., Mennesson, B., Diamond, P. J., Perrin, G., Coudé du Foresto, V., Chagnon, G., van Langevelde, H. J., Ridgway, S., Waters, R., Vlemmings, W., Morel, S., Traub, W., Carleton, N., & Lacasse, M. 2004, *A&A* 414, 275

Cotton, W. D., Vlemmings, W., Mennesson, B., Perrin, G., Coudé du Foresto, V., Chagnon, G., Diamond, P. J., van Langevelde, H. J., Bakker, E., Ridgway, S., Mc Allister, H., Traub, W., & Ragland, S. 2006, *A&A* 456, 339

Deacon, R. M., Chapman, J. M., & Green, A. J. 2004, *ApJS* 155, 595

Deguchi, S. & Iguchi, T. 1976, *PASJ* 28, 307

Diamond, P. J., Kemball, A. J., Junor, W., Zensus, A., Benson, J., & Dhawan, V. 1994, *ApJ* Letters 430, L61

Diamond, P. J. & Kemball, A. J. 2003, *ApJ* 599, 1372

Elitzur, M. 1992, *Astronomical Masers* (Dordrecht:Kluwer)

Etoka, S. & Diamond, P. 2004, *MNRAS* 348, 34

Ford, K. E. S., Neufeld, D. A., Goldsmith, P. F., & Melnick, G. J. 2003, *ApJ* 589, 430

Freytag,B., Steffen, M. & Dorch, B. 2002, *Astron. Nachr.* 323, 213

Genzel, R., Moran, J. M., Lane, A. P., Predmore, C. R., Ho, P. T. P., Hansen, S. S., & Reid, M. J. 1979, *ApJ* Letters 231, L73

Gonidakis, I. 2005, *Ph.D. thesis*, University of Manchester.

Goss, W. M. & Robinson, B. J. 1968, *Ap. Lett.* 2, 81

Green, A. J. 2002, in: V. Migenes & M. J. Reid (eds.), *Cosmic Masers: from Protostars to Blackholes*, Proc. IAU Symposium 206 (San Francisco: ASP), p. 204

Greenhill, L. J., Colomer, F., Moran, J. M., Backer, D. C., Danchi. W. C., & Bester, M. 1995, *ApJ* 449, 365

† http://veraserver.mtk.nao.ac.jp
‡ http://www.alma.nrao.edu
¶ http://www.aoc.nrao.edu/evla

Habing, H. J. 1996, *A&AR* 7, 1.

Hinkle, K. H., Hall, D. N. B. & Ridgway, S. T. 1982, *ApJ* 252, 697

Hinkle, K. H., Lebzelter, T., Scharlach, W. W. G. 1997, *AJ* 114, 2686

Hollis, J. M., Boboltz, D. A., Pedelty, J. A., White, S. M., & Forster, J. R. 2001, *ApJ* Letters 559, L37

Humphreys, E. M. L., Gray, M. D., Yates, J. A., Field, D., Bowen, G. H., & Diamond, P. J. 2002, *A&A* 386, 256

Imai, H., Morris, M., Sahai, R., Hachisuka, K., & Azzollini, F. 2004, *A&A* 420, 265

Kemball, A. J. & Diamond, P. J. 1997, *ApJ* Letters 481, L111

Kurayama, T., Sasao, T. & Kobayashi, H. 2005, *ApJ* Letters 627, L49

Lockett, P. & Elitzur, M. 1991, in: A.D. Haschick & P. T. P. Ho (eds.), *Skyline*, Proc. 3rd Haystack Conf. (Provo:ASP)

Menten, K. M., Philipp, S. D., Güsten, R., Alcolea, J., Polehampton, E. T., & Brünken, S. 2006, *A&A* Letters 454, L107

McIntosh, G. C. 2006, *ApJ* Letters 638, L41

in: V. Migenes & M. J. Reid (eds.), *Cosmic Masers: from Protostars to Blackholes*, Proc. IAU Symposium 206 (San Francisco: ASP)

Miyoshi, M., Matsumoto, K., Kameno, S., Takaba, H., & Lwata, T. 1994, *Nature* 371, 395

Moran, J. M., Ball, J. A., Predmore, C. R., Lane, A. P., Huguenin, G. R., Reid, M. J., & Hansen, S. S. 1979, *ApJ* Letters 231, L67

Morris, M. R., Sahai, R. & Claussen, M. 2003, *Rev. Mex. AA.* 15, 20

Murakaw, K., Yates, J. A., Richards, A. M. S., & Cohen, R. J. 2003, *MNRAS* 344, 1

Pardo, J. R., Alcolea, J., Bujarrabal, V., Colomer, F., del Romero, A., & de Vicente, P. 2004, *A&A* 424, 145

Phillips, R. B., Straughn, A. H., Doeleman, S. S., & Lonsdale, C. J. 2003, *ApJ* Letters 588, L105

Porter, D., Anderson, S. & Woodward, P. 1997, *LCSE report (U.Minnesota)*, http://pinot.lcse.umn.edu/RedGiant.

Reid, M. J. & Menten, K. M. 1997, *ApJ* 476, 327

Reid, M. J. 2002, in: V. Migenes & M.J. Reid (eds.), *Cosmic Masers: from Protostars to Blackholes*, Proc. IAU Symposium 206 (San Francisco: ASP), p. 506

Sevenster, M. N., Chapman, J. M., Habing, H. J., Killeen, N. E. B., & Lindqvist, M. 1997a, *A&AS* 122, 79

Sevenster, M. N., Chapman, J. M., Habing, H. J., Killeen, N. E. B., & Lindqvist, M. 1997b, *A&AS* 124, 509

Sevenster, M. N., van Langevelde, H. J., Moody, R. A., Chapman, J. M., Habing, H. J., & Killeen, N. E. B. 2001, *A&A* 366, 481

Sevenster, M. N. 2002, *AJ* 123, 2788

Soria-Ruiz, R., Alcolea, J., Colomer, F., Bujarrabal, V., Desmurs, J.-F., Marvel, K. B., & Diamond, P.J. 2004, *A&A* 426, 131

Szczerba, R., Szymczak, M., Babkovskaia, N., Poutanen, J., Richards, A. M. S., & Groenewegen, M. A. T. 2006, *A&A* 452, 561

Thorwirth, S., Wyrowski, F., Schilke, P., Menten, K. M., Brünken, S., Müller, H. S. P., & Winnewisser, G. 2003, *ApJ* 586, 338

van Langevelde, H. J., Vlemmings, W., Diamond, P. J., Baudry, A., & Beasley, A. J. 2000, *A&A* 357, 945

Vlemmings, W. H. T., van Langevelde, H. J. & Diamond, P. J. 2002, *A&A* Letters 393, L33

Vlemmings, W. H. T., van Langevelde, H. J., Diamond, P. J., Habing, H. J., & Schilizzi, R. T. 2003, *A&A* 407, 213

Vlemmings, H. T., Diamond, P. J. & Imai, H. 2006, *Nature* 440, 58

Weiner, J., Hale, D. S. S. & Townes, C. H. 2003, *ApJ* 588, 1064

Wittkowski, M., Boboltz, D. A., Driebe, T., & Ohnaka, O. 2005, *Mem. S.A. It.* 76, 457

Yi, J., Booth, R. S., Conway, J. E., & Diamond, P. J. 2005, *A&A* 432, 531

Zijlstra, A. A., Chapman, J. M., te Lintel Hekkert, P., Likkel, L., Comeron, F., Norris, R. P., Molster, F. J., & Cohen, R. J. 2001, *MNRAS* 322, 280

Astrophysical Masers and their Environments
Proceedings IAU Symposium No. 242, 2007
J.M. Chapman & W.A. Baan, eds.

doi:10.1017/S1743921307013075

The Mira star S Ori: SiO maser shells related to the stellar photosphere, the molecular layers, and the dust shell at three epochs

M. Wittkowski[1], D. A. Boboltz[2], K. Ohnaka[3], T. Driebe[3] and M. Scholz[4,5]

[1]European Southern Observatory, Garching, Germany
email:mwittkow@eso.org

[2]United States Naval Observatory, Washington, DC, USA
email:dboboltz@usno.navy.mil

[3]Max-Planck-Insitut für Radioastronomie, Bonn, Germany

[4]Institut für Theoretische Astrophysik, Heidelberg, Germany

[5]Institute of Astronomy, University of Sydney, Australia

Abstract. We present recent results from our project of concurrent radio and infrared interferometry of oxygen-rich Mira stars. Our pilot study on the Mira variable S Ori included coordinated VLBA mapping of the v=1, J=1-0 43.1 GHz and v=2, J=1-0 42.8 GHz SiO maser radiation at one epoch and concurrent near-infrared K-band interferometry to constrain the stellar photospheric diameter. We recently obtained new maps of these SiO maser transitions toward S Ori at three additional epochs with the VLBA and concurrent mid-infrared interferometric data with VLTI/MIDI. The MIDI data are analyzed using recent self-excited dynamic model atmospheres including molecular shells close to continuum-forming layers, which we complement by a radiative transfer model of the dust shell. The modelling of our MIDI data results in phase-dependent continuum photospheric angular diameters. The dust shell can best be modelled with Al_2O_3 grains alone. We find that the mean SiO maser ring radii lie between about 1.9 and 2.4 stellar continuum radii. The maser spots mark the region of the molecular atmospheric layers shortly outward of the steepest decrease of the mid-infrared model intensity profile. Our results suggest that the SiO maser shells are co-located with the Al_2O_3 dust shell near minimum visual phase. Their velocity structure indicates a radial gas expansion. Preliminary results from a study of the Mira star GX Mon suggest a similar picture.

Keywords. masers, radiative transfer, techniques: high angular resolution, techniques: interferometric, stars: AGB and post-AGB, stars: atmospheres, circumstellar matter, stars: fundamental parameters, stars: individual (S Ori, GX Mon), stars: mass loss

1. Introduction

Interferometric observations providing a high spatial resolution are best suited to provide further observational constraints on the regions close to the surface where stellar pulsation and dust formation take place and where the mass-loss is initiated. Hereby, interferometry at different wavelength regions is needed to probe different regions: Near-infrared interferometry is most sensitive to the continuum photosphere and to warm molecular layers lying above the photosphere. Mid-infrared interferometry is most sensitive to the molecular layers lying above the photosphere, which cause the mid-infrared diameter to be significantly larger than the near-infrared diameter, and to the dust shell. Complementary information regarding the circumstellar environment can be obtained by observing the maser radiation that some of the molecules emit. The structure and dynamics of the CSE of Mira variables and other evolved stars can be investigated by mapping SiO maser emission at typically about 2 stellar radii toward these stars using very long baseline interferometry (VLBI) at radio wavelengths.

We have established a program of concurrent infrared interferometry using the VLTI and radio interferometry using the VLBA. The former aims at constraining the photospheric radius, the characteristics of atmospheric molecular layers, and parameters of the dust shell. The latter aims at concurrently mapping the SiO maser emission. Our final goal is a better understanding of the mass-loss process and its connection to stellar pulsation. Our pilot study in Boboltz & Wittkowski (2005) on the Mira variable S Ori included coordinated near-infrared K-band interferometry to constrain the stellar photospheric diameter and VLBA mapping of the SiO maser radiation toward this source. We have now obtained the first multi-epoch study that includes concurrent radio and mid-infrared interferometry of an oxygen-rich Mira star (Wittkowski *et al.* (2007)), the main results of which we outline in the following sections.

2. Relating the maser shell radius to the photospheric radius

Two basic problems arise when attempting to relate the maser shell radius to the photospheric stellar radius.

Firstly, observational results regarding the detailed relationships between the stellar photosphere and the SiO maser ring often suffer from uncertainties inherent in comparing observations of variable stars widely separated in time and stellar phase as discussed in Boboltz & Wittkowski (2005). We overcome this limitation by obtaining concurrent VLTI and VLBA observations. In our recent study, each combined VLTI/VLBA epoch has a width of less than 4% of the stellar variability period.

Secondly, a further uncertainty in comparing photospheric radii to the extensions of the maser shells arises from the complication that near- and mid-infrared intensity profiles of finite bandwidth include a blend of intensities from continuum-forming layers and overlying molecular layers. This effect has often resulted in over-estimated continuum photospheric diameter values. An overestimated photospheric diameter would result in biased relative distances of the dust shell and the maser ring from the stellar surface even if obtained at the same stellar phase and cycle. We employ a detailed comparison of our observations to dynamic model atmospheres to relate the observable quantities to a meaningful continuum photospheric radius and thus to overcome this limitation.

3. Modeling the molecular atmosphere and the dust shell

Mid-infrared photometric and interferometric data of Mira stars are in general sensitive to details of the structure of the stellar atmosphere in which molecular layers of various geometrical and optical characteristics lie above the continuum-forming photosphere, as well as to the surrounding dust shell. We use the P and M model atmosphere series discussed in Ireland *et al.* (2004a), Ireland *et al.* (2004b) to describe the stellar atmosphere including the continuum photosphere and overlying molecular layers, and add an ad-hoc radiative transfer model of the dust shell. The P and M model series are complete self-excited dynamic model atmosphere of Mira stars. We use the Monte Carlo radiative transfer code `mcsim_mpi` by Ohnaka *et al.* (2006) for the radiative transfer model of the dust shell and use the SED of the dust-free atmosphere model as central source. Lorenz-Martins & Pompeia (2000) have investigated the envelopes of 31 oxygen-rich AGB stars using the IRAS LRS spectra. They classify the objects into three groups, whose IRAS LRS spectra can be modeled using either silicate grains, Al_2O_3 grains, or a mix thereof.

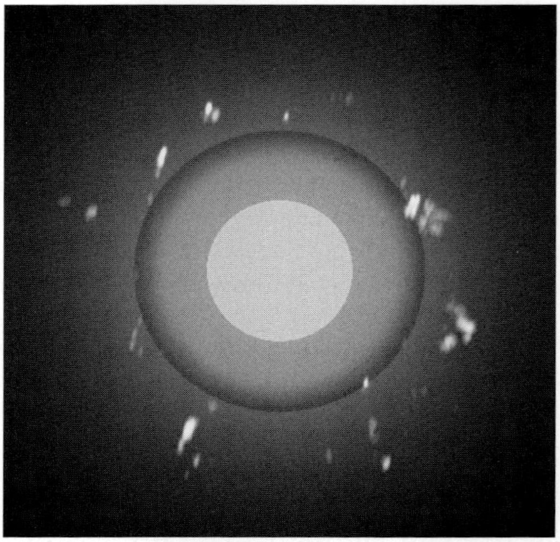

Figure 1. The (red) $v = 2$, $J = 1-0$ (42.8 GHz) and (green) $v = 1$, $J = 1-0$ (43.1 GHz) maser images overlaid onto pseudo-color representations of the infrared intensity for the example of our second epoch from Wittkowski *et al.* (2007). The continuum photosphere is marked by the light blue color. The darker blue shades represent the model intensity profile of the molecular atmosphere seen at mid-infrared wavelengths. The green shades represent the location of the Al_2O_3 dust shell.

We follow their results for the details of the two grain species. Our overall model is thus defined with this set of five independent parameters: (1) choice of the dust-free M model, (2) optical depth of Al_2O_3, (3) optical depth of silicate (4) inner dust shell boundary (5) dust shell density gradient. We use the well-defined continuum photospheric radius at a wavelength of 1.04 μm of the respective dust-free M model to characterize the photospheric radius.

4. VLBA and VLTI measurements

We observed the $v = 1, J = 1 - 0$ (43.1 GHz) and $v = 2, J = 1 - 0$ (42.8 GHz) SiO maser emission associated with S Ori using the 10 stations of the VLBA in January, February, and November 2005. The first two epochs occurred near stellar minimum and were spaced 42 days apart or approximately 10% of the stellar period. The third epoch occurred roughly 9 months later just after maximum on the next stellar cycle. Fig. 1 shows the total intensity images of the two maser transitions for the example of the second epoch compared to the model of our mid-infrared VLTI/MIDI data (described in the next section). In the velocity structure of the masers (not shown here) there appears to be a velocity gradient at all epochs, with masers toward the blue- and red-shifted ends of the spectrum lying closer to the center of the distribution than masers at intermediate velocities. We interpret this effect by a radial expansion of the maser shell with average velocity of about 10 km/sec.

We concurrently obtained mid-infrared interferometry of S Ori with the instrument MIDI at the ESO VLT Interferometer in December 2004, February/March 2005, and November 2005. For the example of our second epoch, Fig. 2 shows the resulting MIDI flux and the MIDI visibility compared to our best-fitting model. For more details on the observations, we refer to Wittkowski *et al.* (2007).

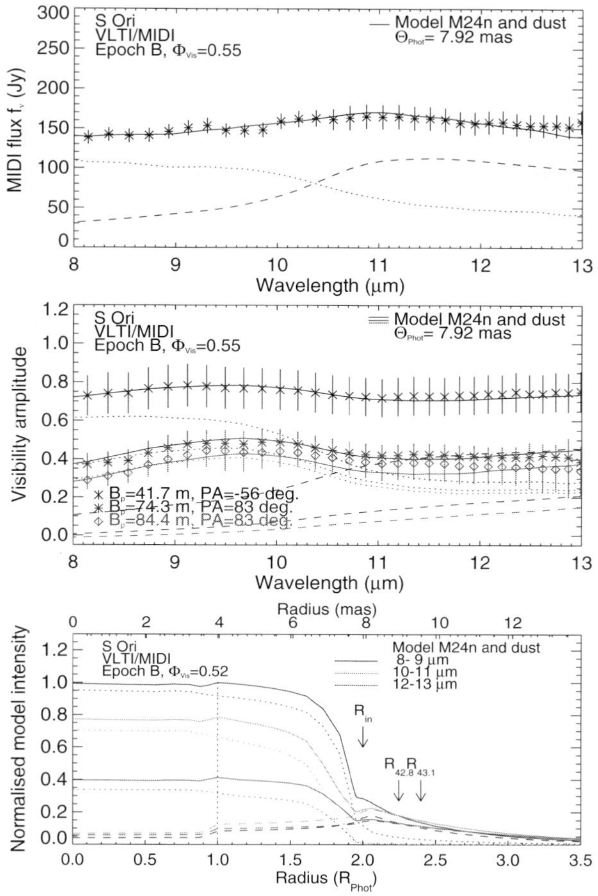

Figure 2. VLTI/MIDI results for the example of our second epoch from Wittkowski *et al.* (2007). Top: MIDI total flux; Middle: MIDI visibility; Bottom: Model intensity profile. The dotted lines denote the contribution from the stellar atmosphere, the dashed line that of the dust shell, and the solid line the overall model.

5. Results

Fig. 3 illustrates the results of our findings for the radial structure of S Ori's circumstellar environment (CSE) at our near-minimum and post-maximum phases which is widely consistent with the scheme of Reid & Menten (1997). The resulting continuum photospheric angular diameters lie between 7.9 mas and 9.7 mas for the different epochs. The dust shell can well be modelled with Al_2O_3 grains alone and our data shows no indication of silicates, consistent with the modelling of the IRAS LRS spectra by Lorenz-Martins & Pompeia (2000). The inner dust shell angular diameters lie between 1.8 and 2.4 times the continuum photospheric diameters for the different epochs. The 43.1 GHz and 42.8 GHz maser spots show the typical structure of partial to full rings with a clumpy distribution. There is no sign of a globally asymmetric gas distribution. The mean distances of the maser spots from the center of their distribution ranges between 1.9 and 2.4 continuum photospheric radii for the different epochs, with the 42.8 GHz masers lying systematically at closer distances compared to 43.1 GHz masers by about 0.1 continuum photospheric radius. This location of the maser spots is mainly consistent with the modeling by Gray

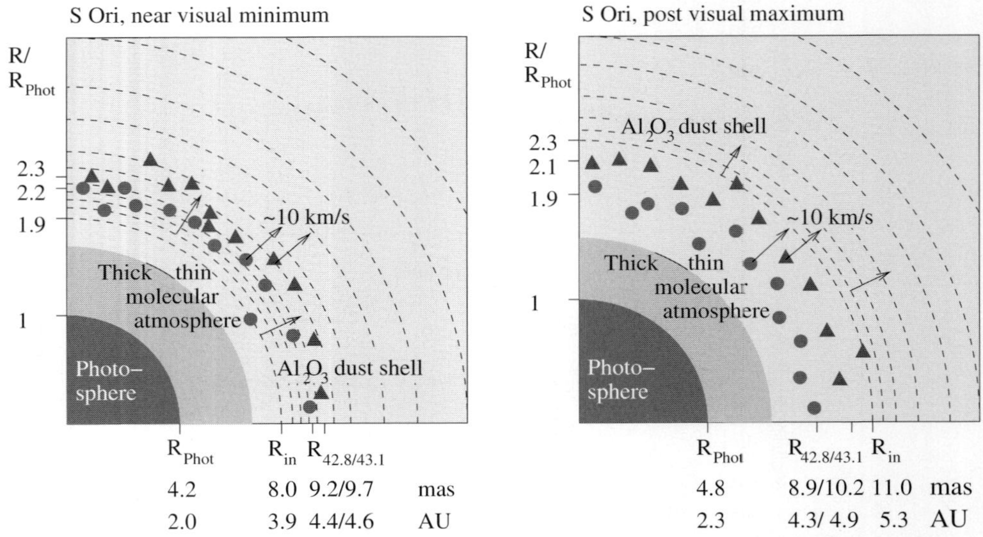

Figure 3. Sketch of the radial structure of S Ori's CSE at (left) near-minimum and (right) post-maximum visual phase as derived in Wittkowski *et al.* (2007). Shown are the locations of (dark gray) the continuum photosphere, (medium dark gray) the at N-band optically thick molecular atmosphere, (light gray) the at N-band optically thin molecular atmosphere, (dashed arcs) the Al_2O_3 dust shell, and (circles/triangles) the 42.8 GHz and 43.1 GHz maser spots.

& Humphreys (2000) and Humphreys *et al.* (2002), while details of the variability and widths of the maser ring radii may differ from the model predictions. Compared to the molecular atmosphere, the maser shells mark the region of the molecular atmosphere shortly outward of the steepest decrease of the mid-infrared model intensity. Compared to the dust shell, the maser spots are co-located with the inner dust shell near stellar minimum. At our post-maximum phases, the maser spots remain at about the same location, while the dust shell has expanded outward.

Preliminary results from a study of the Mira variable GX Mon using the same technique suggest a similar picture, but where the MIDI data indicates dust shells of corundum *and* silicates with the silicate shell being located farther outward compared to the corundum shell (work in preparation).

References

Boboltz, D. A. & Wittkowski, M. 2005, *ApJ* 618, 953
Gray, M. D. & Humphreys, E. M. L. 2000, *New Astron.* 5, 155
Humphreys, E. M. L., Gray, M. D., Yates, J. A., *et al.* 2002, *A&A*, 386, 256
Ireland, M. J., Scholz, M., & Wood, P. R. 2004a, *MNRAS* 352, 318
Ireland, M. J., Scholz, M., Tuthill, P. G., & Wood, P. R. 2004b, *MNRAS* 355, 444
Lorenz-Martins, S. & Pompeia, L. 2000, *MNRAS* 315, 856
Ohnaka, K., Driebe, T., Hofmann, K.-H., *et al.* 2006, *A&A* 45, 1015
Reid, M. J. & Menten, K. M. 1997, *ApJ* 476, 327
Wittkowski, M., Boboltz, D. A., Ohnaka, K., Driebe, T. & Scholz, M. 2007, *A&A* submitted

Astrophysical Masers and their Environments
Proceedings IAU Symposium No. 242, 2007
J.M. Chapman & W.A. Baan, eds.

doi:10.1017/S1743921307013087

The hypergiant masers: episodic mass loss, convection and magnetic fields

Roberta M. Humphreys
Astronomy Department, University of Minnesota, Minneapolis, MN 55455, USA

Abstract. The cool hypergiants are the most luminous known stars in the upper HR Diagram in the apparent temperature range represented by spectral types A to M. Most of the stars in this regime are unstable as evidenced by their high mass loss rates, variability, and in some cases large IR excesses and circumstellar ejecta. We have obtained high resolution multi-wavelength images with HST/WFPC2 of several of the most known evolved cool stars including several well known stellar masers. VX Sgr and S Per were marginally resolved, while NML Cyg has a peculiar asymmetric envelope that has been shaped by its environment. The powerful maser sources IRC+10420 and VY CMa have extensive and complex circumstellar ejecta due to high mass loss episodes apparently driven by large-scale convective activity.

Keywords. supergiants, stars: winds, outflows, stars: activity

1. Introduction

To better understand the evolution of luminous, cool evolved stars and their mass loss mechanism, my collaborators and I have been obtaining high resolution imaging and spectroscopy of several of the stars called "cool hypergiants". These stars lie on or near the empirical upper luminosity boundary in the HR Diagram (Humphreys & Davidson 1994, de Jager 1998). They were selected based on their infrared excess emission, high mass loss rates, molecular emission, and photometric and spectroscopic variability.

The "yellow" hypergiants, ρ Cas, HR 5171a and HR 8752 and the M-type supergiant μ Cep are all point sources, while VX Sgr, S Per, NML Cyg, VY CMa, and IRC+10420 were all resolved and the latter three have circumstellar ejecta. *Note that all five of these stars are strong maser sources.* VX Sgr and S Per have marginally resolved envelopes comparable is size to their dust formation radii. S Per's circumstellar nebula appears elongated in a NE/SW orientation similar to that for its OH and H_2O masers, while VX Sgr is embedded in a spheroidal envelope. NML Cyg, IRC +10420 and VY CMa are discussed in more detail below

2. The post red supergiant – IRC +10420

With its high luminosity of $L \sim 5{\times}10^5\ L_{\odot}$ (Jones *et al.* 1993), extraordinary mass loss rate of $3 - 6 \times 10^{-4}\ M_{\odot}\ \mathrm{yr}^{-1}$ (Knapp & Morris 1985, Oudmaijer *et al.* 1996, Humphreys *et al.* 1997), and intermediate temperature, IRC+10420 is one of the the most important evolved stars in the upper HRD. It is also one of the warmest maser sources known.

Our HST images of IRC+10420 (Humphreys *et al.* 1997) revealed a complex circumstellar environment, with a variety of structures including condensations or knots, ray-like features, and semi–circular arcs or loops within 2" of the star which are evidence for high mass loss ejections in the past few hundred years (Figure 1). One or more distant reflection shells at 5" to 6" from the star were ejected ≈ 3000 yrs ago, possibly when it was still a red supergiant.

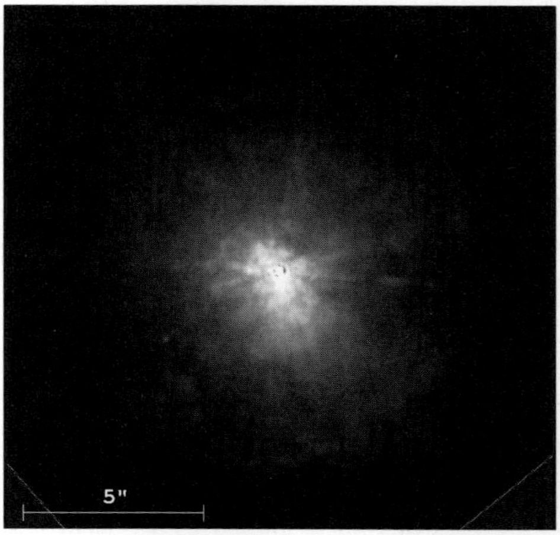

Figure 1. HST/WFPC2 image of IRC +10420 showing its complex inner ejecta.

IRC +10420's OH maser emission is peculiar with a complex and clumpy distribution within 1.5" of the star and demonstrating complex variations in its intensity with time (Lewis *et al.* 1986). Although it is a strong OH source, it is not a known source of H_2O and SiO emission common in many evolved late-type stars. This may be because it is so warm. IRC +10420 may not be of much interest as a source of abundant maser emission, but it is an extremely important star for understanding the final stages of massive star evolution.

3. NML Cyg – Interacting with its environment

The powerful OH/IR source NML Cyg is also near the empirical upper luminosity boundary for red supergiants with a luminosity of 5×10^5 $L_\odot$ and a mass loss rate of 6.4×10^{-5} $M_\odot yr^{-1}$ (Hyland *et al.* 1972, Morris & Jura 1983). It is only $\sim$ 100 pc from the large association Cyg OB2 in the Cygnus X superbubble (Morris & Jura 1983, Knodlseder 2003).

Our HST/WFPC2 images (Schuster *et al.* 2006) show that NML Cyg has a small but very obvious circumstellar nebula with a peculiar asymmetric shape (Figure 2). There are remarkable similarities between the asymmetric envelope that we see and the much more distant (~30" from the star) H II contours around NML Cyg (Habing *et al.* 1982). Morris & Jura (1983) showed that the asymmetric "inverse" H II region was the result of the interaction of a spherically symmetric, expanding wind from NML Cyg and photo-ionization from Lyman continuum photons from the luminous, hot stars in Cyg OB2. Our images show circumstellar material much closer to NML Cyg than the surrounding H II region and coincident with the water masers as well as SiO masers, suggesting that we are likely imaging the photo-dissociation boundaries. Schuster *et al.* have modeled the molecular photodissociation boundary shown superposed on the image of NML Cyg. Note that the overall distribution of the H_2O masers is not only similar in extent to the reflection nebula, but also matches its convex shape. Thus, it appears that the molecular material SE of the star is preferentially shielded from photo-dissociation.

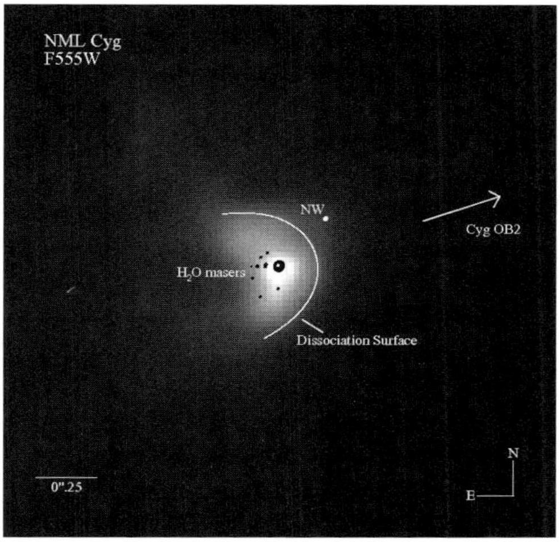

Figure 2. HST/WFPC2 image of NML Cyg with the photodissociation boundary and positions of the H_2O masers superposed.

4. The extreme red hypergiant – VY CMa

The extreme red supergiant and powerful infrared source and OH maser, VY CMa is one of the most luminous and largest evolved stars known. At its distance of 1.5 kpc (Herbig 1972; Lada & Reid 1978; Marvel 1997), VY CMa's luminosity is $\approx 4.3 \times 10^5 L_\odot$. Its very visible asymmetric nebula, 10" across, combined with its high mass loss rate of $4 \times 10^{-4} M_\odot yr^{-1}$ (Danchi *et al.* 1994), makes VY CMa a special case even among the cool hypergiants that define the upper luminosity boundary.

High resolution imaging with HST (FOC, Kastner & Weintraub 1998; WFPC2, Smith *et al.* 2001) and near–IR interferometry (Monnier et al 1999) have revealed VY CMa's complex circumstellar environment. The multi-wavelength HST/WFPC2 images of VY CMa (Smith *et al.* 2001) showed the complexity of detail in its ejecta (Figure 3) including the prominent arc to the northwest, two bright filamentary arcs to the southwest, plus relatively bright clumps of dusty knots near the star, and numerous small arcs throughout the nebula. All of which are evidence for multiple and asymmetric mass loss episodes. The apparent random orientations of the arcs suggested that they were produced by localized ejections, possibly from large-scale convective regions on the star.

The extensive maser (OH. H_2O, SiO), and CO observations, however, have been interpreted as evidence for an axis of symmetry with possible bipolar outflows and a disk-like distribution for the circumstellar material. The OH maser intensity maxima are within 0.5" and are distributed along a NE-SW axis at a position angle of $\sim 50°$ (Bowers *et al.* 1983). If there is a NE-SW polar/rotation axis, then Arcs 1 and 2 could both be part of an associated bipolar outflow depending on the opening angle of the cone, but from different locations on the star corresponding to their spatial separation.

The H_2O maser spots appear to be oriented much more east-west (Richards, Yates & Cohen 1998). The highly polarized SiO maser spots (Shinaga *et al.* 2004) have a 0.2" north-south distribution on the sky, although the SiO emission is bipolar with a mean position angle of 72°. Similarly, diffraction – limited speckle interferometry in the near – infrared also shows a primarily north-south orientation for the dust shell extended 0.2" with a position angle $\sim$ 153 to 176° (Wittkowski, Langer & Weigelt 1998), while near–IR aperture–masked interferometry (Monnier *et al.* 1999) showed a southward extension of the dust emission within 0.1" but no clear disklike or bipolar geometry in the images.

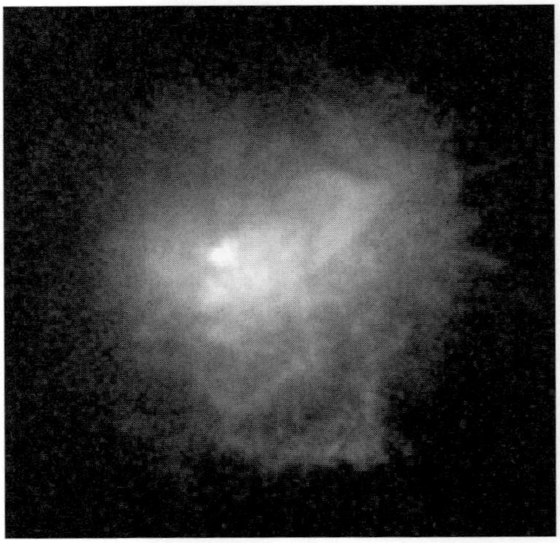

Figure 3. HST/WFPC2 image of showing the prominence-like arcs and clumps of knots in its ejecta.

K-band interferometric measurements however, indicate a bipolar distribution within 0.1" of the star but with the dusty disk oriented east-west (Monnier *et al.* 2004). Recent interferometric millimeter observations of CO and SO (Muller *et al.* 2006) have been modelled as a bipolar outflow in the east-west direction, but with a very wide opening angle ($\sim 120°$) and an expanding shell elongated north-south.

The maser maps and interferometric observations are thus not always consistent with each other. With this lack of strong evidence for a well-defined axis of symmetry or bipolar axis in VY CMa, we conclude that the morphology and kinematics of the numerous arcs and knots are more consistent with *a history of localized mass ejections from active regions on the star not strongly aligned with a presumed axis or equator.*

5. Convective activity and magnetic fields

We have recently obtained second epoch images and polarimetry of VY CMa with the HST (Humphreys *et al.* 2007, Jones *et al.* 2007). The polarization together with the color from the images can yield information on the line of sight distribution of the nebulosity and let us determine the relative locations of the arcs and knots, while the transverse motions combined with the Doppler velocities (Humphreys *et al.* 2005) provide a complete picture of the kinematics of the ejecta including the total space motions and directions of the outflows. Our results show that the arcs and clumps of knots are moving at different velocities, in different directions, and at different angles relative to the plane of the sky and to the star, confirming their origin from eruptions at different times and from physically separate regions on the star, not by a more uniform long-term mass loss. They represent localized, relatively massive (few $\times 10^{-3} M_{\odot}$) ejections. We are suggesting that they are the result of activity associated with large convective regions or starspots, which have now been observed on numerous red giants, AGB stars and supergiants.

Recently, Vlemmings *et al.* (2002, 2005) have estimated the magnetic field strength from the circular polarization of H_2O masers in the ejecta of AGB stars and several evolved supergiants including the strong OH/IR sources VY CMa, VX Sgr, NML Cyg, and S Per. They report magnetic fields in VY CMa of $\sim$ 200mG at distances of 220 AU. Their analysis supports the Zeeman interpretation of the circular polarization of the

SiO masers only a few AU from the surface of the stars (Barvainis *et al.* 1987; Kemball & Diamond 1997). Together with Zeeman splitting of the OH emission far out in the wind at a few thousand AU (Szymczak & Cohen 1997; Masheder *et al.* 1999), these measurements confirm the presence of a magnetic field throughout the ejecta of VY CMa. The OH measurements at distances in the ejecta comparable to the prominence-like arcs, imply a magenetic field at the star of $\sim$ 8000 G. Similarly, circular polarization of the OH emission in IRC +10420 (Nedoluha & Bowers 1992) suggests a field strength of $\sim$ 3000 G at the star.

These values seem too large for a global magnetic field but may be representative of the field associated with the active regions.

Acknowledgements

It is a please to thank my collaborators on these studies, Kris Davidson, Robert Gehrz, Andrew Helton, George Herbig, Terry J. Jones, Massimo Marengo, Gerald Ruch, Michael Schuster, Nathan Smith and George Wallerstein.

References

Barvainis, R., McIntosh, G. & Predmore, C. R. 1987, *Nature* 329, 613
Bowers, P. F., Johnston, K. J., & Spencer, J. H . 1983, *ApJ* 274, 733
Danchi, W. C., Bester, M., Degiacomi, C. G., Greenhill, L. J., & Townes, C.H. 1994, *AJ* 107, 1469
de Jager, C. 1998, *A&AR* 8, 145
Habing, H. J., Goss, W. M., & Winnberg, A. 1982, *A&A* 108, 412
Herbig, G. H. 1972, *ApJ* 172, 375
Humphreys, R. M. & Davisdon, K. 1994, *PASP* 106, 1025
Humphreys, R. M., Smith, N., Davidson, K., Jones, T. J. *et al.* 1997, *AJ* 114, 2778
Humphreys, R. M., Davidson, K., Ruch, G., & Wallerstein, G. 2005, *AJ* 129, 492
Humphreys, R. M., Helton, L. A., & Jones, T. J. 2007, *AJ* in press
Jones, T. J., Humphreys, R. M., Gehrz, R. D. *et al.* 1993, *ApJ* 411, 323
Jones, T. J., Humphreys, R. M., Helton, L. A., Gui, C. & Xiang, H. 2007, *AJ* in press
Kastner, J. H. & Weintraub, D. A. 1998, *AJ* 115, 1592
Kemball, A. J. & Diamond, P. J. 1997, *ApJ* Letters 481, L111
Knapp, G. R. & Morris, M. 1985, *ApJ* 292, 640
Knödlseder, J. 2003, IAU Symp. 212, A Massive Star Odyssey, from Main Sequence to Supernova, eds. K.A. van der Hucht, A. Herrero & C. Esteban, 505
Lada, C. J. & Reid, M. J. 1978, *ApJ* 219, 95
Marvel, K.B. 1997, *PASP* 109, 1286
Masheder, M. R. W. *et al.* 1999, *New Astron* 43, 563
Monnier, J. D., Tuthill, P. G., Lopez, B., Cruzalebes, P. *et al.* 1999, *ApJ* 512, 351
Monnier, J. D. *et al.* 2004, *APJ* 605, 436
Morris, M. & Jura, M. 1983, *APJ* 267, 179
Muller, S., Dinh-V-Trung, Lim, J. Hirano, N., Muthu, C, & Kwok , S. 2007, *ApJ* 656, 1109
Nedoluha, G. E. & Bowers, P. F. 1992, *ApJ* 392, 249
Oudmaijer, R. D., Groenewegen, M. A. T., Matthews, H. E., *et al.* 1996, *MNRAS* 280, 1062
Richards, A. M. S., Yates, J. A., & Cohen, R. J. 1998, *MNRAS*, 299, 319
Shinnaga, H., Moran, J. M., Young, K. H. & Ho, P. T. P. 2004, *ApJ* Letters 616, L47
Schuster, M. T., Humphreys, R. M., & Marengo, M 2006, *AJ* 131, 603
Smith, N., Humphreys, R. M., Davidson, K., *et al.* 2001, *AJ* 121, 1111
Szymczak, M. & Cohen, R. J. 1997, *MNRAS* 288, 945
Vlemmings, W. H. T., Diamond, P. J. & van Langevelde, H. J. 2002, *A&A* 394, 589
Vlemmings, W. H. T., van Langevelde, H. J., & Diamond, P. J. 2005, Mem. Soc. Astr. It. 76, 462
Wittkowski, M., Langer, N., & Weigelt, G. 1998, *A&A* 340, L39

Astrophysical Masers and their Environments
Proceedings IAU Symposium No. 242, 2007
J.M. Chapman & W.A. Baan, eds.

doi:10.1017/S1743921307013099

The 3D morphology of the ejecta surrounding VY Canis Majoris

Terry Jay Jones[1], Roberta M. Humphreys[1] and L. Andrew Helton[1]

[1]Department of Astronomy, University of Minnesota, Minneapolis, MN 55455, USA
email: tjj@astro.umn.edu, roberta@aps.umn.edu, ahelton@astro.umn.edu

Abstract. We use second epoch images taken with WFPC2 on the HST and imaging polarimetry taken with the HST/ACS/HRC to explore the three dimensional structure of the circumstellar dust distribution around the red supergiant VY Canis Majoris. Transverse motions, combined with radial velocities, provide a picture of the kinematics of the ejecta, including the total space motions. The fractional polarization and photometric colors provide an independent method of locating the physical position of the dust along the line-of-sight. Most of the individual arc-like features and clumps seen in the intensity image are also features in the fractional polarization map, and must be distinct geometric objects. The location of these features in the ejecta of VY CMa using kinematics and polarimetry agree well with each other, and strongly suggest they are the result of relatively massive ejections, probably associated with magnetic fields.

Keywords. stars: individual (VY CMa), stars: mass loss, polarization

1. Introduction

VY CMa is one of the most luminous evolved stars known, and is near the empirical upper limit for cool hypergiants (Humphreys & Davidson 1994, de Jager 1998). At a distance of 1.5 kpc (Herbig 1972, Lada & Reed 1978, Marvel 1997), the star has a luminosity of $\sim 4.3\times10^5 L_o$. VY CMa is surrounded by an asymmetric nebula, 10" across, created by a very high mass loss rate of $4 \times 10^{-4} M_o yr^{-1}$ (Danchi *et al.* 1994). This high luminosity and high mass loss rate make VY CMa one of the richest stellar maser sources known. Multi-wavelength HST/WFPC2 images of VY CMa (Smith *et al.* 2001) revealed a complex circumstellar environment dominated by multiple filamentary arcs and knots. The apparent random orientations of the arcs suggested that they were produced by localized ejections, not necessarily aligned with either the stars presumed NE/SW axis (Morris & Bowers 1980, Bowers *et al.* 1983, Richards *et al.* 1998) or its equator. Smith *et al.* (2001), Smith (2004), and Humphreys *et al.* (2005) argue that the circumstellar environment of VY CMa is dominated by complex, distinct arcs and features that were created at different epochs. To further investigate the distribution of material in the ejecta surrounding VY CMa, we use second epoch images taken with WFPC2 on the HST and imaging polarimetry taken with the HST/ACS/HRC.

2. Kinematics

Humphreys *et al.* (2007) combined previous radial velocity measurements with new multi-epoch HST images to determine the total space motion of many of the knots and arcs seen in the nebula surrounding VY CMa. They find that these features are moving at different velocities, in different directions, and at different angles relative to the plane of the sky and to the star, confirming their origin from eruptions at different times and from physically separate regions on the star. The projected motions on the sky of these features are illustrated in Figure 1. Most of these vectors point back to the star, indicative of purely radial motions. Under the assumption their total space motion is purely radial from the star, the addition of the radial velocity observations allows us to locate the features in 3 dimensions around the star.

The principle features and the travel time since their ejection are illustrated in Figure 2. The NW Arc is redshifted with respect to the systemic stellar velocity and stretches out behind the star. Arcs 1 and 2 are blueshifted and travelling towards us relative to the star. The S Arc has two velocity components, one of which shows a redshift, placing the arc behind the star moving away from us.

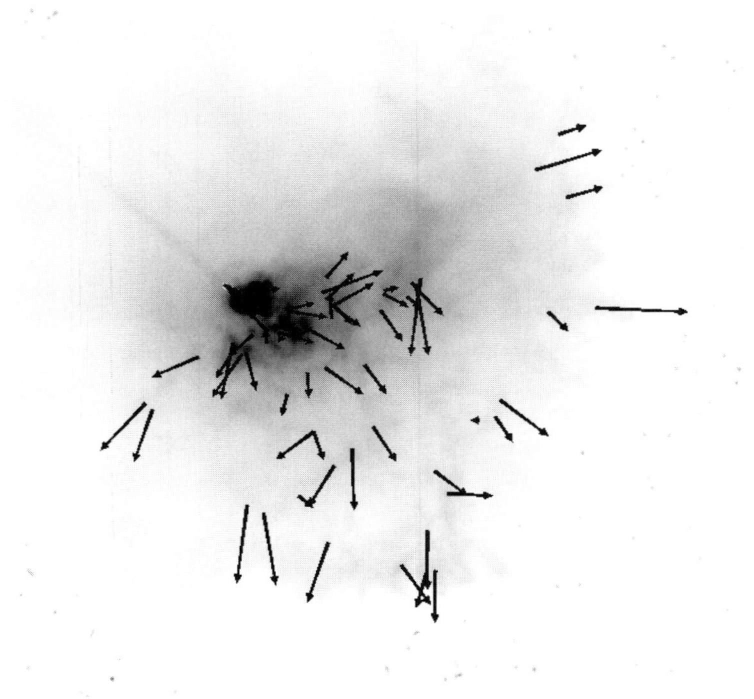

Figure 1. Proper motion vectors for features seen in the F550 image of VY CMa

3. Polarimetry

Polarimetry has proved a powerful tool for studying the geometry of extended circumstellar dust and gas, in particular bipolar systems. For example, Shure *et al.* (1995) were able to accurately fix the tilt of the bipolar outflow axis in the peculiar red giant OH 0739. Schulte-Ladbeck *et al.* (1999) were able to distinguish between different axisymmetric models for the ejecta surrounding Eta Carina. Polarimetry combined with theoretical modeling has been used to further refine the details of bipolar outflows and circumstellar disks (Lowe & Gledhill 2007, Hoffman *et al.* 1997). In the visual images, VY CMa does not appear to have a mass loss nebula with any simple, underlying symmetry like a classic bipolar nebula and the analysis of the radial velocities and proper motions found a more complex distribution of ejecta. Imaging polarimetry can be used as an independent method to locate each of these features in three dimensions with respect to the star, and without reference to any model distribution.

If the scattering is optically thin, we can compute two possible scattering angles from

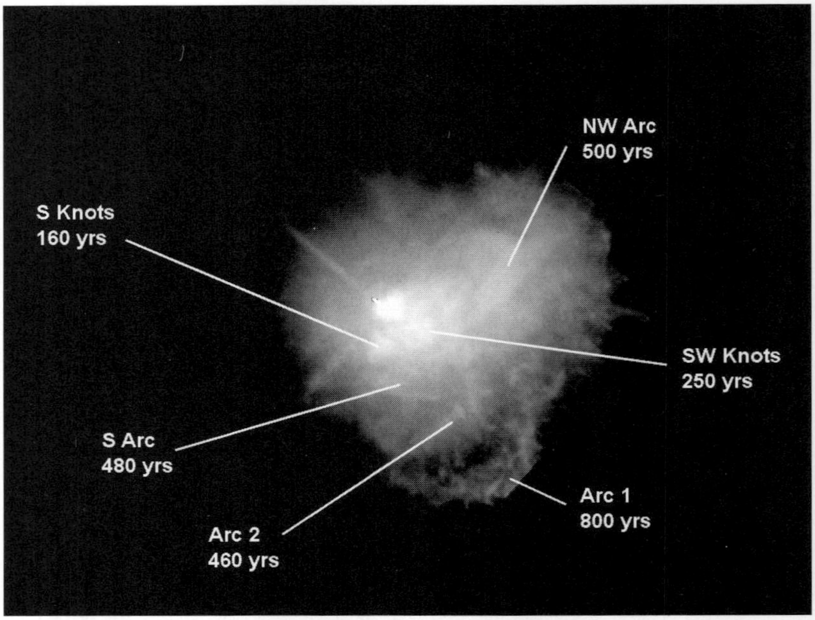

Figure 2. Principle features and an estimate of their age since ejection.

the observed fractional polarization, but we can not determine from the polarization alone whether the dust is at the larger or smaller scattering angle. For fractional polarizations less than the observed maximum, the dust must be either behind or in front of the plane of the sky containing the scattering angle corresponding to maximum polarization. In principle, light scattered off of dust behind the plane of optimum scattering will experience more reddening exiting the nebula than light scattered off of dust in front of that plane. Jones *et al.* (2007) used the color of the nebulosity as a means of breaking the foreground/background degeneracy in locating the dust. This technique can not be expected to extract fine details of the geometry of the dust surrounding VY CMa, only large scale patterns. Any single line of sight through the nebulosity likely intersects dust at a range of scattering angles. Given these limitations, we will concentrate on the bright, distinct regions identified in Figure 2 that are not too close to the star and too optically thick for our technique to work.

A map of fractional polarization along with a total intensity image and a map of the instrumental [547]–[1024] color from Smith *et al.* (2001) is shown in Figure 3. The NW Arc is prominent in the total intensity image and is quite red in the color map. In the fractional polarization map this feature is clearly apparent as more weakly polarized than its surroundings. Arcs 1 and 2 are relatively blue in color and can be seen as moderately polarized features in the fractional polarization map. The highest fractional polarization, reaching 80%, is found in the regions surrounding the NW Arc. The entire NE half of the nebula is faint and has relatively low polarization, varying between 10% and 20% on average.

Most of the individual arc-like features and clumps appear to be separate geometric objects, and do not display any clear association with a simple bipolar outflow geometry. They are distinct both in the fractional polarization map and the intensity image. If these features were just local density enhancements, the fractional polarization would

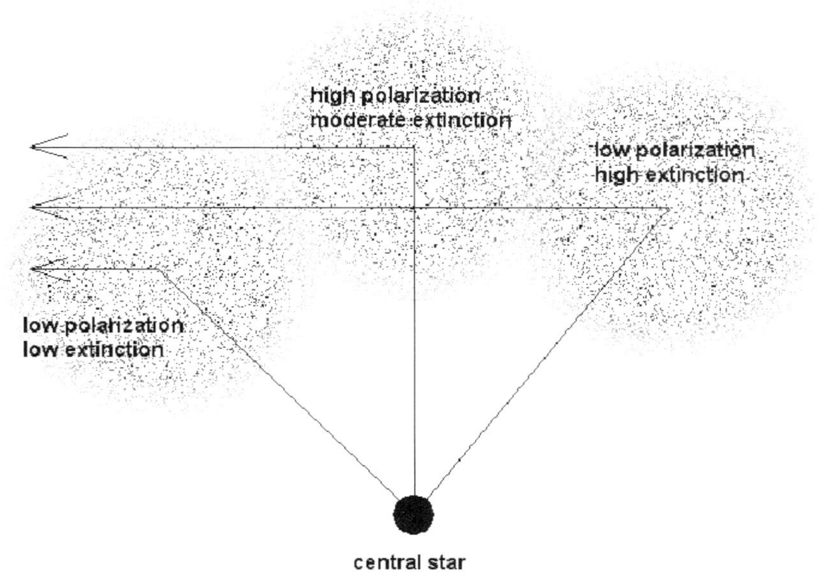

Figure 3. Schematic of the technique to use fractional polarization and color to locate a region of scattering along the line of sight.

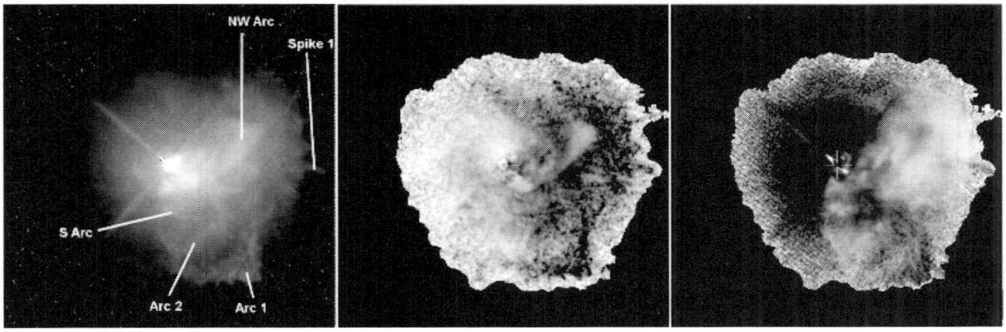

Figure 4. Left: F658 total intensity; Center: [547]-[1024] color; Right: fractional polarization.

not change across strong brightness gradients. In all cases, the locations of these features are consistent with their observed total space motions.

The polarimetry analysis places the NW Arc (which is moderately optically thick in scattering at 658nm) behind the plane of the sky at an angle with respect to the star of about 30°, in good agreement with the kinematic location from Humphreys *et al.* (2007) for features A, D, X, Y in the NW Arc. Arc 1 is in front of the star at an angle of about 30° based on the polarimetry, in good agreement with the value of 35° from the kinematics. Arc 2 is also in front of the star at about 20° according to the polarimetry and the same value based on the kinematics. Humphreys *et al.* (2007) list the S Arc as being located in front of the star, but there are two velocities associated with the S Arc. The redshifted feature would place the S Arc behind the star at an orientation similar to that derived from the polarimetry.

4. Summary

The morphology of the ejecta surrounding VY CMa is more complex than a simple outflow with one axis of symmetry. Most of the individual arc-like features and clumps seen in the HST images appear to be separate geometric objects, and do not display any clear association with a bipolar outflow geometry. They are distinct both in the fractional polarization map and the intensity image. If these features were just local density enhancements, the fractional polarization would not change across strong brightness gradients. Placement of these arcs and knots within the volume surrounding the central star using the observed fractional polarization and colors of the reflection nebulosity is in good agreement with the kinematics, assuming the motions are primarily radial out from the star.

Extrapolation of the space motions of these features back to the star yields different ejection times spanning the last 1000 years or so. The motions are not consistent with a single ejection time or a smooth, continuous outflow. In analogy to the Sun, we suggest these arcs and knots may be associated with very large scale magnetic phenomena at the surface of the star, but on much larger energy and mass scales.

References

Bowers, P. F., Johnston, K. J., & Spencer, J. H. 1983, *ApJ*, 274, 733
Danchi, W. C., Bester, M., Degiacomi, C. G., Greenhill, L. J., & Townes, C. H. 1994, *AJ* 107, 1469
de Jager, C. 1998, *A&AR* 8, 145
Herbig, G. H. 1972, *ApJ*, 172, 375
Hoffman, J. L., Whitney, B. A., Wood, K., & Nordsieck, K. H. 1997, *Bulletin of the American Astronomical Society*, 29, 1279
Humphreys, R. M., & Davidson, K. 1994, *PASP* 106, 1025
Humphreys, R. M., Helton, L. A., & Jones, T. J. 2007, *AJ*, in press.
Humphreys, R. M., Davidson, K., Ruch, G., & Wallerstein, G. 2005, *AJ*, 129, 492
Jones, T. J., Humphreys, R. M., Helton, L. A., Gui, C. & Xiang, H. 2007, *AJ*, in press.
Lada, C. J., & Reid, M. J. 1978, *ApJ*, 219, 95
Lowe, K. T. E., & Gledhill, T. M. 2007, *MNRAS*, 374, 176
Marvel, K. 1997, *PASP*, 109, 1286
Morris, M., & Bowers, P. F. 1980, *AJ*, 85, 724
Richards, A. M. S., Yates, J. A., & Cohen, R. J. 1998, *MNRAS*, 299, 319
Schulte-Ladbeck, R. E., Pasquali, A., Clampin, M., Nota, A., Hillier, D. J., & Lupie, O. L. 1999, *AJ*, 118, 1320
Shure, M., Sellgren, K., Jones, T. J., & Klebe, D. 1995, *AJ*, 109, 721
Smith, N. 2004, *MNRAS*, 349, L31
Smith, N., Humphreys, R. M., Davidson, K., Gehrz, R. D., Schuster, M. T., & Krautter, J. 2001, *AJ* 121, 1111

Astrophysical Masers and their Environments
Proceedings IAU Symposium No. 242, 2007
J.M. Chapman & W.A. Baan, eds.

doi:10.1017/S1743921307013105

Turbulent, steamy red supergiant winds

A. M. S. Richards[1], I. Bains[2], A. Bartkiewicz[3], R. J. Cohen[1], P. J. Diamond[1], S. Etoka[1], M. D. Gray[4], E. E. Lekht[5], M. R. W. Masheder[6], E. Mendoza-Torres[5], K. Murakawa [7], M. Szymczak[3], H. J. van Langevelde[8,9], W. Vlemmings[1], and J. A. Yates[10]

[1]JBO, University of Manchester, SK11 9DL, UK. a.m.s.richards@manchester.ac.uk
[2]Centre for Astrophysics and Supercomputing, Swinburne University of Technology, PO Box 218, Hawthorn, Victoria 3122, Australia.
[3]Centre for Astronomy, Nicolaus Copernicus University, Torun, Poland.
[4]Dept. of Physics, University of Manchester, PO Box 88, Manchester M60 1QD, UK.
[5]INAOE, Apdo Postal 51 y 216, 72840 Tonantzintla, Puebla, Mexico.
[6]Department of Physics, Bristol University, Bristol, BS8 1TL, UK.
[7]Max-Planck-Institut für Radioastronomie, Auf dem Hugel 69, G-53121, Bonn, Germany.
[8]JIVE, Postbus 2, 7990 AA Dwingeloo, The Netherlands.
[9]Sterrewacht Leiden, Leiden University, Postbus 9513, 2300 RA Leiden, The Netherlands.
[10]UCL, Gower Street, London WC1E 6BT, UK.

Abstract. Rapidly-evolving red supergiants (RSG) lose half or more of their mass before ending their lives as supernovae. Masers allow us to study the mass loss from 4 nearby RSG in AU-scale detail using MERLIN and EVN/global VLBI. The water maser clouds are over-dense and over-magnetised with respect to the surrounding wind. In most cases, the brighter an individual maser component is the smaller its apparent (beamed) FWHM appears, as predicted for approximately spherical clouds. Individual water maser features have a typical half-life of 5-10 yr, but comparison with single dish monitoring suggests that the water vapour clouds themselves survive many decades (the water maser shell crossing time), within which the local masers wink on and off. OH mainline masers are found in the tenuous surrounding gas, overlapping the water maser shell, surrounded by OH 1612-MHz masers at a greater distance from the star.

Keywords. masers, stars: mass loss, late-type, supergiants, polarization, techniques: high angular resolution

1. Introduction

Stars more than about 8 times the mass of the Sun live fast and go out in a blaze of glory as a supernova. It is less well known that even before then, these stars contribute up to half of all the dust and a large proportion of light elements which enrich the interstellar medium and hence the next generation of star formation. This was especially important in the early universe owing to the rapid evolution of high-mass stars. Less than one in 200 stars are high-mass at birth, but they typically lose half their mass in the Red Super Giant (RSG) stage, over a few hundred thousand years. The stars lose mass through a combination of pulsation (with irregular periods of 2–5 yr) and radiation pressure on dust. They have no detectable rotation and the simplest "onion" model places spherical SiO, H_2O and OH maser shells at increasing distances from the star. We have studied four RSG, whose vital statistics are given in Table 1.

2. Water maser cloud measurements

We observed the H_2O emission at 22 GHz from all four sources using MERLIN, with a resolution of $\geqslant$ 10 mas. We measured maser emission in each channel by fitting 2D

Table 1. Red Supergiants studied. References not explicitly given are cited in the article listed for each star. The IRAS colour regions (C.R.) are defined in van der Veen & Habing (1988) and the 9.7 μm silicate feature and dust morphologies are discussed in Monnier *et al.* (1999) and references therein. The OH 1612-MHz column gives the shell shape and approximate peak flux density.

RSG	D (kpc)	$M_\star$ ($M_\odot$)	$T_\star$ (K)	IRAS C.R.	Silicate feature	Dust morphology	OH 1612-MHz morphology	log(Jy)	References
S Per	2.2	20	2400	IIIa	emission	compact	biconical	0	RYC99
VX Sgr	1.7	10	2900	VII	emission	compact	~spherical	1	M+03
VY CMa	1.5	50	2700	VII	declining	complex	♡	2	RYC98
NML Cyg	2.0	50	3650	VIb	absorption	multiple shells	elongated, detached	2	RYC96, ED04

References: RYC96 Richards *et al.* (1996); RYC98 Richards *et al.* (1998); RYC99 Richards *et al.* (1999); M+03 Murakawa *et al.* (2003); ED04 Etoka & Diamond (2004).

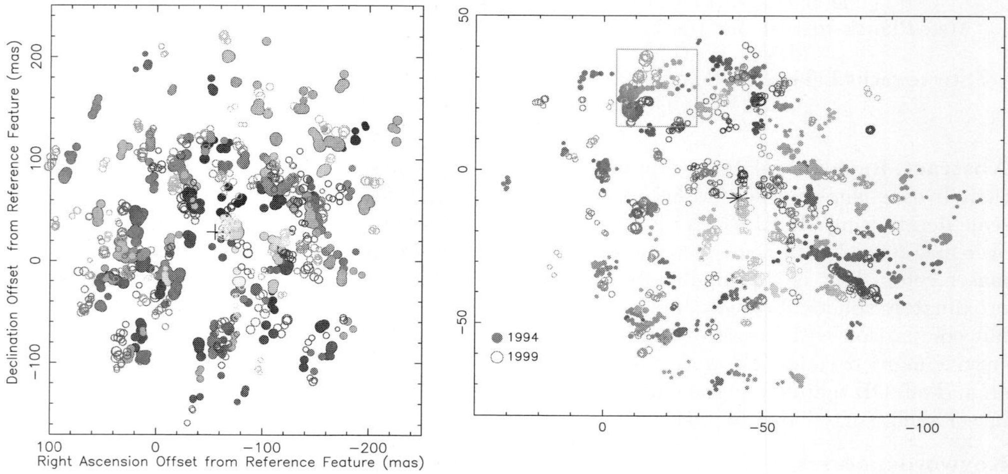

Figure 1. Positions of H_2O maser components around VX Sgr (*left*) and S Per(*right*) at 2 epochs. The darker (paler) symbols represent emission which is more red- (blue)-shifted with respect to $V_\star$ and the size is proportional to log(flux density).

Gaussian components, giving sub-mas position accuracy. We resolved the individual components in S Per and NML Cyg. Fig. 1 shows examples of H_2O maser shells.

2.1. *Water maser shell sizes and densities*

The water masers mostly occupy approximately spherical shells around each star, but with increasing signs of asymmetry, from a mild equatorial density enhancement in S Per to protruding jet-like features in NML Cyg. In each case 22-GHz emission traces strong acceleration away from the star; typically the outflow velocity doubles from 10 to 20 km s^{-1} over 100–200 or more AU, illustrated in Fig. 2. The inner rim of the shell is defined by the number density at which the masers are collisionally quenched, $n_\mathrm{q} \approx 5 \times 10^{15}$ m^{-3}. The inner and outer radii and expansion velocities (r_i, r_o, v_i, v_o) were used to estimate the mass-loss rate $\dot{M}(n_\mathrm{q})$ if the number density of the whole wind was n_q at r_i. These properties are tabulated in Table 2, from which it can be seen that $\dot{M}(n_\mathrm{q})$ is unfeasibly large compared with other estimates or, indeed, RSG mass loss lifetimes.

Table 2. The mass loss rate measured from thermal emission $\dot{M}(\mathrm{CO/IR})$ is compared with $\dot{M}(n_q)$; the latter is unfeasibly large, and an explanation is given in Section 2.2. The remaining columns give the inner and outer maser shell limits, see Table 1 for references.

RSG	$\dot{M}(\mathrm{CO/IR})$ ($\mathrm{M_\oplus\ yr^{-1}}$)	$\dot{M}(n_q)$ ($\mathrm{M_\oplus\ yr^{-1}}$)	H_2O (AU)		OH 1665-MHz (AU)		OH 1667-MHz (AU)	
			r_i	r_o	r_i	r_o	r_i	r_o
S Per	3–60	*930*	55	165	80	80	570/175[2]	570/245[2]
VX Sgr	5–80	*1900*	95	320	130	130	540/1950[3]	540/1950[3]
VY CMa	65–115	*3150*	115	540[1]	225	225	4500	10000
NML Cyg	35–60	*5150*	100	480[1]	2400	3000	3200	5000

Footnotes [1]Including biconical/bipolar features; [2]Observed in 1993/1999 [3]Multiple shells;

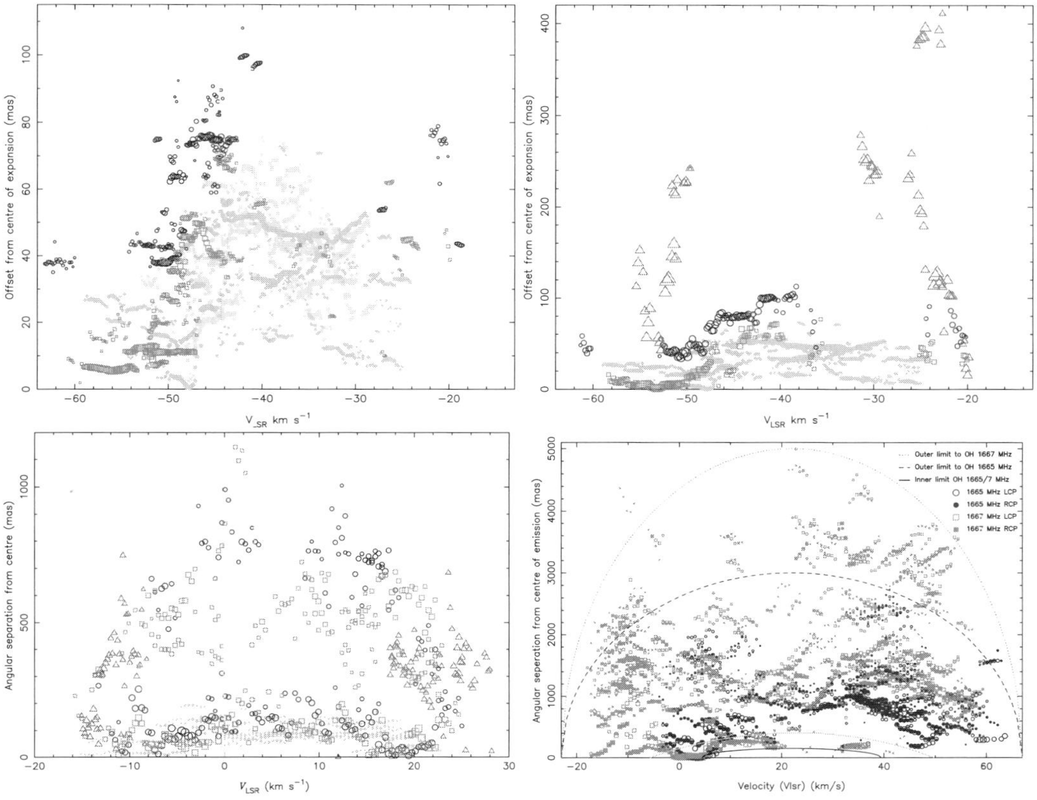

Figure 2. H_2O masers: pale filled circles; OH 1665/1667-MHz masers hollow squares; 1612-MHz masers: triangles. The V_{LSR} axes are centred on $V_\star$. (*top*) S Per. (*bottom*)(*left*) VX Sgr, (*right*) VY CMa (only OH mainlines shown; the outer limit of H_2O emission is shown by the innermost arc).

2.2. *Cloud sizes and survival*

Fig. 1 shows that many maser components form ordered patterns with systematic velocity/position gradients, corresponding to an average cloud size of 20 AU. Some clouds have a total velocity extent 2 or more times the thermal line width and/or a total angular extent greater than the estimated gain length. This shows that the clouds are physically

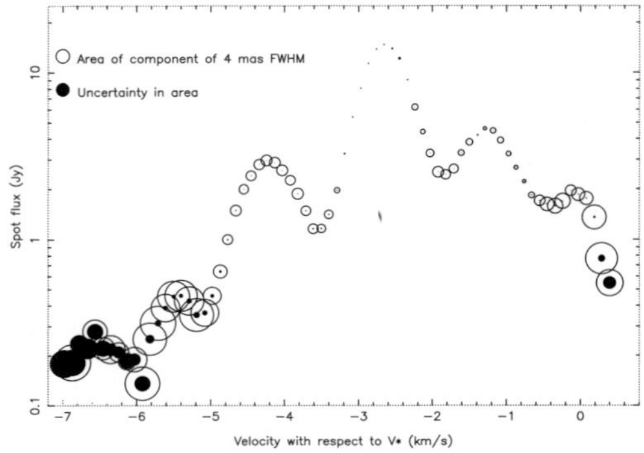

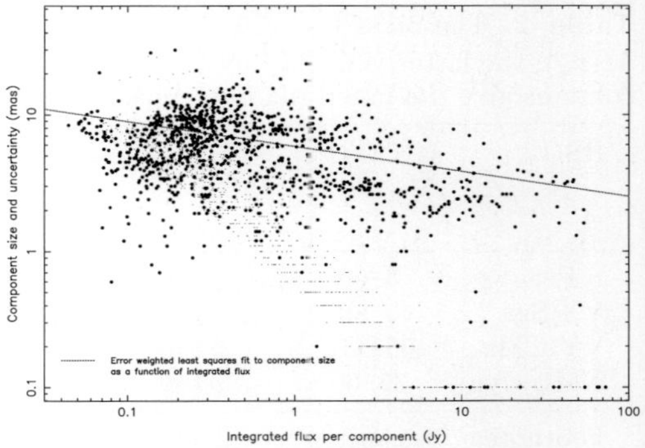

Figure 3. (*left*) Beamed areas of 22-GHz components in the feature of S Per boxed in Fig. 1. (*right*) FWHM of 22-GHz components (large symbols) and uncertainties (small symbols).

bounded, most likely by a density differential (Section 2.1). Each CSE contains about 50–100 H_2O maser clouds (assumed to be on average spherical), giving a volume filling factor of a few percent or less. The crossing time for the H_2O maser shell is $\approx$ 50−80 yr, suggesting that a few clouds appear at r_i per stellar period and inject about 10–20 $M_\oplus$ yr^{-1} into the CSE. This solves the mass-loss-rate problem, since the total mass in H_2O maser clouds is comparable to the mass-loss rate measured from thermal species. The surrounding gas number density needs to be $\sim$1/50 or less of that in the clouds.

Comparison with other epochs of interferometry show that about half the maser clouds survive 5 or more years. Proper motions were measured for VX Sgr, VY CMa and NML Cyg. The velocities and accelerations are consistent with the observed V_{LSR} gradients. It is, however, surprising that the other half of the clouds vanish as the overall shell sizes imply that they survive for many decades. An explanation is provided by single-dish monitoring every few months by Lekht *et al.* (2005) using the Puschino telescope. We identified a bright feature in S Per seen at a very similar velocity and position in the 1994 and 1999 maps (the box in Fig. 1). The spectra show that this feature vanished just a few months after the 1994 MERLIN observations but re-appeared in 1998. This shows that the clouds themselves may survive any length of time, but detectable masers wink on and off, either due to local excitation conditions or to the beaming direction.

2.3. *Direct measurements of maser beaming*

The brightest maser emission is most strongly amplified and this leads to the narrowing of the apparent angular size of a single maser component propagating through an approximately spherical cloud (Elitzur 1992). The FWHM of the H_2O maser components in S Per is inversely proportional to their flux density illustrated in Fig. 3. A similar effect is seen for the OH mainline masers from VX Sgr and for three out of four AGB stars (Bains *et al.* 2003) but U Ori shows a direct relationship for the brightest masers at some epochs only. This could be explained if the masers originate from a shock front and there are other indications that this may be the case for U Ori.

3. OH maser distribution

The 1.6-GHz OH masers around all four RSG have been mapped by MERLIN at $\geqslant$ 120 mas resolution. The 1665 and 1667 MHz mainline masers from S Per and VX Sgr were also imaged using the EVN/global VLBI at $\geqslant$ 7 mas resolution. Fig 2 shows that the OH mainline masers overlap the H_2O maser shells in position and velocity in all CSEs apart from NML Cyg. OH mainline emission close to $V_\star$ is seen between H_2O masers, showing that the apparent overlap is not just a projection effect of the front and back caps. No excited OH masers are detected, so the OH gas must be at a temperature $\leqslant 500$ K and a number density $n \leqslant 10^{14}$ m^{-3}, consistent with the deduction in Section 2.2 that the surrounding gas is 1/50–1/100 less dense than the H_2O clouds.

We used Zeeman splitting of the OH mainline masers to measure a magnetic field strength of ~ 0.3–3 μT. The field strength in the H_2O maser clumps of these stars is ~ 10–400 μT (Vlemmings *et al.* 2002, 2005). In S Per for example, the H_2O magnetic field is ~ 10–$20\times$ stronger than the OH magnetic field at 140 AU from the star. This suggests that the H_2O maser clumps have compressed, frozen-in fields where the magnetic field strength varies as $n^{\sim 0.3-0.5}$ (Mouschovias & Ciolek 1999).

4. Conclusions and future work

We have measured the apparent beamed size of individual 22-GHz H_2O maser components, finding that their FWHM is inversely proportional to the maser intensity. The components form distinct streaks which represent discrete water vapour clouds, typically 20 AU across. These take many decades to cross the H_2O maser shell but the 22-GHz emission from any single cloud may wink on and off several times during this lifetime. The H_2O clouds are 50 times denser than the surrounding gas. This explains how OH mainline masers can emanate at a similar distance from the star, but in more diffuse regions surrounding the H_2O masers, and how they appear to be associated with a much weaker magnetic field than the H_2O masers at a similar distance.

If the size of the clouds is extrapolated back to the stellar surface, allowing for an r^2 expansion in the wind, they would be born with a radius of $\sim 0.1R_\star$. This could result from the ejection of matter associated with a convection cell bringing chemically enriched material to the surface, or with a star spot. Betelgeuse is the only RSG which has been so far resolved in detail; models (Freytag *et al.* 2002) produce structures on compatible scales. In future we will be able to resolve many more AGB and RSGs using *e*-MERLIN, the EVLA, VLBI and ALMA and even track individual clumps as they are ejected from the stellar surface, pass through the SiO maser zone, form dust, produce water masers and pass through OH mainline masers before joining the OH 1612-MHz maser shell.

References

Bains, I., Cohen, R. J., Louridas, A., Richards, A. M. S., Rosa-González, D., & Yates, J. A. 2003, *MNRAS*, 342, 8

Etoka, S., & Diamond, P. 2004, *MNRAS*, 348, 34

Elitzur, M. 1992, *Astronomical Masers*, Kluwer

Freytag, B., Steffen, M., & Dorch, B. 2002, *AN*, 323, 213

Lekht, E. E., Rudnitskij, G. M., Mendoza-Torres, J. E., & Tolmachev, A. M. 2005, *A&A*, 437, 127

Mouschovias, T. C., & Ciolek, G. E. 1999, in: C. J. Lada & N. D. Kylafis (eds.), *The Origin of Stars and Planetary Systems*, NATO ASIC Proc., 540, 305

Monnier, J. D., Geballe, T. R., & Danchi, W. C. 1999, *ApJ*, 521, 261

Murakawa, K., Yates, J. A., Richards, A. M. S., & Cohen, R. J. 2003, *MNRAS*, 344, 1

Richards, A. M. S., Yates, J. A., & Cohen, R. J. 1996, *MNRAS*, 282, 665

Richards, A. M. S., Yates, J. A., & Cohen, R. J. 1998, *MNRAS*, 299, 319

Richards, A. M. S., Yates, J. A., & Cohen, R. J. 1999, *MNRAS*, 306, 954

van der Veen, W. E. C. J. & Habing, H. J. 1988, *A&A*, 184, 125

Vlemmings, W. H. T., Diamond, P. J., & van Langevelde, H. J. 2002, *A&A*, 394, 589

Vlemmings, W. H. T., van Langevelde, H. J., & Diamond, P. J. 2005, *A&A*, 434, 1029

Astrophysical Masers and their Environments
Proceedings IAU Symposium No. 242, 2007
J.M. Chapman & W.A. Baan, eds.

doi:10.1017/S1743921307013117

SiO maser observations of a wide dust-temperature range sample

Jun-ichi Nakashima[1] **and Shuji Deguchi**[2]

[1]Academia Sinica Institute of Astronomy and Astrophysics, P.O. Box 23-141, Taipei 10617, Taiwan
email: junichi@asiaa.sinica.edu.tw

[2]Nobeyama Radio Observatory, National Astronomical Observatory, Minamimaki, Minamisaku, Nagano 384-1305, Japan

Abstract. We present the results of SiO line observations of a sample of known SiO maser sources covering a wide dust-temperature range. The aim of the present research is to investigate the causes of the correlation between infrared colors and SiO maser intensity ratios among different transition lines. We observed in total 75 SiO maser sources with the Nobeyama 45m telescope quasi-simultaneously in the SiO $J = 1{-}0$ $v = 0$, 1, 2, 3, 4 and $J = 2{-}1$ $v = 1$, 2 lines. We also observed the sample in the ^{29}SiO $J = 1{-}0$ $v = 0$ and $J = 2{-}1$ $v = 0$, and ^{30}SiO $J = 1{-}0$ $v = 0$ lines, and the H_2O $6_{1,6}{-}5_{2,3}$ line. As reported in previous papers, we confirmed that the intensity ratios of the SiO $J = 1{-}0$ $v = 2$ to $v = 1$ lines clearly correlate with infrared colors. In addition, we found possible correlation between infrared colors and the intensity ratios of the SiO $J = 1{-}0$ $v = 3$ to $v = 1$&2 lines.

Keywords. masers — stars: AGB and post-AGB — stars: late-type — stars: mass loss — stars: statistics

1. Introduction

An important problem in the studies of SiO masers was that the sample of all known SiO maser sources was considerably biased. Specifically, the dust (effective) temperature of known SiO maser sources, which was calculated from mid-infrared flux densities (such as the IRAS and MSX flux densities), was limited roughly to the range of 250 K $\lesssim T_{\rm dust} \lesssim$ 2000 K. This is because the previous SiO maser surveys have been limited to relatively warm dust-temperature ranges. Consequently a non-negligible number of potential SiO maser sources (especially with a low dust-temperature) have been omitted from the previous SiO maser surveys.

Nyman *et al.* (1993) first realized the importance of SiO maser sources exhibiting a low dust-temperature. They investigated how SiO maser emission behaves in a low dust-temperature range by observing OH/IR stars in the SiO $J = 1{-}0$ $v = 1$&2 and $J = 2{-}1$ $v = 1$ lines. The OH/IR stars often exhibit a low dust-temperature less than $T_{\rm dust} = 250$ K. In their observations the cold objects clearly show a larger intensity ratio of the SiO $J = 1{-}0$ $v = 2$ to $v = 1$ lines. Both collisional and radiative schemes cannot fully explain this observational property of the SiO masers (Bujarrabal 1994; Doel *et al.* 1995). Nyman *et al.* (1993) suggested that an infrared H_2O line ($11_{6,6}$ $\nu_2 = 1 \rightarrow 12_{7,5}$ $\nu_2 = 0$) overlapping with the SiO $J = 0$ $v = 1 \rightarrow J = 1$ $v = 2$ transition might play an important role. However, in the early 1990s the number of cold SiO maser sources (like OH/IR stars) was quite limited, and it was difficult to statistically investigate the relation between infrared colors and the intensity ratios of SiO maser lines.

Nakashima & Deguchi (2003b) recently extended the Nyman *et al.* (1993) study by surveying SiO maser emissions in cold, dusty IRAS sources exhibiting dust temperature

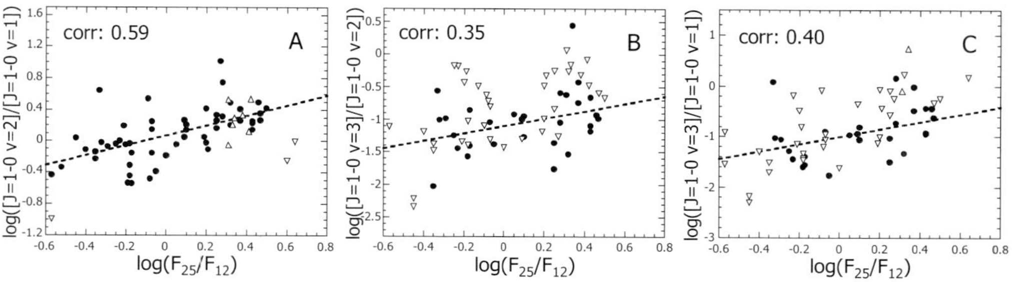

Figure 1. Infrared colors versus intensity ratio of SiO maser lines. The horizontal axes represent infrared colors. F_{25} and F_{12} denote the IRAS flux densities at $\lambda = 25$ and 12 μm, respectively. The filled dots (•), upward triangles (△) and downward triangles (▽) respectively represent the intensity ratios of the SiO maser lines, lower limits of the ratio and upper limits of the ratio. Correlation coefficients are given in the upper-left corners of each panel. The dashed lines are the results of least-square fitting of a first order polynomial.

less than 250 K. They found roughly 40 new SiO maser sources in the cold dusty objects and, in conjunction with the results of another SiO maser survey of relatively warm IRAS objects (Nakashima & Deguchi 2003a), they clearly demonstrated that the intensity ratio of the SiO $J = 1{-}0$ $v = 2$ to $v = 1$ lines increases inversely proportional to the dust temperature. Nakashima & Deguchi (2003b) again suggested that the line overlap with H_2O might explain this correlation if the line overlap becomes stronger with decreasing dust temperature. To further consider this problem, we need to confirm whether properties of the SiO lines other than the $J = 1{-}0$ $v = 1$ and 2 lines are consistent with the existence of H_2O line overlap. In this contributed paper, we present the result of quasi-simultaneous observations in the multiple different SiO rotational lines with the Nobeyama 45m telescope. The main aim of the observation is to check the behavior of SiO maser intensity ratios including lines other than the $J = 1{-}0$ $v = 1$ and 2 lines.

2. Observations and results

The observing targets were selected from Nakashima & Deguchi(2003a, b) and the Nobeyama SiO maser source catalog (Gorny *et al.* in preparation) in terms of the IRAS colors and flux densities. The targets are distributed roughly in the right ascension range between 18^h and 22^h, because the cold SiO maser sources found by Nakashima & Deguchi (2003b) are distributed roughly in this range. We selected the observing targets basically in order of the brightness at $\lambda = 12$ μm, but we also paid attention to the source distribution in the IRAS two-color diagram so that the observing targets continuously cover the entire color range.

SiO line observations with the Nobeyama 45m telescope were made in two separated periods: May 11–19, 2004 and February 15–19, 2006. In the first period we observed a total of 38 objects. The observed SiO transitions in the first period were $J = 1{-}0$ $v = 1$, 2, 3 and $J = 2{-}1$ $v = 1$, 2. We also observed the ^{29}SiO $J = 1{-}0$ $v = 0$ and $J = 2{-}1$ $v = 0$ lines. In addition, we observed 27 objects in the H_2O maser line at 22 GHz ($6_{1,6}{-}5_{2,3}$) as a backup observation under rainy/heavy cloud conditions. In the second period we observed, in total, 53 objects. The observed transitions in the second period were SiO $J = 1{-}0$ $v = 0$, 1, 2, 3, 4, ^{29}SiO $J = 1{-}0$ $v = 0$ and ^{30}SiO $J = 1{-}0$ $v = 0$. The technical details of the observations will be presented in our future paper (Nakashima & Deguchi, in preparation).

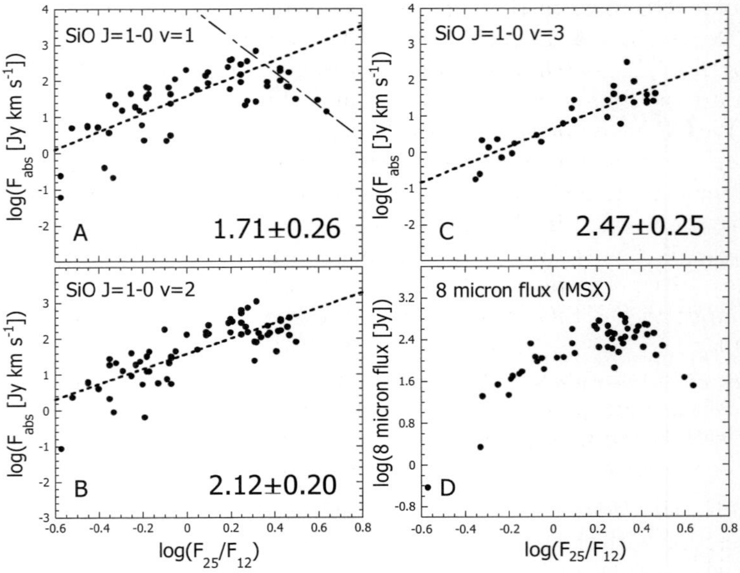

Figure 2. [*Panels A, B and C*]: Relation between infrared colors and absolute intensity of SiO maser lines. The notation of the infrared colors is same with that used in Figure 1. The intensity of SiO maser lines is standardized at the distance of 1 kpc. The thick dashed lines represent the results of least-square-fitting of a first order polynomial. The inclinations of the fitted lines (thick dashed lines) are given at the lower-right corners of each panel with statistical uncertainty. In the panel A, only the data points below $\log(F_{25}/F_{12}) = 0.5$ were fitted by the polynomial. The data points above $\log(F_{25}/F_{12}) = 0.5$ are independently fitted by a first order polynomial, and the results of the fitting is given as the chain line. [*Panel D*]: Relation between 8μm absolute flux density and infrared color. The 8μm flux is standardized at the distance of 1 kpc.

In this paper, we focus on the properties of the SiO $J = 1\text{–}0$ $v = 1$, 2 and 3 lines, in which we detected a large number of objects for statistical analysis. Figure 1 shows the relation between infrared colors and intensity ratios among the SiO maser lines. The line intensities used to calculate the intensity ratios are velocity-integrated intensities. In panel A of Figure 1 we can clearly confirm the positive correlation between the $\log(F_{25}/F_{12})$ color and the intensity ratio of the SiO $J = 1\text{–}0$ $v = 2$ to $v = 1$ lines as reported by Nakashima & Deguchi (2003b). Interestingly, the intensity ratios of the $J = 1\text{–}0$ $v = 3$ to $v = 2$ lines and of the $J = 1\text{–}0$ $v = 3$ to $v = 1$ lines seem to also correlate with the $\log(F_{25}/F_{12})$ color (see, panels B and C) even though the correlation coefficients are slightly smaller than that of panel A.

Figure 2 shows the relations between infrared colors and absolute intensities of the SiO maser lines. The intensity of the SiO maser lines is standardized at the distance of 1 kpc using the luminosity distances. The panels A, B and C of Figure 2 show the relations between the $\log(F_{25}/F_{12})$ color and the absolute intensity of the SiO $J = 1\text{–}0$ $v = 1$, 2 and 3 lines. A notable feature seen in these panels is that the SiO maser absolute intensities undoubtedly correlate with the $\log(F_{25}/F_{12})$ color. Another clear feature is that for higher vibrational transitions, the dashed lines representing the least-square-fitting with a first-order polynomial also become steeper. This tendency is consistent with the correlation seen in Figure 1. In panel A in Figure 2, the values of the absolute intensity of the SiO maser emission seem to maximize at $\log(F_{25}/F_{12}) \sim 0.5$, and the values tend to decrease with increasing color in the red region above $\log(F_{25}/F_{12}) = 0.5$. The $\log(F_{25}/F_{12})$ color of 0.5 corresponds to the boundary between distributions of

AGB and post-AGB stars in the $\log(F_{25}/F_{12})$ color. In fact, panel A in Figure 1 (and Figure 8 in Nakashima & Deguchi 2003b) shows a sudden change in strength of the feature at $\log(F_{25}/F_{12}) \sim 0.5$. No such change is seen in the panels B and C in Figure 2, simply because the SiO $J = 1{-}0$ $v = 2$ and 3 lines have not been detected above $\log(F_{25}/F_{12}) = 0.5$ in the present observations.

A possible reason for the correlation seen in panels A, B and C in Figure 2 is that the energy input to the SiO maser region increases with the infrared colors. To confirm this possibility, in panel D in Figure 2 we plotted the 8μm flux densities as a function of the infrared colors. The values of the 8μm flux densities were taken from the MSX point source catalog. If we rely on the radiative scheme the 8μm flux should well represent the energy input to the SiO maser region, because the $\lambda = 8\mu$m corresponds to the $\Delta v = 1$ SiO transition (e.g., Deguchi & Iguchi 1976). In panel D in Figure 2, the 8μm flux densities are standardized to the distance of 1kpc using the luminosity distance. The distribution of the data points seen in panel D is, in fact, strikingly similar with those seen in panels A, B and C, supporting the idea that the 8μm flux tightly correlates with the SiO maser intensity as suggested by Bujarrabal *et al.* (1987).

3. Discussion

In this section we discuss the possible explanation for the correlation between infrared colors and SiO maser intensity ratios among the $v = 1$, 2 and 3 lines at 43 GHz. One possible explanation is to introduce the line overlap with H_2O ($11_{6,6}$ $\nu_2 = 1 \rightarrow 12_{7,5}$ $\nu_2 = 0$), which has been first suggested by Olofsson *et al.* (1981) to explain the anomalous, weak intensity of the SiO $J = 2{-}1$ $v = 2$ line in oxygen-rich (O-rich) stars. The H_2O line overlaps with the SiO $J = 0$ $v = 1 \rightarrow J = 1$ $v = 2$ transition with a velocity difference of 1 km s^{-1}. With this line overlap, the $J = 1$ $v = 2$ level is overpopulated, and the weakness of the SiO $J = 2{-}1$ $v = 2$ line is explained by this overpopulation. The overpopulation at the $J = 1$ $v = 2$ level is also consistent with the strong intensity of the $J = 1{-}0$ $v = 2$ line. Thus, the correlation between the infrared colors and the intensity ratio of the SiO $J = 1{-}0$ $v = 2$ to $v = 1$ lines may be explained if this line overlap with H_2O becomes stronger with increase of the infrared colors. One problem in this interpretation is that the intensity ratios of the SiO $J = 1{-}0$ $v = 3$ to $v = 1\&2$ lines cannot be explained only with the H_2O $11_{6,6}$ $\nu_2 = 1 \rightarrow 12_{7,5}$ $\nu_2 = 0$ line. However Cho *et al.* (2007) recently reported an interesting detection of the SiO $J = 2{-}1$ $v = 3$ line toward an S-type star, χ Cyg. They also confirmed that the SiO $J = 2{-}1$ $v = 3$ line is weak in O-rich stars. The S-type stars have almost the same amount of oxygen and carbon atoms in their envelopes, and consequently they have few H_2O molecules in the envelopes. These results potentially suggest that another line overlap with H_2O affects the population distribution of SiO in O-rich stars, and Cho *et al.* (2007) have suggested that the H_2O $5_{0,5}$ $\nu_2 = 2 \rightarrow 6_{3,4}$ $\nu_2 = 1$ line overlapping with the SiO $J = 0$ $v = 2 \rightarrow J = 1$ $v = 3$ line (with a velocity difference of about 1.5 km s^{-1}) acts on the population distribution of SiO. Thus, if both H_2O $11_{6,6}$ $\nu_2 = 1 \rightarrow 12_{7,5}$ $\nu_2 = 0$ and $5_{0,5}$ $\nu_2 = 2 \rightarrow 6_{3,4}$ $\nu_2 = 1$ lines become stronger with increasing infrared colors, all correlations between infrared colors and the SiO maser intensity ratios among the $J = 1{-}0$ $v = 1$, 2 and 3 lines might be explained. The line intensity of the H_2O $5_{0,5}$ $\nu_2 = 2 \rightarrow 6_{3,4}$ $\nu_2 = 1$ line is usually weaker than that of the $11_{6,6}$ $\nu_2 = 1 \rightarrow 12_{7,5}$ $\nu_2 = 0$line. This fact also seems to be consistent with the relatively weak intensity of the SiO $J = 1{-}0$ $v = 3$ line.

However, there are some other problems with the explanation of the line overlap with H_2O. First, we have to explain how the H_2O infrared lines overlapping with the SiO lines become stronger with increasing infrared colors. The relative abundance of H_2O

molecules possibly increases with infrared colors, but this is not conclusive. Second, the correlation between infrared colors and the intensity ratios of the SiO $J = 1{-}0$ $v = 2$ to $v = 1$ lines might be explained without line overlap with H_2O. In the envelopes of very cold objects, strong 8μm emission comes from every direction to the SiO masering region, causing ineffective pumping through the SiO $\Delta v = 1$ transition. On the other hand, 4μm emission corresponding to the SiO $\Delta v = 2$ transition is more effectively pumping the SiO population instead of the 8μm emission. These processes might explain the correlation seen in Figure 2 (e.g., Doel *et al.* 1995). Third, a recent theoretical calculation predicted that if we introduce line overlap with H_2O, the spatial distribution of the maser spots cannot be theoretically reproduced (Soria-Ruiz *et al.* 2004). Thus, this problem will remain controversial for some more time.

4. Summary

In this research we observed 75 known SiO maser sources quasi-simultaneously in the SiO $J = 1{-}0$, $v = 0$, 1, 2, 3 and 4 lines, SiO $J = 2{-}1$ $v = 1$ and 2, ^{29}SiO $J = 1{-}0$ $v = 0$ and $J = 2{-}1$ $v = 0$, and ^{30}SiO $J = 1{-}0$ $v = 0$ lines. We also observed the targets in the H_2O $6_{1,6}{-}5_{2,3}$ line under rainy/heavy cloud condition. The sample continuously covers a very wide dust-temperature range from 150 K to 2000 K. The correlation between infrared colors and the intensity ratio of the SiO $J = 1{-}0$ $v = 2$ to $v = 1$ lines is confirmed as reported by Nakashima & Deguchi (2003b). The intensity ratios of the SiO $J = 1{-}0$ $v = 3$ to $v = 1\&2$ lines possibly correlate with infrared colors. The line overlaps with H_2O might explain the correlations between the infrared colors and the SiO maser intensity ratios among the $J = 1{-}0$ $v = 1$, 2 and 3 lines, although there are alternative ways to interpret the phenomena.

Acknowledgements

The present research has been supported by the Academia Sinica Institute of Astronomy & Astrophysics in Taiwan.

References

Bujarrabal, V. 1994, *A&A* 285, 953
Bujarrabal, V., Planesas, P., & del Romero, A. 1987, *A&A* 175, 164
Cho, S.-H., Lee, C. W., & Park, Y.-S. 2007, *ApJ* 657, 482
Deguchi, S., & Iguchi, T. 1976, *PASJ* 28, 307
Doel, R. C., *et al.* 1995, *A&A* 302, 797
Nakashima, J., & Deguchi, S. 2003a, *PASJ* 55, 203
Nakashima, J., & Deguchi, S. 2003b, *PASJ* 55, 229
Nyman, L.-Å., Hall, P. J., & Le Bertre, T. 1993, *A&A* 280, 551
Olofsson, H., Rydbeck, O. E. H., Lane, A. P., & Predmore, C. R. 1981, *ApJ* 247, L81
Olofsson, H., Rydbeck, O. E. H., & Nyman, L.-A. *A&A* 150, 169
Soria-Ruiz, R., *et al.* 2004, *A&A* 426, 131

Astrophysical Masers and their Environments
Proceedings IAU Symposium No. 242, 2007
J.M. Chapman and W.A. Baan, eds.

doi:10.1017/S1743921307013129

A study of the close environments of evolved stars from SiO masers

Valentín Bujarrabal[1], **J. Alcolea**[2], **F. Colomer**[1], **J.-F. Desmurs**[2], **C. Sánchez Contreras**[3], **and R. Soria-Ruiz**[4]

[1]Observatorio Astronómico Nacional, Apartado 112, E-28803 Alcalá de Henares, Spain
email: (v.bujarrabal,f.colomer)@oan.es

[2]Observatorio Astronómico Nacional, Calle Alfonso XII 3, E-28014 Madrid, Spain
email: (j.alcolea,jf.desmurs)@oan.es

[3]DAMIR, IEM, CSIC, Serrano 121, 28006 Madrid, Spain. email: carmen@damir.iem.csic.es

[4]JIVE, Postbus 2, 7990 AA Dwingeloo, The Netherlands. email: soria@jive.nl

Abstract. SiO maser emission allows us to study the innermost circumstellar layers around AGB and post-AGB stars, at a few AU from the stellar photosphere, what is crucial to understand the mass loss process in AGB stars and the jet launching mechanisms in post-AGB stars. Observations of SiO masers are also useful to address the question of the pumping mechanism itself, still under debate. In particular, VLBI observations of such emissions enables the study of the innermost shells with extremely high spatial resolution, equivalent (in nearby stars) to a few 10^{12} cm, about one tenth of the stellar radius. In this contribution, recent results on this topic obtained by our group are summarized.

Keywords. stars: AGB and post-AGB, circumstellar matter, stars: mass loss, stars: variables: long-period variables.

1. General observational properties of SiO maser emission

Maser emission appears in SiO rotational lines, in general within vibrationally excited states. These lines are usually observed in evolved stars, mainly O-rich Mira-type variables. They are also detected in semi-regular and irregular variables, supergiant stars, S-type stars and a few proto-planetary nebulae around post-AGB stars (as well as in a few regions of star formation).

Many SiO maser lines have been detected, up to $J=8$–7, in vibrational states up to $v=4$. But the most intense and better studied transitions, by far, are $v=1,2$ $J=1$–0 (43 GHz frequency, 7 mm wavelength) and $v=1$ $J=2$–1 (86 GHz). The $v=2$ $J=2$–1 line is known to be anomalously weak.

When we compare the emission of different stars, the SiO maser intensity depends on the amplitude of the stellar variability, being the high amplitude pulsators and the most intense emitters, e.g. Alcolea *et al.* (1990). The maser intensity also correlates very well with the IR stellar intensity, particularly at 8 μm, the wavelength of the $\Delta v=1$ fundamental vibrational transition, see Bujarrabal *et al.* (1987).

More results from wide surveys of SiO maser emission can be found in the contribution by Deguchi (2007, these proceedings).

The rare isotopes ^{29}SiO and ^{30}SiO also present weak maser emission, but in them the masers usually appear in rotational transitions within the ground vibrational state. (Also lines from the vibrationally excited states of these species may be masers, but they are even weaker and more rarely detected.)

High resolution mapping of SiO masers, using VLBI techniques, show that the maser emission is in general distributed in rings of spots at a few stellar radii from the photosphere. Very probably, the maser formation is dominated by tangential amplification in the innermost circumstellar shells, in which the material is kept in levitation by shock energy dissipation, before the characteristic circumstellar expansion starts. Finally, polarization measurements show that the maser emission from each spot is linearly polarized, often in the tangential direction. Details on such measurements are discussed in the contribution by A. Kemball (Kemball 2007).

2. The pumping mechanisms of SiO masers from evolved stars

It is generally accepted that the main mechanism responsible for the inversion of the SiO masers is that proposed by Kwan & Scoville (1974). SiO maser emission usually appears in rotational lines within vibrationally excited states, therefore requiring a relatively high excitation degree, since the first vibrationally excited state is at 1800 K from the ground. The population of these $v > 1$ states from the fundamental one is followed by a fast radiative decay through the $\Delta v = 1$ transition. The mechanism by Kwan and Scoville is efficient when this transition is optically thick. Then, the efficient de-excitation probability of a given ro-vibrational component is proportional to $A\beta$, where A is its Einstein coefficient and β is the escape probability. In the optically thick case, $\beta \sim 1/\tau$. We also know that $\tau \underset{\sim}{\propto} Ag_u$, where g_u is the statistical weight of the upper level; note that the variation of τ for the different rotational components of a given vibrational transition is largely dominated by this law in our case, in which a high rotational excitation can be assumed.

Therefore, the efficient probabilities of the different components are essentially proportional to $1/g_u$. This means that the molecules are relatively more trapped in levels with higher J-values, though the effect is vanishingly noticeable for consecutive levels with very high Js. In other words, the appearance of high opacities in the $\Delta v = 1$ de-excitation tends to produce a systematic inversion of the rotational transitions in the upper v-state. A chains of masers so appears that are mutually reinforced, because the molecules decaying to one rotational level strengthen the maser from that level to the lower next one.

The mechanism responsible for the population of the vibrationally excited states, i.e. the final energy source of the maser radiation, is still debated. The $v > 1$ level population could be radiative or collisional. In any case, the excitation of lines lying at least at 1800 K from the ground requires a high excitation environment, and all models place the masers very close to the stellar photosphere.

The collisional pumping mechanism is relatively simple and only requires efficient collisional population of the excited states, because it is known that the collisional population itself does practically not select the rotational levels. A great theoretical and computational effort has been devoted to those models (e.g. Lockett & Elitzur 1992, Humphreys *et al.* 2002), including a detailed modeling of the shock fronts traversing the masing region.

On the other hand, radiative pumping is somewhat more complex and requires that the exciting radiation field is anisotropic (essentially coming from the central star or close surroundings, e.g. Bujarrabal 1994). These conditions naturally lead to the tangential maser amplification and linear polarization deduced from the observations. Radiative models are also supported by the observed correlation of the SiO intensity with the IR stellar continuum. We will discuss below the comparison of the variations of the maser and continuum fluxes during the stellar cycle, found to be remarkably in phase.

3. SiO maser variability

Very soon after their discovery, it was realized that SiO masers are variable, as it is expected from the nature of the emission. Because of the few SiO masers detected in regions of massive star formation, the variability of the SiO maser emission in these type of sources is very scarcely studied. On the contrary, the properties of SiO masers in long period variables are fairly well known thanks to numerous monitoring studies, of which the most complete in many senses is the one carried out at 7 mm ($J=1$–0) with the 13.7 m dish at Yebes (see Pardo *et al.* 2004 with references therein and a full listing of all other monitoring works on SiO maser variability). The Yebes monitoring started in 1984 and ended in 1995, so in the best cases we have data for a period of 11 yr. The typical spacing between observations is 3 weeks. This spacing was set to properly sample the typical time scale of the SiO maser variations in regular (Mira-type) variables, which was known to present periods of about one year. Initially, the monitoring included 14 sources that were only observed in the $v=1$ line. Later more sources and the $v=2$ line were added. Since 1989, 20 long-period variables and Orion A were observed in the two $J=1$–0 transitions. In total the monitoring comprises more than 3500 spectra, taken under an special procedure to ensure a relative calibration better than 10%.

This monitoring represents a major improvement with respect to previous similar works, and in fact it is the only one in which its length and tight spacing allow a time-series analysis of the data, resulting in the first and so far only quantitative estimates of the variability properties of SiO masers. In addition, the data were combined with optical and/or infrared monitoring of the central stars, which were also subject to time-series investigations. These analysis showed that, al least in Mira variables, the three types of emission (SiO, optical and NIR) have the same dominant period of variability, within $\pm 4\%$: the period of the pulsation of the star. In addition, it is also found that SiO maxima happen accurately in phase with NIR maxima, phase lags $\lesssim 0.05$ in absolute value, and with a positive phase lag typically between 0.05 and 0.20 with respect to optical maxima. The presence of regular variability in other SiO maser parameters, such as the mean velocity, was also investigated with no significant results. For the non-Mira variables in our sample, the time-series analysis yielded no clear trends except for GY Aql. This object is considered to be a semi-regular variable but the SiO properties are very similar to those of Mira stars. In fact the SiO period found in our data for this source coincides with the optical period derived from optical data obtained during the last part of our monitoring.

All these results indicate a tight and universal SiO/NIR phase coincidence, making a strong case in favor of radiative pumping of SiO masers, which explains such a correlation in a very natural way. On the other hand, collisional dominated models require a very fine ad hoc tuning on a case by case of the shock front properties, in order to explain such a tight coincidence.

4. The relative distribution of the maser spots for different transitions

4.1. $v=1$ $J=1$–0 *vs.* $v=2$ $J=1$–0

The first attempts to measure the relative positions of two SiO maser lines were performed by Miyoshi *et al.* (1994), who mapped the $v=1,2$ $J=1$–0 lines. These authors concluded that both transitions appeared spatially coincident, but with a relatively poor resolution. It was also concluded that such a coincidence strongly favored collisional pumping. Both collisional and radiative mechanisms require generally different conditions to excite these lines, because they belong to different vibrational states and require different excitation

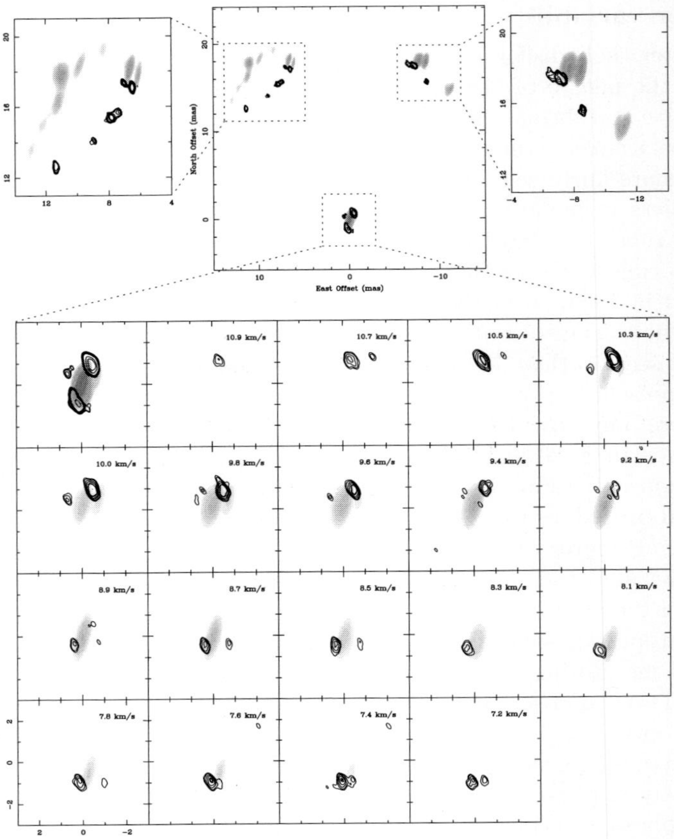

Figure 1. Maps of $v=1$ and 2, $J=1$–0 maser emission (respectively, grey scale and contours). From Soria-Ruiz *et al.* (2004).

degrees. Under the radiative schemes, it is more difficult that both lines are intense simultaneously.

However, observations with much higher resolution (Desmurs *et al.* 2000) have shown that, in fact, the spots of the $v=1$ $J=1$–0 and $v=2$ $J=1$–0 transitions are almost never coincident, although the overall distributions are similar. In general, the $v=2$ 43 GHz transition appears slightly, but systematically closer to the central star, by 1 or 2 mas.

In some cases, one has the impression that each transition avoids the other. We can see an example in Fig. 1, Soria-Ruiz *et al.* (2004), which shows observations of the very red AGB star IRC +10011. In the southern clump, the $v=2$ emission in some way moves around the $v=1$ clump, depending on the velocity, but avoiding a true coincidence.

We see another example of this behavior in Fig. 2, which represents observations of the same source at a later epoch. Other VLBI observations of the relative positions of these lines have been published by Cotton *et al.* (2004, 2006), confirming the above conclusions.

4.2. $v=1$ $J=1$–0 *vs.* $v=1$ $J=2$–1

The $v=1$ $J=2$–1 transition is also a strong maser, but VLBI data are scarce, because of its relatively high frequency, 86 GHz.

The first maps of this line have revealed a surprising result. The emitting spots at 86 GHz show a completely different distribution than those of $v=1$ $J=1$–0, see Soria-Ruiz *et al.* (2004, 2005) and Fig. 2. Here, we cannot say that there are small shifts

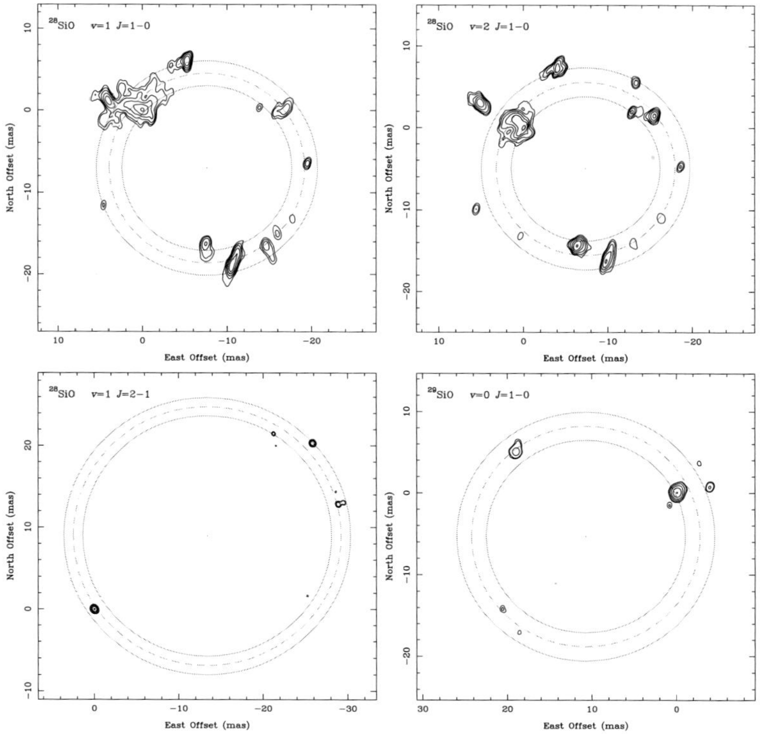

Figure 2. Integrated intensity maps for SiO maser emissions in IRC +10011. Circles denote the ring fitting of the masering regions. From Soria-Ruiz *et al.* (2005).

between the spots: the distributions of both lines are clearly different. In the few cases observed, the $v = 1$ $J = 2$–1 emission is always placed significantly outwards with respect to the 43 GHz rings.

This relative distributions were completely unexpected because, as mentioned (Sect 2), the $v = 1$, $J = 1$–0 and $J = 2$–1 transitions belong to the same rotational ladder and require very similar pumping conditions. In fact we expected the formation of chains of masers in strong coupling. This result is then in clear contradiction with the standard, usually accepted theory; we will see (Sect. 5) that the presence of line overlap may explain such a behavior.

4.3. ^{29}SiO, $v = 0$ $J = 1$–0

The results from the maps of the ^{29}SiO, $v = 0$ $J = 1$–0 masers are also surprising. This line (the most intense maser line of the rare isotopes) requires a very low excitation. In fact, the standard models for this maser, place it quite far from the star, yielding in particular a dominant radial amplification (Robinson & Van Blerkom 1981, Deguchi & Nguyen-Q-Rieu 1983).

However, the observed distribution of the ^{29}SiO, $v = 0$ $J = 1$–0 spots is similar to that of the other masers; see Fig 2, Soria-Ruiz *et al.* (2005). One would say that the distribution of this line is *too similar* to those of the other masers we have mentioned, although identical to none of them.

It has been proposed that line overlap effects could also explain these observations.

5. Effects on the maser pumping of line overlap

Line overlap occurs when two lines have almost identical frequencies, such that, taking into account the local or macroscopic velocity dispersions, both line profiles significantly overlap. Then, the photon field available to one of the lines is affected by the emission or absorption by the other. The anomalously weak maser in the $v=2$ $J=2$–1 transition could be explained by the overlap between the H_2O $v_2=0, 12_{7,5}$ – $v_2=1, 11_{6,6}$ and the ^{28}SiO $v=2$, $J=1$ – $v=1$, $J=0$ ro-vibrational components (Olofsson *et al.* 1985). The H_2O line would increase the number of photons available for the SiO component, leading to a relative overpopulation of the $v=2$ $J=1$ level and quenching the (otherwise intense) $v=2$ $J=2$–1 maser, provided that the water component is emitting a sufficiently high number of photons.

This overlapping pair could also explain some of the observations presented above, see details in Soria-Ruiz *et al.* (2005), Bujarrabal *et al.* (1996). The overpopulation of the $v=2$ $J=1$ level appears at the same time as a relative under-population of the $v=1$ $J=0$ level, therefore establishing a strong coupling between the $v=1$ $J=1$–0 and $v=2$ $J=1$–0 lines. Both emissions would then appear in relatively similar regions. Meanwhile, the $v=1$ $J=2$–1 one would now require different conditions than those for the 43 GHz transitions. Detailed calculations presented by Soria-Ruiz *et al.* (2005) explain, at least qualitatively, all the observational properties of the relative spatial distributions of these lines.

Line overlap could also explain the main properties of the ^{29}SiO, $v=0$ $J=1$–0 maser emission. In this case, two ro-vibrational components of SiO would overlap: ^{28}SiO $v=2$, $J=4$ – $v=1$, $J=3$ and ^{29}SiO $v=1$, $J=1$ – $v=0$,$J=0$. The strong ^{28}SiO line would excite the ^{29}SiO ro-vibrational component, yielding an under-population of the ^{29}SiO $v=0$ $J=0$ level. This would explain the $v=0$ $J=1$–0 maser, without invoking other mechanisms. The fact that strong vibrational emission of ^{28}SiO is required also explains that the ^{29}SiO maser only appears in high-excitation regions, even if the ^{29}SiO line itself is close to the ground level.

We note that overlap has been proposed to explain other maser lines of ^{29}SiO and ^{30}SiO, but theoretical results are not always conclusive.

6. SiO masers in proto-planetary nebulae (PPNe)

SiO masers have also been detected around a few post-AGB stars. The best studied one, by far, is OH 231.8+4.2. This source is a peculiar PPNe, an elongated bipolar nebula surrounding a double star.

SiO masers in OH 231.8+4.2 show a characteristic three-peak structure that has persisted during more than ten years, in spite of changes in the relative intensity or shape of the peaks (Sánchez Contreras *et al.*, private communication; Fig. 3). It is improbable that this profile is just due to a random clump distribution, since in ten years clumps moving at the measured velocities should in general leave the small emitting region (see mapping data below). The three-peak profile must be a property of some stable structure, like a rotating disk. This profile shape is in fact expected from Keplerian disks (see Deguchi 1993, Elitzur 2007), independently of the exact density distribution in them.

Recent VLBI mapping of SiO maser emission in OH 231.8+4.2 (Sánchez Contreras *et al.* 2002, Desmurs *et al.* 2007a,b) shows that this emission comes from an elongated structure in the center of the nebula, perpendicular to the symmetry axis. Although only two peaks were detected, the derived spatial distribution and velocities are consistent with emission coming from a rotating structure.

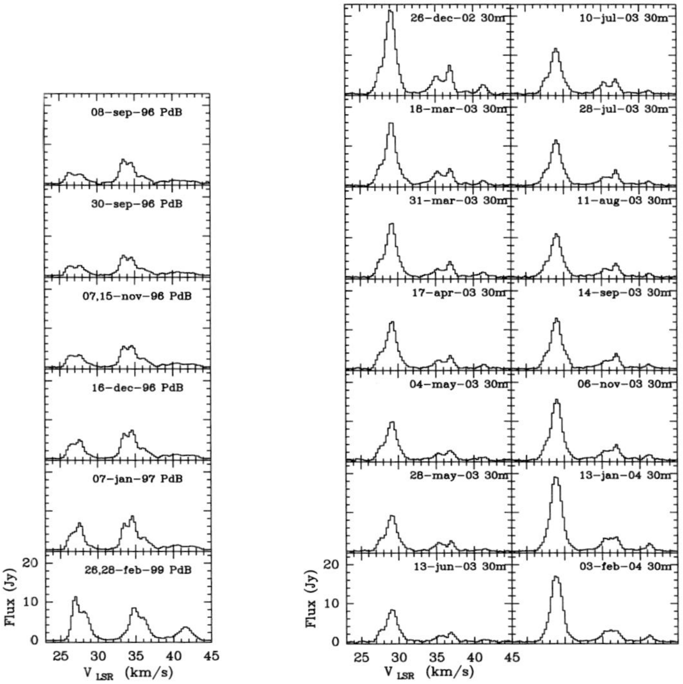

Figure 3. Time variation of the $v = 1$ $J = 2$–1 maser emission in OH 231.8+4.2. Data from the Plateau de Bure (PdB) and 30m telescopes are presented.

The combination of both the single-dish observations and the VLBI mapping argue in favor of that SiO masers in OH 231.8+4.2 identify a very small disk (diameter $\sim 10^{13}$ cm), kept in rotation by gravitation around a stellar mass of about 1 $M_\odot$.

The detection of central Keplerian disks in PPNe is particularly important. The presence of jets in post-AGB stars is thought to be the leading mechanism in the shaping of planetary nebulae: a phenomenon that involves an impressive shock dynamics, which accelerates, in just a few hundred years, a gas mass of $\sim 1/2$ $M_\odot$ up to almost 100 km s^{-1}. It is usually assumed that re-accretion of circumstellar material, through the formation of a rotating disk, is the most probable mechanism to explain such ejections. (See e.g. Bujarrabal *et al.* 2001, Frank & Blackman 2004.) However, these disks have been only well detected to date in one PPNe, the Red Rectangle, from thermal CO emission. In other sources, the disks could be too small to be mapped in thermal emission, with resolutions $\sim 1''$. The reason is probably that the requirement of angular momentum to form big enough disks, that must come from a close stellar or substellar companion, is too strong and only satisfied in relatively massive stellar systems. VLBI observations (of maser emission) can attain much higher resolutions and then detect much smaller disks, for which the angular momentum problem is much less severe. We think that the detection of rotating disks around post-AGB stars from SiO maser data can be a key observational tool to disentangle the actual association of Keplerian disks with jets in these objects.

Acknowledgements

We would like to acknowledge partial support from the Ministerio de Educación y Ciencia, grants AYA2003-7584 and ESP2003-04957.

References

Alcolea, J., Bujarrabal, V. & Gómez-González 1990, *A&A* 231, 431
Bujarrabal, V., Planesas, P. & Del Romero, A. 1987, *A&A* 175, 164
Bujarrabal, V. 1994, *A&A* 285, 953
Bujarrabal, V., Alcolea, J., Sánchez Contreras, C. & Colomer, F. 1996, *A&A* 314, 883
Bujarrabal, V., Castro-Carrizo, A., Alcolea, J. & Sánchez Contreras, C. 2001, *A&A* 377, 868
Cotton, W. D., Mennesson, B., Diamond, P. J., *et al.* 2004, *ApJ* 414, 275
Cotton, W. D., Vlemmings, W., Mennesson, B., *et al.* 2006, *ApJ* 456, 339
Deguchi, S., Good, J., Fan, Y. *et al.* 1983, *ApJ* 264, L65
Deguchi, S. & Nguyen-Q-Rieu 1983, *A&A* 117, 314
Deguchi, S. 2007, in: J. Chapman & W. Baan (eds.), *Astrophysical Masers and their Environments*, Proceedings IAU Symp. No 242 (Cambridge University Press)
Desmurs, J.-F., Bujarrabal, V., Colomer, F., & Alcolea, J. 2000 *A&A* 360, 189
Desmurs, J.-F., Alcolea, J., Bujarrabal, V., Sánchez Contreras, C., & Colomer, F. 2007a, *A&A*, in press
Desmurs, J.-F. 2007b in: J. Chapman & W. Baan (eds.), *Astrophysical Masers and their Environments*, Proceedings IAU Symp. No 242 (Cambridge University Press)
Elitzur, M. 2007 in: J. Chapman & W. Baan (eds.), *Astrophysical Masers and their Environments*, Proceedings IAU Symp. No 242 (Cambridge University Press)
Frank, A. & Blackman, E. G. 2004, *ApJ* 614, 737
Humphreys, E. M. L., Gray, M. D., Yates, J. A. *et al.* 2002, *A&A* 386, 256
Kemball, A. 2007 in: J. Chapman & W. Baan (eds.), *Astrophysical Masers and their Environments*, Proceedings IAU Symp. No 242 (Cambridge University Press)
Kwan, J. Y. & Scoville, N. 1974 *ApJ* 194, L97
Lockett, P. & Elitzur, M. 1992, *ApJ* 394, 704
Miyoshi, M., Matsumoto, K., Kameno, S., Takaba, H., & Iwata T. 1994, *Nature* 371, 395
Olofsson, H., Rydbeck, O. E. H. & Nyman, L.-A. 1985, *A&A* 150, 169
Pardo, J. R., Alcolea, J., Bujarrabal, V. *et al.* 2004, *A&A* 424, 145
Robinson S. E. & Van Blerkom, D. J. 1981, *ApJ* 249, 566
Sánchez Contreras, C., Desmurs, J.-F., Bujarrabal, V., Alcolea, J., & Colomer, F. 2002 *A&A* 385, L1
Soria-Ruiz, R., Alcolea, J., Colomer, F., *et al.* 2004, *A&A* 426, 131
Soria-Ruiz, R., Colomer, F., Alcolea, J., *et al.* 2005, *A&A* 432, L39

Astrophysical Masers and their Environments
Proceedings IAU Symposium No. 242, 2007
J.M. Chapman & W.A. Baan, eds.

doi:10.1017/S1743921307013130

Stellar molecular jets traced by maser emission

Hiroshi Imai

Department of Physics, Faculty of Science, Kagoshima University, Japan
email: hiroimai@sci.kagoshima-u.ac.jp

Abstract. Highly collimated jets found in AGB and post-AGB stars are expected to play an important role for shaping planetary nebulae. Recent VLBI observations of H_2O maser sources have revealed that some of the spatio-kinematical structures of H_2O maser sources exhibit stellar jets with extremely spatially and kinematically high collimation. Such stellar H_2O maser jets, so-called "water fountains", have been identified in about 10 sources to date. Here we review recent VLBI observations and relevant observational results of the water fountain sources. They have revealed a typical dynamical age and the detailed kinematical structures of the water fountains, the possibility of the existence of "equatorial flows", and the evolutionary status of the host stars. The location and kinematics of one of the water fountain sources in the Galaxy is also revealed and shown here.

Keywords. masers, stars:AGB and post-AGB, mass loss, winds, outflows

1. Introduction

A stellar jet appearing at the final stage of stellar evolution is a manifestation of the important phenomena that form asymmetric planetary nebulae (PNe). Based on the large incidence of bipolar and multipolar optical morphologies and the common presence of point-symmetry found in a sample of young PNe, Sahai & Trauger (1998) concluded that collimated jets have been operational in a large fraction of PNe, and expected to be one of the major factors shaping these objects. Similar morphologies have been found in a survey of pre-PNe (Sahai 2004), leading to the current paradigm that such a collimated jet had already been launched at an earlier phase of stellar evolution – the asymptotic giant branch (AGB) phase - and is frequently observable in molecular line emission. Subsequently an optically visible PN is seen at the post-AGB phase.

There are candidate sources that show us the earliest stage of stellar jet emergence. "Water fountains" are the most promising candidates; they exhibit extremely-high-velocity flows traced by H_2O maser emission. The outflow velocity is much higher than the typical expansion velocity of circumstellar envelopes of AGB stars found in 1612 MHz OH masers (10–25 km s^{-1}, e.g. te Lintel Hekkert *et al.* 1989). The H_2O maser emission exhibits high collimation in both its morphology and kinematics. This suggests that high density gas is supplied from the vicinity of the stellar surface to the jet tips where H_2O maser emission is excited (e.g., Elitzur *et al.* 1989). Because VLBI observations have been made for some of these water fountains, it is possible to undertake statistical analysis to estimate typical values of their physical parameters, which leads to understanding the final evolution of the circumstellar envelopes and the formation mechanism of stellar jets.

Note that high precision astrometry of H_2O masers in the water fountains provides great opportunity to directly estimate the locations and kinematics of the water fountains

Table 1. Parameters of "water fountain" sources.

IRAS name	Other name	Length (arcsec)	$\Delta V_{\rm los}$† (km s^{-1})	$t_{\rm jet}$‡ (year)	Distance (kpc)	Reference¶
15445−5449	OH 326.5−0.4		91			4
15544−5332	OH 325.8−0.3		40			4
16342−3814	OH 344.1+5.8	2.4	240	∼100	2	3, 15, 16, 17, 18
16552−3050	GLMP 498		170			19
18043−2116	OH 0.9−0.4		204			4
18139−1816	OH 12.8−0.9	0.11	55	70	?	1, 2, 10
18286−0959		0.24	200	∼15	3.1	5, 11
18450−0148	W43A, OH31.0+0.0	0.8	190	50	2.6	6, 8, 13, 15,16, 20
18460−0151	OH 31.0−0.2	0.11	290	∼5	6.8	5, 11
18596+0315	OH 37.1−0.8		59			4
19134+2131		0.15	100	40	8.0	12, 14, 16

†Full range of the line-of-sight velocities of maser emission.
‡Dynamical age of the jet.
¶1: Baud *et al.* 1979, 2: Boboltz & Marvel 2005, 3: Claussen *et al.* 2004 4: Deacon *et al.* 2007, 5: Deguchi *et al.* 2007, 6: Diamond *et al.* 1985, 7: Engels 2002, 8: Genzel & Downes 1977, 9: Engels *et al.* 1986, 10: Gómez *et al.* 1994, 11: Imai *et al.* 2007 in preparation, 12: Imai, Sahai, & Morris 2007, 13: Imai *et al.* 2005, 14: Imai *et al.* 2004, 15] Imai *et al.* 2002, 16: Likkel *et al.* 1992, 17: Morris *et al.* 2003, 18: Sahai *et al.* 1999, 19: Suárez, Gómez, & Morata 2007, 20: Vlemmings *et al.* 2006

in the Galaxy, leading to estimates of masses and the birth places of the stars driving these water fountains.

2. Lifetime and timing of the water fountain

Table 1 gives physical parameters of the water fountain jets derived from single-dish and interferometric observations. Lengths of stellar jets traced by H_2O maser emission range over 6000 AU (in IRAS 16342−3814, Morris *et al.* 2003, Claussen *et al.* 2004), which is still equal to or shorter than the radius of a circumstellar envelope of an AGB star (e.g., Elitzur 1992). On the other hand, the flow velocities $V_{\rm flow} \sim \Delta V_{\rm los}/2$ are sometimes over 100 km s^{-1}. These give dynamical ages of jets shorter than 100 years, only ∼5 years in an extreme case (Imai *et al.* , in preparation). In the time interval of 100 years, a stellar jet traveling with a typical water fountain velocity of 100 km s^{-1} reaches ∼2000 AU from the central star, which corresponds to the typical size of an OH maser shell in a circumstellar envelope around an OH/IR star, at the final stage of energetic mass loss. The H_2O maser emission is excited near the tip of the stellar jet where high compression takes place in the shock at the interface between the jet and the ambient circumstellar gas. But because the ambient gas density declines with radius, the maser emission may be quenched at the distance where the post-shock gas density falls below a critical value of $n_{\rm H_2} \sim 10^6$ cm^{-3} (Elitzur *et al.* 1989).

Note that while optical lobes are observed in some water fountains (IRAS 16342−3814, and IRAS 19134+2131), the optical emission is not detected in others (W43A and OH12.8−0.9). This suggests that the presence of water fountains may help in the identification of new pre-PNe, that may be optically invisible due to heavy circumstellar extinction. The optical lobes in IRAS 19134+2131 and IRAS 16342−3814 coexist with the H_2O maser emission (Sahai *et al.* 1999, Morris *et al.* 2003, Imai, Sahai, & Morris 2007). It is expected that the lobes have been recently formed along the collimated fast jet, in which the circumstellar envelope around the evolved star has been penetrated by the jet and starlight. These sources may be at a similar stage in the transition from an AGB to post-AGB star.

Does the size of a circumstellar envelope determine the lifetime of a water fountain? We already know that H_2O masers can survive until the photoionization of the circumstellar envelope starts to form a PN, such as K3−35 (Miranda *et al.* 2001). In K3−35, H_2O maser emission is seen at the tips of the radio lobes as well as in the inner and denser part of the circumstellar envelope where the UV radiation from the star has not yet penetrated. However, the maser emission at the lobe tips of K3-35 is only weakly collimated, both morphologically and kinematically, implying that an initially collimated jet traceable by H_2O maser emission is being destroyed. The dynamical age of the radio lobes is estimated to be 800 yr (Miranda *et al.* 2001), which may correspond to an upper limit to the lifetime of a water fountain source but it is one order of magnitude higher than the dynamical age estimated using VLBI observations of the maser spatio-kinematical structures.

Here we speculate that the collimated jet may drill through the circumstellar envelope formed during the AGB phase and the tip of the jet may reach the outer surface of the envelope before the onset of photoionization of the envelope. The masers lying beyond the optical lobes, such as those in IRAS 16342−3814, might soon be quenched within a few decades or centuries owing to the decline of the gas density in the pre-shock gas as the jet propagates outward. This in turn leads to a loss of collimation and ultimately to a loss of the maser inversion as the post-shock gas density declines.

3. Equatorial flows in the water fountain sources

Highly collimated fast stellar jets as described above are expected to be driven by magneto-hydrodynamical processes (e.g., Blackman *et al.* 2001, Vlemmings *et al.* 2006). However it is unclear whether the stellar object driving the jet is a single star or a binary system; one is an AGB star providing material to another through an accretion disk. Astrometry for the H_2O and 1612 MHz OH maser emission will provide an important clue to find each of the locations of the driving sources of the collimated jet and the AGB flow. Only W43A has SiO maser emission (Nakashima & Deguchi 2003, Imai *et al.* 2005), which provides a better opportunity of exactly finding the location of the AGB star.

Figure 1 compares the locations of the dynamical centers of the H_2O maser jet and the SiO maser flow using a common extragalactic position reference. Although astrometric accuracy is poorer in the north–south direction because W43A is located very close to the celestial equator, the dynamical centers coincide within 4 mas (10 AU at 2.6 kpc). The SiO maser flow is well modeled by a biconically expanding flow with a low velocity ($\sim$20 km s^{-1}) and a large opening angle (Imai *et al.* 2005). These implies that both of the collimated jet and the biconical flow are driven in the region within 10 AU.

Note that the spatio-kinematical structure of the SiO masers does not exhibit any rotation or contraction as suggested when mass accretion occurs from a disk. Figure 2 shows relative proper motions of the W43A H_2O masers. We found a maser feature that is located close to the dynamical center and has a Doppler velocity close to the systemic velocity of W43A ($V_{\rm LSR}$ =34 km s^{-1}) and a proper motion corresponding to a speed of $\sim$30 km s^{-1}. The velocity vector is pointed to the different direction from the jet major axis. The low velocity maser feature is expected to be associated with an "equatorial" flow, rather than an accretion disk. Such equatorial flows have recently been recognized in other water fountain sources, IRAS 18286−0959 and IRAS 18460−0151, as well (Deguchi *et al.* 2007). Figure 3 and 4 shows the spatio-kinematics of IRAS 18460−0151 and IRAS 18286−0959, respectively. The former source has the fastest jet ever observed with a speed over 300 km s^{-1}. This source also has an equatorial flow with a velocity of $\sim$30 km s^{-1} in the region close to the midpoint of the most blue-shifted and red-shifted

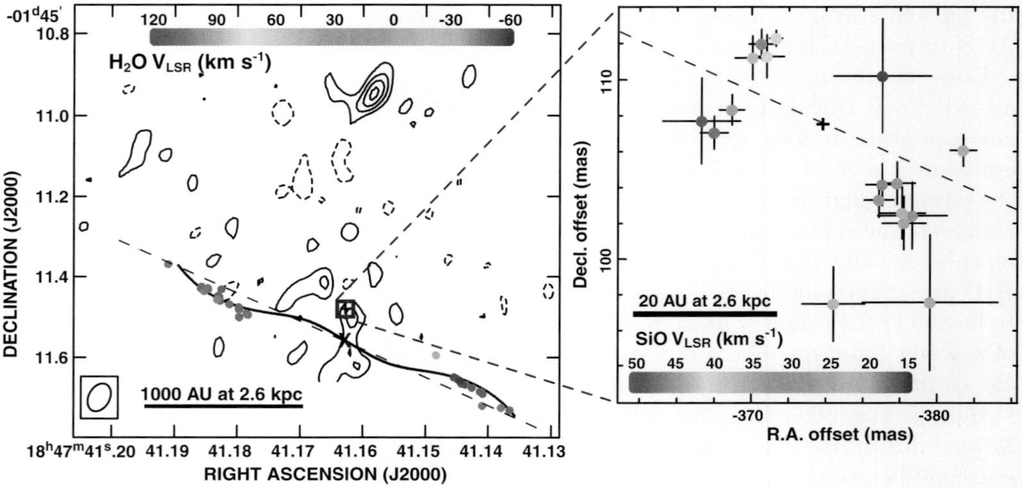

Figure 1. H_2O, SiO (v =1, J =1–0) maser emission, and 7 mm continuum emission in W43A. Different colors denote the maser feature line-of-sight velocities, according to the color scale. *Left* Spatio-kinematics of the H_2O maser features observed on 2002 April 3. A dotted line shows a pattern of a precessing jet appearing in the year 2002.3, which is modeled by Imai *et al.* 2002. A symbol "X" indicates the dynamical center of the modeled jet. A thin dashed line indicates the axis direction of the H_2O maser jet. *Right* Spatio-kinematics of the SiO maser features. Horizontal and vertical bars of individual maser features indicate uncertainties of feature positions in R.A. and declination, respectively. A plus symbol indicates a roughly-estimated location of the dynamical center of the biconically-expanding flow found in the SiO maser kinematics. Roughly a flux density of a spot is inversely proportional to the error bar of the spot.

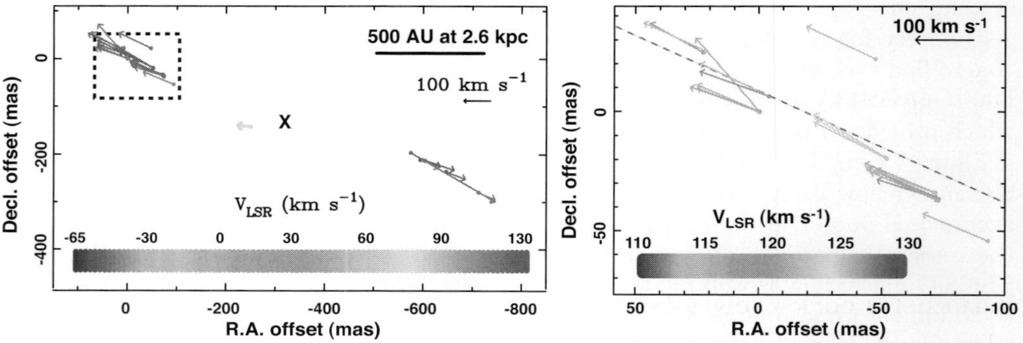

Figure 2. Relative proper motions of H_2O masers in W43A measured during 2003–2004 with a mean proper motion subtracted. A cross indicates the location of the dynamical center of the precessing bipolar jet. *Left* the whole distribution of maser features. *Right* zoom-up view of the feature distribution in the area displayed in a dashed-line box in the left panel.

maser features forming the high velocity jet. The latter source has many velocity maser components, only the most blue-shifted and red-shifted maser features are clearly separated from each other and each of them is distributed in a small area. These imply the coexistence of a collimated fast flow and a slow outflow associated with either an AGB envelope or an equatorial flow driven by the fast flow as its sub-product. The existence of such equatorial flows may support a hypothesis that a single star can create both the fast collimated and the slow equatorial flows when the star is collapsing into a white dwarf (H. Washimi and S. Miyaji, in private communication).

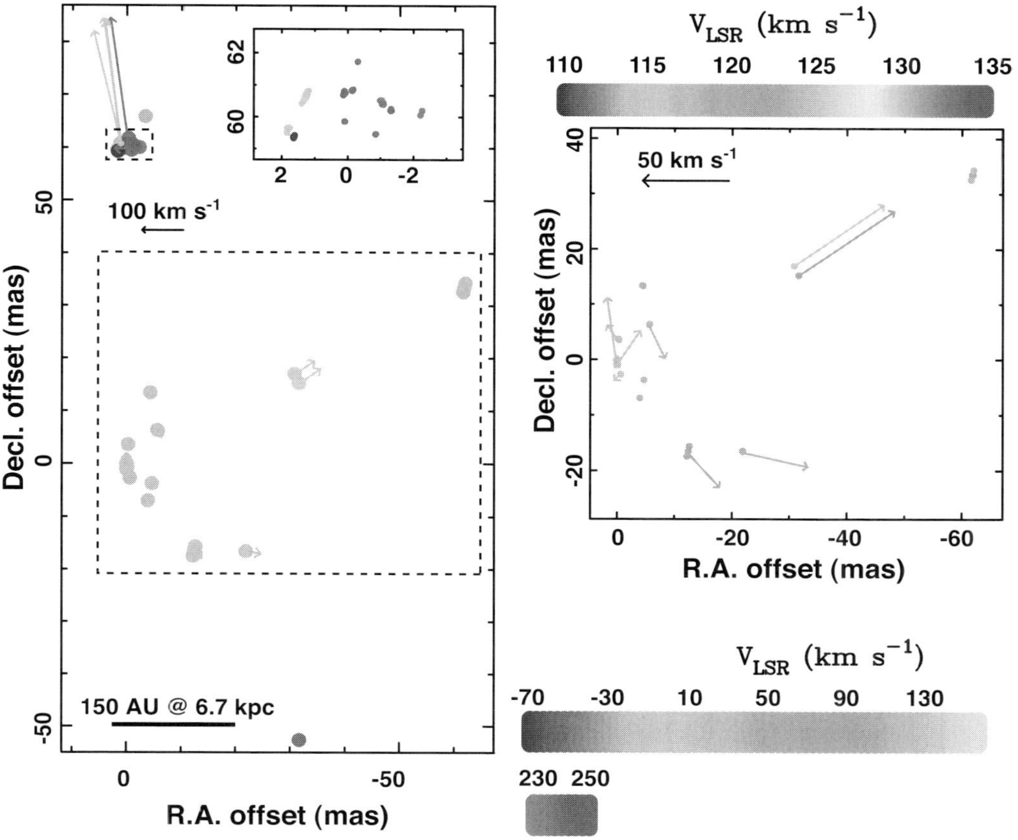

Figure 3. Distribution (filled circle) and relative proper motions (arrows) of H_2O masers in IRAS 18460−0151 during 2006 September–November. *Left* Overview of the distribution. *Right* Zoom-up view of the area displayed with a dashed line box in the left panel. Different colors denote the features' line-of-sight velocities, according to the color scale above this panel. *Relative* proper motions of maser features are measured with respect to the maser feature located at the origin and biased in the north–west direction by $\sim$30 km s^{-1}. Color renditions of the figures can be found in the online version of this paper.

4. Ballistic corkscrew jets

The morphology of water fountains may provide a unique clue of the ejection of the material in the jets, leading to a characterization of the central stars. Some of the water fountain jets exhibit spiral patterns. In particular, such a pattern is clearly seen in the distributions of H_2O masers in W43A (Imai *et al.* 2002, Imai *et al.* 2005). On the other hand, a "corkscrew jet" is found in optical emission in IRAS 16342−3814 (Sahai *et al.* 2005). Recently we found twin spiral patterns of the H_2O maser distribution in W43A (see figure 1), one of the spiral is seen in the axisymetrically opposite side of another (dotted line) with respect to the jet major axis (thin dashed line). However, curved motions of the maser features are not recognized. The right panels of figure 2 and figure 5 show details of the proper motions of the H_2O masers found in W43A and IRAS 19134+2131, respectively. In each source, we find an alignment of maser features whose direction has an offset from the major axis of the jet. On the other hand, proper motion vectors of maser features are still parallel to the jet axis. These suggest that the corkscrew jets found in these sources are formed by ballistic motions of maser features from the driving source of

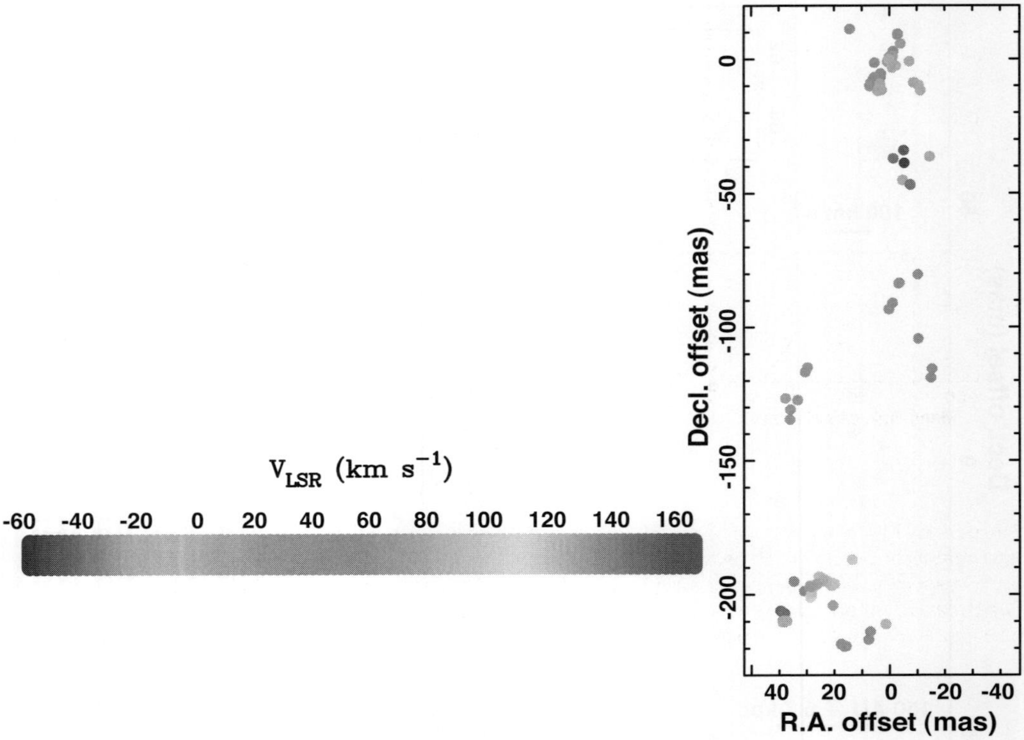

Figure 4. Same as figure 3, but for IRAS 18286−0959 observed on 2006 September 14.

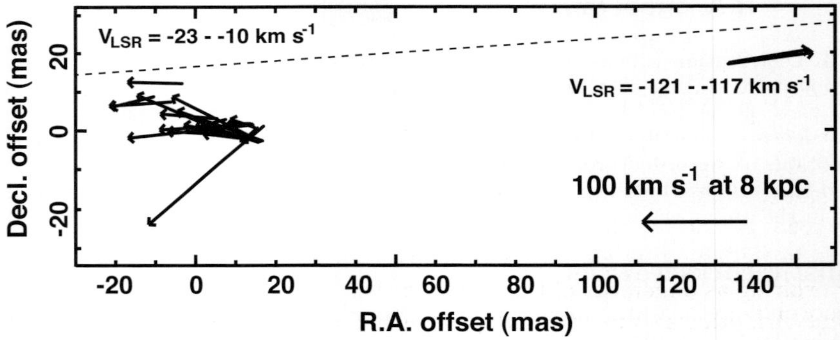

Figure 5. H_2O maser feature distribution in IRAS 19134+2131 (roots of arrows) and relative proper motion vectors of these masers (arrows) with respect to a position reference maser feature with a mean proper motion subtracted.

the jets, but whose directions are precessing with time. The alignments of maser features may be formed at the shock fronts interfaced with the ambient circumstellar envelopes.

5. Location and motion of the water fountain in the Galaxy

The *absolute* coordinates of H_2O masers associated with IRAS 19134+2131 (hereafter I19134) have been measured with respect to the extragalactic position reference source J1925+2106 for a span over one year (Imai, Sahai, & Morris 2007). We therefore fit the temporal variation of the mean position of three maser features to a kinematical model

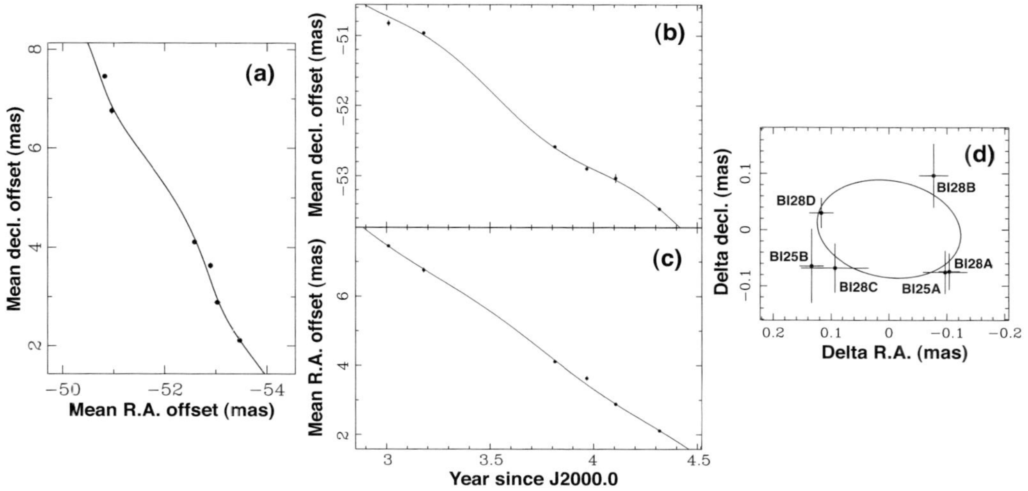

Figure 6. Mean motion of three H_2O maser features in IRAS 19134+2131 with respect to the extragalactic position reference source J1925+2106. (*a*) Mean R.A. and decl. offsets on the sky with respect to the phase-tracking center. (*b*) and (*c*) Mean R.A. and decl. offsets against time. (*d*) Relative mean offsets with a mean proper motion subtracted. The observation code is shown alongside each measured position.

consisting of an annual parallax and a constant velocity mean secular motion. Figure 6 shows the fitting result. We obtain a best-fit annual parallax, corresponding to a distance to I19134 of $D = 8.0^{+0.9}_{-0.7}$ kpc. On the basis of the inferred heliocentric distance and the systemic secular maser motion, we estimate the location of I19134 in the Galaxy to be $[R, \theta, z] = [7.4^{+0.4}_{-0.3}$ kpc, $62°\pm5°$, $0.65^{+0.07}_{-0.06}$ kpc], and the 3-D velocity vector in Galactic cylindrical coordinates to be, $[V_R, V_\theta, V_z] = [3^{+53}_{-46}, 125^{+20}_{-28}, 8^{+48}_{-39})$ [km s^{-1}].

The estimated location is close to the Perseus spiral arm within ~1 kpc (c.f., Nakanishi & Sofue 2006). Note that the estimated Galactic rotation velocity of I19134 (~125 km s^{-1}) is much slower than that implied by the Galactic rotation curve (~220 km s^{-1}, e.g. Clemens 1985, Dehnen & Binney 1998). It implies that the I19134 orbit must have large excursions in radius although I19134 is orbiting along a circle at present. I19134 is apparently orbiting as a member of the Galactic "thick" component rather than the "thin" component. Adopting a source velocity perpendicular to the Galactic plane V_z =8 km s^{-1}, we find that the present Galactic altitude gives a lower limit to the travel time of I19134 from the Galactic plane to the present position to be $\sim 2.4 \times 10^7$ yr. This suggests that the stellar object of I19134 has a lifetime of at least this travel time and a stellar mass smaller than ~8.6 $M_\odot$ (here we adopt a stellar lifetime to be $t_* \propto M_*/L_* \propto M_*^{-2.8}$) if the central star of I19134 was born in the Galactic plane as a member of the Galactic "thin disk" component. The statistical study of Galactic PNe by Manchado 2004 shows that the Galactic scale height of bipolar PNe such as I19134 is $< z >$=100 pc, suggesting that their progenitors should be high mass stars. Although the mass of I19134's progenitor is therefore expected to be near the upper mass limit estimated above, the present galactic altitude of I19134 and its anomalously slow galactic rotation velocity are quite odd. These may imply an abnormal environment during the birth of I19134's central star. However, the estimated mass of I19134 still has a big uncertainty due to the poor accuracy of the vertical velocity, V_z.

Acknowledgements The present work have been done in collaborations with S. Deguchi, P. J. Diamond, S. Kwok, A. Miyazaki, M. Morris, J. Nakashima, K. Obara, T. Omodaka, R. Sahai, T. Sasao, and W. H. T. Vlemmings. The NRAO VLBA and VLA are facilities of the National Science Foundation are operated under a cooperative agreement with Associated Universities, Inc. The EVN is a joint facility of European, Chinese, South African and other radio astronomy institutes funded by their national research councils. H.I. was supported by an IAU Grant for traveling to the symposium and a Grant-in-Aid for Scientific Research from the Japan Society for the Promotion of Science (18740109).

References

Baud, B., Habing, H. J., Mattheus, H. E., & Winnberg, A. 1979, *A&AS*, 36, 193
Blackman, E. G., Frank, A., Markiel, J. A., Thomas, J. H., & Van Horn, H. M. 2001, *Nature*, 409, 485
Boboltz, D. A., & Marvel, K. B. 2005, *ApJ*, 627, L45
Claussen, M., Sahai, R., & Morris, M. 2004, in ASP Conf. Ser. 313, Asymmetrical Planetary Nebulae III: Winds, Structure and the Thunderbird, ed. M. Meixner *et al.* (San Francisco: ASP), 331
Clemens, D. P. 1985, *ApJ*, 295, 422
Deacon, R. M., Chapman, J. M., Green, A. J., & Sevenster, M. N. 2007, *ApJ*, 658, 1096
Deguchi, S., Nakashima, J, & Kwok, S. 2007, *ApJ*, in preparation
Dehnen, W., & Binney, J. 1998, *MNRAS*, 294, 429
Diamond, P. J., Norris, R. P., Rowland, P. R., Booth, R. S., & Nyman, L.-Å. 1985, *MNRAS*, 211, 1
Engels, D. 2002, *A&A*, 388, 252
Engels, D., Schmid-Burgk, J., & Walmsley, C. M. 1986, *A&A*, 67, 129
Elitzur, M. 1992, in Astronomical Masers (Dordrecht: Kluwer)
Elitzur, M., Hollenbach, D. J., & McKee, C. F. 1989, *ApJ*, 346, 983
Genzel, R., & Downes, D. 1977, *A&As*, 30, 145
Gómez, Y., Rodríguez, L. F., Contréras, M. E., & Moran, J. M. 1994, Rev. Mex. AA, 28, 97
Imai, H., Sahai, R., & Morris, M. 2007, *ApJ*, submitted
Imai, H., Nakashima, J., Diamond, P. J., Miyazaki, A., & Deguchi, S. 2005 *ApJ*, 622, L125
Imai, H., Morris, M., Sahai, R., Hachisuka, K., & Azzollini, J. R. F. 2004, *A&A*, 420, 265
Imai, H., Obara, K., Diamond, P. J., Omodaka, T., & Sasao, T. 2002, *Nature*, 417, 829
Likkel, L., Morris, M., & Maddalena, R. J. 1992, *A&A*, 256, 581
te Lintel Hekkert, P., Versteege-Hensel, H. A., Habing, H. J., & Wiertz, M. 1989, *A&AS*, 78, 399
Manchado, A. 2004, in ASP Conf. Ser. 313, Asymmetrical Planetary Nebulae III: Winds, Structure and the Thunderbird, ed. M. Meixner *et al.* (San Francisco: ASP), 3
Miranda, L. F., Gómez, Y., Anglada, G., & Torrelles, J. M. 2001, *Nature*, 414, 284
Morris, M. R., Sahai, R., & Claussen, M. 2003, Rev. Mex. AA Conf., 15, 20
Nakanishi, H., & Sofue, Y. 2006, *PASJ*, 58, 847
Nakashima, J., & Deguchi, S. 2003, *PASJ*, 55, 229
Sahai, R., Le Mignant, D., Sánchez Contreras, C., Campbell, R. D., & Chaffee, F. H. 2005, *ApJL*, 622, L53
Sahai, R. 2004, in ASP Conf. Ser. 313, Asymmetrical Planetary Nebulae III: Winds, Structure and the Thunderbird, ed. M. Meixner *et al.* (San Francisco: ASP), 141
Sahai, R., & Nyman, L.-Å. 2000, *ApJ*, 537, L145
Sahai, R., te Lintel Hellert, P., Morris, M., Zijlstra, A., & Likkel, L. 1999, *ApJ*, 514, L115
Sahai, R., & Trauger, J. 1998, *AJ*, 116, 1357
Suárez, O., Gómez, J. F., & Morata, O. 2007, *A&A*, in press (astro-ph/0703502)
Vlemmings, W. H. T., Diamond, P. J., & Imai, H. 2006, *Nature*, 440, 58

Astrophysical Masers and their Environments
Proceedings IAU Symposium No. 242, 2007
J.M. Chapman & W.A. Baan, eds.

doi:10.1017/S1743921307013142

A search for water maser emission from post-AGB stars

J. M. Chapman[1], R. M. Deacon[2], A. J. Green[2], and M. Cohen[3]

[1] Australia Telescope National Facility, PO Box 76, Epping, NSW 1710, Australia
[2] School of Physics A29, University of Sydney, NSW 2006, Australia
[3] Radio Astronomy Laboratory, University of California, Berkeley, CA 94720, USA

Abstract. We have used the Tidbinbilla 70-m antenna to search for 22 GHz H_2O maser emission from a sample of 85 evolved stars. 21 detections were made. Of these 15 were from massive AGB stars. High-velocity H_2O maser emission was detected from five sources, of which four are post-AGB stars. Three of the high-velocity sources, b292 (IRAS 18043–2116), d46 (IRAS 15445–5449), and d62 (IRAS 15544–5332) were new discoveries. d46 is also a source of non-thermal radio continuum emission. The high-velocity H_2O maser emission and the radio continuum from post-AGB stars are probably associated with shocks that form from wind-wind interactions.

Keywords. masers, stars: AGB and post-AGB, evolution, mass-loss, radio continuum

Introduction

Post-asymptotic giant branch (post-AGB) stars are in a short-lived evolutionary stage between the AGB and planetary nebulae. During this stage the regular long-period pulsations of AGB stars cease and the stars change from losing mass in slow dense winds, with wind expansion velocies of typically 5–15 km s^{-1}, to losing mass in hot low-density winds with wind velocities of several hundreds kilometres per second.

Masers provide an excellent means of studying mass-loss from AGB and post-AGB stars. OH maser emission at 1612 MHz has been detected from the outer circumstellar envelopes of more than 1500 AGB stars. In most cases the OH 1612 MHz spectra are double-peaked, consistent with spherical, or near-spherical outflows. In contrast, planetary nebulae often show complex and diverse morphologies with elliptical, bipolar and filamentary structures observed. It is likely that changes in the envelope morphologies begin during the early post-AGB stage.

We are studying the maser properties of a well-defined sample of 85 evolved stars selected from the OH 1612 MHz Galactic Plane survey of Sevenster *et al.* (2001 and references therein). The sources were selected to be post-AGB candidate stars on the basis of their IRAS and/or MSX far-infrared colours. The OH 1612, 1665 and 1667 MHz properties of the sources, and the full selection criteria are discussed by Deacon *et al.* (2004). Here we discuss the results of a search for H_2O maser emission.

Figure 1 shows the MSX colour-colour plot for the sources in our sample with good quality MSX fluxes. Following Sevenster (2002a,b) the plot is divided into four quadrants. Post-AGB stars are located in quadrants I (late-type or older sources) and IV (early-type or younger sources). The sources seen in quadrant III are known as 'LI' stars and have an excess of emission at 60 μm emission compared to AGB stars with little or no OH 1612 MHz maser emission. On the standard IRAS [12–25], [25–60] colour-colour plot, they are seen to the *left* of the well-known evolutionary track for AGB stars (van der Veen & Habing 1988). Our sample included 30 LI sources and 55 sources with the MSX colours of post-AGB stars.

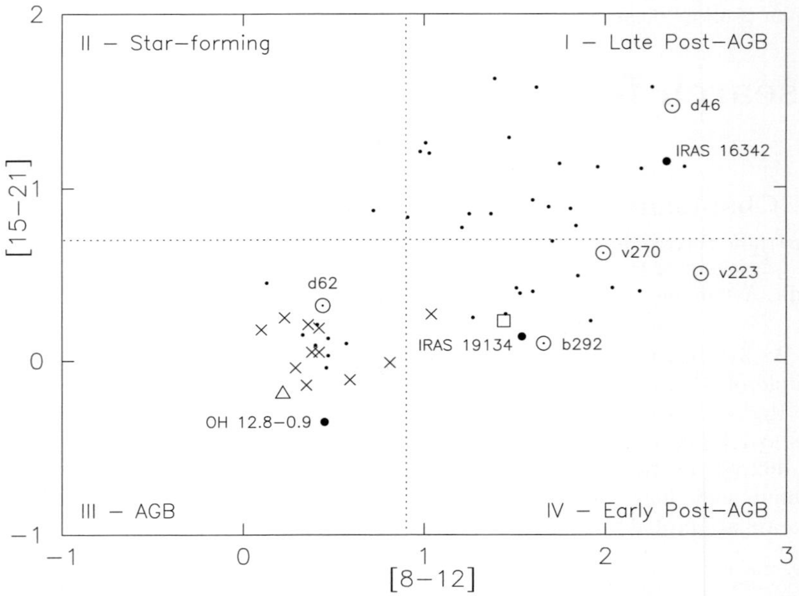

Figure 1. MSX colour-colour plot for sources in the sample. Sources with no detected H_2O emission are shown as small dots. Sources with regular H_2O maser profiles are shown as crosses. The dotted circles indicate five high-velocity H_2O maser sources detected with the Tidbinbilla antenna. Three high-velocity sources with previously published detections (IRAS 19134+2131, IRAS 16342–3814 and OH 12.8–0.9) are shown as large dots (Deacon *et al.* 2007). All the high-velocity H_2O maser sources are labelled.

Observations at a rest frequency of 22.235 GHz were taken between 2003 May and 2004 April with the Tidbinbilla 70-m DSS 43 antenna. Each of the 85 sources was observed for a total on-source integration time of 7 minutes, using a bandpass of 16 MHz and 8193 channels, corresponding to a total velocity coverage of $\sim$ 200 km s^{-1}. The spectra were Hanning smoothed to give a velocity resolution of 0.18 km s^{-1}. The antenna had a pointing accuracy of about 7 arcseconds. The one-sigma noise levels in the spectra were between 40 and 120 mJy. Repeated observations were taken in a few cases to confirm weak detections.

The LI sources

We detected H_2O maser emission from 21 of the 85 sources in the sample. Of these, 15 detections were from the LI sources. In 14/15 cases, the H_2O maser emission was detected at velocities that were equal to, or within, the OH 1612 MHz velocity ranges. A typical example is shown in Figure 2. Such double-peaked H_2O profiles are characteristic of AGB stars with high mass-loss rates where the H_2O molecules are located closer to the central stars than the OH molecules. The OH 1612 MHz spectra of the LI sources are almost invariably doubled-peaked indicating spherical or near-spherical circumstellar envelopes. For one LI source in the present sample, d62 (IRAS 15544–5332), the 22 GHz maser spectrum showed a single narrow peak located *outside* of the OH velocity range, indicating a possible high-velocity outflow.

The evolutionary status of LI stars as AGB or post-AGB stars has been uncertain. From their average Galactic latitude of 0.46 degrees, Sevenster (2002a,b) determined that they are massive stars with an average mass of about 4 $M_\odot$. From their Galactic

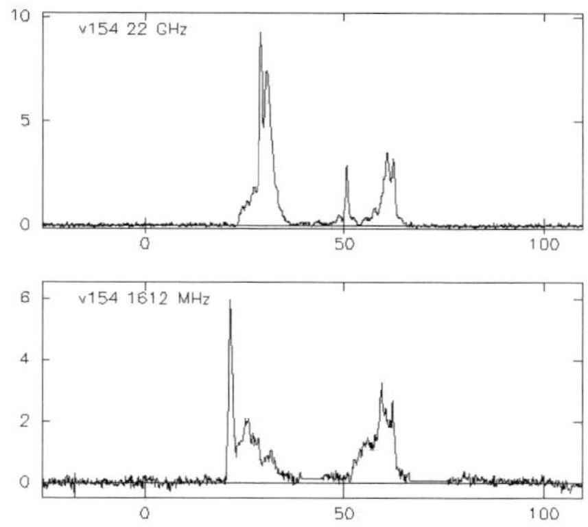

Figure 2. 22 GHz H_2O (top) and OH 1612 MHz (bottom) maser spectra for the LI source, v154 (IRAS 18327–0715). The stellar velocity, determined from the OH 1612 MHz profile is 40.5 km s^{-1}.

distribution and their OH and H_2O properties we infer they are massive stars with high mass-loss rates that are still AGB stars. As such they are the precursors to the most massive post-AGB stars and planetary nebulae.

Cohen *et al.* (2006) have discussed the radio properties of a remarkable source, V1018 Sco, where a faint ionised nebula is seen centred on an AGB star with a pulsation period of $\sim$ 1460 days. This source has MSX colours similar to the LI sources in our sample. Radio continuum emission has been detected from two locations from within the nebula. The stronger continuum feature (their feature A) has a spectral index of –0.8 between 3 and 13 cm, consistent with non-thermal synchroton radio emission.

High-velocity water sources

We detected H_2O maser emission at velocities outside of the OH velocity range from five sources. One source, d62, is an LI star while the other four are post-AGB stars (Figure 1).

Two sources, v223 (W43A) and v270 (IRAS 18596+0315) were previously known. W43A belongs to a rare group of sources known as 'water fountains'. In these sources H_2O maser emission is detected over a large velocity range from collimated bipolar jets. The jets have very short dynamical ages of typically 100 years. From polarisation observations of the H_2O maser emission, Vlemmings, Diamond & Imai (2006) have shown that the jets in W43A are likely to be magnetically collimated. Our 22 GHz spectrum of v270 (not shown) shows a symmetric H_2O maser profile with emission over two fairly narrow velocity ranges, each approximately 30 km s^{-1} from the stellar velocity, at twice the expansion velocity of the OH 1612 MHz maser emission. We consider that v270 is a likely water fountain source, although high angular resolution images are needed to confirm this.

b292, a new water fountain source

We discovered extremely broad H_2O maser emission from the source b292 (IRAS 18043–2116). As shown in Figure 3, numerous emission features were detected over velocities between -20 and 185 km s^{-1}. It is possible that other features are present beyond the range of observed velocities. The stellar velocity is known from the OH 1612 MHz spectrum to be $\sim$ 83 km s^{-1}. The strongest H_2O emission was detected at velocities between $\sim$ 95 and 120 km s^{-1}, corresponding to the far side of the star. b292 is the

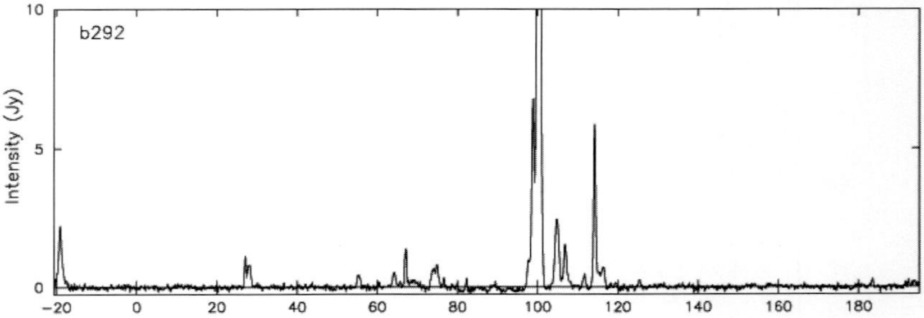

Figure 3. H_2O maser spectrum of b292 (IRAS 18093–2116) obtained with the Tidbinbilla 70-m antenna in 2004 March. The spectrum is truncated in intensity to show the weak features.

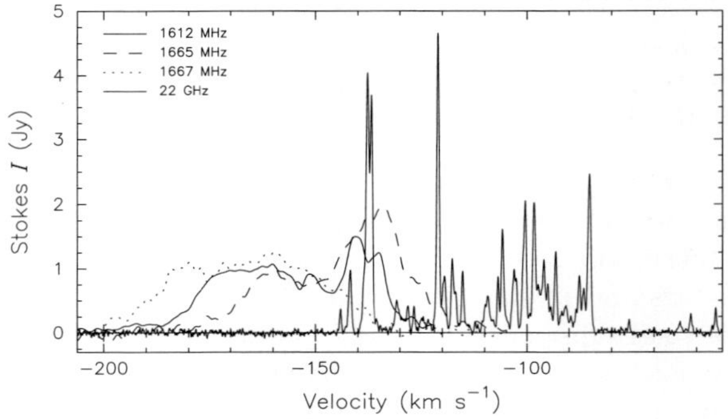

Figure 4. OH 1612, 16655, and 1667 MHz maser spectra of the post-AGB star d46 (IRAS 15445–5449), are shown together with the Tidbinbilla 22GHz H_2O maser spectrum (from Deacon *et al.* 2007).

only post-AGB star to have been detected in the OH 1720 MHz transition (Sevenster & Chapman 2001) and this emission is thought to arise behind a shock front associated with a slow-wind – fast-wind collision. From the H_2O maser spectrum, the OH 1720 MHz detection and the stellar classification as a young post-AGB star (Figure 1) we classify this as a water fountain source.

The peculiar source d46

Figure 4 shows the Tidbinbilla H_2O maser spectrum for d46 together with OH spectra from Deacon *et al.* (2004). This source has unusual maser profiles with emission detected over a total velocity range of approximately 150 km s^{-1}. The OH profiles at 1612, 1665 and 1667 MHz are all broad and similar in shape. However, the OH 1665 MHz emission is red-shifted relative to the OH 1612 MHz emission while the 1667 MHz emission is blue-shifted. The stellar velocity is unknown but the OH spectra suggest a systemic velocity near -150 km s^{-1}. The 22 GHz H_2O spectrum is red-shifted by about 50 km s^{-1} relative to the OH maser emission with numerous narrow features detected at velocities between -150 and -50 km s^{-1}. As with b292, other features outside of the observed velocity range could be present. The MSX colours indicate that d46 is a more highly evolved post-AGB star than b292 (Figure 1) while the complex and asymmetric maser profiles indicate a highly disturbed outflow.

Figure 5 shows a radio continuum spectrum of d46 from observations at 3, 6 and 13 cm taken with the Australia Telescope Compact Array in 1998. The radio continuum spectrum of the LI star V1018 Sco at wavelengths between 3 and 36 cm is also shown (Feature A, see Cohen *et al.* 2006). For both sources the radio continuum spectrum has a

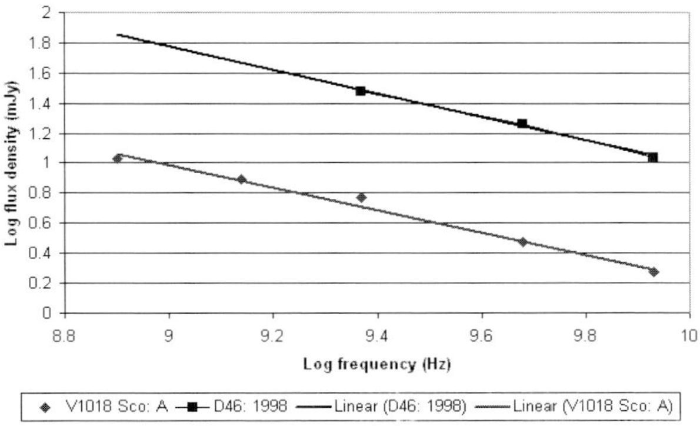

Figure 5. Radio continuum spectra for d46 (top) and the LI source V1018 Sco (bottom). In both cases a linear fit to the data gives a spectral index of –0.8 (adapted from Cohen *et al.* 2006).

spectral index of $\alpha = -0.8$ (defined as $S_\nu \propto \nu^\alpha$), consistent with synchroton radio emission. We interpret this as evidence for shocks due to a wind-wind interaction between the hot wind that turns on as the star leaves the AGB, and the AGB circumstellar envelope.

Discussion and conclusions

From this survey of 85 sources, we detected two different types of H_2O maser emission. AGB-type H_2O maser profiles were detected from 50% of the 30 LI sources in our sample. This is consistent with an identification of these objects as massive AGB stars. We also detected five sources with high-velocity H_2O emission. Four of these were post-AGB stars. We did *not* detect any 'regular' H_2O maser profiles from sources with very red IRAS or MSX colours. We surmise that the H_2O maser emission that is seen during the AGB evolutionary stage, disappears as a star evolves away from the AGB. However, a new source of H_2O maser emission, associated with high-velocity jets, may appear. The detection of high-velocity H_2O emission in two LI sources (d62 and OH12.8–0.9, see Figure 1) suggests that for some massive stars the jet outflows may start before the star has left the AGB.

References

Cohen, M., Chapman, J. M., Deacon, R. M., Sault, R. J., Parker, Q. A., & Green, A. J. 2006, *MNRAS* 369, 189

Deacon, R. M., Chapman, J. M. & Green, A. J. 2004, *ApJS* 155, 595

Deacon, R. M., Chapman, J. M., Green, A. J., & Sevenster, M. N. 2007, *ApJ* 658, 1096

Sevenster, M. N., van Langevelde, H. J., Moody, R. A., Chapman, J. M., Habing, H. J., & Killeen, N. E. B. 2001, *A&A*, 366, 481

Sevenster, M. N. & Chapman, J. M. 2001, *ApJ*, 546, L119

Sevenster, M. N., 2002a, *AJ*, 123, 2772

Sevenster, M. N., 2002b, *AJ*, 123, 2788

van der Veen, W. E. C. J. & Habing, H. J. 1988, *A&A*, 194, 125

Vlemmings, W. H. T., Diamond, P. J. & Imai, H. 2006, *Nature*, 400, 58

Astrophysical Masers and their Environments
Proceedings IAU Symposium No. 242, 2007
J.M. Chapman & W.A. Baan, eds.

doi:10.1017/S1743921307013154

Maser emission in planetary nebulae

Yolanda Gómez

Centro de Radioastronomía y Astrofísica, UNAM, A.P. 3-72, C.P. 58090, México
email: y.gomez@astrosmo.unam.mx

Abstract. Stars at the top of the asymptotic giant branch (AGB) can exhibit maser emission from molecules like SiO, H_2O and OH. These masers appear in general stratified in the envelope, with the SiO masers close to the central star and the OH masers farther out in the envelope. As the star evolves to the planetary nebula (PN) phase, mass-loss stops and ionization of the envelope begins, making the masers disappear progressively. The OH masers in PNe can be present in the envelope for periods of ~1000 years but the H_2O masers can survive only hundreds of years. Then, H_2O maser emission is not expected in PNe and its detection suggests that these objects are in a very particular moment of its evolution in the transition from AGB to PNe. We discuss the unambiguous detection of H_2O maser emission in two planetary nebulae: K 3-35 and IRAS 17347-3139. The water-vapor masers in these PNe are tracing disk-like structures around the core and in the case of K3-35 the masers were also found at the tip of its bipolar lobes. Kinematic modeling of the H_2O masers in both PNe suggest the existence of a rotating and expanding disk. Both PNe exhibit a bipolar morphology and in the particular case of K 3-35 the OH masers are highly polarized close to the core in a disk-like structure. All these observational results are consistent with the models where rotation and magnetic fields have been proposed to explain the asymmetries observed in planetary nebulae.

Keywords. Planetary Nebulae, masers, magnetic fields, individual(K 3-35, IRAS 17347-3139

1. Introduction

Planetary nebulae (PNe) represent the final stage of evolution for intermediate mass stars (1-8 $M_\odot$). A PN consists of an expanding gaseous shell of highly ionized gas that has been ejected by the star during the end of the AGB evolution, and a central star that will end as a white dwarf. The study of PN is very important not only because our Sun will end as a PN but also because these objects expel heavy elements like carbon and oxygen producing an enrichment of the interstellar medium. The lifetime of a PNe is about 10^4 years and it is expected that there are 10^4-10^5 PNe in our galaxy, from which only ~1500 have been reported (Kohoutek 2001). PNe are characterized by exhibiting a variety of shapes (e.g. Balick & Frank 2002). Several observational studies have revealed that most of the PNe (~75%; Manchado 2004) exhibit asymmetric morphologies that go from elliptical to very collimated outflows. One of the open questions in this field is to understand the origin of these asymmetries.

To answer this question we should review the formation process itself in the transition between the AGB and PNe stages. After completion of the H and He core burning, the star evolves to the AGB. At the top of the AGB phase, the star is surrounded by an expanding envelope of gas and dust (Caswell 1974). The study of the star in this evolutionary phase is better pursued with radio and infrared telescopes and the presence of neutral atomic and molecular gas in the envelope of the star is common (e.g Mufson, Lyon & Marionni 1975; Huggins *et al.* 1996; Josselin & Bachiller 2003; Bujarrabal 2006). For AGB stars with oxygen-rich envelopes it is possible to detect maser emission of one or more molecules, such as SiO, H_2O and OH that appear stratified in the envelope

Table 1. OH and H_2O maser emission in young PNe

PNe	OH	H_2O	Morphology	References
NGC 6302	YES	NO	bipolar	Payne *et al.* (1988)
OH 349.36-0.20	YES	NO	?	Zijlstra *et al.* (1989)
IRAS 17207-2855	YES	NO	?	Zijlstra *et al.* (1989)
PK 356+2° 1	YES	NO	?	Zijlstra *et al.* (1989)
IRAS 17347-3139	YES	YES	bipolar	de Gregorio-Monsalvo *et al.* (2004)
IRAS 17371-2747	YES	NO	?	Zijlstra *et al.* (1989)
IRAS 17375-2759	YES	NO	?	Zijlstra *et al.* (1989)
IRAS 17375-3000	YES	NO	?	Zijlstra *et al.* (1989)
OH0.9+1.3	YES	NO	?	Zijlstra *et al.* (1989)
IRAS 17443-2949	YES	YES	?	Zijlstra *et al.* (1989); Suárez *et al.* (2007)
IRAS 17580-3111	YES	YES[a]	?	Zijlstra *et al.* (1989); Suárez *et al.* (2007)
IRAS 18061-2505	NO	YES[b]	?	Suárez *et al.* (2007)
IC 4997	YES	NO	bipolar	Tamura & Kazes (1989)
K 3-35	YES	YES	bipolar	Engels *et al.* (1989); Miranda *et al.* (2001)
Vy2-2	YES	NO	shell	Davis, Seaquist & Purton (1979)
M1-92	YES	NO	bipolar	Lépine & Rieu (1974)

[a] Probably not a PN; [b] No OH maser emission (Suárez, Gómez & Morata 2007).

(e.g. Reid & Moran 1981; Elitzur 1992). As the slow and massive mass loss of the late AGB phase stops, the gravity contracts the core heating it up, and the star enters its PN phase, from this moment the maser conditions will disappear in short timescales (Lewis 1989, Gómez, Moran & Rodríguez 1990). In particular, the water molecules are expected to disappear in a timescale of decades, and only OH masers seem to persist for a considerable time ($\sim$1000 yr; Kwok 1993). In this way H_2O masers are not expected to be present in a regular PN. However, two PNe (K3-35 and IRAS 17347-3139; Miranda *et al.* 2001; de Gregorio-Monsalvo *et al.* 2004) have been found to harbor H_2O and OH maser emission, suggesting that these objects can be in an early stage of their evolution as PNe. Recently, two more PNe have been reported to exhibit H_2O maser emission and interferometric observations are required to confirm the association (Suárez, Gómez & Morata 2007). In this talk we will present details about the kinematics and polarization of the maser emission in the two confirmed PNe: K3-35 and IRAS 17347-3139.

2. Young PNe with OH and H_2O masers

Modeling of the OH maser emission in late AGB stars has shown that the typically observed double peak profile corresponds, in general, to the blueshifted and redshifted maser components which are coming mainly from the front and the back parts of the envelope, respectively (Reid *et al.* 1977). We mentioned that after the mass loss stops the OH maser can survive around $\sim 10^3$ years, then it may be possible to find OH maser emission in very young PNe. In the particular case when an ionized region is present, which is the case for young PNe, the redshifted OH maser peak component coming from the back, may not be seen because the emission is absorbed by the ionized gas, and only the blueshifted OH peak will be present (Rodríguez, Gómez & García-Barreto 1985). Then a very young PNe will have a compact ionized inner part while the outer part of the envelope remains neutral, possibly with OH maser emission. It is possible to appreciate in several young OH-PNe, with known LSR velocities, that the blueshifted OH maser components are the dominant ones (Shepherd *et al.* 1990). For example, for

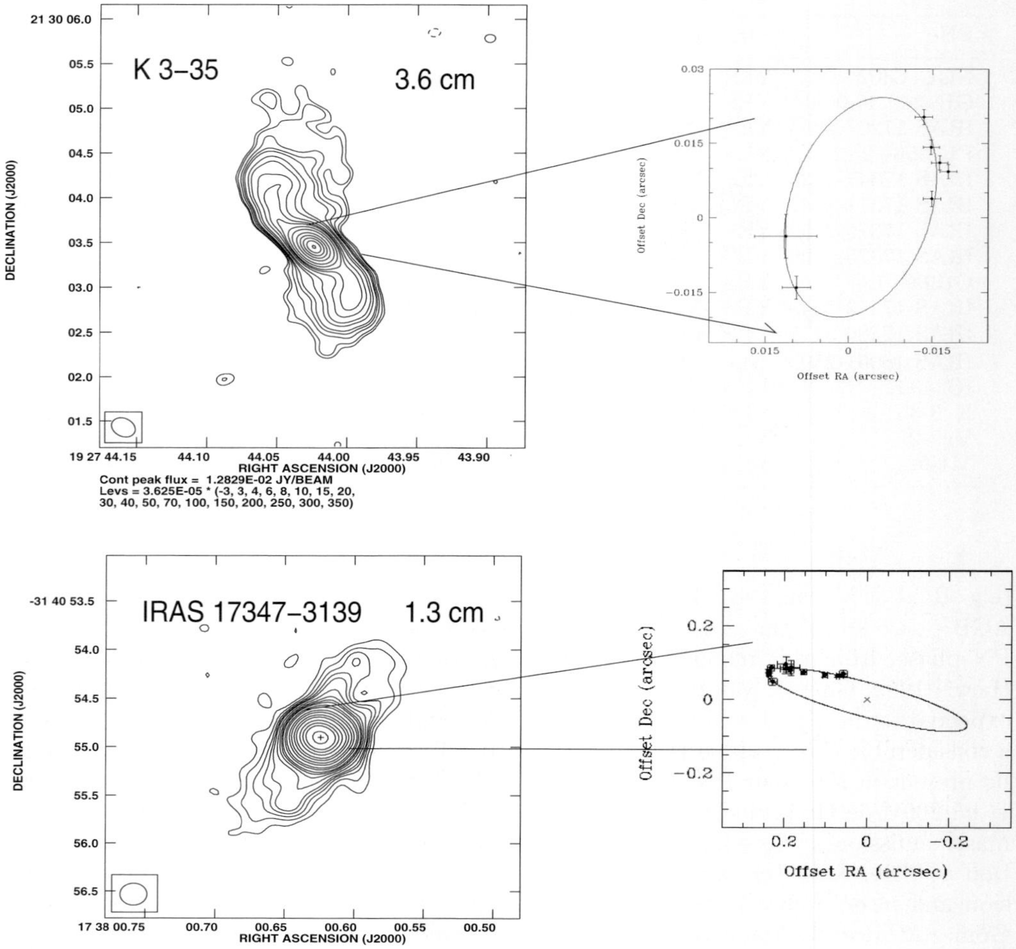

Figure 1. Left: Contour images of the radio continuum observed with the VLA in K 3-35 (Miranda *et al.* 2001) and IRAS 1737-3139 (Tafoya *et al.* 2007b). Right: The respective water maser spots with a disk fit (Uscanga *et al.* 2007).

NGC 6302 the ionized gas has a V_{LSR} (He 121α)$= -29$ km s^{-1} (Gómez, Rodríguez & García-Barreto 1987), while the strong OH 1612 MHz maser emission is blueshifted, arising from a velocity of -40 km s^{-1} (Payne, Phillips & Terzian 1988). Several surveys of OH maser emission have being carried out toward PNe (e.g. Johansson *et al.* 1977; Caswell *et al.* 1981; Payne, Phillips & Terzian 1988; Bowers & Knapp 1989; Zijlstra *et al.* 1989; Sevenster *et al.* 2001) showing that it is uncommon to observe OH maser emission in PNe, with the detection of very few objects (see Table 1). Then, if it is difficult to detect OH maser emission, the detection of H_2O masers in PNe is expected to be more rare, however our group have confirmed the detection of two H_2O-PNe (Miranda *et al.* 2001; de Gregorio-Monsalvo *et al.* 2004) and another two H_2O-PNe have been recently detected with the Robledo de Chavela Antenna (Suárez, Gómez & Morata 2007). In Table 1 we list the detections of H_2O maser emission in PNe.

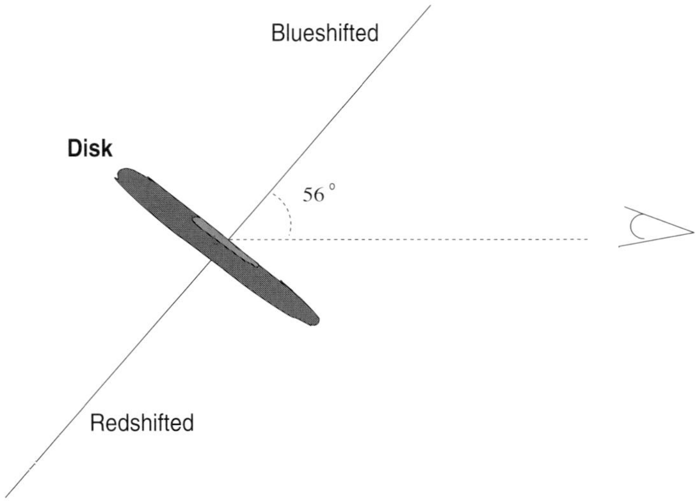

Figure 2. Schematic diagram for the inclined disk model in K 3-35.

3. K 3-35 and IRAS 17347-3139

The first confirmed cases of PNe where H_2O maser emission is present and coexistent with OH maser and radio continuum emission are: K 3-35 (Miranda *et al.* 2001) and IRAS 17347-3139 (de Gregorio-Monsalvo *et al.* 2004). Both H_2O-PNe are emission line nebulae that exhibit bipolar morphologies at radio wavelengths (K 3-35: Aaquist 1993; Miranda *et al.* 2001, IRAS 17347: Tafoya *et al.* 2007b) and the H_2O masers are arising from the core of the sources in a torus-like structure (see figure 1). In the particular case of K 3-35, there is also H_2O maser emission arising from the tips of the bipolar lobes at 5000 AU from the central star (Miranda *et al.* 2001; Gómez *et al.* 2003). Bipolar morphologies in these H_2O-PNe are also appreciated from optical/IR images (Sánchez-Contreras *et al.* 2006). The systemic LSR velocities for these two H_2O-PNe, obtained from CO observations, are 24±3 km s^{-1} for K 3-35 (Tafoya *et al.* 2007a) and −55 km s^{-1} for IRAS 17347-3139 (Uscanga *et al.* 2007). The LSR velocities of the H_2O maser components in these PNe appear blueshifted, <24 km s^{-1} for K 3-35 (Miranda *et al.* 2001; Gómez *et al.* 2003; Tafoya *et al.* 2007a) and $\geqslant -65$ km s^{-1} for IRAS 17347-3139 (de Gregorio-Monsalvo *et al.* 2004; Uscanga *et al.* 2007). In general, young PNe have a very high molecular to ionized gas mass ratio (Huggins *et al.* 1996; Josselin & Bachiller 2003). In the case of K 3-35, the ratio of molecular to ionized mass is ∼2 and the recent detection of HCO^+ emission suggests the presence of very high density molecular gas (∼10^5 cm^{-3}) that could be protecting the H_2O molecules from the dissociating radiation of the central star (Tafoya *et al.* 2007a). The bipolar morphology, and the presence of large amounts of neutral molecular gas not only suggest that these objects are young, but also that they have followed a similar evolutionary track corresponding to massive progenitors.

4. Kinematics of the H_2O masers

Figure 1 shows radio continuum contour images and the H_2O maser spot positions for K 3-35 and IRAS 17347-3139. Modeling of the H_2O masers was made for a homogeneous disk-like structure fitting each maser spot velocity with an expanding and rotating disk.

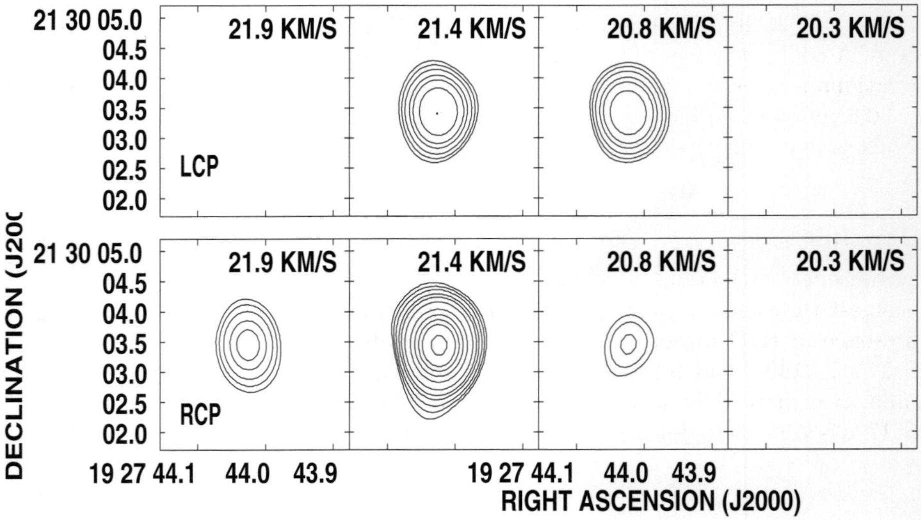

Figure 3. Contour images of OH 1720 MHz emission from K 3-35. Top panels are for the left circular plarization (LCP) and the bottom panels are for the right circular polarization (RCP) maser emission. The LSR velocities are indicated in the top right-hand corner of each image. Contours are 5, 7, 10, 15, 20, 30, 50, 70, 90, 110 and 150 times 8 and 9 mJy beam^{-1}, the *rms* noise of the LCP and RCP images, respectively.

For K 3-35 the model resulted in expanding and rotating velocity values of 1.1 km s^{-1} and 3.6 km s^{-1}, respectively. The inclination angle of the disk, between the line-of-sight and the major axis of the outflow, was also derived (56°) (see figure 2), which is in agreement with having the north lobe of the outflow blueshifted and the south redshifted. The estimated radius of the ring is $0\farcs023$ ($\sim$ 100 AU). For IRAS 17347-3139 it was more difficult to do the modeling since the H_2O masers are interpreted as tracing only one edge of the equatorial disk. However, in both cases the orientation of the major axis of the rings appears perpendicular to the corresponding outflows, supporting a disk model interpretation (Uscanga *et al.* 2007).

5. Equatorial magnetic field in K 3-35

The detection of magnetic fields in PNe is a key parameter to understand the generation of jets and bipolar structures in PNe (see Vlemmings; these proceedings). K 3-35 is one of the few PNe where magnetic fields have been detected (Gómez *et al.* 2005). In particular, the OH 1665 MHz maser emission toward K 3-35, was found to be very highly polarized (>50%), suggesting the presence of a magnetic field (Miranda *et al.* 2001). Since the 1665 MHz maser spots seems to be tracing an elongated structure perpendicular to the bipolar outflow, it was proposed that they could be tracing an equatorial magnetic field. New VLA observations of the four OH maser transitions confirmed the strong circular polarization of the OH 1665 MHz emission and assuming that at least one Zeeman pair is present, we derived a magnetic field in the line-of-sight of 0.14 mG at a radius of 250 AU. In addition, the OH 1720 MHz transition was clearly detected. A single OH 1720 MHz feature was previously reported by te Lintel Hekkert (1991) toward K 3-35. Our VLA OH observations confirmed the presence of a single feature in the spectrum of the 1720 MHz and also proved that it is associated with the H_2O-PNe, since the peak position of the maser feature coincides with the peak of the radio continuum emission (Gómez *et al.* 2005). It is the only one PNe where the OH 1720 MHz maser transition has been detected. A contour image of the OH 1720 MHz from K 3-35 in its left and right circular polarizations is shown in figure 4. Assuming that a Zeeman pair is present, a magnetic field of 0.8 mG was derived at a radius of 120 AU. The strength of the magnetic field in K 3-35 is in agreement with the values derived toward evolved stars.

6. Conclusions

The presence of H_2O and OH maser emission in PNe is not common, and their detection suggest that these objects are in a very early phase of their evolution. In particular the presence of H_2O maser emission has been confirmed only in two PNe (K 3-35 and IRAS 17347-3139), and from a recent H_2O survey, made with the Robledo de Chavela Antenna, two more PNe seem to be detected. The H_2O maser positions in K 3-35 and IRAS 17347-3139 show kinematics consistent with masers located in rings. From modeling, it is derived that in K 3-35 there is a rotating and expanding disk perpendicular to the bipolar outflow. There is evidence of magnetic fields in K 3-35 that could help to constrain the models for jets and bipolar morphologies in PNe.

Acknowledgements

YG acknowledge financial support from PAPIIT-UNAM (IN100407) and CONACyT, México.

References

Aaquist, O. B. 1993 *A&A* 267 260
Balick, B. & Frank, A. 2002 *ARA&A* 40, 439
Bujarrabal, V. 2006 *Planetary Nebulae in our Galaxy, IAU Proceedings* 234, 193
Bowers, P. F. & Knapp, G. R. 1989 *ApJ* 347, 325
Caswell, J. L. 1974, *IAU Proceedings. Ed: F. J. Kerr and S. C. Simonson* 60, 423
Caswell, J. L., Haynes, R. F., Goss, W. M., & Mebold, U. 1981 *AuJPh* 34, 333
Davis, L. E., Seaquist, E. R., & Purton, C. R 1979, *ApJ* 230, 434
de Gregorio-Monsalvo, I., Gómez, Y., Anglada, G., Cesaroni, R., Miranda, L. F., Gómez, J. F., & Torrelles, J. M. 2004, *ApJ* 601, 921
Elitzur, M. 1992, *ARA&A* 30, 75
Gómez, Y., Miranda, L. F., Anglada, G. & Torrelles, J. M. 2003, *Planetary Nebulae, IAU Proceedings* 209, 263
Gómez, Y., Moran , J. M., & Rodríguez, L. F. 1990, *RevMexAA* 20, 55
Gómez, Y., Rodríguez, L. F., & García-Barreto, J. A. 1987, *RevMexAA* 14, 560
Gómez, Y., Tafoya, D., Anglada, G., *et al.* 2005 *Memorie della Societa Astronomica Italiana* 76, 472
Huggins, P. J., Bachiller, R., Cox, P. & Forrveille, T. 1996, *A&A* 315, 284
Johansson, L. E. B., Andersson, C., Goss, W. M., & Winnberg, A. 1977, *A&AS* 28 199
Josselin, E. & Bachiller, R. 2003, *A&A* 397, 659
Kohoutek, L. 2001, *A&A* 378, 843
Kwok, S. 1993, *ARA&A* 31, 63
Lépine, J. R. D. & Rieu, N. Q. 1974, *A&A* 36, 469
Lewis, B. M. 1989, *ApJ* 33, 234
Manchado, A. 2004, *ASP Conference Proceedings* 313, 3
Miranda, L. F., Gómez, Y., Anglada, G. & Torrelles, J. M. 2001, *Nature* 414, 284
Mufson, S. L., Lyon, J., Marionni, P. A. 1975, *ApJ* (Letters) 201, L85
Payne, H. E., Phillips, J. A., & Terzian, Y. 1988 *ApJ* 326, 368

Reid, M. J., Muhleman, D. O., Moran, J. M., Johnston, K. J., & Schwartz, P. 1977, *ApJ* 214, 60
Reid, M. J. & Moran, J. M. 1981, *ARA&A* 19, 231
Rodríguez, L. F., Gómez, Y. & García-Barreto, J. A. 1985, *RevMexAA* 11, 139
Sánchez-Contreras, C., Le Mignant, D., Sahai, R., Chaffee, F. H., & Morrism M. 2006, *Planetary Nebulae in our Galaxy, IAU Proceedings* 234, 71
Sevenster, M. N., van Langevelde, H. J., Moody, R. A., Chapman, J. M., Habing, H. J., & Killeen, N. E. B. 2001, *A&A* 366, 481
Shepherd, M. C., Cohen, R. J., Gaylard, M. J., & West, M. E. 1990 *Nature* 344, 6266
Suárez, O., Gómez, J. F., & Morata, O. 2007, *A&A* in press
Tafoya, D., Gómez, Y., Anglada, G. *et al.* 2007a, *AJ* 133, 364
Tafoya, D., *et al.* 2007b, *in preparation*
Tamura, S. & Kazes, I. 1989 *Planetary Nebulae, IAU Proceedings* 131, 209
te Lintel Hekkert, P. 1991 *A&A* 248, 209
Uscanga, L. *et al.* 2007 *in preparation*
Zijlstra, A. A. *et al.* 1989, *A&A* 217, 157

Astrophysical Masers and their Environments
Proceedings IAU Symposium No. 242, 2007
J.M. Chapman & W.A. Baan, eds.

doi:10.1017/S1743921307013166

OH (1720 MHz) masers: signposts of SNR/molecular cloud interactions

Crystal L. Brogan

National Radio Astronomy Observatory, 520 Edgemont Rd, Charlottesville, VA, 22903, USA
email:cbrogan@nrao.edu

Abstract. Over the last decade it has been demonstrated that supernova remnant (SNR) OH (1720 MHz) masers are unique tracers of SNR/molecular cloud interactions. Here I briefly review the current state of our observational understanding of these masers including results from recent MERLIN and VLBA full polarization studies of the masers in W28, W44, and W51C. Some of our findings include that (1) in accordance with theory, the linear polarization position angles are either parallel or perpendicular to the plane-of-sky magnetic field determined through other observations; (2) the maser spot sizes are fairly large ($\sim 10^{14}$ cm) and exhibit a core/halo morphology; and (3) while the magnetic field strengths do increase slightly with higher resolution, this effect can be completely explained by spectral/spatial blending.

Keywords. Masers,(ISM:) supernova remnants

1. Introduction

Although "anomalous" OH (1720 MHz) emission toward a supernova remnant (SNR) was first identified in 1968 (Goss & Robinson 1968), this phenomenon was essentially ignored until Frail, Goss, & Slysh (1993) conclusively identified the emission toward W28 as arising from a maser process, and suggested that these masers may trace sites of SNR/molecular cloud interactions. This suggestion was supported by the earlier work of Elitzur(1976) that showed that OH (1720 MHz) masers can be collisionally pumped under the conditions present in such interactions. A number of surveys have subsequently been carried out toward most of the known SNRs in the Galaxy, as well as toward the Galactic Center region by Frail *et al.* (1996), Green *et al.* (1997), Yusef-Zadeh *et al.* (1996), Yusef-Zadeh *et al.* (1999), Koralesky *et al.* (1998). From these surveys OH (1720 MHz) masers have been found in $\sim$ 20 SNRs, or 10% of the known SNRs in our Galaxy. Interestingly, all of the OH (1720 MHz) SNRs are located towards the inner Galaxy ($l = \pm 50°$) where most of the molecular gas resides except one (IC443). SNR OH (1720 MHz) masers exhibit a number of characteristics that are distinct from OH (1720 MHz) masers observed in H II regions. For example, the SNR variety spot sizes tend to be several orders of magnitude larger (Claussen *et al.* 2002, Hoffner *et al.* 2003), have narrow simple line profiles, and have significantly weaker linear and circular polarization percentages of $\sim$ 10%. SNR OH (1720 MHz) masers have not been observed to be accompanied by any other maser species (see for example Claussen *et al.* 1999b).

More detailed analysis of the pumping mechanism for SNR OH (1720 MHz) masers suggest that they form in the post-shock molecular gas behind C-type shocks, and that a sufficient column of OH with the necessary velocity coherence can only be achieved if the shock is moving (more or less) transverse to the line of sight (Lockett *et al.* 1999; Wardle 1999). The X-rays from the SNR are a critical ingredient in these theories, as they induce the conversion of the most abundant species formed in post C-shock gas H_2O, to OH with about a 1% efficiency. These theories suggest that the collisional pump is only efficient

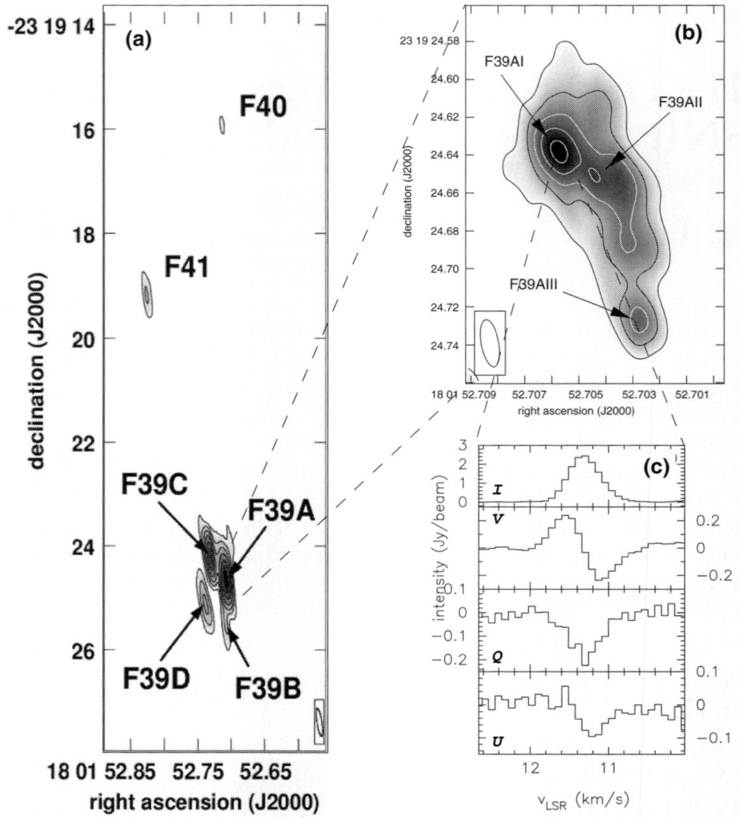

Figure 1. (a) MERLIN image of the W28 maser regions F39-41 with 550 × 100mas resolution (P.A. °). (b) VLBA image of the F39A masers with 25 × 9 mas resolution (P.A. 10°). (c) VLBA full polarization spectra of region F39AI. These figures are from Hoffman *et al.* (2005a).

over a very limited range of densities $\sim 1 \times 10^{(4-5)}$ cm^{-3} and temperatures in the range 50 K $< T <$ 125 K. Hence, observations of this OH maser line can indeed serve as a diagnostic probe of SNR/molecular cloud interactions. Moreover, since OH has a strong Zeeman coefficient at 1720 MHz (0.65 Hz μG^{-1}) it can also provide a powerful probe of the magnetic field strength in the post-shock molecular gas. Assuming the standard thermal Zeeman equation is appropriate for these masers, magnetic field strengths in the 0.2 – 5 mG range have been observed using the Very Large Array (VLA) (Claussen *et al.* 1997, Koralesky *et al.* 1998, Yusef-Zadeh *et al.* 1999, Brogan *et al.* 2000).

Although circa 2000, much had been learned about the distribution and general physical conditions accompanying OH (1720 MHz) masers using instruments like the Very Large Array (VLA) and Australia Telescope Compact Array (ATCA), a deeper understanding would require full polarization data obtained at high spatial resolution. Some of the key remaining questions include: (1) What are the detailed properties of the polarization, and what do they tell us about the strength and orientation of the 3-D magnetic field? (2) How is the maser flux distributed on small size scales and what are the true brightness temperatures? (3) Does the magnetic field strength increase when observed with higher resolution as might be expected if the magnetic fields are tangled on larger sizescales?

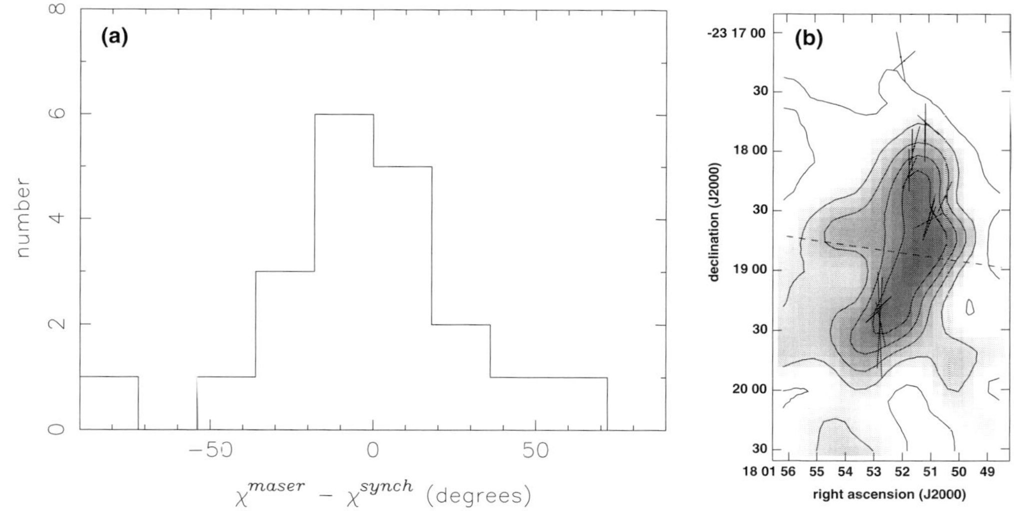

Figure 2. (a) Difference between χ^{synch} and χ^{maser} position angles. (b) Comparison between molecular shock morphology traced by CO(3–2) (in greyscale and contours) and the magnetic field direction defined by χ^{maser} rotated by 90° as suggested by similarity in $\chi^{synch} = 80°$ (dashed line) and χ^{maser}. These figures are from Hoffman *et al.* (2005a).

2. MERLIN and VLBA studies of OH (1720 MHz) maser sources

2.1. *SNR: W28*

The W28 SNR is located at a distance of $\sim$ 2.5 kpc (Velazquez *et al.* 2002) and is partially encircled by shocked molecular gas on its northern and eastern sides. Several regions of OH (1720 MHz) maser emission are coincident with the shocked molecular gas, with the strongest masers located toward the NE in regions denoted E and F (Claussen *et al.* 1997). Figure 1a shows the morphology of the W28 masers located in region F39-F41 observed with MERLIN, and Figure 1b shows the morphology of the F39A maser observed with the VLBA from the recent study by Hoffman *et al.* 2005a. Comparison between the total maser flux densities recovered from the VLA (Claussen *et al.* 1997) vs. MERLIN (Hoffman *et al.* 2005a) data yield a range of results. For example, essentially all of the VLA maser flux density from regions E31 and F39 is recovered in the MERLIN images. In contrast, many of the other maser spots have MERLIN flux densities that are 50%–75% less than measured with the VLA. These results suggest that at least some of the masers have structure on scales $> 3''$ (7,500 AU). A similar comparison between the MERLIN and VLBA flux densities reveal that the VLBA recovers 20%−85% of the MERLIN flux density. Overall these results suggest a core/halo morphology for SNR OH (1720 MHz) masers, but with a wide range of cutoff scalesizes (also see Yusef-Zadeh *et al.* 2003, and J. Hewitt *et al.* these proceedings).

VLBA spectral line profiles for all four Stokes parameters for the F39AI maser are shown in Figure 1c. The percent linear polarization for the MERLIN data ranges from 2%−22%, with an average value of 6% (over 20 distinct measurements). The measured Zeeman magnetic field strength B_θ ranges from 0.2 to 1.2 mG, assuming the thermal Zeeman equation (with no θ dependence). Comparison between the MERLIN data and the few spots with sufficient sensitivity for analysis show no trends in B_θ with increased resolution, beyond that expected from spatial blending of spots with slightly different v_{lsr}. For example, the VLBA line center velocities observed for the F39AI, II, III maser spots

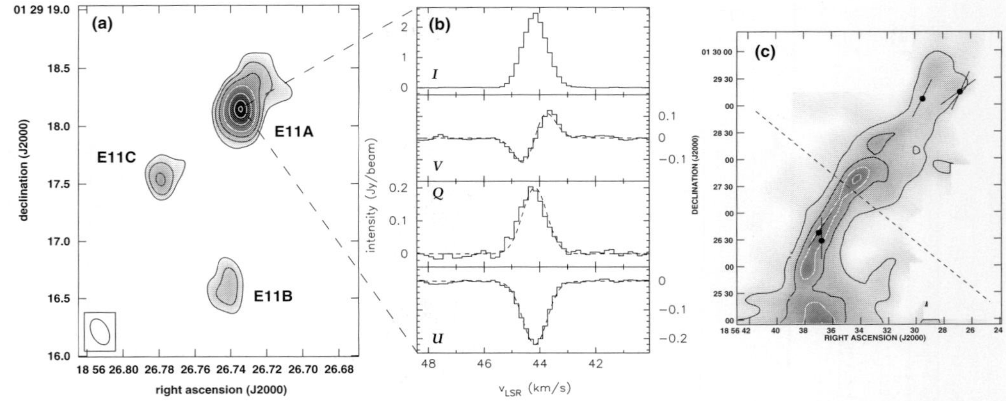

Figure 3. (a) MERLIN image. (b) MERLIN full polarization spectra. (c) Comparison between molecular shock morphology traced by CO(3–2) (in greyscale and contours) and the magnetic field direction defined by χ^{maser} (solid black lines). The dotted black line shows the position angle of $\chi^{synch} = 55°$. These figures are from Hoffman *et al.* (2005b)

differ by only 0.1 km s^{-1}, a factor of 14 times smaller than the range observed across the F39A,B,C,D masers at MERLIN scales. The low velocity dispersion observed at VLBA scales is an indication that this resolution ($\sim$ 25 AU, 4×10^{14} cm) is approaching the intrinsic scalesize of OH (1720 MHz) maser spots.

The position angle of the synchrotron linear polarization χ^{synch} has been measured on $2'$ scales by Dickel & Milne(1976) to be $\sim 80°$. The Faraday rotation in the vicinity of the masers is estimated by these authors to be less than $\sim 20°$, it is also notable that χ^{synch} is quite uniform throughout the eastern and NE parts of the SNR. Figure 2a shows the difference between χ^{synch} and the maser linear polarization position angles χ^{maser}. This distribution is strongly peaked at a difference of 0°. Since synchrotron linear polarization is perpendicular to the plane of sky direction of the magnetic field ($B_\perp$), this comparison suggests that the maser linear polarization for this source is also perpendicular (with the caveat that the scalesizes of the observations are quite different). A comparison between the morphology of the shocked molecular filament in which the masers are immersed (Frail & Mitchell 1998) and the χ^{maser} *rotated by* 90° is shown in Figure 2b (a line indicating the *unrotated* χ^{synch} is also shown). The evident agreement with the rotated χ^{maser} suggests that for W28 $B_\perp$ is aligned with the long axis of the filament in agreement with expectation since only the component of the field parallel to the shock front will be compressed and amplified.

2.2. *SNR: W44*

The W44 SNR is located at a distance of 3.2 kpc (see for example Wolszczan *et al.* 1991) similar to W28 and has its strongest OH (1720 MHz) maser emission on its northeastern side. Figure 3a shows a MERLIN image of the W44 E11 region (maser spot E11A is also detected with the VLBA, but is unresolved). These data are from the recent MERLIN and VLBA study by Hoffman *et al.* (2005b). Similar to W28, the flux density measured for one maser group (C8) in the MERLIN data includes all of the VLA flux density observed by Claussen *et al.* (1997), but the MERLIN data are missing more than half the VLA flux density for most of the other maser spots. The VLBA data contain 50%–70% of the MERLIN flux density. This result provides another indication that SNR OH (1720 MHz) masers have a core/halo morphology.

Figure 3b shows MERLIN spectra of all four Stokes polarization parameters for the

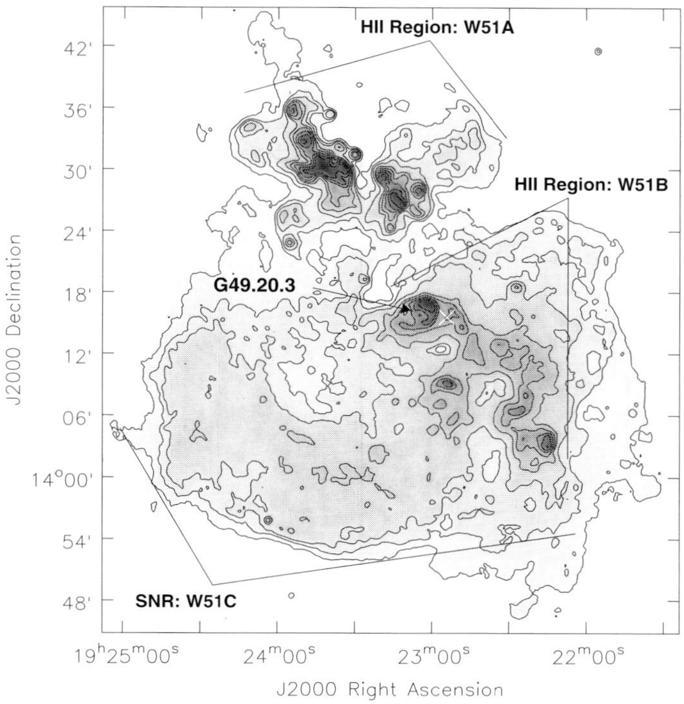

Figure 4. VLA 90cm image (greyscale and contours) of the W51 complex with 34″ resolution. The contour levels are 20, 60, 90, 150, 210, 300, 400, and 500 mJy beam^{-1}. The sources corresponding to W51 A, B, and C; as well as the ultracompact H II region G49.2-0.3 are labeled. The position of the OH (1720 MHz) masers is indicated by a white x symbol (Brogan et al. in prep.).

E11A maser. The magnetic field strengths measured from the MERLIN data are similar to the VLA measurements by Claussen *et al.* (1997), after accounting for a factor of two error in the Zeeman equation used by these authors. Unfortunately, the low signal-to-noise of the VLBA detections preclude measurement of the Zeeman magnetic field strengths. The percentage of linear polarization in the MERLIN data range from $7\% - 14\%$, similar to W28. Figure 3c shows the orientations of χ^{maser} compared to the linear filament of shocked CO(3–2) emission observed by Frail & Mitchell (1998). Both the filament and χ^{maser} are approximately perpendicular to the orientation of $\chi^{synch} \sim 55^\circ$ as observed by Dickel & Milne 1976 with $2'$ resolution. Like the case of W28, Dickel & Milne(1976) do not find evidence for significant Faraday rotation in the vicinity of the masers. Thus, the direction of the plane-of-sky magnetic field implied by the χ^{synch} is still parallel to the long axis of the filament. However, instead of χ^{maser} and χ^{synch} being along a similar position angle as with W28, in W44 the two position angles are perpendicular to each other, suggesting that χ^{maser} is parallel to $B_\perp$ in this case.

2.3. *W51B/C region*

W51 is composed of two large H II region complexes (W51A, and W51B), as well as the SNR W51C (see Figure 4). W51 is located near the Sagittarius arm tangent point at $\sim l = 49^\circ$ and $\sim b = 0.3^\circ$ and thus precise distances for these sources have been difficult to determine due to the severe velocity crowding in this direction, though the range is likely between 4-7 kpc. Soft X-ray absorption toward the W51B region compared to the

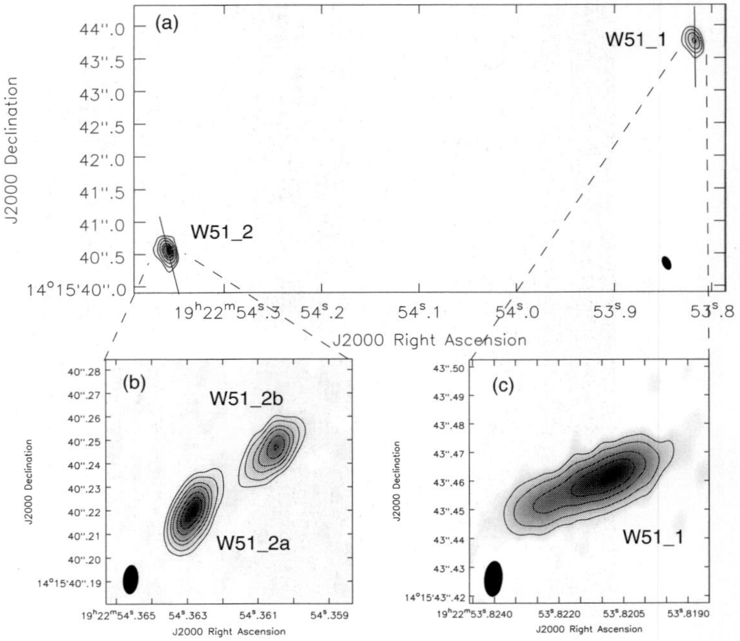

Figure 5. (a) MERLIN integrated intensity image (greyscale and contours) of maser spots W51C_1 and W51C_2. The contour levels are 70 (3σ), 200, 500, 1100, & 1700 mJy beam^{-1} km s^{-1}. The beam with a size of 225 $\times$ 126 mas P.A.= 23.0° is shown in the lower right. The straight solid lines indicate the linear polarization position angles. (b) VLBA integrated intensity image (greyscale and contours) of maser spots W51C_2a and W51C_2b. The contour levels are 50 (3σ), 100, 200, 300, 500, 700, & 900 mJy beam^{-1} km s^{-1}. (c) VLBA integrated intensity image (greyscale and contours) of maser spots W51C_1. The contour levels are 25 (3σ), 50, 100, 150, & 200 mJy beam^{-1} km s^{-1}. The beam with a size of 12.5 $\times$ 6.3 mas P.A.= −5.2° is shown in the lower left (Brogan et al. in prep.).

rest of the W51C SNR suggests that W51C lies behind W51B (Koo, Kim, & Seward 1995). The OH (1720 MHz) masers reported by Green et al. (1997) are located $\sim 2'$ west of the W51B UCH II region G49.2-0.3. Despite this proximity, VLA observations of the W51 OH (1720 MHz) masers by Brogan et al. (2000) demonstrated that these masers share all the hallmarks of SNR OH (1720 MHz) masers, and like Green et al., conclude that they are most likely excited by the W51C SNR (the only known SNR in the region) and that the requisite molecular gas may be associated with the W51B H II region. The SNR OH (1720 MHz) masers in this region are denoted W51C_1 and W51C_2 to clearly distinguish them from the H II region OH (1720 MHz) masers in W51A (though it is not certain they are associated with W51C).

MERLIN and VLBA images of the W51C OH (1720 MHz) masers from Brogan et al. (2007, in preparation) are shown in Figure 5a,b,c. The B_θ magnetic field strengths measured for W51C_1 using MERLIN and the VLBA are both ~ 1.5 mG, and coincidentally, this maser does not show significant substructure on VLBA scales. In contrast the MERLIN B_θ for W51C_2 (1.9 ± 0.1 mG) is intermediate between the values measured for the two VLBA maser spots, W51C_2a and W51C_2b: 1.7 ± 0.1 and 2.2 ± 0.1, respectively, that comprise W51C_2. As was true for W28, this difference is entirely consistent with the combined spectral/spatial blending present at the MERLIN resolution. The χ^{maser} measured from the MERLIN full polarization data are also shown on Figure 5a.

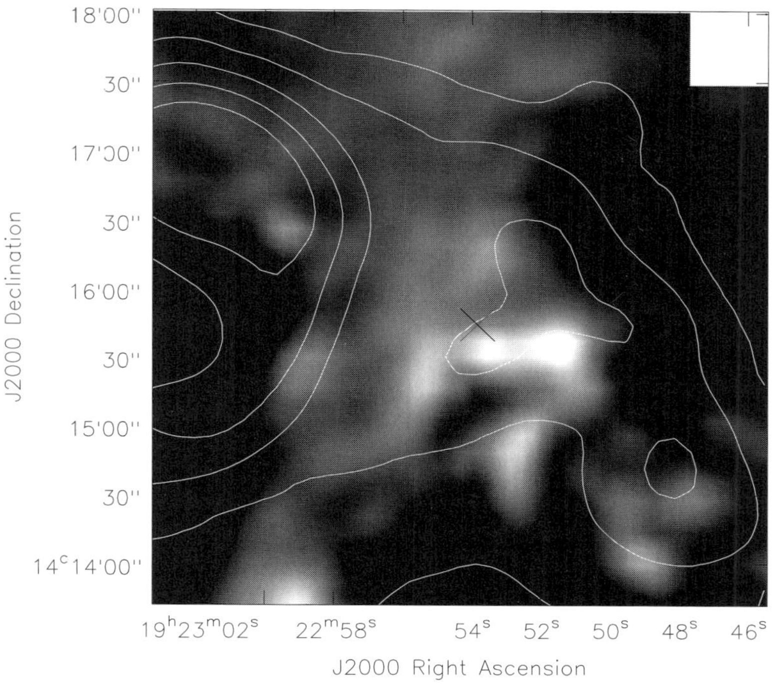

Figure 6. Greyscale of the shocked CO(3–2) integrated intensity with the 90cm continuum contours from Fig. 4 superposed in white. The positions of the OH (1720 MHz) masers are marked by the black × symbol (Brogan et al. in prep.).

From a diverse set of archival and new multi-wavelength data we have determined that (1) there is a 20 and 90cm continuum enhancement at the position of the maser spots with an angular size of $\sim 1.5'$; (2) this enhancement is encircled by a ring of shocked CO(3–2) emission (Figure 6); and (3) there is weak, hard X-ray emission at the same location (Koo *et al.* 2005). These data suggest two possibilities: either the continuum emission, shock, and X-rays are due to W51C (which lies behind W51B) or along this crowded line of sight there is a previously unknown small supernova remnant in front of (or co-distant with) the W51B H II regions. As will be explained in more detail in Brogan *et al.* (in prep.), we favor the second explanation. Comparison of the North-South orientation of the ring of shocked CO(3–2) and 90cm continuum emission in the vicinity of the masers (Fig. 6, note the orientation of the filament in this region is clearer in the channel maps, not shown) and the orientation of the χ^{maser} (Fig. 5) suggest that similar to the case of W44, the linear polarization vectors of the OH (1720 MHz) W51C masers are aligned with the shocked filament, and by association with the direction of the plane-of-sky component of the magnetic field (note this is a bit clearer in the channel maps, not shown).

3. Implications

Maser theories suggest that χ^{maser} will either be parallel or perpendicular to the magnetic field direction in the plane-of-the-sky ($B_\perp$) (see e.g. Elitzur 1998, Watson & Wyld 2001). Furthermore, these theories also show that for the case where $\chi^{maser} \parallel B_\perp$ the angle to the line-of-sight component of the magnetic field ($B_\parallel$) is $< 55°$ and that $B_\parallel > 55°$ for the case where $\chi^{maser} \perp B_\perp$. The fact that some sources fall into the $\chi^{maser} \| B_\perp$

case and some the $\chi^{maser} \perp B_\perp$ breaks the uncomfortable coincidence pointed out by Brogan *et al.* (2000) that since all SNR OH (1720 MHz) masers have linear polarization percentages of $\sim$ 10% they could all have essentially the same $B\|$ which is extremely unlikely. The work of Watson & Wyld (2001) further suggests that the value measured for B_θ is also a function of the degree of maser saturation, though the specific J transition appropriate for OH (1720 MHz) masers has not yet been modeled. The brightness temperatures of these masers $T_b \sim 10^{8-10}$ K, along with the lack of significant variability, are consistent with at least a moderate degree of saturation. For moderate saturation, Watson & Wyld (2001) find that at least for the J = 1–0 and 2–1 transitions the Zeeman equation $V = (Z/2)BdI/d\nu$ (Z = the Zeeman coefficient, and B is the total field strength) with no assumed θ dependence is approximately correct, i.e. $B \approx B_\theta$.

From these data we have also learned that SNR OH (1720 MHz) masers exhibit a core/halo morphology with 50%-80% resolved out by MERLIN and then another 50%-80% by the VLBA. The comparatively large intrinsic scalesizes measured for these masers ($\sim 10^{14-15}$ cm) is in agreement with collisional pump theories (Wardle 1999, Lockett *et al.* 1999). We also find that the small increases observed in the magnetic field strengths with decreasing resolution can be fully explained by spatial/spectral blending effects, and do not appear indicative of tangled B fields.

References

Brogan, C. L., Frail, D. A., Goss, W. M., & Troland, T. H. 2000, *ApJ*, 537, 875
Claussen, M., Frail, D. A., Goss, W. M., & Gaume, R. A. 1997, *ApJ*, 489, 143
Claussen, M. J., Frail, D. A., Goss, W. M., & Desai, K. 1999a, *ApJ*, 522, 349
Claussen, M. J., Goss, W. M., Frail, D. A., & Seta, M. 1999b, *AJ*, 117, 1387
Claussen, M. J., Goss, W. M., Desai, K. M., & Brogan, C. L. 2002, *ApJ*, 580, 909
Dickel, J. R., & Milne, D. K. 1976, Australian Journal of Physics, 29, 435
Elitzur, M. 1976, *ApJ*, 203, 124
Elitzur, M. 1998, *ApJ*, 504, 390
Frail, D. A., Goss, W. M., Reynoso, E. M., Giacani, E. B., Green, A. J., & Otrupcek, R. 1996, *AJ*, 111, 1651
Frail, D. A., Goss, W. M., & Slysh, V. I. 1994, ApJ, 424, L111
Frail, D. A., & Mitchell, G. F. 1998, *ApJ*, 508, 690
Goss, W. M. & Robinson, B. J. 1968, *ApL*, 2, 81
Green, A. J., Frail, D. A., Goss, W. M., & Otrupcek, R. 1997, *AJ*, 114, 2058
Hoffman, I. M., Goss, W. M., Brogan, C. L., Claussen, M. J., & Richards, A. M. S. 2003, *ApJ*, 583, 272
Hoffman, I. M., Goss, W. M., Brogan, C. L., & Claussen, M. J. 2005a, *ApJ*, 620, 257
Hoffman, I. M., Goss, W. M., Brogan, C. L., & Claussen, M. J. 2005b, *ApJ*, 627, 803
Koo, B. -C., Kim, K. -T., & Seward, F. D. 1995, *ApJ*, 447, 211
Koo, B.-C., Lee, J.-J., Seward, F. D., & Moon, D.-S. 2005, *ApJ*, 633, 946
Koralesky, B., Frail, D. A., Goss, W. M., Claussen, M. J., & Green, A. J. 1998, *AJ*, 116, 1323
Lockett, P., Gauthier, E., & Elitzur, M. 1999, *ApJ*, 511, 235
Velázquez, P. F., Dubner, G. M., Goss, W. M., & Green, A. J. 2002, *AJ*, 124, 2145
Wardle, M., 1999, *ApJ*, 525, L101
Watson, W. D., & Wyld, H. W. 2001, *ApJ*, 558, L55
Wolszczan, A., Cordes, J. M., & Dewey, R. J. 1991, *ApJ*, 372, L99
Yusef-Zadeh, F., Goss, W. M., Roberts, D. A., Robinson, B., & Frail, D. A. 1999, *ApJ*, 527, 172
Yusef-Zadeh, F., Roberts, D. A., Goss, W. M., Frail, D. A., & Green, A. J. 1996, *ApJ*, 466, L25
Yusef-Zadeh, F., Wardle, M., & Roberts, D. A. 2003, *ApJ*, 583, 267

Astrophysical Masers and their Environments
Proceedings IAU Symposium No. 242, 2007
J.M. Chapman & W.A. Baan, eds.

doi:10.1017/S1743921307013178

Extended OH(1720 MHz) maser emission from supernova remnants

J. W. Hewitt[1], F. Yusef-Zadeh[1], M. Wardle[2] and D. A. Roberts[1]

[1]Department of Physics and Astronomy, Northwestern University, Evanston, IL 60608, USA
[2]Department of Physics, Macquarie University, Sydney, NSW 2109, Australia

Abstract. Compact OH(1720 MHz) masers have proven to be excellent signposts for the interaction of supernova remnants with adjacent molecular clouds. Less appreciated has been the weak, extended OH(1720 MHz) emission which accompanies strong compact maser sources. Recent single-dish and interferometric observations reveal the majority of maser-emitting supernova remnants(SNRs) have accompanying regions of extended maser emission. Enhanced OH abundance created by the passing shock is observed both as maser emission and absorption against the strong background of the remnant. Modeling the observed OH profiles gives an estimate of the physical conditions in which weak, extended maser emission arises. I will discuss how we can realize the utility of this extended maser emission, particularly the potential to measure the strength of the post-shock magnetic field via Zeeman splitting over these large-scales.

Keywords. supernova remnants, masers, ISM: magnetic fields

1. Introduction

OH(1720 MHz) masers are an excellent indicator of a SNR shocking adjacent molecular material (see review by Brogan, elsewhere in these proceedings). The study of bright compact maser spots has successfully measured post-shock magnetic fields (e.g. Claussen *et al.* 1997) and resolved the sizes and geometries of masering clumps in a number of remnants (Hoffman 2004). Additionally, it has become increasingly apparent that weak, extended OH(1720 MHz) maser emission is also present on parsec scales in SNRs W28, G359.1-0.5, G357.7+0.3 and G357.7-0.1 (Yusef-Zadeh *et al.* 1999, 2003). This extended emission is well correlated with shocked molecular tracers (Lazendic *et al.* 2002) and OH absorption in the other ground-state transitions. Here we discuss the frequency of extended maser emission in known maser-emitting SNRs as well as its utility in tracing the larger extent of the SNR/MC interaction and the potential to map the magnetic field via sensitive Zeeman measurements.

2. Evidence of extended OH(1720 MHz) emission

To better determine the prevalence of extended OH(1720 MHz) emission we have used the 100m Green Bank Telescope(GBT) to map 15 maser-emitting remnants. At 1.7 GHz the GBT has a beamwidth of $7\rlap{.}'2$, so pointings are spaced by $3\rlap{.}'3$ to satisfy Nyquist sampling criterion and fully recover all flux from these sources. All four ground-state transitions of OH (1612, 1665, 1667, 1720 MHz) are observed simultaneously permitting OH absorption to be studied against the bright background of these remnants. Observations obtain a velocity resolution of $0.28\,\mathrm{km\,s^{-1}}$ and a sensitivity of $\sim$20 mJy channel^{-1} comparable to radio synthesis observations which first detected these masers (Frail *et al.* 1994, Frail *et al.* 1996, Green *et al.* 1997, Koralesky *et al.* 1998).

Table 1. Integrated line intensity of OH(1720 MHz) emission

SNR	Name	Velocity Range (km s^{-1})	GBT 0th Moment (Jy km s^{-1})	VLA 0th Moment (Jy km s^{-1})	Ref.
6.4–0.1	W28 A,B	+4.8,+6.3	15.31	7.826	C97
	W28 C,D	+7.2,+14.1	11.57	6.980	C97
	W28 E,F	+8.6,+15.9	273.2	176.0	C97
16.7+0.1		+19.9	0.411	0.245	G97
21.8–0.6	Kes 69	+69.8	0.130	0.093	G97
31.9+0.0	3C391	+105.3	0.200	0.063	F96
		+109.8	0.255	0.077	F96
32.8–0.1	Kes 78	+86.2	0.523	0.510	K98
34.7–0.4	W44 A	+42.9	7.640	1.288	C97
	W44 B,C	+44.7,+46.7	17.91	2.314	C97
	W44 D,E,F	+43.7,+46.9	48.91	33.41	C97
49.2–0.7	W51C	+67.4,+74.3	7.815	8.308	B00
189.1+3.0	IC 443 G	–4.6,0.5	3.916	3.168	C97
	IC 443 D	–9.9,–4.9	0.291	–	–
	IC 443 B	–12.6,–4.9	0.223	–	–
348.5+0.1	CTB 37A	–21.4	0.333	0.383	B00
		–63.5,–66.2	1.671	1.702	B00
349.7+0.2		+14.3,+16.7	3.528	2.153	B00
357.7–0.1	Tornado	–12.8,–14.7	3.011	0.590	F99
357.7+0.3	Square	–34.1,–37.2	9.426	1.117	F99

References: F96=Frail *et al.* (1996); C97=Claussen *et al.* (1997); G97=Green *et al.* (1997); K98=Koralesky *et al.* (1998); F99=Yusef-Zadeh *et al.* (1999); B00=Brogan *et al.* (2000).

It is clear that significant flux from 1720 MHz masers is missed by previous observations. A comparison of integrated maser line intensity between GBT and published VLA measurements is given in Table 1. We give the integrated line intensity† as it is a robust measure of maser emission less sensitive to the variable channel widths if the narrow maser line is not fully resolved. Absolute flux calibration for the GBT is estimated to be better than 10%. Enhancements exceeding this uncertainty are interpreted as evidence for low gain, extended maser emission which is resolved out due to the interferometer's lack of sensitivity to large spatial scales and low brightness temperatures.

It is unlikely that the observed flux enhancement is a result of time-variable masering that has brightened significantly since VLA observations were conducted. Goss & Robinson (1968) first detected OH (1720 MHz) maser emission in W28 and no time variability has yet been reported for that source or any other (Frail *et al.* 1994).

Thus extended OH(1720 MHz) maser emission appears to be a common phenomenon, detected in nine of twelve remnants considered here‡. Even more striking is our detection of new extended maser components in IC 443 that appear without associated compact maser spots (Hewitt *et al.* 2006). Previously only masers associated with shocked molecular clump G had been detected. Narrow maser emission associated with clumps B and D is observed with the GBT yet no compact maser spots was detected toward these clumps down to 5 mJy beam^{-1} channel^{-1} with high-resolution VLA observations. These new

† The integrated line intensity $\gamma = S_p \frac{2(\Delta \mathrm{v})}{\sqrt{\pi/ln(2)}}$ is measured in units of Jy km s^{-1} where S_p is the peak flux density and Δv is the full-width at half maximum as fit by a Gaussian. This is also commonly referred to as the "zeroeth moment" of a Gaussian.

‡ We note that for three SNRs located near the Galactic Center (Sgr A East, Sgr D SNR and G1.4-0.1) absorption at 1720 MHz by intervening gas makes our measurements of maser intensity uncertain, and will be considered in a future work.

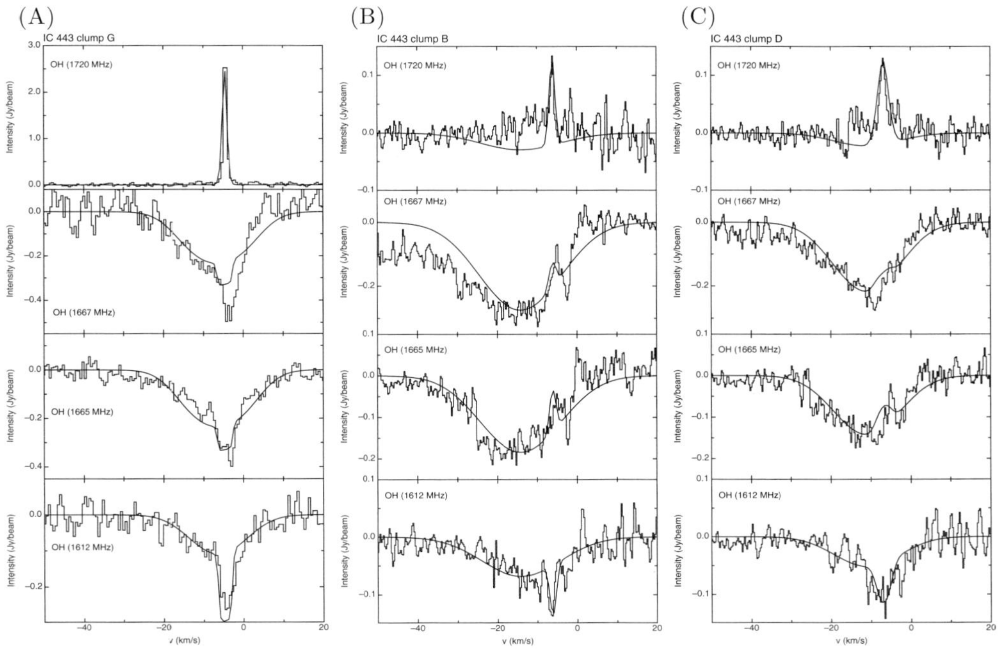

Figure 1. Spectra of all four ground-state transitions of OH from pointings taken toward three clumps with detected 1720 MHz emission. The solid line shows the best-fit model for the observed spectrum. [**A**] Clump G shows a strong, narrow emission feature is seen at 1720 MHz, whereas broad absorption is seen at the other transitions of OH, indicating a transverse shock. [**B**] Clump B shows a narrow maser feature at 1720 MHz with broad, blue-shifted absorption. [**C**] Clump D also shows weak maser emission and broad absorption.

masers are observed at a brightness level some 20 times lower than the bright masers in clump G but with narrow line widths($\sim$ 2 km s^{-1}) and low brightness temperatures ($T_B \leqslant$ 2500 K) consistent with the properties of extended masers. Modeling discussed in the next section supports the idea that in clump G conditions are sufficient to obtain strong inversion of the 1720 MHz line, while in clumps B and D conditions are less ideal.

3. OH modeling and abundance

OH pumping models indicate that the 1720 MHz line can be weakly inverted over a broad range of conditions provided the FIR continuum is not strong. Theoretical considerations give strict constraints on the conditions necessary for SNR OH (1720 MHz) masers: moderate temperatures of 50–125 K, densities of 10^5 cm^{-3}, and OH column densities of the order 10^{16} to 10^{17} cm^{-2}(Lockett *et al.* 1999). Observations of all four OH ground-state transitions permits detailed modeling of the post-shock gas in which masers arise. Here we discuss modeling of masers detected toward IC 443 to illustrate the conditions for weak, extended OH(1720 MHz) emission.

A simple two-component fit is sufficient to model line profiles of shocked gas towards both compact (G) and extended (B and D) masers in IC 443. As shown in figure 1 we include components fit to both narrow maser emission and broad absorption arising from the shocked gas. Each component is characterized by its line-of-sight OH column density N_{OH}, kinetic temperature T_k, beam filling factor f, and mean and FWHM velocities, v_0 and Δv. For further details of the modeling procedure see Hewitt *et al.* (2006).

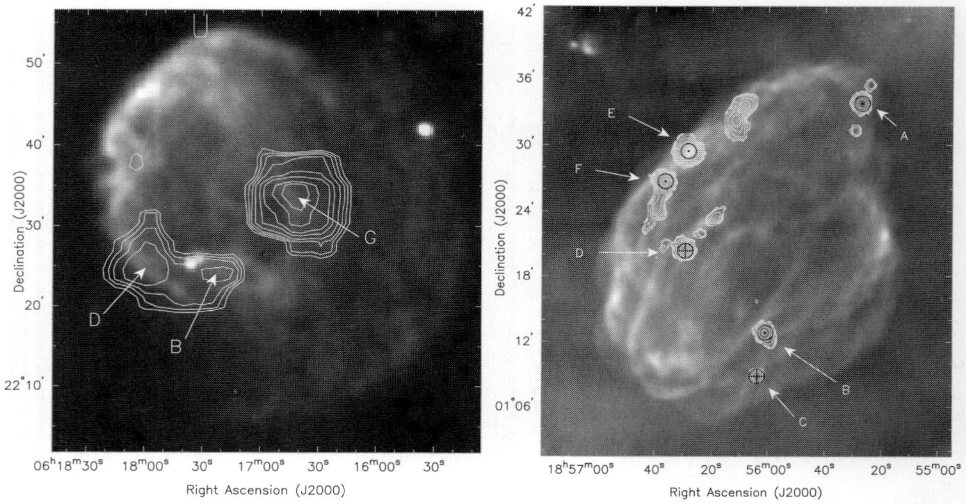

Figure 2. **[Left]** Greyscale image of 327 MHz continuum from IC 443. Contours of the integrated intensity of OH(1720 MHz) emission detected by the GBT are shown with maser features B, D and G identified. **[Right]** Greyscale of 1.4 GHz continuum from W44 (Giancani *et al.* 1997). Contours of integrated maser intensity detected with preliminary VLA observations are superposed with maser features A-F identified. Here the line of sight magnetic field determined for each maser is denoted by symbols ($\odot$ = toward observer [negative field] and $\oplus$ = away from observer [positive field]).

The geometry of the shock can be roughly inferred from absorption profiles. For clump G the absorption is largely symmetric about the maser velocity indicating the shock is propagating transversely across the plane of the sky. Toward weak maser features in clumps B and D blue-shifted absorption features extending out to large negative velocities clearly indicate the shock is propagating with a significant inclination toward the observer's line of sight. Maser features appear only at velocities systemic to the remnant which is likely due to sufficient velocity coherence only being obtained in sufficiently transversely shocked clumps. Weak maser emission is seen extending along the southern shocked ridge of IC443 in figure 2. The observed line profiles for clumps B and D are best fit by T_k = 35 K, $N_{OH} \approx 10^{16}$ cm^{-2} and a low filling factor of order 10%. A comparable column is determined in the broad, blue-shifted absorption component but with a filling factor near unity. Bright maser clump G is fit by T_k = 60 K and $N_{OH} \approx 10^{17}$ cm^{-2}, physical conditions which yield a stronger 1720 MHz line inversion. In IC 443 it appears that weak maser emission results from physical conditions and a shock geometry which are less optimal than for bright compact masers.

To obtain the high OH columns observed it had been expected that C-type shock waves would very efficiently convert the bulk of gas-phase oxygen into water, a few percent of which could be dissociated into OH just behind the shock (Wardle 1999). Interestingly, SWAS observations of shocked water give columns of only a few$\times 10^{13}$ cm^{-2}, yielding a post-shock water abundance of between 10^{-8} and 10^{-7} (Snell *et al.* 2005). This is orders of magnitude lower than the post-shock abundance of OH of a few $\times 10^{-6}$ inferred from our observations (Hewitt *et al.* 2006). This apparent lack of water is not accounted for in any present theory for the production of OH(1720 MHz) masers. As the oxygen chemistry is well known, resolving this riddle of missing water likely lies in a better understanding of the kinds of shocks present and the freeze-out of water ice onto grain surfaces.

4. Zeeman mapping of extended maser emission

As the distribution of weak, extended maser emission appears to be a widespread tracer of the large-scale shock distribution, Zeeman splitting measurements of this emission allow us to probe the structure of the magnetic field behind the shock. As an example, results of sensitive Zeeman measurements of W44 with the VLA is given in figure 2 (Yusef-Zadeh, Roberts & Hewitt 2007, in preparation). Here the line of sight magnetic field measured from fitting the Zeeman effect on the stokes V signal gives field strengths from -828 to $+864$ μG. The direction of the field with respect to the observer is seen to reverse direction on scales of a few parsecs. Extended OH(1720 MHz) emission is clearly seen as an arc connecting compact maser spots D, E and F. Near-IR H_2 and Spitzer 4.5μm images show shocked filaments corresponding to the observed extended maser emission (Reach *et al.* 2006). Given W44's relatively nearby location, strong continuum and large angular size it is particularly well suited for sensitive Zeeman measurements of the weak, extended maser emission. Typically observations of OH(1720 MHz) masers are done with snapshots or highly spaced configurations that are not sensitive to such extended structures. Sensitive Zeeman measurements of the extended maser component are an exciting direction of future study that hold the promise of resolving the structure of the magnetic field in SNR shock interactions.

Acknowledgements

We thank Ron Maddalena, Jim Braatz and the GBT operators for their help throughout these observations. Support for this work was provided by the NSF through award GSSP06-0009 from the NRAO. JWH acknowledges partial support from an IAU Grant.

References

Brogan, C. L., Frail, D. A., Goss, W. M., & Troland, T. H. 2000, ApJ, 537, 875
Claussen, M. J., Frail, D. A., Goss, W. M., & Gaume, R. A. 1997, ApJ, 489, 143
Frail, D. A., Goss, W. M., & Slysh, V. I. 1994, ApJL, 424, L111
Frail, D. A., Goss, W. M., Reynoso, E. M., Giacani, E. B., *et al.* 1996, AJ, 111, 1651
Giacani, E. B., Dubner, G. M., Kassim, N. E., Frail, D. A., *et al.* 1997, AJ, 113, 1379
Goss, W. M., & Robinson, B. J. 1968, AP Letters, 2, 81
Green, A. J., Frail, D. A., Goss, W. M., & Otrupcek, R. 1997, AJ, 114, 2058
Hewitt, J. W., Yusef-Zadeh, F., Wardle, M., *et al.* 2006, ApJ, 652, 1288
Hoffman, I. M., Goss, W. M., Brogan, C. L., *et al.* 2003, ApJ, 583, 272
Hoffman, I. M. 2004, Ph.D. Thesis,
Koralesky, B., Frail, D. A., Goss, W. M., Claussen, M. J., & Green, A. J. 1998, AJ, 116, 1323
Lazendic, J. S., Wardle, M., Burton, M. G., Yusef-Zadeh, F., *et al.* 2002, MNRAS, 331, 537
Lockett, P., Gauthier, E., & Elitzur, M. 1999, ApJ, 511, 235
Reach, W. T., *et al.* 2006, AJ, 131, 1479
Snell, R. L., Hollenbach, D., Howe, J. E., Neufeld, D. A., *et al.* 2005, ApJ, 620, 758
Wardle, M. 1999, ApJL, 525, L101
Yusef-Zadeh, F., Goss, W. M., Roberts, D. A., Robinson, B., & Frail, D. A. 1999, ApJ, 527, 172
Yusef-Zadeh, F., Wardle, M., & Roberts, D. A. 2003, ApJ, 583, 267

Astrophysical Masers and their Environments
Proceedings IAU Symposium No. 242, 2007
J.M. Chapman & W.A. Baan, eds.

doi:10.1017/S174392130701318X

VLBI studies of SiO masers around VX Sagittarii

X. Chen[1,2], Z.-Q. Shen[1] and D.-R. Jiang[1]

[1]Shanghai Astronomical Observatory, 80 Nandan Road, Shanghai 200030, PR China
email: chenxi, zshen, djiang@shao.ac.cn

[2] Department of Earth Sciences, National Taiwan Normal University, 88 Sec. 4, Ting-Chou Rd., Taipei 116, Taiwan

Abstract. Observations of 43 GHz ^{28}SiO ($v=1$, $J=1$–0) masers in the circumstellar envelope of the M-type semi-regular variable star VX Sgr were performed using the VLBA at 3 epochs during 1999 April-May. By tracing 42 matched maser features appearing in all the three epochs, we determined the contraction of an SiO maser shell toward VX Sgr at a velocity of about 4 km s^{-1} at a distance of 1.7 kpc to VX Sgr. We also report on some preliminary results from our first epoch of simultaneous VLBA observations of three 7 mm SiO masers toward VX Sgr.

Keywords. circumstellar matter — masers — stars: individual (VX Sgr)

1. VLBA observations

In order to study the kinematics of the extended atmosphere around VX Sagittarii (VX Sgr), we observed the 43 GHz ^{28}SiO ($v=1$ $J=1$–0) maser emission toward VX Sgr at 3 epochs (1999 April 24, May 23 and May 31) using the Very Long Baseline Array (VLBA). Furthermore, recently we have been performing the simultaneous VLBA observations of three 7 mm SiO masers (^{28}SiO $v=1$ $J=1$–0 (43.122 GHz), ^{28}SiO $v=2$ $J=1$–0 (42.820 GHz), and ^{29}SiO $v=0$ $J=1$–0 (42.879 GHz)) toward VX Sgr at multiple epochs to test the pumping mechanism of SiO maser emission. Here, we will present some preliminary results from the first simultaneous observation of three 7 mm lines on 2006 July 16 as well as results from three-epoch observations in 1999. All data were correlated with the FX correlator in Socorro, New Mexico (USA). The correlator output data has a spectral resolution corresponding to a velocity resolution of 0.22 km s^{-1}.

2. Results and discussions

2.1. *Contraction of SiO maser shell observed in 1999 April-May*

By identifying 42 matched common maser features among all the three epochs in 1999, we were able to estimate their proper motions (left panel in Fig. 1). Analysis of the pairwise separation (right panel in Fig. 1) of the 42 matched features suggests a contraction of the SiO maser shell toward VX Sgr at a velocity of about 4 km s^{-1} at a distance of 1.7 kpc to VX Sgr. This contraction speed can be easily obtained through the gravitational infall. This is the first direct evidence for an inward motion of the SiO maser shell around a red supergiant variable star. Interestingly, the optical phase at which the SiO maser shell around the red supergiant contracts is nearly the same as that seen in Mira variables: VX Sgr (ϕ=0.75-0.80; Chen et al. 2006), R Aqr ($\phi=$0.78-0.04; Boboltz, Diamond & Kemball 1997), TX Cam (ϕ=0.50-0.65; Diamond & Kemball 2003). This infers that the contraction of the SiO maser shell would occur during an optical stellar phase of 0.5-1, which agrees with the theoretical kinematical model results of Humphreys *et al.* (2002).

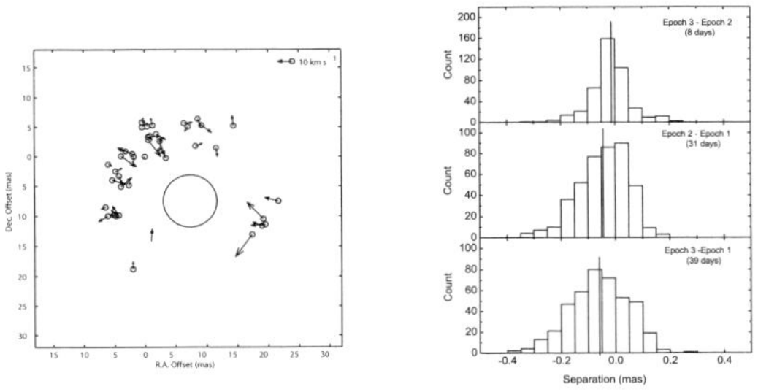

Figure 1. Left panel: The distribution of 42 proper motion vectors (Chen *et al.* 2007). Right panel: Histograms of pairwise component separations with bold line indicating the mean inward shift (Chen *et al.* 2006).

2.2. *Simultaneous observations of three 7 mm SiO masers on 2006 July 16*

The ^{29}SiO (v=0, J=1–0) maser emission was not detected, while two other 7 mm ^{28}SiO lines were clearly seen (left panel in Fig. 2). The fitted ring diameter of the ^{28}SiO (v=2, J=1–0) masers is smaller than that of the ^{28}SiO (v=1, J=1–0) masers (see right panel in Fig. 2), i.e., the ^{28}SiO (v=2, J=1–0) masers are a bit closer to the star. But this is still not adequate to constraint the dominant pumping mechanism for SiO maser emission, because both the radiative and the revised collisional pumping models can explain the systematically smaller radii for ^{28}SiO (v=2, J=1–0) masers. However, the modified collisional pumping model predicts an evolution of radii over the stellar pulsation period. Therefore, our on-going multi-epoch simultaneous VLBA observations of these lines will be critical to further distinguish between the two models.

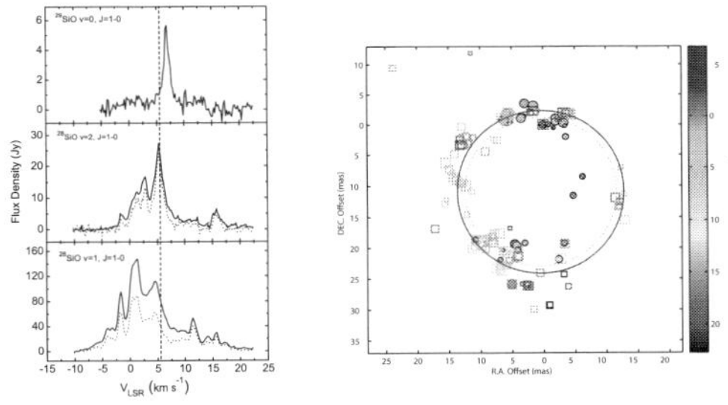

Figure 2. Left panel: The spectra of cross power imaged by VLBI (dotted line) and total power (solid line) obtained from MK antenna. Right panel: The distribution of ^{28}SiO maser features at v=2, J=1–0 transition (filled circle) aligned with that v=1, J=1–0 transition (open square).

Acknowledgement This work was supported in part by the National Natural Science Foundation of China (grants 10573029, 10625314, and 10633010) and the Knowledge Innovation Program of the Chinese Academy of Sciences (Grant No. KJCX2-YW-T03), and sponsored by Program of Shanghai Subject Chief Scientist (06XD14024).

References

Boboltz, D. A., Diamond, P. J., & Kemball, A. J. 1997, *ApJ*, 487, L147
Chen, X., Shen, Z.-Q., Imai, H., *et al.* 2006, *ApJ*, 640, 982
Chen, X., Shen, Z.-Q., & Xu, Y. 2007, *ChJAA*, (in press)
Diamond, P. J., & Kemball, A. J. 2003, *ApJ*, 599, 1372
Humphreys, E. M. L., Gray, M. D., Yates, J. A., *et al.* 2002, *A&A*, 386, 256

Astrophysical Masers and their Environments
Proceedings IAU Symposium No. 242, 2007
J.M. Chapman & W.A. Baan, eds.

doi:10.1017/S1743921307013191

Observations of SiO J=2–1 and J=3–2 masers towards evolved stars with the TRAO 14 m telescope

Se-Hyung Cho

Korean VLBI Network, Korea Astronomy and Space Science Institute, P.O. Box 88, Yonsei University, 134 Sinchon, Seodaemun, Seoul 120-749, Korea
email: cho@kasi.re.kr

Abstract. We present the results of survey and time-monitoring observations of SiO J=2–1 and J=3–2 masers towards evolved stars with the 14 m radio telescope at Taeduk Radio Astronomy Observatory (TRAO) from 1995 February to 2001 February. The first detection of SiO v=3, J=2–1 maser emission toward S-type Mira variable χ Cyg is also presented.

Keywords. circumstellar matter, masers, evolved stars, variables

1. Introduction

The SiO masers associated with evolved stars are a good probe for investigating the physical processes in the extended atmosphere of these stars. Nowadays, SiO maser emission has been detected in more than 500 evolved stars. However, these observations were mostly focused on J=1-0 and J=2-1 transitions. No systematic observational studies of J=3-2 transition have yet been performed. Therefore, we have performed J=3-2 observations in parallel with J=2-1 observations using the TRAO telescope.

2. Observations

The survey observations of SiO v=0, 1, 2, J=3–2 lines for 42 evolved stars (known SiO J=2-1 maser sources) were performed in 1995, 1996 February (Cho *et al.* 1998), and 1997, 1998 March We used a 124–174 GHz double-sideband Schottky-barrier diode mixer receiver and a 85–116 GHz single-sideband SIS receiver (Han *et al.* 1996). The half-power beamwidths and the aperture efficiencies were about 50″, 37% at 86 GHz and 40″, 33% at 128 GHz, respectively. For simultaneous time-monitoring observations of J=3–2 and J=2–1 masers from 1999 to 2001 (Kang *et al.* 2006), we employed 100/150 GHz dual-channel SIS receiver (Park *et al.* 1999). The corrected antenna temperature $T_A{}^*$ is related to the flux density as 1 K $\approx$ 54 Jy at 86 GHz and 1 K $\approx$ 62 Jy at 128 GHz, respectively. A 256 channel x 250 kHz filter bank was used as a main spectrometer. Integration time was 30–60 minutes to achieve 0.05 K rms at 3 σ level.

3. Results

The SiO v=1, J=3–2 maser was detected in 30 evolved stars for the J=3-2 survey observations from 1995 to 1998. The v=2, J=3–2 maser was detected in 9 stars, unlike the anomalously weak emission seen previously in the rare v=2, J=2–1 maser (Cho *et al.* 1998). For the simultaneous time-monitoring observations of J=2–1 and J=3–2 masers towards 10 evolved stars, the line profile and peak velocity variations of SiO v=1, J=3–2

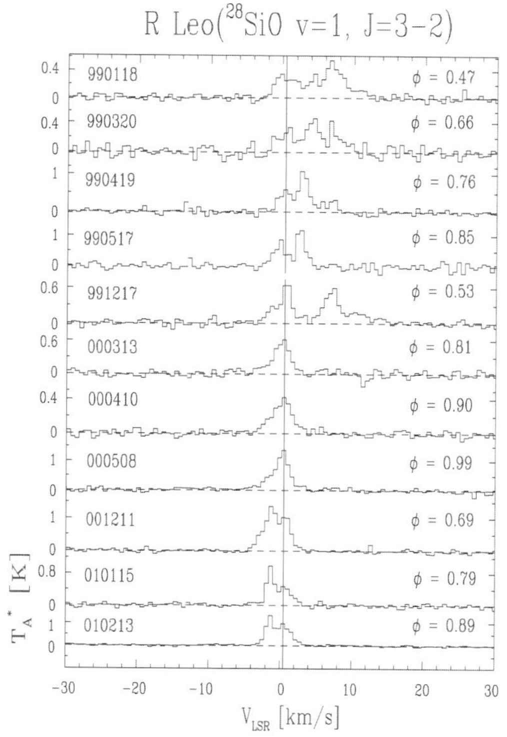

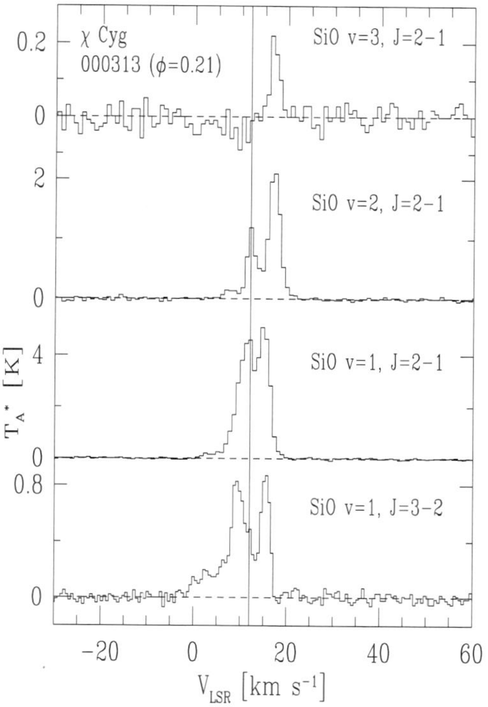

Figure 1. Sample spectra of SiO J=2-1 and J=3-2 masers obtained from time-monitoring observations (Kang *et al.* 2006)

Figure 2. Spectra of χ Cyg obtained from simultaneous observations of SiO J=2-1 and J=3-2 masers (Cho, Lee, & Park 2007)

maser with pulsation phase were different from those of the v=1, J=2–1 maser (Fig. 1). The peak and integrated antenna temperature (PT and IT) ratios between rotational ladders are averaged to be PT(v=1, J=3–2)/PT(v=1, J=2–1) $\approx$ 0.29, and IT(v=1, J=3–2)/IT(v=1, J=2–1) $\approx$ 0.21, respectively (Kang *et al.* 2006). In addition, SiO v=3, J=2–1 maser emission was detected for the first time toward S-type Mira variable χ Cyg as a single spike (Fig. 2). These SiO J=2–1 and J=3–2 results by the 14 m telescope will be a fundamental basis for one key science area for the main observational bands of the Korean VLBI Network (KVN).

References

Cho, S.-H., Chung, H. S., Kim, H. R. & Oh, B. Y. 1998, *ApJS* 115, 277

Cho, S.-H., Lee, C. W., & Park, Y. S. 2007, *ApJ* 657, 482

Han, S.-T., Lee, C. H., Kim, H.-R., Park, D.-C. 1996, *Int. J. Infrared Millimeter Waves* 17, 105

Kang, J., Cho, S.-H., Kim, H.-G., Chung, H. S., Kim, H. R., Roh, D.-G., Lee, C. W., & Kim, S.-J. 2006, *ApJS* 165, 360

Park, J. A., Han, S. T., Kim, T. S., Kim, H. R., Chung, H. S., Cho, S.-H., Lee, C. H., & Yang, J. 1999, *Int. J. Infrared Millimeter Waves* 20, 1769

Astrophysical Masers and their Environments
Proceedings IAU Symposium No. 242, 2007
J.M. Chapman & W.A. Baan, eds.

doi:10.1017/S1743921307013208

A database of circumstellar OH masers

Dieter Engels and Felix Bunzel

Hamburger Sternwarte, Gojenbergsweg 112, D–21029 Hamburg, Germany
email: dengels,fbunzel@hs.uni-hamburg.de

Abstract. We present a new database of circumstellar OH masers at 1612, 1665, and 1667 MHz. The database contains 10774 observations and 2274 stars with OH maser emission detected. The database contains flux densities and velocities of the two strongest maser peaks, the expansion velocity of the shell and the radial velocity of the star. Access to the database is possible over the Web (www.hs.uni-hamburg.de/maserdb), allowing cone searches for individual objects and lists of objects. Object selection is possible on the base of flux densities and velocities.

Keywords. masers, catalogs, stars: AGB and post-AGB, stars: late-type

1. Introduction

Since the discovery of masers 40 years ago several thousand observations have been made to detect masers in circumstellar shells. Comfortable tools to access these observations are lacking. The last catalogs listing 1612 MHz OH masers in AGB stars ("OH/IR stars") were published by te Lintel Hekkert *et al.* (1989) and Benson *et al.* (1990). The te Lintel Hekkert *et al.* catalog contains detected OH masers with their flux densities and velocities and covers the literature until 1984. The Benson *et al.* catalog lists references to OH maser observations and covers the years until 1989. The number of detected 1612 OH Masers listed are 439 and 713 respectively. Since then the number of detected masers has been almost tripled.

2. The database

The search of the literature started with the contents of the te Lintel Hekkert *et al.* catalog. The additional literature search covers the years 1984 – 2006. The database is considered to be (almost) complete for 1612 MHz detections, but contains no non-detections published prior to 1984. For the main lines only measurements published after 1984 are contained. Only galactic masers were taken into account.

The database contains the coordinates of the maser, as well as flux densities and velocities of the two strongest peaks. For the 1612 MHz masers these are usually the outermost peaks, from which the radial velocity of the stars and the expansion velocity of the circumstellar shells are calculated. For more complex spectra, as usually seen for main-line masers, radial and expansion velocities are calculated from the outermost peaks listed in the reference paper. Modifications to the original data have been logged on the reference pages. A page compiling statistical data (galactic distribution, velocity distributions etc.) gives an overview on the content of the database.

Database search is possible according to frequencies, coordinates, flux densities, and velocities.

Acknowledgements

This work is supported by the Deutsche Forschungsgemeinschaft under grant number En 176/33-1.

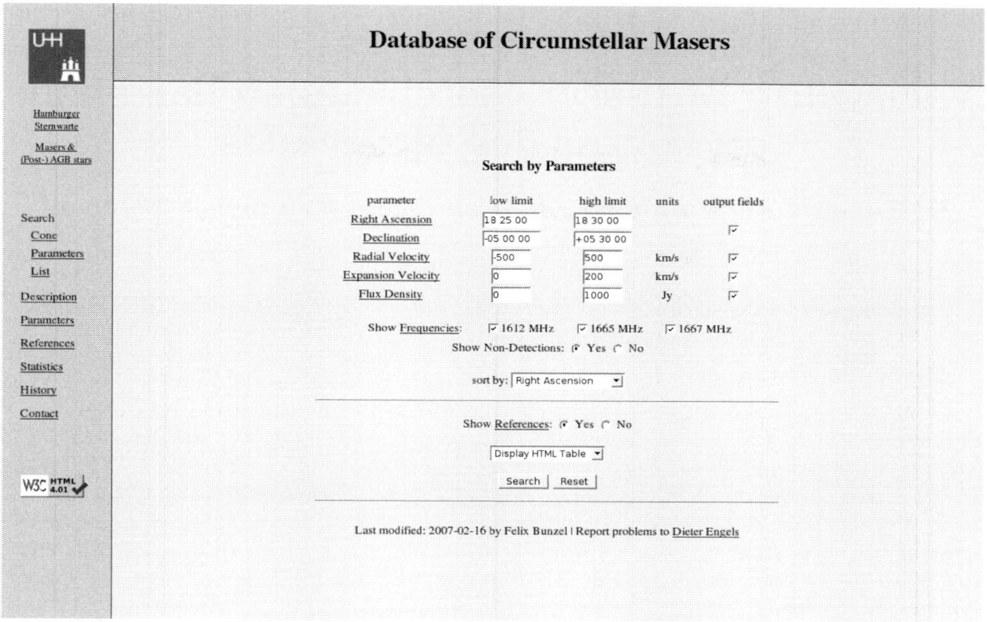

Figure 1. Input mask. **Adress: www.hs.uni-hamburg.de/maserdb**

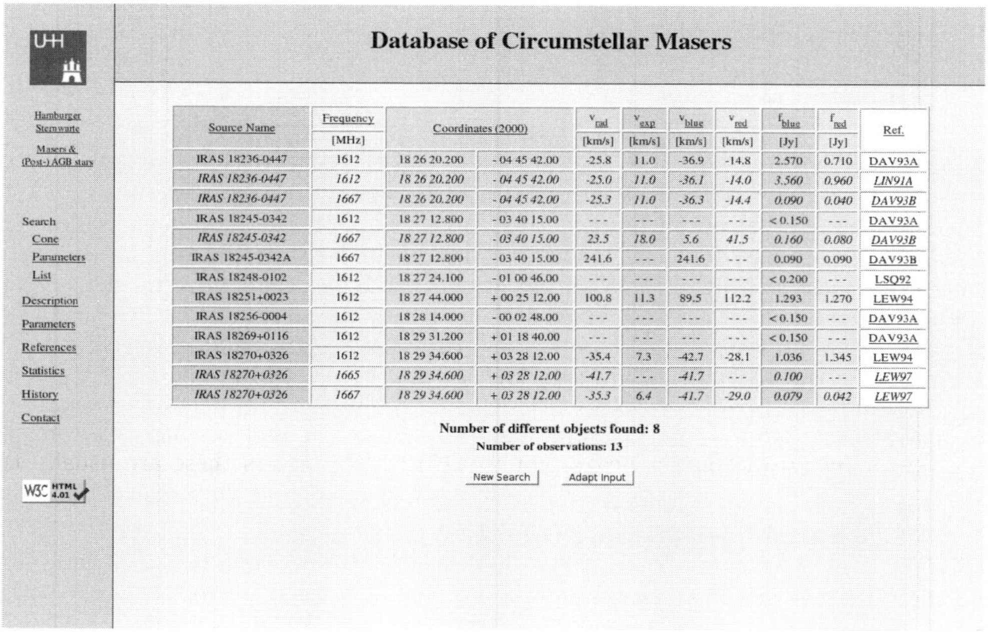

Source Name	Frequency [MHz]	Coordinates (2000)		v_{rad} [km/s]	v_{exp} [km/s]	v_{blue} [km/s]	v_{red} [km/s]	f_{blue} [Jy]	f_{red} [Jy]	Ref.
IRAS 18236-0447	1612	18 26 20.200	- 04 45 42.00	-25.8	11.0	-36.9	-14.8	2.570	0.710	DAV93A
IRAS 18236-0447	*1612*	*18 26 20.200*	*- 04 45 42.00*	*-25.0*	*11.0*	*-36.1*	*-14.0*	*3.560*	*0.960*	*LIN91A*
IRAS 18236-0447	*1667*	*18 26 20.200*	*- 04 45 42.00*	*-25.3*	*11.0*	*-36.3*	*-14.4*	*0.090*	*0.040*	*DAV93B*
IRAS 18245-0342	1612	18 27 12.800	- 03 40 15.00	- - -	- - -	- - -	- - -	< 0.150	- - -	DAV93A
IRAS 18245-0342	*1667*	*18 27 12.800*	*- 03 40 15.00*	*23.5*	*18.0*	*5.6*	*41.5*	*0.160*	*0.080*	*DAV93B*
IRAS 18245-0342A	1667	18 27 12.800	- 03 40 15.00	241.6	- - -	241.6	- - -	0.090	0.090	DAV93B
IRAS 18248-0102	1612	18 27 24.100	- 01 00 46.00	- - -	- - -	- - -	- - -	< 0.200	- - -	LSQ92
IRAS 18251+0023	1612	18 27 44.000	+ 00 25 12.00	100.8	11.3	89.5	112.2	1.293	1.270	LEW94
IRAS 18256-0004	1612	18 28 14.000	- 00 02 48.00	- - -	- - -	- - -	- - -	< 0.150	- - -	DAV93A
IRAS 18269+0116	1612	18 29 31.200	+ 01 18 40.00	- - -	- - -	- - -	- - -	< 0.150	- - -	DAV93A
IRAS 18270+0326	1612	18 29 34.600	+ 03 28 12.00	-35.4	7.3	-42.7	-28.1	1.036	1.345	LEW94
IRAS 18270+0326	*1665*	*18 29 34.600*	*+ 03 28 12.00*	*-41.7*	*- - -*	*-41.7*	*- - -*	*0.100*	*- - -*	*LEW97*
IRAS 18270+0326	*1667*	*18 29 34.600*	*+ 03 28 12.00*	*-35.3*	*6.4*	*-41.7*	*-29.0*	*0.079*	*0.042*	*LEW97*

Figure 2. Sample output

References

Benson, P. J., Little-Marenin, I. R., Woods, T. C., Attridge, J. M., Blais, K. A., Rudolph, D. B., Rubiera, M. E., & Keefe, H. L. 1990, *ApJS* 74, 911

te Lintel Hekkert, P., Versteege-Hensel, H. A., Habing, H. J., & Wiertz, M. 1989, *A&AS* 78, 399

Astrophysical Masers and their Environments
Proceedings IAU Symposium No. 242, 2007
J.M. Chapman & W.A. Baan, eds.

doi:10.1017/S174392130701321X

Variability of masers in circumstellar shells on timescales of decades

D. Engels[1], A. Winnberg[2], J. Brand[3] and F. Jiménez-Esteban[4]

[1]Hamburger Sternwarte, Gojenbergsweg 112, D–21029 Hamburg, Germany
email: dengels@hs.uni-hamburg.de

[2]Onsala Rymdobservatorium, Observatorievägen, S–43992 Onsala,at 1612 MHz Sweden

[3]INAF – Istituto di Radioastronomia, Via Gobetti 101, I–40129 Bologna, Italy

[4]Observatorio Astronomico Nacional, Apartado 112, E–28803 Alcala de Henares, Spain

Abstract. Based on a 15–20 year long monitoring program of H_2O masers and new observations of OH masers discovered 35 years ago, we studied the variability of maser emission in AGB stars over timescales of decades. The H_2O maser features in the semi-regular variable stars RX Boo and SV Peg, and of the Mira stars U Her and R Cas, showed strong fluctuations superposed in case of the Mira stars on regular variations due to the pulsation of the stars. The spatial distribution of the emission regions in RX Boo and U Her showed deviations from spherical symmetry, which remained unchanged over >10 years. We conclude that the spatial asymmetry is determined by the underlying asymmetry of the mass loss process. There is no evidence that mass loss rates or the wind geometry in these stars have varied over the last 20 years.
In 2005 we re-detected at 1612 MHz the 114 OH/IR stars discovered before 1978, implying an OH maser lifetime of >2800 years (1σ).

Keywords. masers, stars: AGB and post-AGB, stars: late-type

Maser emission from OH, H_2O and SiO molecules is frequently observed for stars on the Asymptotic Giant Branch (AGB). Interestingly, a large fraction of oxygen-rich AGB stars do not exhibit detectable masers or emit only in one or two of these molecules. The reasons for the absence of masers among apparently similar stars (e.g. similar distances, luminosities, mass loss rates) is not known. Recently the idea that the (non-)detection of masers is a temporary effect has gained support by the disappearance of five 1612 MHz OH masers, when re-observed 12 years after their first detection (Lewis 2002). Lewis concluded that the mean lifetime of 1612 OH emission is a mere 100–400 years, implying that masers in general might be short-lived phenomena. Long-term monitoring programs and new observations of known masers should then uncover more stars which do no longer have masers. This question can be addressed with our systematic study of the variability of H_2O maser properties of several AGB stars. We also made a simple test to determine the lower limit of the lifetime of 1612 OH masers in OH/IR stars.

The 22 GHz H_2O maser lines of the semi-regular variable RX Boo were observed with the Effelsberg and Medicina radiotelescopes between 1987 and 2007, complemented by five maps obtained 1990–1995 with the VLA (Winnberg *et al.* 2007). The maser was strongly variable with peak flux densities varying between a few Jy and >1000 Jy (Fig. 1a). In each of the VLA maps we identified 10–15 unresolved maser spots. Spots common in consecutive maps were used to align the maps taken in 1990–92 (Fig. 1b). Most of the emission comes from an incomplete ring of inner radius 91 mas (15 AU), showing a velocity gradient in a NW–SE direction. We modelled the maser region as a shell with thickness 14–35 AU ($\approx 4-12R_\star$), which is only partially complete. The wind crossing time is 18 years. Combining our 5 maps with one obtained in 1985 (Bowers *et al.* 1993)

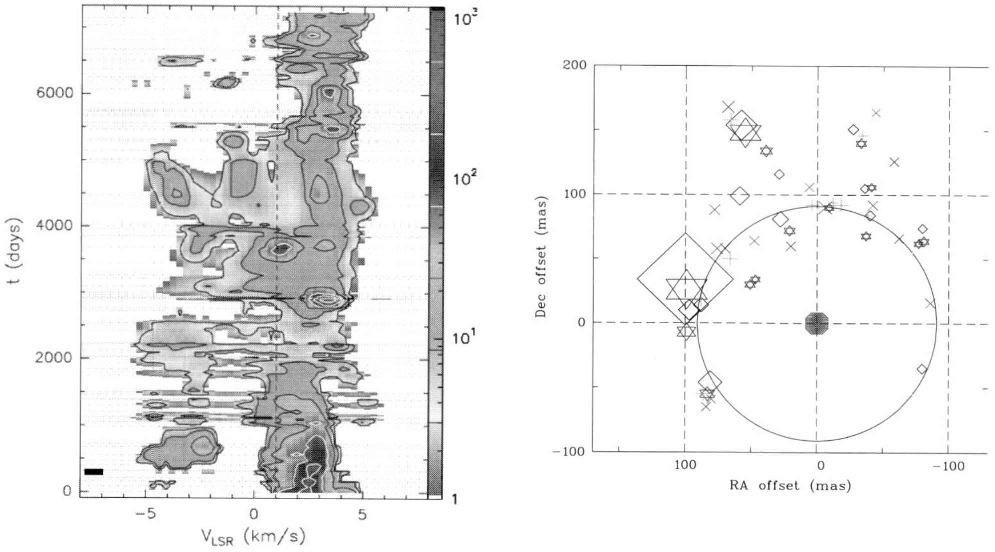

Figure 1. *Left:* Flux density variations of the H_2O masers of RX Boo from single-dish observations. The time axis starts on 1987 March 30. The dashed horizontal lines mark the actual observations and flux densities in between were interpolated in steps of 4 days. The dashed vertical line marks the stellar systemic velocity. *Right:* All maser components identified from four VLA maps taken between 1990 and 1992. The component sizes are scaled according to their strengths and the observing epochs are marked by different symbols and colors. The assumed location of the star is shown at the origin.

we find that structure and velocity gradient remained stable over at least 11 years, while the maser line profiles varied strongly. This implies that the spatial distribution observed at a given epoch is not accidental due to the random process of maser cloud formation, but that the mass loss process in RX Boo itself is not spherically symmetric. The results for RX Boo are in agreement with the properties of the H_2O maser variability that we have observed in the Mira stars U Her (Engels *et al.* 1999) and R Cas (Brand *et al.* 2002).

To assess the stability of the 1612 OH maser emission in OH/IR stars we re-observed in 2005 at Effelsberg a sample of 116 OH masers discovered before 1978 (Baud *et al.* 1981). We rediscovered all OH/IR stars (N=114) in the sample (Engels & Jiménez-Esteban 2007). Probabilities that all masers are re-detected for lifetimes T are: 1% for T= 700 years, 10% for T=1400 years, and 32% for T=2800 years. The short lifetimes found by Lewis (2002) are therefore excluded for this sample. However, the stability of the OH masers depends probably on gas density, and hence on mass loss rates. Lewis found 'dead OH/IR stars' only among the blue objects of his sample, which have a much lower mass loss rate than the typical star of our sample. The two results may therefore not be contradictory.

References

Baud, B., Habing, H. J., Matthews, H. E., & Winnberg, A. 1981, *A&AS* 95, 156
Bowers, P. F., Claussen, M. J. & Johnston, K. J. 1993, *AJ* 105, 284
Brand, J., Baldacci, L. & Engels, D. 2002 in V. Migenes, M. Reid (eds.), IAU Symp. 206, p. 202
Engels, D., *et al.* 1999 in T. Le Bertre *et al.* (eds.), IAU Symp. 191, p. 373
Engels, D. & Jiménez-Esteban, F. 2007, *A&A* submitted
Lewis, B. M. 2002, *ApJ* 576, 445
Winnberg, A., Engels, D., Brand, J., *et al.* 2007, in preparation

Astrophysical Masers and their Environments
Proceedings IAU Symposium No. 242, 2007
J.M. Chapman & W.A. Baan, eds.

doi:10.1017/S1743921307013221

Infrared OH absorption lines in 1612 MHz OH maser sources

J. H. He[1,2], P. S. Chen[1], R. Szczerba[3] and A. M. Sobolev[4]

[1]National Astronomical Observatories/Yunnan Observatory, CAS, PR China
[2]Academia Sinica, Institute of Astronomy and Astrophysics, Taiwan, China.
email: jhhe@asiaa.sinica.edu.tw
[3]N. Copernicus Astronomical Center, Rabianska 8, 87-100 Torun, Poland
[4]Astronomical Observatory, Ural State University, Lenin Street 51, Ekaterinburg, Russia

Abstract. ISO mid-infrared absorption data helps to distinguish between radiative and collisional pumping mechanisms in 1612 MHz OH masers in various environments. Archive data of OH absorption at 34.6 and 53.3 μm shows different behavior and different pumping rates for different types of maser sources.

1. Mid-Infrared absorption in 1612 MHz OH masers

Mid-infrared spectroscopy by the Infrared Space Observatory (ISO) first provided a way to directly distinguish between infrared pumping and collisional pumping in OH masers in various astrophysical environments using the presence of IR absorption lines.

The ISO archive spectra for all known Galactic OH 1612 MHz masers have been searches for OH related absorption. Absorption at 34.6 μm is detected by the ISO Short Wavelength Spectrometer (SWS) in only 5 sources (out of 87) while at 53.3 μm absorption is detected by the ISO Long wavelength Spectrometer (LWS) in 11 sources (out of 48). The non-detection of the two lines in most of the sources is due to the low spectral resolution and/or sensitivity of the ISO instruments. However, we found respectively 16 and 7 sources with spectra of good enough quality that do not show the 34.6 and 53.3 μm absorption lines (He *et al.* 2005, He & Chen 2004) possibly due to clumpy shell or limb-filling emission.

The spectra of the detected lines are reproduced in Fig. 1 (He *et al.* 2005, He & Chen 2004). Panel A shows the five 34.6 μm line profiles while Panels B and C present the 53.3 μm line profiles of the 11 sources. The spectra of the 53.3 μm spectra in panels B and C are grouped according to the following classification: *RaBe* – redshifted absorption and blueshifted infilling emission, *Asymm* – asymmetrical profile, *GCMega* – symmetrical profile of GC or megamasers, and *BaRe* – blueshifted absorption and redshifted infilling emission. The weak infilling emission features are found to be ubiquitous: blueshifted *RaBe* infilling in stellar OH masers, and redshifted *Asymm* or *BaRe* infilling in interstellar OH masers. Infilling is not present in the symmetric *GCMega* profiles of GC and megamaser sources. The shifted infilling emission is possibly due to emission in only one of the 53.3 μm doublet components. The alternative of a velocity shift of about 560 km/s ($\approx 0.1\,\mu$m) from the nominal line center is unreasonable in such sources. However, it is not clear how such doublet asymmetry arises in the two types of OH masers. The OH maser pumping routes tracing technique developed by M. Gray (2007, these proceedings) may provide an answer.

The exact pump rate may be defined as: *the integrated photon flux of the blue+red OH maser peaks divided by the integrated 35+53 μm absorption photon flux.* The pump rate was calculated for six sources with both infrared OH absorption lines (at least tentatively)

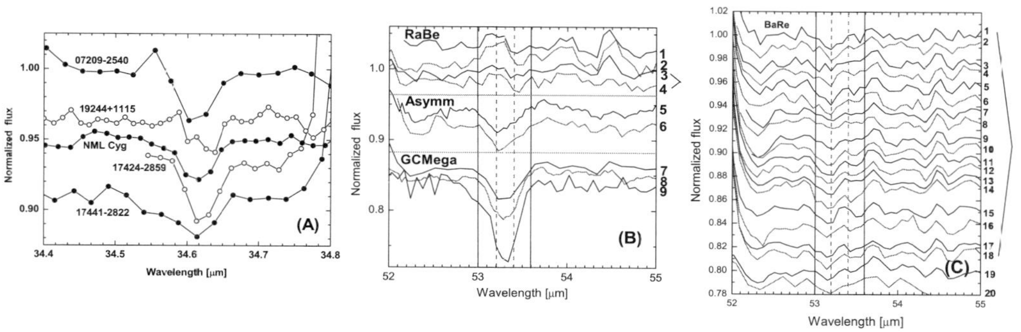

Figure 1. Infrared OH absorption line spectra at 34.6 μm (A) and 53.3 μm (B & C). The classification *RaBe*, *Asymm*, *GCMega*, *BaRe* is explained in the text. The vertical lines mark the range of the line while the two dashed lines mark the blue and red shifted components respectively. IRAS names associated with the spectra: Panel B (1) 03507+1115 (AGB), (2) 07209-2540 (RSG), (3,4) NML Cyg (RSG), (5) 17424-2852 (H II), (6) 17430-2848 (cluster), (7) 17424-2859 (GC), (8) 17441-2822 (GC), (9) Arp 220 (Galaxy); Panel C (1-18) 20255+3712 (H II), (19) 06053-0622 (H II), (20) 17574-2403 (H II).

detected. The pump rates of three stellar OH maser sources (two red supergiants 07209-2540 and NML Cyg and one yellow hypergiant 19244+1115) are very similar (0.079, 0.041 and 0.054, respectively), while the two GC sources ($17424-2859$ and $17441-2822$) have much smaller rates (3.56×10^{-6} and 1.56×10^{-5}, respectively). The pump rate of the interacting galaxy Arp 220 of 0.32 is much larger than any of the considered sources. The similar pump rates of stellar masers suggest radiative pumping of the stellar masers, since the infrared absorption is proportional to maser emission (Elitzur *et al.* 1976). The varying (small) pump rate of GC sources may be due to interstellar absorption by non-maser OH cloudlets (He 2005). The large pump rate of Arp 220 may result from strong beaming (A.M. Sobolev and W.A. Baan – private communication).

2. Conclusions

Improved mid-infrared spectrometry is needed to resolve the weak 34.6 and 53.3 μm lines in many galactic OH 1612 MHz maser sources. A limited sample of detections shows ubiquitous infilling emission in the 53.3 μm absorption line profiles of galactic OH masers with blueshifted features in stellar masers and redshifted features in interstellar OH masers. Infilling emission features can be interpreted as emission of a single 53.3 μm doublet component. Similar pump rates among stellar OH 1612 MHz masers demonstrate an effective radiative pumping mechanism. Observational and theoretical studies are needed to confirm/interpret the weak line features and *true* non-detection sources.

Acknowledgments This work was partly supported by the Chinese National Science Foundation under Grant No. 10433030 and 10503011.

References

Elitzur, M., Goldreich, P., & Scoville, N. 1976, *ApJ* 205, 384
Gray, M. 2007, *IAUS* 242
He, J. H. 2005, *New Astronomy* 10, 283
He, J. H., Chen, P. S. 2004, *New Astronomy* 9, 545
He, J. H., Szczerba, R., Chen, P. S., & Sobolev, A. M. 2005, *A&A* 434, 201

Astrophysical Masers and their Environments
Proceedings IAU Symposium No. 242, 2007
J.M. Chapman & W.A. Baan, eds.

doi:10.1017/S1743921307013233

VLBI phase-referencing observations of SiO masers towards R Aquarii

Ryuichi Kamohara and the VERA team

Mizusawa VERA Observatory, National Astronomical Observatory of Japan,
VERA Observatory, NAOJ, Mitaka, Tokyo, JAPAN
email: ryuichi.kamohara@nao.ac.jp

Abstract. We show a phase-referenced image of the SiO maser emissions towards the Mira variable R Aqr. These data have been obtained using the VLBI Exploration of Radio Astrometry (VERA). The proper motion we have obtained is different from that obtained using the HST. The proper motion from 2004 to 2005 is not different from that of 1991 to 2005 and supports the current orbital elements.

Keywords. SiO maser, late-type star, symbiotic Binary

1. Motivation

R Aqr is one of the brightest SiO maser sources and located close to the bright extra-galactic source J2348-1631. The star is a symbiotic stellar system, but the orbital motion has not been detected directly. This project tries to detect it and estimate its orbital elements using the VERA. We also have evaluated the performance of VERA at 43 GHz.

2. Observations and results

The $v = 1, 2$ $J = 1\text{-}0$ SiO maser emissions towards R Aqr and the continuum source, J2348-1631, were observed simultaneously with 4 antennas of NAOJ VERA on 2004 December 23 and 2005 December 24. The SiO maser emission data was recorded in a 8 MHz band, and the continuum source data with a 224 MHz bandwidth. Data analysis was done in the NRAO AIPS package. The resultant synthesized beam for the image is 0.94×0.41 mas with a major-axis position angle of $-27.6°$.

We have obtained the spatial distributions of the SiO maser emission for both epochs. We also obtained the stellar position by circular fitting. The positions of the star relative to the tracking center ($23^h 43^m 49^s.4616 - 15^d 17' 04''.202$ (J2000)) are listed in Table 1.

Table 1. Stellar position relative to a tracking center

Date	Relat. RA (mas)	Relat. DEC (mas)	Diam. (mas)
23 Dec. 2004	136.7 (1.8)	−132.7 (0.7)	19.2
24 Dec. 2005	167.1 (0.6)	−162 (2.0)	14.5

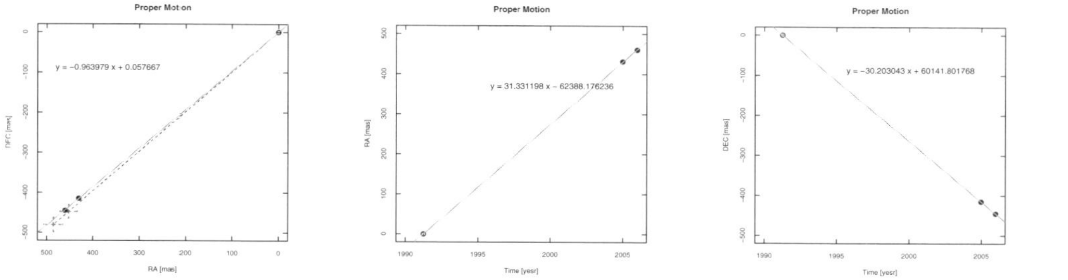

Figure 1. The proper motion of R Aqr. The origin of these figures is ($\alpha_{J2000} = 23^{h}43^{m}49^{s}.440$, $\delta_{J2000} = -15°17'03''.9$). Cross-dotted marks show the positions that are estimated from the proper motion reported by HST at each epochs. *Left*: Dec vs. RA *Center*: RA vs. time *Right*: Dec vs. time.

3. Discussion

3.1. *Proper motion*

We compare the position obtained from our observations with the astrometric position (Figure 1) identified by Hipparcos on 1991.25 (Perryman *et al.* 1997). Figure 1 also shows the positions estimated from the proper motion which is reported by the Hipparcos in each epochs. The positions we have obtained are different from the estimated positions by about 40 mas. We find a linear variation of our positions and the Hipparcos astrometric position at 1991.25. We have obtained a linear fit yielding a proper motion of $(\mu_\alpha, \mu_\delta) = (31.33 \pm 0.12, -30.20 \pm 0.12)[\text{mas yr}^{-1}]$. The differences of proper motions between out observation and the HST observation are -1.65 mas for RA, and 2.41 mas for Dec.

3.2. *Orbital motion*

From 2004.98 to 2005.98, the star move 29.3±2.0 mas in RA and -30.0 ±2.1 mas in Declination. This motion is consistent with the proper motion between 1991.25 and 2005.98 within the error. We could not detect an orbital motion for R Aqr. Hollis *et al.* (1997) proposed orbital elements of this system of $(P, a, e, T, i, \Omega, \omega, K) = (44\ \text{yr}, 2.54 \times 10^{14}\text{cm}, 0.8, 2442100.0\ \text{JD}, 70°, \sim 90°, -90°, 7.2\ \text{km s}^{-1})$. The position of R Aqr shifted 23.0 mas to the east and 0.9 mas to the south. The motion is consistent with -1.56 mas yr^{-1} in RA and 0.06 mas yr^{-1} in Dec. Positional shifts due to the orbital motion between 2004.98 and 2005.98 are -1.5 mas in RA and 0.6 mas in Dec. The difference of the orbital motion between these epochs is so small that we think we could not detect the motion in the observation.

4. Summary

We have observed SiO maser emissions towards the symbiotic star R Aqr using the VERA. We have obtained a proper motion of the star from 1991.25 to 2005.98. The proper motion is different from that reported from the HST observation. Proper motions between 1991.25 and 2005.98 and between 2004.98 and 2005.98 coincide with each other within the error. This result supports the orbital elements proposed by Hollis *et al.* (1997).

References

Hollis, J. M., Pedelty, J. A., & Lyon, R. G. 1997, *ApJ* (Letters) 482, L85
Perryman, M. A. C., *et al.* 1997, *A&A* (Letters) 323, L49

Astrophysical Masers and their Environments
Proceedings IAU Symposium No. 242, 2007
J.M. Chapman & W.A. Baan, eds.

doi:10.1017/S1743921307013245

Full polarization VLBA maps of a proto-planetary nebula

Kim McAlpine[1], A. J. Kemball[2] and J. L. Jonas[1]

[1]Department of Physics and Electronics, Rhodes University, Grahamstown, 6139, South Africa

[2]National Center for Supercomputing Applications, University of Illinois at Urbana-Champaign, 605 East Springfield Avenue, Champaign, IL 61820, USA

Abstract. The morphological evolution of stars from the asymptotic giant branch (AGB) to the planetary nebula (PN) phases presents an intriguing problem in stellar research. Planetary nebulae show bright-rims, well-defined shell-like structures and a large proportion of them are axisymmetric. In contrast their progenitor AGB stars are largely spherically symmetric (Sahai, 2004). Studies of proto-planetary nebulae, which are objects in transition between these two phases of stellar evolution, offer insight into the mechanisms which are responsible for the onset of axisymmetry. This poster presents the first polarization VLBI observations of the 1612 MHz OH maser emission from the proto-planetary nebula candidate OH O.9+1.3. The morphology, kinematics and polarization properties of these masers are discussed.

Keywords. proto-planetary nebulae, OH masers

1. Introduction

The period immediately following the asymptotic giant branch (AGB) is one of the least understood phases of stellar evolution and there remain many open questions regarding the transformation of AGB stars into planetary nebulae. Many promising models detailing the morphological evolution of late-type stars invoke the role of stellar magnetic fields (e.g. Matt *et al.* 2004; García-Segura *et al.* 2005). Maser polarization measurements can be of considerable use in establishing the morphology and magnitude of such fields. Thus the polarization VLBI maps of OH 0.9+1.3 presented in this work were used to investigate the morphology and kinematics of this source with a view to gaining insight into the relative dynamic influence of magnetic fields and the development of asymmetry in post-AGB objects.

2. Observations

The source was observed with the VLBA on the 24th and 25th of October 1994. The total observing run lasted 12 hours and included interleaved observations of continuum calibrators. The observations were done in spectral line mode and the data was recorded in dual circular polarization in two overlapping spectral windows of 250 kHz each.

3. Results

The spectrum of OH 0.9+1.3 consists of several strong, $\sim$ 50 Jy, blue-shifted peaks at ~ -124 km.s^{-1} and a considerably weaker red-shifted emission feature of 600 mJy at -93 km.s^{-1}. Contour maps of the Stokes I emission of both the blue- and red-shifted emission from this source are presented in Figure 1.

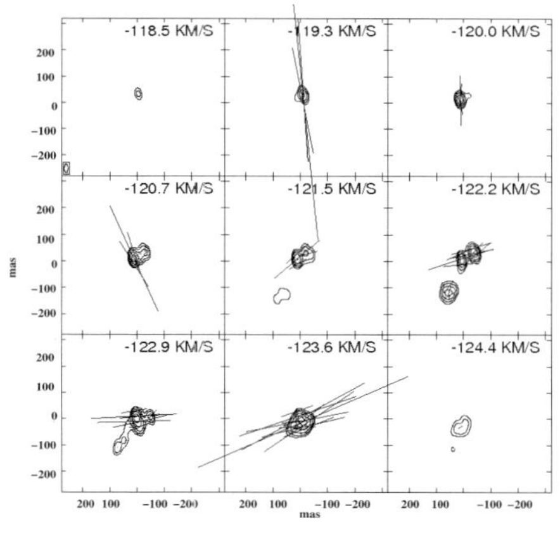

(a)

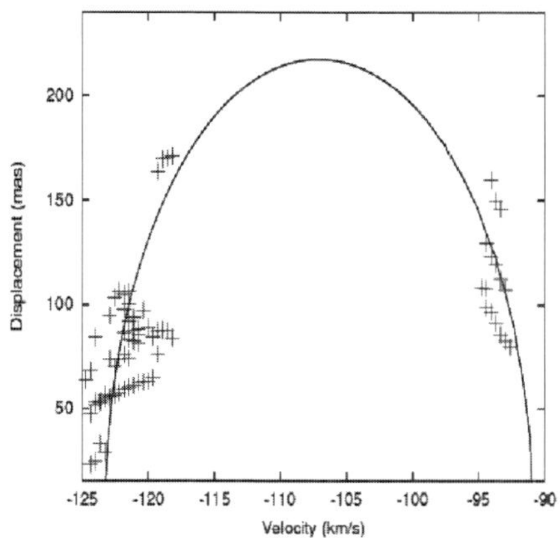

(b)

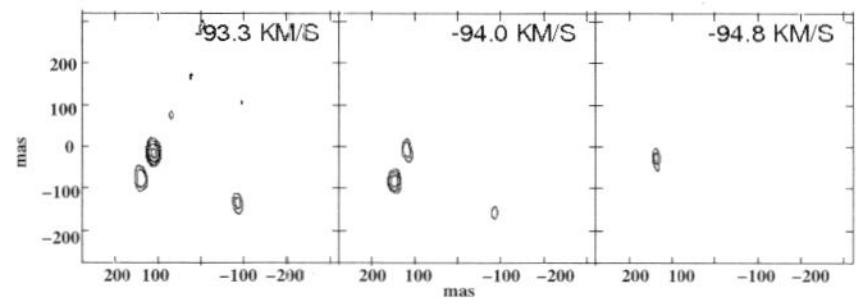

(c)

Figure 1. a) Contour maps of blue-shifted emission from OH 0.9+1.3. Contours are plotted at 0.368 ×(-1,1,2,4,8,16,32) Jy/Beam. The magnitude and angle of the linear polarization vectors are shown by the length and direction of the vectors; here 1 mas = 2.00 mJy/Beam of polarized emission. b) An expanding shell model fit to the maser components. The solid line indicates the model and the crosses represent the displacement of the maser components relative to the assumed stellar position. c) Contour maps of red-shifted emission from OH O.9+1.3. Contours are plotted at 19.3 ×(-1,1,2,4,8,16,32) mJy/Beam.

4. Conclusion

The total intensity maser maps demonstrated a considerable degree of asymmetry with the blue- and red-shifted emission located in spatially distinct regions of the envelope. The morphology of this source is not consistent with the standard symmetric thin-shell model and an attempt to fit the traditional OH/IR kinematic model of a simple expanding shell to the maser components was found to be unsatisfactory. The fractional linear and circular polarizations of the maser components were derived from the Stokes parameter maps. In all except one of the components the total fractional polarization was found to be low ($< 15\%$). The mean fractional linear and circular polarization were calculated to be 5.5% and 7.1% respectively. The absence of an identifiable Zeeman pair in the Stokes V map prohibited the estimation of the magnetic field in this source.

Acknowledgements

I would like to acknowledge the financial assistance of Rhodes University, in the form of a Henderson award, and SKA South Africa during the course of the research.

References

Sahai, R. 2004, *ASP CS* 313, 141
García-Segura, G., López, J. A. & Franco, J. 2005 *ApJ* 618, 919
Matt, S., Frank, A. & Blackman, E. G. 2004 *ASP CS* 313, 449

Astrophysical Masers and their Environments
Proceedings IAU Symposium No. 242, 2007
J.M. Chapman & W.A. Baan, eds.

doi:10.1017/S1743921307013257

VLBI monitoring of Mira variables with VERA

A. Nakagawa[1], T. Omodaka[1], K. M. Shibata[2], T. Kurayama[2], H. Imai[1], S. Kameno[1], M. Tsushima[1], M. Shintani[1], N. Matsumoto[1], M. Matsui[1], S. Oizumi[1], T. Yasuda[1], Y. Arao[1] and the VERA project[2]

[1]Faculty of science, Kagoshima University, 1-21-35, Korimoto, Kagoshima, 890-0065, JAPAN
email:nakagawa@astro.sci.kagoshima-u.ac.jp

[2]Mizusawa VERA observatory, National Astronomical Observatory of Japan, 2-21-1 Osawa, Mitaka, Tokyo, 181-8588, JAPAN

Abstract. We have started a VLBI monitoring program for Asymptotic Giant Branch (AGB) stars at 22 and 43 GHz as part of a project of the VLBI Exploration of Radio Astrometry (VERA) for precisely obtaining the period-luminosity (PL) relation of Galactic Mira variables. Using accurate distances measured with VERA, we reveal PL relation in the Galaxy based on the absolute magnitudes of the sources. We selected the sources for VLBI monitoring so that they have a good coverage of various pulsation periods. Photometry in the infrared J, H, and K bands for over 600 AGB stars has also started since 2003 with the 1m telescope of Kagoshima University to obtain the pulsation periods and magnitudes. Current analysis of the phase referencing VLBI observations of S Crt shows that the parallax of 2.3 ± 0.2 milliarcsec (mas) corresponds to a distance of 435^{+41}_{-35} pc. From the infrared monitoring data, pulsation periods and magnitudes in K band for 248 sources were obtained.

Keywords. astrometry, masers, AGB and post-AGB, interferometric, photometric

1. Introduction

Mira variables are stars pulsating with periods in the range 100 to 1000 days and showing rapid mass loss before ejecting their outer layers as planetary nebula shells. Although a narrow PL relation in the Large Magellanic Cloud (LMC) was reported by van Leeuwen *et al.* (1997), the same relation for the Galactic Miras have not been precisely obtained (Feast *et al.* 1989). Such uncertainty arises from the ambiguities of absolute magnitudes suffering directly from inaccurate distances of the objects. Using absolute magnitudes derived from accurate distance measurements with VERA, we can precisely investigate PL relation of the Galactic Miras. Furthermore, this program is important in a sense that once we have provided the calibration for the PL relation as for use as a relative distance estimator, we can convert the relative distance measurements to an absolute scale.

2. Observations in radio and infrared

Single dish flux monitoring of ~500 sources at 22 and 43 GHz for grasping the radio bright state are in progress and one third of these samples are observed monthly at 22 GHz. This monitoring is used to select sources that are bright enough to give successful detections of VLBI fringes. The unique dual beam system installed in VERA allows simultaneous observations of target maser and extragalactic reference sources.

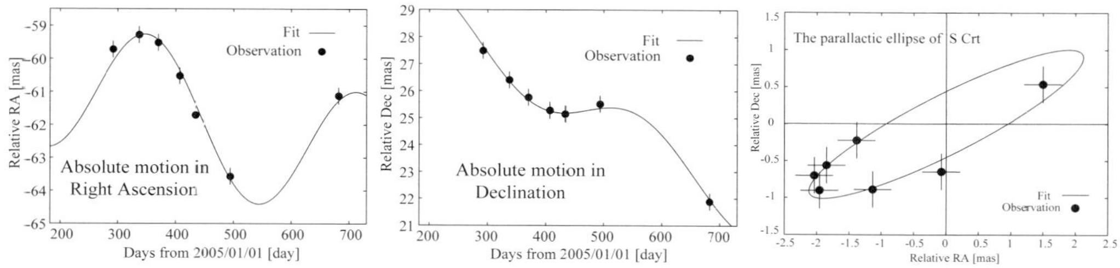

Figure 1. Position offset of the 34.7 $\mathrm{km s^{-1}}$ maser spot in S Crt in right ascension and declination are presented with filled circles in left and middle panels, respectively. Solid lines show a best fit model to the measured positions. The right panel is the parallactic ellipse of S Crt.

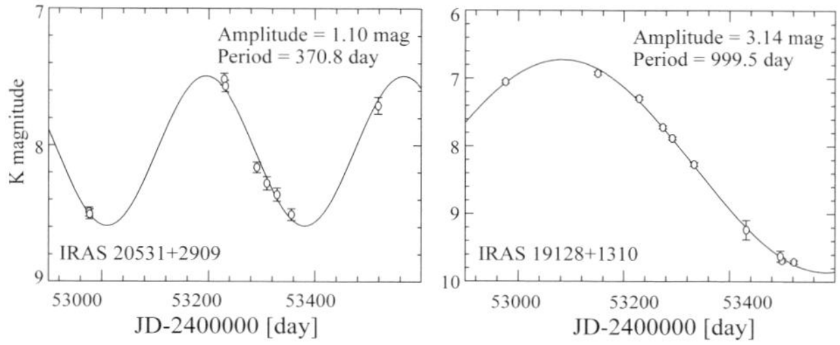

Figure 2. Light curves in the infrared (2.2 μm) of IRAS 20531+2909 (left) and IRAS 19128+1310 (right) indicate a pulsation period of 370.8 and 999.5 days, respectively.

Photometries in the infrared J, H, and K bands (λ=1.2, 1.6, and 2.2 μm, respectively) are also in progress. The sources for photometry are selected mainly from the IIIa, IIIb, IV, and VIb regions of the two-color diagram of the IRAS point source catalog (van der Veen & Habing 1988), where sources have a high mass loss ratio and are expected to be accompanied by H_2O/SiO masers.

3. Results

Multi-epoch VERA observations revealed the absolute motion of a maser spot in S Crt using the extragalactic continuum source J1147−0724 as a reference (see figure 1). The measured parallax of 2.3±0.2 mas corresponds to the distance of 435^{+41}_{-35} pc. Zenith atmospheric delay residuals, the structure of the maser, and model errors account well for the positional errors in our analysis. Our distance measurement of S Crt is larger than previous results of 285 pc (Patel *et al.* 1992) and 420 pc (Bowers & Johnston 1994) estimated from the PL relation. We are conducting monthly phase referencing VLBI monitoring for the following sources, AP Lyn, R UMa, Z Pup, SY Scl, and WX Psc.

Figure 2 represents examples of the light curves obtained with our infrared monitoring. We have obtained pulsation periods ($\leqslant$1000 days) and magnitudes for more than 200 sources at the present time and continuing work will find the sources with even longer periods. We are planning radio surveys for these sources to find new VLBI targets.

References

Bowers, P. F., & Johnston, K. J. 1994, *ApJS* 92, 189
Feast, M. W., Glass, I. S., Whitelock, P. A., & Catchpole, R. M. 1989, *MNRAS* 241, 375
Patel, N. A., Joseph, A., & Ganesan, R. 1992, *JA&A* 13, 241
van der Veen, W. E. C. J., & Habing, H. J. 1988, *A&A* 194, 125
van Leeuwen, F., Feast, M. W., Whitelock, P. A., & Yudin, B. 1997, *MNRAS* 287, 955

Astrophysical Masers and their Environments
Proceedings IAU Symposium No. 242, 2007
J.M. Chapman & W.A. Baan, eds.

doi:10.1017/S1743921307013269

A VLBI polarization study of SiO masers towards VY CMa

L. L. Richter[1], A. J. Kemball[2] and J. L. Jonas[1]

[1]Physics and Electronics Department, Rhodes University, Grahamstown, 6140, South Africa

[2]National Center for Supercomputing Applications, University of Illinois at Urbana-Champaign, 1025 W. Clark Street, Urbana, IL 61801, USA

Abstract. Maser emission from the SiO molecule has been widely observed in the near-circumstellar envelopes of late-type, evolved stars. VLBI images can resolve individual SiO maser spots, providing information about the kinematics and magnetic field in the extended atmospheres of these stars. This poster presents full polarization images of several SiO maser lines towards the supergiant star VY CMa. VY CMa is a particularly strong SiO maser source and allows observations of a wide range of maser transitions. We discuss implications of these observations for VY CMa morphology, polarization, and pumping models.

Keywords. stars: late-type, polarization

1. Introduction

SiO maser emission has been widely observed in the near-circumstellar envelopes of late-type evolved stars. VLBI images can resolve individual SiO maser spots around these stars, providing information about the kinematics and magnetic field close to the stellar surface at high spatial resolution. Comparison of the emission from different SiO maser lines may provide insight into the pumping mechanisms providing the population inversion necessary for the masers and also holds promise in discriminating between competing SiO maser polarization models.

2. Observations

The source VY CMa was observed with the Very Long Baseline Array (VLBA) on 2 December 1994 (epoch 1) and 20 and 23 December 2003 (epoch 2). The data were reduced and imaged with AIPS following the technique described by Kemball & Diamond (1995). Several SiO maser lines were observed, three of which are shown in Figure 1. Unfortunately there was very little data for each of the epoch 2 lines and the signal to noise ratio is lower than it is in epoch 1.

3. Discussion

The intensity of the imaged maser emission in epoch 1 was significantly higher than the intensity of any of the epoch 2 lines. The epoch 2 v=1 J=1-0 and v=2 J=1-0 emission has a very similar spatial distribution. These maps in Figure 1 were aligned by correlating the images, channel by channel. There is no obvious trend in the direction of the polarization vectors between epochs.

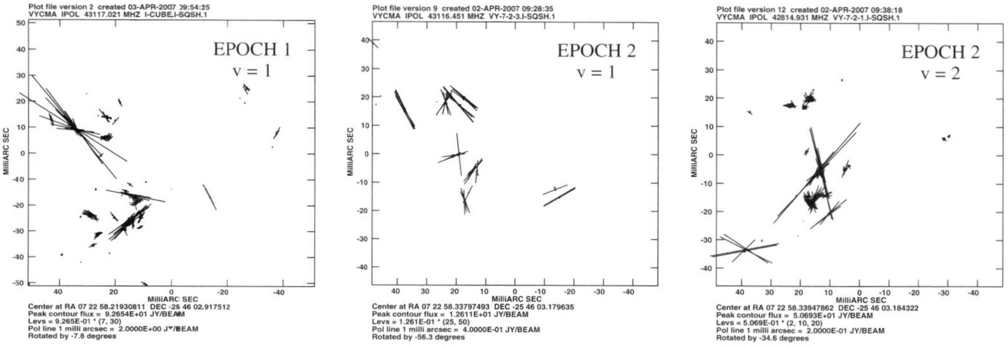

Figure 1. 43 GHz J=1-0 SiO masers towards VY CMa. The intensity is given by the contours and the magnitude and direction of the linear polarisation are represented the vectors.

3.1. *Pumping models*

The primary pumping mechanism driving SiO maser emission remains an issue of dispute in the literature. Models of masers in the circumstellar environment make use of both radiative pumping (e.g. Bujarrabal 1994) and collisional pumping (e.g. Lockett & Elitzur 1992; Doel *et al.* 1995).

One means to identify the primary pumping mechanism at work is to compare the spatial distribution of different vibrationally-excited SiO maser lines. The spatial coincidence of the rotational transitions in different vibrational lines, such as the v=1 and v=2 J=1-0 lines, would argue against a radiative pumping model (Bujarrabal 1994; Doel *et al.* 1995). Kinematic models of the SiO maser emission, using collisional pumping, predict that the v=1 J=1-0 emission should lie further from the star than the v=2 J=1-0 emission, in a thicker shell (Gray & Humphreys 2000).

In the epoch 2 maps above, the v=1 J=1-0 and v=2 J=1-0 maps have many overlapping features and a similar overall distribution. The overlap of several features in the aligned v=2 J=1-0 and v=1 J=1-0 maps argues against purely radiative pumping. Where the v=1 and v=2 features overlap, the v=2 features generally extend inwards further than the v=1 features.

We cannot draw definitive conclusions from comparisons of different rotational lines at just one epoch. Further observations in this area are continuing.

Acknowledgements

I gratefully acknowledge the support of the NRF, the Fuchs Foundation and the Rhodes University Physics Department.

References

Bujarrabal, V. 1994, *A&A* 285, 953

Doel, R. C., Gray, M. D., Humphreys, E. M. L., Braithwaite, M. F. & Field, D. 1995, *A&A* 302, 797

Gray, M. D. & Humphreys, E. M. L. 2000, *New Astron.* 5, 155

Kemball, A. J., Diamond, P. J. & Cotton, W. D. 1995, *A&AS* 110, 383

Lockett, P. & Elitzur, M. 1992, *ApJ*, 399, 704

Astrophysical Masers and their Environments
Proceedings IAU Symposium No. 242, 2007
J.M. Chapman & W.A. Baan, eds.

doi:10.1017/S1743921307013270

Twenty-six-year monitoring of water masers

G. M. Rudnitskij[1], M. I. Pashchenko[1], V. F. Esipov[1], V. A. Samodurov[2], I. A. Subaev[2], A. M. Tolmachev[2] and E. E. Lekht[1,3]

[1]Sternberg Astronomical Institute, Moscow State University, 13 Universitetskij prospekt, Moscow, 119992 Russia
email: gmr@sai.msu.ru

[2]Pushchino Radio Astronomy Observatory, Astrospace Center of the Lebedev Institute of Physics, Russian Academy of Sciences, Pushchino, Moscow Region, 142290 Russia
email: sam@prao.ru

[3]Instituto Nacional de Astrofísica, Óptica y Electrónica, Luis Enrique Erro No. 1, Apdo Postal 51 y 216, 72840 Tonantzintla, Puebla, México
email: lekht@inaoep.mx

Abstract. Since 1980 variability of a sample of H_2O maser sources has been monitored on the 22-metre radio telescope in Pushchino, Russia. The interval between successive observational sessions is 1–2 months. The sample includes 125 maser sources in star-forming regions (SFR) and late-type variable stars. Twenty-six-year time series of H_2O line profiles have detected flares and velocity drift of spectral features. Very fast variations in the H_2O maser flux ($\Delta t \lesssim 1$ hour) have been detected in several SFR sources, in particular, W33B. Variations of circumstellar H_2O masers in late-type stars correlate with visual light curves with a time lag of 0.3–0.4P (P is the star's period). Exceptionally strong H_2O maser flares were recorded in SFR sources (Sgr B2 and others) and in the stars W Hya, R Cas and U Ori. Models for H_2O maser variability are reviewed. For stellar masers shock-wave excitation of H_2O line variability is discussed.

Keywords. molecular processes, radio lines: ISM, radio lines: stars, masers, surveys, stars: formation, stars: AGB and post-AGB, circumstellar matter, shock waves

---

Since 1980 our team has been monitoring cosmic maser sources radiating in the 6_{16}–5_{23} rotational line of the water vapour molecule at $\lambda = 1.35$ cm (Lekht *et al.* 1982; Berulis *et al.* 1983). The observations are carried out on the 22-metre radio telescope of the Pushchino Radio Astronomy Observatory. The sensitivity is about 10 Jy, the velocity resolution is 0.081 km/s. The interval between successive observational sessions is 1–2 months.

The sample includes 125 maser sources of which 65 are in star-forming regions (Ori A, W75N, G43.8−0.1, NGC 2071 etc.) and the rest are in circumstellar envelopes of late-type variable stars (U Ori, RS Vir, RT Vir, U Her, VY CMa, VX Sgr etc.). For most sources H_2O line profiles have been measured for more than 200 epochs between 1980 and 2006.

For masers in star-forming regions the long time series have revealed some previously unknown regularities in the variations of the H_2O line profiles (flares and velocity drift of individual spectral features). Figure 1 (*left*) presents examples of strong maser flares observed in the SFR source Sgr B2 (Ramírez Hernández *et al.* 2005). In some SFR sources (S255, NGC 7538) models of rotating proto-planetary discs and/or bipolar outflows are applicable (Lekht *et al.* 2004). Velocity drifts of maser features may reflect orbital motion of proto-planetary condensations. Dedicated observations allowed us to detect very fast variations in the maser flux of some sources (in particular, Sgr B2 and W33B) on a timescale shorter than 1 hour (Ramírez Hernández *et al.* 2005; Samodurov *et al.* 2007).

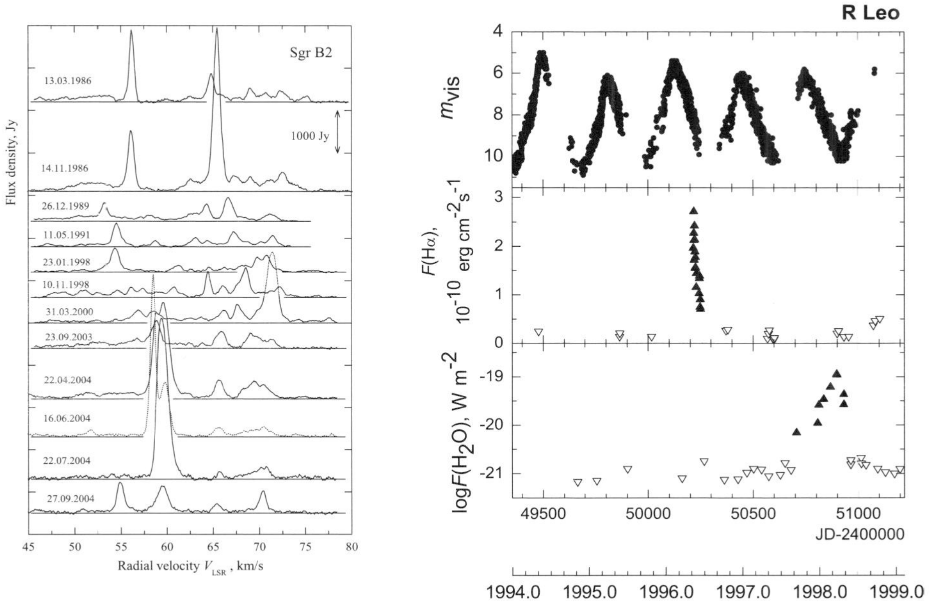

Figure 1. Flares of the H_2O maser sources Sgr B2 and R Leo.

Variations of circumstellar H_2O masers in late-type stars (U Ori, S CrB, U Her, RS Vir and others) correlate with the visual light curves with a time lag of 0.3–0.4P (P is the star's period). Exceptionally strong H_2O maser flares were noted in W Hya, R Cas and U Ori. Parallel optical spectroscopy of the stars has been carried out since 1994. Flares of the Hα line emission recorded in R Leo, U Aur, R Cas and R LMi were followed 1.5–2 years later by corresponding flares of the H_2O masers. Figure 1 (*right*) shows an example of such a flare in the Mira-type star R Leo (Esipov *et al.* 1999). These phenomena are interpreted as a consequence of propagation of a shock wave driven by stellar pulsation. A strong shock propagating in the inner layers of a circumstellar envelope produces optical emission lines; further out, as the shock becomes weaker, it excites the H_2O maser.

Acknowledgements

This work is supported by the Russian Foundation for Basic Research, project codes 06-02-16806 and 07-02-08014. The RT-22 telescope is sponsored by the Ministry of Education and Science of the Russian Federation, facility registration number 01-10.

References

Berulis, I. I., Lekht, E. E., Pashchenko, M. I. & Rudnitskij, G. M. 1983, *Soviet Astron.* 27, 179

Esipov, V. F., Pashchenko, M. I., Rudnitskij, G. M. & Fomin, S. V. 1999, *Astron. Lett.* 25, 672

Lekht E. E., Pashchenko, M. I., Rudnitskij, G. M. & Sorochenko, R. L. 1982, *Soviet Astron.* 26, 168

Lekht, E. E., Munitsyn, V. A. & Tolmachev A. M. 2004, *Astron. Rep.* 48, 200

Ramírez Hernández, O., Lekht, E. E. & Tolmachev, A. M. 2005, *Astron. Rep.* 49, 777

Samodurov, V. A., Rudnitskij, G. M., Tolmachev, A. M., Subaev, I. A., Lekht, E. E., Pashchenko, M. I. & Logvinenko, S. V. 2007, *Astron. Astrophys. Trans.*, in press

Astrophysical Masers and their Environments
Proceedings IAU Symposium No. 242, 2007
J.M. Chapman & W.A. Baan, eds.

doi:10.1017/S1743921307013282

SiO maser emission in oxygen AGB stars

R. Soria-Ruiz[1], F. Colomer[2], J. Alcolea[3], V. Bujarrabal[2], and J.-F. Desmurs[3]

[1]Joint Institute for VLBI in Europe, Postbus 2, 7990 AA Dwingeloo, The Netherlands
email: soria@jive.nl

[2]Observatorio Astronómico Nacional, Apartado 112, E-28803, Alcalá de Henares, Spain
[3]Observatorio Astronómico Nacional, Alfonso XII 3, E-28014, Madrid, Spain

Abstract. We present high resolution observations of SiO masers in the late-type star R Cas as part of a multi-line study of SiO circumstellar masers in oxygen-rich AGB stars. We used the NRAO Very Long Baseline Array (VLBA) to map the 43 GHz maser emission of ^{28}SiO. We study the relative spatial distribution of the different maser lines and discuss the relevance of line overlaps in the SiO pumping theory.

Keywords. technique: interferometric – stars: circumstellar matter – stars: AGB

1. Introduction

SiO maser emission is mainly detected in the circumstellar envelopes around evolved stars, which are formed due to the mass loss process occurring during the AGB phase. This emission is produced in the innermost shells of these envelopes, at a few stellar radii from the central star. It is well established that the SiO maser emission is distributed in ring structures composed of many compact spots with very high brightness temperatures, therefore being ideal targets for VLBI studies. Maser amplification is present in ^{28}SiO, ^{29}SiO and ^{30}SiO in many rotational lines extensively studied with single-dish telescopes. From these observations, we know the strong correlation between the SiO maser emission and the IR 8μm stellar radiation or its time variability (Pardo *et al.* 2004).

2. Observations and results

High resolution observations of the ^{28}SiO v=1 and v=2 J=1–0 maser transitions were carried out in January 2005 with the NRAO† VLBA in R Cas, obtaining an angular resolution of $\sim$0.17 *mas*. The calibration was done following the standard procedures for spectral line VLBI.

When comparing the two maps, the v=2 emission is produced in an inner part of the envelope and the spatial spot distribution is similar (see Fig. 1). This shift between the masing regions is consistent with previous results in other O-type stars (see e.g. Desmurs *et al.* 2000 and Soria-Ruiz *et al.* 2007) and in R Cas (Yi *et al.* 2006).

3. Discussion

Currently, two types of pumping mechanisms are considered as the primary mechanism responsible for the amplification: the stellar radiation at 8 μm in radiative models (Bujarrabal 1994), or the collisions with the H_2 molecules in the collisional models

† The National Radio Astronomy Observatory is a facility of the National Science Foundation operated under cooperative agreement by Associated Universities, Inc.

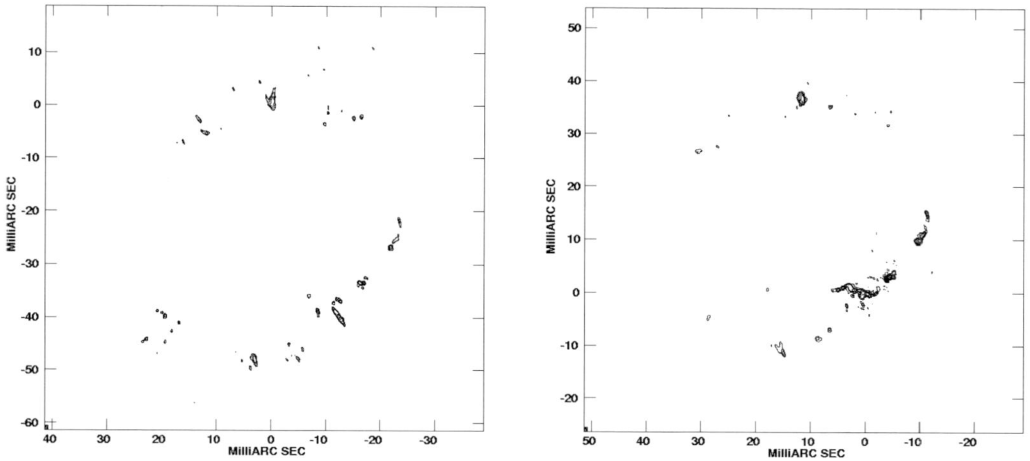

Figure 1. ^{28}SiO v=1 (left) and v=2 (right) J=1–0 maser emission in R Cas. The figure shows the integrated intensity maps in Jy beam^{-1} units. For the ^{28}SiO v=1 J=1–0 map, the peak intensity is 7.55 Jy beam^{-1} and the contour levels are -8, 8, 10, 20, 40 and 80% of the peak. For the ^{28}SiO v=2 J=1–0 map, the peak intensity is 6.20 Jy beam^{-1} and the contour levels are -3, 3, 6, 9, 18, 36 and 72% of the peak.

(Humphreys *et al.* 2002). In all cases, the inversion of the levels in the maser transition is produced when the Δv=1 de-excitation is opaque. Because of the nature of the inversion mechanism, the physical conditions for the arising of SiO masers within the same v-state are similar regardless of the model. They occur when the corresponding $v \rightarrow v$–1 lines become optically thick, in a similar way for the different J levels. On the contrary, for masers in different vibrationally excited estates, the conditions must be naturally different. For this reason, the emissions in the v=1 and v=2 vibrationally states, which are separated about 1800 K, are expected to be produced in different regions of the envelope. However, the v=1 and v=2 J=1–0 lines present relatively similar, although not identical, distributions, while that of the v=1 J=2–1 line (Soria-Ruiz *et al.* 2004) is clearly different; these results are obviously in contradiction with the theoretical predictions. On the other hand, further calculations of the excitation of the SiO molecule in AGB stars have shown that the conditions under which different maser transitions occur, change when the line overlap between infrared lines of H_2O and ^{28}SiO is introduced in the pumping models and could explain the observed relative distributions (Bujarrabal *et al.* 1996, Soria-Ruiz *et al.* 2004).

Finally, although our new maps support the relevance of line overlaps in the SiO maser pumping in O–rich shells, it would be desirable to perform similar studies in a larger sample including all types of long-period variable stars, namely, Mira-type, semi-regular and irregular variables, as well as supergiant stars.

References

Bujarrabal, V. 1994, *A&A*, 285, 953
Bujarrabal, V., Alcolea, J., Sánchez Contreras, C., *et al.* 1996, *A&A*, 314, 883
Desmurs, J.-F., Bujarrabal, V., Colomer, F., & Alcolea, J. 2000, *A&A*, 360, 189
Humphreys, E. M. L., Gray, M. D., Yates, J. A., *et al.* 2002, *A&A*, 386, 256
Pardo, J. R., Alcolea, J., Bujarrabal, V., *et al.* 2004, *A&A*, 424, 145
Soria-Ruiz, R., Alcolea, J., Colomer, F., *et al.* 2004, *A&A*, 426, 131
Soria-Ruiz, R., Alcolea, J., Colomer, F., *et al.* 2007, *A&A*, in press
Yi, J., Booth, R., & Conway, J., 8th European VLBI Network Symposium, Torun, 2006

Astrophysical Masers and their Environments
Proceedings IAU Symposium No. 242, 2007
J.M. Chapman & W.A. Baan, eds.

doi:10.1017/S1743921307013294

Maser emission towards the young planetary nebula IRAS 17347-3139

D. Tafoya[1,2], Y. Gómez[2], J. F. Gómez[3], I. de Gregorio-Monsalvo[3], L. Uscanga[2], G. Anglada[4] and J. M. Torelles[5]

[1]Harvard-Smithsonian Center for Astrophysics, 60 Garden Street, Cambridge, MA 02138, USA

[2]Centro de Radioastronomía y Astrofísica, UNAM Apartado Postal 3-72 (Xangari), 58089 Morelia, Michoacán, México

[3]Laboratorio de Astrofísica Espacial y Física Fundamental INTA, Spain

[4]Instituto de Astrofísica de Andalucía, CSIC Apartado 3004, E-18080, Granada, Spain

[5]Instituto de Ciencias del Espacio, Consejo Superior de Investigaciones Científicas-Instit D'Estudis Espacials de Catalunya, (CSIC-IEEC), Gran Capità 2, E-08034 Barcelona, Spain

Abstract. We have re-analyzed observations of the water maser emission in IRAS 17347-3139 carried out previously and compared them with new higher angular resolution and more sensitive radio-continuum observations to explain the shift between the position of the peak of the radio-continuum with respect to the center of the ring-like distribution. From our analysis, we found that the water maser emission is not distributed in a closed ring-like structure, but in what appears to be a segment of an ellipse centered in the position of the peak of the radio-continuum. These results can explain the shift between the radio-continuum and the water maser emission. We also present interferometric observations of the OH maser emission towards IRAS 17347-3139 from which we have found that the OH maser emission detected previously is not associated with this source. Furthermore, since our observations are more sensitive, we have detected for the first time weak OH maser emission associated with IRAS 17347-3139.

Keywords. planetary nebula: individual (IRAS 17347-3139) — masers — radio continuum: stars

1. Introduction

IRAS 17347-3139 is one of the two planetary nebula in which water maser emission is known to be present (de Gregorio-Monsalvo *et al.* 2004). Their VLA observations revealed that the water maser emission is distributed in what seems to be a ring-like structure perpendicular to the lobes observed in the near-IR images. Interestingly, it was also found that the peak of the radio-continuum emission was located not in the geometrical center of such a distribution but towards one of its edges. One suggestion proposed to explain such displacement with the radio-continuum emission being associated with one source, and the water maser emission with a secondary source. However, due to the lack of angular resolution of the observations, it was not possible to arrive at a definitive conclusion. On the other hand, Zijlstra *et al.* (1989), and more recently Szymczak & Gérard (2004), reported the detection of OH maser emission associated with IRAS 17347-3139. However, since the observations were carried out with a single dish and have low angular resolution, the association of such emission with this source was not completely confirmed. In order to understand how the maser emission is distributed in this source, we decided to carry out higher angular resolution observations.

2. Observations and results

On 2005 January 27, we used the VLA in the hybrid configuration, BnA, to carry out high angular observations of the radio continuum at wavelengths 3.6, 1.3 and 0.7 cm, as well as the four OH maser transitions towards the planetary nebula IRAS 17347-3139. The angular resolution ranges from $\simeq 3'.5$ to $\simeq 0'.1$ at λ=18 and 0.7 cm, respectively. In addition, we retrieved the water maser emission data of this source from the VLA archive to re-analyze them and compare them with our observations.

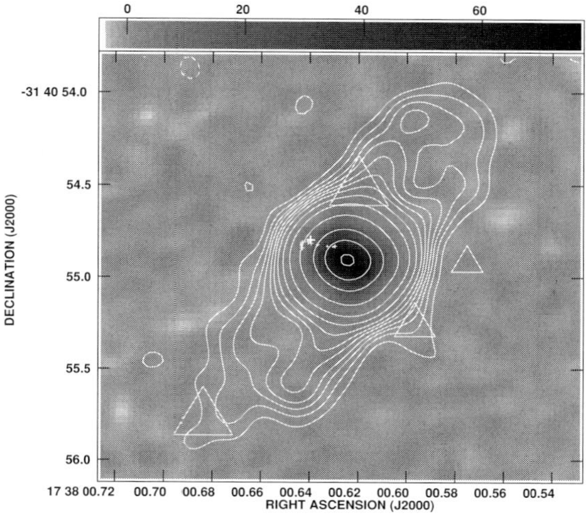

Figure 1. VLA image of IRAS 17347-3139. White contours: λ=1.3 cm, the contours are -3, 3, 5, 7, 9, 12, 15, 20, 30, 50, 100, 200, 300 and 400 times 2.5×10^{-4} Jy/beam. Grey scale: λ=0.7 cm. For the water and OH masers, the position of the emission from the different channels are shown in this image as plus signs and triangles, respectively.

In Figure 1, we show an image of the radio-continuum emission of IRAS 17347-3139 at λ =1.3 and 0.7 cm in contours and grey scale, respectively. The water maser emission is shown as plus signs towards the north-east of the peak of the continuum emission. From our analysis, we have found that the masers are distributed in what appears to be a segment of an ellipse of size $\simeq 0''.1 \times 0''.5$ centered at the peak of the continuum. This result, along with the image at λ =0.7 cm, suggests that the water maser emission is arising in a circumstellar high density torus-like structure.

Finally, from the four OH maser transitions, we have detected emission of the 1612 MHz line at three velocity ranges: i) from 5 to -24 km s^{-1}, ii) $\simeq -70$ km s^{-1} and iii) from -90 to -123 km s^{-1}. The emission in the first range is located $\simeq 11'$ south of the position of IRAS 17347-3139, and it has been associated with the OH/IR star OH 356.65-0.15 (Bowers & Knapp 1989). The emission in the third range is located $2'.4$ to the north-east of the position of IRAS 17347-3139, probably too far to be associated to this source. The only emission that coincides with the position of IRAS 17347-3139 is at $v_{LSR} \simeq -70\,\mathrm{km\,s^{-1}}$, which is detected for the first time with our observations and is shown as triangles in Figure 1.

References

Bowers P. F. & Knapp G. R., ApJ, 347, 325

de Gregorio-Monsalvo, I., Gómez, Y., Anglada, G., Cesaroni, R., Miranda, L. F., Gómez, J. F. & Torrelles, J. M. 2004, ApJ, 601, 921

Szymczak & Gérard, 2004, A&A, 423, 209

Zijlstra, A. A., Te Lintel Hekkert, P., Pottasch, S. R., Caswell, J. L., Ratag, M., & Habing, H. J., 1989, A&A, 217, 157

Astrophysical Masers and their Environments
Proceedings IAU Symposium No. 242, 2007
J.M. Chapman & W.A. Baan, eds.

doi:10.1017/S1743921307013300

Collisional excitation of OH(6049 MHz) masers in supernova remnant – molecular cloud interactions

Mark Wardle

Department of Physics, Macquarie University, Sydney, NSW 2109, Australia
email: wardle@physics.mq.edu.au

Abstract. OH (1720 MHz) masers serve as indicators of SNR – molecular cloud interaction sites. These masers are collisionally excited in warm (50–100 K) shocked gas with densities of order $10^5\,\mathrm{cm}^{-3}$ when the OH column density is in the range 10^{16}–$10^{17}\,\mathrm{cm}^{-2}$. Here I present excitation calculations which show that when the OH column density exceeds $10^{17}\,\mathrm{cm}^{-2}$ at similar densities and temperatures, the inversion of the 1720 MHz line switches off and instead the 6049 MHz transition in the first excited rotational state of OH becomes inverted. This line may serve as a complementary signal of warm, shocked gas when the OH column density is large.

Keywords. supernova remnants, molecular processes, radiation mechanisms: nonthermal, radio lines: ISM

The OH excitation calculations discussed here include the 32 lowest energy levels, with transition wavelengths and A values from Detombes *et al.* (1977) and Brown *et al.* (1982), and rates for collisional de-excitation by H_2 kindly provided by Alison Offer (private communication). Following Lockett & Elitzur (1989) and Lockett, Gauthier & Elitzur (1999), I adopt a uniform slab model for the masing medium. Radiative transfer is approximated using escape probabilities based on the mean optical depth, including the effects of line overlap. The slab is parametrised by density n_H, temperature T, column N_OH, velocity width, and a radiation field contributed by the CMB and warm dust.

Figs. 1 and 2 show results for $T = 50\,\mathrm{K}$, assuming small velocity gradients within the slab, and no far-infrared radiation from dust. These conditions are broadly consistent with warm gas that is cooling off behind a shock wave driven into a molecular cloud by an adjacent supernova remnant. Fig. 1 shows the maser optical depth through the slab for the 1720 MHz satellite line in the ground rotational state and its analogues in the first (6049 MHz) and second (4765 MHz) excited rotational states. At low OH column densities the 1720 MHz line is inverted, peaking in the range 10^{16}–$10^{17}\,\mathrm{cm}^{-2}$. At about $10^{17}\,\mathrm{cm}^{-2}$ the optical depth of the slab to the 1720 MHz line suppresses the inversion. Meanwhile the inversion in the 6049 MHz line grows, peaking at $N_\mathrm{OH} \approx 3 \times 10^{17}\,\mathrm{cm}^{-2}$ (similar results were found by Pavlakis & Kylafis 2000). The 4765 MHz line peaks at $N_\mathrm{OH} \sim 10^{18}\,\mathrm{cm}^{-2}$. The effect of varying n_H and N_OH is explored in Fig. 2. As expected, masing in the 1720 MHz line is strongest for $n_\mathrm{H} \sim 10^5\,\mathrm{cm}^{-3}$ and $N_\mathrm{OH} \sim 10^{16.5}\,\mathrm{cm}^{-2}$. The inversion of the 6049 MHz line requires column densities in excess of $10^{17}\,\mathrm{cm}^{-2}$, while inversion at 4765 MHz (not shown) requires an OH column a few times higher still.

OH (1720 MHz) masers are a signature of SNR – molecular cloud interactions (Frail *et al.* 1994; Wardle & Yusef-Zadeh 2002). Masing at 6049 MHz may take over this role when the OH column density exceeds $10^{17}\,\mathrm{cm}^{-2}$, too high for 1720 MHz masers to exist. Note, however, that the OH column density typically produced in SNR–cloud interactions is uncertain. Existing models of OH production rely on UV or X-ray dissociation of the water produced in molecular shocks, and yield column densities $\lesssim 10^{16}\,\mathrm{cm}^{-2}$ (Lockett

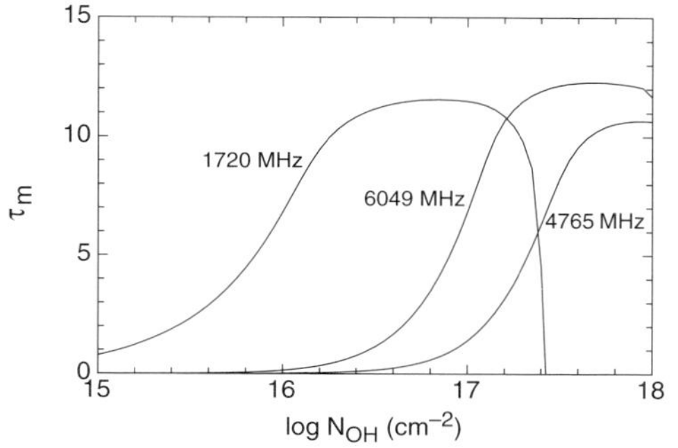

Figure 1. Maser optical depth, τ_m ($I_\nu \propto \exp(\tau_m)$) of the 1720 MHz OH line and its analogues at 6049 and 4765 MHz, as a function of OH column density for T = 50 K and $n_H = 10^5 \mathrm{cm}^{-3}$.

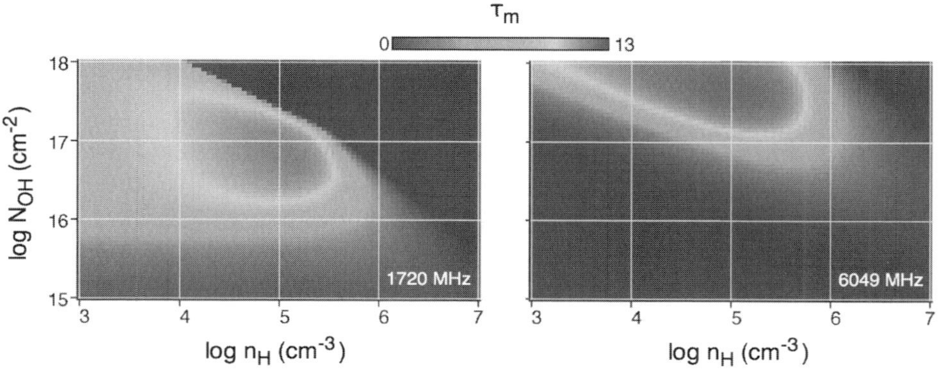

Figure 2. Maser optical depth as a function of n_H and N_{OH} for T = 50 K for the 1720 MHz (left) and 6049 MHz (right) lines of OH.

et al. 1999; Wardle 1999). As the dissociation rate of OH is approximately half that of H_2O, these models predict $OH/H_2O \lesssim 2$, and typically much less. This conflicts with recent absorption measurements in IC 443, which find that $N_{H_2O} \sim 10^{14-15} \mathrm{cm}^{-2}$ (Snell *et al.* 2005) and $N_{OH} \sim 10^{16-17} \mathrm{cm}^{-2}$ (Hewitt *et al.* 2006). Preliminary analysis of a survey for 6049 MHz masers towards southern SNRs has not yielded detections to date (see McDonnell, Vaughan & Wardle, elsewhere in these proceedings).

References

Brown, J. M., Schubert, J. E., Evenson, K. M., & Radford, H. E. 1982, *ApJ*, 258, 899

Destombes, J. L., Marliere, C., Baudry, A., & Brillet, J. 1977, *AA*, 60, 55

Frail, D. A., Goss, W. M., Reynoso, E. M., Giacani, E. B., Green, A. J., & Otrupcek, R. 1996, *AJ*, 111, 1651

Hewitt, J. W., Yusef-Zadeh, F., Wardle, M., Roberts, D. A., & Kassim, N. E. 2006, *ApJ*, 652, 1288

Lockett, P., & Elitzur, M. 1989, *ApJ*, 344, 525

Lockett, P., Gauthier, E., & Elitzur, M. 1999, *ApJ*, 511, 235

Pavlakis, K. G., & Kylafis, N. D. 2000, *ApJ*, 534, 770

Snell, R. L., *et al.* 2005, *ApJ*, 620, 758

Wardle, M. 1999, *ApJ*, 525, L101

Wardle, M. & Yusef-Zadeh, F. 2002, *Science*, 296, 2350

Astrophysical Masers and their Environments
Proceedings IAU Symposium No. 242, 2007
J.M. Chapman & W.A. Baan, eds.

doi:10.1017/S1743921307013312

VLBA H_2O and SiO maser observations in the pPN OH 231.8+4.2

J.-F. Desmurs[1], **C. Sánchez Contreras**[2], **V. Bujarrabal**[3], **J. Alcolea**[1], **and F. Colomer**[3]

[1]Observatorio Astronómico Nacional, Madrid, Spain
email: desmurs@oan.es

[2]Departamento de Astrofísica Molecular e infrarroja, Instituto de Estructura de la Materia, CSIC, Madrid, Spain

[3]Observatorio Astronómico Nacional, Alcalá de Henares, Spain

Abstract. We report high-resolution observations of H_2O, $6_{1,6}$–$5_{2,3}$ and ^{28}SiO v=2, J=1-0 maser emissions obtained with the Very Long Baseline Array in the bipolar pre-planetary nebulae OH 231.8+4.2 (see Desmurs *et al.*, 2007). A phase referencing technique was used to recover the absolute position of both emissions. We detected two groups of water vapour emission oriented nearly north-south. SiO masers are tentatively found to be placed between the two H_2O maser emitting regions.

Keywords. Maser, stars: AGB and post-AGB, radio lines: stars

1. Introduction

OH 231.8+4.2 is a well studied bipolar pre-planetary nebula (pPN) at a distance of ~1500 pc. The central source is a binary system formed by an M9-10 III Mira variable (i.e. an AGB star) and a A0 main sequence companion (Sánchez Contreras *et al.*, 2004). The nebula shows all the signs of post-AGB evolution: fast bipolar outflows with velocities ~200–400 km/s, shock-excited gas and shock-induced chemistry. The presence of a late-type star in the core of a bipolar post-AGB nebula like OH 231.8 is very unusual since the central stars of pPNe are typically hotter, with spectral types from B to K. This enables the survival of SiO and H_2O masers in the Mira's vicinity, where the agents governing non-spherical mass ejections are expected to operate.

2. Results and discussion

We performed VLBA observations of the H_2O $6_{1,6}$–$5_{2,3}$ and SiO v=1 and v=2, J=1–0 maser emission in the pPN OH 231.8 using the phase referencing technique to measure the relative position of the H_2O and SiO maser spots (see Desmurs *et al.* 2007).

H_2O maser emission arises in two groups of compact spots separated by ~ 60 mas (80 AU) along the north-south direction, i.e. roughly along the nebular symmetry axis. A clear velocity gradient is observed indicating an expansive kinematics in this region (Fig. 1). The deprojected expansion velocity of the spots of 10-15 km/s for an inclination of the nebular axis with respect of the plane of the sky of i~36°(Shure *et al.* 1995) is in good agreement with the usual velocities found in AGB stars (e.g. Loup *et al.*, 1993). However, the total velocity dispersion in OH 231.8 of ~20 km/s is larger than values obtained from H_2O maser profile in most AGB stars (e.g. Colomer *et al.*, 2000). The SiO masers arise in a torus like structure around the Mira placed approximately in the middle of both H_2O maser-emitting areas.

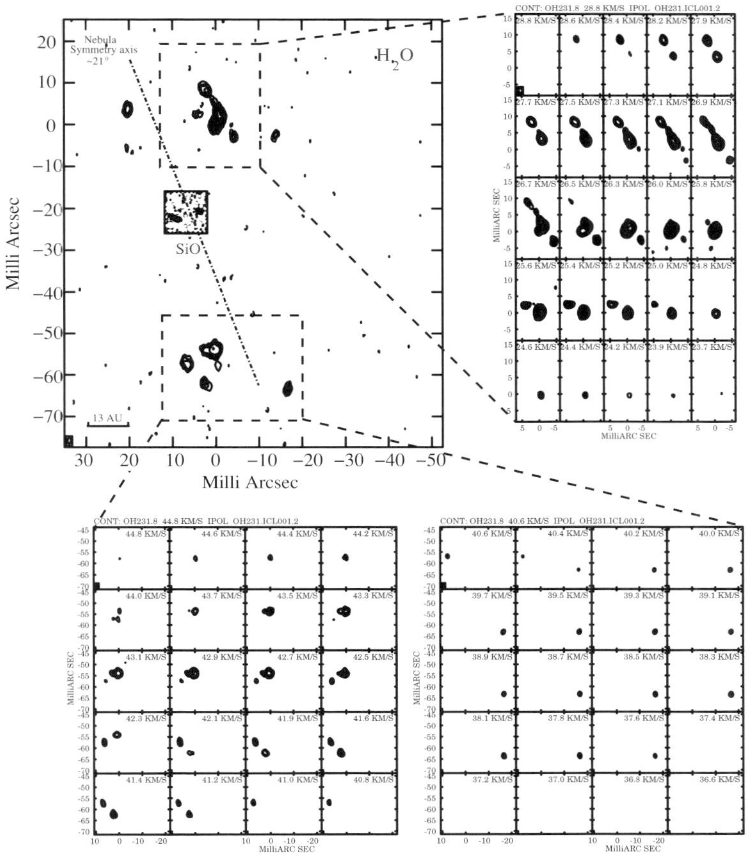

Figure 1. Integrated and velocity H_2O map (SiO map from Sánchez Contreras *et al.*, 2002).

The distribution of the H_2O and SiO masers in OH 231.8 is different from the ring-like chain of spots typically found in AGB stars. The distribution of the H_2O and SiO masers suggests that the equatorial regions of the inner circumstellar envelope of OH 231.8 have been strongly modified. In particular, an equatorial density enhancement, e.g. induced by the presence of the companion, could have quenched and intensified the H_2O and SiO masers respectively in these equatorial regions. Alternatively, the H_2O masers could trace the base of a relatively slow bipolar outflow.

References

Colomer, F., Reid, M. J., Menten, K. M., & Bujarrabal, V. 2000, *A&A* 355, 979

Desmurs, J.-F., Bujarrabal, V., Colomer, F., & Alcolea, J. 2000, *A&A* 360, 189

Desmurs, J.-F., Alcolea, J., Bujarrabal, V., Sánchez Contreras C., & J.Colomer, F. 2007, *A&A* in press

Loup, C., Forveille, T., Omont, A., & Paul, J. F. 1993, *A&AS* 99, 291

Sánchez Contreras, C., Desmurs, J.-F., Bujarrabal, V., Colomer, F., & Alcolea, J. 2002, *ApJ* 385, L1

Sánchez Contreras, C., Gil de Paz, A., & Sahai, R. 2004, *ApJ* 616, 519

Shure, M., Sellgren, K., Jones, T. J., & Klebe, D. 1995, *AJ* 109,721

Astrophysical Masers and their Environments
Proceedings IAU Symposium No. 242, 2007
J.M. Chapman & W.A. Baan, eds.

doi:10.1017/S1743921307013324

A collimated jet and an infalling-rotating disk in G192.16−3.84 traced by H_2O maser emission

Hiroshi Imai[1], Toshihiro Omodaka[1], Tomoya Hirota[2], Tomofumi Umemoto[3], Kazuo Sorai[4], Tetsuro Kondo[5] and the VERA collaboration[2]

[1]Department of Physics, Faculty of Science, Kagoshima University, Japan
email: hiroimai@sci.kagoshima-u.ac.jp

[2]Mizusawa VERA Observatory Mitaka Office, National Astronomical Observatory, Japan

[3]Nobeyama Radio Observatory, National Astronomical Observatory, Japan

[4]Division of Physics, Graduate School of Science, Hokkaido University, Japan

[5]Kashima Space Research Center, National Institute of Information and Communication Technology, Japan

Abstract. We present H_2O masers associated with the massive-star forming region G192.16−3.84 observed with the Japan VLBI network since the year 2005, The spatio-kinematical structure of the maser feature clusters has well persisted since previous observations, in which the masers are associated with two young stellar objects (YSOs) separated by ∼1200 AU and expected to be associated with a highly-collimated bipolar jet and an infalling-rotating disk in the northern and southern YSOs, respectively. We estimated a jet speed of ∼100 km s^{-1} and re-estimated a dynamical age of the whole jet to be 5.6×10^4 years. The spatial distribution of maser Doppler velocities found during the previous and present observations, relative proper motions of H_2O maser features in the southern cluster found in the present observations, a relative bulk motion between the two maser clusters are well explained by a model of an infalling–rotating disk with a radius of ∼1000 AU and a central stellar mass of ∼8 $M_\odot$.

Keywords. masers, stars:mass accretion; individual (G192.16−3.84), ISM:jets and outflows

---

The massive-star forming region G192.16−3.84 (hereafter G192) has massive YSO candidates, one of which is a B2-type star with a giant Herbig-Haro bipolar outflow and a possible circumstellar disk (e.g., Shepherd, & Kurtz 1999, Shepherd*et al.* 2004). H_2O masers are associated with both YSOs. The northern and southern H_2O maser clusters exhibit alignments of maser features, respectively, parallel and perpendicular to the bipolar outflow, which is traced by ^{12}CO emission and projected roughly in the east–west direction, parallel to the Herbig-Haro objects. On the basis of the spatial distributions of H_2O masers and $C^{18}O$ emission tracing high density gas, it is suggested that the southern H_2O maser features are associated with a flattened rotating gas torus around the B2-type star, in which the maser velocities are roughly consistent with those in a Keplerian disk.

The VLBI observations for G192.16−3.84 were made at three epochs during 2005 March–June using five or six telescopes of the Japanese VLBI Network, four 20 m telescopes of the VLBI Exploration of Radio Astrometry (VERA), the 45 m telescope of Nobeyama Radio Observatory (NRO), and the 34 m telescope of the National Institute of Information and Communication Technology (NiCT). These observations aimed to elucidate the 3D velocity field of H_2O masers in term of *relative* proper motions of the masers. The astrometric VLBI observations have been made using only the VERA

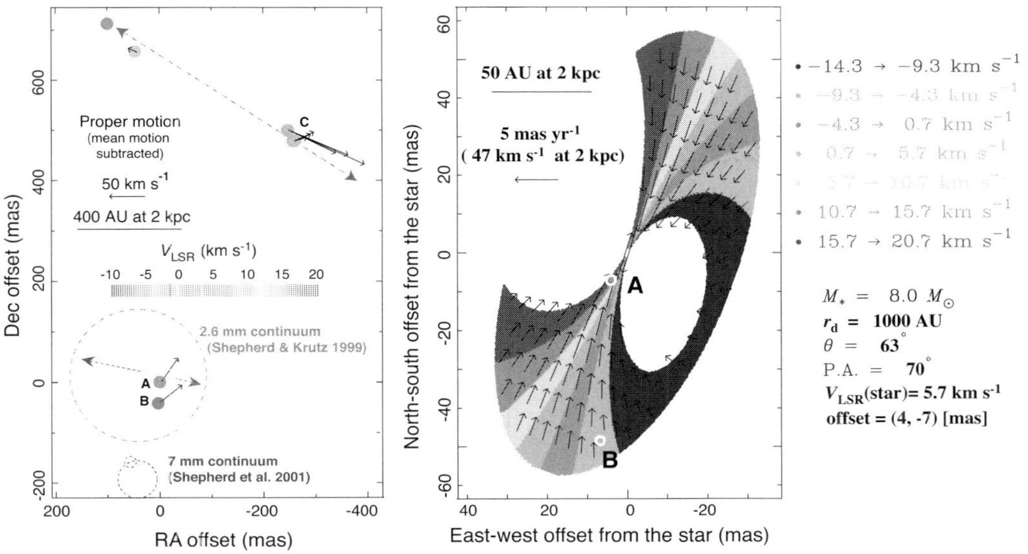

Figure 1. *Left*: Doppler velocity distribution (colored filled circle) and relative proper motions (black arrow) of the H_2O masers in G192.16–3.84. The displayed proper motion vector is the original vector subtracted by a velocity bias (see Imai*et al.* 2006 in detail). Axes of the northern jet and the outer outflow proposed by Shepherd*et al.* 2004 are indicated by dashed arrows. *Right*: Calculated LSR-velocity distribution for masing gas clumps in the modelled thin disk. The H_2O maser features A and B (filled circles, see also the left panel) are indicated after shifting the offset coordinates by the value shown in the panel. Arrows indicate the proper motion of a gas clump located at the root of the arrow.

telescopes equipped with the dual beam system since 2006 October to find an annual parallax and absolute proper motions of the masers. Figure 1 shows the spatial distribution and relative proper motions of maser features. The VERA astrometry enables us to put maser features in the absolute coordinate system at every observing epoch since 2005 October. We found that the dynamical center of the northern Herbig-Haro bipolar outflow is located around the feature *C*, where maser proper motions in two opposite directions are found with respect to the extragalactic position reference source J0603+1742. The southern maser features *A* and *B* were moving towards the northern YSO with a speed of $\sim$40 km s^{-1}. Such fast motions and large Doppler velocity offsets of the southern maser features from the systemic velocity (V_{LSR} =5.7 km s^{-1}) are well explained by a model, in which maser gas clumps are impinging onto a thin gas disk around the southern YSO found in previous molecular line and continuum emission (e.g., Shepherd*et al.* 2004).

Acknowledgements NRO and VERA observatory are branches of the National Astronomical Observatory, an interuniversity research institute operated by the Ministry of Education, Culture, Sports, Science and Technology. H. I./T. H. and H.I. were supported by Grant-in-Aid for Scientific Research from Japan Society for the Promotion of Science (16540224, 18740109, respectively).

References

Imai, H., Omodaka, T., Hirota, T., Umemoto, T., Sorai, K., & Kondo, T. 2006, *PASJ*, 58, 883
Shepherd, D. S. *et al.* 2004 *ApJ*, 614, 211
Shepherd, D. S., Claussen, M. J., & Kurtz, S. E. 2001, *Science*, 292, 1513
Shepherd, D. S., & Kurtz, S. E. 1999, *ApJ*, 523, 690

Astrophysical Masers and their Environments
Proceedings IAU Symposium No. 242, 2007
J.M. Chapman & W.A. Baan, eds.

doi:10.1017/S1743921307013336

Improving the parallaxes of OH bearing Miras

W. H. T. Vlemmings[1,2] **and H. J. van Langevelde**[3,4]

[1]Argelander Institute for Astronomy, University of Bonn, Auf dem Hügel 71, 53121 Bonn, Germany

[2]Jodrell Bank Observatory, Univ. of Manchester, Macclesfield, Cheshire SK11 9DL, U.K. email: Wouter.Vlemmings@manchester.ac.uk

[3]Joint Institute for VLBI in Europe, Postbus 2, 7990 AA Dwingeloo, The Netherlands

[4]Sterrewacht Leiden, Postbus 9513, 2300 RA Leiden, The Netherlands

Abstract. We have carried out observations with the Very Long Baseline Array (VLBA) to measure the parallaxes of Mira variables and are able to improve the distance estimates significantly for a fraction of our sample (U Her, S CrB and RR Aql). This is predominantly because we have enhanced our technique by making use of nearby, in-beam calibrators. Additionally, the observing conditions have improved during the current solar minimum. The distances of these stars are of fundamental importance for studying the physical properties of Asymptotic Giant Branch (AGB) stars with high mass loss.

Keywords. astrometry, stars: AGB, masers

1. Introduction

Astrometric observations of the circumstellar OH masers can be used to determine the proper motion and parallax of maser bearing stars (van Langevelde *et al.* 2000; Vlemmings *et al.* 2003, hereafter V03). This will for instance allow the inclusion of the more extreme Mira stars in studies of the fundamental properties of these stars, like the pulsation and mass-loss mechanism. In order to use the maser positions to monitor the stellar trajectory, an assumption has to be made about the motion of the masers with respect to the star. In V03, it was shown that for several stars the brightest, most blue-shifted circumstellar maser spot corresponds to the *Amplified Stellar Image*. However, not all stars show such a maser spot, depending on for instance an inhomogeneous distribution of the masing gas (Vlemmings 2002). Still the observations in V03 indicate that even without an amplified stellar image, Very Long Baseline Interferometry astrometry of OH masers yields accurate results for sources up to $\sim 1-2$ kpc.

2. Observations

The 1665 and 1667 GHz OH masers of S CrB, U Her and RR Aql were observed using the NRAO† VLBA for 6 epochs between February 2004 and March 2005. For each of our sources we observed a regular nodding phase-reference calibrator at up to $\sim 4.5°$ from the target source as well as an in-beam calibrator source (except for U Her, where no compact source was available). The in-beam calibrators are within $30'$ of the maser source, inside the 1.6 GHz primary beam of the VLBA telescopes, and are used to calibrate the remaining differential phase fluctuations between the nodding calibrator and the maser

† The National Radio Astronomy Observatory (NRAO) is a facility of the National Science Foundation operated under cooperative agreement by Associated Universities, Inc.

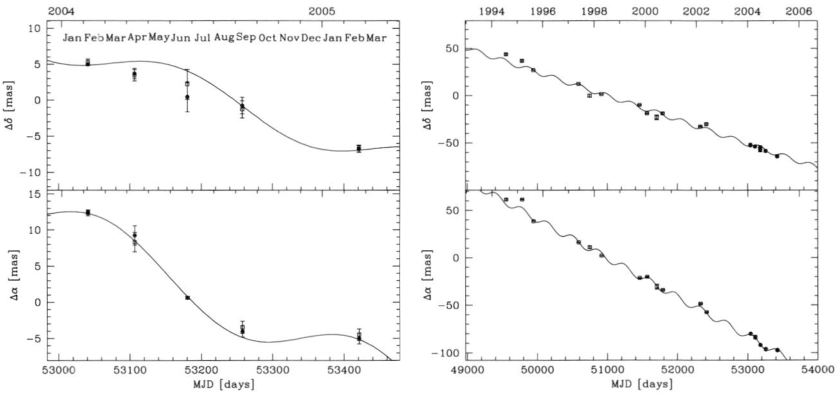

Figure 1. The position of the most blue-shifted 1665 and 1667 MHz maser spots of U Her that have been found to correspond to the stellar image. Drawn is the best fitting parallax and proper motion trajectory combining the results on both maser transitions for the 5 recent epochs of observations (left), and combining the new observations with the 12 previous epochs (right).

Table 1. Improved VLBI astrometry

Source	Period (days)	VLBI parallax (mas)	Hipparcos parallax (mas)	d_{P-L} (pc)	$d_{\rm VLBI}$ (pc)
S CrB	360	2.39 ± 0.17	2.40 ± 1.17	470	418^{+21}_{-18}
U Her	406	3.76 ± 0.27	1.88 ± 1.31	380	266^{+32}_{-28}
RR Aql	394	1.58 ± 0.40	2.48 ± 2.57	540	633^{+214}_{-128}

line of sight. As it has been shown that the astrometric errors are proportional to the target-calibrator separation (Chatterjee *et al.* 2004), in-beam calibration significantly improves the maser astrometry. Additional improvements over the results in V03 were due to the improved conditions during the current solar minimum.

3. Results

Fig. 1 shows the results of our observations and model fitting for U Her. The complete results for to parallax measurements of our sources, presented in Vlemmings & van Langevelde (2007), is given in Table 1, which also contains the revised Hipparcos parallax (Knapp *et al.* 2003) and the Period-Luminosity distance using the P-L relation from Whitelock & Feast (2000).

References

Chatterjee, S., Cordes, J. M., Vlemmings, W. H. T., Arzoumanian, Z., Goss, W. M., & Lazio, T. J. W. 2004, *ApJ*, 604, 339

Knapp, G.R., Pourbaix, D., Platais, I., & Jorissen, A., 2003, *A&A*, 403, 994

van Langevelde, H. J., Vlemmings, W., Diamond, P. J., Baudry, A., & Beasley, A. J. 2000, *A&A*, 357, 945

Vlemmings, W. 2002, Ph.D. Thesis, Leiden University

Vlemmings, W. H. T., van Langevelde, H. J., Diamond, P. J., Habing, H. J., & Schilizzi, R. T. 2003, *A&A*, 407, 213

Vlemmings, W. H. T. & van Langevelde, H. J., 2007, *A&A*, submitted

Whitelock, P., & Feast, M., 2000, *MNRAS*, 319, 759

Astrophysical Masers and their Environments
Proceedings IAU Symposium No. 242, 2007
J.M Chapman & W.A. Baan, eds.

doi:10.1017/S1743921307013348

Discrete survey of 5-cm OH emission from planetary and proto-planetary nebulae

J.-F. Desmurs[1], A. Baudry[2], P. Sivagnanam[3], C. Henkel[4] and A. M. S. Richards[5]

[1]Observatorio Astronómico Nacional, Madrid, Spain
email: desmurs@oan.es

[2]LAB - Observatoire Aquitain des Sciences de l'Univers, Floirac, France

[3]Observatoire de Paris, France

[4]MPIfR, Bonn, Germany

[5]Jodrell Bank Observatory, UK

Abstract. We have undertaken a comprehensive search for the first excited state of OH emission (J=5/2, $^2\Pi_{3/2}$) in Planetary and Proto- Planetary Nebulae. With the Effelsberg telescope, we confirm the detection of Vy 2–2 and we present one new detection in the pPN K3−35. This detection has been confirmed by subsequent observations made at 6035 MHz with the MERLIN interferometer. This is the first detection of 6 GHz OH maser emission from a post-AGB star.

Keywords. Maser, stars: AGB and post-AGB, radio lines: stars

1. Introduction

PNe evolve from the envelopes of AGB stars, through the very short ($\sim$ 1000 yr) phase of protoplanetary nebula (pPN). During this phase, the nebular morphology and kinematics are dramatically altered: the spherical, slowly expanding AGB envelope becomes a PN with, usually, axial symmetry and high axial velocities. Sensitive observations of Planetary Nebulae in the 5 cm lines of OH are motivated by several reasons. The 6 GHz observations probe highly excited inner layers and provide important information; excitation models are complex and still uncertain, and predictions on the main pumping routes leading to OH inversion of the ground state will be constrained by observations of the $J = 5/2$ state.

2. Results and discussion

With the 100m MPIfR radiotelescope at Effelsberg, we have undertaken a complete discrete source survey for the first excited state of OH emission (J =5/2, $^2\Pi_{3/2}$) in 31 PN and pPN above a declination of –20 degrees. It is based on previous 18 cm OH surveys in which we have selected all sources exhibiting both 1612 and/or 1665/1667 MHz emission. We reached an average noise value of about $3\sigma \sim$30 mJy for a spectral resolution of 0.29 km/s. We confirm the detection of Vy 2−2 (see Desmurs *et al.* 2002) and we detected one new source, the young PN K 3−35, as a single 1 km/s-wide component around 21 km/s LSR velocity. This is the first detection of 6 GHz OH maser emission from a post-AGB star.

To confirm the 6035 MHz maser emission detected with Effelsberg in K 3−35, we made new observations using the MERLIN interferometer in the ground state ($^2\Pi_{3/2}$, $J = 3/2$) and the first excited level ($^2\Pi_{3/2}$, $J = 5/2$). Figure 1 presents our map of the 6035 MHz

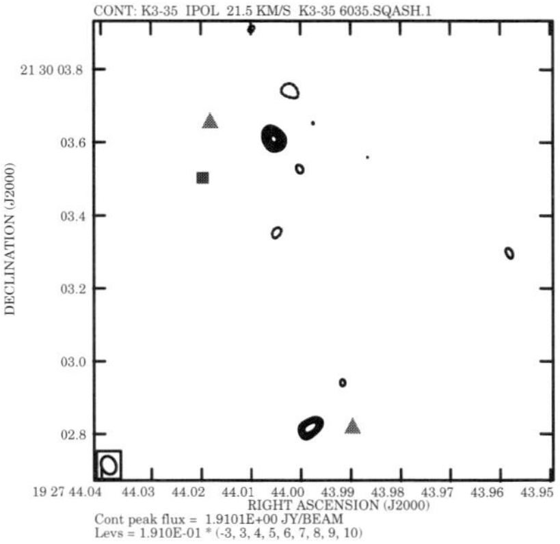

Figure 1. Map of the OH 6035 MHz maser distribution (line contours) compared to the 1612 (blue square) and 1720 MHz (red triangle) maser emission centers.

maser emission (see line contours). The 18-cm positions (position uncertainty is about 40 mas) derived from our interferometric observations are shown for the 1720MHz (red triangle) and the 1612 MHz (blue square). Other 18-cm OH components detected with the Nançay radio telescope around 9 and 18 km/s did not show any 6 GHz counterpart or 18-cm compact emission in the MERLIN data.

K 3–35, is considered to be a very young PN (see OH maser observations by Zijlstra *et al.* 1989) situated in the constellation Vulpecula. VLA observations by Miranda *et al.* (2001) at 22 GHz reveal the shape of a bipolar structure typical of Proto Planetary Nebulae and allow the authors to conclude that this object has just begun its transformation into a planetary nebula. K 3–35 has been detected at 18 cm in the 1612, 1667, and 1720 MHz OH lines (Engels *et al.* 1985, te Linkel Hekkert 1990, Gómez *et al.* 2006). Our interferometric observations of K 3–35, show the quasi spatial coincidence of the 18 cm and 5 cm maser emissions.

The absence of detectable excited emission at 5 cm in pPNe (except in Vy 2–2 and K 3–35) tends to argue in favour of a pumping scheme based on the absorption of 35 and 53 μm photons. In the case of Vy 2–2 the ionisation shell from where the maser emission seems to originate, may present physical conditions (shock, higher temperature and density) similar to those prevailing in HII regions. We note that 1720 MHz masers are exceptionally rare around evolved stars (but have been observed in many star forming regions). A hybrid pumping model applying to both OH/IR stars and HII regions or even a pumping scheme similar to OH maser emission in massive star forming regions may be successful.

References

Desmurs, J.-F., Baudry, A., Sivagnanam, P., & Henkel, C. 2002, *A&A* 394, 975

Engels, D., Schmid-Burgk, J., Walmsley, C. M., & Winnberg, A. 1985, *A&A* 148, 344

Gómez, Y., Tafoya, D., Anglada, G., Franco-Hernandez, R., Torrelles, J. M., & Miranda, L. F. 2006 *RMxAC* 26, 22

Miranda, L. F., Gómez, Y. Anglada, G., & Torrelles, J. M. 2001 *Nature* 414, 284

te Linkel Hekkert 1990, PhD Thesis

Zijlstra, A. A., te Linkel Hekkert, P., Pottasch, S. R., Caswell, J. L., Ratag, M., & Habing H. J., 1989, *A&A* 217, 157

Astrophysical Masers and their Environments
Proceedings IAU Symposium No. 242, 2007
J.M. Chapman & W.A. Baan, eds.

doi:10.1017/S174392130701335X

Outburst in OH17.7-2.0

M. Szymczak[1], E. Gérard[2] and P. Wolak[1]

[1]Toruń Centre for Astronomy, Nicolaus Copernicus University, Poland
email: msz@astro.uni.torun.pl

[2]GEPI, Observatoire de Paris, France
email: Eric.Gerard@obspm.fr

Abstract. We present OH 1612 MHz maser observations towards the pre-planetary nebula OH17.7−2.0 taken with the Nançay radio telescope in 2002−2007. A bursting feature at 73 km s^{-1} grew by a factor of 11 in 740 days in left-hand circular polarization intensity, reaching a maximum in mid-2005. The OH flux density of features at neighbouring velocities exhibited a rise by a factor of 5−7. During the burst, the degree of circular polarization increased from 7 to 83%. The blue-shifted emission showed weak variations of 7−10% in amplitude with the exception of the bluest feature for which the OH flux decreased with the rate of ∼4.4 Jy year^{-1}. Very unusual OH outbursts of individual features and remarkable changes in the shape of the maser profile appear to be linked with recent evolutionary changes in the detached shell.

Keywords. masers, stars: AGB and post-AGB, stars: individual: OH17.7-2.0, radio lines: stars

---

In 2003 the OH 1612 MHz maser line profile of OH17.7−2.0 underwent a very unusual change; the intensity of the red-shifted emission at velocity ∼73 km s^{-1} increased nearly threefold over a period of ∼430 days (Szymczak & Gérard 2005). The follow-up VLBI observations revealed a bipolar morphology of the 1612 MHz maser envelope with the bursting polarized emission coming from two unresolved components of brightness temperature up to 10^{11} K located at the edge of the biconical region 2500 AU from the central star (Szymczak, Bartkiewicz & Gérard 2006). Optical and NIR images confirm a bipolar morphology with two lobes separated by a dark waist (Sánchez Contreras *et al.* 2007). These observational facts bring further arguments that OH17.7−2.0 is in an intermediate evolutionary stage from asymptotic giant branch to planetary nebula.

After the detection of an outburst of the ∼73 km s^{-1} feature at 1612 MHz, we continued to monitor the target in a more systematic manner (typically twice a month) with the Nançay radio telescope. The 1612, 1665 and 1667 MHz lines were observed in full polarization mode. Here, we report the 1612 MHz data taken with a velocity resolution of 0.14 km s^{-1} over a period of 5 years. The calibration uncertainty of the absolute flux density scale was 4−5%.

Figure 1 shows the behaviour of 8 selected maser features in left- (LHC) and right-hand circular (RHC) polarization. Spectacular changes in OH flux appeared in the red-shifted ($V > 61.6$ km s^{-1}) emission. A slow rise of the 73.0 km s^{-1} feature at LHC was seen; the polarized emission increased by a factor of 11 in 740 days. The maximum occurred near mid-2005 followed by a gradual decline since. The degree of circular polarization increased up to 83%. A similar behaviour was observed for the neighbouring feature at 72.8 km s^{-1}; the LHC flux density increased by a factor of about 5. The maxima of both features coincide very well. The variations of the other neighbouring feature at 73.2 km s^{-1} differ considerably; the LHC flux density increased by a factor of 7 in 420 days followed by a high level plateau seen over about one year. A decline of this feature began 125 days prior to the maximum of the strongest feature. The RHC flux density variations of these three flaring features generally followed those in LHC but their amplitudes were

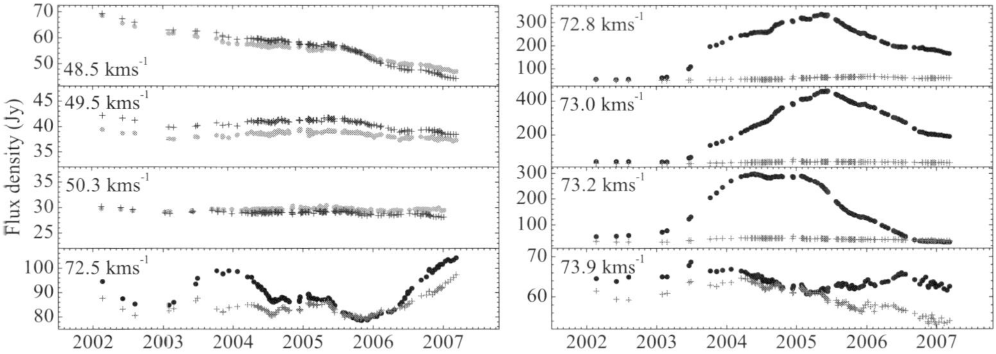

Figure 1. Flux density variations of the selected features in the OH 1612 MHz maser spectrum of OH17.7−2.0. Each panel is labelled with the velocity of maser feature. The dots and crosses mark the left- and right-hand circular polarizations, respectively. Note in particular the burst of polarized emission at velocity ~73 km s^{-1}.

typically lower than 30%. The other red-shifted features showed small amplitude (~25%) variations and their degrees of circular polarization were typically lower than 12%.

The blue-shifted (V <61.6 km s^{-1}) emission did not exhibit dramatic changes. It is usually weakly polarized; the degree of circular polarization was always lower than 8%. Features at 50.3 and 49.5 km s^{-1} showed OH amplitude variations of 7 and 12% respectively. However, the bluest feature of 48.5 km s^{-1} experienced a decline with a rate of about 4.4 Jy year^{-1} (Fig. 1).

We noted small amplitude variations in the OH flux of the blue-shifted emission (for instance the 49.5 km s^{-1} feature) and the RHC flux of the red-shifted emission. It is intriguing that the onset of the outburst coincided with a minimum of the OH flux while a maximum slightly preceded the outburst maximum. This suggests that there is a quiet OH emission emission that still exhibits regular variations with a period ⩾1200 days.

Our monitoring of OH17.7−2.0 documented well the behaviour of the OH 1612 MHz maser emission over a period of 5 years. The most striking result, not seen from the data reported in Szymczak & Gérard (2005), is a gradual decline of the bluest maser feature and evidence for periodic changes of several blue-shifted features as well as the RHC flux of red-shifted features. This seems to be consistent with a scenario in which the outburst emission arises in a very confined region of interaction of the jet-like outflow from the central star with the detached OH shell (Szymczak, Bartkiewicz & Gérard 2006), while the OH maser from the overall AGB shell still varies in a periodic manner as observed in OH/IR objects (Herman & Habing 1985). The OH maser emission from OH17.7−2.0 appears to be a good probe of recent evolutionary changes in the detached shell, which are presumably induced by bipolar winds. The present case study of OH 17.7-2.0 demonstrates the need for both long-term monitoring of well selected OH sources and prompt follow-up VLBI observations of outstanding events if one wants to understand the complex physics of circumstellar OH masers.

References

Herman, J. & Habing, H. J. 1985, *A&AS* 59, 523

Sánchez Contreras, C., Le Mignant, D., Sahai, *et al.* 2007, *ApJ* 656, 1150

Szymczak, M., Bartkiewicz, A. & Gérard, E. 2006, *Proc. 8th EVN Symposium* (submitted)

Szymczak, M. & Gérard, E. 2005, *A&A* 433, L29

Astrophysical Masers and their Environments
Proceedings IAU Symposium No. 242, 2007
J.M. Chapman & W.A. Baan, eds.

doi:10.1017/S1743921307013361

Masers and Galactic structure: Micro-arcsecond astrometry with the VLBA

M. J. Reid[1], A. Brunthaler[2], K. M. Menten[2], Xu Ye[2], Zheng Xing-Wu[3] and L. Moscadelli[4]

[1]Harvard-Smithsonian CfA, Cambridge, MA, USA
email: reid@cfa.harvard.edu

[2]MPIfR, Bonn, Germany

[3]Nanjing University, Nanjing, China

[4]Arcetri Observatory, Firenze, Italy

Abstract. Astrometric observations with the VLBA with accuracies approaching ~ 10 μas are being conducted in order to better understand the Galaxy. The location of Sgr A* on infrared images can be determined with an accuracy of a few mas, using stars with SiO maser emission as a calibration grid for infrared images. The apparent proper motion of Sgr A*, which is dominated by the effects of the orbit of the Sun around the Galactic center, has been measured with high accuracy. This measurement strongly constrains Θ_0/R_0 and offers a dynamical definition of the Galactic plane with Sgr A* at its origin. The intrinsic motion of Sgr A* is very small and comparable to that expected for a supermassive black hole. When combined with infrared results, this provides overwhelming evidence that Sgr A* is a supermassive black hole. Finally, we are engaged in a large project to map the spiral structure and kinematics of the Galaxy. Preliminary trigonometric parallaxes, obtained with the VLBA, to eight massive star forming regions are presented.

Keywords. Galactic structure, supermassive black holes, SiO masers, methanol masers, astrometry, VLBA, parallaxes, proper motions

---

1. Where is Sgr A*(IR)?

Sgr A*, the supermassive black hole that anchors the center of the Galaxy, is a strong radio source but quite dim at infrared wavelengths. This makes it difficult to locate Sgr A* on infrared images: Sgr A*(IR). Also, the central stellar cluster is quite dense, and these stars move considerably from year to year, further complicating locating Sgr A*(IR). In order to transfer the highly accurate radio position of Sgr A* to infrared images, one needs a grid of stars near Sgr A* visible at both radio and infrared wavelengths. Fortunately, this is available as Mira-like variables found in the Galactic center are both strong infrared sources and have circumstellar masers that are strong radio sources.

Menten *et al.* (1997) provided the first highly accurate position of Sgr A*(IR) relative to stars in the Galactic center. We have been continuing this effort in order to find more SiO maser stars and to improve upon the accuracy of their position and proper motions. Fig. 1, from Reid *et al.* (2007), shows a composite spectrum of SiO v=1, J= $1-0$ maser emission from 15 stars within $\approx 45''$ of Sgr A*. This spectrum was taken with the National Radio Astronomy Observatory's † VLA on three observing runs between 1998 and 2006.

† The National Radio Astronomy Observatory is operated by Associated Universities, Inc., under a cooperative agreement with the National Science Foundation.

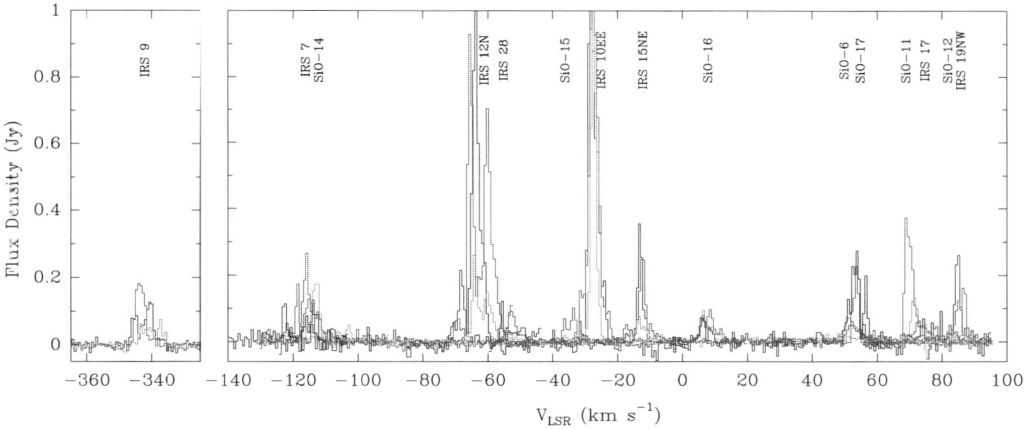

Figure 1. Composite spectrum of stellar SiO masers detected with the VLA in 1998, 2000, and 2006. Stars are identified at the top. Taken from Reid *et al.* (2007).

These masers are variable, as expected from observations of Mira-like variables in the solar neighborhood.

We have been measuring the positions of these stars, some for nearly a decade. Not all masers are detected at all epochs, and our position and motion accuracies depend on the strength of the masers and on the time span of observations with detections. A typical position accuracy is ≈ 1 mas and proper motion accuracy is ~ 0.3 mas y^{-1}. The locations of the nine maser stars within $20''$ of Sgr A* are shown in Fig. 2.

The maser stars provide a critical calibration "grid" for the measurement of stellar orbits in the central milli-pc with infrared techniques. Since the stars move rapidly, continued monitoring is needed to avoid long extrapolations using measured proper motions. Currently, the SiO spectra are pieced together by splitting the observing time among about ten 3-MHz bands, in order to achieve adequate spectral resolution. The upcoming EVLA capabilities of broad bandwidth and high spectral resolution will greatly enhance future observations.

One star, IRS 9, has a very large 3-D motion of 370 km s^{-1} at a projected distance of 0.33 pc from Sgr A*. If bound to the central pc, this requires an excess of nearly 10^6 $M_\odot$ above that attributed to Sgr A* and the observed stellar cluster. Possible explanation for this anomaly include 1) that IRS 9 is not bound to the central parsec, but is on a highly elliptical orbit that takes it out to ≈ 10 pc; 2) that a cluster of dark stellar remnants accounts for the missing mass; and/or 3) that the galactocentric distance R_0 is considerably greater than 8 kpc. These and other possibilities are discussed in detail by Reid *et al.* (2007).

2. The proper motion of Sgr A*

We have been measuring the proper motion of the compact radio source Sgr A* since 1995. In Fig. 3 we show the expected *apparent* motion for Sgr A*, relative to two background quasars, on a 90 cm wavelength image of the Galactic center. The apparent motion stems from our moving vantage point near the Sun, which orbits the Galactic center about every 225 My.

Were Sgr A* a stellar mass object, for example a mass transfer binary with a compact object such as a neutron star or black hole, it would be moving at speeds of many thousands of km s^{-1}, as observed for stars in the deep gravitational potential. However,

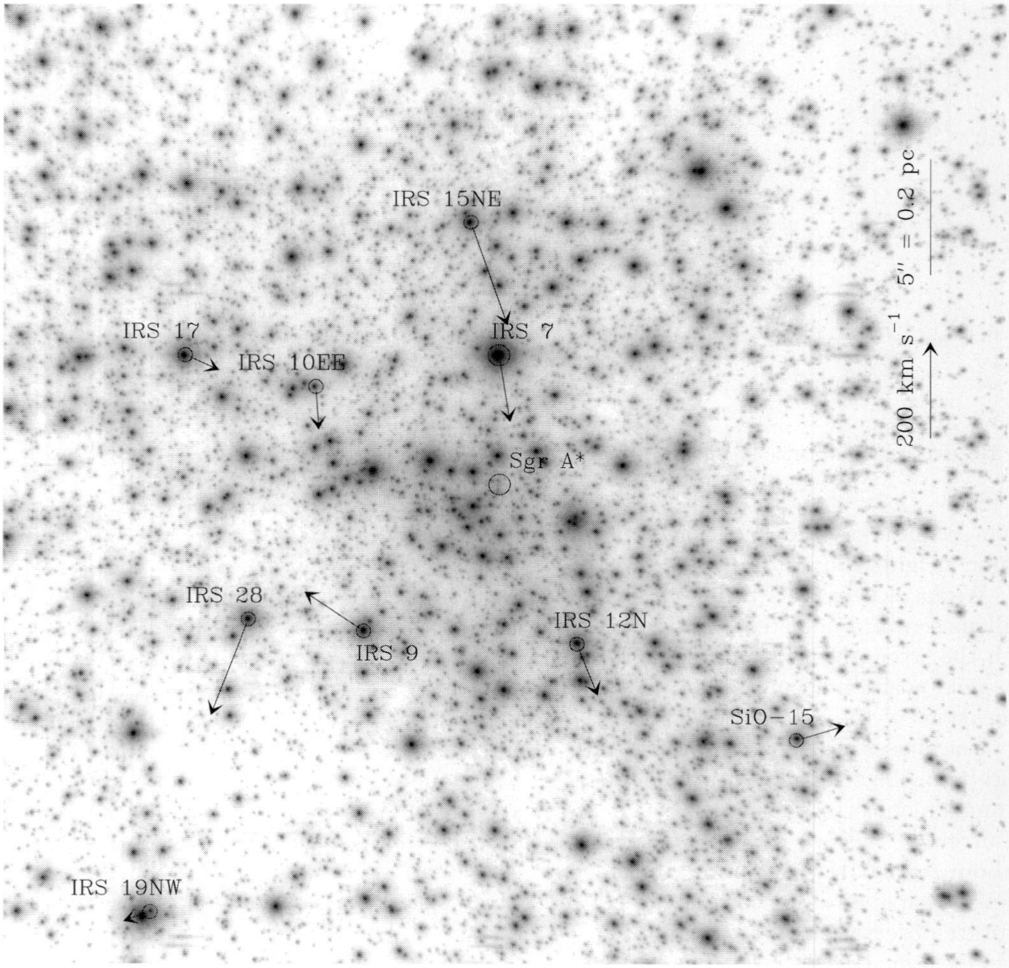

Figure 2. Infrared (2.2 μm wavelength) image of the central 20″ of the Galactic center from Reid *et al.* (2007). North is up and East is to the left. Stellar SiO masers are circled and their proper motions indicated by arrows. Linear and motion scales are at the right side of the figure. Sgr A*, a very weak infrared source, is circled at the center.

for a supermassive black hole at the Galactic center, one expects a very small motion of ~ 0.3 km s^{-1} owing to small random perturbations from the $\sim 10^6$ stars (and perhaps upwards of 10^4 stellar remnants) within Sgr A*'s gravitational sphere of influence. With 8 years of VLBA data, we find that the true motion of Sgr A* in the direction perpendicular to the Galactic plane is indeed very small: -0.4 ± 0.9 km s^{-1}(Reid & Brunthaler (2004)).

The motion upper limit for Sgr A* can be directly transformed to a mass lower limit of $\sim 4 \times 10^5$ M$_\odot$. Combining this mass lower limit with an upper limit of ≈ 0.5 AU for the intrinsic (non-scatter broadened) size of Sgr A* (see Bower *et al.* (2004) and references therein), yields a mass density $> 7 \times 10^{21}$ M$_\odot$ pc^{-3}. This provides overwhelming evidence that Sgr A* is a supermassive black hole.

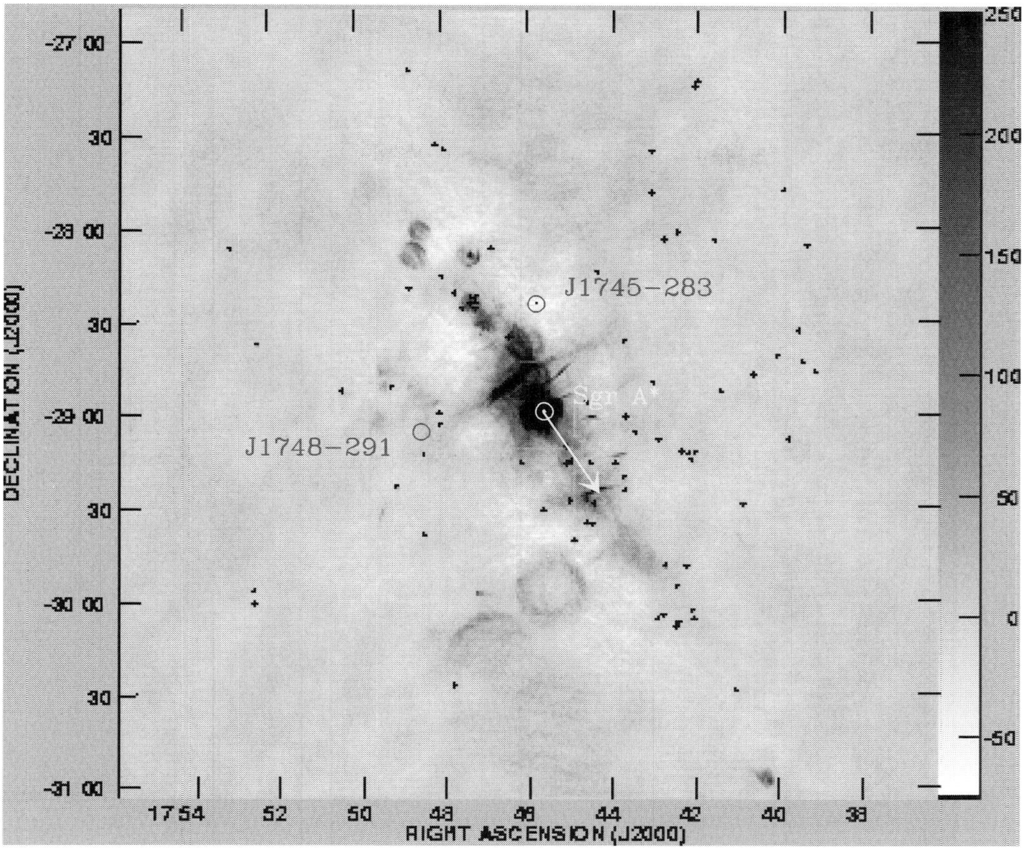

Figure 3. A schematic overlay of the location and expected direction of motion of Sgr A* on a 90-cm wavelength image of the Galactic center region (LaRosa *et al.* (2000)). Also shown are the locations of two compact extragalactic radio sources used as references for Sgr A*'s motion.

Since the apparent proper motion of Sgr A* is totally dominated by the orbit of the Sun about the Galactic center, one could use this to provide a dynamical definition of the plane of the Galaxy. The current IAU definition of the Galactic plane is based primarily on the distribution of HI, with the Sun defining the zero of Galactic longitude and latitude. The Sun is now known to be $\sim$ 10 pc above the plane. The IAU definition of the Galactic plane, places Sgr A* $\approx$ 6 pc below the plane. A better definition of the Galactic plane could come from the direction of the apparent proper motion of Sgr A* (corrected for the small peculiar motion of the Sun perpendicular to the plane), with Sgr A*'s position defining the zero of Galactic latitude. Of course, since the Sun orbits the Galactic center, the position of Sgr A* changes by about 6 mas y^{-1}; thus the zero of Galactic longitude would become a function of time.

The accuracy of proper motions improves with the time span as $t^{-3/2}$ (for uniform time sampling). Our current data span 8 years; observations in 2007 will increase the time span to 11 years and should reduce our motion uncertainty by a factor of 0.6.

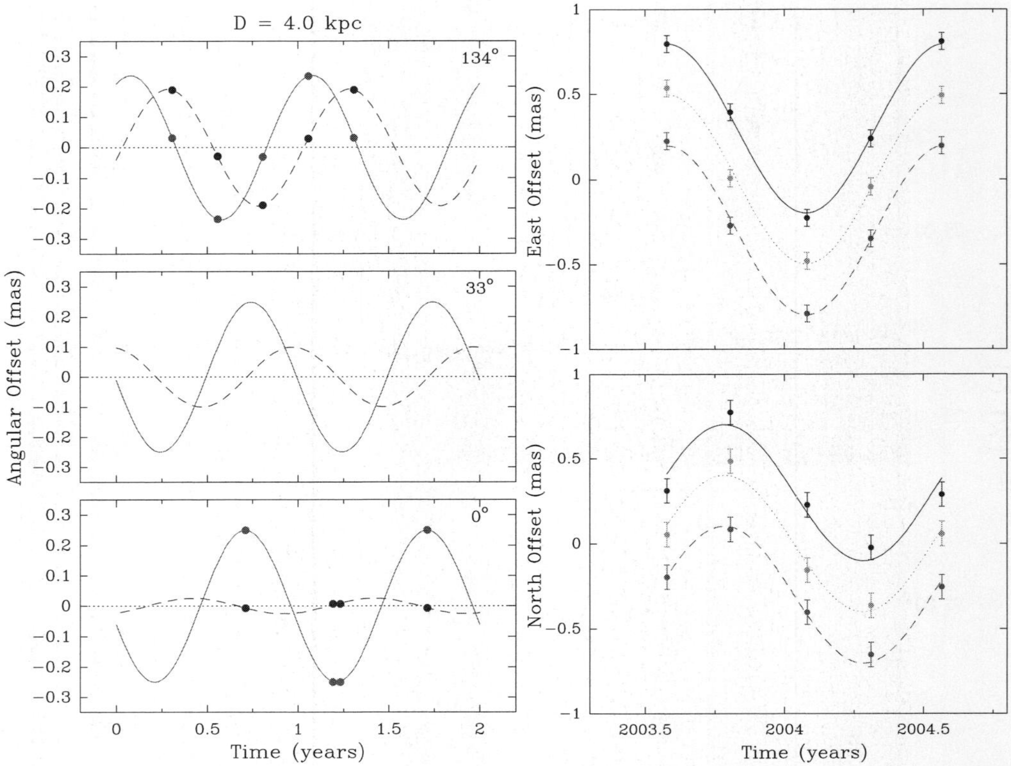

Figure 4. *Left panels:* Parallax signatures for three hypothetical sources at a distance of 4 kpc at the Galactic longitudes indicated in the upper right-hand corners. Solid lines and dashed lines indicate the expected Eastward and Northward positions as a function of time after removing any proper motion. Near optimum time samplings are indicated for two cases. *Right panels:* Real parallax data from Xu, Reid, Zheng & Menten (2006): VLBA positions of a 12.2 GHz methanol maser spot of W3OH (at Galactic longitude 134°) measured relative to three extragalactic sources (offset for clarity), after removing the fitted proper motion.

3. Parallax and proper motions of star forming masers

We are conducting a large program with the VLBA to map the spiral structure of the Milky Way and to define its rotational kinematics and mass profile. This involves measuring trigonometric parallaxes and proper motions to methanol (and some other) masers in massive star forming regions. We are observing primarily in the 12.2 GHz methanol transition as that falls in one of the VLBA bands.

Examples of expected parallax signatures are shown in the left-hand panel of Fig. 4. Note that at Galactic longitude 134° the Earth's orbit appears almost face-on as viewed by, eg, W3OH. For this case, one expects the amplitude of the Dec. parallax sinusoid to be nearly as large as the R.A. sinusoid. Sampling quarterly for one year, as shown, is a nearly optimum. However, at Galactic longitude 0° the Earth's orbit appears almost edge-on as viewed by, eg, Sgr B2. For this case, the Dec. parallax sinusoid has a very low amplitude (and the VLBA beam is also bigger in this direction) and one can optimize the observations by measuring only the R.A. component at 6 month intervals as shown in the figure.

In the right-hand panels of Fig. 4, we show the 12.2 GHz maser parallax data for W3OH by Xu, Reid, Zheng & Menten (2006) with proper motions removed. These data

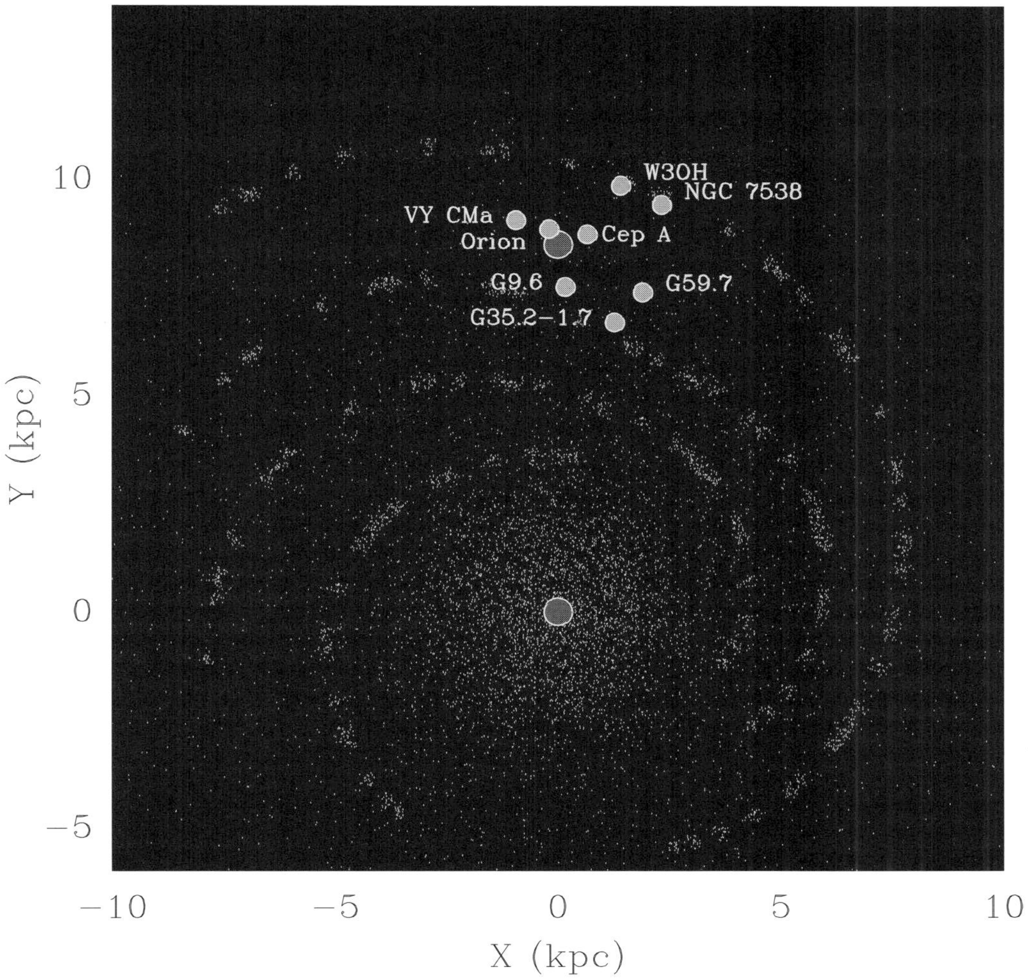

Figure 5. Schematic model of the Milky Way with spiral arms from Taylor & Cordes (1993), with locations of massive star forming regions for which we have preliminary trigonometric parallaxes. The position of the Sun (0,8) and Sgr A* (0,0) kpc are also indicated.

yield a parallax of 0.512 ± 0.010 mas (ie, a distance of 1.95 ± 0.04 kpc). A similar parallax of 0.489 ± 0.017 mas, based on H_2O masers in W3OH, has been measured by Hachisuka *et al.*(2006). This resolves a long-standing problem of the discrepancy of a factor of two between (associated O-star) luminosity-based distances of 2.2 kpc by Humphreys (1978) and kinematic distances near 4.3 kpc for this region of the Perseus spiral arm. The luminosity distance is consistent with our parallax and the kinematic distance is incorrect. We also measured the proper motion of W3OH, which reveals a large non-circular component of 22 km s^{-1} inward and slower than Galactic rotation (ie, roughly toward the Sun). This explains the large kinematic distance anomaly, but raises the question of how a massive star forming region could acquire such a large non-circular velocity component.

We are in the process of measuring parallaxes and proper motions for a dozen 12.2 GHz methanol maser sources, as well as some other star forming regions at other frequencies. While the observations are not complete, preliminary parallaxes for some sources are shown in Fig. 5. We have accurate distances to massive star forming regions in

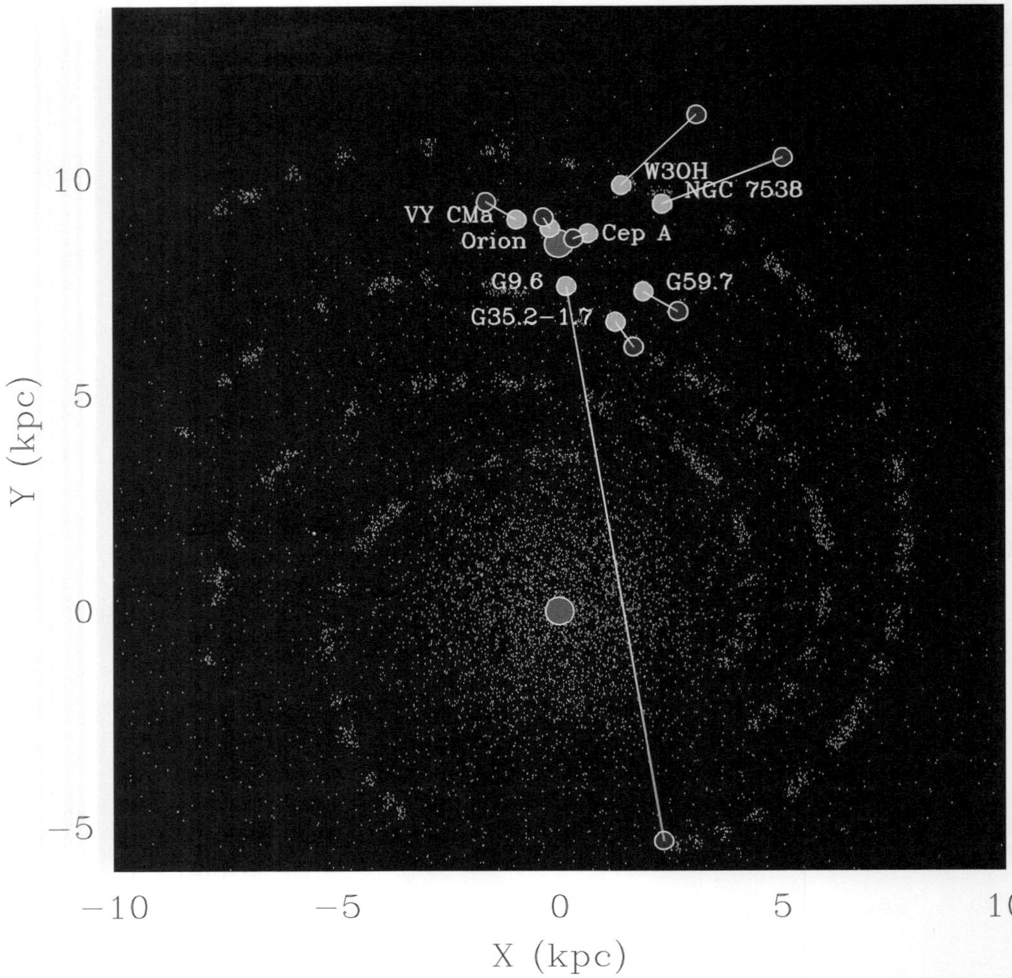

Figure 6. Comparison of trigonometric parallaxes (light cyan) with kinematic distances (dark blue), assuming $R_0 = 8.5$ kpc and a flat rotation curve with a local velocity of $\Theta_0 = 220$ km s^{-1}. For all sources, except Cep A, the kinematic distances are larger than the trigonometric parallaxes.

the Perseus, Local and Sagittarius arms. All sources, with the possible exception of G59.7+0.1, lie close to these spiral arms. With more measurements we plan to map the locations of the major spiral arms of the Galaxy.

In Fig. 6 we compare kinematic and trigonometric parallax distances. As for W3OH, we find the Perseus arm source NGC 7538 also has a large kinematic anomaly. The source G9.6+0.2 is generally thought to be at its far kinematic distance, well past the Galactic center. However, our preliminary (3-epoch) parallax indicates that it is nearby in the Sagittarius arm. This resolves the problem of the extreme luminosity of its 6.7 GHz methanol maser, which was based on the far kinematic distance. Our measurements indicate that G9.6+0.2 is among the nearest high-mass star forming regions in the Galaxy.

Most sources appear to have trigonometric parallax distances that are nearer than their kinematic distances. Decreasing the galactocentric distance R_0 and/or increasing the rotational velocity of the solar neighborhood Θ_0, from the adopted IAU values, would reduce the kinematic distances and bring them into better agreement with the trigonometric parallaxes. We note that the proper motion of Sgr A* by Reid & Brunthaler (2004), after correcting for Solar Motion, implies $\Theta_0/R_0 = 29.45 \pm 0.15$ km s^{-1} kpc^{-1}. If $R_0 = 8.0$ kpc, this requires $\Theta_0 = 236$ km s^{-1}; using these Galactic parameters would reduce most kinematic distances by roughly 10-15%.

We eagerly await more parallaxes from the VLBA and VERA.

References

Bower, G. C. Falcke, H. Herrnstein, R. M, Zhao, J.-H., Goss, W. M, & Backer, D. C. 2004, *Science* 304, 704
Hachisuka, K. *et al.*2006, *ApJ*, 645, 337
Humphreys, R. M. 1978, *ApJS* 38, 309
LaRosa, T. N., Kassim, N. E., Lazio, T. Joseph W., & Hyman, S. D. 2000, *AJ*, 119, 207
Menten, K. M., Reid, M. J, Eckart, A., & Genzel R. 1997, *ApJ* 475, L111
Reid, M. J., Menten, K. M., Trippe, S, Ott, T., & Genzel, R. 2007, *ApJ*, 659, 378
Reid, M. J., & Brunthaler, A. 2004, *ApJ* 616, 872
Taylor, J. H., & Cordes, J. M. 1993,*ApJ* 411, 674
Xu, Y., Reid, M. J., Zheng, X. W., & Menten, K. M. 2006, *Science*, 311, 54

Astrophysical Masers and their Environments
Proceedings IAU Symposium No. 242, 2007
J.M. Chapman & W.A. Baan, eds.

doi:10.1017/S1743921307013373

Do maser stars trace a rotating Galactic bar?

H. J. Habing

Leiden Observatory, the Netherlands
email: habing@strw.leidenuniv.nl

Abstract. I have analyzed the radial velocities in two large samples of maser stars that cover the Galactic plane from $l = +40$ deg to $l = -40$ deg through the Galactic Centre. I compare a diagram of longitude versus radial velocity as observed with a diagram obtained from the calculation of orbits in a simple, but realistic potential containing a weak, rotating bar. Almost all stars belong to the Galactic disk but there is a small sample with high velocities close to the Galactic Centre and in the so-called forbidden quadrants. These velocities cannot be explained by a combination of pure Galactic rotation and a modest dispersion in velocity. An acceptable explanation is one in which the stars of this small sample move up and down a rotating Galactic bar.

1. The problem

Maser stars are all late AGB-stars at the centre of a sphere of matter continuously ejected by the star. The formation of dust in the envelope makes the stars very faint at wavelengths below 1 μm and strong in the infrared. In the extreme cases the peak of the SED can be at 20 to 30 μm. Stellar maser emission is seen in the 1612 MHz line of OH, in several lines of SiO and in a 22 GHz line of H_2O. The line emission makes it possible to measure very precisely the radial velocity of each star. The OH and SiO maser emission is stable in time, but stellar H_2O-masers vary strongly and irregularly; they may remain undetectable during long periods and then suddenly flare. For that reason H_2O-maser stars will not be discussed here. Stars with maser emission at 1612 MHz will be called OH/IR stars. Infrared emission provides the non-thermal excitation of the molecules and appears to be well-understood. Recent spectroscopic studies of OH/IR stars at red wavelengths have shown the presence of s-type elements and thus the occurrence of Hot Bottom Burning (Garcia-Hernandez *et al.*, 2006).

Over the last 25 years several radio astronomical surveys have been made for maser stars in the Milky Way in the Northern and the Southern skies. The results have provided me with a data base that contains almost 800 OH/IR stars detected in a 6°-wide strip along the Galactic equator over the longitude range from $-45°$ to $+40°$ and a velocity range from -300 to $+300$ km s^{-1}. At all longitudes there is a strong concentration to the Galactic plane. The longitude distribution shows a clear concentration around the Galactic Centre (GC) proving that our sensitivity is good enough to reach beyond the GC. From earlier surveys we know that the number of detections drops considerably in the anti-centre direction; maser stars are confined to the inner Galaxy. Stars with infrared excess do in general not show this confinement and the scarcity of maser stars in the anti-centre may be due to a lower metallicity the outer Galaxy.

In Figure 1 I show two lV-diagrams of radial velocity as a function of longitude-see the legend. The x-axis and the pair of inclined lines bracket the bulk of the data; these curves represent the maximum rotational velocity at positive longitudes and the minimum at negative longitudes, if the gas had a constant circular velocity of 200 km s^{-1}.

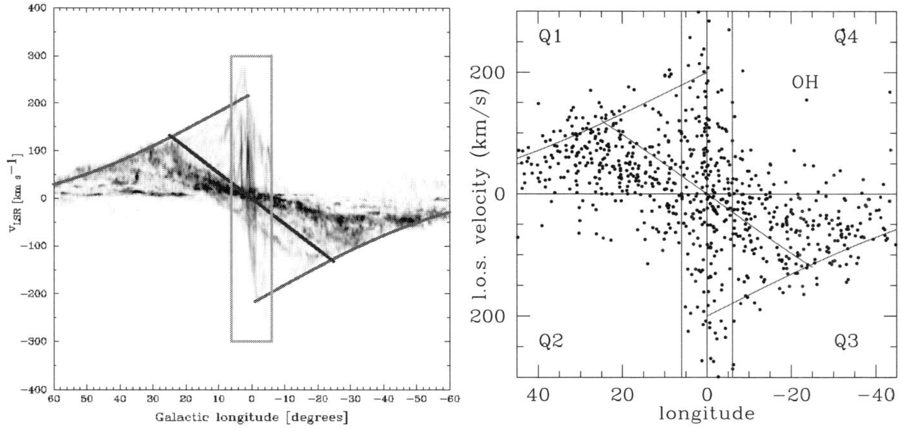

Figure 1. The left diagram shows the intensity of the interstellar CO-line in the Galactic plane as a function of longitude and latitude. For the meaning of the lines see the text.The diagram on the right shows the distribution in longitude and velocity of OH maser stars. The inclined curves and the transversal and vertical lines have the same meaning as in the diagram on the left.

The transversal line through the centre indicates the upper limit if there were no CO-emitting gas inside 3 kpc from the Centre and the box marks the innermost part of the diagram, i.e. the area between $-6°$ and $+6°$. Comparing the two diagrams in Fig. 1 we find the following:

- Almost all CO-emission is confined by the inclined curves and the transversal line,
- A significant number of maser stars cross limits set by Galactic rotation- see the number of stars outside the limits observed by the CO-gas,
- Within 6° from the GC one finds little gas at "forbidden" velocities, i.e. rectangles defined by: ($V \leqslant 0°$, $l \geqslant 0°$) and ($V \geqslant 0°$, $l \leqslant 0°$) but many stars.

The question that we want to answer is: what kinds of stellar orbits will explain the distribution of the stars in the right hand diagram of Fig.1?

2. Building a model

The starting point for our study was the significant number of stars with high and forbidden velocities in the area limited by the box in the left diagram of Fig. 1. Could these velocities be explained by stars falling into the GC from the molecular ring at 3 kpc from the GC? Such falling in produces velocities high enough, but the lV-diagram is wrong. The next question is: what high and forbidden velocities would they have if these stars were falling in inside a rotating bar? To explore this suggestion we selected some areas in the lV-diagrams and counted stars in each area. We then used a simple potential for a 2D galaxy with a rotating Galactic bar, we calculated orbits for different initial conditions and derived the time spent in each area. By making linear combinations of different orbits we then attempted to reproduce the observed numbers of stars in each area.

The potential we used has been given by Binney and Tremaine (1987). It has the following form:

$$\Phi = \frac{1}{2} V_0^2 \, ln\{r_0^2 + x^2 + (1+\varepsilon)y^2\}$$

where x is measured along the bar and y perpendicular to the bar; ε measures the strength

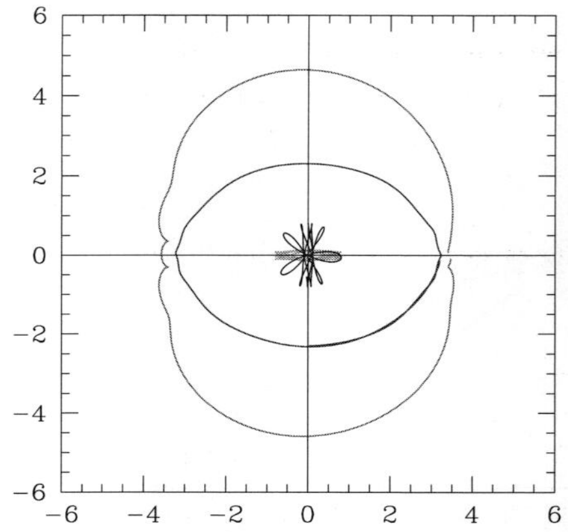

Figure 2. This figure shows two orbits of stars in the Galactic plane. The orbits start at almost the same position on the x-axis and with almost identical velocities and yet they cross the y-axis at very different locations. See text for explanation.

of the bar. It is assumed that the bar rotates with angular speed Ω. The two equations for the motion are now given by:

$$\frac{d^2x}{dt^2} = -\frac{\partial\Phi}{\partial x} + \Omega^2\ x + 2\ \Omega\frac{dy}{dt}$$

and by

$$\frac{d^2y}{dt^2} = -\frac{\partial\Phi}{\partial y} + \Omega^2\ y - 2\ \Omega\frac{dx}{dt}.$$

The free parameters of the model were chosen as: $v_0 = 200\mathrm{km\,s^{-1}}$, $r_0 = 0.15$ kpc, $\varepsilon = 0.11,\ and\Omega = -60\ \mathrm{km\,s^{-1}kpc^{-1}}$.

3. Properties of selected orbits

We started the calculation of the orbits at the x-axis assigning the star a velocity in the y-direction equal to f times the local circular velocity; f was always chosen in the interval $(0, 1)$. Pure circular orbits no longer exist because the potential is not circularly symmetric. However, the deviation from circular symmetry of Φ is small and quasi-circular orbits, or rather orbits limited to a narrow range in distance are always present, inside as well as outside of co-rotation. This is, however, not true around co-rotation at $x = 3.33$ kpc where $x\ \Omega = v_0$, the co-rotation distance.

Fig. 2 shows two orbits in the plane of the Galaxy that have almost the same initial conditions. They both start at the x-axis just inside and just outside the co-rotation distance ($x = 3.33$ kpc and with the local rotational velocity, i.e. just below and just above 200 $\mathrm{km\,s^{-1}}$. In spite of the very small differences in initial conditions they cross the y-axis at distances that differ by almost a factor 2. There is an explanation for this effect: stars just inside the bar will go faster than the bar and are slowed down by the bar, whereas stars outside co-rotation will go slower than the bar and the bar will thus accelerate the two stars.This speeding-up and slowing-down explains why the two orbits pass the y-axis at such different distances. It also implies another property: if the majority of maser stars is on almost circular orbits than we do not expect to find stars in between the two larger orbits in Fig. 2.

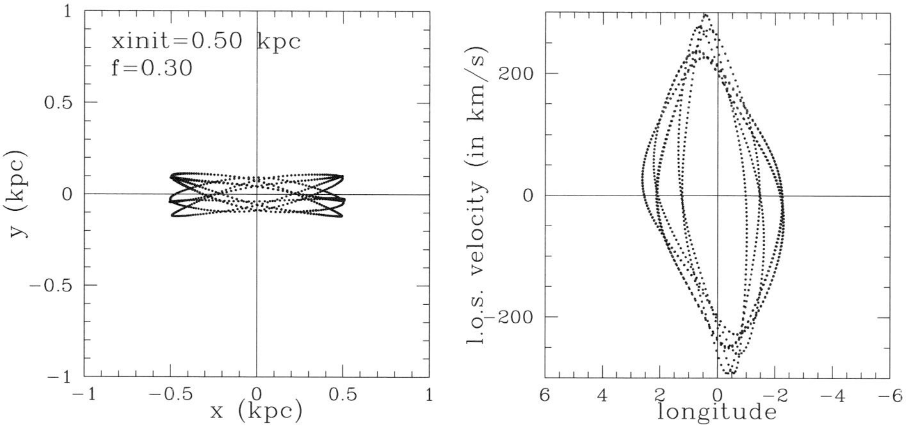

Figure 3. The left diagram shows an orbit in the Galactic plane as described in the text. The right diagram shows the same orbit in lV-space.

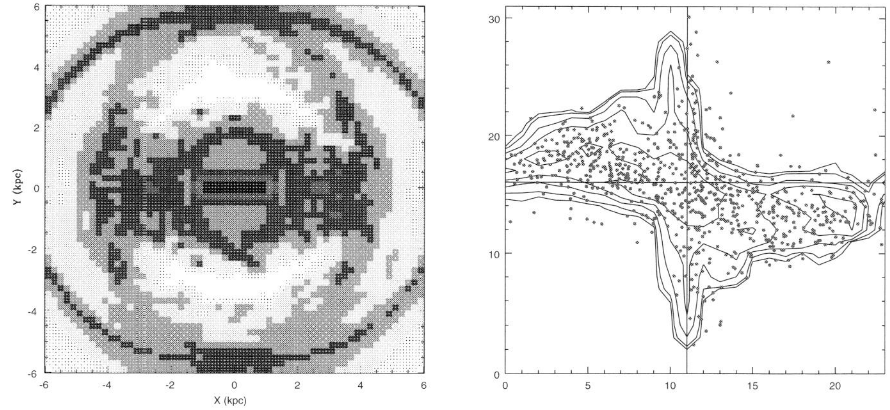

Figure 4. The left diagram shows the distribution of OH maser stars in the Galactic plane in a frame that rotates with angular velocity $\Omega = -60$ km s^{-1}. The oval at the centre indicates the highest density; it is the rotating bar. The diagram on the right shows the individual OH maser stars overlaid over the phase space distribution in longitude and velocity.

Figure 3 shows an interesting type of orbits. They again set forth from the x-axis but with a velocity in the y-direction that is only 0.3 times the circular velocity. Figure 3 (left diagram) shows this particular orbit in the frame of the (rotating) bar: the star moves up and down along the bar. In doing this it reaches very high and forbidden in the lV-diagram: it is the type of orbit that we looked for at the beginning of this article.

4. Final model

We calculated numerically many orbits and found that a set of initial conditions, i.e. of initial values $(x, \dot{x} = 0, y, \dot{y})$ can be obtained that predict rather well the numbers in the selected areas. This set specifies the distribution of Galactic orbits of the maser stars. As expected, most stars are on almost disk-like orbits but there is a significant number of stars on bar-like orbits such as in Fig. 3.

The distribution of initial conditions equals the so-called phase space distribution. Knowing the orbits one can calculate the distribution of the maser stars in the inner Milky Way Galaxy. Figure 4 (left diagram) shows the distribution in the plane of our Galaxy. One easily recognizes the Galactic bar and also the two empty, ear-like areas centred at the corotation point at the y-axis. Fig. 4 (right diagram) shows the velocity distribution belonging to the model with the OH maser velocities superposed.

These results have been published by Habing *et al.* (2006).

Acknowledgement

I thank my collaborators Maartje Sevenster, Maria Messineo, Glenn van der Ven and Koen Kuijken

References

Binney J., & Tremaine S. 1987, Galactic Dynamics (Princeton University Press)

Garcia-Hernandez, D. A., Garcia-Lario, P., Plez, B., Manchado, A., D'Antona, F., Lub, J., & Habing, H. J. 2007, A&A 462, 711

Habing, H., Sevenster, M., Messineo, M., van der Ven, G., & Kuijken, K. 2006, A&A, 458, 151

Astrophysical Masers and their Environments
Proceedings IAU Symposium No. 242, 2007
J. M. Chapman & W. A. Baan, eds.

doi:10.1017/S1743921307013385

Galactic rotation measurements based on H_2O maser astrometry with VERA

Mareki Honma[1,2], Takeshi Bushimata[1,3], Yoon Kyung Choi[1,4], Tomoya Hirota[1], Hiroshi Imai[5], Kenzaburo Iwadate[6], Takaaki Jike[6], Osamu Kameya[2,6], Ryuichi Kamohara[1], Yukitoshi Kan-ya[1,7], Noriyuki Kawaguchi[1,2], Masachika Kijima[1,2], Hideyuki Kobayashi[1,3,4,6], Seisuke Kuji[6], Tomoharu Kurayama[1], Seiji Manabe[2,6], Takeshi Miyaji[1,3], Takumi Nagayama[5], Akiharu Nakagawa[5], Chung Sik Oh[1,4], Toshihiro Omodaka[5], Tomoaki Oyama[1], Satoshi Sakai[6], Katsuhisa Sato[6], Tetsuo Sasao[8,9], Katsunori M. Shibata[1,3], Motonobu Shintani[5], Hiroshi Suda[4,6], Yoshiaki Tamura[6], Miyuki Tsushima[5] and Kazuyoshi Yamashita[1,2]

email: honmamr@cc.nao.ac.jp

[1]Mizusawa VERA Observatory, NAOJ, Mitaka, 181-8588 Tokyo, Japan
[2]Graduate University for Advanced Studies, Mitaka, Tokyo 181-8588
[3]Space VLBI Project, NAOJ, Mitaka, Tokyo 181-8588
[4]Department of Astronomy, University of Tokyo, Bunkyo, Tokyo 113-8654
[5]Faculty of Science, Kagoshima University, Korimoto, Kagoshima, Kagoshima 890-0065
[6]Mizusawa VERA observatory, NAOJ, Mizusawa, 023-0861, JAPAN
[7]Department of Astronomy, Yonsei University, Seoul, 120-749, Republic of Korea
[8]Ajou University, Suwon 442-749, Republic of Korea
[9]Korean VLBI Network, KASI, Seoul, 120-749, Republic of Korea

Abstract. We present results of astrometric observations of S269 H_2O maser performed with VERA (VLBI Exploration of Radio Astrometry). We have monitored the positions of S269 H_2O masers for 1 year and successfully detected its parallax to be 189 ± 8 micro-arcsecond. This corresponds to a source distance of $5.28^{+0.24}_{-0.22}$ kpc, and is the smallest parallax (and thus the largest distance) that has ever been measured by means of annual parallax. Proper motions of S269 H_2O maser were also measured and used to determine the Galactic rotation velocity at the position of S269. Our measurements show that the Galactic rotation velocity at S269 is the same to that at the Sun within 3%, indicating that the Galactic rotation curve is flat out to R$\sim$13 kpc.

Keywords. ISM:star forming regions — ISM:individual(Sharpless 269) — masers (H_2O) — VERA

1. Introduction

VLBI astrometry of maser sources provide unique tools to explore the 3-D structure of the Galaxy. For instance, recent maser observations with VLBA proved that kpc-scale astrometry is possible with phase-referencing technic (Kurayama *et al.* 2005; Xu *et al.* 2006, Hachisuka *et al.* 2006). VERA (VLBI Exploration of Radio Astrometry) is a new Japanese VLBI array dedicated to VLBI astrometry (Honma *et al.* 2000, Kobayashi *et al.* 2003). With its dual-beam system, the VERA telescope has a capability to observe simultaneously a target and a reference source within 2.2 degrees. This dual-beam system allows us to cancel out tropospheric fluctuations effectively and to achieve an astrometry accuracy of 10 μas-level (Honma *et al.* 2003). With the accuracy at the 10 μas, VERA aims to precisely locate hundreds of maser sources and to explore the 3-D structure and dynamics of the Galaxy. The VERA projected was funded in 1999, and the construction of the array was started in 2000. At the time of the last maser meeting in Brazil in 2001, the array was under construction, and the construction status was reported at the meeting (Honma 2002). The VERA array was completed in 2002, and regular observations have been started since the fall of 2003. Here we present initial results of high precision astrometry with VERA toward the Galactic star forming region S269, in particular to do the parallax measurements and to constrain the Galactic rotation velocity in the outer Galaxy.

2. Astrometry of S269 H_2O maser

H_2O masers in the Galactic star forming region Sharpless 269 (S269) were regularly monitored with VERA since November of 2004. In this paper, we present the data of 6 epochs spanning nearly one year. At each epoch, the S269 H_2O maser at 22 GHz (rest frequency of 22.235080 GHz, H_2O 6_{16}-5_{23} transition) and an ICRF calibrator source J0613+1306 were simultaneously observed in dual-beam mode for nearly 9 hours. The separation angle between the maser and the reference source is 0.73 degree. Details of observations and data reductions are presented in Honma *et al.* (2007).

Figure 1(a) and 1(b) show the total-power spectrum of the S269 H_2O maser and the maser spot distribution on DOY (day-of-year) 073 in 2005. Six maser spots were detected in the velocity range from 19.0 to 20.1 km s^{-1}, and these maser spots are aligned in the east-west direction with a scale of 0.4 mas. Positions in Figure 1(b) are the residuals to the tracking center positions of the maser and the reference sources, which are taken to be (α, δ)=(06h14m37.08s, +13d49′36.7″) for S269, and (06h13m57.692764s, +13d06′45.40116″) for J0613+1306, both in J2000 coordinates. The absolute position of the S269 H_2O maser is consistent with the position of S269 IRS2w, which is the most luminous infrared star in the S269 YSO cluster (Jiang *et al.* 2003).

The positional variations of the brightest maser spot are shown in Figure 1(c) (for east-west offset, X) and figure 1(d) (for north-south offset, Y). The east-west position offsets clearly show a systematic sinusoidal modulation with a period of 1 yr, that is the parallax of S269. In this paper we use only the east-west component to determine the parallax of S269, partly because the Y-direction error is large and also because S269 is near the ecliptic and the parallax ellipse is highly elongated in the $X-$direction, making the contribution of Y to the parallax determination small. Based on least-squares fit to positions of the three brightest maser spots, the best estimate of the parallax π is obtained to be 189 $\pm$ 8 μas. This corresponds to a source distance of $5.28^{+0.24}_{-0.22}$ kpc, being the largest distance ever measured by means of trigonometric parallax. The distance to S269 is found to be slightly larger than previous estimates, which claimed a distance of $\sim$4 kpc (Moffat *et al.* 1979).

In addition to the source parallax, mean proper motions of the brightest three spots are obtained as (μ_X, μ_Y)=(-0.422 ± 0.010, -0.121 ± 0.042) mas yr^{-1}, respectively. Using the solar motion based on the HIPPARCOS satellite data (Dehnen & Binney 1998), the proper motions in the Galactic coordinate are obtained as (μ_l, μ_b)=(-0.184 ± 0.032, -0.149 ± 0.029) mas yr^{-1}. With the source distance of 5.28 kpc, these proper motions correspond to the velocity vector of (v_l, v_b)=(-4.60 ± 0.81, -3.72 ± 0.72) km s^{-1}.

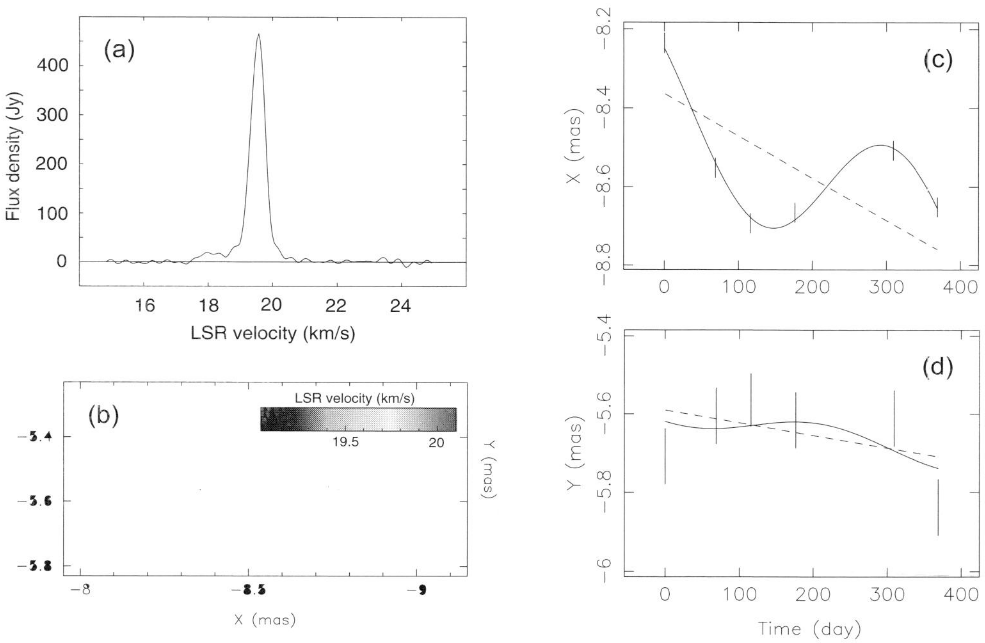

Figure 1. (a, top left) Auto-correlation spectrum of S 269 H_2O maser emissions taken on DOY 073 in 2005. (b, bottom left) maser spot distribution obtained with VERA on DOY 073 in 2005. The coordinates are with respect to the tracking center position, which is (06h14m37.08s, +13d49'36.7") in J2000. (c, top right) Positional variations of the brightest maser spot at $V_{\rm LSR}$=19.6 km $^{-1}$ for X (east-west) direction. (d, bottom right) the same plot for Y (north–south) direction. The time origin is the first epoch of our observations, which is DOY 323 in 2004. Solid curves are best-fit results with parallax and proper motions, and dashed lines show the best-fit proper motions.

3. Galactic rotation measurements

Here we use the proper motions and parallax obtained above to constrain the Galactic rotation velocity at the position of S269. The two velocity components $(v_l, v_b)=(-4.60 \pm 0.81, -3.72 \pm 0.72)$ km s^{-1}, are remarkably small compared to the rotation speed of the Galaxy, which is an order of $\sim$ 200 km s^{-1}. The small proper motion perpendicular to the Galactic plane ($v_b = -3.72 \pm 0.72$ km s^{-1}) implies that S269 is basically on its galactic rotation and also that the H_2O maser proper motions truly reflect the systemic motion of S269. Then, given that S269 is located in the anti-center region ($l = 196°$), the small value of v_l indicates that the Galactic rotation velocity at the Sun and that at S269 are close to each other and proper motions were cancelled out in relative proper motion measurements. If a source is in perfect Galactic rotation, tangential velocity with respect to the LSR observer can be written as,

$$v_l = \left(\frac{\Theta}{R} - \frac{\Theta_0}{R_0}\right) R_0 \cos l - \frac{\Theta}{R} D, \tag{3.1}$$

Using the above equation, the Galactic rotation velocity at the position of S269 is obtained as $\Theta = 1.00(\pm 0.03) \times \Theta_0$. Thus, the rotation velocity at the S269 (Θ) must be the same with Θ_0 within 3% level. This is the strongest constraint of the rotation velocity in the outer galaxy ever obtained.

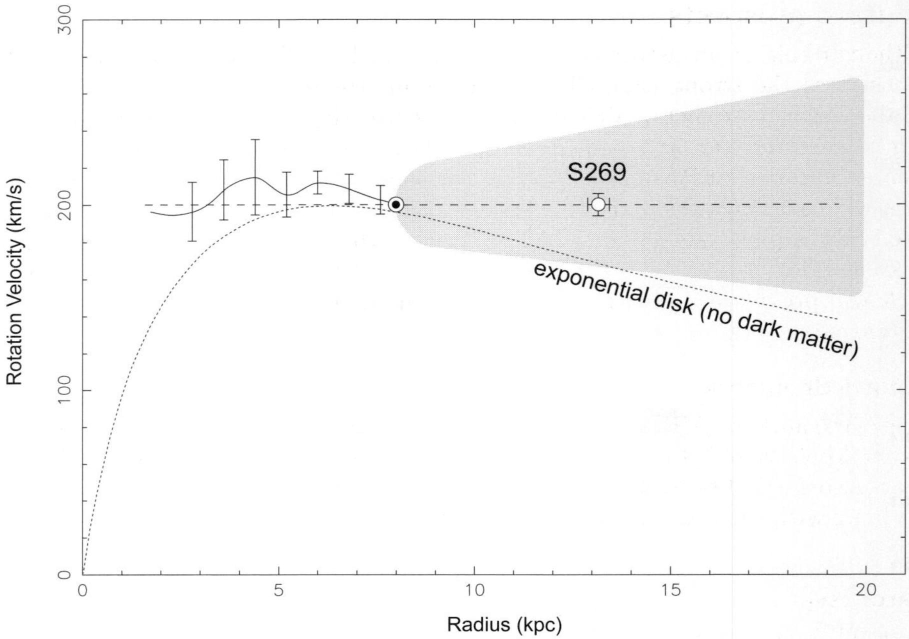

Figure 2. Rotation curve of the Galaxy obtained in the previous studies as well as the result from S269 using Galactic constants of R_0=8 kpc and Θ_0=200 km $^{-1}$. The dashed line is the flat rotation curve with Θ=200 km $^{-1}$, and the shadowed area shows the possible range of outer rotation curves in previous studies (Honma & Sofue 1997). Points at $R \leqslant 8$ kpc are inner rotation curves determined from the tangential velocities of Galactic HI gases (Honma & Sofue 1997), with a smoothed fit (thick curve). The dotted curve is the rotation curve for an exponential disk, corresponding to a constant mass-to-light ratio disk without dark matter. The discrepancy between the observed point for S269 and the exponential disk is evident, demonstrating the existence of large amount of dark matter in the outer region of the Galaxy.

In previous works, the Galactic rotation curve has an uncertainty up to 100 km s^{-1} in the outer region if one includes the strong dependence on the Galactic constant Θ_0 (Honma & Sofue 1997). This situation is summarized in Figure 2, showing the area of uncertainty in previous studies. The point for S269 determined in this study is also shown in figure 2. The coincidence of rotation velocities at the Sun and at S269 confirms the idea that the rotation curve of the outer Galaxy is basically flat, as widely known for other spiral galaxies (e.g., Rubin *et al.* 1980, Sofue & Rubin 2001). In disk galaxies like the Galaxy, optical surface brightness obeys an exponential law, i.e., $I(R) = I_0 \exp(-R/h)$, where R is the radius and h is the disk scale length. Assuming that the surface brightness traces the mass density (i.e., constant mass-to-light ratio), one can calculate the rotation curve of the optical disk without dark matter (Freeman 1970). In Figure 2, we also showed such a rotation curve for the Galaxy's disk, assuming the disk scale length of $h = 3$ kpc and the maximum rotation velocity of 200 km s^{-1}. Our measurements of S269 clearly rules out such a no-dark-matter case. At the position of S269, the rotation velocity of the exponential disk ($V_{\rm exp}$) is smaller than $V_{\rm obs}$ by 16%. The ratio of these two velocities gives $(V_{\rm exp}/V_{\rm obs})^2 = 0.70$, and hence, assuming a spherical mass distribution (where an enclosed mass can be estimated as $M \sim RV^2/G$), at least $\sim$ 30% of the enclosed mass within 13 kpc must be composed of dark matter.

4. Future prospects

Although this is an astrometric measurements for only one source, the S269 case demonstrates the strong capability of VERA in studying the Galactic rotation based on Galaxy-scale astrometry. VERA has been conducting monitoring programs of tens of maser sources per year, and some interesting astrometric results are already obtained, as presented in other VERA-related papers in this proceeding volume. In the next decade, VERA will continue astrometry of nearly 1000 H_2O and SiO maser sources in the Galaxy. There is also a plan to install 6.7 GHz receivers so that $\sim$500 of 6.7 GHz methanol maser sources are also used to trace the structure of the Galaxy. Hopefully, in the near future, VERA will provide a dramatic revolution in our understanding the Galaxy's structure and dynamics.

Acknowledgements

One of the authors (MH) acknowledges financial support from grant-in-aid (No.16740120) from the Ministry of Education, Culture, Sports, Science and Technology (MEXT). The authors also would like to thank all the supporting staff at Mizusawa VERA observatory for helping with the observations.

References

Dehnen, W. & Binney, J. 1998, MNRAS, 298, 387
Freeman, K. C. 1970, ApJ 160, 811
Hachisuka, K., *et al.* 2006, ApJ 645, 337
Honma, M. & Sofue, Y. 1997, PASJ 49, 453
Honma, M., Kawaguchi, N. & Sasao, T. 2000, in Proc. SPIE Vol.4015 Radio Telescope, ed H. R. Buthcer, 624
Honma, M. 2002, in IAU Symposium Vol. 206 Cosmic MASERs: FromProtostars to Black holes, ed V. Migenes and M. J. Reid, 501
Honma, M., *et al.* 2003, PASJ, 55, L57
Honma, M., *et al.* 2007, submitted to PASJ
Jiang, Z., *et al.* 2003, ApJ 596, 1064
Kobayashi, H., *et al.* 2003, in ASP Conf. Ser. 306, New Technologies in VLBI, ed. Y. C. Minh, 367
Kurayama, T., Sasao, T., & Kobayashi, H. 2005, ApJ, 627, L49
Moffat, A. F. J., Jackson, P. D., & Fitzgerald, M. P. 1979, A&AS 38, 197
Rubin, V. C., Thonnard, N. T., & Ford, W. K. Jr. 1980, ApJ 238, 471
Sofue, Y., Rubin, V., 2001, ARA&A, 39, 137
Xu, Y., Reid, M. J., Zheng, X. W., & Menten, K. M., 2006, Science 311, 54

Astrophysical Masers and their Environments
Proceedings IAU Symposium No. 242, 2007
J.M. Chapman & W.A. Baan, eds.

doi:10.1017/S1743921307013397

Masers as probes of massive star formation in the nuclear disk

F. Yusef-Zadeh[1], R. G. Arendt[2], C. O. Heinke[1], J. L. Hinz[3], J. W. Hewitt[1], P. Pratap[4], S. V. Ramirez[5], G. H. Rieke[3], D. A. Roberts[1], S. R. Stolovy[5], M. Wardle[6] and B. A. Whitney[7]

[1]Department of Physics and Astronomy, Northwestern University, Evanston, IL 60208 USA
email: zadeh@northwestern.edu, j-hewitt@northwestern.edu, cheinke@northwestern.edu, doug-roberts@northwestern.edu

[2]CRESST/UMBC/GSFC, Code 665, Greenbelt, MD 20771 USA
email: arendt@milkyway.gsfc.nasa.gov

[3]Steward Observatory, University of Arizona, 933 N. Cherry Ave., Tucson, AZ 85721 USA
email: grieke@as.arizona.edu, jhinz@as.arizona.edu

[4]MIT Haystack Observatory, Westford, MA 01886 USA
email: ppratap@haystack.mit.edu

[5]IPAC, Cal Tech, Pasadena, CA 91125 USA
email: solange@ipac.caltech.edu, stolovy@ipac.caltech.edu

[6]Department of Physics, Macquarie University, Sydney NSW 2109, Australia
email: wardle@ics.mq.edu.au

[7]Space Science Institute, 4750 Walnut Street, Suite 205, Boulder, CO 80301 USA
email: bwhitney@spacescience.org

Abstract. OH(1720 MHz) and methanol masers are now recognized to be excellent probes of the interactions of supernova remnants with molecular clouds and tracers of massive star formation, respectively. To better understand the nature of star formation activity in the central region of the Galaxy, we have used these two classes of masers combined with the IRAC and MIPS data to study prominent sites of ongoing star formation in the nuclear disk. The nuclear disk is characterized by massive GMCs with elevated gas temperatures, compared to their dust temperatures. We note an association between methanol masers and a class of mid-infrared "green sources". These highly embedded YSOs show enhanced 4.5μm emission due to excited molecular lines.

The distribution of methanol masers and supernova remnants suggest a low efficiency of star formation (with the exception of Sgr B2), which we believe is due to an enhanced flux of cosmic ray electrons impacting molecular clouds in the nuclear disk. We also highlight the importance of cosmic rays in their ability to heat molecular clouds, and thus increase the gas temperature.

Keywords. Masers, supernova remnants, cosmic rays.

1. Introduction

Understanding the processes occurring in the nuclear disk of our own Galaxy is interesting not only for insight into our own Milky Way Galaxy, but also because it is the closest galactic nucleus. This important region of the Galaxy hosts several sources of energetic activity, and is the site of massive molecular clouds with pockets of past and present massive star formation. However, the study of the nuclear disk of our galaxy has led to a number of paradoxes, puzzles and discrepancies. One of the best paradoxical examples is related to the nature of the compact radio source Sgr A* at the dynamical center of the Galaxy. Although there is compelling evidence that Sgr A* is a massive black hole (Schödel *et al.* 2003; Ghez *et al.* 2003), its X-ray luminosity is orders of magnitude lower than expected. Another paradox is related to the expectation that the formation of stars near Sgr A* should be forbidden due to the extreme tidal forces exerted by Sgr A*; however, many young stars are distributed in a disk orbiting Sgr A* (Paumard *et al.* 2006). Another outstanding puzzle is the origin of nonthermal filaments in the Galactic nucleus, which has not been understood since their discovery (see Yusef-Zadeh, Hewitt & Cotton 2004; Nord *et al.* 2004).

On a larger scale, the nature of on-going star formation in the nuclear region remains unclear. Although there is a high concentration of dense molecular clouds distributed in this region, the star formation rate has shown extreme values. On the one hand, the massive star forming region Sgr B2 points to the closest example of starburst activity in our Galaxy, whereas the quiescent giant molecular cloud (GMC) G0.25+0.01 (Lis & Carlstrom 1994; Lis & Menten 1998) appears to contain highly inefficient star formation. This region contains some of the most spectacular molecular clouds with elevated gas temperatures but no obvious source of heating of the clouds. In the last two decades, studies of this region have indicated that the molecular gas is warm, ranging between 75 to 200 K (e.g., Hüttemeister *et al.* 1993). However, studies based on far-IR observations have revealed dust temperature to be $<\sim$30K (Odenwald & Fazio 1984; Cox & Laureijs 1989; Pierce-Price *et al.* 2000). This discrepancy between the gas and dust temperatures in the central region of the Galaxy is puzzling.

Using SNR masers and methanol masers, we place constraints on the low efficiency of star formation in the Galactic nucleus. We then argue that the low efficiency of star formation and the elevated gas temperature with respect to dust temperature can be explained by the enhanced energy density of cosmic rays interacting with molecular clouds in the nuclear disk. The interstellar medium of the nuclear disk is distributed in the region with an extent $-1^0 < l < 2^0$ and $-0.5^0 < b < 0.5^0$ and is characterized by a strong concentration of molecular gas in the so-called "Central Molecular Zone" with a radius of ~200 pc (Morris & Serabyn 1996). Physical conditions in this region are extreme, characterized by high velocity dispersion (~20 km/s), high density ($\sim 10^4$ cm^{-3}) and high column density ($\sim 10^{23}$ cm^{-2}).

2. Supernova remnant masers

The high number density as well as the high column density of warm gas in the nuclear disk are similar to the restricted physical conditions required to produce OH(1720 MHz) masers (Lockett, Gauthier & Elitzur 1999). At the site of C-type shocked gas, OH abundance is expected to be enhanced sufficiently to detect OH(1720 MHz) maser emission throughout the nuclear disk (Wardle 1999). With this in mind, we carried out an unbiased, snapshot survey of the Galactic nucleus to search for SNR masers. When this survey is combined with other targeted surveys, a total of six SNRs are detected toward roughly the inner $8^0 \times 1^0$ (Yusef-Zadeh *et al.* 1995, 1996, Frail *et al.* 1996). However, only four SNR masers G0.0+0.0 (Sgr A East), Sgr D (G1.13-0.1), G1.4-0.1 & G359.1-0.5 are detected toward the nuclear disk. The only known SNRs with no maser counterparts in the nuclear disk are G0.9+0.1 and G0.33+0.01. Here we discuss SNR masers associated with the SNRs Sgr A East and G359.1-0.5.

2.1. *SNR G0.0+0.0*

The most well known interacting SNR in the nuclear disk is Sgr A East (SNR G0.0+0.0) projected toward Sgr A*. The kinematics of OH(1720 MHz) masers indicate two populations. The first kinematic population has velocities ranging between 30 and 60 km s^{-1} distributed around the shell of Sgr A East. Recent sensitive observations of OH masers toward Sgr A East have discovered a number of additional maser spots (see the references

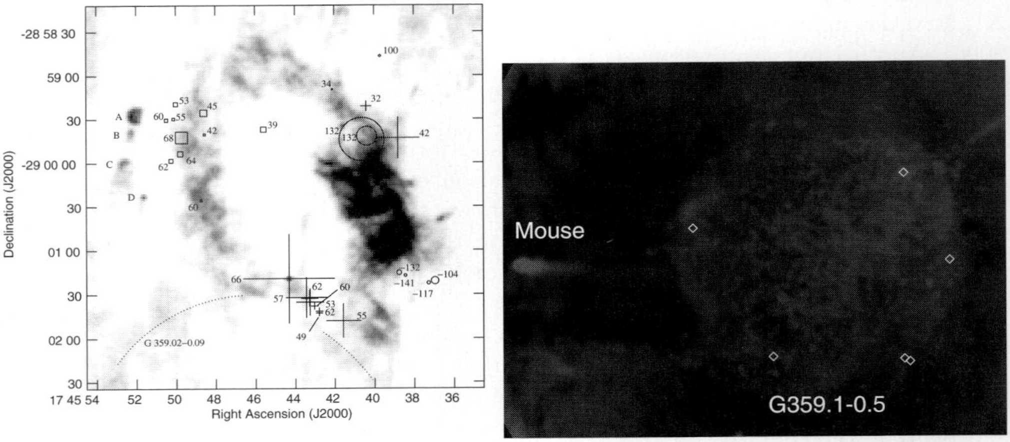

Figure 1. (*a*) A grayscale image of the continuum emission from Sgr A East and West at 20cm. The distribution of OH(1720 MHz) maser spots with their V_{LSR} velocities are superimposed on the nonthermal Sgr A East shell and the thermal spiral-like structure of Sgr A West (taken from Figure 1 of Pihlström and Sjouwerman 2006). (*b*) A 20cm continuum image of G359.1-0.5 (red color) shows the shell-like structure of this remnant. The interior of the shell is represented by the distributions of X-ray (blue color). The diamond points represent the positions of maser spots. The elongated feature to the east of the figure corresponds to radio emission from the pulsar wind nebula known as the Mouse. A colour rendition of this figure can be found in the on-line version.

within Pihlstrom and Sjouwerman 1996; Karlsson *et al.* 2003). These measurements established that Sgr A East is interacting with molecular gas associated with the 50 km s^{-1} GMC. A number of weak OH(1720 MHz) masers to the northeast of Sgr A East coincide with a large number of shocked molecular clumps, as probed by CS (1-0) line observation with NMA (Tsuboi *et al.* 2006), The broad line emission from molecular gas provides additional support for shocked 50 km s^{-1} clouds as a result of the expansion of the SN shock. The line profiles of the CS line emission are asymmetrically blue shifted which implies that the far side of the the 50 km s^{-1} GMC is being shocked by the expansion of the Sgr A East shell located behind the 50 km s^{-1} cloud (Tsuboi *et al.* 2006).

The second kinematic component shows highly blue- and red-shifted velocities at $\sim \pm 132$ km s^{-1}(Yusef-Zadeh *et al.* 2002; Pihlström and Sjouwerman 2006). These high velocity masers are distributed symmetrically at the tangent points of the lobes of the circum-nuclear ring orbiting Sgr A*. OH(1720 MHz) masers are excellent indicators of the systemic motion of the molecular gas that is being shocked. The low velocity dispersion noted between the highly blue- and red-shifted masers places a constraint on the average motion of the enclosed material within the ring of $V_{LSR} \sim 0$ km s^{-1}. The origin of the masers associated with the circum-nuclear ring remains less clear than those of Sgr A East. On the one hand, it is possible that the expansion of Sgr A East runs into the circum-nuclear ring and produces the observed OH (1720 MHz) masers. This is consistent with the fact that Sgr A East lies behind Sgr A* (Pedlar *et al.* 1989). In this picture, Sgr A East lies behind both the 50 km s^{-1}molecular cloud and the circum-nuclear molecular ring and is responsible for production of OH(1720 MHz) masers in both clouds. On the other hand, these high velocity maser spots may originate within the ring by alignment of molecular clumps, resulting in amplification along the line of sight (Pihlström & Sjouwerman 2006). Future study of these masers should clarify their relationship to the circum-nuclear molecular ring which is known to be orbiting Sgr A*.

2.2. *SNR G359.1-0.5*

G359.1-0.5 shows several masers surrounding the remnant at a velocity of -5 km s^{-1} and a low velocity dispersion. Such low velocity maser features are unusual if the remnant is interacting with molecular gas in the nuclear disk. This low velocity is in contrast to the kinematics of a CO shell suggested to surround this remnant, which shows velocities ranging between -60 and -150 km s^{-1} (Uchida *et al.* 1992). Several more recent studies such as the detections of a bar of H_2 emission at low velocities and the linear polarization from the brightest maser spot as well as the relatively low rotation measure of the remnant itself indicate that G359.1-0.5 is unlikely be located in the Galactic nucleus. More recently, an XMM observation of thermal X-ray gas within the remnant supports a low column density to the remnant, consistent with it being a foreground source (Heinke *et al.* 2007, in preparation), as described below. These observations indicate clearly that OH(1720 MHz) maser emission is associated with a foreground remnant. This implies that OH(1720 MHz) masers are powerful indicators of the systemic motion of the interacting remnant especially in the confused region of the nuclear disk.

The overall fit to the X-ray emission from G359.1-0.5 places a strong constraint on the low-energy absorption, $N_H = 2.6 \pm 0.2 \times 10^{22}$ cm^{-2} (Heinke *et al.* 2007, in preparation). The pulsar wind nebula known as the Mouse lies only 22 arcminutes from the center of the remnant, and pulsar dispersion measurements infer a distance to the Mouse between 3 and 5.5 kpc (requiring the Mouse to be within the molecular ring at 3-5 kpc distance, but not beyond the spiral arm 3 kpc from the Galactic center), giving a nominal value of 5 kpc. The N_H to the Mouse is also estimated from X-ray measurements to be $2.60{\pm}0.09 \times 10^{22}$ cm^{-2} (Mori *et al.* 2004; Gaensler *et al.* 2004). Our measured N_H value is consistent with the Mouse's N_H, suggesting that G359.1-0.5 is located at a similar distance and is likely to be the origin of the Mouse.

2.3. *Star formation rate from known SNRs*

If we remove G359.1–0.5 from the list of remnants interacting with molecular gas in the nuclear disk, then 3/5 or ~60% of all known remnants in this region are interacting with molecular gas in the nuclear disk. This is in contrast to the fraction of interacting remnants in the disk of the Galaxy, 10%. This implies a uniform distribution of molecular gas throughout the nuclear disk.

Given the number of known SNRs and assuming a residence time of $2.5{\times}10^4$ yrs, the SN rate is estimated to be one per 5000 years. This implies that the rate of formation of massive stars is 10^{-2} $M_{\odot}yr^{-1}$ (Condon 1992). This is consistent with the far-IR measurements inferring a paucity of massive stars in the nuclear disk (Odenwald and Fazio 2004). This estimate is, however, quite uncertain since the residence time of a remnant expanding into a dense medium is unknown. If the residence time is shorter than the above estimate, the star formation rate estimate increases, accordingly.

3. Methanol masers

It is now well established that methanol masers are signposts of ongoing massive star formation throughout the Galaxy. A survey of the inner 2 degrees of the Galactic center showed a total of 23 class II methanol masers at 6.7 GHz (Caswell 1996). This class of masers is recognized to be radiatively pumped, unlike class I methanol masers, known to be collisionally pumped (e.g., Menten 1991). There is a strong correlation between the 6.7 GHz methanol masers and infrared dark clouds as revealed in the 8μm Spitzer observations of the Galactic center (Stolovy *et al.* 2006). The infrared dark clouds are likely progenitors of high mass stars, and the masers indicate locations where star formation

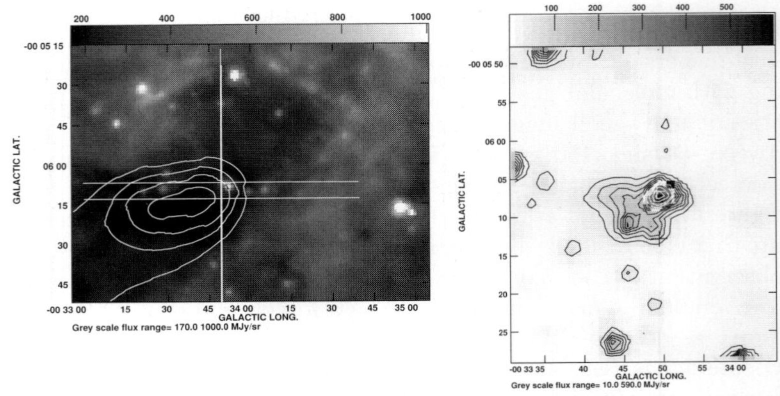

Figure 2. (*a*) Contours of 850μm emission (Pierce-Price *et al.* 2000) are superimposed on the 8μm image of Sgr C. (*b*) Contours of 4.5μm emission based on Channel 2 of IRAC data. The crosses in (a) and (b) represent methanol maser distributions associated with Sgr C.

is currently taking place. Closer examination of the kinematics of the 6.7 GHz methanol masers indicates that 11 maser spots arise from Sgr B2, two from Sgr C, and 3-5 from the quiescent dust ridge G0.25-0.01. Given that there is a large reservoir of dense molecular gas, up to 10^8 $M_\odot$, distributed in the nuclear disk, there appears to be a relative paucity of methanol masers (above 1 Jy) in the nuclear disk (Caswell 1996). With the exception of Sgr B2, the rarity of bright 6.7 GHz masers suggests low star formation activity in the nuclear disk. The inner 30 pcs of the Galactic center show no evidence of any class II methanol masers. Remarkably, this inner region is known to have a large concentration of dense giant molecular clouds associated with the 20, 40 and -30 km s^{-1} molecular clouds (e.g., Güsten *et al.* 1981). As described below, this region is also known to be coincident with strong nonthermal emission from Sgr A East, its halo and the radio continuum arc near l$\sim$ 0.2^0.

3.1. *"Green" sources*

To determine if any of the methanol masers have a stellar counterpart in the mid-IR, we examined the color of stars in the nuclear disk. One of the most powerful methods of detecting molecular line emission in mid-IR wavelengths using IRAC images is to look for excess emission at 4.5μm ("green" in 3-color images combining 3.6, 4.5 and 8 μm emission). Using this color selection criterion, a recent IRAC survey of the Galactic plane has identified a number of interacting supernova remnants showing excess emission at 4.5μm, indicating shocked, excited CO and H_2 molecular line emission (Reach *et al.* 2006). Also, the "green nebulosity" has been detected in DR21, a massive star formation site, in which the excess 4.5μm color is considered to be due shocked molecular outflow (Marston *et al.* 2004; Smith *et al.* 2006).

Using mid-IR data, we have examined the color of stars and dust clouds in star forming sites throughout the nuclear disk. The infrared images of the nuclear disk are generated from the IRAC Galactic center survey mosaics (Stolovy *et al.* 2006). A quantified indication of excess 4.5 μm emission is obtained by constructing the ratio $I(4.5)/[1.2*I(3.6)*I(5.8)]^{(0.5)}$. This essentially is the ratio of the actual 4.5 μm intensity to that determined by a power–law interpolation of the 3.6 and 5.8 μm intensities. It was found empirically that a slight ($\sim$3%) modification to the interpolation coefficients helps emphasize sources with unusually strong 4.5 μm emission. In a map of this ratio across the entire Galactic center survey, most point sources exhibit a uniform ratio that is not very different from the background. In regions of high extinction (e.g. IR dark clouds, and Sgr B2), this ratio rises because the reddening reduces the 3.6 μm intensity. However, there were of order 100 sources, located in regions of both high and low extinction, which exhibit ratios distinctly higher than those of other sources in their vicinity (within several arc minutes). These include 33 sources that were visually selected as "green" sources in 3-color images. Yet the ratio also clearly identifies a number of interesting sources that do not stand out in the 3-color images, such as 3 planetary nebulae. There are thirty green sources suggesting YSOs embedded within infrared dark clouds. A correlation of methanol masers in the nuclear disk with the green sources has led to the surprising result that about 1/3 of green sources have maser counterparts, 1/3 have no maser counterparts, and the rest have not yet been observed. Given the limited sensitivity of 1 Jy in the maser survey by Caswell (1996), it is possible that many of the green sources have weak maser counterparts or that they signify early sites of massive star forming regions.

Although the $\sim 2''$ spatial resolution of IRAC images being different than the $2'' \times 4''$ resolution of methanol maser observations (Caswell 1996), we have mapped the data with $1''$ pixels. The green sources are generally seen to be extended on a scale $\sim 10''$ and the maser sources fall within the extent of the 4.5μm emission. An example in which the relationship between a green source and a methanol maser can be seen is in the Sgr C complex region. This complex is known to consist of an evolved, extended HII region and a prominent nonthermal radio filament running perpendicular to the Galactic plane (Liszt & Spiker 1995; Yusef-Zadeh, Hewitt & Cotton 2004). At the core of the HII region, an elongated infrared dark cloud is noted in which two maser sources and a green source are embedded. Figure 2a shows contours of submm emission at 850μm (Pierce-Price *et al.* 2000) superposed on an IRAC image of this region at 8μm. 4.5μm contours of the bright source which shows prominently in a three-color image of Sgr C are shown in Figure 2b. The crosses show the positions of methanol maser emission at 6.7 GHz. As pointed out earlier both these maser sources could be associated with the green source in Sgr C. We have also mapped 44-GHz methanol maser emission from the green source using Haystack Observatory. This transition of methanol maser emission, which is known to be collisionally pumped, is identified on a scale of $1'$ oriented along the extent of the major axis of the green source. This is also consistent with a picture in which the green source is associated with the maser emission source in Sgr C (a more detailed account of this source will be given elsewhere).

We have modelled the SED of the green source G359.436-0.102 in Sgr C and have estimated the mass and luminosity of the central protostar coincident with the norther maser source by fitting the SED from a large grid of 2-D YSO models (Robitaille *et al.* 2006, 2007). The models include a disk and an accreting envelope with bipolar cavities. Figure 3a shows the best fitted models to the data using IRAC, MIPS and submm data accounting for 30 magnitudes of visual extinction. Due to the lower resolution of submm data, we show upper limits for the peak flux densities at 850 and 450 μm (Pierce-Price *et al.* 2000) at the position of the green source. Different curves of the model fits correspond to the variation of the observed apertures. These model fits all seem to be consistent with the SED of the source. Figure 3b shows the derived mass and luminosity of the green source based on the well-fit SED models. This fit of the SED is consistent with the identification of the green source with a massive protostar, associated with the methanol masers. The derived parameters of protostars using green sources with methanol maser counterparts should be useful indicators of the evolutionary phase of these sources. In addition, further study of the green/maser sources may also address whether linear distribution of maser spots come from circumstellar disks or outflows (De

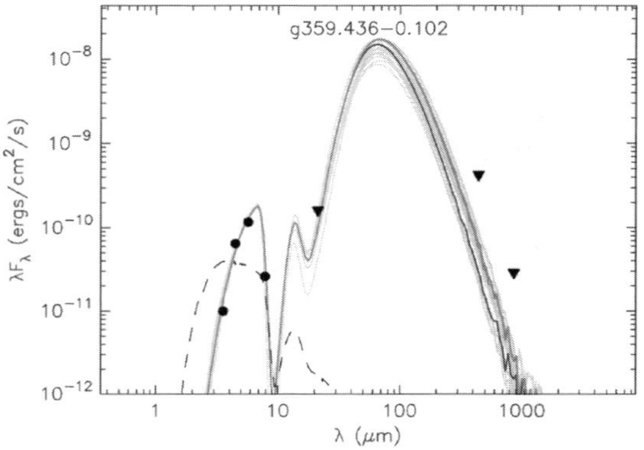

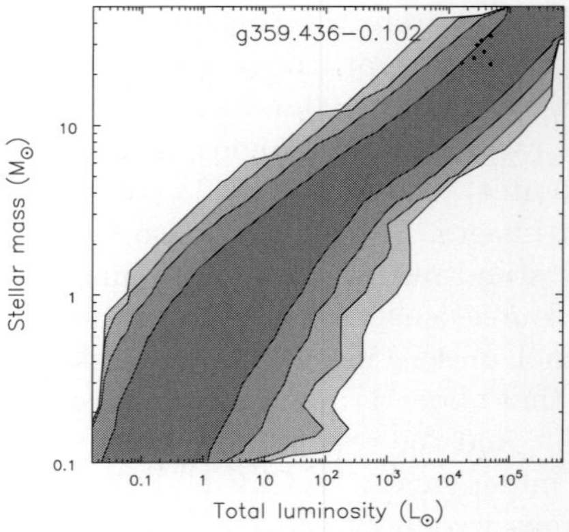

Figure 3. (*a*) A preliminary SED of the northern maser G359.436-0.102 in Sgr C and the fitted model with different aperture sizes. (*b*) The grey scale shows distribution of stellar mass and luminosity from a grid of 20,000 YSO radiative transfer models. The black dots show the mass and luminosities of the well-fit SEDs (those in Figure 3a).

Buizer *et al.* 2005). Furthermore, green sources found based on IRAC observations can provide an extensive target list for future methanol maser studies.

4. The role of cosmic rays

We believe that the low on-going star formation activity inferred from the estimate of the supernova rate and the rarity of methanol masers can be explained by enhanced cosmic ray flux in the nuclear disk. High cosmic-ray fluxes in molecular clouds affect star formation by heating the gas and increasing its ionization fraction (Yusef-Zadeh, Wardle & Roy 2007a). Higher cloud temperatures increase the Jeans mass, potentially changing the IMF, while high ionization increases magnetic coupling to the cloud material, reducing ambipolar diffusion and increasing the time taken for gravitationally unstable cores to contract to the point that they overwhelm their magnetic support.

Several studies have recently indicated the importance of cosmic rays in the nuclear disk. One is from the detection of low frequency 74 MHz radio emission from the central region of the Galaxy, indicating enhanced cosmic rays with cosmic ray electron density of $\sim$7 eV cm^{-3} (LaRosa *et al.* 2005). The other is strong H_3^+ absorption lines detected along several lines of sight towards the Galactic center (Oka *et al.* 2005). These absorption lines are much stronger than those seen in the Galactic disk. These measurements infer that the ionization rate in the inner 30pc of the Galaxy is more two orders of magnitude higher than that that in the Galactic disk. The high ionization rate is more likely to be due to high cosmic ray flux as revealed by high synchrotron emissivity in raido wavelengths than due to diffuse hot X-ray emitting gas. Lastly, the large scale distribution of Kα Iron emission line at 6.4 keV associated with a number of cloud complexes in the nuclear disk has been correlated with the distribution of nonthermal radio emission from this region (Yusef-Zadeh *et al.* 2007b). The energy density of the cosmic rays required to explain the 6.4 keV X-ray emission from the inner $2^0 \times 0.5^0$ of the molecular nuclear disk ranges between 20 and 10^3 eV cm^{-3}.

Acknowledgements

We thank Y. Pihlström for providing us with Figure 1a. C.O.H. is supported by the Lindheimer Postdoctoral Fellowship.

References

Caswell, J. L. 1996, *MNRAS* 283, 606

Candon, J. J. 1992, *ARA&A* 30, 575

Cox, P., & Laureijs, R. 1989, The Center of the Galaxy, 136, 121

De Buizer, J. M., Radomski, J. T., Telesco, C. M., & Piña, R. K. 2005, ApJ Supp, 156, 179

Frail, D. A. et al. 1996, *AJ* 111, 1651

Gaensler, B. M. et al. 2004, ApJ, 616, 383

Ghez, A. M., et al. 2003, ApJ, 586, L127

Gusten, R., & Downes, D. 1981, A&A, 99, 27

Hüttemeister, S. et al. 1993, *A&A* 280, 255

Karlsson et al. 2003, *A&A* 403, 1011

LaRosa, T. N. et al. 2005, *ApJ* 626, L23

Lis, D. C.& Carlstrom, J. E. 1994, *ApJ* 424, 189

Lis, D. C.& Menten, K. 1998, *ApJ* 507, 794

Liszt, H.S., & Spiker, R.W. 1995, *ApJS* 98, 259

Lockett, P., Gauthier, E. & Elitzur, M. 1999, *ApJ* 511, 235

Marston et al. 2004, *ApJS* 154, 333

Menten, K. M. 1991, ApJ, 380, L75

Morris, M. & Serabyn, E. 1996, *ARA&A* 34, 645

Mori, H. et al. 2004, 35th COSPAR Scientific Assembly, 35, 3705

Nord et al. 2004, *AJ* 128, 1646

Odenwald, S. F., & Fazio, G. G. 1984, ApJ, 283, 601

Oka, T. et al. 2005, *ApJ* 632, 882

Paumard, T. et al. 2006, *ApJ* 643, 1011

Pedlar et al. 1989, *ApJ* 342, 769

Pierce-Price, D. et al. 2000, *ApJ* 545, L121

Pihlström, Y. M. & Sjouwerman, L. O. 2006, *JPhCS* 54, 77

Reach, W. T. 2006, *AJ* 131, 1479

Robitaille, T. P., Whitney, B. A., Indebetouw, R., Wood, K., & Denzmore, 2006, *ApJS* 167, 256

Robitaille, T. P., Whitney, B. A., Indebetouw, R. & Wood, K. 2007, *ApJS* (in press, astro-ph/0612690)

Schödel, R., Ott, T., Genzel, R., Eckart, A., Mouawad, N., & Alexander, T. 2003, ApJ, 596, 1015

Smith, H., Hora, J. L., Marengo, M. & Pipher, J. L. 2006, *ApJ* 645, 1264

Stolovy, S. et al. 2006, *JPhCS* 54, 176

Tsuboi, M., Okumura, S. K. & Miyazaki, A. 2006, *JPhCS* 54, 16

Uchida et al. 1992, *ApJ* 398, 128

Wardle, M. 1999, *ApJ* 525, L101

Yusef-Zadeh, F., Roberts, D. A., Bower, G., Wardle, M., & Goss, W. M. 2002, Cosmic Masers: From Proto-Stars to Black Holes, 206, 212

Yusef-Zadeh, F., Hewitt, J. W., & Cotton, W. 2004, *ApJS* 155, 421

Yusef-Zadeh, F., Muno, M., Wardle, M. & Lis, D. C. 2007b, *ApJ* 656, 847

Yusef-Zadeh, F., Uchida, K. I. & Roberts, D. 1995, *Science* 270, 1801

Yusef-Zadeh, F., Roberts, D. A., Goss, W. M., Frail, D. A., & Green, A. J. 1996, ApJ, 466, L25

Yusef-Zadeh, F.; Roberts, D. A.; Goss, W. M.; Frail, D. A.; Green, A. J. 1999, *ApJ* 512, 230

Yusef-Zadeh, F., Wardle, M. & Roy, S. 2007a, *ApJ* submitted

Astrophysical Masers and their Environments
Proceedings IAU Symposium No. 242, 2007
J.M Chapman & W.A. Baan, eds.

doi:10.1017/S1743921307013403

The distance to G59.7+0.1

Y. Xu[1,2]†, M. J. Reid[3], K. M. Menten[1], A. Brunthaler[1], X. W. Zheng[4] and L. Moscadelli[5]

[1]Max-Planck-Institute für Radioastronomie, Bonn, Germany
email: xuye@mpifr-bonn.mpg.de

[2]Purple Mountain Observatory, Nanjing, China

[3]Harvard-Smithsonian Center for Astrophysics, USA

[4]Nanjing University, China

[5]Arcetri Obs., Firenze, Italy

Abstract. We have measured the distance to the high-mass star-forming region G59.7+0.1, which harbors the far-infra-red source IRAS 19410+2336. The distance is 2.20 ± 0.11 kpc and was determined by triangulation using Very Long Baseline Array (VLBA) observations of 12.2 GHz methanol masers phase-referenced to two compact extragalactic radio sources.

Keywords. ISM: molecules, Galaxy: kinematics and dynamics, Galaxy: structure.

---

1. Introduction

We are carrying out a large project to study the spiral structure and kinematics of the Milky Way (Xu *et al.* 2006). Here we report our results for G59.7+0.1, a very active star-forming region (Fig.1, left), which contains millimeter-wavelength dust continuum emission on multiple scales (Beuther & Schilke 2004), including a dense hot core with an associated methanol maser emission and a weak centimeter-wavelength continuum sourcse (Beuther *et al.* 2002).

2. Observations and Results

We have conducted phase-referenced observations of G59.7+0.1 with the VLBA over 7 epochs in order to measure its relative position (using its methanol masers as astrometric targets) with respect to extragalactic radio sources. In Fig. 1 (right) we show an image from the the first epoch of a maser spot and of each of the background radio sources, which were phase-referenced relative to the strong maser emission. The QSOs as well as the maser spot are compact and appear dominated by a single component.

In order to provide the data needed to measure the parallax and proper motion, we fit Gaussian brightness distributions to the brightest maser spot and the two background radio sources for all seven epochs. In Fig. 2 we plot the positions of the maser spot relative to the two background radio sources. The results are:
Parallax: 0.454 ± 0.022 mas (Distance: 2.20 ± 0.11 kpc)
Proper motion in R.A.: -1.63 ± 0.04 mas y^{-1}
Proper motion in Decl.: -5.12 ± 0.1 mas y^{-1}

The 3-D motion of G59.7+0.1 at its position in the Galaxy, assuming $R_0 = 8.5$ kpc and $\Theta_0 = 220$ km s^{-1}, is as follows:
In the direction of Galactic rotation: 215 ± 10 km s^{-1};
Radial toward the Galactic Center: $+9 \pm 3$ km s^{-1}

† Present address: Auf dem Hügel 69 D-53121 Bonn, Germany.

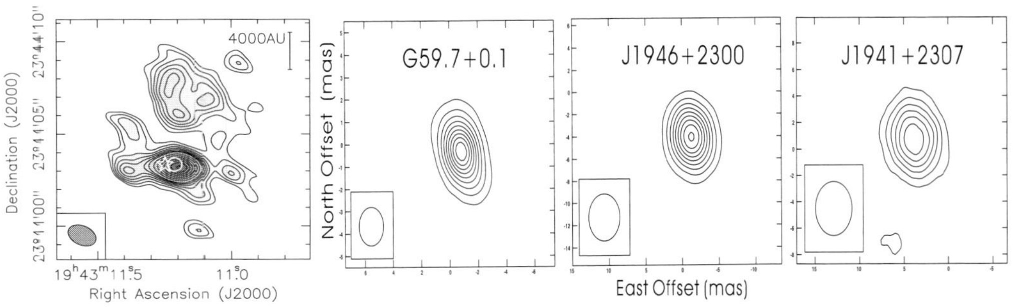

Figure 1. Left: The contours represent 230 GHz mm continuum emission imaged with the IRAM Plateau de Bure interferometer by Beuther & Schilke (2004). The circle marks the position of a compact, weak (1 mJy) 3.6 cm radio continuum source (from Beuther *et al.* 2002) and the star that of the CH_3OH maser. The restoring beam of the interferometer observation is represented in the lower left part of the figure. Right: Synthesized images of the maser (left) and two background radio sources from the first epoch. The contour levels are multiples of 150 mJy for G59.7+0.1, 2 mJy for J1941+2307 and 10 mJy for J1946+2300, with the zero contour suppressed. All sources appear dominated by a single compact component.

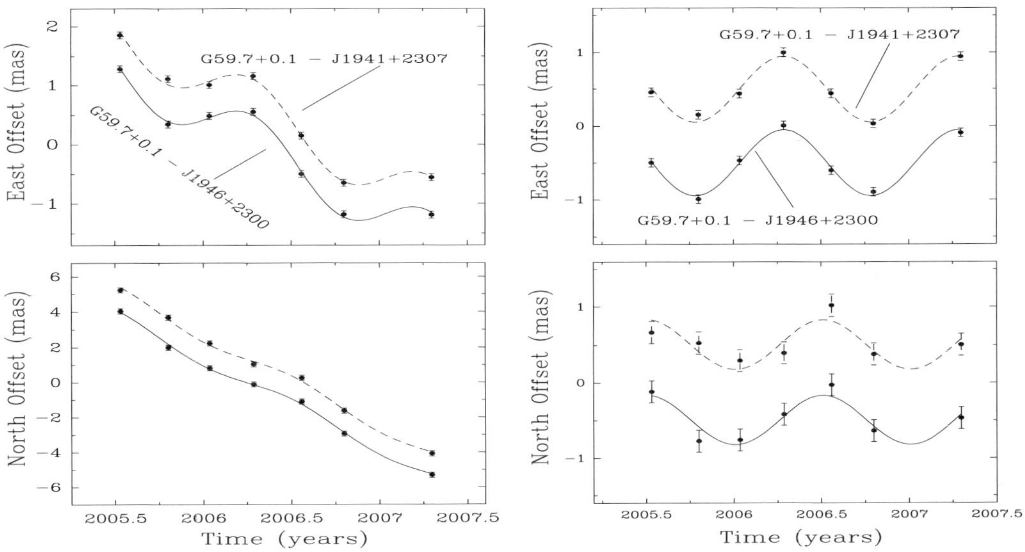

Figure 2. Position versus time for the reference maser spot at V_{LSR} 27.4 km s^{-1} relative to two background radio sources (offset for clarity). The left and right panels indicate the positions of the maser spot before and after removing the fitted proper motion. The top and bottom panels show the eastward and northward offsets.

Toward North Galactic Pole: -6 ± 0.5 km s^{-1}.

This work is supported by NSFC grant 10673024.

References

H. Beuther & P. Schilke, 2004, *Scince*, 303, 1167

H. Beuther *et al.* 2002, *A&A* 390, 289

Y. Xu, M. J. Reid, X. W., & Zheng, K. M. Menten 2006, *Scince*, 311, 54

Astrophysical Masers and their Environment
Proceedings IAU Symposium No. 242, 2007
J.M. Chapman & W.A. Baan, eds.

doi:10.1017/S1743921307013415

TCS-CAIN: NIR survey of the Galactic plane

C. González-Fernández[1], A. Cabrera-Lavers[1,2], F. Garzón[1,3], P. L. Hammersley[1], M. López-Corredoira[1] and B. Vicente[1]

[1]Instituto de Astrofísica de Canarias, E-38205 La Laguna, Tenerife, Spain
[2]GTC Project Office, E-38205 La Laguna, Tenerife, Spain
[3]Departamento de Astrofísica, Universidad de La Laguna, E-38205 La Laguna, Tenerife, Spain

Abstract. We present TCS-CAIN, a NIR survey of the Galactic plane, recently made public at the Instituto de Astrofísica de Canarias, and some results derived from it: with star counts derived directly the structure of the inner the Milky Way can be dissected, obtaining also estimations for the extinction toward the inner MW.

Keywords. Surveys, Galaxy: structure, Galaxy: bulge, Galaxy: stellar content, Stars: distances

1. Introduction: the catalogue

TCS-CAIN is a deep multi-color NIR survey. Mapping selected areas along the Galactic plane, it reaches deeper than 2MASS or DENIS. The main scientific goal of TCS-CAIN is to study the structure of the Milky Way, mainly the bulge/bar and the disk, so the fields to observe have been chosen to sample well enough the $0° < l < 35°$ zone and to give enough coverage both in l and b of the disk.

The body of the catalogue comprises about 500 fields, $4.25' \times 4.25'$ each, obtained using JHKs photometry in the TCS Telescope (1.5m, Tenerife, Spain) yielding 10 million source points, with limiting magnitudes of 17 (J), 16.5 (H) and 15.2 (Ks), and with a photometric precision of 0.1 mag (against 2MASS data). The spatial resolution is $1'$, with astrometric errors of $0.15'$ for the position of the point sources.

2. Results: the bar

Hammersley *et al.* (2000), studying red giant star counts, spotted the presence of an overdensity from l=27° to l=5° at different magnitudes. Assuming an extinction law, the magnitude of this overdensity can be translated into distances along different lines of sight, rendering a elongated feature that runs from l=5° to l=27° with a position angle of $43°\pm7°$ and a half-length of 4kpc, that may be interpreted as a bar.

This structure can be traced further making use of the red clump stars. Since their luminosity function is rather narrow, they present a well-defined clump in the color-magnitude diagrams (CMD). It is possible then to extract star density and interstellar extinction by isolating this clump on a CMD (López-Corredoira *et al.* (2002) for details).

We can obtain a magnitude estimate for the red clump stars, which can be translated into a distance with the estimate of the extinction. Since TCS-CAIN has a good coverage of the inner parts of the Milky Way, we can apply this analysis to several fields in various lines of sight, and compare them with similar studies, as can be seen in Fig. 1, left panel. Clear traces of a long structure appear along the plane, reaching almost l=30°, and with an angle of $43°\pm3°$, in concordance with that proposed by Hammersley *et al.* (2000).

3. Extinction towards the inner Milky Way

With TCS-CAIN DCM's it is possible to obtain an extinction estimate along the desired line of sight, without any prior assumption of the stellar or interstellar matter

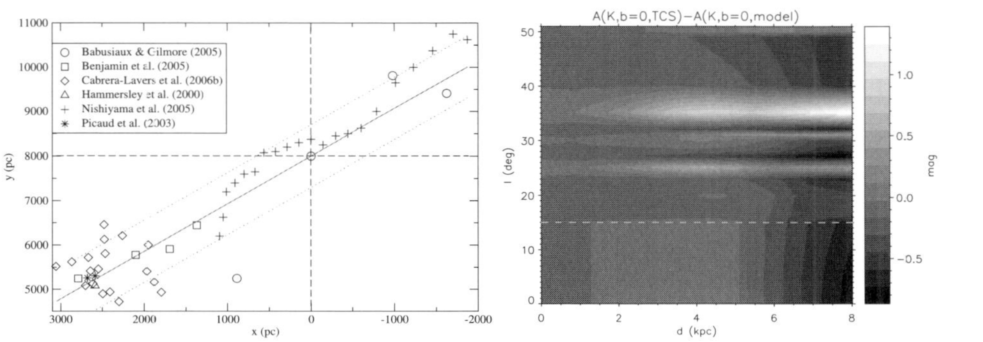

Figure 1. Left: Distance estimations for the red clump giants. Data are taken from Nishiyama *et al.* (2005), Babusiaux & Gilmore (2005), Benjamin *et al.* (2005), Picaud *et al.* (2003), López-Corredoira *et al.* (2007) and Hammersley *et al.* (2000). A feature with an angle of 43° and a width of 1 kpc is marked. Right: The difference in magnitude between the modelled values in Deguchi *et al.* (2004) and the ones derived with TCS-CAIN. The white line denotes the lower Galactic longitude limit of our present calculations.

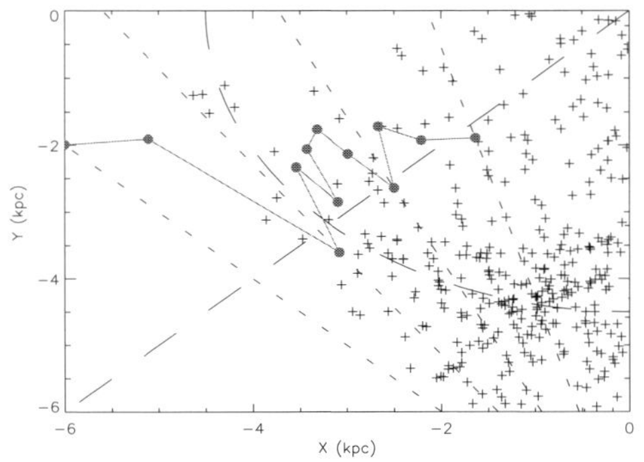

Figure 2. New distance estimation for the Deguchi *et al.* (2004) data, based in our derived extinctions. Red line and dots denote the completeness limit of our calculations.

distribution and thus, among other things, reproduce the patchy nature of the dust distribution at certain Galactic longitudes. Differences between models and our direct measurements can add up to 2 mag in A_K. With this extinction, it is possible to derive a new distance to the Deguchi *et al.* (2004) masers (see Fig. 2).

References

Babusiaux, C., & Gilmore, G. 2005, *MNRAS*, 358, 1309

Benjamin, R. A., *et al.* 2005, *ApJ*, 630, L149

Cabrera-Lavers, A., Garzón, F., Hammersley, P. L., Vicente, B., & González-Fernández, C. 2006, *A&A*, 453, 371

Deguchi, S., *et al.* 2004, *PASJ*, 56, 765

Hammersley, P. L., Garzón, F., Mahoney, T. J., López-Corredoira, M., & Torres, M. A. P. 2000, *MNRAS*, 317, L45

López-Corredoira, M., Cabrera-Lavers, A., Garzón, F., & Hammersley, P. L. 2002, *A&A*, 394, 883

López-Corredoira, M., Cabrera-Lavers, A., Mahoney, T. J., Hammersley, P. L., Garzón, F., & González-Fernández, C. 2007, *AJ*, 133, 154

Nishiyama, S., *et al.* 2005, *ApJ*, 621, L105

Picaud, S., Cabrera-Lavers, A., & Garzón, F. 2003, *A&A*, 408, 141

Astrophysical Masers and their Environments
Proceedings IAU Symposium No. 242, 2007
J.M. Chapman & W.A. Baan, eds.

doi:10.1017/S1743921307013427

Measurements of annual parallaxes and proper motions of the red supergiant S Per

Yoshiharu Asaki[1], Shuji Deguchi[2], Hirishi Imai[3], Kazuya Hachisuka[4], Makoto Miyoshi[5] and Mareki Honma[5]

[1]ISAS, 3-1-1 Yoshinodai, Sagamihara, 229-8510, JAPAN
email: asaki@vsop.isas.jaxa.jp

[2]Nobeyama Radio Observatory, Nobeyama, Minamimaki, Minamisaku, 384-1305, JAPAN

[3]Kagoshima University, 1-21-24 Korimoto, Kagoshima, 890-8580, JAPAN

[4]Shanghai Astronomical Observatory, Chinese Academy of Sciences, Shanghai, 200030, CHINA

[5]National Astronomical Observatory, 2-21-1, Osawa, Mitaka, 181-8588, JAPAN

Abstract. VLBI phase-referencing monitoring of water vapor masers around the red supergiant, S Per, was conducted over four years. We successfully obtained proper motions and an annual parallax of the masers and determined the distance to S Per of 2.51 ± 0.09 kpc. The proper motion of the star itself was inferred from the maser proper motions, and it was -0.38 and -1.54 mas/yr for right ascension and declination, respectively. Assuming the distance from the sun to the Galactic center, R_0, of 8.5 kpc and the rotation velocity around the sun, Θ_0, of 220 km/s, the Galactic rotation velocity around S Per is 200 km/s.

Keywords. Astrometry, Masers, Supergiants.

1. Introduction

VLBI has a capability of determining positions of water masers relative to a phase-referencing calibrator with sub-milliarcsecond accuracy at short wavelengths (Hachisuka *et al.* (2006)). There are a number of stellar water maser sources in the circumstellar envelopes (CSE) of evolved stars, which are observable with VLBI at 22 GHz. The red supergiant star S Per is an interesting target for VLBI astrometry. Marvel (1996) estimated the distance to be 2.3 ± 0.5 kpc from the relative proper motions of the water maser features fitted to a symmetrically expanding outflow model. Phase-referencing VLBI for the star can directly detect absolute position changes of the maser features on the sky, which are very useful for investigating the annual parallax and proper motions. In addition, S Per is in the outer Galaxy so that it is also interesting for the study of the Galaxy's rotation curve. The Doppler shift of the maser emissions, and the transverse speed can be estimated from the proper motion and the distance determined from the annual parallax.

We have conducted VLBI phase-referencing observations of S Per relative to a closely located continuum source, KR143, with the VLBA at 22 GHz over seven epochs (four years) since Nov 21st, 2000. We have also conducted phase-referencing observations of KR143 and ICRF 0244+624 at 15 GHz in the last four epochs to check the positional stability of KR143, which was selected from the 6-cm northern sky catalogue by Becker *et al.* (1991). Here we show the results of the annual parallax and proper motions of the water masers as well as the distance to S Per and the three-dimensional motion. The observed pairs of sources in our monitor program are listed in Table 1 with their separation angles and the observing frequencies.

Table 1. Observed pairs of the sources in our monitoring program.

Target	Positional reference	Separation angle	Observation frequency
S Per	KR143	0.22°	22.2 GHz
KR143	ICRF 0244+624	4.8°	15.3 GHz

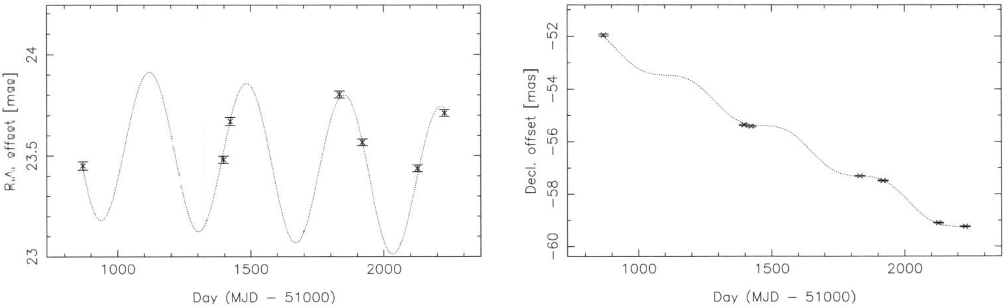

Figure 1. Maser feature motion relative to KR143. Dotted lines represent the best fit annual parallax and proper motion. The error bar is the root-square-sum of the position error of the maser feature and the assumed astrometric accuracy of 15μarcsec.

2. Results

The position of KR143 relative to ICRF 0244+624 was obtained with astrometric accuracy of 0.1 mas. If we take VLBI errors into consideration (the hydrostatic tropospheric delay error, the EOP error, and so on), this astrometric accuracy is appropriate for the 4.8° separate pair of sources. The obtained KR143 positions at the four epochs do not show an indication of the proper motion and/or annual parallax. We then used KR143 as position reference for the phase-referencing astrometry of S Per. The astrometric accuracy of S Per is assumed to be 15 μarcsec for right ascension and declination.

A positional change of one of the maser features with the systemic velocity of −42.0 km/s is shown in Figure 1. The maser structure including this feature shows an arc-like structure expanding with time. This maser feature could be identified through all epochs. We fitted the motion of the maser feature to the simple model for annual parallax and proper motion. We estimated nine annual parallaxes. The averaged value and the standard deviation is 0.399 ± 0.014 mas (2.51 ± 0.09 kpc).

In the first three epochs, we identified 69 maser features. The proper motions is combined kinematics of the CSE and the central star. We fitted the proper motions to the expanding outflow model to obtain the proper motion of the central star, and the result is −0.38 and −1.54 mas/yr for right ascension and declination, respectively. This proper motion is consistent with the *Hipparcos* result within its accuracy of 2 mas/yr. Assuming R_0 of 8.5 kpc and Θ_0 of 220 km/s, the distance from the Galactic center to S Per is 10.4 kpc. The estimated proper motion of the central star indicates that the Galactic rotation velocity around S Per is 200 km/s.

References

Becker, R. H., White, R. L., & Edwards, A. L. 1991, *ApJS* 75, 1

Hachisuka, K., Brunthaler, A., Menten, K. M., Reid, M. J., Imai, H., Hagiwara, Y., Miyoshi, M., Horiuchi, S., & Sasao, T. 2006, *ApJ* 645, 337

Marvel, K. B. 1996, PhD thesis, New Mexico State Univ., New Mexico.

Photos: Miller Goss, Montage: Vicki Drazenovic

Astrophysical Masers and their Environments
Proceedings IAU Symposium No. 242, 2007
J.M. Chapman & W.A. Baan, eds.

doi:10.1017/S1743921307013439

Masers in AGN environments

L. J. Greenhill

Harvard-Smithsonian Center for Astrophysics, 60 Garden St, Cambridge, MA 02138 USA
email: greenhill@cfa.harvard.edu

Abstract. Galactic nuclei are well known sources of OH and H_2O maser emission. It appears that intense star formation in ultra-luminous infrared galaxies drives most OH sources. In contrast, nuclear activity appears to drive most H_2O sources. When H_2O emission originates in accretion disk structures, constrained geometry and dynamics enable robust interpretation of spectroscopic and imaging data. The principal science includes study of AGN geometry at parsec and sub-parsec radii and measurement of geometric distances in the Hubble Flow. New high accuracy estimates of the "Hubble constant, $H_\circ$" obtained from maser distances may enable new substantively improved constraints on fundamental cosmological parameters (e.g., dark energy).

Keywords. masers, accretion, galaxies: active, galaxies: Seyfert, cosmological parameters, distance scale, radio lines: galaxies

1. OH and H_2O masers in AGN

Molecular emission is widespread among active galactic nuclei (e.g., Henkel *et al.* 1990, García-Burillo *et al.* 2007, and references therein). However, maser action is comparatively rare, having been detected only for OH and H_2O.

In general, OH maser sources lie in luminous infrared galaxies. Seven have been resolved on scales $\lesssim 100$ pc. Most are associated at least in part with inclined circumnuclear disks, inferred from position-velocity gradients or limb brightened structures (see review by Pihlström 2005b and case studies by Momjian *et al.* 2006; Pihlström *et al.* 2001, 2005a; Richards *et al.* 2005; Rovilos *et al.* 2003; Trotter *et al.* 1997; Yates *et al.* 2000). Estimated radii are 10^1 to 10^2 pc and dynamical masses are 10^6 to $10^8 M_\odot$. However, connections between these structures and nuclear activity have yet to be demonstrated. Observational hurdles include separating star forming and AGN processes, and identifying direct links between the OH structures and accretion or outflow (e.g., Vignali *et al.* 2005; Klöckner & Baan 2005; Baan & Klöckner 2006).

In contrast, connections between structures traced by H_2O masers and nuclear activity are reasonably well established. Water maser host galaxies are in nearly every case readily identified as AGN; luminosity from star formation is secondary. (A few H_2O masers are seen in star forming galaxies – Henkel *et al.* 2005; Greenhill 2002a, and references therein – and five H_2O hosts exhibit OH emission as well (Gallimore *et al.* 1996; Tarchi *et al.* 2007) – but these constitute a small minority.) Where H_2O emission has been mapped, and the underlying structures have been resolved, they are typically found to lie a few tenths to a few parsecs from the central engines, and to exhibit dynamical signatures characteristic of accretion and outflow.

In light of greater understanding of the association between H_2O masers and AGN processes, the remainder of this review will focus on H_2O, leaving the complex and rapidly evolving story of OH to treatment elsewhere in this volume.

2. History

Early interpretations of extragalactic H_2O maser emission in nuclei (mistakenly) focused on star formation as driver. It was a familiar context. The NGC 4945 H_2O maser was the first discovered in a nucleus (Table 1), prior to identification of an AGN therein (Dos Santos & Lépine 1979). The second nuclear H_2O source was in a well-known Seyfert galaxy, Circinus, but this too was attributed to star formation (Gardner & Whiteoak 1982). Claussen & Lo (1986) delivered the first strong challenge, with evidence that nuclear H_2O maser emission could be compact on scales of 3-10 pc and in close proximity to central engines. In light of the proposed unification paradigm for AGN, Claussen & Lo conjectured that the emission arose from warm gas in the now canonical torus. Nearly a decade later, studies of NGC 4258 were the first to demonstrate that the emission in fact originates from annuli in the still more compact, relatively thin disks at radii on the order of 0.1 to 1 pc (Miyoshi *et al.* 1995; Greenhill *et al.* 1995). Two additional contexts for maser emission were identified later: (1) downstream from central engines in association with radio jets (Gallimore *et al.* 1996; Claussen *et al.* 1998; Sawada-Satoh, this meeting) and (2) wide-angle outflows of AGN narrow-line regions (Greenhill *et al.* 2003a).

Table 1. Historical Perspective

Date [a]	Event	Target	Reference
11/76	First extragalactic λ1cm H_2O maser	M 33	1
09/77	First H_2O maser in a galactic nucleus	NGC 4945[b]	2
04/82	First H_2O maser in a known AGN	Circinus	3
09/83	Detection of compact structure with VLA & VLBI	NGC1068/4258	4
10/84		NGC 4258	5
	—— *First Lull in Detection of New Masers* ——		
01/92	High-velocity emission discovered	NGC 4258	6
03/93	Narrow-band surveys target optical AGN		7,8
10/94	Demonstration that masers can trace accretion disks	NGC 4258	9
04/94			10
08/95	Demonstration that masers can trace radio jets	NGC 1068	11
11/95		NGC 1052	12
06/97	Demonstration that masers can trace AGN winds		13
	—— *Second Lull in Detection of New Masers* ——		
12/01	Wide-band surveys begin at DSN 70m antennas		14,15
04/02	Wide-band surveys begin at GBT 100m antenna		16
12/03	HST program begins to X-calibrate Cepheid/maser distances		17
01/05	First maser discovered in a quasar ($z \sim 0.7$)	J0804+3607	18
03/05	First submm H_2O maser discovered in an AGN	NGC 3079	19
11/06	Water Maser Cosmology Project begins (NRAO/CfA/MPI)		20

[a] Observing dates associated with events in column 2.
[b] Lépine & Dos Santos (1977) reported an earlier detection possibly toward the nucleus of NGC 253, but independent studies failed to confirm it. See review by Greenhill (2002a).
References: (1) Churchwell *et al.* (1977), (2) Dos Santos & Lépine (1979), (3) Gardner & Whiteoak (1982), (4) Claussen & Lo (1986), (5) Claussen *et al.* (1988), (6) Nakai, Inoue, & Miyoshi (1993), (7) Braatz *et al.* (1996), (8) Greenhill *et al.* (2002b), (9) Greenhill *et al.* (1995), (10) Miyoshi et al (1995), (11) Gallimore *et al.* (1996), (12) Claussen *et al.* (1998), (13) Greenhill *et al.* (2003a), (14) Greenhill *et al.* (2003b), (15) Kondratko *et al.* (2006a), (16) Braatz *et al.* (2004), (17) Macri *et al.* (2006), (18) Barvainis & Antonucci (2005), (19) Humphreys *et al.* (2005), and (20) Braatz *et al.* (this volume).

3. Surveys

Water maser sources in nuclei are rare. The historical record of search programs is full of hypothesized correlations and physical arguments intended to identify selection criteria that would enable detection rates $\gg$ a few percent. Though small samples have achieved detection rates as large as $\sim$ 50% (Henkel *et al.* 2005), no large survey (N$>$ 10^2) has exceeded 1-10%. The most consistently successful sample is optically identified AGN, as first pursued in the mid-1990s (e.g., Braatz *et al.* 1996). Nearly all masers lie in Seyfert 2 objects and Low Ionization Nuclear Emission Regions (LINERs), though some are known to lie in ostensibly inactive galaxies and transition objects between Seyfert 1 and 2. None are known to lie in "pole-on" active systems. The incidence rate of maser emission among known Seyfert-2 objects and LINERs is on the order of 20% for $cz < 5000$ km s^{-1}, within which past surveys are believed to be reasonably complete. The incidence rate drops with increasing velocity, probably because of incompleteness and sensitivity limitations (Braatz *et al.* 2004; Kondratko *et al.* 2006a). Today, *there are* $\sim$ 90 *known masers in AGN (published and unpublished) with* $cz < 20000$ *km* s^{-1} (Figure 1), plus the maser seen toward quasar J0804+3607 at $z \sim 0.66$ (Barvainis & Antonucci 2005). For the time being, the internal structure and physical nature of this maser is not known.

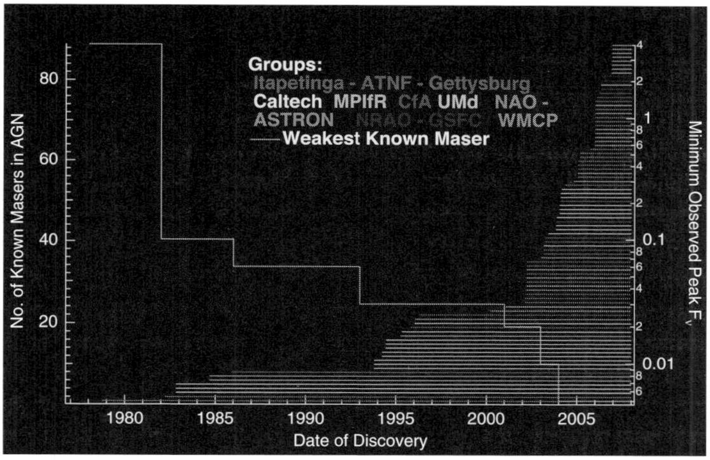

Figure 1. Growth in the number of H_2O masers believed to lie in AGN (*horizontal bars*). Color coding associates discoveries with lead institutions. The evolution in the peak flux density of the weakest known maser (*down-sloping curve*) marks gradual improvement in observing systems.

Because of low detection rates, maser surveys often test stamina, whether comprising shallow integrations on a large number of objects or deep integrations on a small number. The largest surveys have been made with the Green Bank Telescope, NASA Deep Space Network, and Effelsberg 100-m (e.g., Braatz *et al.* 2004; Greenhill *et al.* 2003b; Braatz *et al.* 1996), with important work on smaller samples also completed at Nobeyama (e.g., Nakai *et al.* 1995). As of now, $>$ 2000 galaxies outside the Local Group had been targeted since the early work by Henkel *et al.* (1984) and Claussen *et al.* (1984).† Another $\sim$660 ostensibly normal galaxies have been observed as well (Braatz, Greenhill, unpublished).

The broadest bandwidth maser source known covers $\sim$2600 km s^{-1} (Figure 2), centered within $\sim$100 km s^{-1} of the host galaxy systemic velocity (V_{sys}), and the weakest known peak source flux densities are on the order of 10 mJy, which reflects the detection

† See www.cfa.harvard.edu/wmcp/surveys/survey.html for cumulative lists of nondetections.

thresholds of existing observing systems (Figure 1). Because it is impossible to know *a priori* the breadth or velocity of the peak maser emission from any particular galaxy, successful observing systems combine high sensitivity and broad bandwidth (e.g., Greenhill *et al.* 2003b). Preponderance of narrow-band observing systems up to 2001, posed an obstacle. Today, the majority of survey targets have been observed with systems capable of detecting source with peak fluxes on the order of 10 mJy over many thousand $\mathrm{km\ s^{-1}}$.

Commonplace qualitative designations are: "kilomaser," "megamaser," and "gigamaser." These are used loosely in the literature (e.g., the so-called gigamaser in TXS 2226-182 is apparently only 10× more luminous than the common megamaser, e.g., NGC 3079), and classification is based on apparent integrated luminosity relative to that observed in local high-mass star forming regions.

Prefixes will not be used here because of implicit ambiguity. Specifically, maser emission is always beamed, and apparent luminosity (L_{H_2O}) is computed assuming isotropic emission. It may exceed true luminosity by orders of magnitude. The ratio between the two, the beam solid angle, is determined at least in part by source geometry and may differ widely from source to source (e.g., a thin accretion disk beaming into a plane; a thick fragmented disk beaming emission into a squat cylinder; a foreground cloud amplifies jet emission along a pencil beam). Consequently, correlations between various quantities and apparent maser luminosity may be anticipated to exhibit substantial scatter (e.g., L_{H_2O} vs. L(2-10 keV), Kondratko *et al.* 2006b; L_{H_2O} vs. X-ray absorption column density, Zhang *et al.* 2006). Prospectively, the discovery of any tight correlations with L_{H_2O} would be notable, suggesting commonalities among underlying physical systems. However, absent this, qualitative (loose) correlations are still valuable, in particular an apparent transition at $L_{H_2O} = 1 - 10\,L_\odot$ between excitation by star formation and by nuclear activity (Henkel *et al.* 2005).

4. Spectroscopy

Because VLBI follow-up is time consuming, working physical classification of maser systems is often made based on the morphology of spectra.

• **High-velocity** – distinct red and blue-shifted emission complexes approximately symmetrically offset from V_{sys} ($|\mathrm{V} - \mathrm{V_{sys}}| \gtrsim 200\ \mathrm{km\ s^{-1}}$). Perceived degrees of symmetry vary from source to source depending on the density of Doppler components. Emission close to V_{sys} may also be visible, offset by $\lesssim 10\%$ of the total source breadth. Spectra that include high-velocity and systemic emission complexes are distinctive indicators of emission from highly inclined accretion disks (Figure 2).

• **Low-velocity/broad** – emission broader than $\sim$100 $\mathrm{km\ s^{-1}}$, peaking within $\sim$300 $\mathrm{km\,s^{-1}}$ of V_{sys} (Figure 3). These masers are circumstantially associated with jet activity based on, thus far, a small sample.

• **Low-velocity/narrow** – emission comprising a small number of narrow Doppler components within $\sim$100 $\mathrm{km\ s^{-1}}$ of V_{sys}. The types of the underlying physical systems in these cases are uncertain.

• **Ambiguous** – Doppler components distributed in velocity, including what may be high-velocity emission, but without apparent symmetry (Figure 3). The number of components may be limited, which is one element that makes pattern matching difficult.

All "high-velocity" maser systems that have been studied thus far with very long baseline interferometry (VLBI) trace highly inclined accretion disks or disk-structures whose rotation curves suggest central dynamical masses of 10^6 to $10^8\ M_\odot$. These include NGC 4258 (Miyoshi *et al.* 1995), NGC 1068 (Greenhill & Gwinn 1997), Circinus (Greenhill *et al.* 2003a), NGC 3079 (Trotter *et al.* 1998; Yamauchi *et al.* 2004; Kondratko

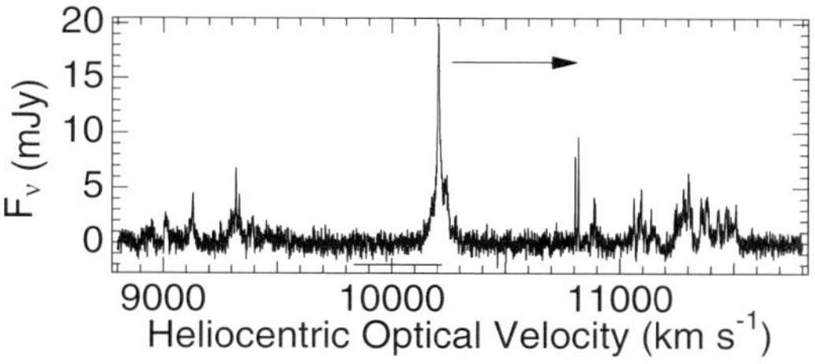

Figure 2. Distinct red and blue-shifted emission complexes, symmetry, and secular drift (arrow) in the UGC 9618B maser. All three are spectroscopic signatures of emission from edge-on disks. A bar along the velocity axis indicates the range of published V_{sys} (Kondratko *et al.* 2006b).

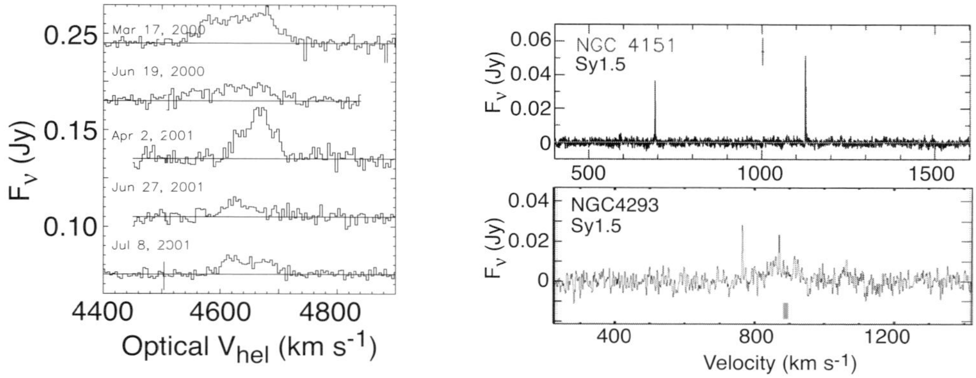

Figure 3. (*left*)—The Mrk348 maser is associated with jet activity and dominated by broad emission features, an empirical signature of the origin (Peck *et al.* 2003). (*right top*)–A maser with ambiguous classification; LSR/optical velocity (Braatz *et al.* 2004). (*right bottom*)—a low-velocity/narrow maser; heliocentric/radio velocity(Kondratko *et al.* 2006a). Bars mark V_{sys}.

et al. 2004), NGC 3393 (Kondratko *et al.* 2007), IC 2560 (Ishihara *et al.* 2001; Greenhill *et al.* in prep.), and NGC 6323 and UGC 3789 (Braatz, this volume). Three broad low-velocity systems that have been mapped are offset from the central parsec and are seen toward radio jets (NGC 1068, Gallimore *et al.* 1996, 2004; NGC 1052, Claussen *et al.* 1998; Mrk 348, Peck *et al.* 2003).

Supplemental classifications are also used to clarify uncertain identifications.

• **Accelerating** – long-term redward secular drift of Doppler components close to V_{sys}. This has been linked to orbital centripetal acceleration within inclined disks.

• **Symmetric** – emission extending $\gtrsim 200\,\mathrm{km\,s^{-1}}$ (symmetrically) to either side of V_{sys}, though perhaps without separable emission complexes. This also marks disks.

• **Low-luminosity** – apparent luminosities $\lesssim 10\,L_\odot$. Low luminosity often appears to be associated with maser action driven by star formation.

For instance, lacking high-velocity emission, the NGC 2639 maser classification would be "Ambiguous" but for observed secular drift that suggests the maser emission originates in a disk (Wilson *et al.* 1995). The NGC 1068 maser exhibits chiefly plateau emission, but symmetry in maximum velocity offsets of $\pm 330\,\mathrm{km\,s^{-1}}$ is a directly attributable to origin in a disk (Greenhill & Gwinn 1997). Lastly, the M 82 maser spectrum exhibits just a few Doppler components near V_{sys}, but low apparent luminosity suggests origin in star formation (Ho *et al.* 1987), as confirmed by direct imaging (Hagiwara 2007). These are "static" examples. In contrast, spectra may develop defining characteristics due to natural source variability; an "ambiguous" maser with no high-velocity emission at one epoch may display symmetric high-velocity emission at a later epoch (e.g., NGC 3735, Kondratko *et al.*, in prep). Such cases reinforce the importance of monitoring in spectroscopic studies.

5. Mapping

Water maser emission traces underlying dense, warm gas. The details of amplification in any individual parcel of gas may be difficult to quantify, but nonetheless each observable maser spot samples line-of-sight velocity on the sky, marks the sky position of a stationary point along the line of sight in the line-of-sight velocity field, and points to volumes where energy is deposited into the molecular gas. Necessarily, sampling is incomplete because maser emission is beamed (i.e., any one observer will see only a fraction of masers in a given source). However, inferences from the distribution of maser positions and velocities, coupled with *a priori* constraints on geometry (e.g., largely planar, point-symmetric, rotating structures, with circular or nearly circular orbits), enable construction of robust models and meaningful 3D deprojections.

For a nearly edge-on, differentially rotating, well ordered, thin ($h/r \lesssim 0.1$), flat disk that is heated by irradiation broadly across its surface or internally via viscous dissipation, velocity coherence favors H_2O maser emission: (1) in a narrow sector on the near/far side, and (2) close to the disk-diameter perpendicular to the line of sight (the "midline"). The former gives rise to spectral features near the systemic velocity (Figure 2). The latter gives rise to high-velocity emission that traces the rotation curve of the disk when mapped with interferometers. NGC 4258 is the archetype in this respect (Herrnstein *et al.* 2005; Moran, this volume).

Emission loci for thick disks ($h/r \gtrsim 0.1$) are comparatively difficult to predict. NGC 3079 is a good example (Kondratko *et al.* 2004). Red and blue-shifted emission are segregated on opposite sides of the central engine, but otherwise the maser structure is clumped with local velocity dispersions $> 10\%$ of the rotation speed. Between the two extremes are the relatively well-ordered mid-range cases of Circinus, which is somewhat clumpy but Keplerian to within modelling errors (Greenhill *et al.* 2003b), and NGC 1068, which is moderately clumpy and distinctly sub-Keplerian (Greenhill & Gwinn 1997; Lodato & Bertin 2003).

Warps also complicate prediction of emission loci. Maser photons are beamed close to disk planes, (probably) with narrower beam angles for thinner disks. Warps broaden the solid angle into which emission is beamed but and separate the loci where line-of-sight velocity is coherent and where the disk is seen edge-on. Despite variation of $\sim$0.1 rad in Euler angles over a 2:1 range of radius, the high-velocity emission loci in NGC 4258 are dictated by velocity coherence rather than local tangency of the line of sight to the disk. This could be attributed to finite disk thickness, where in the limiting case of a paper-thin disk, the reverse would be true.

6. Key science

Key science directly enabled by study of disk masers follows two principal threads (Figure 4): study of AGN structure (e.g., Morganti *et al.* 2004) and "maser cosmology," i.e., characterization of cosmological parameters via estimation of accurate geometric distances to maser host galaxies (Greenhill 2004; also Braatz, this volume). Notably, the latter depends on the former in that estimation of distances requires clean 3D dynamical models of maser disks, obtained via fitting to maser positions and line-of-sight velocities.

AGN Structure– Very long baseline study of disk masers is the only means by which structures < 1 pc from massive black holes can be directly mapped. A 1 pc diameter annulus of maser emission at a distance of 100 Mpc subtends on the order of 10 interferometer

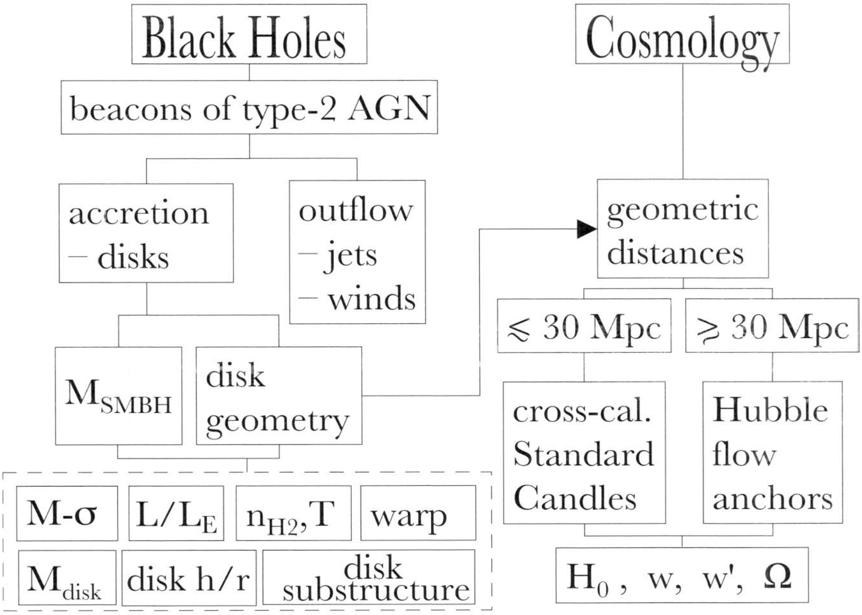

Figure 4. Two threads in the science that may be accomplished through study of H_2O masers in AGN. Water masers are quite often beacons of activity in galactic nuclei. Related science impacts studies of AGN physics and refinement of the extragalactic distance scale, with ramifications for estimation of cosmological parameters.

beams; a maser that is 2000 km s^{-1} across can be resolved into at least as many spectral channels with VLB correlators. Mapping disks in type-2 AGN can be particularly valuable because obscuration complicates the spectroscopic and timing analyses of central engines at other wavebands, and high inclinations superpose reflected, scattered, and partially obscured emission from the central engines and surrounding media with a range of temperatures. *Effectively, the AGN that are the most difficult to study with existing instruments at other wavebands are amenable to detailed radio study when maser emission is present.* The contrast is perhaps strongest for AGN with Compton thick obscuration, for which disk masers have a strong preference (Figure 5).

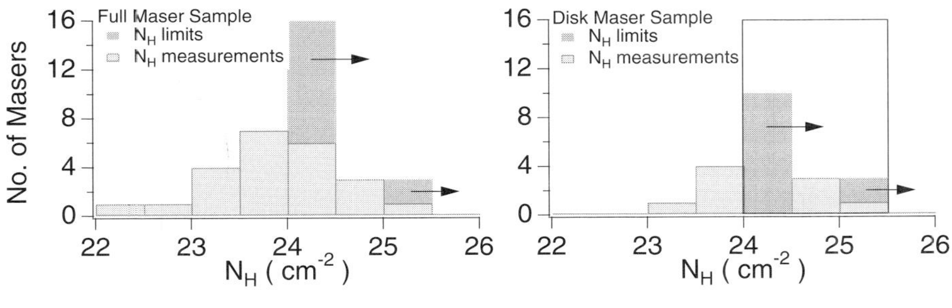

Figure 5. Relatively high incidence of disk borne maser emission among heavily obscured AGN. *(left)* – Histogram of known X-ray absorption columns for maser AGN. *(right)* – Subset believed to originate in disks, based on evidence from spectroscopy or mapping. Columns are Compton thick in 16 of 21 cases (heavy-lined box). See Madejski *et al.* (2006) and Zhang *et al.* (2006).

Modelling VLB data in principal enables accurate measurement of central engine dynamical masses, disk masses, Eddington luminosities, and descriptive ratios such as L_{Edd}/L_{bol}. Dynamical masses may be used to refine $M_\bullet - \sigma$ and $M_\bullet - M_{gal}$ relations (e.g., Ferrarese & Merritt 2000; Ferrarese *et al.* 2006) that address the question: what are the roles of massive black holes in galaxy formation? As well, detailed maps of disk structure may be used to assess the balance among central engine gravity, disk self-gravity, luminosity, and dissipation (e.g., Tilak *et al.* 2008). Independent estimation of disk systemic velocities is also possible, enabling refinement of galactic systemic velocities, or estimation of central engine drift (cf. Reid & Brunthaler 2004). In the case of warped disks, orientations are known to change by 0.1-1 rad over radii of 0.1-1 pc. Misalignment between central engine principle axes and reservoirs of accreting material on parsec and larger scales can be significant (e.g., NGC 3079; Kondratko *et al.* 2004). Quantitative characterization of warps enables assessment of torque origins and disk stability (e.g., Maloney *et al.* 1996). This also enables assessment of warped disks as the origin of obscuring columns in at least some, with ramifications for the unification paradigm. The suggestion is borne out for NGC 4258 (Fruscione *et al.* 2005; Herrnstein *et al.* 2005), and circumstantial evidence for Circinus supports the hypothesis that shadowing by warps may in some cases bound outflows and photoionized regions (Greenhill *et al.* 2003a).

Cosmology– The Standard Model is flat and comprises radiation, baryonic matter, non-baryonic matter, and dark energy. It has been defined largely by analyses of Cosmic Microwave Background fluctuations and supernova distances (e.g., Spergel *et al.* 2007). Nevertheless, estimates of geometry and the physical nature of dark energy remain too uncertain to distinguish among contending theories without substantive assumptions; external constraints are required, as from S-Z effect or lensing. High-accuracy measurement of the Hubble constant, $H_\circ$, would be another and arguably a more direct one. Accuracy of $\sim$1% would enable strong constraint on the dark energy equation of state (Hu 2005; Macri *et al.* 2006). The present best independent estimate of $H_\circ$ is perhaps accurate to 10%, though systematic errors are difficult to quantify (Argon *et al.* 2007).

A sample of a few to a few tens of anchor galaxies with measured geometric distances could support direct estimation of $H_\circ$, probably with higher accuracy than is possible with traditional Standard Candles, considering intrinsic and systematic uncertainties. All distance estimates for N anchors being equally good, uncertainties in $H_\circ$ would scale as $N^{-0.5}$ because systematic uncertainties for disk models among galaxies depend chiefly on individual emission distributions and are uncorrelated. In fact, focus on a few very good cases, e.g., NGC 4258-like galaxies in the Hubble flow, might yield better results overall than a larger sample of more middling cases, though the distribution in cz matters.

In principal, two geometric distance estimates may be obtained for each maser galaxy, one via analysis of centripetal acceleration, and one via analysis of proper motion (Herrnstein *et al.* 1999). Both observables are greatest for masers close to the systemic velocity. Measurement of acceleration is easier and less resource intensive. (Proper motions are a very small fraction of a beamwidth for earth-bound baselines; 1000 km s^{-1} subtends $\sim 2\mu$as yr^{-1} at $\sim$100 Mpc, or $\sim$1% of a beam at λ1 cm. For an orbital radius of 1 pc, this speed corresponds to a $\sim$1 km s^{-1} yr^{-1} drift or about one line width per year.) Nonetheless, VLB study is requisite to establish a disk rotation curve and model from which to estimate central mass divided by distance and deprojected orbital radii for systemic emission. The latter are distinct from the radii of high-velocity emission inferred from rotation curves and in first order analyses may be inferred from the rate of change of line-of-sight velocity as a function of position along the disk plane.

The basic methodology to measure the geometric distance to an anchor galaxy involves several VLB epochs, coupled with single-dish monitoring of spectra over a time interval that includes the VLB observations (to enable cross-referencing). Considering time variability of Doppler components over weeks to months, VLB epochs need to be closely spaced in time (in the event more than one is required to achieve adequate sensitivity). However, widely spaced clusters of VLB epochs are also useful in cases where the velocity pattern of Doppler components evolves (i.e., new lines appearing at new velocities). Repeated VLB observation may broaden the fraction of the disk that can be traced (e.g., by sampling the midline over a larger range of radii), thus improving model quality.

VLB modelling efforts thus far suggest the best and highest priority masers for estimation of extragalactic distances would be those that conform to five basic criteria:

- red *and* blue-shifted high-velocity emission –
  - many high-velocity lines sampling rotation curves over a broad range of radius;
- emission close to the systemic velocity –
  - many lines near systemic that display measurable centripetal accelerations;
  - measurable position-velocity gradients from which radii may be inferred;
- thin disk structure –
  - emission divisible by eye into narrow Doppler components;
  - Keplerian rotation (i.e., little evidence for fragmentation or other perturbation);
  - a rich emission distribution that enables robust modelling, including independent estimation of disk inclination, systemic velocity, and dynamical center location;
- peculiar motion that is small with respect to total recessional velocity –
  - origin in the Hubble Flow;
- a line strong enough for VLB self-calibration *or* a strong calibrator $\ll 1^\circ$ away.

References

Argon, A. L., Greenhill, L., Reid, M., Moran, J., & Humphreys, E. 2007, *ApJ* 659, 1040

Baan, W. A. & Klöckner, H.-R. 2006, *A&A* 449, 559

Barvainis, R. & Antonucci, R. R. J. 2005, *ApJ* 628, 89

Braatz, J. A., Wilson, A. S., & Henkel, C. 1996, *ApJS* 106, 51

Braatz, J. A., Henkel, C., Greenhill, L. J., Moran, J. M., & Wilson, A. S. 2004, *ApJ* 617, L29

Churchwell, E., Witzel, A., Pauliny-Toth, I., Sieber, W., Huchtmeier, W., Roland, J. 1977, *A&A* 54, 969

Claussen, M. J., Heiligman, G. M., & Lo,K.-Y. 1984, *Nature* 310, 298

Claussen, M. J. & Lo, K.-Y. 1986, *ApJ* 308, 592

Claussen, M. J., Reid, M. J., Schneps, M. H., Lo, K.-Y., Moran, J. M., & Güsten, R. 1988, in: M. J. Reid & J. M. Moran (eds.), *The Impact of VLBI on Astrophysics and Geophysics*, Proceedings of the 129th IAU Symposium (Dordrecht: Kluwer) vol. 129, p. 231

Claussen, M. J., Diamond, P. J., Braatz, J. A., Wilson, A. S., Henkel, C. 1998, *ApJ* 500, L129

Dos Santos, P. M. & Lépine, J. R. D. 1979, *Nature* 278, 34

Ferrarese, L., *et al.* 2006, *ApJ* 644, L21

Ferrarese, L. & Merritt, D. 2000, *ApJ* 539, L9

Fruscione, A., Greenhill, L. J., Filippenko, A. V., Moran, J. M., Herrnstein, J. R., Galle, E. 2005, *ApJ* 624, 103

Gallimore, J. F., Baum, S. A., O'Dea, C. P., Brinks, E., Pedlar, A. 1996, *ApJ* 462, 740

Gallimore, J. F., Baum, S. A. & O'Dea, C. P. 2004, *ApJ* 613, 794

García-Burillo, S., Combes, F., Usero, A., Graciá-Carpio, J. 2007, *New Astron.* 51, 160

Gardner, F. F. & Whiteoak, J. B 1982, *MNRAS* 201, P13

Greenhill, L. J., Jiang, D., Moran, J., Reid, M., Lo, K.-Y., Claussen, M. 1995, *ApJ* 440, 619

Greenhill, L. J. & Gwinn, C. R. 1997, *Ap&SS* 248, 261

Greenhill, L. J. 2002a, in: V. Migenes and M. J. Reid (eds.), *Cosmic Masers: from Proto-Stars to Black Holes*, Proceedings of IAU Symposium 206, (San Francisco: ASP) vol. 206, p. 381

Greenhill, L. J., *et al.* 2002b, *ApJ* 565, 836
Greenhill, L. J., *et al.* 2003a, *ApJ* 590, 162
Greenhill, L. J., Kondratko, P. T., Lovell, J. E. J., Kuiper T. B. H., Moran, J. M., Jauncey, D. L., Baines, G. A. 2003b, *ApJ* 582, L11
Greenhill, L. J. 2004, *New Astron.* 48, 1079
Hagiwara, Y. 2007 *AJ* 133, 1176
Henkel, C., Güsten, R., Downes, D., Thum, C., Wilson, T., Biermann, P. 1984, *A&A* 141, L1
Henkel, C., Whiteoak, J. B., Nyman, L.-A., Harju, J. 1990, *A&A* 230, L5
Henkel, C., Peck, A. B., Tarchi, A., Nagar, N. M., Braatz, J. A., Castangia, P., Moscadelli, L. 2005, *A&A* 436, 75
Herrnstein, J. R., *et al.* 1999, *Nature* 400, 539
Herrnstein, J. R., Moran, J. M., Greenhill, L. J., & Trotter, A. S. 2005, *ApJ* 629, 719
Ho, P. T. P., Martin, R. N. Henkel, C., Turner, J. L. 1987, *ApJ* 320, 663
Hu, W. 2005, in: S. Wolff and T. Lauer (eds.) *Observing Dark Energy* (San Francisco: ASP) vol. 339, p. 215
Humphreys, E. M. L., *et al.* 2005, *ApJ* 634, L133
Ishihara, Y., Nakai, N., Iyomoto, N., Makishima, K., Diamond, P. Hall, P. 2001, *PASJ* 53, 215
Klöckner, H.-R. & Baan, W. A. 2005, *Ap&SS* 295, 277
Kondratko, P. T., Greenhill, L. J. & Moran, J. M., 2004, *ApJ* 618, 618
Kondratko, P. T., *et al.* 2006a, *ApJ* 638, 100
Kondratko, P. T., Greenhill, L. J., & Moran, J. M. 2006b, *ApJ* 652, 136
Kondratko, P. T., Greenhill, L. J., & Moran, J. M. 2007, *ApJ* submitted
Lépine, J. R. D. & Dos Santos, P. M. 1977, *Nature* 270, 501
Lodato, G. & Bertin, G. 2003, *A&A* 398, 517
Macri, L., Stanek, K., Bersier, D., Greenhill, L., & Reid, M., 2006, *ApJ* 652, 1133
Madejski, G., Done, C., Życki, P. T., & Greenhill, L. 2006, *ApJ*, 636, 75
Maloney, P. R., Begelman, M. C. & Pringle, J. E. 1996, *ApJ* 472, 582
Miyoshi, M., Moran, J., Herrnstein, J., Greenhill, L., Nakai, N., Diamond, P., & Inoue, M. 1995, *Nature* 373, 127
Momjian, E., Romney, J. D., Carilli, C. L., Troland, T. H. 2006, *ApJ* 653, 1172
Morganti, R., Greenhill, L. J., Peck, A. B., Jones, D. L., Henkel, C. 2004, *New Astron.* 48, 1195
Nakai, N., Inoue, M. & Miyoshi, M. 1993, *Nature* 361, 45
Nakai, N., Inoue, M., Miyazawa, K., Miyoshi, M., Hall, P. 1995, *PASJ* 47, 771
Peck, A. B., Henkel, C., Ulvestad, J. S., Brunthaler, A., Falcke, H., Elitzur, M., Menten, K. M., Gallimore, J. F. 2003, *ApJ* 590, 149
Pihlström Y. M., Conway J., Booth R., Diamond P., Polatidis A. 2001, *A&A* 377, 413
Pihlström, Y. M., Baan, W. A., Darling, J., Klöckner, H.-R. 2005a, *ApJ* 618, 705
Pihlström, Y. M. 2005b, in: J. Romney & M. Reid (eds.), *Future Directions in High Resolution Astronomy*, (San Francisco: ASP) vol. 340, p. 224
Reid, M. J. & Brunthaler, A. 2004, *ApJ* 616, 872
Richards, A. M. S., *et al.* 2005, *MNRAS* 364, 353
Rovilos, E., Diamond, P., Lonsdale, C., Lonsdale, C., Smith, H. 2003, *MNRAS* 342, 373
Spergel, D. N., et al. 2007, *ApJS* 170, 377
Tarchi, A., Castangia, P., Henkel, C., Menten, K. M. 2007, *New Astron. Revs* 51, 67
Tilak, A., Greenhill, L. J., Done, C., Madejski, G. 2008, *ApJ* submitted
Trotter, A. S., Greenhill, L., Moran, J., Reid, M., Irwin, J., Lo, K.-Y. 1998, *ApJ* 495, 740
Trotter, A. S., Moran, J., Greenhill, L., Zheng, X., Gwinn, C. 1997, *ApJ* (Letters) 485, L79
Vignali, C., Brandt, W. N., Comastri, A., Darling, J. 2005, *MNRAS* 364, 99
Wilson, A. S., Braatz, J. A., & Henkel, C. 1995, *ApJ* 455, L127
Yamauchi, A., Nakai, N., Sato, N., Diamond, P. 2004, *PASJ* 56, 605
Yates J. A., Richards A. M. S., Wright M. M., Collett J. L., Gray M. D., Field D., Cohen R. J. 2000, *MNRAS* 317, 28
Zhang, J. S., Henkel, C., Kadler, M., Greenhill, L. J., Nagar, N., Wilson, A. S., Braatz, J. A. 2006, *A&A* 450, 933

Astrophysical Masers and their Environments
Proceedings IAU Symposium No. 242, 2007
J.M. Chapman & W.A. Baan, eds.

doi:10.1017/S1743921307013440

The structure of the accretion disk in NGC 4258

James M. Moran, Elizabeth Humphreys, Lincoln Greenhill, Mark Reid and Alice Argon

Harvard-Smithsonian Center for Astrophysics†
email: greenhill@cfa.harvard.edu

Abstract. A wealth of new information about the structure of the maser disk in NGC 4258 has been obtained from a series of 18 VLBA observations spanning three years, as well as from 32 additional epochs of spectral monitoring data from 1994 to the present, acquired with the VLA, Effelsberg, and GBT. The warp of the disk has been defined precisely. The thickness of the maser disk has been measured to be 12 micro-arcseconds (FWHM), which is slightly smaller than previously quoted upper limits. Under the assumption that the masers trace the true vertical distribution of material in the disk, from the condition of hydrostatic equilibrium the sound speed is 1.5 km s^{-1}, corresponding to a thermal temperature of 600K. The accelerations of the high velocity maser components have been accurately measured for many features on both the blue and red side of the spectrum. The azimuthal offsets of these masers from the midline (the line through the disk in the plane of the sky) and derived projected offsets from the midline based on the warp model correspond well with the measured offsets. This result suggests that the masers are well described as discrete clumps of masing gas, which accurately trace the Keplerian motion of the disk. However, we have continued to search for evidence of apparent motions caused by "phase effects." This work provides the foundation for refining the estimate of the distance to NGC 4258 through measurements of feature acceleration and proper motion. The refined estimate of this distance is expected to be announced in the near future.

Keywords. NGC 4258, galaxies: active, accretion disks, masers, galaxies: distances and redshifts, black hole physics

1. Introduction

A group based largely at the CfA has been working for the past decade to better understand the properties of the remarkable accretion disk in NGC 4258 that is traced by its water vapor maser emission (for a general review of its properties, see Moran, Greenhill & Herrnstein, 1999). In addition to the recent work reported here, important results have been obtained by a series of research projects by Harvard graduate students: Ann Bragg (acceleration analysis; Bragg *et al.*, 2000), James Herrnstein (warp analysis; Herrnstein 1997, Herrnstein *et al.* 1997, 1998, 2005 and precise distance estimate; Herrnstein *et al.* 1999), Maryam Modjaz (magnetic field upper limit; Modjaz *et al.* 2005), and Adam Trotter (warp analysis; Herrnstein *et al.*, 2005). There are two principal foci of the current work: (1) to improve the understanding of the physics of the accretion disk, and (2) to model precisely the dynamics to allow for the determination of an accurate distance to the galaxy, and to guide the modelling of other galaxies that have more limited maser data. Here, we describe some of the refinements to the basic model. The distance determined from the accelerations and proper motions of the maser features under the assumption of circular orbits is 7.2 ± 0.5 Mpc (Herrnstein *et al.*, 1999; compare cepheid distance of

† Present address: 60 Garden St, MA-42, Cambridge, MA 02138 USA

7.5 ±0.25 Mpc for an LMC distance of 50 kpc by Macri *et al.*, 2007). At this (maser distance) 1 mas = 0.035 pc = 1.1 × 10^{17} cm. For convenience, most radii are given in angular units in this paper.

The principal dataset is a series of 18 VLB experiments (VLBA only, plus VLBA augmented by large dishes) undertaken over a period of three years from March 1997 to August 2000. These data have been systematically analyzed, described in the literature (Argon *et al.*, 2007) and the tabulated results made available in the on-line version of the paper. In all, a total of about 14,000 features (defined here as velocity channels) were imaged and their positions, intensities, line widths and accelerations reported (3,967 features in the redshifted group of features, 267 in the blue group, and 10,018 in the systemic group). A critical contribution of this dataset is the measurement of many new features in the blueshifted group, which are much weaker and underrepresented compared to the red group. (A possible explanation for persistent weakness of the blue masers with respect to the redshifted masers is that, because of the warp and orientation of the disk, line of sight to the blue masers passes through the foreground disk and may suffer substantial attenuation in an ionized layer of gas.) The measurement of these masers will provide a much more accurate determination of parameters (e.g., center of symmetry) (Humphreys *et al.*, 2007a). In addition, we also analyzed spectra from the VLA, GBT and Effelsberg to obtain feature accelerations.

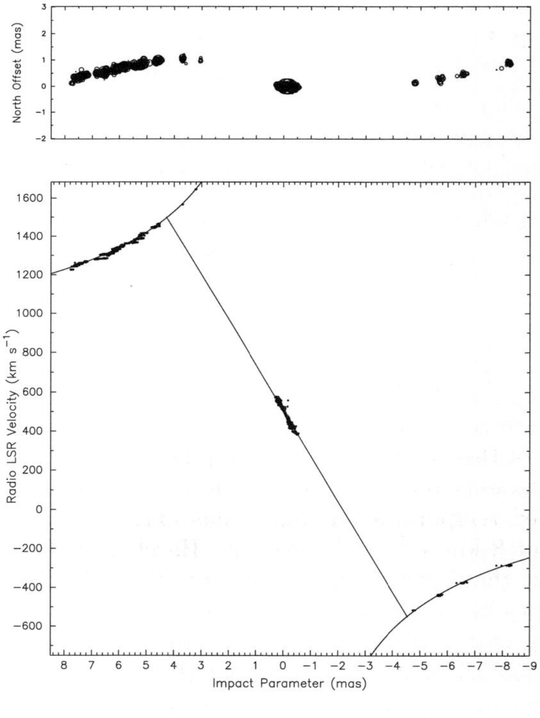

Figure 1. (top) Image of the masers in NGC 4258 (1 mas = 0.035 pc). (bottom) Position-velocity diagram of the masers. Note that the masers at 1556 and 1647 km s^{-1} lie inside the point where the line through the systemic features intersects the Keplerian curve, which indicates that they lie at smaller radii than the systemic features. From Argon *et al.* (2007).

2. The warp of the disk

Figure 1 shows the image of the masers and the position-velocity (PV) diagram computed along the major axis of the maser distribution. Since the accelerations of the high velocity features are close to zero, these masers lie along the so-called "midline" (the line through the disk from the black hole which is in the plane perpendicular to the line of sight). The anti-symmetric distribution of masers is readily understood as a "position-angle" warp. The high velocity portions of the PV diagram follow Kepler's law to an astonishing accuracy of about 1 percent. However, the velocities show a slight flattening, which can be thought of as systematic deviations from the Keplerian prescription by about 8 km s^{-1}. The most reasonable explanation of this effect is that the disk has a component of "inclination-angle warp." The warp is fully characterized by these two components (i.e., α is the position angle of the disk measured West from North, i is the inclination angle measured from North), and has been parameterized in the form (in degrees):

$$\alpha = 66 + 5.0r - 0.22r^2 \tag{2.1}$$
$$\phi = 107 - 2.29r \tag{2.2}$$

where r is the radius in mas, by Herrnstein *et al.* (2005) based on pre-1997 data. Hence, the disk is defined by y $= r$ cos α and y $= r$ sin i in the plane of the sky and along the line of sight, respectively. This parameterization reveals three important characteristics of the disk as we view it. Firstly, there is a point 7.4 mas from the center of the disk where the inclination crosses 90°, i.e., the disk is seen edge on at this radius. Inside this radius, the front of the disk tips down; outside, it tips up. Hence, the extreme outer portions of the high velocity emission are seen towards the underside of the disk. Secondly, the warp creates a shallow "bowl" on the front side of the maser that has a minimum point at a radius of 3.9 mas along the line of sight from the black hole. Finally, along the line of sight, the inclination angle becomes 90° at a radius of 7.4 mas. The gas at this location may be responsible for the thermal absorption of the x-ray emission (Fruscione *et al.*, 2005).

3. Acceleration of the systemic features

The positions, velocities and accelerations of the systemic features are shown in Figure 2. The spectrum of the systemic maser group is very crowded with features, which makes it difficult to track them individually because of changes in lineshape and intensity and because of limited temporal sampling. It has long been known that the systemic features drift at about 8 km s^{-1} yr^{-1} (see, e.g., Haschick & Baan 1990 and Greenhill *et al.*, 1995), traversing the observed masing region in about 10 years (from this, the beam angle of the masers has been deduced to be about 8°). Humphreys *et al.* (2007b) have found a global solution for a set of Gaussian components, each of which is assumed to be drifting at a constant velocity. They show that there is an approximate linear change in the acceleration with position in the systemic features (see Figure 2-h). A simple explanation of this tendency may be that the masing region has a fixed dependence of the radius of the masing region as a function of position. The line-of-sight component of the acceleration is given by

$$a = \frac{GM}{r^2} \cos\theta, \tag{3.1}$$

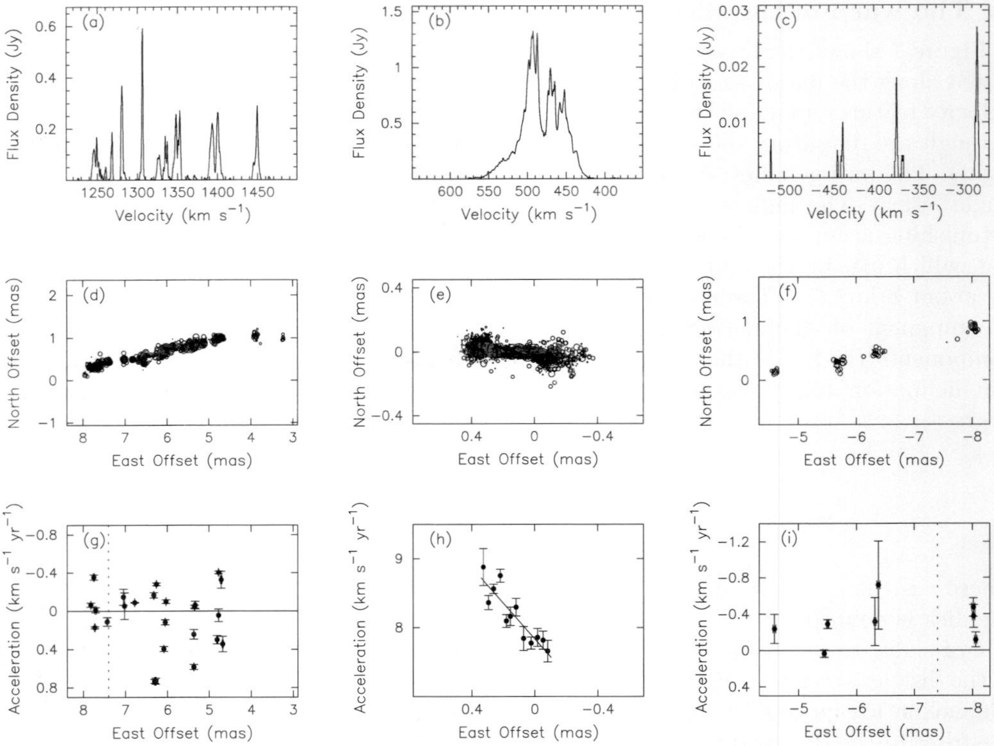

Figure 2. Spectra (top row), positions (middle) and accelerations (bottom) of the redshifted (left column), systemic (middle), and blueshifted masers (right). Note that the acceleration data for the systemic masers has been binned. The offset distance is measured from the position of the black hole. Note changes in acceleration scales for different velocity groups. The apparent periodicities in velocities (panels a and c) and positions (panels d and f) of red and blueshifted features may indicate spiral structure as suggested by Maoz (1995). Adapted from Argon *et al.* (2007) and Humphreys *et al.* (2007b).

where G is the gravitational constant, M is the black hole mass, r is the radius, and θ is the azimuth angle measured from the line of sight eastward. Hence, the change in acceleration from 7.7 to 8.8 km s^{-1} yr^{-1} may be due to a change in radius from 4.0 to 4.2 mas (the azimuth angles are close to zero). Note that the measurements of acceleration from earlier epochs seem to show this same dependency, which suggests that this pattern does not move with the disk rotation. However, there is a more intriguing explanation for this dependency. Perhaps the accretion disk contains a spiral density wave, with a pattern speed less than the rotation speed, which because of its overdensity provides an extra component of acceleration. A model with a spiral arm mass of about 10^4 $M_\odot$ (about 15 percent of the maximum allowable disk mass) could account for the trend in acceleration with position (Humphreys *et al.*, 2007b). If this model is correct, changes in the acceleration might allow the pattern speed to be measured. Finally, it is conceivable, though theoretically improbable (see, e.g., Statler *et al.*, 2001), that the maser orbits are confocal ellipses. Tight limits on any eccentricity in the orbits have been placed by Humphreys *et al.* (2007b).

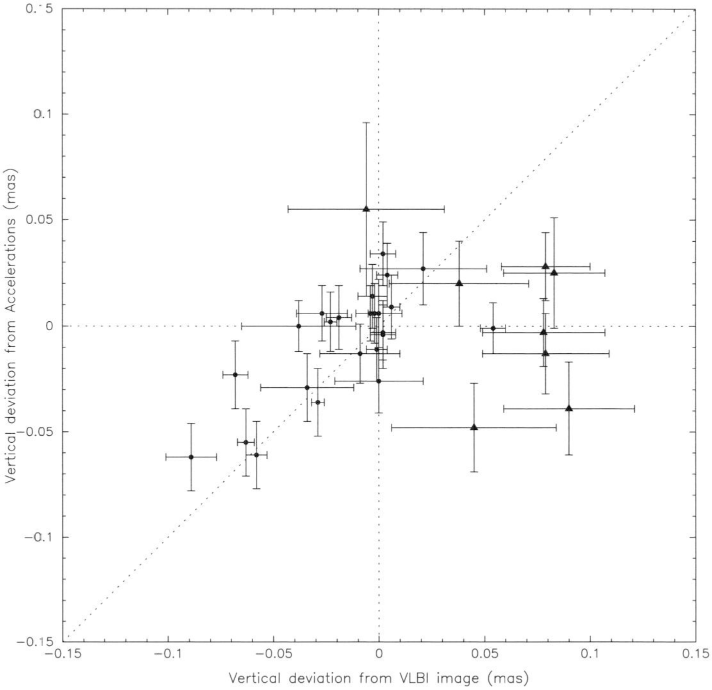

Figure 3. Comparison of expected projected vertical positions based on the azimuth angles calculated from accelerations and projection based on the warp versus the observed projected vertical positions. From Humphreys *et al.* (2007b).

4. Acceleration of the high velocity features

The accelerations of the high velocity features (see Figure 2) are about a factor of twenty smaller than those of the systemic masers. The origin of these accelerations is probably the result of a purely geometric effect. That is, masers lying on the midline will have exactly zero acceleration, while those off the midline will have accelerations (see Eq. 3.1) of $a \sim GM(90-\theta)/r^2$. With this interpretation, the spread of about $\pm$ 1.0 km s^{-1} yr^{-1} leads to a range of inferred azimuth angles of $\pm$ 10 degrees. If these azimuth angles are correct and the parameters of the warp model are accurate, then there should be a strong correlation between the offset of the masers from the local midplane as calculated from the azimuth angle and the warp, and the measured position offsets. Figure 3 shows that these position estimates are highly correlated. The correlation is not perfect and is weakest for the blue features. However, the warp model parameters are largely determined by the redshifted features. Furthermore, the finite thickness of the disk of about 12 μas (see section 5 below) will degrade the correlation. We consider this correlation to be strong evidence that the masers are accurately modelled as ballistic clumps and that the warp, as parameterized in Eq. 2.1, has been reasonably well modelled.

A model of a spiral density wave moving through the accretion disk could create apparent accelerations, i.e., the radial motion of the spiral density wave causes the excitation point in the masing gas to move outward in time causing a decrease in velocity on the redshifted side of the disk as a function of time and a blue positive drift in velocity on the blueshifted side. The magnitude of the effect is $a_{app} = \pm 0.05(\theta_p/2.5°)$ km s^{-1}yr^{-1}, where θ_p is the pitch angle, the positive sign refers to the blueshifted masers and the negative sign to the redshifted ones (Maoz & McKee 1998). This effect may be present, but it is much smaller than, and masked by, the geometric effect. Note, however, that the mean value of the accelerations seems to be slightly positive on the red side and significantly negative on the blue side, the opposite of what is expected for a trailing spiral. Our conclusion is contrary to that of Yamauchi *et al.* (2005), but their analysis is based on only a single blueshifted acceleration.

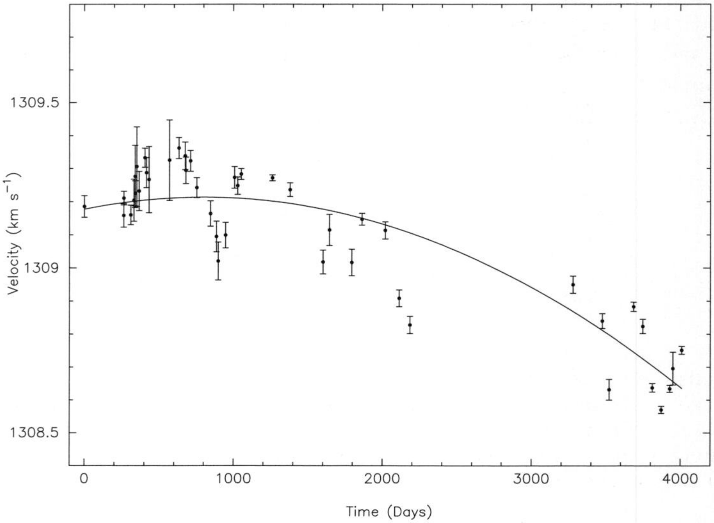

Figure 4. The velocity drift of the "1309" km s^{-1} feature versus time. The curve has the shape of the 1830 year period expected for this feature. Updated from Humphreys *et al.* (2004).

Over a long period of time the accelerations of the maser features should show deviations from a constant value. That is, the line of sight velocity should be given by a sine wave of about 1,830 years, that is,

$$v = \sqrt{\frac{GM}{r}} \cos[\Omega_r (t - t_o)], \tag{4.1}$$

where Ω_r is the orbital frequency at radius r, and t_o is the time when the maser crosses the midline. The feature at "1309" km s^{-1} has been particularly long lasting, and we have tracked its velocity over a period of more than 12 years now (see Figure 4). The expected velocity function for a central mass of $3.8 \times 10^6 M_\odot$ is shown. There is not enough of a time line to prove convincingly that the curvature is real. The confirmation of sinusoidal variation would be another powerful confirmation of the robustness of the simple dynamical model.

5. The disk thickness

The first VLBI data suggested that the disk thickness was less than about 10 μas (Moran *et al.*, 1995). The statistical distribution of the maser features from the midline can be most reliably estimated by examining the features at azimuth angles near the center of the bowl where changes in radial distance do not cause vertical projected displacements. In Figure 5 we show a histogram of these projected vertical displacements. The distribution is nearly a Gaussian function with a width (FWHM) of 12 micro-arcseconds. If this represents the real thickness of a disk that is in hydrostatic equilibrium, then the sound speed is about 1.5 km s^{-1} and the temperature is about 600K (Argon *et al.*,

2007). Note that this may be a slight overestimate since there may be some effects of projection and measurement noise included. This temperature is quite appropriate for the excitation of water maser emission. The temperature could be lower if part of the support originates from magnetic pressure. For magnetic pressure to completely support this disk would require a magnetic field of about 100 mG, which is about the current upper limit on its measurement (Modjaz *et al.*, 2005). Note, however, that it has been suggested that the masers form in a skin on a thicker disk. It might be possible to test this model if, say, the redshifted masers are generated on the top side of the disk and the blue on the bottom.

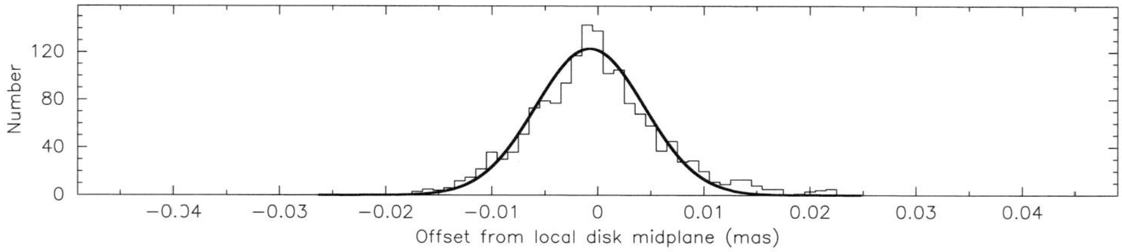

Figure 5. A histogram of the vertical positions of the maser features in the systemic group stronger than 250 mJy between velocities of 485 and 510 km s^{-1}, corresponding to azimuth angles of 2 to 6 degrees, where the masers originate near the bottom of the bowl. The average rms measurement error in the relative positions is about 1 μas, which is negligible for this analysis. From Argon *et al.* (2007).

6. Prospects for distance measurements

For a simple flat disk, where the high velocity features lie on the midline and the systemic features are at a fixed radius, the distance can be estimated by the formula

$$D = \frac{v_H^2}{a\ b_H\ \sin i}, \tag{6.1}$$

where v_H and b_H define the point in the PV diagram (see Figure 1) where the line through the systemic features intersects the Keplerian curve of the high velocity features, a is the line-of-sight acceleration of the systemic features, and i is the inclination of the disk. The slope of the systemic features in the position-velocity diagram, $\Omega = dv/db = \sqrt{GM/r^3}$ so $v_H = \Omega b_H$. (See Greenhill 2004 for an alternate formulation). The angular radius of the systemic features can be estimated from Ω in the case of NGC 4258 where the inclination is nearly 90°. This is the only way to determine the radius of the systemic features when the disk is nearly edge-on. Hence, the measurements of a and Ω provide the critical link between angular and linear scales. In the case of NGC 4258, the accelerations vary from 7.7 to 8.8 km s^{-1}yr^{-1}. To first order, an accurate distance can be obtained if the local slope of the PV diagram is used with the corresponding acceleration and inclination. Proper motions of the systemic features determined from VLBI measurements can also be used to estimate the distance independently of the accelerations. However, in the case of NGC 4258, the change in velocity in a year's time is about 5 times the feature linewidth for a systemic feature, whereas the proper motion is only about 30 μas, or one-tenth of the typical angular resolution. However a clear demonstration of the correlation between the proper motions and accelerations would be an important confirmation of the simple dynamical model. Clearly, to obtain a distance estimate of accuracy of better than about 5 percent for a warped disk that has a range of accelerations among its systemic features and has effects due to spiral structure requires sophisticated modelling. This work is underway (see Humphreys *et al.*, 2007a).

References

Argon, A. L., Greenhill, L. J., Reid, M. J., Moran J. M., & Humphreys, E. M. L. 2007, *ApJ* 659, 1040

Bragg, A. E., Greenhill, L. J., Moran, J. M., Henkel, C. 2000, *ApJ* 535, 73

Fruscione, A., Greenhill, L. J., Filippenko, A. V., Moran, J. M., Herrnstein, J. R., & Galle, E. 2005, *ApJ* 624, 103

Greenhill, L. J. 2004, *New Astr. Rev.* 48, 1079

Greenhill, L. J., Henkel, C., Becker, R., Wilson, T. L., & Wouterloot, J. G. A. 1995, *A&A* 304, 21

Haschick, A. D. & Baan, W. A. 1990, *ApJ* (Letters) 355, L23

Herrnstein, J. R., Moran, J. M., Greenhill, L. J., & Trotter, A. S. 2005, *ApJ* 629, 719

Herrnstein, J. R., Moran, J. M., Greenhill, L. J., Diamond, P. J., Inoue, M., Nakai, N., Miyoshi, M., Henkel, C., Riess, A. 1999, *Nature* 400, 539

Herrnstein, J. R., Greenhill, L. J., Moran, J. M., Diamond, P. J., Inoue, M., Nakai, N., & Miyoshi, M. 1998, *ApJ* (Letters) 497, L69

Herrnstein, J. R., Moran, J. M., Greenhill, L. J., Diamond, P. J., Nakai, N., & Inoue, M. 1997, *ApJ* (Letters) 475, L17

Herrnstein, J. R. 1997, PhD thesis, Harvard University

Humphreys, E. M. L., *et al.*, 2007a, in preparation

Humphreys, E. M. L., Reid, M. J., Greenhill, L. J., Moran, J. M., & Argon, A. L. 2007b, *ApJ*, submitted

Humphreys, E. M. L., Argon, A. L., Greenhill, L. J., & Reid, M. J., Moran, J. M. 2004, *Ap&SS* 295, 285

Macri, L. M., Stanek, K. Z., Bersier, D., & Greenhill, L. J., Reid, M. J. 2007, *ApJ* 652, 1133

Maoz, E. & McKee, C. F. 1998, *ApJ* 494, 218

Maoz, E. 1995, *ApJ* (Letters) 455, L131

Modjaz, M., Moran, J. M., Kondratko, P. T., Greenhill, L. J. 2005, *ApJ* 626, 104

Moran, J. M., Greenhill, L. J., & Herrnstein, J. R. 1999, *JA&A* 20, 165

Moran, J. M., Greenhill, L. J., Herrnstein, J. R., Diamond, P., Miyoshi, M., Nakai, N., & Inoue, M. 1995, *Proc. Nat. Acad. Sci.* 92, 11427

Statler, T. S. 2001, *AJ* 122, 2257

Yamauchi, A., Sato, N., Hirota, A., & Nakai, N. 2005, *PASJ* 57, 861

Astrophysical Masers and their Environments
Proceedings IAU Symposium No. 242, 2007
J.M Chapman & W.A Baan, eds.

doi:10.1017/S1743921307013452

Precision cosmology with H_2O megamasers: progress in measuring distances to galaxies in the Hubble flow

J. Braatz[1], L. Greenhill[2], M. Reid[2], J. Condon[1], C. Henkel[3] and K.-Y. Lo[1]

[1]National Radio Astronomy Observatory, Charlottesville, VA, USA
email: jbraatz@nrao.edu

[2]Harvard-Smithsonian Center for Astrophysics, Cambridge, MA, USA

[3]Max-Planck-Institut für Radioastronomie, Bonn, Germany

Abstract. A determination of the Hubble Constant (H_0) to better than 3% would be the best complement to cosmic microwave background (CMB) data to constrain the equation of state of Dark Energy. Water vapor megamasers provide perhaps the best opportunity for measuring direct distances to galaxies out to about 200 Mpc. We have formed a team of astronomers in the Megamaser Cosmology Project to pursue the ambitious goal of making a precise measurement of H_0 by measuring such distances using the techniques pioneered on the disk maser in NGC 4258 by Herrnstein *et al* (1999). In recent surveys we have made significant progress identifying new maser systems analogous to that in NGC 4258, but more distant. Once the appropriate candidates are identified, two types of observations are necessary to ultimately measure a distance: single-dish monitoring to measure the acceleration of gas in the disk, and sensitive VLBI imaging in order to measure the angular size of the disk, measure the rotation curve, and model radial displacement of the systemic maser features. We have recently obtained preliminary VLBI maps of the masers in two systems, NGC 6323 and UGC 3789. The maser disks in both galaxies were discovered and monitored with the Green Bank Telescope (GBT) and subsequently imaged with the High Sensitivity Array (VLBA + GBT + Effelsberg). In this contribution we present a map of the maser distribution in one of those systems, NGC 6323. The map demonstrates that pc-scale maser disks as distant as $\sim$ 100 Mpc can be imaged with existing telescopes. Results on UGC 3789 will be presented in a later publication.

1. The megamaser cosmology project

The Megamaser Cosmology Project (MCP) aims to make a precise measurement of the Hubble Constant, an important complement to CMB data for constraining the nature of Dark Energy, the geometry (flatness) of the universe, and the fraction of the critical density contributed by matter. The measurement technique to be used by the MCP is based on observations of 22 GHz water vapor maser emission from AGNs with edge-on, nuclear accretion disks. The technique was first applied by Herrnstein *et al.* (1999) to determine the distance to NGC 4258. Because maser distances are independent of standard candle assumptions, they are not subject to the systematic uncertainties that currently limit the H_0 measurement based on standard candle techniques to about 10% precision.

In order to place a meaningful constraint on cosmological models, a measurement of H_0 to 3% or better is necessary. The precision in H_0 that can be achieved by measurements of maser disks depends on the quality of the individual measurements, the number of galaxies that can be measured, their distance distribution, and their distribution on the sky. A reasonable distribution on the sky is necessary to smooth over any local

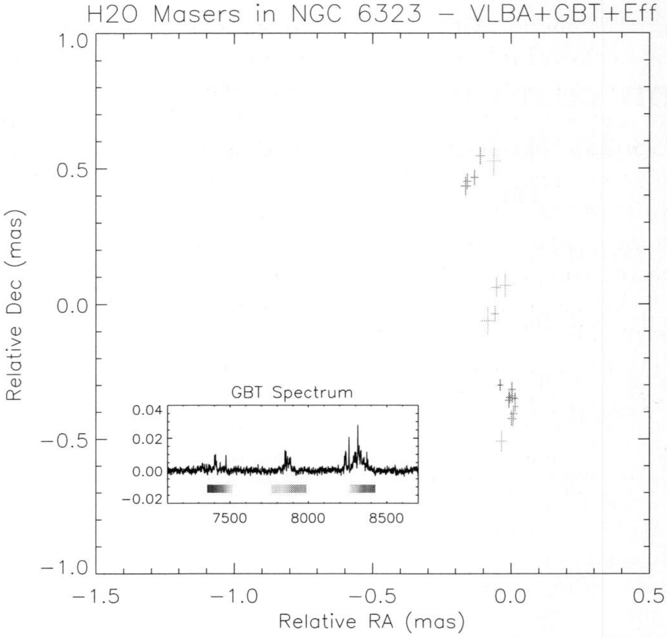

Figure 1. This preliminary VLBI map shows the distribution of maser emission toward the nucleus of the Seyfert 2 galaxy NGC 6323. The crosses in the figure represent the position of maser emission in each channel detected in the VLBI map, with the size of the cross proportional to the measured position uncertainty. The northernmost cluster of features are blueshifted with respect to the systemic recession velocity, the southernmost cluster of features are redshifted, and the features near the center of the map are near the recession velocity. The map is consistent with maser emission from an edge-on, sub-pc accretion disk in rotation about a central black hole with mass $\sim 10^7$ solar masses.

anomalies in the Hubble expansion rate. We could achieve an overall 3% precision in H_0, for example, by measuring the distances to 10 galaxies if each distance could be measured to 10% precision, assuming the individual distance measurements are uncorrelated.

NGC 4258 is the only galaxy with a 10% or better distance determination but it is too close to constrain H_0 directly. However, we have discovered H_2O masers in galaxies well into the Hubble flow (e.g. Braatz *et al.* 2004, Kondratko *et al.* 2006). Altogether, there are currently about 100 galaxies detected in water vapor maser emission. Of those, about one third show some evidence of disk origin. Currently about six of these galaxies are promising candidates for maser distance measurement. More are needed, and surveys are continuing to find additional candidates.

2. The maser disk in NGC 6323

An important milestone to the Megamaser Cosmology Project is the demonstration that maser disks in the Hubble flow can be imaged with high fidelity, given the faintness of the masers (a few tens of mJy at peak) and the need to apply external phase calibration techniques. We have recently made VLBI observations of several galaxies that have single-dish spectra consistent with edge-on disk maser emission. In Figure 1 we show a preliminary map of the maser distribution in the nucleus of one of these galaxies, NGC 6323, which has a recession velocity of 7772 km s^{-1} corresponding to a distance of

110 Mpc for $H_0 = 72$ km s^{-1} Mpc^{-1}. The map was observed with the High Sensitivity Array, consisting of the VLBA, the GBT, and the Effelsberg 100 m telescope. A single track across the sky was used for this preliminary map. The maser system in the disk spans an angular radius of $\sim$0.5 mas and is aligned nearly N–S. The disk shows differential rotation, with orbital velocities increasing at smaller radii. The map demonstrates that it is possible to image maser disks at cosmological distances with high fidelity.

To ultimately determine a distance to the galaxy, it is necessary to measure an acceleration in the systemic features using long-term monitoring. Preliminary GBT monitoring of the maser spectrum in NGC 6323 reveals features with accelerations between 1.8 and 3.0 km s^{-1} yr^{-1}. A careful analysis of the monitoring data will improve these estimates and will be presented in a later publication.

In addition to measuring an acceleration and a rotation curve, a robust distance determination also depends on a measurement of the position-velocity gradient of the systemic features, which determines the radial velocity structure in the disk. The preliminary VLBI map shown in Figure 1 does not have sufficient sensitivity to make such a measurement, but more sensitive, multi-track VLBI observations of the maser in NGC 6323 are underway.

References

Braatz, J. A., Henkel, C., Greenhill, L. J., Moran, J. M., & Wilson, A. S. 2004, *ApJ*, 617, L29

Herrnstein, J. R., Moran, J. M., Greenhill, L. J., Diamond, P. J., Inoue, M., Nakai, N., Miyoshi, M., Henkel, C., & Riess, A. 1999, *Nature* 400, 539

Kondratko, P. T., Greenhill, L. J., & Moran, J. M. 2006, *ApJ*, 652, 136

Astrophysical Masers and their Environments
Proceedings IAU Symposium No. 242, 2007
J.M. Chapman and W.A. Baan, eds.

doi:10.1017/S1743921307013464

An atlas of extragalactic water vapor masers

J. Braatz[1], P. Kondratko[2], L. Greenhill[2], J. Condon[1], C. Henkel[3], N. Gugliucci[4], L. Hao[5], M. Reid[2], J. Moran[2] and K.-Y. Lo[1]

[1]National Radio Astronomy Observatory, Charlottesville, VA, USA
email: jbraatz@nrao.edu

[2]Harvard-Smithsonian Center for Astrophysics, Cambridge, MA, USA

[3]Max-Planck-Institut für Radioastronomie, Bonn, Germany

[4]Dept. of Astronomy, University of Virginia, Charlottesville, VA, USA

[4]Dept. of Astronomy, Cornell University, Ithaca, NY, USA

Abstract. We present an atlas of extragalactic water vapor masers. As of 2007, one hundred galaxies have been detected as sources of water vapor maser emission, two thirds of them discovered since 2003. Extragalactic water masers fall in at least three categories: those associated with nuclear jets or winds, those in starbursts or star-forming regions, and those in AGN accretion disks. While all maser systems offer the possibility of unique investigations into their physical environments, it is the disk masers that have been most aggressively sought because of their potential for use as precision distance indicators. Type 2 Seyfert and LINER galaxies are hosts to such disk masers. In single-dish spectra, disk masers are often revealed by the presence of high-velocity emission features (defined roughly by having anomalous velocities in excess of 200 km s^{-1}). About one third of the extragalactic water masers detected to date show evidence of disk origin. Only a few galaxies are currently identified as jet-type or star-forming type. The remaining systems show only a few narrow doppler components, usually near the systemic velocity, and are difficult to categorize. These unclassified systems are detected toward AGNs and are also possibly associated with disks or winds near the nucleus. Detection rates in large maser surveys are typically 5% or less, but the observing efficiency and sensitivity of the Green Bank Telescope (GBT) allow for short integration times (typically 10 minutes to detect a narrow 30 mJy line) so many galaxies can be searched. Recently, GBT surveys targeting type 2 Seyfert galaxies identified by the SDSS have been the most productive, identifying 17 systems in surveys observed during 2006.

---

1. About the maser atlas

The atlas consists of single-dish spectra of 22 GHz water vapor maser emission detected toward the nucleus of each galaxy. Most of the observations were taken with the GBT, and cover at least 200 MHz bandwidth and 24 kHz channel spacing, corresponding to about 2700 km s^{-1} velocity coverage and 0.33 km s^{-1} channels. Typical rms noise levels of GBT spectra are 2–5 mJy per channel. Spectra of southern sources were observed either at Parkes or Tidbinbilla. The on-line atlas will be updated to reflect new discoveries and includes a list of undetected sources as well. The full atlas and catalogs are available at the sites:

http://www.nrao.edu/~jbraatz
and
http://www.cfa.harvard.edu/wmcp.

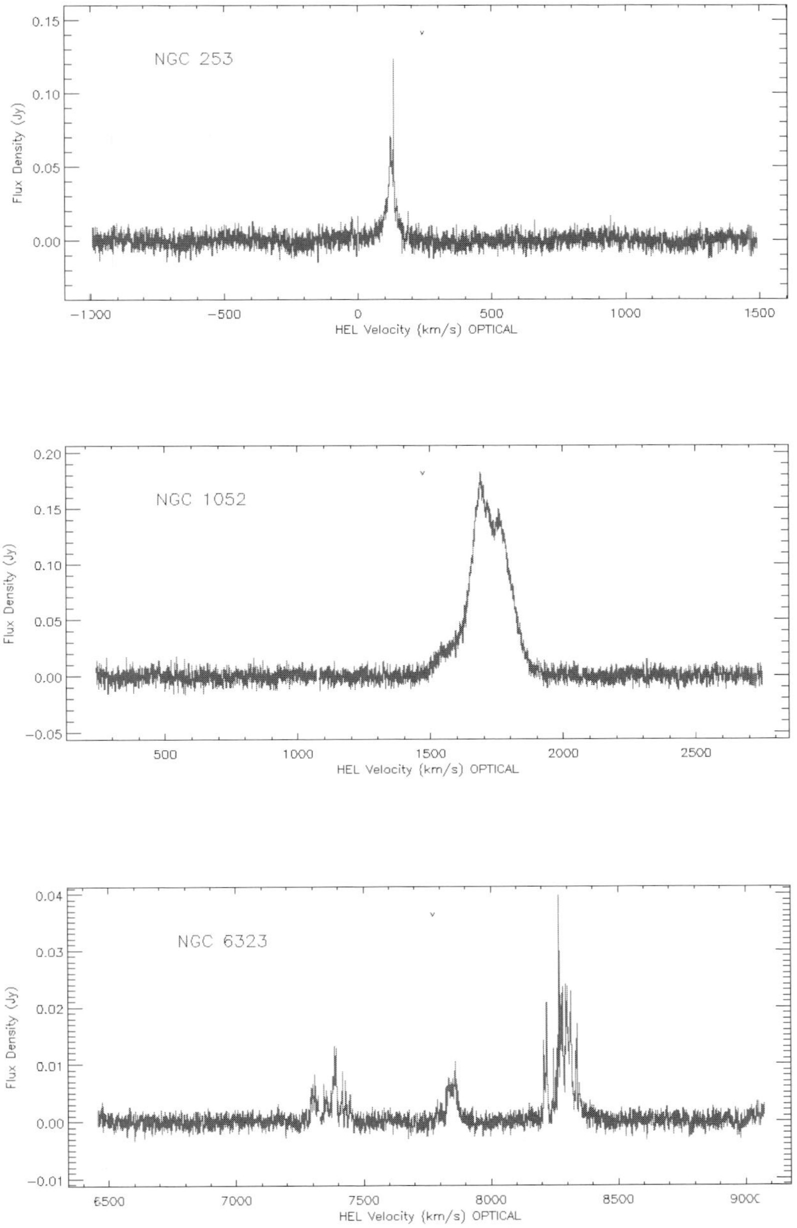

Figure 1. The atlas consists of spectra of nearly 100 extragalactic maser systems. Here we present three representative examples showing GBT spectra of a starburst maser (NGC 253, top), a jet maser (NGC 1052, middle) and a disk maser (NGC 6323, bottom). The complete atlas can be accessed from http://www.nrao.edu/∼jbraatz.

Astrophysical Masers and their Environments
Proceedings IAU Symposium No. 242, 2007
J.M Chapman & W.A. Baan, eds.

doi:10.1017/S1743921307013476

Investigating the nature of low-luminosity extragalactic H_2O masers

Paola Castangia[1,2], A. Tarchi[1], C. Henkel[2] and K. M. Menten[2]

[1]INAF-Osservatorio Astronomico di Cagliari, Loc. Poggio dei Pini, Strada 54, 09012 Capoterra (CA), Italy
email: pcastang@ca.astro.it

[2]Max-Planck-Institut für Radioastronomie, Auf dem Hügel 69, 53121 Bonn, Germany

Abstract. We have searched for 22 GHz water maser emission in a sample of FIR bright galaxies and detected two new kilomaser sources. The newly detected masers have been promptly followed-up using interferometric observations to derive positions and constraints on the size and brightness temperature of the emitting spots. Here we report results related to the newly detected kilomasers, also including the well-known kilomaser source in in NGC 253. These are discussed within the framework of the kilomaser/megamaser dichotomy.

Keywords. masers, galaxies: active, galaxies: starburst

---

1. Introduction

At present, it is still a tantalizing puzzle whether the extragalactic kilomasers ($L_{H_2O} < 10\,L_\odot$) belong to the same family as the Galactic masers associated with star formation or if some of them belong instead to a hypothetical weak tail of the megamasers ($L_{H_2O} > 10\,L_\odot$) associated with Active Galactic Nuclei. The small number of kilomasers found so far does not allow us to draw statistically definite conclusions on their nature. In order to extend the number of such sources, we have searched for 22 GHz water maser emission in a sample of galaxies with declination $\delta > -30°$ and IRAS point source flux densities of $S_{100\mu m} > 30$ Jy, with the Effelsberg 100-m telescope. The survey yields two new kilomaser sources with $L_{H_2O} \sim 1\,L_\odot$ in the nearby starburst galaxy NGC 3556 (Henkel *et al.* 2005) and in the merging system NGC 520 (Castangia 2006). Here we report the results of the interferometric follow-ups of the newly detected kilomasers, also including the prominent kilomaser source in NGC 253, which is also part of the sample. All the interferometric observations have been performed with the Very Large Array (VLA).

2. Results and discussion

All three extragalactic maser spots discussed below are found to be associated with star-forming activity. The H_2O emission in **NGC 520** coincides, within the relative positional uncertainty of $\sim 0.''1$ ($\sim$ 15 pc), with a continuum source associated with a compact non-thermal radio object that is probably a supernova remnant or a radio supernova (source D in Fig.1, lower panel). The small linewidth ($\sim 0.6\,\mathrm{km\,s^{-1}}$) and the time variability (70% decrease of the flux density within 24 days, see Fig. 1, upper panel) of the maser resembles that of the two kilomasers in IC 10 (Henkel *et al.* 1986) and IC 342 (Tarchi *et al.* 2002), which are also associated with star formation activity, reinforcing our interpretation. The kilomaser in **NGC 3556** is somewhat resolved both in space and velocity, indicating that we are likely observing a collection of weak masers, the brightest of which has an isotropic luminosity of $\sim 0.3\,L_\odot$. This maser is coincident (within the

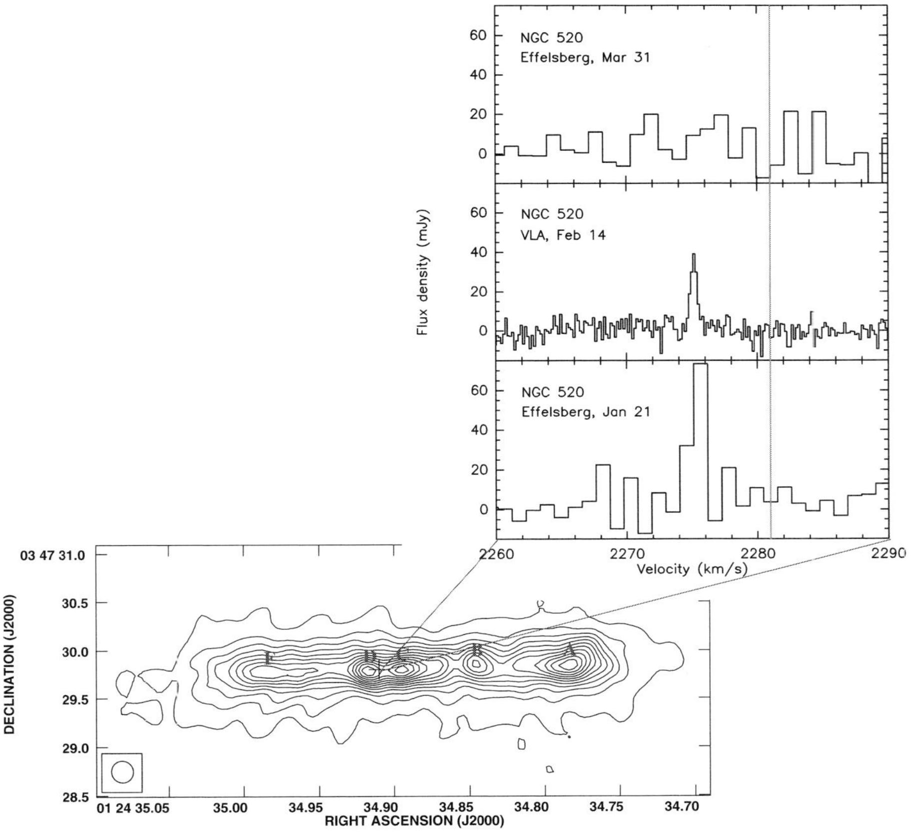

Figure 1. *Upper panel*: H_2O kilomaser spectra toward NGC 520, observed between January and March 2005. Channel spacings are 1.1 km s^{-1} (lower and upper panels) and 0.3 km s^{-1}. Pointing position is α_{2000} = 01^h 24^m 35.1^s and δ_{2000} = +03° 47′ 33″. The vertical line indicates the systemic velocity V_{sys}=2281 km s^{-1} (NED). *Lower panel*: VLA A-array radio continuum map of the nuclear region of NGC 520 at 8 GHz, with labels indicating the compact sources. Contour levels are: –1, 1, 2 ,3...16×0.1 mJy beam^{-1}. The cross marks the position of the maser spot.

relative position errors of ~0.″1, ≡6 pc) with a compact continuum source for which a multi-frequency study rules out an AGN nature. In **NGC 253**, the main H_2O maser component is associated with the source TH4 (for the nomenclature, see e. g. Ulvestad & Antonucci 1997), which is confidently a SNR, confirming the interpretation of Hofner *et al.* (2006). Our results, therefore, favour the conclusion that kilomasers and megamasers form two distinct well separated family of extragalactic H_2O masers.

References

Castangia, P. 2006, PhD Thesis, *"Water masers in active galaxies"*, Univ. of Cagliari

Henkel, C., Peck, A. B., Tarchi, A. Nagar, N.M, Braatz, J. A., Castangia, P., & Moscadelli, L. 2005, *A&A* 436, 75

Henkel, C., Wouterloot, J. G. A., & Bally, J. 1986 *A&A* 155, 193

Hofner, P., Baan, W. A., & Takano, S. 2006 *AJ* 119, 1592

Tarchi, A., Henkel, C., Peck, A. B., & Menten, K. M. 2002 *A&A* 385, 1049

Ulvestad, J. S., & Antonucci, R. R. J. 1997 *ApJ* 488, 621

Astrophysical Masers and their Environments
Proceedings IAU Symposium No. 242, 2007
J.M. Chapman & W.A. Baan, eds.

doi:10.1017/S1743921307013488

Positional coincidence between an H_2O maser and a plasma torus in NGC 1052

S. Sawada-Satoh[1]†, S. Kameno[2], K. Nakamura[2], D. Namikawa[2], K. M. Shibata[3] and M. Inoue[3]

[1]ASIAA, P.O. Box 23-141, Taipei 10617, TAIWAN
email: sss@yamaguchi-u.ac.jp

[2]Kagoshima University, Korimoto, Kagoshima, Kagoshima 890-0065 JAPAN

[3]NAOJ, Osawa, Mitaka, Tokyo 181-8588 JAPAN

Abstract. We present VLBA observation towards the nucleus of a nearby radio galaxy NGC 1052. In NGC 1052, two-sided jet structure and a dense plasma circumnuclear torus with a radius of 0.7 pc have been found around the central mass. It emits a H_2O megamaser, which is redshifted with respect to the systemic velocity of the galaxy (1491 km s^{-1}) with a large velocity width of 100 km s^{-1} (FWHM). The maser gas is found at the inner jet components of both the approaching and receding jets. The maser gas is positionally coincident with a plasma torus. The maser gas in NGC 1052 could be explained as a circumnuclear torus or disk, as found for the nucleus of NGC 4258.

Keywords. galaxies: active, galaxies: individual(NGC 1052), galaxies: H_2O maser

1. Free-free absorption opacity distribution

The continuum images at 15, 22 and 43 GHz show the two-sided jet structure which consists of several components. After restoring with a same beam size, we obtained spectral indices along the jets. The spectral indices indicate that most parts of the two-side jet structure have optically thin spectra at 15–43 GHz except its inner edge part; a steeply inverted spectrum ($\alpha = 3.2$; $S \propto \nu^{-\alpha}$) is revealed between B and C3. The spectrum index exceeds the theoretical limit for synchrotron self-absorption ($\alpha = 2.5$). The highly inverted spectrum of the inner edge of the jets implies that synchrotron emission is obscured through free-free absorption (FFA) by the foreground dense plasma, and it is consistent with past multi-frequency observations (Kameno *et al.* 2001, Vermeulen *et al.* 2003 and Kadler *et al.* 2004).

Fitting the continuum spectrum at 15–43 GHz to FFA model ($S_\nu \propto \nu^{-\alpha} \exp(-\tau\nu^{-2.1})$), we obtained FFA opacity (τ) distributions along the jet axis (figure 1), which reveals that a high opacity ($\tau > 1000$) is found in the inner edge. It implies that the dense cold plasma covers ~2 mas in the inner edge of the jets, where the central engine is supposed to exist.

2. H_2O maser emission

In our observation, significant maser emission within the velocity range of 1550–1850 km s^{-1} was detected. The maser spots consists of two clusters; the east cluster and the west cluster are located at Component B and C3, respectively. The velocity range of the east cluster is 1550–1850 km s^{-1}, which is the same as the whole velocity width

† Present address: Yamaguchi University, Yoshida, Yamaguchi, Yamaguchi 753-8512 JAPAN

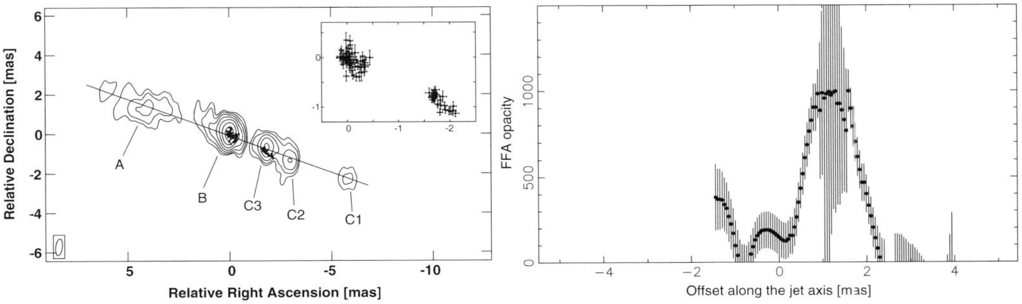

Figure 1. *Left*: Relative distributions of H_2O maser spots (full circle) with respect to the continuum image at 22 GHz (contour). Continuum jet features are labelled by A, B, C3, C2 and C1, along the jet axis (solid line). The inset box shows the detailed distributions of maser spots. *Right*: Free-free absorption opacity along the jet axis.

of the H_2O maser profile. The maser spots with a velocity of 1550–1700 $km\,s^{-1}$ in the east cluster are distributed within 0.1 pc, although the masers with a velocity of 1700–1850 $km\,s^{-1}$ are concentrated within 0.02 pc on the peak of Component B. On the other hand, the west cluster is detected with a velocity range of 1550–1750 $km\,s^{-1}$ along the NE-SW direction or the jet axis. It shows a velocity gradient of $\sim$ 200 $km\,s^{-1}\,mas^{-1}$ along this direction. We note that Claussen *et al.* (1998) showed a velocity gradient along the east-west direction ($\sim$ 100 $km\,s^{-1}\,mas^{-1}$) in the maser cluster of the west jet component.

3. Discussions

The maser gas is located where FFA opacity is large, or where the plasma torus covers the foreground of the inner edge of the jets. The positional coincidence between the plasma and the H_2O masers supports the subparsec-scale torus model, which consists of dense plasma, molecular gas, and the X-ray dissociation region (XDR: Maloney 2002) described by Kameno *et al.* (2005). The other possible explanation is the jet model. The jet propagation hits the molecular gas in the AGN region. The molecular gas within or at the front shock of the jet emits the maser. The west maser cluster shows some velocity gradient along the jet axis. This trend is also seen in Mrk348 whose H_2O maser has been explained an interaction between a jet and a molecular cloud (Peck *et al.* 2003). The west cluster is seen at the receding jet, and the velocity of the west cluster is all redshifted.

Acknowledgements

The VLBA is operated by the NRAO, a facility of the National Science Foundation operated under cooperative agreement by Associated Universities, Inc.

References

Claussen, M. J.,Diamond, P. J., Braatz, J. A., Wilson, A. S., & Henkel, C. 1998, *ApJ* 500, L129

Kadler, M., Ros, E., Lobanov, A. P., Falcke, H., & Zensus, J. A. 2004, *A&A* 426, 481

Kameno, S., Sawada-Satoh, S., Inoue, M., Shen Z.-Q., & Wajima, K. 2001, *PASJ* 53, 169

Kameno, S., Nakai, N., Sawada-Satoh, S., Sato N., & Haba, A. 2005, *ApJ* 620, 145

Maloney, P. R. 2002, *PASA* 19, 401

Peck, A. B., Henkel, C., Ulvestad, J. S., Brunthaler, A., Falcke, H., Elitzur, M., Menten, K. M., Gallimore, J. F. 2002, *ApJ* 590, 149

Vermeulen, R. C., Ros, E., Kellermann, K. I., Cohen, M. H., Zensus, J. A., & van Langevelde, H. J. 2003, *A&A* 401, 113

Astrophysical Masers and their Environments
Proceedings IAU Symposium No. 242, 2007
J.M. Chapman & W.A. Baan, eds.

doi:10.1017/S174392130701349X

The nuclear environment of the water megamaser radiogalaxy 3C 403

Andrea Tarchi[1,2], A. Brunthaler[3], C. Henkel[3], K. M. Menten[3], J. A. Braatz[4] and A. Weiss[3]

[1]INAF-Osservatorio Astronomico di Cagliari, Loc. Poggio dei Pini, Strada 54 09012 Capoterra (CA), Italy; email: atarchi@ca.astro.it

[2] INAF-Istituto di Radioastronomia, Via Gobetti 101, 40129 Bologna, Italy

[3] Max-Planck-Institut für Radioastronomie, Auf dem Hügel 69, D-53121 Bonn, Germany

[4] National Radio Astronomy Observatory, 520 Edgemont Road, Charlottesville, VA 22903

Abstract. So far, few direct sub-arcsecond resolution studies exist of the dense material that, according to the standard unified scheme, should obscure the central engines of radio galaxies. In the following, observations are presented that highlight the nuclear environment of the prototypical X-shaped FR II galaxy 3C 403. To date, it is the only powerful radio galaxy known to host a water megamaser.

Keywords. masers, galaxies: individual (3C 403), galaxies: active

1. Introduction

The standard unified scheme requires the presence of high column densities of absorbing material obscuring the central engines of active galactic nuclei. To date, few direct sub-arcsecond resolution studies of this gas have been performed toward radio galaxies. Tarchi *et al.* (2003) reported the detection of luminous maser emission from the FRII radio galaxy, 3C 403, providing a suitable target for such investigations. 3C 403 is located at a distance of ∼230 Mpc. Optically, it has a normal elliptical structure (Govoni *et al.* 2000), while at radio frequencies it has a remarkable X-shape with "wings" of ∼ 100 kpc stretching toward the NW, SW, NE, and SE (e.g. Black *et al.* 1992; their Fig. 13). To better investigate the nuclear environment of this galaxy, the maser detection of Tarchi *et al.* (2003) has been followed-up by us through a number of radio observations at different frequencies. A summary of the main results obtained is reported in the following section.

2. Results and discussion

Our 22 GHz VLA A-array spectral line and continuum maps suggest that the positions of line emission from the maser features and continuum emission from the core are aligned within 0.″1 (110 pc; Fig. 1 and left panel of Fig. 2). The 5 GHz VLBI continuum observations (Fig. 2, right panel) allows analysis of the spatial structure of the nuclear environment at the parsec scale, which supports a situation where a nuclear core, represented by the dominant central radio continuum component, is accompanied by a jet and counterjet directed toward the western and eastern large scale lobes of the galaxy. In view of the statistical properties of the sample of known megamaser galaxies, an association between the maser emission and an accretion disk seems to be the most plausible scenario, with the two main velocity components (Fig. 1) representing the two tangentially seen parts of the masing disk. However, the small velocity difference between

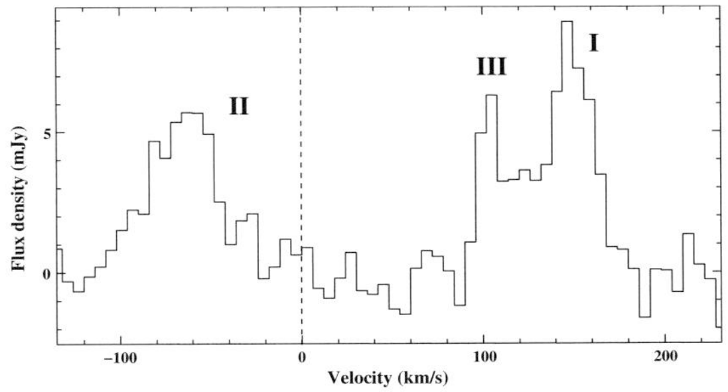

Figure 1. The main water maser features in 3C 403 observed with the VLA A-array on July 23, 2003. The dashed vertical line marks the nominal systemic velocity of the galaxy, $V_{sys} = cz = 17688\,\mathrm{km\,s^{-1}}$.

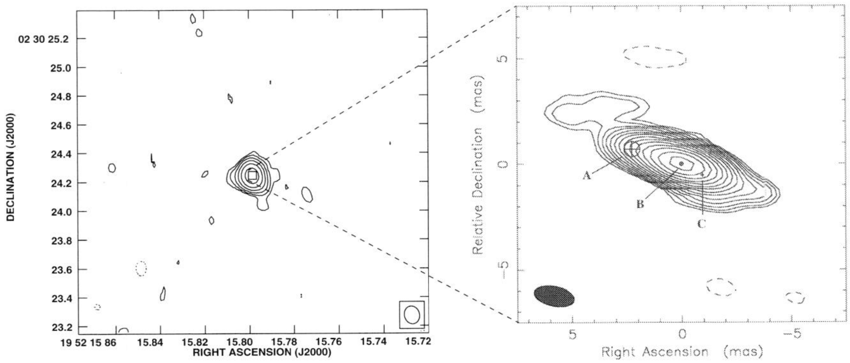

Figure 2. *Left panel:* A naturally weighted 22 GHz VLA radio continuum image (resolution: 0.″1, corresponding to 110 pc) of the nuclear region of 3C 403 made using channels free of line emission. The peak is at 40 mJy beam^{-1}. The contours are (−1, 1, 2, 4, ..., 32) × 0.75 mJy beam^{-1}. *Right panel:* Full resolution EVN image of the nucleus of 3C 403 at λ=6 cm. The contours start at 0.2 mJy and increase with a factor of 1.5. The beam size of the observations is 1.87 mas×0.94 mas at a position angle of 76.°9.

the two features makes this interpretation questionable. Investigations of J=1-0 and 2-1 transitions of CO using the IRAM 30-m telescope at Pico Veleta in order to detect the thermal emission from a spatially extended molecular environment were unsuccessful. Our ongoing single-dish 1.3 cm monitoring campaign of the maser lines performed with the Effelsberg and the GBT radio telescopes indicates strong variations in the intensities of the two main maser components with (isotropic) luminosities sometimes surpassing 1000 solar luminosities. Typical timescales of the variation are of order 1–2 years.

References

Black, A. R. S., Baum, S. A., Leahy, J. P., Perley, R. A., Riley, J. M., & Scheuer, P. A. G. 1992, *MNRAS* 256, 186

Govoni, F., Falomo, R., Fasano, G., & Scarpa, R. 2000, *A&A* 353, 507

Tarchi, A., Henkel, C., Chiaberge, M., & Menten, K. M. 2003, *A&A* (Letters) 407, L33

Astrophysical Masers and their Environments
Proceedings IAU Symposium No. 242, 2007
J.M. Chapman & W.A. Baan, eds.

doi:10.1017/S1743921307013506

PV diagrams for the maser emission from a Keplerian ring

L. Uscanga[1], J. Cantó [2] and A. C. Raga[3]

[1]Centro de Radioastronomía y Astrofísica, UNAM, Ap. 3-72, 58089 Morelia, Mich., Mexico
email: l.uscanga@astrosmo.unam.mx

[2]Instituto de Astronomía, UNAM, Ap. 70-264, 04510 México, D.F., Mexico

[3]Instituto de Ciencias Nucleares, UNAM, Ap. 70-543, 04510 México, D.F., Mexico

Abstract. We study the maser emission arising from a thin, planar, gaseous ring in Keplerian rotation around a central mass when it is observed edge-on. Assuming that the absorption coefficient within the amplifying ring is a decreasing function of distance from the central mass (i.e., $\kappa = \kappa_0 r^{-q}$), we calculated position-velocity (PV) diagrams for the most intense maser features using different values of the exponent q. We found that, depending on the value of q, these diagrams can be qualitatively different.

Keywords. galaxies: individual (NGC 4258) — galaxies: nuclei — masers

1. Introduction

The spatial distribution and line-of-sight velocities of the systemic and high-velocity water masers in NGC 4258 trace a thin molecular ring in Keplerian rotation around a massive black hole of 3.6×10^7 M$_\odot$ seen nearly edge-on (Miyoshi *et al.* 1995). The PV diagram for the maser emission shows distinct Keplerian orbits defined by the high-velocity maser emission, as well as a line traced by the systemic maser emission. The systemic masers lie within a narrow range of radii, on the near side inner edge of the ring, while the high-velocity masers are located near the ring diameter. Previously, Watson & Wallin (1994) showed that the maser emission from a rapidly rotating, thin Keplerian ring viewed edge-on can reproduce the general features of the observed 22 GHz radiation of this galaxy. However, their assumption of a uniform absorption coefficient within the amplifying ring results in a PV diagram for the most intense masers that is qualitatively different from the observed one. Here we show that this discrepancy can be resolved if the absorption coefficient decreases with distance from the central mass.

2. Model and results

We consider a masering Keplerian ring seen edge-on. The masering gas is located between R_0 and R, the inner and outer radii of the ring, respectively. We assume that the absorption coefficient is $\kappa = \kappa_0 r^{-q}$. Distances are measured in units of R, and velocities are measured in units of $v_{\rm out}$, the rotation velocity at the outer edge of the ring. For the case of an unsaturated maser and neglecting the spontaneous emission, the intensity of the maser radiation from a line of sight with impact parameter y at a velocity v_r is

$$I(v_r, y) = I_0 e^{\tau(v_r, y)}, \tag{2.1}$$

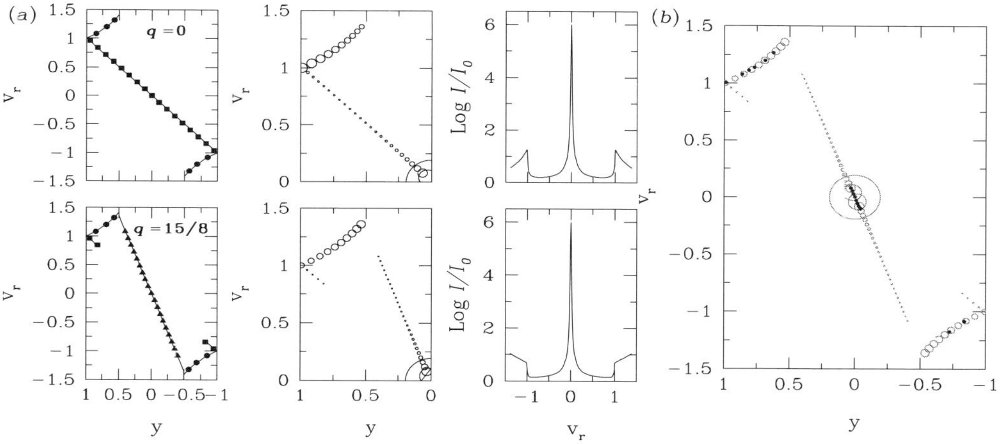

Figure 1. (*a*) Results of model with $q = 0, 15/8$. *Left panels*: PV diagrams for the maser emission peak. The filled squares, circles, and triangles represent the strongest maser emission which is coming from the outer edge, mid-line, and inner edge of the ring, respectively. The straight lines or curves represent the velocity dependencies of the regions where this emission arises. *Central panels*: PV diagrams for the maser emission peak, only positive velocities are shown. The radii of the open circles are proportional to the maximum maser intensity at each position and velocity. *Right panels*: the logarithm of the ratio between the maximum intensity and the background intensity as a function of the velocity. (*b*) Comparison between the calculated PV diagram for the maser emission peak using $q = 15/8$ and the PV diagram delineated by the water masers observed in NGC 4258. The dots represent the observed maser spots (Miyoshi *et al.* 1995). We subtracted the ring systemic velocity of 476 km s^{-1} from the observed LSR velocity of the maser spots in order to compare the observed PV diagram with the modelled one. The positions and velocities are in units of the outer radius of the ring (8 mas) and the rotation velocity at the outer edge of the ring (770 km s^{-1}), respectively (Uscanga *et al.* 2007).

where the optical depth or gain along the line of sight is given by

$$\tau(v_r, y) = \kappa_0 \int (x^2 + y^2)^{-q/2} \exp\left[\frac{-(v - v_r)^2}{\Delta v_D^2}\right] dx. \tag{2.2}$$

We numerically solved equations (2.1) and (2.2) and calculated the y-positions (impact parameters) of maximum maser intensity for each specific value of v_r. When we found two local maxima, we kept both. Then we constructed the PV diagrams using the positions of the observer's line of sight with maximum emission at each velocity (Figure 1).

3. Conclusions

A comparison with the PV diagram observed for the water maser emission in the nucleus of the galaxy NGC 4258 indicates that the exponent q must be higher than 1 to have significant emission coming from the inner edge of the amplifying ring and hence explain the form of the PV diagram. According to our simple model, this implies that the absorption coefficient must depend on the radius of the ring as a decreasing function.

References

Miyoshi, M., Moran. J., Herrnstein, J., *et al.* 1995, *Nature*, 373, 127

Uscanga, L., Cantó, J., & Raga, A. C. 2007, *ApJ*, in press

Watson, W. D., & Wallin, B. K. 1994, *ApJ* (Letters), 432, L35

Astrophysical Masers and their Environments
Proceedings IAU Symposium No. 242, 2007
J.M. Chapman & W.A. Baan, eds.

doi:10.1017/S1743921307013518

Water-vapor maser disk at the nucleus of the Seyfert 2 IC 2560

Aya Yamauchi[1], Naomasa Nakai[1], Yuko Ishihara[2], Philip Diamond[3] and Naoko Sato[4]

[1]University of Tsukuba, 1-1-1 Ten-nodai, Tsukuba, Ibaraki 305-8571, Japan
email: yamauchi@physics.px.tsukuba.ac.jp

[2]Koriyama City Fureai Science Center, 2-11-1 Ekimae, Koriyama, Fukushima 963-8002, Japan

[3]Jodrell Bank Observatory, University of Manchester, Macclesfield, Cheshire, SK11 9DL, UK

[4]Wakayama University, 930 Sakaedani, Wakayama, Wakayama 640-8510, Japan

Abstract. We observed the H_2O maser at the nucleus of the Seyfert 2, IC 2560, using the VLBA and the phased VLA. The systemic, red-shifted and blue-shifted maser features and a continuum component have been detected. We propose a maser disk in the nuclear region. The systemic and red-shifted features are emitted from a nearly edge-on disk with the position angle of $PA = -47°$. The thickness is $2H < 0.025$ pc. The binding mass is $3.5 \times 10^6 M_\odot$. Assuming the Keplerian rotation, the radii at the disk are $r = 0.087 - 0.335$ pc and the rotation velocities are 213–418 km s^{-1}. The mean density within the inner radius is $1.3 \times 10^9 M_\odot$ pc^{-3}, suggesting a massive black hole at the center. A continuum component is considered as a jet ejected from the nucleus, with an angle of 70° from the disk. The blue-shifted maser feature is located on the continuum component, being interpreted to be a 'jet maser'.

Keywords. galaxies:active, galaxies:individual (IC 2560), galaxies:nuclei, galaxies: Seyfert

---

IC 2560 is a Seyfert 2 glaxy at a distance of 26 Mpc (Aaronson *et al.* 1989) and its systemic velocity is $V_{\rm LSR,rad} = 2876$ km s^{-1} (Strauss *et al.* 1992). We observed the systemic maser features and continuum emission of IC 2560 on 1998 January 2 and the systemic and high-velocity features on 2000 April 10, using the VLBA and the phased VLA. The synthesized beams were 2.2×1.8 mas (1998) and 1.00×0.36 mas (2000).

Fig.1(a) shows the distribution of H_2O masers in a coordinate system relative to the position of the strongest maser feature at $V_{\rm LSR} = 2876$ km s^{-1}. The systemic features (2867–2890 km s^{-1}) are located around $(\Delta RA, \Delta Decl) \approx (0, 0)$ mas. The peak intensity is 0.145 Jy beam^{-1}. A red-shifted feature (3201 km s^{-1}) is detected at $(\Delta RA, \Delta Decl) = (-0.84, 0.77)$ mas. Its angular separation from the strongest systemic feature is 1.14 ± 0.03 mas (0.144 ± 0.004 pc), and its peak intensity is 0.062 Jy beam^{-1}. A blue-shifted feature (2656 km s^{-1}) is detected at $(\Delta RA, \Delta Decl) = (-1.57, 0.13)$ mas. Its angular separation from the strongest feature is 1.58 ± 0.47 mas (0.199 ± 0.006 pc), and its peak intensity is 0.030 Jy beam^{-1}. 22-GHz continuum emission is detected with a peak flux density of 1.8 ± 0.3 mJy beam^{-1}. Assuming that the peak position of the 2876 km s^{-1} maser feature observed in 1998 is the same as the position observed in 2000, the continuum map is overlaid on the maser map, as seen in Fig.1(a). The peak position of the continuum component is at $(\Delta RA, \Delta Decl) = (-1.40, -0.72)$ mas $= (-0.176, -0.090)$ pc.

The spatial distribution of the systemic and red-shifted features and the continuum component strongly suggest that we are seeing an edge-on maser disk rotating around $(\Delta RA, \Delta Decl) \approx (0, 0)$ mas, and a jet ejected from the centre, perpendicular to the maser disk, as is observed in NGC 4258 (Miyoshi *et al.* 1995, M95). The blue-shifted feature near the continuum component clearly does not belong to the disk but is probably a

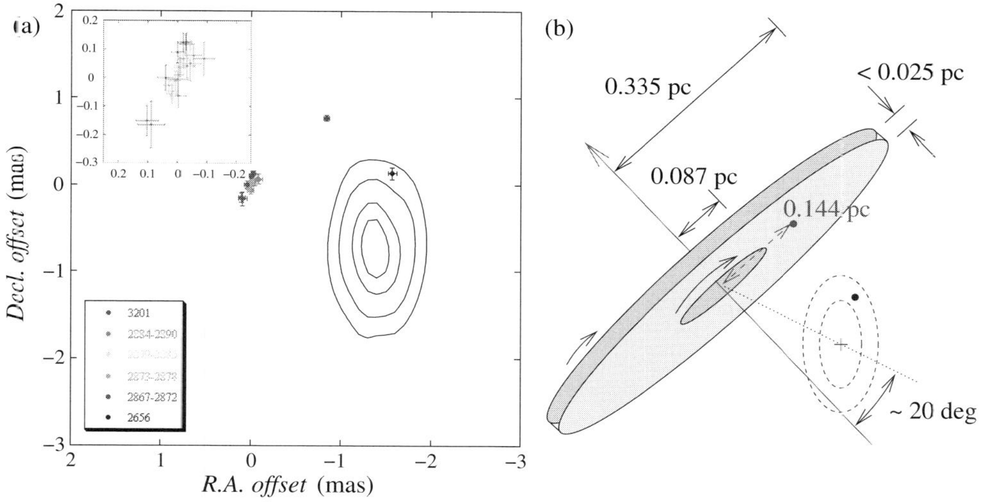

Figure 1. (a) Distributions of maser emission and continuum component in the nuclear region. The contour levels are 3, 4, 5, and 5.5σ (1σ = 0.31 mJy beam^{-1}). (b) Model for the maser disk.

'jet maser' which may be pumped by collisions between a molecular gas and the jet and stimulated by continuum radiation of the jet as seen in NGC 1068 (Gallimore *et al.* 2001). A least-squares fit to the systemic and red-shifted features gives the position angle of the disk, $PA = -47°$. The distribution of the systemic features measured perpendicularly to the PA indicates that the thickness of the disk is $2H < 0.20$ mas (0.025 pc). The red-shifted feature we detected lies at a radius of $r = 1.14 \pm 0.17$ mas (0.144 ± 0.021 pc) and rotates with a velocity of $V_{\rm rot} = 325$ km s^{-1}, which means that the mass inside the radius is $(3.5 \pm 0.5) \times 10^6 M_\odot$, assuming a spherical distribution of the central matter.

In single-dish observations, more blue- and red-shifted maser features are detected. Assuming that the blue-shifted features at $V_{\rm LSR} = 2458$–2556 km s^{-1} and the red-shifted features at $V_{\rm LSR} = 3089$–3222 km s^{-1} are located in the disk, the rotation velocities are $V_{\rm rot} = 213$–418 km s^{-1}. If the rotation is Keplerian ($V_{\rm rot} \propto r^{-1/2}$) as in NGC 4258 (M95), the disk radii can be estimated to be $r = 0.087$–0.335 pc. The mean volume density inside the inner radius is $\rho = (1.3 \pm 0.6) \times 10^9 M_\odot$ pc^{-3} which is comparable with that of NGC 4258 ($3.4 \times 10^9 M_\odot$ pc^{-3}, M95) and thus suggests the existence (Ishihara *et al.* 2001) of a massive black hole of $M_{\rm BH} \approx 3.5 \times 10^6 M_\odot$ at the nucleus of IC 2560.

Acknowledgements

A. Y. thanks the IAU Grant and the Hayakawa Fund in Astronomical Society of Japan.

References

Aaronson, M., Bothun, G. D., Cornell, M. E., Dawe, J. A., Dickens, R. J., Hall, P. J., Sheng, H. M., Huchra, J. P., Lucey, J. R., Mould, J. R., Murray, J. D., Schmmer, R. A., & Wrighy, A. E. 1989, *ApJ* 338, 654

Gallimore, J. F., Henkel, C., Baum, S. A., Glass, I. S., Claussen, M. J., Prieto, M. A., & Von Kap-Herr, A. 2001, *ApJ* 556, 694

Ishihara, Y., Nakai, N., Iyomoto, N., Makishima, K., Diamond, P., & Hall, P. 2001, *PASJ* 53, 215

Miyoshi, M., Moran, J., Herrnstein, J., Greenhill, L., Nakai, N., Diamond, P., & Inoue, M. 1995, *Nature* 373, 127

Strauss, M. A., Huchra, J. P., Davis, M., Yahil, A., Fisher, K. B., & Tonry, J. 1992 *ApJS*, 83, 29

Astrophysical Masers and their Environments
Proceedings IAU Symposium No. 242, 2007
J.M. Chapman & W.A. Baan, eds.

doi:10.1017/S174392130701352X

Statistical properties of extragalactic H_2O maser sources

Jiang-Shui, Zhang[1] **and Christian Henkel**[2]

[1]Center for Astrophysics, Guangzhou university, Guangzhou 510006, China
email: jszhang365@yahoo.com.cn

[2]Max-Planck-Institut für Radioastronomie, Auf dem Hügel 69, D-53121 Bonn, Germany
email: p220hen@mpifr-bonn.mpg.de

Abstract. Recently, with the development of sensitive maser receiver and wide-band spectrometer, maser emission were detected toward more and more galaxies. 69 galaxies beyond the Magellanic Clouds has been reported with H_2O maser emission (to 2006.12). The statistical properties of the sample of extragalactic H_2O masers are not well explored. Here we collect observation data of all these detected extragalactic H_2O maser sources and give a statistical analysis to investigate the statistical properties of the sample.

Keywords. H_2O masers, statistical properties

1. The sample

We collect all published data on extragalactic H_2O maser sources (69, to the end of 2006) and their observational parameters at other wavelengths. Some of these parameters reflect the properties of the host galaxies, include the distance, infrared luminosity, the dust temperature, the apparent B magnitude, and the absolute B magnitude. Others parameters reflect the nuclear properties, include the AGN type, the isotropic H_2O maser luminosity, the emission line flux and luminosity, the central black hole mass, the nuclear X-ray emission luminosity, the X-ray absorbing column density, and the radio flux and luminosity.

2. Statistical analysis

2.1. *Distributions of some parameters*

Based on the literature and archival data, X-ray absorbing column densities are compiled for the 69 galaxies with reported maser sources. We obtained X-ray column densities for 38 maser sources, 36 from the literature and 2 from processing archival data. Fig. 2 shows the distribution of their column density. More than 80% (22/26) of the megamaser sources are heavily obscured ($N_H > 10^{23}$ cm^{-2}) and half are Compton-thick ($N_H \geqslant 10^{24}$ cm^{-2}). Among the 13 known kilomaser sources ($L_{H_2O} < 10\,L_\odot$), however, only NGC 2273 and NGC 5194 are Compton-thick. At a confidence level of at least 99.4%, the column density distributions of kilomaser and megamaser sources are different (see details in Zhang *et al.*. 2006).

2.2. *Possible correlations*

A correlation would be most convincing if there were a strong dominant X-ray source known to coincide with both the nuclear engine of the parent galaxy and the observed H_2O maser. In Fig. 1, the H_2O maser isotropic luminosities are plotted against their X-ray

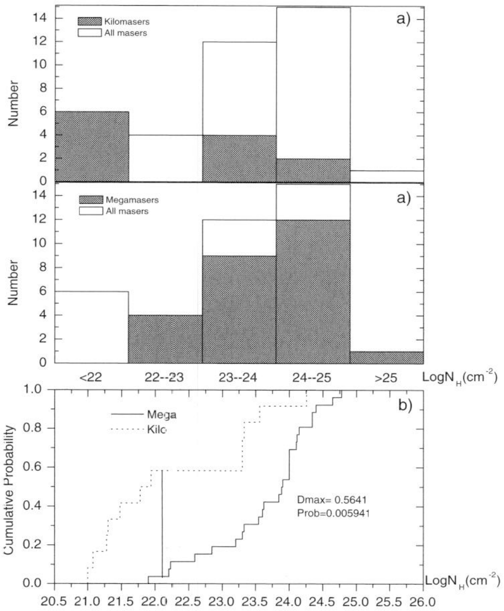

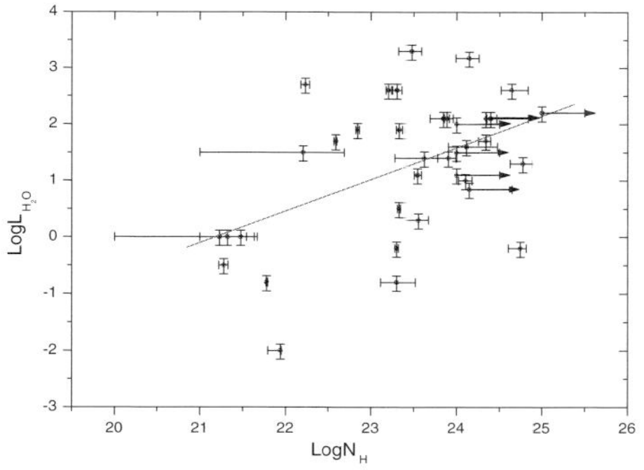

Figure 1. Isotropic maser luminosities (in $L_\odot$) versus X-ray absorbing column densities N_H (in cm^{-2}), including a linear fit. Taken from Zhang *et al.* (2006).

Figure 2. a. Histograms showing the number of H_2O masers (kilomasers: L_{H_2O}<10 $L_\odot$; megamasers: L_{H_2O}>10 $L_\odot$) as a function of column density determined from X-ray spectroscopy (upper panels). **b.** Kolmogorov-Smirnov test for the kilomaser and megamaser samples (lower panel). Taken from Zhang *et al.* (2006).

absorbing column densities. A trend of rising column density with rising maser luminosity may be apparent, but the scatter is large and the correlation is weak. The megamaser subsample (not displayed) shows no trend at all. Presumably a clumpy cloud structure in the cirumnuclear environment, diverging positions between maser and nuclear sources, and occasional amplification of a background radio continuum source are sufficiently decoupling the X-ray column densities and the H_2O maser properties not to show a clear correlation.

In addition, the nuclear 2−10 keV X-ray luminosity and the H_2O isotropic luminosity was discussed in Kondratko *et al.* (2006, see their Figure 3). The correlation between FIR luminosity and H_2O isotropic luminosity are confirmed (see Henkel *et al* . 2005). Other possible correlations are also investigated (Zhang *et al.* in preparation).

References

Henkel, C., Peck, A. B., Tarchi, A., *et al.*. 2005, *A&A* 436, 75

Kondratko, P. T., Greenhill, L. J., & Moran, J. M. 2006, *ApJ* 652, 136

Zhang, J. S., Henkel, C., Kadler, M., Greenhill, L. J., Nagar, N., Wilson, A. S., & Braatz, J. A. 2006, *A&A* 450, 933

Photo: Peter Hajottofi

The Todd River

The Todd river in Alice Springs is usually dry but on the last day of the meeting it rained and the river flooded.

Astrophysical Masers and their Environments
Proceedings IAU Symposium No. 242, 2007
J.M. Chapman & W.A. Baan, eds.

doi:10.1017/S1743921307013531

Masers in starburst galaxies

Jeremy Darling

Center for Astrophysics and Space Astronomy, Department of Astrophysical and Planetary Sciences, University of Colorado, 389 UCB, Boulder, CO 80309-0389
email: jdarling@origins.colorado.edu

Abstract. Masers in starburst galaxies are outstanding probes of a range of phenomena related to galaxy and black hole evolution, and offer unique high brightness temperature illumination that can be used to probe small scales in the host galaxy and in our own. But we require a deeper understanding of the galaxy-scale maser phenomenon if we wish to employ starburst galaxy masers as probes using the next generation of radio telescopes. This review summarizes what is known about the different flavors of masers in starburst galaxies and the setting and structure of OH megamasers. The question of which galaxies produce megamasers and which do not is critical to our understanding of the megamaser phenomenon, and recent studies of HCN and H_2CO are particularly instructive. Constraints on the lifetime of OH megamasers and the predictability of OH megamaser line properties are critical issues to address in the near future. It is also time to begin the next wave of OH megamaser surveys at higher redshifts and to finally employ them as probes of starbursts, massive black holes, galaxy evolution, and intervening media.

Keywords. masers, galaxies: starburst, galaxies: nuclei, galaxies: magnetic fields, galaxies: ISM, galaxies: interactions, galaxies: evolution, radio lines: galaxies

---

1. The woeful inadequacy of extragalactic observations

Extragalactic maser research is severely hampered by the vast distances between galaxies. While Galactic masers can be observed at AU scales and commonly show flux densities of many — even thousands— of Jy, when we move from kpc to Mpc distances, resolution decreases 1000-fold and sensitivity drops by a million. For the nearest megamaser galaxies, the resolution is another factor of 100 lower, and the sensitivity goes down by 10^4. Thus, AU scales become pc scales in megamaser galaxies and 1 Jy becomes 0.1 nJy. Thanks to distance dilution, Galactic-style masers have been detected only in the Local Group. Enhanced maser emission that is thought to follow a similar mechanism as Galactic masers, the "kilomaser," has been observed in nearby galaxies, such as M82. But for most galaxies in the local universe, modern facilities can only detect "megamasers," masers that do not have Galactic analogs and are extremely rare.

Thanks to distance dilution, extragalactic maser research is information-poor: there are only a few well-mapped systems (pc-scale), only a few hundred megamaser galaxies detected (none beyond $z = 1$), and there are only a few maser species found to date. Unfortunately, the state of the art in extragalactic maser research treats most megamasers and their host galaxies as point sources, and the samples include only the most exceptional systems in the local universe. Megamasers, however, are very bright extragalactic signposts that we need to learn how to read: if we know what these extraordinary phenomena signify (or conversely how they form), then we can employ them as probes of galaxy evolution, star formation, cosmology, and massive black holes.

This review focuses on masers in starburst galaxies (H_2O masers associated with active galactic nuclei are reviewed by Greenhill in this volume), which are mostly OH megamasers (OH MMs; § 2). While examining the setting (§ 3) and structure (§ 4) of OH MMs is instructive, the question of which galaxies produce megamasers and which do not is crucial to understanding the underlying mechanisms behind the megamaser phenomenon in starbursts (§ 5). Examination of other molecular species, both masing and thermal, associated with starburst galaxies is highly instructive in this case. New questions arise (§ 7), but the future looks very promising both in theory and observation for understanding megamasers and employing them as probes of all manner of extragalactic phenomena (§ 8).

2. Extragalactic maser flavors

It is important to remember that *all* masers observed in the Galaxy occur in other galaxies and that the dearth of observed maser species is simply a limitation in sensitivity. Currently, we cannot usually detect the aggregate maser emission from common molecules in external galaxies, much less detect individual maser regions. It is likely that upcoming advances in instrumentation will reveal Galactic-style masing in nearby galaxies outside of the Local Group, and several groups have already detected OH and H_2O kilomasers (e.g., Castangia *et al.*, this meeting).

Currently, observations of extragalactic masers are mostly limited to megamasers, a markedly different phenomenon in scale and setting from Galactic masers or kilomasers. To date, only OH, H_2O, and H_2CO (formaldehyde) have been detected as megamasers. Searches for other common masers, such as CH, SiO, and CH_3OH (methanol), have produced no detections.

In starburst galaxies, the most common megamaser by far is OH. H_2CO megamasers have been detected in a few cases, most notably in the first OH MM discovered and archetypical ultraluminous infrared galaxy, Arp 220 (Baan, Wood, & Haschick 1982; Baan, Güsten, & Haschick 1986), but at about 1% of the intensity of OH MMs (see, e.g., Araya, Baan, & Hofner 2004). We therefore focus primarily on OH MMs in this review.

3. Setting of OH megamasers

3.1. *OH megamaser hosts*

OH MMs are produced in the extreme starbursts occurring during galaxy mergers, apparently favoring late-stage mergers with closely separated nuclei, often in a common envelope. The hosts of OH MMs are (ultra)luminous IR galaxies ([U]LIRGs), and the OH MM fraction is a strong function of the IR luminosity and may depend weakly on far-IR color, favoring "warmer" systems (Darling & Giovanelli 2002a). ULIRGs are roughly as common in the local universe as quasars, and since OH MMs appear in only a fraction of (U)LIRGs, they are exceedingly rare. Large volumes of space ($\sim$1 Gpc^3) have been surveyed in order to detect $\sim$ 100 OH MMs.

It is incorrect to treat OH MM hosts as point sources. Figure 1 shows the illustrative case of IRAS 06487+2208: this OH MM is a multiple-merger system with two optically bright nuclei, two radio continuum sources (only one of which corresponds to an optical nucleus), and an unresolved maser source associated with an optically unremarkable component. The maser source is associated with strong star formation, as expected. Typically, statistical studies of OH MM properties, such as radio continuum, IR luminosity, or optical spectral classification, treat the OH MM and host galaxy as a point source because sub-arcsec angular resolution is not possible. But one should always bear in mind that most OH MM samples are contaminated by non-masing nuclei, and some wavelengths may not even detect the regions responsible for the maser action.

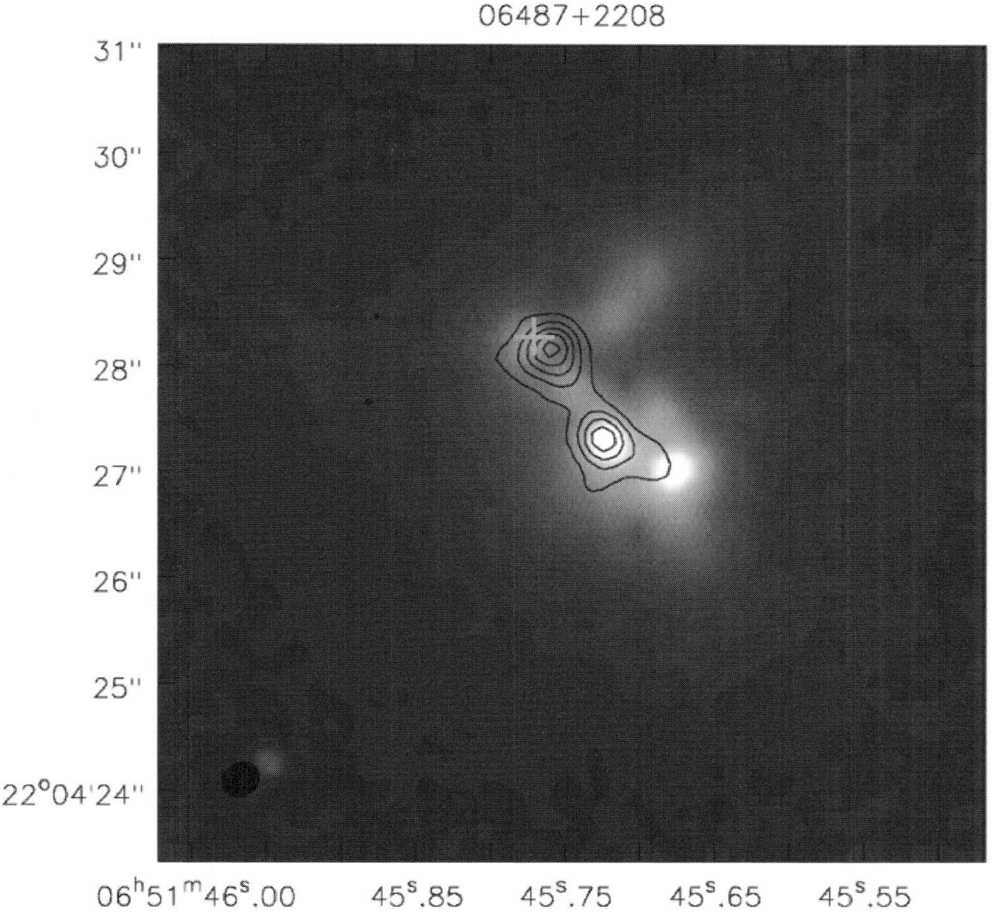

Figure 1. IRAS 06487+2208 ($z = 0.1437$). The image is from the Hubble Space Telescope (WFPC2 F814W filter: $\lambda = 7940$ Å, $\Delta\lambda = 1531$ Å), revealing sub-arcsec structure. The contours are 5 GHz radio continuum from the VLA, and the cross marks the location of the unresolved OH megamaser emission. The offset between the OH emission and the upper continuum peak should not be construed as significant. Coordinates are J2000.

3.2. *OH megamasers and binary massive black holes*

The evolution of massive black holes in galaxies is merger- and accretion-driven, and both processes are expected to spike during galaxy merging events, so OH MM studies give us a window on this interesting episode in the life of a massive black hole. Of great interest is the emission of gravitational waves by hard binary black holes with sub-pc separation (Begelman, Blandford, & Rees 1980), but the demographics of this population is unknown. Whereas massive galaxies are all thought to harbor massive black holes, whereas a merger has as its inevitable evolutionary outcome a hard binary (i.e., the stalling problem can be surmounted, and neither galaxy already contains a binary massive black hole), and whereas OH MMs occur during a specific stage in merging, therefore OH MMs sample the population of gravitational wave-emitting binary massive black holes with the same lifetime as the OH MM lifetime, as shown in Figure 2. Surveying the population of OH MMs versus redshift thus samples this population of massive black

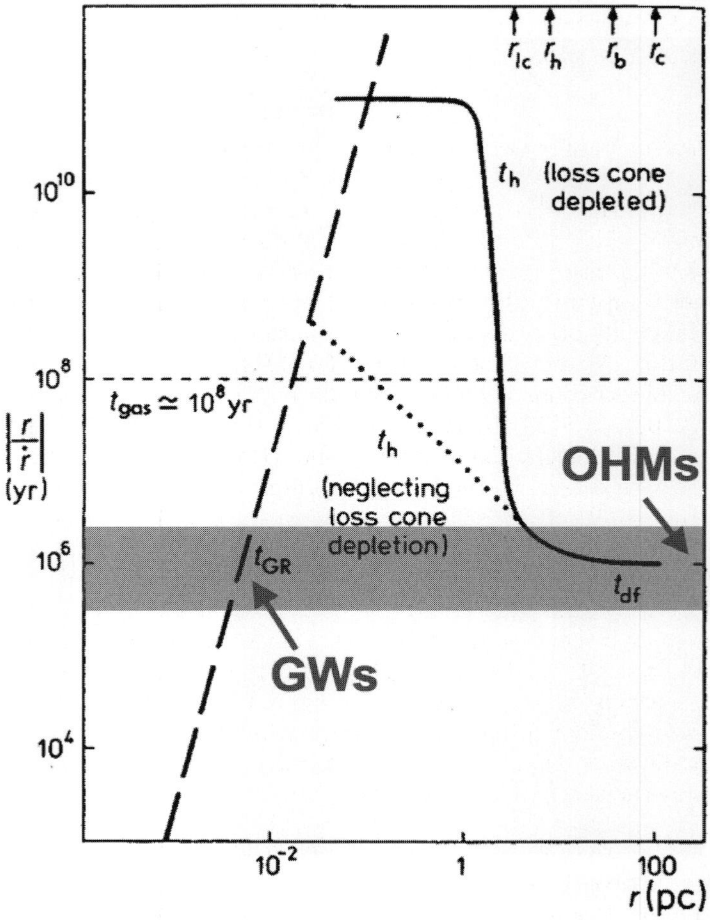

Figure 2. Evolutionary sequence of binary massive black holes from Begelman, Blandford, & Rees (1980). The dashed line indicates the gravitational wave-emitting track leading to coalescence, and the solid line indicates the evolution of the pair by dynamical friction. The arrows and shaded region have been added to emphasize the characteristic timescale sampled by OH MM surveys.

hole binaries. Since the evolution of binaries on the gravitational wave-emitting track is well-defined, OH MM surveys will furthermore provide a census of a large fraction of the hard binary massive black hole population. The missing elements for this type of study are the OH MM lifetime and surveys for redshifted OH MMs (see § 8).

4. Structure of OH megamasers

4.1. *Theory*

There is an encouraging convergence of the theory behind OH MM formation thanks to work by, among others, Parra *et al.* (2005; this volume) and Lockett & Elitzur (2007; this volume). There are, however, areas where observations can confront the theory and make it robust. Observers need to expand the sample of OH MMs with 18 cm satellite line measurements because Lockett & Elitzur (2007) make firm predictions in this area. It will be interesting to see if observed constraints on maser brightness temperatures, particularly those at high redshift showing variability (§ 4.3), can be reproduced by models. There are also new hints of a critical H_2 density for OH MM formation that should be explored with maser models (§ 5).

4.2. *Clues from VLBI*

The fundamental units of maser action, unimaginatively called "spots" (the author proposes a new term: "mason"), are notoriously difficult to resolve, even with very long baseline interferometric (VLBI) observations. The rarity of OH megamasers places them at large distances, which compounds the resolution problem. Addressing such basic issues as where OH MMs are produced in merging systems, their spatial extent, how the emission geometries relate to line profiles and merger kinematics, and the relationships between OH MMs and optical spectral types of nuclei, are exceedingly difficult with modern instrumentation. A handful of the closest OH MMs have been studied in some detail: the most recent observations include III Zw 35 by Pihlström *et al.* (2001); Arp 220 by Rovilos *et al.* (2003); Mrk 231 by Klöckner, Baan, & Garrett (2003); Mrk 273 by Yates *et al.* (2000) and Klöckner & Baan (2004); and *IRAS* 17208−0014 by Momjian *et al.* (2006). The sample size remains small and the results of these studies generally indicate that many of the details of OH MMs remain hidden from view by their small sizes and low fluxes.

VLBI studies do appear to have converged on some aspects: OH is observed in 1–200 pc structures; CO and HI are more extended and mostly exterior to OH; a large fraction of the sample shows toroidal OH MM regions with thickness of the same order as the height and velocity dispersion roughly equal to rotation (e.g., III Zw 35); and line widths remain large on the smallest observed scales. But the small VLBI sample remains extremely diverse, ranging from nearly pure starbursts to Sey 1 AGN. Of particular richness are recent observations of the OH, both in emission and absorption, in Arp 220 toward radio supernovae (Lonsdale *et al.* this volume).

4.3. *Variability in OH megamasers*

Variability was discovered in OH MMs by Darling & Giovanelli (2002b) with properties consistent with refractive interstellar scintillation (ISS). Further study of a sample of OH MMs over long timescales is underway (Darling 2007b). Variability in these spectral line sources offers a powerful super-VLBI probe of OH MM structure because not all spectral line components vary, and those that do vary do not always vary together. There is a critical threshold angular size scale above which the scintillation of radio waves induced by the Galactic free electrons washes out and below which the observed scintillation can be quite strong and rapid. Applying an ISS model to the observed variability allows us to segregate maser component size scales based on the presence or absence of variation and the magnitude of fluctuation and to segregate positions of masing regions on the sky based on correlation (or not) of variations in scintillating components. For OH MMs, one can obtain sub-milliarcsecond "resolution" from this approach and significantly constrain the brightness temperature of OH MM spots.

An interesting case study is IRAS 02524+2026 at $z = 0.18$, which has been observed as part of an OH MM monitoring campaign at Arecibo (Darling 2007b). This OH MM shows strong variability in many spectral components in both of the "main" 18 cm OH lines (Figure 3). The rms spectrum significantly understates the day-to-day variation seen in this source. Preliminary analysis suggests that variable features are smaller than ~ 1 pc (0.3 mas), which implies a brightness temperature $T_b > 8 \times 10^{11}$ K. We observe spectral components that vary together, implying coincidence on the sky, and components that are uncorrelated, implying some minimum separation on the sky.

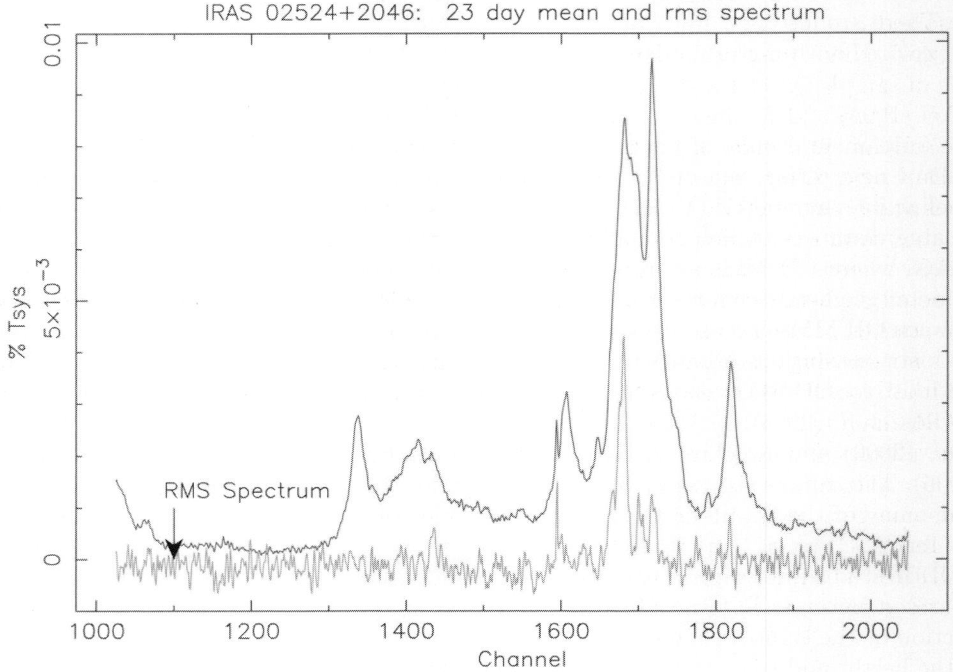

Figure 3. IRAS 02524+2046 ($z = 0.18$) observed at Arecibo in a variability monitoring campaign. The red spectrum is the 23-day mean and the green spectrum is the rms over the same interval. Both of the "main" 18 cm OH lines show variation.

4.4. *Magnetic fields*

Robishaw (2006; this volume) has detected Zeeman splitting in several OH MM galaxies, from Arp 220, which shows many polarized line components, to IRAS 12032+1707 at $z = 0.217$. IRAS 12032+1707 shows variability in at least two line components (Darling 2007b), one of which corresponds with the polarized component identified by Robishaw (personal communication). This is an exciting new result: the variable feature, if it is due to ISS, is smaller than about 1 pc across. What does this imply about the magnetic field strength and density across the maser emission region?

It is likely that coupled with VLBI observations, modelling, and variability studies, detections of Zeeman splitting in OH MMs will provide new insight into OH MM physics and the associated ISM in distant interacting starburst galaxies.

5. Who mases?

5.1. *Searches for differences*

Lo (2005) posed a key question critical to our understanding of OH megamasers: why do 80% of luminous IR galaxies show no OH MM activity? To reframe the question: take two merging systems with similar global IR and radio continuum properties in the same morphological stage of merging; one shows OH MM emission while the other does not. Why? What is the difference between the two systems? Perhaps there is no difference and the fraction of OH MMs among mergers simply reflects beaming. Or perhaps OH MM activity depends on small-scale conditions that are relatively decoupled from global properties of mergers (remember that in most studies, OH MM samples are completely unresolved).

Several groups have investigated this issue with studies of IR, radio, and optical properties of samples of OH MMs and non-masing systems. For example: Baan, Haschick, & Henkel (1992) and Darling & Giovanelli (2002a) have investigated the OH MM fraction in (U)LIRGs versus star formation rate and IR color; Baan, Salzer, & LeWinter (1998) and Darling & Giovanelli (2006) used optical spectral classification to distinguish populations and quantify AGN fraction in OH MM hosts; Darling & Giovanelli (2002a) and to a much greater extent Baan & Klöckner (2006) studied radio and IR properties vis-a-vis the AGN versus starburst contributions to OH MM activity; and Vignali *et al.* (2005) conducted an X-ray study of the contribution of AGN to OH MM hosts. While some of these studies pointed to some small differences in statistical samples of OH MM hosts versus non-masing systems, they could not identify in a case-by-case basis which systems would harbor OH MMs and which would not based on any observable quantity except the OH line itself.

These types of studies are complicated by the fact that in most cases, multiple-nucleus merging systems are not resolved: masing nuclei are contaminated by non-masing nuclei within mergers, and the optically dominant nucleus is often not the masing nucleus (Figure 1). So even if there were global properties of starburst nuclei that are responsible for OH MM production, it is likely that these properties will not be identifiable without sub-arcsecond resolution in the IR and optical regimes.

5.2. *Lessons from HCN*

Let us diverge from OH for a moment to discuss recent studies of HCN in nearby starburst galaxies. HCN is a high density tracer of molecular gas ($n(H_2) \gtrsim 3 \times 10^4$ cm^{-3}), shown by Gao & Solomon (2004) to correlate in a linear fashion with star formation (proxied by IR luminosity), even in ULIRGs where the CO line saturates. Gao & Solomon (2004) show that the population of CO-saturated ULIRGs have an extreme dense gas fraction, as measured by $L_{\rm HCN}/L_{\rm CO}$. Darling (2007a) show that while OH absorbers in the Gao & Solomon sample appear nearly uniformly distributed in $L_{\rm IR}$ and $L_{\rm HCN}$, OH MMs represent the *majority* of the CO-saturated population. It appears that OH MMs form the bulk of the high dense gas fraction starbursts and that we can at last observe a quantity ($L_{\rm HCN}/L_{\rm CO}$) that is highly predictive of OH MM activity. This is consistent with the modelling work by Parra *et al.* (2005) showing that the critical component for OH MM formation is cloud-cloud overlap; a galaxy-scale high dense gas fraction provides the required overlap of many dense clouds. Moreover, OH MMs are thus signposts of the most extreme star formation in the local universe: surveys for OH MMs will now provide more information about the detected host galaxies and their mode of star formation. These issues are discussed in detail in Darling (2007a).

5.3. *Lessons from H_2CO*

One more divergence to examine formaldehyde (H_2CO). The most commonly observed line is the ground state cm transition of ortho-H_2CO at 4.8 GHz. Empirically, there are two flavors of extragalactic formaldehyde seen in starburst galaxies: in "normal" starbursts, this line is seen in absorption ostensibly against the starburst radio continuum, but in OH MM galaxies, this line appears in emission. To date, *all* confirmed extragalactic H_2CO masers are also OH megamasers and *all* H_2CO absorbers are also OH absorbers (Araya *et al.* 2004; Mangum *et al.* 2007; Darling & Mangum 2007). Recent modelling of H_2CO lines and line ratios by Mangum *et al.* (2007) shows that this flip in the ground state line can be explained entirely by collisional excitation. The excitation of H_2CO cm lines thus depends critically on $n(H_2)$, and the flip occurs at roughly $n(H_2) \sim 10^{5.6}$ cm^{-3}. If this correlation between OH and H_2CO holds, then the notion of a critical density for OH MM formation plus the galaxy-wide molecular gas concentration suggested by HCN studies gives us new insight into the OH MM process and allows us to predict the presence or absence of OH MMs based on the fundamental properties of a given starburst. The new insight into OH MMs provided by H_2CO is discussed in detail in Darling & Mangum (2007).

6. An emerging picture?

Pulling all of these threads in the OH MM story together, we can weave a first-order tapestry of the setting and process of OH MM formation. During interaction and merging of two (or more) massive gas-rich galaxies:

(1) Tidal torques move gas into radial orbits, depositing a large fraction (of order half) of the gas into the nuclear regions (inner ~kpc) and throwing much of the remaining gas into tidal tails.

(2) There is (often, but perhaps not always) a stage of extreme dense gas fraction in the nuclear starburst, presumably due to radially infalling gas.

(3) During this high dense gas fraction stage, the fraction of clumps with the requisite density range for OH masing is strongly enhanced, providing enough cloud-cloud overlap for stochastic maser amplification.

(4) Maser emission, while beamed, is stochastic, implying that the OH MM would be viewed as such from most directions.

It is appropriate to think of OH MMs as a transient phenomenon during major mergers, which is why constraining the lifetime of OH MMs is critical to employing them as tracers of extreme stages of star formation. Given the OH MM fraction in starbursts, constraints on the OH MM lifetime will indicate the fraction of mergers that go through the extreme starburst stage. Clearly understanding OH MMs goes beyond the maser itself and has fundamental applications to galaxy evolution and the formation of stars in bursts.

7. New questions

While we may now be able to predict which galaxies show OH MM activity, several questions remain before OH MMs can be employed as effective tracers of galaxy and black hole evolution:

• Can the OH MM line luminosity and profile can be predicted? If we can find a connection to the OH MM line properties, then OH MMs become a very powerful probe of the detailed conditions of star formation at cosmological distances.

• What is the OH/H_2CO megamaser lifetime? Constraints on the OH MM lifetime will provide a connection to the star formation history in giant ellipticals and a connection to gravitational waves produced by the binary massive black hole population.

• What is the physics of OH MM activity? Do OH and H_2CO arise in the same regions? Why is there a density threshold for OH MMs? Models are key, given new observational inputs.

8. The future

It is time to break the OH MM redshift record held by IRAS 14070+0525 at $z = 0.265$ (Baan *et al.* 1992) and to employ OH MMs as tracers of galaxy, black hole, and star formation evolution to their full potential, particularly as we gain a deeper understanding of what OH MMs signify. There are barriers and boons to OH MM searches at high(er) redshifts. Barriers include radio frequency interference (RFI) — which will be disastrous once the conversion to digital television is complete — limited receiver coverage below 1100 MHz particularly on interferometers that can spatially filter RFI, and the simple rarity of OH MMs in the universe. But many of these barriers can be overcome by the boons of increased star formation and merging activity in the past, telescope sensitivity adequate for detections of known OH MMs out to $z \sim 1$, and the fact that roughly half of known OH MMs are quasar-like in the sense that they are "overluminous" compared to a typical flux-limited sample. That is, they could be detected at many times their actual redshift in the short integration times of the Arecibo OH MM survey (Darling & Giovanelli 2002c). If a survey can reach a 5σ sensitivity of 1 mJy in ~ 10 km s^{-1} channels, then OH gigamasers can be detected out to $z \sim 1$ and OH MMs with $L_{\rm OH} > 10^3 L_\odot$ can be detected up to $z = 0.5$ (Darling & Giovanelli 2002c). An obvious sample for targeted observations is the submm galaxies, which are known to have extreme star formation rates and ample molecular gas.

It is also essential to constrain the lifetime of OH MMs in order to understand the role and significance of the OH MM/extreme starburst phase in major galaxy mergers. The OH MM lifetime has implications for nearly all applications of OH MMs as signposts of mergers including the merging history of galaxies, the stellar mass created in mergers, the growth of massive black holes, and the gravitational wave-producing population of massive black hole binaries.

Acknowledgements

The author would like to thank M. Elitzur, P. Lockett, T. Robishaw, and J. Mangum for useful discussions. The author's participation in this meeting was supported by an IAU grant.

References

Araya, E., Baan, W. A., & Hofner, P. 2004, *ApJS* 154, 541
Baan, W. A., Wood, P. A., & Haschick, A. D. 1982, *ApJ* 260, L52
Baan, W. A., Güsten, R., & Haschick, A. D. 1986, *ApJ* 305, 830
Baan, W. A., Haschick, A. D., & Henkel, C. 1992, *AJ* 103, 728
Baan, W. A., Rhoads, J., Fisher, K., Altschuler, D. R., & Haschick, A. D. 1992, *ApJ* 396, L99
Baan, W. A., Salzer, J. J., & LeWinter, R. D. 1998, *ApJ* 509, 633
Baan, W. A. & Klöckner, H.-R. 2006, *A&A* 449, 559
Begelman, M. C., Blandford, R. D., & Rees, M. J. 1980, *Nature* 287 307
Darling, J. & Giovanelli, R. 2002a, *AJ* 124, 100
Darling, J. & Giovanelli, R. 2002b, *ApJ* 569, L87
Darling, J. & Giovanelli, R. 2002c, *ApJ* 572, 810
Darling, J. & Giovanelli, R. 2006, *AJ* 132, 2596
Darling, J. 2007a, in preparation
Darling, J. 2007b, in preparation
Darling, J. & Mangum, J. G. 2007, in preparation
Gao, Y. & Solomon, P. M. 2004, *ApJ* 606, 271
Klöckner, H.-R., Baan, W. A., & Garrett, M. A. 2003, *Nature* 421, 821
Klöckner, H.-R. & Baan, W. A. 2004, *A&A* 419, 887
Lo, K. Y. 2005, *ARAA* 43, 625
Lockett, P. & Elitzur, M. 2007, in preparation
Mangum, J. G., Darling, J., Menten, K. & Henkel, C. 2007, in preparation
Momjian, E., Romney, J. D., Carilli, C. L., & Troland, T. H. 2006, *ApJ* 653, 1172
Parra, R., Conway, J. E., Elitzur, M., & Pihlström. Y. M. 2005, *A&A*, 443, 383

Pihlström, Y. M., Conway, J. E., Booth, R. S., Diamond, P. J., & Polatidis, A. G. 2001, *A&A* 377, 413

Robishaw, T. 2006, *BAAS* 209, #198.06

Rovilos, E., Diamond, P., Lonsdale, C. J., Lonsdale, C. J., & Smith, H. E. 2003, *MNRAS* 342, 373

Vignali, C. Brandt, W. N., Comastri, A., & Darling, J. 2005, *mnras* 364, 99

Yates, J. A., Richards, A. M. S., Wright, M. M., Collett, J. L., Gray, M. D., Field, D., & Cohen, R. J. 2000, *MNRAS* 317, 28

Astrophysical Masers and their Environments
Proceedings IAU Symposium No. 242, 2007
J.M. Chapman & W.A. Baan, eds.

doi:10.1017/S1743921307013543

OH main line masers in the M 82 starburst

M. K. Argo, A. Pedlar, T. W. B. Muxlow and R. J. Beswick

Jodrell Bank Observatory, University of Manchester, Macclesfield, Cheshire SK11 9DL, UK

Abstract. A study of the distribution of OH gas in the central region of the nearby active starburst galaxy M82 has confirmed two previously known bright masers and revealed several new main line masers. Three of these are seen only at 1665 MHz, one is detected only at 1667 MHz, while the rest are detected in both lines. Observations covering both the 1665 and 1667 MHz lines, conducted with both the Very Large Array (VLA) and the Multi-Element Radio Linked Interferometer Network (MERLIN), have been used to accurately measure the positions and velocities of these features. This has allowed a comparison with catalogued continuum features in the starburst such as HII regions and supernova remnants, as well as known water and satellite line OH masers. Most of the main line masers appear to be associated with known HII regions although the two detected only at 1665 MHz are seen along the same line of sight as known supernova remnants.

Keywords. masers – galaxies: individual: M82 – galaxies: ISM – galaxies: starburst

1. Introduction

M82 is one of the closest, and therefore best studied, starburst galaxies and radio observations of this galaxy are numerous. An OH maser was first detected in M82 using the Effelsberg telescope by Nguyen-Q-Rieu *et al.* (1976). Using the VLA, Weliachew *et al.* (1984) detected two main line maser regions in the galaxy and noted that, although both spots are an order of magnitude brighter than the most luminous Galactic maser, the emission from those in M82 may be coming from a number of smaller masers within each spot. The resolution of these observations however was not sufficient to detect any velocity structure or spatial extension in either masing region.

Subsequent observations have resulted in further detections of masers at other frequencies. A total of six OH features were detected using the 1612 and 1720 MHz satellite lines by Seaquist *et al.* (1997). Four features were seen in emission at 1720 MHz, two were seen in absorption at the same frequency, while only one feature was detected in emission at 1612 MHz. Several H_2O masers have also been detected in M82 at 22 GHz (Baudry & Brouillet 1996; Hagiwara 2005).

Recently MERLIN and the VLA have been used to probe the OH absorption across the central starburst region in M82 in order to provide a comparison of the distributions of molecular (OH) and atomic (HI) gas with similar resolution. As well as deep absorption features, eleven main line OH masers have been detected to a limit of 3σ, nine of which are new detections.

Unless otherwise stated, all maser positions are given in J2000 coordinates† in the format aa$''$aaa bb^{s}bb corresponding to $09^{h}55^{m}$aa^{s}aaa and $+69^{\circ}40'$bb$''$bb respectively. At the distance of M82 (3.2 Mpc; Freedman *et al.* 1994), one arcsecond corresponds to a linear distance of 15.5 parsecs.

† Note that positions of the continuum features in M82 are quoted in B1950 coordinates following the naming convention of Kronberg & Wilkinson (1975).

Telescope used	Date of observation	Angular resolution	Bandwidth	Velocity resolution
MERLIN	1995 Nov	$0\farcs2$	8 MHz	23 km s^{-1}
MERLIN	1997 May	$0\farcs2$	2 MHz	1.4 km s^{-1}
VLA	2002 Apr/May	$1\farcs4$	6.3 MHz	18 km s^{-1}

Table 1. Summary of the observations of OH masers in M82 discussed here.

ID (J2000)	1995 S_{1665}	1995 S_{1667}	1997 S_{1665}	1997 S_{1667}	2002 S_{1665}	2002 S_{1667}	Nearest feature (B1950)	Nearest feature (J2000)
53.62+50.1	<2.6	<2.6	<17	<18	2.44	2.49	44.93+63.9 (HII)	53.65+50.0
53.11+48.0	5.57	<2.6	<17	<18	1.86	<0.77	44.43+61.8 (SNR)	53.14+47.8
52.71+45.8	<2.6	<2.6	<17	<18	5.02	<0.77	44.01+59.6 (SNR)	52.73+45.7
51.87+48.3	<2.6	<2.6	<17	<18	1.71	<0.77	-	-
51.23+44.5	<2.6	<2.6	<17	<18	4.85	5.93	42.48+58.4 (HII)	51.22+44.5
50.97+45.3	8.35	7.83	<17	23.0	13.0	18.5	42.21+59.2 (HII)	50.95+45.2
50.51+45.6	<2.6	<2.6	<17	<18	1.85	1.28	-	-
50.36+44.2	7.01	29.9	<17	57.2	10.7	50.5	41.64+57.9 (HII)	50.39+44.1
50.02+45.8	4.30	5.09	<17	<18	<0.77	<0.77	41.29+59.7 (SNR)	50.06+45.9
49.71+44.4	<2.6	5.29	<17	<18	<0.77	2.34	40.96+57.9? (HII)	49.71+44.1
48.45+42.1	4.15	6.94	<17	<18	2.64	5.97	39.68+55.6 (HII)	48.44+41.9

Table 2. All masers detected in the data-sets discussed here. Maser IDs are constructed from the VLA 2002 position with the exception of 50.02+45.8 which was only detected in the MERLIN 1995 observations. Fluxes are given for each maser in each line in which is was detected in each epoch (lower limits are the 3σ noise levels in each dataset), along with the closest continuum feature from McDonald *et al.* (2002) given in both B1950 and J2000 coordinates.

2. Observations

Data from the MERLIN archives, as well as more recent VLA observations, have been used to investigate the maser population in M82. A summary of the observations discussed here is given in Table 1. Some of these observations were carried out with the primary aim of investigating the molecular gas distribution across the M82 starburst via OH absorption in order to provide a comparison with that seen in HI for example. The low velocity resolution which resulted from this setup is clearly not ideal for maser studies, however useful results were obtained from these data.

3. The masers

A total of eleven main line OH masers were detected in the observations outlined above, nine of which are new detections. Table 2 lists the masers and their properties while Figure 1 shows the velocities of each maser along with the OH absorption on a position-velocity diagram constructed using the VLA 2002 data.

In the MERLIN 1995 observations, four of the six masers are detected in both main lines, while 53.11+48.0 is detected only at 1665 MHz and 49.71+44.4 is present only at 1667 MHz. In 1997, the two masers detected were seen only at 1667 MHz. In 2002 six of ten are seen in both lines, 53.11+48.0, 52.71+45.8 and 51.87+48.3 are seen only at 1665 MHz, while 49.71+44.4 is again seen only at 1667 MHz.

Of those which appear to be associated with HII regions, most are brighter at 1667 MHz than at 1665 MHz. The only exception to this is 50.97+45.3 which is reversed only in the 1995 observations. The difference in magnitude between the two lines is, however, less

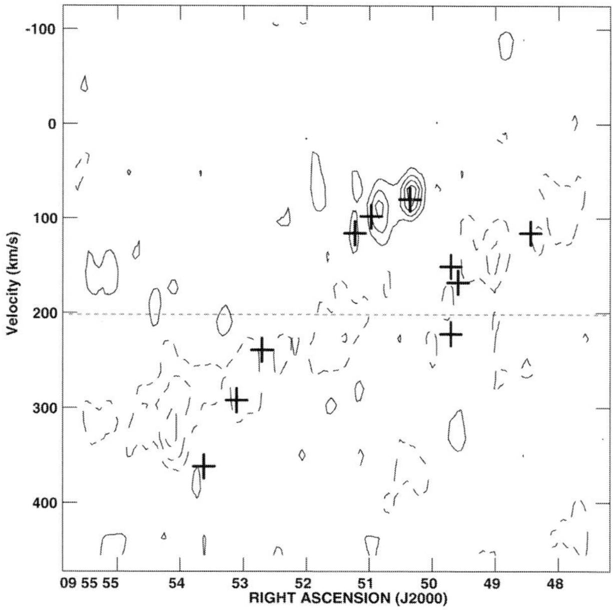

Figure 1. A comparison between the maser velocities and the absorption. Crosses represent measurements of maser velocities from the 2002 data. The contours represent the OH absorption (and emission) from the VLA 2002 dataset and the dashed line represents the systemic velocity of M82.

than 1σ. Two of the masers, 53.11+48.0 and 52.71+45.8, appear to be associated with supernova remnants. These two masers are both detected only at 1665 MHz, if they are detected at all, with no emission apparent at 1667 MHz.

Any variability between epochs could be due either to variability of a few strong sources in a particular spot, the creation/destruction of many weaker maser sources within a spot, or a combination of both. Whether the masers are varying between epochs is not trivial to determine since each observation was carried out using a different setup.

4. Two interesting sources

49.71+44.4: Despite the velocity resolution being more suited to absorption studies in the VLA 2002 and MERLIN 1995 observations, some evidence of velocity structure was seen in this maser source. Two components were observed in the 1667 MHz line in both observations. The spectra from these two observations is shown in Figure 2. Preliminary results from a further VLA dataset obtained in 2006 at higher velocity resolution also show two well-separated peaks in the 1667 MHz line with one of these further split into several components.

50.97+45.3: The only maser to show evidence of spatial structure in any of the observations examined to date. The VLA 2002 data shows an extension to the south west at a position angle almost aligned with the direction of the peculiar continuum source 41.95+57.5. The higher spatial resolution 1995 MERLIN observations show an unresolved component at the position of the peak seen in the 2002 data, as well as a 3σ component located to the SW along the same position angle as the extension seen with the VLA in 2002, see Figure 3.

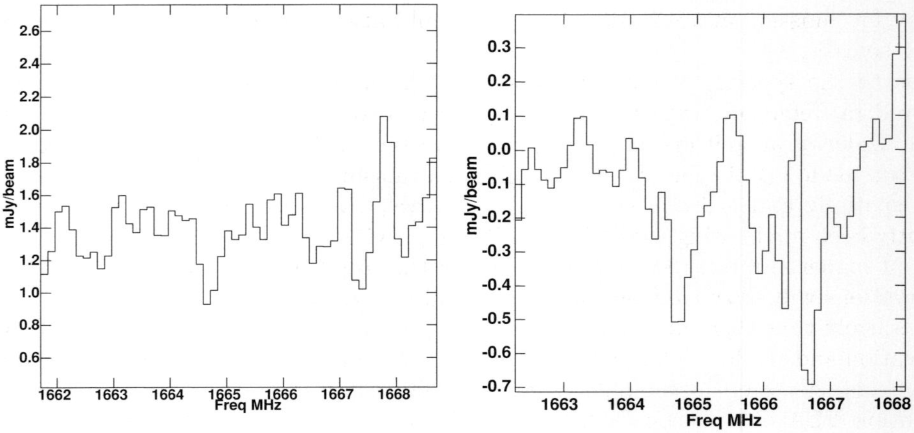

Figure 2. Spectra of 49.71+44.4 from the 1995 MERLIN observations (left) and 2002 VLA observations (right). Two components are detected at 1667 MHz separated by of order 70 km s^{-1}. A component is also visible in the absorption at 1665 MHz in the VLA 2002 data.

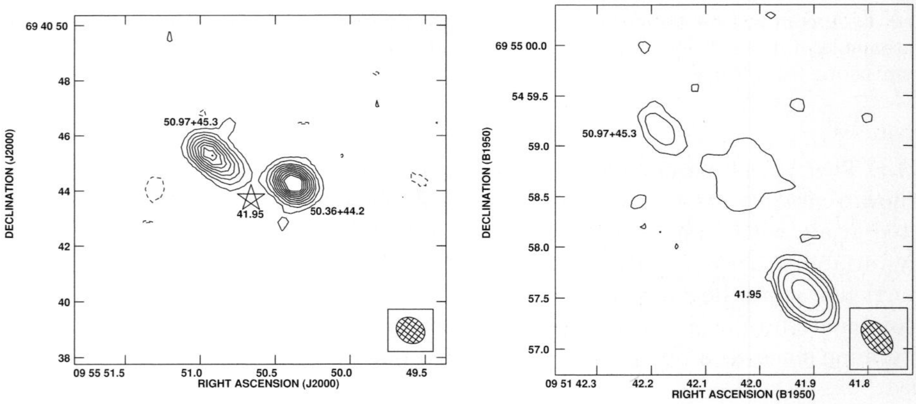

Figure 3. Images from the 2002 VLA observations (left) and 1995 MERLIN observations (right). Both maps show the maser at 50.97+45.3 relative to the continuum source 41.95 and show the extension of the maser, see main text. Note that the VLA data have been continuum subtracted while the MERLIN data have not.

5. Some conclusions

Nine new OH main line masers have been detected in the central starburst region of the nearby galaxy M82. On a position-velocity plot of the galaxy, most of the masers lie along the same axis as the gas. Four of the masers are blue-shifted from the main distribution and could possibly be on the edge of an expanding shell or orbiting within a bar.

The apparent association of most maser features with HII regions is not surprising given the results of surveys carried out in our own Galaxy, although the comparative strength of the 1667 MHz line is more characteristic of late-type stars than HII regions. The association of two masers with known supernova remnants, while unusual, is not unique. Higher resolution imaging of these sources will help determine whether these associations are real or just chance close alignments along a line of sight. A more sensitive survey for emission at 1612 and 1720 MHz would also help determine the nature of the maser sources.

Due to the spectral resolution of the 2002 VLA observations the masers were not resolved in frequency, with most only visible in two channels. As the individual spots are an order of magnitude more luminous than typical Galactic masers, it is likely that they are made up of more than one individual masing region, so would have structure both spatially and in frequency. In order to investigate this further, observations have recently been made using the EVN in order to provide higher spatial resolution, and the VLA at higher spectral resolution. It is hoped that further investigation using these two observations will show evidence of both spatial and velocity structure.

It is likely that there are more main line OH masers in M82 but, due to the depth of absorption and the low velocity resolution, faint or narrow masers could be buried to the extent that they are undetectable in the VLA 2002 observations. The recent high spectral resolution VLA observations will have a gain in sensitivity over the 2002 observations which should enable the detection of weaker masers.

Acknowledgements

MERLIN is run by the University of Manchester as a National Facility on behalf of PPARC. The VLA is operated by the National Radio Astronomy Observatory, a facility of the National Science Foundation operated under cooperative agreement by Associated Universities, Inc.

References

Baudry, A. & Brouillet, N. 1996, *A&A* 316, 188
Caswell, J. L., Haynes, R. F. & Goss, W. M. 1980 *Australian Journal of Physics* 33, 639
Freedman, W. L., Hughes, S. M., Madore, B. F., Mould, J. R., Lee, M. G., Stetson, P., Kennicutt, R. C., Turner, A., Ferrarese, L., Ford, H., Graham, J. A., Hill, R., Hoessel, J. G., Huchra, J. & Illingworth, G. D. 1994, *ApJ* 427, 628
Hagiwara, Y. 2005, *Ap&SS* 295, 125
Kronberg, P. P. & Wilkinson, P. N. 1975, *ApJ* 200, 430
McDonald, A. R., Muxlow, T. W. B., Wills, K. A., Pedlar, A. & Beswick, R. J. 2002, *MNRAS* 334, 912
Nguyen-Q-Rieu, Mebold, U., Winnberg, A., Guibert, J. & Booth, R. 1976, *A&A* 52, 467
Seaquist, E. R., Frayer, D. T. & Frail, D. A. 1997, *ApJ* 487, L131
Shen, J. & Lo, K. Y. 1995, *ApJL* 445, L99
Weliachew, L., Fomalont, E. B. & Greisen, E. W. 1984, *A&A* 137, 355
Wills, K. A., Das, M., Pedlar, A., Muxlow, T. W. B. & Robinson, T. G. 2000, *MNRAS* 316, 33
Wills, K. A., Pedlar, A. & Muxlow, T. W. B. 2002, *MNRAS* 331, 313

Astrophysical Masers and their Environments
Proceedings IAU Symposium No. 242, 2007
J.M. Chapman & W.A. Baan, eds.

doi:10.1017/S1743921307013555

OH emission and absorption associated with supernovae in Arp 220

Colin J. Lonsdale[1], Katherine R. de Kleer[1], Philip J. Diamond[2], Hannah Thrall[2], Carol J. Lonsdale[3] and Harding E. Smith[4]

[1]MIT Haystack Observatory, Westford, MA 01886
[2]Jodrell Bank Observatory, Macclesfield, SK11 9DL, UK
[3]Infrared Processing and Analysis Center, MS 100-22, Pasadena, CA 91125, USA
[4]CASS, U.C. San Diego, 9500 Gilman Dr., La Jolla CA 92093-0424, USA

Abstract. We present parsec-resolution spectral-line VLBI data for two epochs separated by 15 months as a precise new probe of the innermost regions of the nearby Ultraluminous Infrared Galaxy (ULIRG) Arp 220. This galaxy hosts a powerful starburst, with an associated supernova (SN) rate of order 4/yr. An extensive population of compact continuum sources interpreted as radio supernovae (RSNe) and young supernova remnants (SNR) has been imaged. We show here that many of the supernova-related radio continuum point sources exhibit clear evidence of OH absorption or maser emission in the intervening gas, and as such provide us with a sampling of conditions along very narrow and specific lines of sight through the nuclear environment. The OH gas along these lines of sight exhibits velocity dispersions of up to several tens of km/sec, and that in some cases, multiple distinct concentrations of masing gas at different radial velocities can be discerned. There is evidence for variability in the OH properties on ~1yr timescales. Our results are discussed in the context of the overall OH megamaser properties of Arp 220.

Keywords. galaxies: active — infrared: galaxies — radio lines: galaxies — masers

1. Introduction

The galaxy Arp 220 (IC 4553) is the closest Ultraluminous Infrared galaxy (ULIRG), and has consequently been the subject of intense multi-wavelength study. A persistent challenge for all these studies, however, is the fact that the source of the extreme luminosity of the system lies deep in the innermost regions of the twin galaxy nuclei, where optical depths are extreme. To overcome this problem, we have therefore been exploiting the technique of Very Long Baseline Interferometry (VLBI) to probe the nuclei of Arp 220 with parsec-scale linear resolution. Such studies have revealed clusters of weak continuum point sources identified as radio supernovae (RSNe) and young supernova remnants (SNR), tracing a powerful ongoing nuclear starburst (Smith *et al.* 1998, Rovilos *et al.* 2005, Parra *et al.*, 2007). Lonsdale *et al.* (2006) catalogued 49 continuum sources, including 4 new sources in a 12-month interval.

Arp 220 is also a powerful OH megamaser galaxy, and indeed the megamaser phenomenon was originally discovered by Baan, Wood and Haschick (1982) in this galaxy. A series of VLBI studies capable of spatially resolving the emission has shown the presence of considerable variety in the properties of different maser components in the source (Diamond *et al* 1989, Lonsdale *et al* 1998, Rovilos *et al* 2003). In particular there are now known to be two observationally distinct components of the maser emission, one that is diffuse on ~100pc scales and exhibits relatively modest 1667MHz/1665MHz line ratios, and a second that appears on much smaller size scales (<1 to 10 pc) and features extremely high 1667MHz/1665MHz line ratios.

These observations prompted speculation that diffuse and compact maser components arise in gas clouds with physically different properties. The diffuse component was thought to arise in a screen of OH clouds, each supporting low-gain unsaturated masing, while the compact component, which exhibits relatively high gain, was hypothesized to originate in physically small clouds, presenting a challenge to radiative pumping models driven by the far infrared radiation field of the galaxy nucleus. More recently, and based substantially on VLBI observations of another OH megamaser galaxy IIIZw35 (Diamond *et al.* 1999, Pihlström *et al.* 2001), an alternative explanation for compact, bright OH megamaser components has been explored (Parra *et al.* 2005). In this interpretation, only one class of OH cloud exists, and compact features arise due to chance superposition of multiple such clouds along our line of sight.

Consideration of this model demonstrates that in a clumpy medium the effects of cloud superposition along our line of sight can be significant, and must be accounted for in any complete megamaser model. The cloud overlap model of Parra *et al.* (2005) succeeds in IIIZw35, primarily because the compact features in this source are all presumed to be of very small angular extent. The situation with the compact masers in Arp 220 is, however, quite different in that the structures span a range of scale sizes, and there are clearly coherent structures of order 10pc in extent (Lonsdale *et al.* 1998), suggesting that there are coherent physical associations of OH gas on these scale sizes.

Clearly, new observational constraints on the nature of the OH masing medium are needed. We describe observations of maser amplification and OH absorption of unresolved, sub-parsec-scale RSN continuum sources. These data constitute a probe of the OH gas along a very narrow, highly specific line of sight between us and the RSN, and as such allow us to dissect the OH-bearing medium with far greater precision than has hitherto been possible.

2. Observations and results

We observed Arp 220 on 9 November 2003 and 6 March, 2005 with an array comprising the 10 VLBA antennas, the phased VLA, the GBT, Arecibo, and 5 antennas of the European VLBI Network (EVN). Due to interference problems, data from the phased VLA had to be discarded.

Each observation covered a period of almost 21 hours. The dual-polarization spectral-line data spanned 8 MHz centered at 1637 MHz, thereby covering the redshifted frequencies of both the 1665 MHz and 1667 MHz main OH lines, with sensitivity to departures from the systemic velocity of at least ± 400 km s^{-1} . The data were processed in the VLBA correlator in Socorro with 512 channels across the 8 MHz band, yielding 15.625 kHz spectral resolution ($\sim$2.8 km s^{-1} velocity resolution).

A continuum image and a spectral line image cube, with precise spatial alignment via phase referencing on a strong maser feature, were generated using the AIPS software package. Spectra were extracted from the cube at the locations of all 49 continuum point sources, and candidate spectral features were noted both in absorption and emission.

In order to develop confidence in candidate spectral features, we adopted a strategy of examining the degree to which the purported spectral feature was localized to, and thus presumed to be associated with, the continuum source. We examined spectra from locations immediately surrounding the continuum source, in order to verify that all such locations were consistent with the absence of spectral features of any kind, whether or not they match the properties of the candidate feature. We also constructed an image of the region surrounding the source, integrated over the spectral range of the candidate feature, in order to verify that a well-defined, unresolved peak (or hole in the case of absorption features) was present at the target location. These tests guard against specific image pathologies in both position and frequency.

For candidate spectral features that displayed signal-free noise-like comparison spectra at surrounding locations, and which showed a well-defined unresolved peak in the integrated image, we deemed the detection secure.

Both emission and absorption features are easier to detect in association with the stronger continuum sources. The ∼1mJy source W30 (Lonsdale *et al.* 2006) shows both types of spectral feature with high signal to noise ratio, and is shown in Figure 1 as an illustration of the phenomena under study. This particular source displays multiple velocity systems with strong maser emission in both 1665 MHz and 1667 MHz main lines of OH, broad emission line velocity widths, and a broad absorption trough.

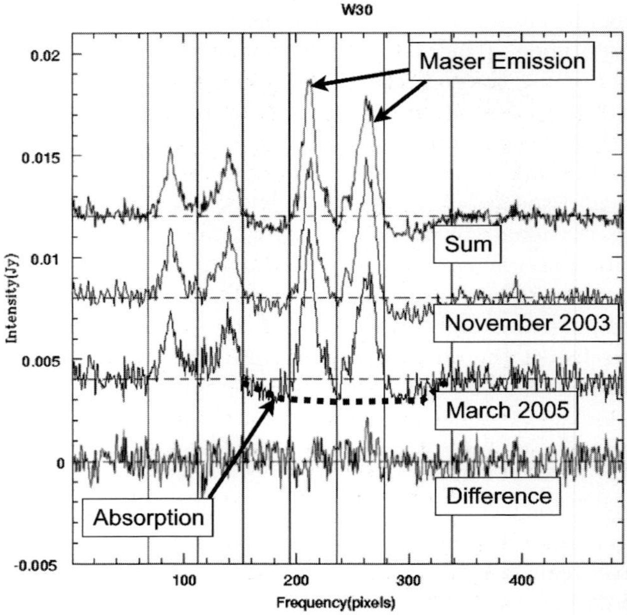

Figure 1. Spectra of the supernova-related continuum source W30 in Arp 220. The two center spectra are the two observing epochs, while the top and bottom spectra are the sum and difference respectively of the two epochs. The spectra have been offset vertically for clarity, and the horizontal dashed lines show the zero flux level for each spectrum. Each "pixel" on the frequency axis corresponds to 15.625 kHz, which is about 2.8 km/sec. The heavy dashed line shows an apparent, very broad absorption trough, several hundred km/sec wide. The vertical lines delineate subjectively separate features for feature verification and parameter extraction purposes. Two distinct velocity systems are present, in each of the main OH lines.

A total of 16 out of 49 supernova-related continuum point sources in Arp 220 have been found to exhibit detectable OH spectral features, and a total of ∼30 discrete spectral features have been identified, roughly 20 in emission and 10 in absorption. A handful of spectral features also exhibited strong variability between the two observing epochs. Typical inferred optical depths, both in emission and absorption, for the detected features are of order 1 or 2. Most of the 49 continuum sources are too weak to allow detection of absorption, and would require maser amplification factors of up to 100 to be detectable in OH emission. Therefore, the 30% detection rate should be regarded as a lower limit to the fraction of sources with OH optical depths in either emission or absorption found to be typical of the detected sources.

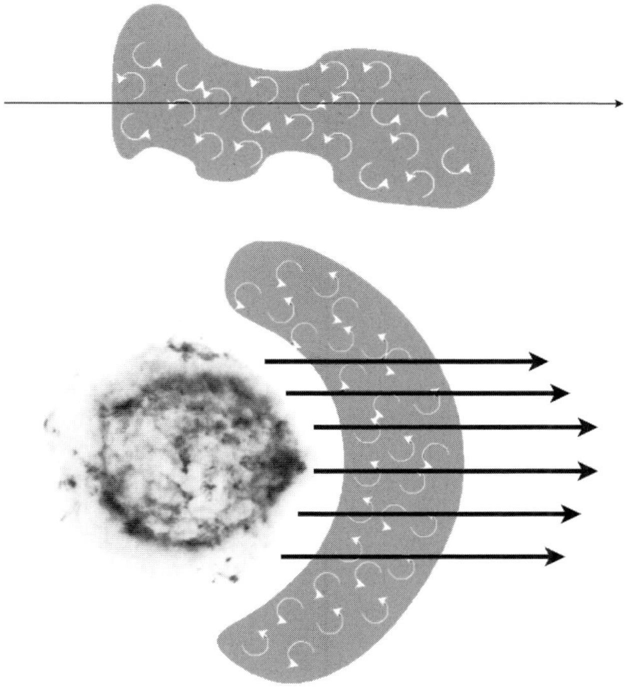

Figure 2. Illustration of two possibilities for generation of spectrally broad OH features associated with a physically compact continuum source. The upper cartoon shows a single, very narrow line of sight through a turbulent OH-bearing medium. The observed velocity width arises from the velocity dispersion along the line of sight. This picture implies that the particular lines of sight picked out by the supernova locations are not special, and that OH optical depths observed in association with the continuum sources are typical of all sight lines. The lower cartoon shows how a turbulent circumstellar OH-bearing medium might be sampled in many different locations by continuum emission from a synchrotron source (the RSN or SNR) extended on comparable scales. In this case the observed velocity width could arise from systematic velocity gradients within the OH-bearing shell, perhaps induced by the massive star progenitor or by the explosion itself.

3. Discussion and conclusions

The compact continuum sources in Arp 220 arise from starburst-related supernova activity, and are likely a mixture of radio supernovae (RSN) and young supernova remnants (SNR), the distinction being whether the ejected material from the explosion is still confined to the pre-existing progenitor stellar wind, or whether it has reached the undisturbed interstellar medium (Lonsdale *et al.* 2006, Parra *et al.* 2007). As such they are expected to be very compact, with linear extents ranging from a few light weeks for the youngest sources to perhaps as much as a parsec for the oldest remnants. Absorption or maser emission seen against these sources therefore represents sampling of a correspondingly narrow cylinder of material in the interstellar medium of Arp 220.

Given the extremely narrow cylinder being sampled, the observed velocity width of the spectral features is surprising, and implies either moderate velocity gradients along the line of sight, or extreme velocity gradients in the transverse direction. These possibilities are illustrated schematically in Figure 2.

In the scenario where the velocity gradients are longitudinal along the line of sight, the OH gas is necessarily distant from the associated supernova. This implies that the continuum supernova sources are randomly sampling the ISM of Arp 220, and that the amplification and absorption optical depths are typical for all sightlines in the region sampled by the supernovae. Consequently, all continuum emission in the region, whether supernova-related or not, should experience similar amplification and absorption, leading to predictable global OH line properties for Arp 220, given sufficient knowledge of the continuum emission distribution. The supernovae account for only a few percent of the total continuum emission in Arp 220. To the extent that discrete spectral features correspond to discrete OH-bearing clouds in the ISM, these observations can also yield constraints on the cloud parameters, in particular the internal velocity dispersion of the clouds. Further analysis may lead to useful constraints on the cloud overlap model described by Parra *et al.* (2005).

An interesting alternative explanation for the observed features is illustrated in the lower cartoon of Figure 2. It is plausible that the immediate vicinity of each supernova is an atypical environment in which OH emission and absorption is enhanced, most likely due to the disturbed circumstellar environment of the progenitor star. Conceivably, a shell of turbulent, OH-bearing gas exists which will be illuminated by the post-explosion synchrotron source. In this picture, strong velocity gradients may occur transversely across the continuum source, despite the small physical extent.

Distinguishing between these possibilities requires further work. It is worth noting, however, that line variability on $\sim$1 year timescales would be a natural consequence of the local shell scenario, because the line properties result from different lines of sight, and those lines of sight evolve on such timescales for sources of the observed age.

Further monitoring of the continuum sources and associated OH features in Arp 220 is ongoing.

Acknowledgements

CJL acknowledges support from NSF grant AST-0352953 to the MIT Haystack Observatory. KdK acknowledges support through the REU program at Haystack Observatory, NSF grant AST-0138508.

References

Baan, W.A., Wood, P.A.D., & Haschick, A. D. 1982, *ApJ* 260, 49

Diamond, P.J., Norris, R.P., Baan, W.A., Booth, R.S. 1989, *ApJ* (Letters) 340, L49

Diamond, P.J., Lonsdale, Colin J., Lonsdale, Carol J., Smith, H.E. 1999, *ApJ* 511, 178

Lonsdale, C.J., Diamond, P.J., Smith, H.E. & Lonsdale, C.J. 1998, *ApJ* (Letters) 493, L13

Lonsdale, C.J., Diamond, P.J., Thrall, H., Lonsdale, C.J. & Smith, H.E. 2006, *ApJ*, 647, 185

Parra, R., Conway, J.E., Elitzur, M. & Pihlstrom, Y.M. 2005, *A&A* 443, 383

Parra, R., Conway, J.E., Diamond, P.J., Thrall, H., Lonsdale, Colin J., Lonsdale, Carol J. & Smith, H.E. 2007, *ApJ* 659, 314

Pihlström, Y.M., Conway, J.E., Booth, R.S., Diamond, P.J., & Polatidis, A.G. 2001, *A&A* 377, 413

Rovilos, E., Diamond, P.J., Lonsdale, C.J., Lonsdale, C.J. & Smith, H.E. 2003, *MNRAS* 342, 373

Rovilos, E., Diamond, P.J., Lonsdale, C.J., Smith, H.E. & Lonsdale, C.J. 2005, *MNRAS* 359, 827

Smith, H.E., Lonsdale, C.J., Lonsdale, C.J., & Diamond, P.J. 1998, *ApJ* (Letters) 493, L17

Astrophysical Masers and their Environments
Proceedings IAU Symposium No. 242, 2007
J.M. Chapman & W.A. Baan, eds.

doi:10.1017/S1743921307013567

Arp 220 – IC 4553/4: understanding the system and diagnosing the ISM

Willem A. Baan

ASTRON, Oude Hoogeveensedijk 4, 7991PD Dwingeloo, The Netherlands
email: baan@astron.nl

Abstract. Arp 220 is a nearby system in final stages of galaxy merger with powerful ongoing star-formation at and surrounding the two nuclei. Arp 220 was detected in HI absorption and OH Megamaser emission and later recognized as the nearest ultra-luminous infrared galaxy also showing powerful molecular and X-ray emissions. In this paper we review the available radio and mm-wave observational data of Arp 220 in order to obtain an integrated picture of the dense interstellar medium that forms the location of the powerful star-formation at the two nuclei.

Keywords. galaxies: ISM – galaxies: nuclei – ISM: molecules – ISM: masers – galaxies: starburst, – galaxies: OH Megamaser

1. OH MM and ULIRGs

The galaxy IC 4335, later referred to as Arp 220, was recognized early as a merging and strongly interacting system with a blue nucleus by Arp (1966) and Nilson (1973). The detection of strong HI absorption against the radio continuum of the nuclear starburst (Mirabel 1982) led to a search for corresponding OH absorption, which resulted in the detection of powerful OH maser emission later classified as OH MegaMaser emission (OH MM) (Baan *et al.* 1982). The extraordinary properties of Arp 220 were confirmed by the IRAS far-infrared (FIR) data (Sanders *et al.* 1988), which made Arp 220 the prototype Ultra-Luminous FIR Galaxy (ULIRG). The enhanced emission line strength of the molecular constituents were recognized and the CO lines were used to map the nuclear regions of ULIRGs (Solomon *et al.* 1997; Downes & Solomon 1998; Sakamoto *et al.* 1999). More recently high-density molecular tracers have been studied in order to diagnose the molecular medium (Aalto *et al.* 2007; Baan *et al.* 2007; Graciá-Carpio *et al.* 2006; Greve *et al.* 2007).

Arp 220 experiences a powerful burst of starformation at each of the nuclei triggered by the ongoing merger, which results in the FIR prominence and multiple radio SNRs at each of the nuclei (Lonsdale *et al.* 2006) and SB-related hard X-ray emission (Iwasawa *et al.* 2005). Arp 220 is a recent example of a short-lived burst of assembly and nuclear activity that are common at redshifts of 2.5 and that defines the characteristics of massive galaxies in the nearby universe. Arp 220 is a template for high redshift active galaxies.

2. Dynamics of Arp 220 and where is what?

The two nuclei of Arp 220 are separated by only 0.97″ (365 pc) as determined from their radio positions. They appear not yet severely deformed at this advanced stage of interaction and have a relatively small velocity difference. Therefore, the two mass centers are still sufficiently far away from each other and need to be one behind the other. Only the high-density tracer emission lines accurately trace the nuclear ISM of the two nuclei

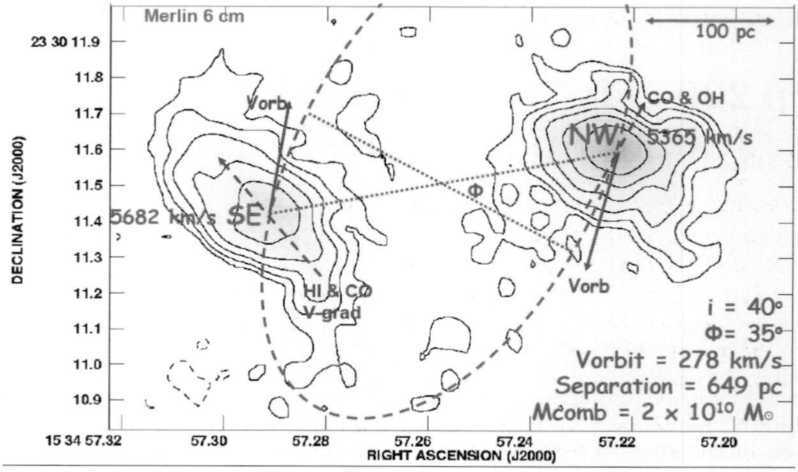

Figure 1. Orbital dynamics of Arp 220. The north-western nucleus is located in front of the south-eastern nucleus. The systemic velocities of the two equal-mass nuclei are indicated and their direction of motion in an orbital plane inclined by 40°. The orbital velocity (V_{orb}) is $278\,\mathrm{km\,s^{-1}}$. The combined mass of the nuclei is based on an observed stellar and gas mass. The directions of the observed velocity gradients of the CO and OH molecular gas in the nuclei are indicated. The MERLIN 6cm continuum map is from Baan, Cohen & Kl''ockner (2007).

having systemic velocities of $5365\,\mathrm{km\,s^{-1}}$ for the western nucleus and $5682\ \mathrm{km\,s^{-1}}$ for the eastern nucleus (see section 6).

The systemic velocities of the nuclear regions obtained from the high-density tracer emission lines may be used for dynamic modelling of the system. Similar models were presented earlier using H_2CO (1-0) and CO (2-1) molecular data (Baan & Haschick 1995; Downes & Solomon 1998; Scoville *et al.* 1997). Reconsideration of the available continuum, CO, HI absorption, and OH and H_2CO emission data gives slightly different picture of the dynamics of the system (Baan, Cohen & Kl''ockner 2007). The eastern nucleus lies behind the western nucleus, which would result in the observed larger column densities in front of the eastern nucleus (Iwasawa *et al.* 2005) and a relatively lower velocity for the western nucleus. Using the CO (2-1) signature, the orbital plane is inclined by 40° (Scoville *et al.* 1997). The combined mass of lower-density molecular gas in CO, the high-density gas in tracers such as HCN, and the stellar component is of the order of 0.6 – 2.0×10^{10} $M_\odot$ (Sakamoto *et al.* 1999; Greve *et al.* 2007). The dynamics of the system may follow from two equal-mass nuclei with a spatial separation of 650 pc (Fig. 1).

Near-infrared imaging of Arp 220 using the NICMOS instrument (1.6 μm) onboard Hubble ST confirm the presence of high column densities in the system (Scoville *et al.* 2000). The visual attenuation is 17 and 24 magnitudes for the East and West nuclei assuming a young stellar population and 13 and 20 magnitudes for an old population (note: one magnitude corresponds to 1.8 x $10^{21}\,\mathrm{cm^{-2}}$ H). The NIR emission actually peaks north of each of the nuclei; there are areas where virtually no radiation can escape, even at 2.2 μm. Reversing the symmetry of the model could place the eastern nucleus in front of the western nucleus, but would not explain the presence of OH emitting gas at the velocity of the western nucleus in front of the eastern nucleus (see section 4).

The hard X-ray (3 – 7 KeV) signature of Arp 220 as observed by Chandra shows two separate (extended) nuclei and there is an extended nebula at 2 keV (Iwasawa *et al.* 2005). The hard X-ray continuum is due to the integrated X-ray emission from binary systems and is a factor 10 too low for the observed SFR. There are plasma features associated with the ionized gas and a Fe Kα feature at 6.5 keV. The obscuration of the X-rays required $N_H = 10^{22-23}$ cm^{-2} (Note: $N_H = 10^{22}$ cm^{-2} is enough for keV absorption).

3. Modelling the starburst

The power generation for the FIR luminosity of $10^{12.2}$ L$_\odot$ is dominated by starburst activity for more than 60% in the radio and more than 90% in the near-infrared (Baan & Kl′′ockner 2006; Genzel *et al.* 1998). The Arp 220 starburst is a scaled-up version of the M 82 activity where also Type II supernova and remnants of massive progenitor stars interact with the ISM. Modelling of the embedded starbursts suggests a star-formation rate SFR =26 $L_{IR,11}$ M$_\odot$ yr^{-1} = 340 M$_\odot$ yr^{-1} and a supernova rate SNR =0.2 $L_{IR,11}$ yr^{-1} = 2.8 yr^{-1} for Arp 220 (Elson, Fall & Freeman 1989). Lonsdale *et al.* (2006) have observed a SNR of 4±2 yr^{-1} and find the rate in the western nucleus to be three times higher than in the eastern nucleus. The mass injection rate for the nuclei of Arp 220 is estimated at 53 M$_\odot$ yr^{-1}. Recent evaluation of the radio data suggests a top-heavy initial mass function (IMF) or a short starburst with a duration of 3×10^5 yr with a SFR $\approx$ 10^3 M$_\odot$ yr^{-1} (Parra *et al.* 2007; Lonsdale *et al.* 2006).

The starburst in Arp 220 must be close to its peak in activity. Much of the molecular ISM appears still intact but in an excited state as evidenced from the warm high-density molecular material (see section 6 below). The line ratios of prominent high-density tracers suggest that the molecular ISM is (or is at the verge of) being depleted and destroyed by the ongoing star-formation process (see section 6 and Loenen, Baan & Spaans 2007). There is evidence that destructive mechanical feedback due to supernovae and high-velocity outflows seen in OH emission are affecting the molecular ISM (section 7). Mechanical feedback due to a violent starburst and radiative feedback on the molecular chemistry could cut off the ongoing star formation, leaving a single (first) generation of stars with a top-heavy IMF.

4. Imaging the OH emission lines

Knowledge of the systemic velocities of the high-density gas at the two nuclei (see section 6) also helps to understand the complex OH groundstate line emission signature. Two pairs of (1665 and 1667 MHz) OH emission lines are offset by 317 km s^{-1} and have a 351 km s^{-1} separation within the pair. Therefore, triple emission lines are seen: the 1667 MHz West line, a convolved 1667 MHz East and 1665 MHz West line, and a 1665 MHz East line. However, also emission in the 1667 MHz line at the systemic velocity of West is seen at the East location (see VLA-A and VLBI data by Baan & Haschick 1987; Rovilos *et al.* 2003). The gas responsible for this 1667 MHz emission at the eastern nucleus cannot be foreground gas trailing from the western nucleus. This is also not solved by placing the eastern nucleus in the foreground in the dynamical picture of Fig. 1. The nature of all emissions and further verification of the dynamical scenario for the two nuclei may be deduced from a study of the lower-gain diffuse emission lines.

In addition to the three emission components due to two line pairs, there is a blue-shifted component that is evidence for outflow in the nuclear region (Baan, Haschick & Henkel 1989). The blue-shifted wing of the 1667 MHz line in Arp 220 extends by 1145 km s^{-1}, while the 1720 MHz line extends by 950 km s^{-1}, although this second outflow wing may be less-pronounced (Ghosh & Salter 2007, personal communication).

The OH emission in Arp 220 at both nuclei encompasses the high-resolution radio continuum structure and the high-density ISM where the star-formation is taking place.

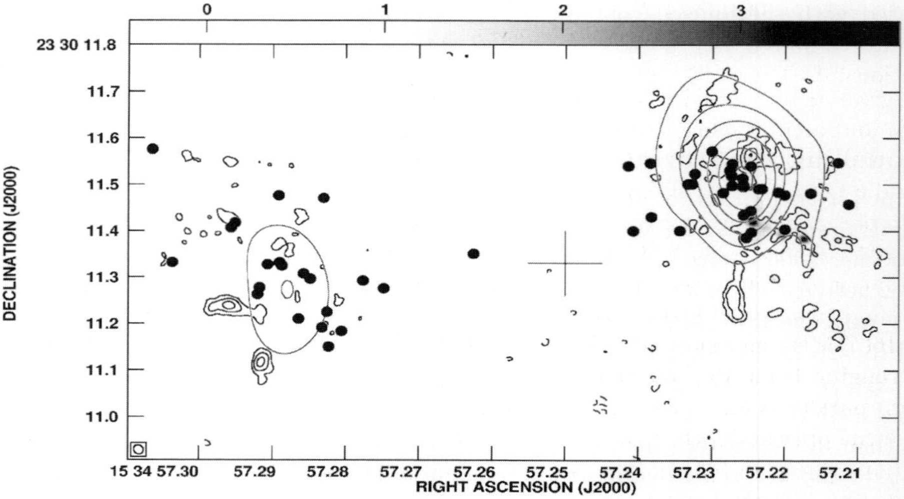

Figure 2. The OH emission at the two nuclei of Arp 220. The high-resolution EVN map of the 1667 MHz emission from Rovilos *et al.* (2003) has been overlayed with the map of supernovae in the two nuclei from Parra *et al.* (2007). Continuum contours indicate the location of the radio nuclei. The superposition is a best effort. ($0.1'' = 35$ pc).

At MERLIN resolution ($0.17'' = 61$ pc) the 1667 MHz OH emission at the eastern nucleus is centrally peaked and has an extent of 170 pc along the velocity gradient (Rovilos *et al.* 2003), which agrees with the extent of the high-brightness components found at EVN resolution ($0.016''$) (see Fig. 2). Similarly the emission at the Western nucleus shows a extended ($1.2'' = 420$ pc) double (torus-like) structure straddling the continuum nucleus (Fig. 2; Rovilos *et al.* 2003) with clumpy high-brightness components only in the central $0.6''$ region. At both nuclei the elongated/edge-on distribution of high-brightness OH components extends beyond the starburst-related cluster of supernova remnants in the continuum (Rovilos *et al.* 2005; Lonsdale *et al.* 2006; Parra *et al.* 2007). The OH velocity gradients at both nuclei are similar to those in the CO (2-1) data (see Fig. 1), which would confirm the presence of orbiting counter-rotating disks (Sakamoto *et al.* 1999).

The OH characteristics suggest that the emissions at both nuclei originate both in the nuclear ISM and in an edge-on torus/disk structure that surrounds the nuclear ISM with the ongoing star formation. At the Eastern component the amplifying medium of the major emission component is the (foreground) frontal section of the torus at the systemic velocity. The double structure of the Western component defines the two edge-sections of a similar thick torus with radius of about 60 pc. The high-brightness components emission components primarily trace structures in the nuclear region and result from spatial superpositions of clouds and (edge-on) shocked regions associated with the star formation. OH emission is also seen in front of individual SNR remnants (see Lonsdale *et al.*, these proceedings). The edge-on torus may serve for further amplification. Successful modelling of observed OH emission structures has been done for Mrk 273 and IIIZw 35 (Parra *et al.*, these proceedings; Kl″ockner 2004; Parra *et al.* 2005).

5. Clumpy low-gain amplification for OH

The standard model for masers of low-gain amplification of radio continuum by foreground molecular material has been suggested in order to explain the extremely powerful emissions in OH MM (Baan 1985, 1989; Henkel & Wilson 1990). The spatial distribution of the background radio continuum and the availability of a pumping agent will allow a foreground screen to selectively amplify the distributed continuum. Assuming a uniform screen gives a first order estimate of the global amplification. A more realistic non-uniform and clumpy nuclear ISM will give non-uniform amplification in high-resolution data. Emission is found in all OH ground state transitions in Arp 220, while the known 5 GHz and 6 GHz lines are all in absorption in both OH MMs and absorbers (Henkel *et al.* 1987; Impellizzeri *et al.* 2005). An amplification scenario is relevant for extragalactic and also galactic emissions of OH, H_2CO, H_2O, and CH_3OH.

Evaluation of the high-resolution data suggests a gain factor of 4 to 5 relative to the background ($\tau = 1.4 - 1.6$) for the extended components that account for the one third of the 1667 MHz emission (Rovilos *et al.* 2003). The compact components have a gain factor ranging from 46 – 65 at the southern part of the western nucleus to 125 at the northern part ($\tau = 3.8 - 4.8$). Clump superpositions or edge-on (shocked) shells need to increase the optical depth by a factor up to three.

Pumping scenarios for the OH molecules (Henkel *et al.* 1987; Randell *et al.* 1995; Lockett & Elitzur 2007; Parra *et al.* 2005, and Elitzur, these proceedings) provide specific requirements for the physics of the ISM of the galactic nuclei. The efficiency of the radiative pumping using the 53 and 35 μm FIR lines puts requirements on the FIR-SED and requires a threshold temperature for the dust of at least 50 K, whereas Arp 220 has 61 K. Typically clouds of 1 pc should have an opacity of one and a linewidth of 10 km s^{-1} for line-overlap conditions. Pumping can be achieved across a density range of n(H_2) = 10^3 to few 10^4 cm^{-3}. For $T_{ex}(OH)$ approximately constant, the line ratios are a function of optical depth and give an optical depth of order 2.0, which is consistent with observations. Simulations of the cloud superposition and the amplification in a torus-like structure has been done by Kl''ockner (2004) and Parra *et al.* (2005).

6. The high-density ISM

Diagnostics of the heavily obscured nuclear ISM can also be done by interpreting molecular tracers to determine densities, temperatures and the molecular chemistry. Accurate estimates of global (integrated) properties can be obtained with LVG (large velocity gradient) radiation transfer studies using line emission along the energy ladder of the molecule. In addition, the chemical evolution of molecular species combined with their radiative transfer can be obtained by modelling X-ray and Photon (UV) dominated regions (XDR and PDR) (Meijerink, Spaans & Israel 2006, 2007, also Spaans *et al.* and Loenen *et al.*, these proceedings).

Many molecular transitions have been detected in Arp 220 and in similar nearby (U)LIRGs like NGC 6240, and Mrk 231. Single-dish spectra obtained with the IRAM 30 m and SEST 15 m telescopes of some of the tracer lines in Arp 220 are presented in Figure 3 (Baan *et al.* 2007). Two features can be seen in Fig. 3 with systemic velocities of 5365 and 5683 km s^{-1} for the western and eastern nuclei (Baan *et al.* 2007). Higher transitions have been detected for CO, CS, CN, HCN, HNC, and HCO^+ (Aalto *et al.* 2007; Graciá-Carpio *et al.* 2006; Greve *et al.* 2007; Wiedner *et al.* 2002). Recent maps of the prominent HCN (4-3) line at 270 GHz display localized emission with a spatial extent similar to that of the cluster of discrete radio sources (Wiedner *et al.* 2007).

The molecular ISM in ULIRGs is expected to have extreme excitation conditions where the physical environment and the molecular chemistry have been affected by the radiation fields and the feedback from the star-formation process. The environment is very different from those in Galactic SF regions. The interpretation of extragalactic phenomena requires

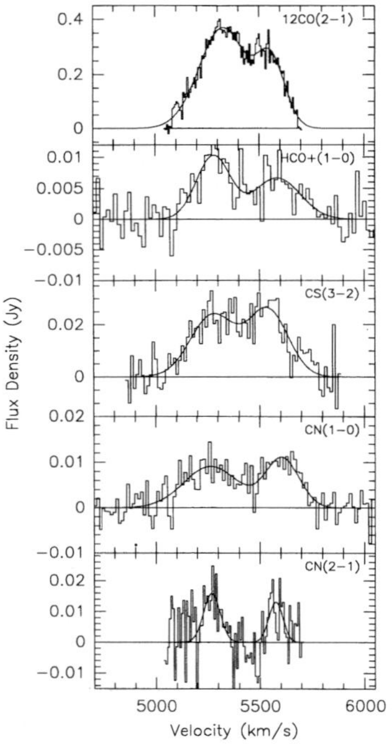

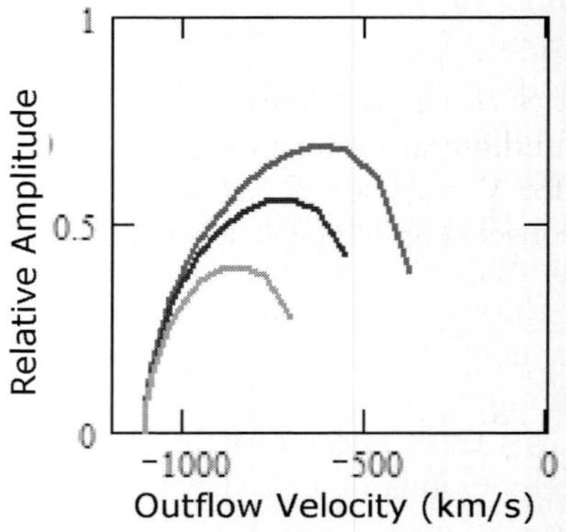

Figure 3. Spectra of emission lines of high-density tracer molecules in Arp 220. From top to bottom the emission line spectra of CO (2-1), HCO^+ (1-0), CS (3-2), CN (1-0), and CN (2-1). Data taken from Baan *et al.* (2007).

Figure 4. Simulations of the amplified OH emission from outflows driven by supernova remnants. The three curves result from different covering factors for the spherical layers of molecular gas relative to the radio continuum generated by the SNR. The chosen terminal velocity is as observed in Arp 220.

an extrapolation of well-resolved Galactic phenomena to integrated and unresolved nuclear environments. High-density components may be used to discriminate between PDR and XDR excitation conditions (Baan *et al.* 2007). In particular, the HNC/HCN ratio has values less than unity in PDRs and values larger than unity in XDRs. Similarly the ratios of HCO^+ with HCN and HNC discriminate between the gas density due to different critical densities of the molecules. The intensity ratio of a high-density tracer molecule with that of a lower density tracer, such as CO (1-0) that is more extended and is less affected by the star-formation activity, shows that the high-density medium varies strongly during the lifetime of the activity (Baan *et al.* 2007). The extreme molecular ratios for Arp 220 and those of other OH MM suggest extreme PDR-like conditions for the nuclear ISM (Loenen *et al.* these proceedings and Loenen, Baan & Spaans 2007).

LVG simulations suggest a two-phase ISM for the combined nuclei of Arp 220 with a diffuse component with $n(H_2) = 2 \times 10^3$ cm^{-3} and $T_k = 40 - 60$ K and dense components with $n(H_2) = 10^{4-6}$ cm^{-3} and $50 - 70$ K (Greve *et al.* 2007). The molecular gas is contained within self-gravitating and virialized clouds with mass $0.6 - 2.0 \times 10^{10}$ $M_\odot$. The considerably higher pressure in the warm high-density medium as compared with the low-density medium will gradually cause disruption of the medium.

7. Signature of nuclear outflows

Arp 220 is one of the OH MM galaxies showing blueshifted outflow components for their 1667 MHz emission lines (Baan, Haschick & Henkel 1989; Baan 2007). The observed outflow velocities vary with L_{FIR} and are likely associated with the shocked shells surrounding supernova remnants. A continuing high-intensity starburst as found in such ULIRGs would eventually result in merging the localized outflows into a large-scale blowout or nuclear superwind Heckman, Armus & Miley (1990). The OH outflow emission results from a superposition shocked SNR shells propagating into the dense ISM that amplify the embedded radio continuum. A simple simulation of a superposition of superposed spherical outflowing shells amplifying a spherical continuum structure provides the correct shape and signature of the outflow (see Figure 4). The observed high velocities of the outflows suggest that J-shock chemistry is responsible for the molecular abundances. Molecules are destroyed and reassembled in J-shocks for V >50 km s^{-1} while for V >300 km s^{-1} this reassembly process happens at a slower rate. Using an energy conserving scenario for the mechanical energy (see Elson, Fall & Freeman 1989), we find that: $N_{SN}\ R^2_{kpc}\ V^3_{100}\ n_{ISM} \approx 23\ L_{IR,11}$, which balances the total energy of the outflows from N_{SN} SNRs with outflow velocity V (in units 100 km s^{-1}) and with radius R (in kpc) and the injected energy L_{IR} (in units of $10^1 1\ L_\odot$). Using $L_{IR} = 1.6 \times 10^{12}\ L_\odot$ for Arp 220, an observed number of $N_{SN} \approx 100$, an observed maximum velocity V(max) =1150 km s^{-1}, and a typical SNR radius =1 pc, we find that the mean ambient density of he pre-shock material $n_{ISM} = 1.5 \times 10^3$ cm^{-3}, which is in an acceptable range. The interpretation of these outflows clearly requires more simulations as it provides further evidence of the eruptive nature of the star-forming environment.

8. Formaldehyde emission

Formaldehyde emission has been verified in Arp 220 as well as two other OH MM, and an H_2O MM using the Arecibo and Effelsberg telescopes (Baan, Haschick & Uglesich 1993; Araya, Baan & Hofner 2004). At low resolution the formaldehyde in Arp 220 peaks at the western nucleus and has weak emission towards the eastern nucleus and the connecting structure (Baan & Haschick 1995). Recent MERLIN data shows an elongated emission region of 80 pc at the western nucleus (resolution 0.04" = 15 pc) centered at 5325 km s^{-1}, which is 40 km s^{-1} below the systemic velocity of the western nucleus (Baan, Cohen & Kl″ockner 2007). The velocity gradient of H_2CO aligns with the weak gradient of the OH emission.

The diagnostics of this emission is still limited by the understanding of the pumping mechanism. Because the formaldehyde emission at lower resolution mimics the 2.2 μm emission, the IR radiation field could play a role in the pumping. A first pumping model proposed for the excitation of the extragalactic formaldehyde was based on pumping by the non-thermal radio continuum, which gives small (few percent) optical depths for densities of 10^{2-5} cm^{-3} (Baan, G″usten & Haschick 1986). While this model could operate in extragalactic environments, it poses a problem for small scale Galactic environments. Evaluation of collisional and other pumping schemes for Galactic and extragalactic environments shows that optical depths at the relevant density range are still relatively small and may not yet satisfy maser requirements under Galactic conditions (see Araya, Baan & Hofner 2004; Baan, Haschick & Uglesich 1993; Araya *et al.* 2007, and Darling, these proceedings). The high- and low-resolution formaldehyde data suggests that emission may be tracing high- and low-density gas components in a larger (foreground) gas structure that is not confined to the site of the star-formation itself.

9. Conclusions

Radio observational tools contribute significantly to the understanding of the nuclear medium of Arp 220. Arp 220 is dynamically advanced such that the two nuclei have developed a common gas/stellar structure in advance of the merger. The current dynamical picture can be determined on the basis of the observed velocities of the two nuclear ISM components. Placing the eastern nucleus, with the highest obscuration, behind the western nucleus on a (first approximation) circular orbit explains the large-scale characteristics of the system but it does not yet explain OH emission at discrepant velocities and the extended continuum structure between the nuclei (see Rovilos *et al.* 2003).

The observed molecular components in Arp 220 sample a density range from 10^2 to 10^6 cm^{-3}. The low-density medium with 10^{2-3} cm^{-3} is traced by the extended diffuse and clumpy CO components. The CO (2-1) transition traces the upper end of this range in nuclear regions enveloping the highest density medium. The high-density constituents have densities in the range 10^4 to 10^6 cm^{-3} depending on the critical density of the molecules. LVG simulations suggest that the temperature of the low-density tracer emissions ranges from 40-60 K, while those of the higher density components suggest 50–70 K. This overpressure of the high-density components results from radiative and mechanical feedback of the star-formation process. The integrated dust temperature from the FIR SED is about T_d = 40 K, while the local conditions in Arp 220 with T_d >60 K provide the required pumping photons for the OH emission.

Both nuclear regions harbor a powerful and rapidly evolving starburst at the point destroying its own ISM. The star-formation rate is on the order of 340 $M_\odot$ yr^{-1} and occurs predominantly in the western nucleus. The ISM in these nuclei is still mostly intact but there is evidence of radiative and mechanical feedback at work. The overpressure of the high-density molecular constituents and the observed OH outflows will in time redistribute and homogenize the nuclear ISM. Radiative feedback also results in a PDR dominated molecular chemistry. The effects of feedback on the molecular medium and the star-formation process would lead to a top-heavy (truncated) IMF and lead to the early termination of the star-formation. There is no evidence for the presence of an AGN in the nuclear regions and any associated XDR conditions. In addition, the observed emission due to X-ray binaries is too weak for the level of star-formation activity. The nuclei of Arp 220 have become the prototype of an extreme nuclear environment with dominant PDR conditions.

The FIR-pumped OH emission extends beyond the nuclear activity regions and shows diffuse structures as well as clumpy and shell-like components with densities of 10^{2-4} cm^{-3} (see Fig. 2 Rovilos *et al.* 2003). The diffuse and clumpy H_2CO can be collisionally or radiatively pumped for densities of 10^{3-5} cm^{-3}. The SNR-driven shells seen in the OH emission are not yet well-constrained but ambient densities of 10^3 cm^{-3} are consistent. The diffuse and the high-brightness OH components cover regions with sizes of 1 to 100 pc, and result from a distributed clumpy molecular medium contained in two edge-on thick tori with a radius of about 60 pc that surround the nuclear ISM at each nucleus. The highest amplifying gains result from cloud superpositions and edge-on shells (also seen as outflows) in the nuclear ISM. The OH and H_2CO emission structures trace intermediate-density components surrounding the highest density ISM structures co-located with the sites of star-formation. The special conditions leading to OH and H_2CO MM activity may occur during a significant part of the duration of the starburst.

The high-resolution VLBI data on the OH and H_2CO emissions and the data for high-density molecular tracers provide a complementary picture of the nuclear ISM and the ongoing SB activity in Arp 220. While the emission of the high-density tracer molecules presents the integrated properties of the nuclear ISM and is mostly unresolved at the two nuclei, the high-resolution data on the OH and H_2CO emissions provides sufficient structural detail of the medium surrounding the sites of star formation. The interpretation of integrated properties depends on our ability to extrapolate from the understanding of Galactic environments to the extreme case of unresolved nuclear environments in starburst nuclei.

References

Aalto, S., Spaans, M., Wiedner, M. C. & H″uttemeister, S. 2007, *A&A*, 464, 193
Araya, E., Baan, W. A. & Hofner, P. 2004, *ApJS*, 154, 541
Araya, E., *et al.* 2007, *ApJ*, submitted
Arp, H. 1966, *ApJS*, 14, 1
Baan, W. A. 1985, *Nature*, 315, 26
Baan, W. A. 1989, *ApJ*, 338, 804
Baan, W. A. 2007, *New Astron. Rev.*, 51, 149
Baan, W. A., Cohen, R. J. & Kl″ockner, H.-R. 2007, in preparation
Baan, W. A. & Haschick, A. D. 1987, *ApJ*, 318, 139
Baan, W. A. & Haschick, A. D. 1995, *ApJ*, 454, 745
Baan, W. A., Haschick, A. D. & Henkel, C. 1989, *ApJ*, 346, 680
Baan, W. A., Haschick, A. D. & Uglesich, R. 1993, *ApJ*, 415, 140
Baan, W. A., Henkel, C., Loenen, A. F., Baudry, A. & Wiklind, T. 2007, *A&A*, in the press
Baan, W. A., G″usten, R. & Haschick, A. D. 1986,
Baan, W. A. & Kl″ockner, H.-R. 2006, *A&A*, 449, 559
Baan, W. A., Wood, P. A. D. & Haschick, A. D. 1982, *ApJL*, 260, L49
Downes, D. & Solomon, P. 1998, *ApJ*, 507, 615
Elson, R. A. W., Fall, S. M. & Freeman, K. C. 1989, *ApJ*, 336, 734
Genzel, R. *et al.* 1998, *ApJ*, 498, 579
Greve,T., Papadopoulos, P., Gao, Y. & Radford, S. 2007, *ApJ*, submitted
Graciá-Carpio, J., García-Burillo, S., Planesas, P. & Colina, L. 2006, *ApJ*, 640, L135
Heckman, T., Armus, L. & Miley, G. 1990, *ApJS*, 74, 833
Henkel, C. & Wilson, T. L. 1990, *A&A*, 229, 431
Henkel, C., G″usten, R. & Baan, W. A. 1987, *A&A*, 185, 14
Impellizzeri, C. M. V., Roy, A. L., Henkel, C. *et al.* 2005, *Astron. Nachr.*, 326, 544
Iwasawa, K., Sanders, D. B., Evans, A. S., *et al.* 2005, *MNRAS*, 357, 565
Kl″ockner, H.-R. 2004, PhD thesis, University of Groningen
Lockett, P. & Elitzur, M. 2007, in preparation
Loenen, A. F., Baan, W. A. & Spaans, M. 2007, in preparation
Lonsdale, C. J., Diamond, P. J., Thrall, H., Smith, H. E. & Lonsdale, C. J. 2006, *ApJ*, 647, 185
Meijerink, R., Spaans, M. & Israel, F. 2006, *ApJL*, 650, L103
Meijerink, R., Spaans, M. & Israel, F. 2007, *A&A*, 461, 793
Mirabel, F. I. 1982, *ApJ*, 260, 75
Nilson, P. 1973, *Nova Acta Regiae Soc. Sci. Upsaliensis*, Ser. V, 0
Parra, R., Conway, J. E., Elitzur, M. & Pihlstr″om, Y. M. 2005, *A&A*, 443, 383
Parra, R., Conway, J. E., Diamond, P. J., *et al.* 2007, *ApJ*, 659, 314
Randell, J., Field, D., Jones, K. N., Yates, J. A. & Gray, M. D. 1995, *A&A*, 300, 659
Rovilos, E., Diamond, P. J., Lonsdale, C. J., *et al.* 2003, *MNRAS*, 342, 373
Rovilos, E., Diamond, P. J., Lonsdale, C.J. *et al.* 2005, *MNRAS*, 359, 827
Sakamoto, K., Scoville, N. Z., Yun, M.S., *et al.* 1999, *ApJ*, 514, 68
Sanders, D. B., Soifer, B. T., Elias, J. H. *et al.* 1988, *ApJ*, 325, 74
Scoville, N. Z., Evans, A. S., Thompson, R., *et al.* 2000, *ApJS*, 119, 991
Scoville, N. Z., Yun, M.S., & Bryant, P. M. 1997, *ApJ*, 484, 702
Solomon, P., Downes, D., Radford, S. J. E. & Barrett, J. W. 1997, *ApJ*, 478, 144
Wiedner, M. C., Wilson, C. D., Harrison, A., *et al.* 2002, *ApJ*, 581, 229
Wiedner, M., Aalto, S., Volgenau, N. H. *et al.* 2007, in preparation

Astrophysical Masers and their Environments
Proceedings IAU Symposium No. 242, 2007
J.M. Chapman & W.A. Baan, eds.

doi:10.1017/S1743921307013579

OH megamasers as extragalactic diagnostics

Ylva M. Pihlström

Department of Physics and Astronomy, 800 Yale Blvd. NE, 1 University of New Mexico, Albuquerque, NM 87131, USA
email: ylva@unm.edu

Abstract. Luminous, extragalactic OH masers are excellent tracers of merger triggered, intense star formation in Ultra Luminous Infrared Galaxies (ULIRGs). As such, they are tracers of high density regions in the centers of active galaxies. From high resolution imaging of the 1667 MHz emission in a small sample of OH megamaser galaxies, the general consensus is that most of the maser emission arises in thick, circumnuclear structures. Here we summarize the current work on OH megamasers, and present the preliminary results on VLA observations of the 1720 MHz satellite line in III Zw 35, Mrk 231 and Arp 220. We also briefly mention ongoing work on investigating the nature of the high molecular gas density environment in a sample of ULIRGs, using CO (J=3–2) observations. A subset of the ULIRG sample harbors OH megamaser emission, and is therefore suited for comparing gas properties between ULIRGs with and without OH megamaser emission.

Keywords. masers, galaxies:starburst, radio lines: galaxies, instrumentation: high angular resolution

---

1. Introduction

The extremely luminous 1667 MHz OH maser emission detected in extragalactic sources has been shown to be associated with the most central regions of Ultraluminous Infrared Galaxies (ULIRGs). ULIRGs are known to be merging systems (e.g. Clements *et al.* 1996), and as a result they display very high far-infrared (FIR) to optical ratios and $L_{FIR} > 10^{11}$ $L_{\odot}$. ULIRGs can be linked to a number of fundamental research areas like the formation of elliptical galaxies, the formation of quasars and the high-z proto-galaxies detected by SCUBA and MAMBO. For instance, studies of ULIRGs allow investigation of the possible evolutionary relationship between Active Galactic Nuclei (AGN) and ULIRGs (the starburst-AGN connection), since both AGN and starburst activity are suggested to be the result of mergers. Other examples of specific research topics include galaxy evolution scenarios, and how star formation and AGN activity might affect the host galaxy and the intergalactic medium.

Given that the OH megamasers reside in merging systems, they could serve as tracers of merger activity. Ultimately, the detection of OH megamasers over a wide range of redshifts could help estimating the merger rate across the cosmic time (Darling & Giovanelli 2002b). Several authors suggest the plausibility of detecting bright OH megamasers at high redshifts (Briggs 1998), and even 'gigamasers' back to the epoch of re-ionization (Darling & Giovanelli 2002b). A future deep OH megamaser survey could perhaps distinguish between different galaxy evolution models. Darling & Giovanelli (2002b) note that even a non-detection could be useful, since maybe then our current assumption that the ULIRGs are the local versions of the high-z submillimeter sources should be re-examined. By these examples it is clear that the OH megamaser emission could serve as an important extragalactic diagnostic tool. Darling (these proceedings) presents a more detailed review of OH megamasers and their properties. Here we will summarize the

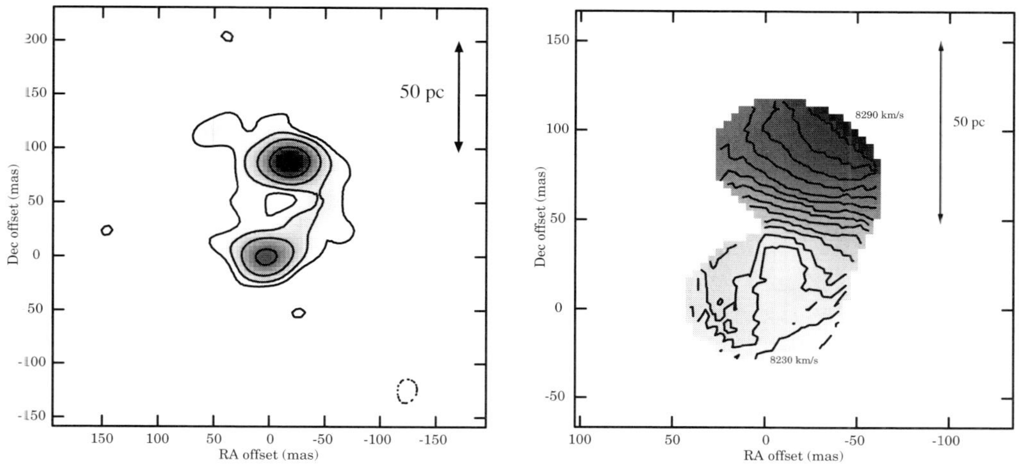

Figure 1. Results from EVN and MERLIN observations of the 1667 MHz OH maser emission in III Zw 35. *a)* Contour map of the integrated maser emission. The beam size is 33 × 29 mas, and plotted contours are -1, 1, 2, 4 and 8 times the 3σ rms noise of 2.7 mJy/beam. *b)* The corresponding velocity field. The gray-scale is between 8224 and 8300 km s^{-1}. Contours are from 8230 km s^{-1} and increasing by 5 km s^{-1} up to 8290 km s^{-1}. The beam size is 43×36 mas; at this resolution the weaker Eastern side is blanked out due to low SNR.

current understanding of the OH megamasers derived mostly from observations of the 1667 MHz main line (Sect. 2). Thereafter we present preliminary results of observations of the much weaker 1720 MHz satellite line in a few of these systems (Sect. 3). Finally, in Sect. 4 we briefly discuss ongoing work on submillimeter data of a sample of ULIRGs, with the aim of investigating any differences in the physical properties or distribution of dense gas in ULIRGs with and without OH megamaser emission.

2. The OH 1667 MHz line

The best observationally studied OH transition is the 1667 MHz main line, which is usually stronger than the 1665 MHz main line with a value often reasonably well in agreement with the expected 9:5 ratio assuming thermodynamic equilibrium. Single dish spectra display line widths ranging from 10 to 1000 km s^{-1}, often with multiple peaks. The emission is generally unresolved at VLA angular scales ($\sim 1.4''$), but can be resolved using long baseline arrays such as the VLBA, EVN and MERLIN. Hitherto a handful of sources have been mapped at high resolution in the 1667 MHz line, and in all cases the emission is centered at one (or both) of the merger system nuclei. Including information from the velocity distribution, there is an emerging general picture of the OH megamasers, with the bulk of the emission being diffuse and distributed in ~ 100 parsec scale circumnuclear structures. Typical enclosed masses are of the order of 10^7 $M_\odot$. Examples include III Zw 35 (Pihlström *et al.* 2001, see Fig. 1), Mrk 231 (Richards *et al.* 2005, Klöckner, Baan & Garrett 2003), Mrk 273 (Yates *et al.* 2000) and the eastern nucleus of Arp 220 (Rovilos *et al.* 2003).

To this general picture of the OH megamasers, however, we have to add the presence of extremely compact sources. At the highest angular resolution, the 1667 MHz emission is most often resolved, and in some sources (e.g. III Zw 35 and Arp 220; Diamond *et al.* 1999, Lonsdale *et al.* 1998) part of the emission is confined to very compact (< 1 pc) scales. More interestingly, the spectra still display line widths exceeding several tens of km s^{-1}, possibly indicating high turbulence. Recent amplification modelling have invoked a clumpy OH medium in order to explain the observed maser characteristics (Parra *et al.* 2005). This model shows that compact masers might be the result of lines of sights intersecting several smaller clouds, with resulting maser emission characteristics very different from what is expected in the case of a smooth medium. Other explanations for the compact emission are also likely, in particular very high sensitivity observation of the OH lines in Arp 220 strongly suggest a connection of the compact masers with individual radio supernovae (Lonsdale, these proceedings).

3. The 1720 MHz line

Assuming thermodynamic equilibrium, the expected line ratio of the 1667 and 1720 MHz transitions is 9, thus a much weaker 1720 MHz line. As a consequence, extragalactic 1720 MHz OH has not been studied in detail, and we have little information about this emission. After a few 1720 MHz detections in early searches for emission in both the main and satellite lines (1612, 1665, 1667 and 1720 MHz) of OH in extragalactic sources (e.g. Baan, Haschick & Henkel 1992), subsequent surveys concentrated on finding the brighter 1667 and 1665 MHz lines using a more limited bandwidth. Examples of sources with 1720 MHz emission include Mrk 231, Arp 220, III Zw 35 and IRAS 17208-0014 (Baan, Haschick & Henkel 1989, Baan & Haschick 1987, Baan, Haschick & Henkel 1992, Martin *et al.* 1989). The peak flux densities of the 1720 MHz lines are typically a few mJy, with 1667/1720 MHz line ratios in the single dish detections that span from about 20 to 100. Similar to the 1667 MHz lines, the 1720 MHz lines are wide. The velocities are comparable, but not always fully corresponding to the 1667 MHz line velocities.

Due to their weak flux densities, no high angular resolution studies have been performed and little is known about the origin and position of these masers. In recent modelling of OH megamaser pumping by Lockett & Elitzur (2007), a weaker 1720 MHz line is predicted to be associated with the 1667 MHz line. If this is the case, a spatial coincidence is expected between the two lines. Another possible origin of the 1720 MHz emission could be in the region of shocks occurring between the merging nuclei in the ULIRG host. Images of H_2 $v = 1-0$ S(1) 2.12μm emission in NGC 6240 and Arp 220 show that the H_2 emission peaks between the two merging nuclei (van der Werf 1993, van der Werf 1999, van der Werf *et al.* 2000). The authors point out that this spatial position, offset from the starbursting nuclei, argues against stellar associated H_2 excitations mechanism such as UV radiation or shocks from supernova remnants. Instead, the excitation mechanism could be due to slow shocks in the dissipating interstellar medium. Such slow shocks could produce similar pumping environments to what is found for Galactic 1720 MHz masers. If this is the case, the extragalactic 1720 MHz emission should be spatially offset from the 1667 MHz emission.

To determine the spatial position of the 1720 MHz emission and thereby constrain its excitation mechanism and origin, we performed a VLA A-array study of the line in three of the sources with published 1720 MHz single dish spectra (III Zw 35, Arp 220 and Mrk 231). The details of the observations will be presented elsewhere, here we report on the preliminary results.

III Zw 35 is a merger with two optical nuclei separated by 9″. The 1667 MHz emission is associated with the northern nucleus in this source, as is also the infrared and radio continuum emission (Chapman *et al.* 1990). In Fig. 2 we plot the VLA 1720 MHz emission in grey scale on top of the radio continuum. The position of the 1720 MHz is coinciding with the 1667 MHz, in agreement with a starburst origin for the 1720 MHz emission. Within the errors, we recover all flux from the single dish spectra (Fig. 3).

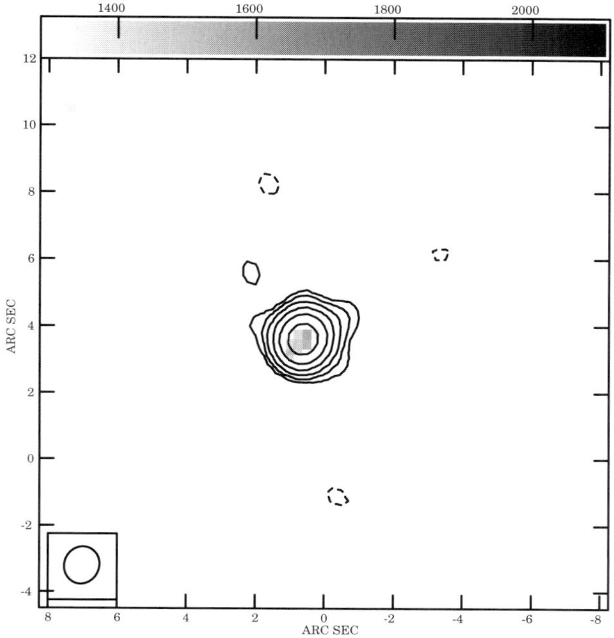

Figure 2. Results from VLA observations of the 1720 MHz OH maser emission in III Zw 35. The contours represents the 1.6 GHz continuum emission, and the plotted contours are −1, 1, 2, 4, 8, 16 and 32 times the 3σ rms noise of 0.7 mJy/beam. The grey scale shows the position of the 1720 MHz maser emission, with the grey scale ranging between 1.3 and 2.1 mJy beam^{-1}.

Mrk 231 displays a similar behavior, with the 1720 MHz emission overlapping with the 1667 MHz. Finally, in Arp 220 the 1720 MHz emission appears to be associated with the Eastern nucleus only, perhaps indicating different density conditions between the two nuclei (1667 MHz emission is present in both of the nuclei of Arp 220).

In summary, extragalactic 1720 MHz OH masers are clearly associated with the starburst regions in mergers. Moreover, 1667/1720 MHz line ratios appear to follow the predictions by the new pumping models of OH megamaser (Lockett & Elitzur 2007).

4. The OH megamaser environment

As mentioned in the introduction, OH megamasers are hosted by ULIRGs. Further, the fraction of OH megamasers in ULIRGs increases with far-infrared luminosity and they are preferentially found in ULIRGs with warm IR color. In general, there is also a higher concentration of dense gas in ULIRGs with higher far-infrared luminosity, for example observed in HCN (Gao & Solomon 2004). Given that molecules form and destroy due to the environment, and that the conditions that are required for OH masers to occur have a limited range, it is likely that the OH megamasers occur during a specific state, or stage of the merger. Thus, to pinpoint the required conditions for OH megamasers, we need to understand the physical connection between galaxy mergers and star formation. In particular, we need a better handle of the time evolution of this process.

A collaboration† has been started with the aim of investigating the dense and warm molecular gas in ULIRGs. The properties and distribution of warm and dense

† The collaboration consists of C. Wilson (McMaster, PI), G. Petitpas, A. Peck, M. Krips (SMA), T.J. Cox (CfA), D. Iono (NAOJ), B. Warren (McMaster), A. Baker (Rutgers), M. Yun

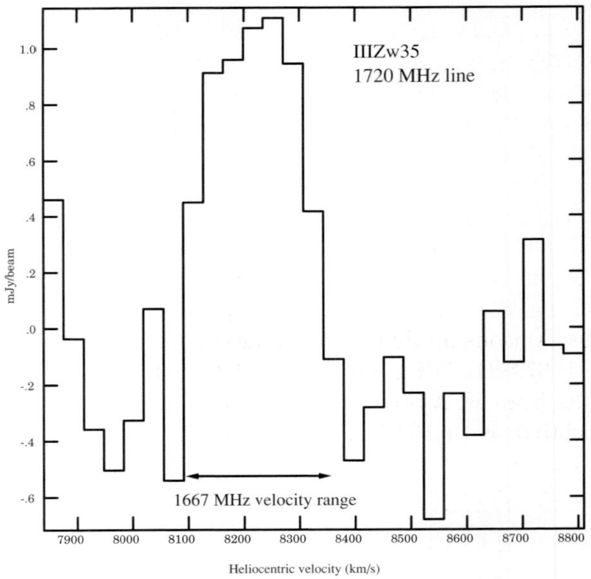

Figure 3. Integrated spectrum of the 1720 MHz OH maser emission in III Zw 35. Within the error bars, all flux is recovered, and the velocity extent of the 1720 and the 1667 MHz lines are in excellent agreement.

molecular gas in ULIRGs could provide key information to understand these processes, since molecular gas is the fuel for star formation. To address this, we selected a sample of 13 ULIRGs, five of which harbors OH megamasers. The sample is observed using the Submillimeter Array (SMA‡), which consists of 6 8m antennas, with a maximum baseline of 500m. SMA has dual frequency operation at 230, 345 and 690 GHz, and with 2 GHz bandwidth we can cover 2600 km s^{-1} at 230 GHz. The five main science goals of the project are as follows:

- To determine the distribution, kinematics and physical conditions of dense molecular gas in ULIRGs
- To determine the spatial distribution of dust in ULIRGs
- To constrain the origin of OH megamasers
- To determine how gas properties might change as the interaction progresses
- To compare properties of dense gas in ULIRGs with high-z submillimeter sources

The first results of this survey are reported on by Wilson *et al.* 2007, and will not be repeated here. However, we note one interesting result that could have important effects on the possibility of high-z submillimeter sources being able to produce 'gigamasers'. Four of the brightest galaxies in the sample were compared to eight high-z submillimeter galaxies that have been mapped in the CO (J=3–2) line. Preliminary results indicate that the molecular gas surface density may be up to an order lower in the high-z sample, than in the local ULIRG sample. This result, however, remain to be confirmed by using our complete sample.

(UMass), Y. Pihlström (UNM), C. Mihos (Case Western Reserve University), S. Matsushita, P. Ho (ASIAA), M. Juvela (Helsinki Observatory) & L. Armus (IPAC)

‡ The Submillimeter Array is a joint project between the Smithsonian Astrophysical Observatory and the Academia Sinica Institute of Astronomy and Astrophysics and is funded by the Smithsonian Institution and the Academia Sinica.

5. Conclusions

OH megamasers are tracers of dense star forming regions in (ULIRGs). High resolution imaging of the 1667 MHz emission in a small sample of OH megamaser galaxies yields a general picture where most of the maser emission arises in thick, circumnuclear structures. Velocity gradients of the 1667 MHz OH emission can be used to estimate the enclosed mass in the nuclei of the merging galaxies, and is typically found to be of the order of $\sim 10^7 M_\odot$ within a few 100 parsecs. The large scales over which the relatively diffuse emission is distributed makes these sources less suited for constraining black hole masses (in contrast to the compact, parsec-scale, thin disk emission traced by H_2O megamasers). In addition to the thick torus, a few OH megamaser galaxies also display extremely compact emission that could be the result of amplification in a clumpy, turbulent medium. There are also observations implying that some compact emission is associated with radio supernovae. The 1667 MHz OH main line is the brightest one, and thus the best studied. Preliminary results from VLA observations indicate that the 1720 MHz line emission also is starburst associated, with 1667/1720 MHz line ratios appear close to those predicted by pumping models assuming more or less the same excitation temperatures for the main as well as the satellite lines.

References

Baan, W. A., & Haschick, A. D. 1987, *ApJ* 318, 139

Baan, W. A., Haschick, A. D., & Henkel, C. 1989, *ApJ* 346, 680

Baan, W. A., Haschick, A., & Henkel, C. 1992, *AJ* 103, 728

Briggs, F. H. 1998, *A&A* 336, 815

Chapman, J. M., Staveley-Smith, L., Axon, D. J., Unger, S. W., Cohen, R. J., Pedlar, A., & Davies, R. D. 1990, *MNRAS* 244, 281

Clements, D. L., Sutherland, W. J., McMahon, R. G., & Saunders, W. 1996, *MNRAS* 279, 477

Darling, J., & Giovanelli, R. 2002, *AJ*, 124, 100

Darling, J., & Giovanelli, R. 2002, *ApJ*, 572, 810

Diamond, P. J., Lonsdale, C. J., Lonsdale, C. J., & Smith, H. E. 1999, *ApJ* 511, 178

Gao, Y., & Solomon, P. M. 2004, *ApJSS* 152, 63

Klöckner, H-R., Baan, W. A., & Garrett, M. A. 2003, *Nature*, 421, 821

Lockett, P., & Elitzur, M. 2007, *in prep*

Lonsdale, C.J., Lonsdale, C. J., Diamond, P. J., & Smith, H. J. 1998, *ApJL* 493, L13

Martin, J.M., Bottinelli, L., Dennefeld, M., Gougenheim, L., & Le Squeren, A. M. Williams, J. A. 1989, *A&A*, 208, 39

Parra, R., Conway, J. E., Elitzur, M., & Pihlström, Y. M. 2005, *A&A* 443, 383

Pihlström, Y. M., Conway, J. E., Booth, R. S., Diamond, P. J., & Polatidis, A. G. 2001, *A&A* 377, 413

Richards, A. M. S., Knapen, J. H., Yates, J. A., Cohen, R. J., Collett, J. L., wright, M. M., Gray, M. D., & Fields, D. 2005, *MNRAS* 364, 353

Rovilos, E., Diamond, P. J., Lonsdale, C. J., Lonsdale, C. J., and Smith, H. E. 2003, *MNRAS* 342, 373

van der Werf, P. P., Genzel, R., Krabbe, A., Blietz, M., Lutz, D., Drapatz, S., Ward, M. J., & Forbes, D. A. 1993, *ApJ*, 405, 522

van der Werf, P. P. 1999, in 'Galaxy Interactions at Low and High Redshift, Proceedings of IAU Symposium #186', *IAUS* 186, 303

van der Werf, P. P. 2000, in 'Molecular Hydrogen in Space', *CUP* p307

Wilson, C. D., Warren, B. E., Petitpas, G., Peck, A., Krips, M., Iono, D., Baker, A., Yun, M., Pihlström, Y. M., Mihos, C., Matsushita, S., Ho, P. T. P., Juvela, M., Cox, T. J., & Armus, L. 2007, in proceedings of 'Science with ALMA: a new era for astrophysics'

Yates, J. A., Richards, A. M. S., Wright, M. M., Collett, J. L., Gray, M. D., Field, D., & Cohen, R. J. 2000, *MNRAS* 447, L95

Astrophysical Masers and their Environments
Proceedings IAU Symposium No. 242, 2007
J.M. Chapman & W.A. Baan

doi:10.1017/S1743921307013580

The irradiated ISM of ULIRGs

M. Spaans[1], R. Meijerink[2], F. P. Israel[3], A. F. Loenen[1,4] and W. A. Baan[4]

[1]Kapteyn Astronomical Institute, P.O. Box 800, 9700 AV Groningen, The Netherlands
email: spaans@astro.rug.nl

[2]Astronomy Department, University of California, Berkeley, CA 94720, United States

[3]Leiden Observatory, P.O. Box 9513, 2300 RA Leiden, The Netherlands

[4]ASTRON, P.O. Box 2, 7990 AA Dwingeloo, The Netherlands

Abstract. The nuclei of ULIRGs harbor massive young stars, an accreting central black hole, or both. Results are presented for molecular gas that is exposed to X-rays (1–100 keV, XDRs) and far-ultraviolet radiation (6–13.6 eV, PDRs). Attention is paid to species like HCO^+, HCN, HNC, OH, H_2O and CO. Line ratios of HCN/HCO^+ and HNC/HCN discriminate between PDRs and XDRs. Very high J (> 10) CO lines, observable with HIFI/Herschel, discriminate very well between XDRs and PDRs. In XDRs, it is easy to produce large abundances of warm ($T > 100$ k) H_2O and OH. In PDRs, only OH is produced similarly well.

Keywords. ISM: molecules – galaxies: evolution

1. Introduction

The power that emanates from (ultra-)luminous infrared galaxies is believed to stem from active star formation and/or an accretion disk around a central super-massive black hole (e.g. Sanders & Mirabel 1996). The unambiguous identification of the central energy source, or the relative contributions from stars and an active galactic nucleus (AGN), remains a major challenge in the study of active galaxy centers (Aalto *et al.* 2007). Molecular lines are ideal in this to penetrate deep into the large column density regions of active galaxies (Israel 2005; Baan & Klöckner 2005). The interested reader is referred to Meijerink & Spaans (2005) and Meijerink, Spaans & Israel (2006, 2007) for details on Photon-Dominated Region (PDR) and X-ray Dominated Region (XDR) physics. PDRs refer to the presence of a starburst that produces photons with energies of 6–13.6 eV, while XDRs indicate an accreting black hole with photon energies of 1–100 keV (Maloney *et al.* 1996; Lepp & Dalgarno 1996). The most important thing to realize is that a 1 keV photon penetrates a hydrogen column of about 10^{22} cm^{-2}, while a UV photon (10 eV) is absorbed by dust after about 1 mag of visual extinction. This is a consequence of the fact that X-ray absorption cross sections scale roughly like energy^{-3}, allowing deep penetration of X-rays into interstellar clouds.

2. Results

All presented models are plane-parallel slabs and are parameterized by a constant density and an impinging UV (in multiples of $G_0 = 1.6 \times 10^{-3}$ erg s^{-1} cm^{-2}) or X-ray (energy$^{-0.9}$) radiation field F_X in erg s^{-1} cm^{-2}. Note that a flux of $F_X = 100$ erg s^{-1} cm^{-2} corresponds to a 10^{44} erg s^{-1} Seyfert nucleus at 100 pc from an interstellar cloud.

It is clear from figures 1, 2, 3 and 4 that OH is easily formed in PDRs and XDRs, while warm water is present mostly in XDRs. In PDRs, OH is the dissociation product of

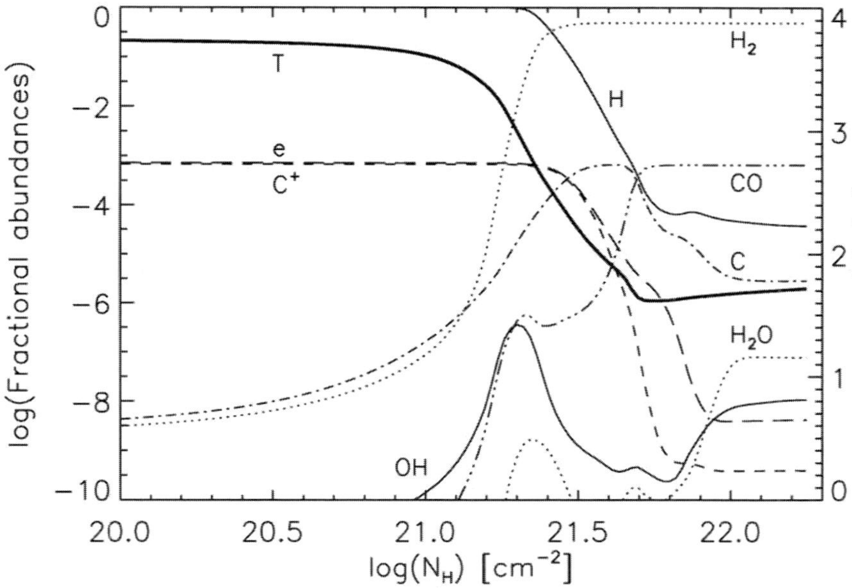

Figure 1. Depth dependence of a few important chemical species in the PDR, $n = 10^5$ cm^{-3} and $10^{3.5} G_0$. Note the clear stratification.

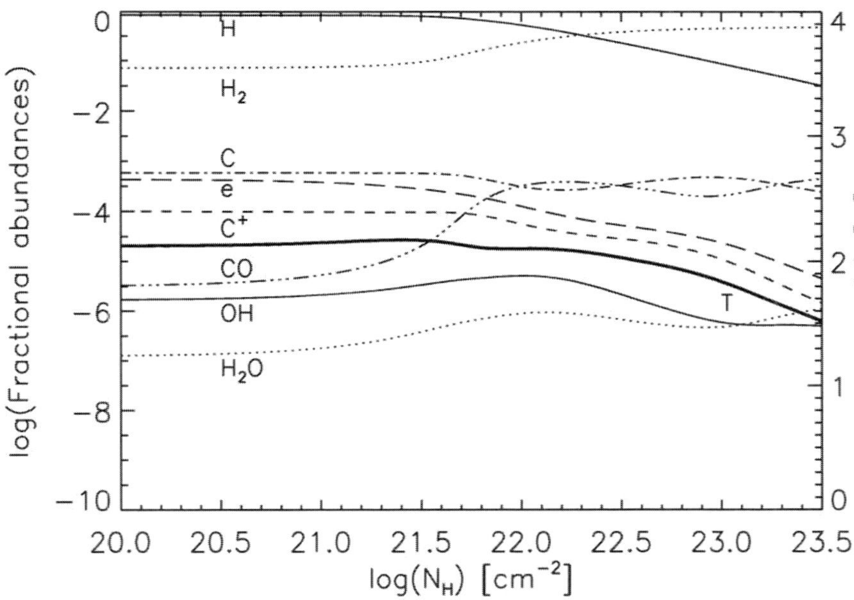

Figure 2. Depth dependence of a few important chemical species in the XDR, $n = 10^5$ cm^{-3} and $F_X = 5$ erg s^{-1} cm^{-2}. Note the different column density scale and warm water.

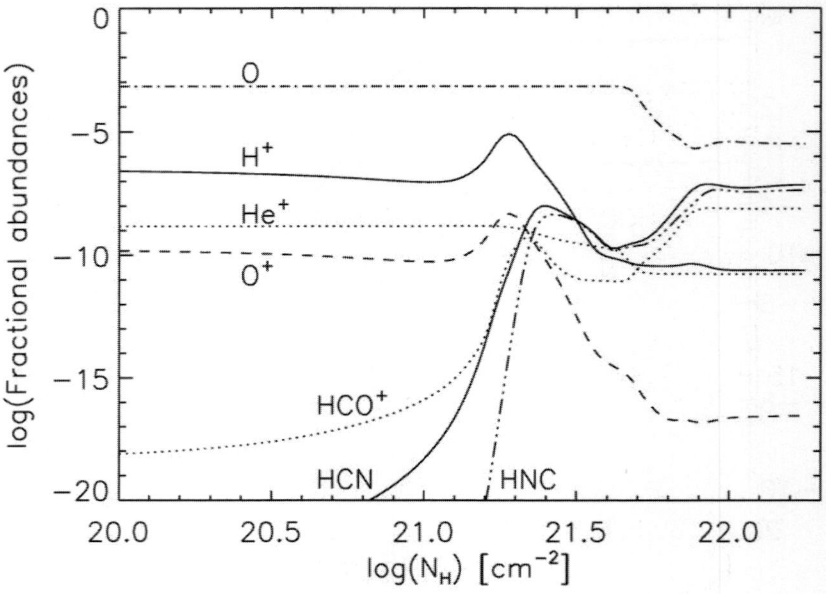

Figure 3. Depth dependence of a few important species in the PDR, same as figure 1.

water, while it is formed indirectly through charge transfer of H^+ and O, rapid reactions with H_2 to OH^+ and H_2O^+, followed by dissociative recombination or directly through $O+H_2 \rightarrow OH+H$. The latter reaction is driven efficiently above 200 k. Additional reactions of H_2O^+ with molecular hydrogen lead to H_3O^+ which dissociatively recombines to, among others, water or OH. Note that an elevated cosmic ray ionization rate also leads to larger abundances of H_3O^+, but not to the same degree as in XDRs (Meijerink *et al.* 2006). In both PDRs and XDRs, vibrationally excited H_2 is also present through, respectively, UV pumping and thermal collisions with electrons. This significantly lowers the effective energy barrier of the $O+H_2$ reaction. In XDRs, an internal UV radiation field is created by collisional excitation of H and H_2 that leads to Lyman α and Lyman-Werner photons through radiative decay. However, this UV field is not strong enough to dissociate water as efficiently as in PDRs. At the high temperatures reached in XDRs, because photo-ionization heating is more efficient than photo-electric heating by dust grains, the endo-ergic reaction $OH+H_2$ also leads to water. Of course, shocks are similarly capable of producing OH and H_2O.

Typically, the HCO^+ lines are stronger in XDRs than in PDRs by a factor of at least three. This is a direct consequence of the higher ionization degree in XDRs (Meijerink & Spaans 2005), leading to an enhanced HCO^+ formation rate. Depending on the incident radiation field, HCN or HCO^+ is more abundant at the PDR edge of the cloud. At sufficiently large column and densities, the HCN(1-0)/HCO^+(1-0) ratio becomes larger than 1. In the XDR models, HCO^+ is chemically less abundant than HCN for very large H_X/n (Meijerink & Spaans 2005). However, for larger columns HCO^+ always becomes more abundant than HCN (Fig. 10 in Meijerink & Spaans 2005). These effects are summarized in figure 2, 4 and 5, see also Loenen *et al.* (this volume).

The critical densities of HCN and HNC are almost identical, so the only differences in ratios can be due to differences in the abundances. It turns out that the HNC/HCN

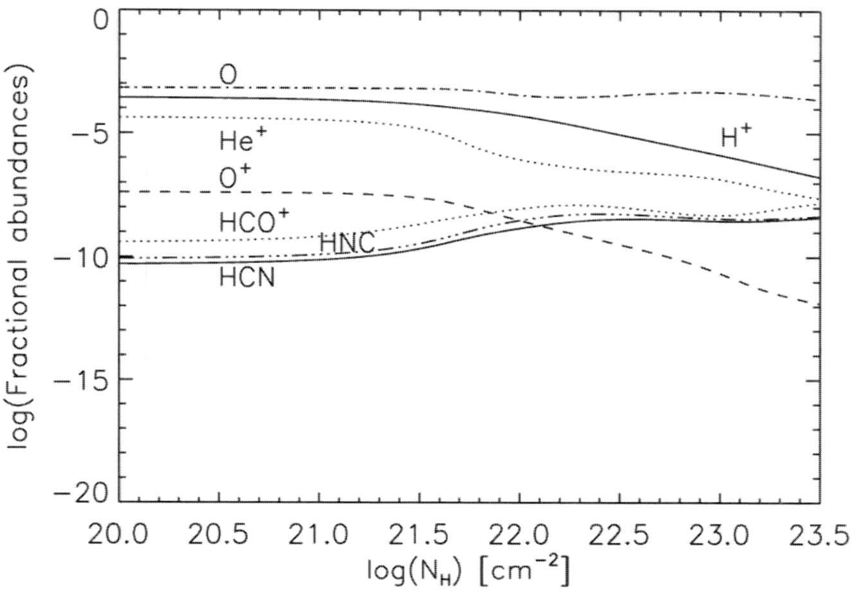

Figure 4. Depth dependence of a few important species in the XDR, same as figure 2. Note the enhancement of HNC with respect to HCN (factor 2).

column density ratio is quite close to that of HNC(1-0)/HCN(1-0) line intensity ratio. In PDRs, HCN is more abundant in the radical region, but deeper into the cloud the ratio becomes about one. In XDRs, HCN is more abundant in the highly ionized part of the cloud. However, HNC is equally or even more abundant than HCN deep into the cloud. All this can be seen in figures 2 and 4.

In XDRs at high densities CO is present all throughout the cloud, even when F_X/n is large. This is a direct consequence of the fact that X-rays do not lead to strong dissociation of CO and thus C^+, C and CO typically co-exist in an XDR at elevated (> 100 K) temperatures. Such warm CO gas produces emission originating from high rotational transitions. Contrary to XDRs, most CO in PDRs is produced after the H/H_2 transition and has on average much lower temperatures ($T \sim 20-50$ K). Future missions such as Herschel/HIFI will be able to distinguish between PDRs and XDRs by observing high rotational transitions such as CO(16-15), see figure 6.

3. Future work

In the future, observatories like Herschel and ALMA will revolutionize our understanding of the nuclear activity in (maser) galaxies by resolving the central regions of these systems spatially. In this light, it is important to stress that starburst activity typically occupies a larger fraction of an active galaxy than accretion onto a central black hole. Consequently, single dish observations are likely to be dominated by a PDR signal even when an XDR is present. This is particularly true for the very high J CO lines and emphasizes the need for high spatial resolution. ALMA can detect and resolve very high J CO emission from ULIRGs at redshifts beyond eight. Additional molecules that allow

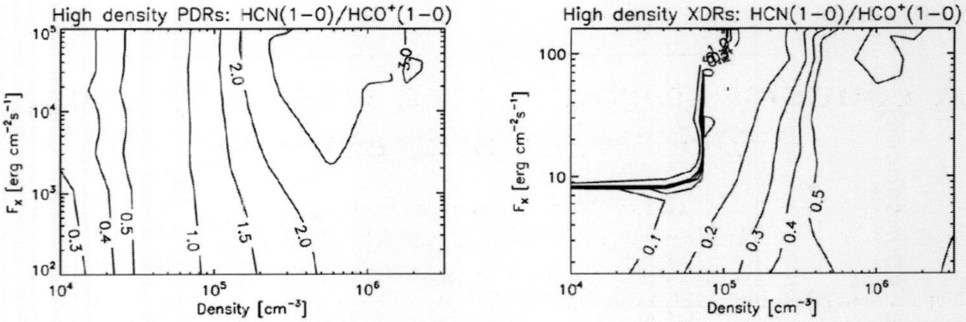

Figure 5. The HCN(1-0)/HCO^+ (1-0) line intensity ratio for a grid of PDR (left) and XDR (right) models.

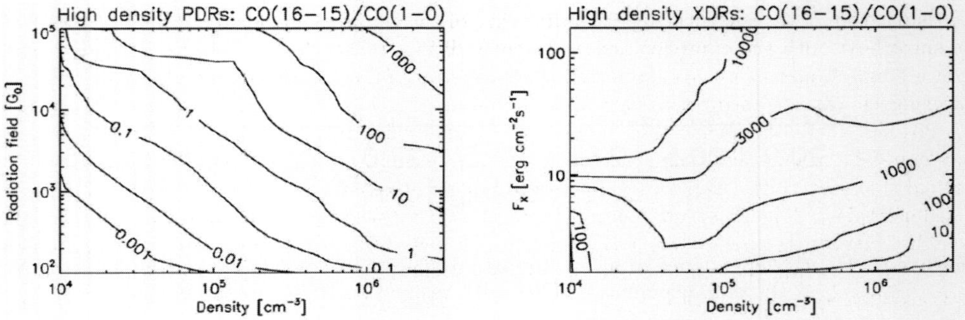

Figure 6. The CO(16-15)/CO(1-0) line intensity ratio for a grid of PDR (left) and XDR (right) models.

one to probe an accreting black buried inside ULIRGs, i.e. that are direct tracers of X-ray irradiation, are CO^+, CH^+ and H_3^+.

Acknowledgements

We are grateful to Dieter Poelman and Juan Pablo Pérez-Beaupuits for discussions.

References

Aalto, S., Spaans, M., Wiedner, M. C., & Hüttemeister, S., 2007, *A&A* 464, 193
Baan, W. A. & Klöckner, H.-R., 2005, *Ap&SS*, 295, 263
Israel, F.P., 2005, *Ap&SS*, 295, 171
Lepp, S. & Dalgarno, A., 1996, *A&A*, 306, L21
Maloney, P. R., Hollenbach, D. J., & Tielens, A. G. G. M., 1996, *ApJ*, 466, 561
Meijerink, R. & Spaans, M., 2005, *A&A*, 436, 397
Meijerink, R., Spaans, M., & Israel, F. P., 2006, *ApJL*, 650, L103
Meijerink, R., Spaans, M., & Israel, F. P., 2007, *A&A*, 461, 793
Sanders, D. B. & Mirabel, I. F., 1996, *ARA&A*, 34, 749

Astrophysical Masers and their Environments
Proceedings IAU Symposium No. 242, 2007
J.M. Chapman & W.A. Baan, eds.

doi:10.1017/S1743921307013592

A compact starburst ring traced by clumpy OH megamaser emission

Rodrigo Parra[1], **John E. Conway**[1], **Moshe Elitzur**[2] **and Ylva M. Pihlström**[3]

[1]Onsala Space Observatory, S 43992, Onsala, Sweden
[2]Department of Physics and Astronomy, University of Kentucky, Lexington, KY 40506, USA
[3]Department of Physics and Astronomy, UNM, 800 Yale Blvd NE, Albuquerque, NM 87131, USA

Abstract. We model the OH megamaser emission from the luminous infrared galaxy IIIZw35 as arising from a narrow rotating starburst ring of radius 22 pc enclosing a mass of 7×10^6 $M_\odot$. We show how both the compact and apparently diffuse maser emission from this ring can arise from a single phase of unsaturated maser clouds amplifying background radio continuum. The masering clouds are estimated to have a diameter of <0.7 pc and internal velocity dispersion of $\sim$20 km s^{-1}. We find that the clouds are neither self-gravitating nor pressure confined, and they could be magnetically confined or freely expanding. Their dispersal lifetimes may set the vertical thickness of the ring. For an estimated internal density of 3×10^3 cm^{-3}, cloud masses are of order 24 $M_\odot$. The observed spectral features and velocity gradients indicate that the clouds must be outflowing and escaping the nucleus. The cloud mass outflow rate is estimated to be 0.8 $M_\odot$ yr^{-1}, while the star formation rate is $\sim$19 $M_\odot$ yr^{-1}. Associated ionised gas, possibly generated from dissipated clouds, provides free-free absorption along the source axis, explaining the observed East-West asymmetries. We show that the clumpiness of a maser medium can have a dramatic effect on what is observed even in a relatively low gain OH megamaser. Specifically, in IIIZw35 our clumpy maser model naturally explains the large line to continuum ratios, the large 1667MHz:1665MHz line ratios and the wide velocity dispersions seen in the compact maser spots. Other astrophysical masers showing both compact and apparently diffuse emission might be explained by similar clumpy structures.

1. Introduction

Extra-galactic OH megamaser (OH MM) emission is generally associated with compact (<100 parsec) starburst activity in the centres of IR luminous galaxies. Observations of this maser emission provide a unique method of studying the structure and kinematics of galactic nuclei at parsec (pc) resolution without dust obscuration effects. Measurements of velocity gradients and line widths (e.g. Pihlström *et al.* 2001) already provide important constraints on stellar mass densities and turbulent velocities in IR luminous galaxies. Potentially OH MM can also tell us about the size, density and temperature of molecular clouds in the central ISM of starburst galaxies. However, to accomplish these goals a better understanding of the OH MM phenomenon is required.

In the standard model (see Baan 1989; Lonsdale 2002), OH MM emission is generated by low gain ($|\tau| \lesssim 2$) unsaturated amplification of background continuum by a foreground OH amplifying medium. Although this medium is comprised of discrete OH clouds it is implicitly assumed that there are many such clouds and they individually have very low gains. These clouds therefore form an effective *gas* in which statistical fluctuations in cloud number between different lines of sight are unimportant and so the maser opacity varies slowly across the source. Given these assumptions the amplifying medium is often described as a *diffuse screen.*

Consistent with the standard model, early VLA and MERLIN observations showed that OH maser and continuum emission overlapped. However, VLBI observations of Arp 220 resulted in the detection of both compact continuum (Diamond *et al.* 1999; Smith *et al.* 1998) and compact OH maser emission (Lonsdale *et al.* 1994, 1998). Remarkably, these bright maser spots were not spatially coincident with the continuum spots, and some displayed extreme line-to-continuum ratios (LCR) in excess of 800 (Lonsdale *et al.* 1998). These observations were clearly inconsistent with diffuse screen models. Slightly less extreme compact maser emission was subsequently detected in other sources (Trotter *et al.* 1997; Diamond *et al.* 1999; Klöckner *et al.* 2003; Klöckner & Baan 2004). These same sources also contain diffuse maser components which account for between 50% and 90% of the total maser flux density. It has been suggested that the compact masers occur in saturated, perhaps collisionally pumped regions, while the diffuse component comes from an extended unsaturated, radiatively pumped screen fully consistent with the standard model (Lonsdale *et al.* 1998; Diamond *et al.* 1999).

One of the clearest cases of an OH MM showing both compact and diffuse maser emission is the LIRG IIIZw35. Two groups of compact masers were detected in VLBI observations (Trotter *et al.* 1997; Diamond *et al.* 1999) recovering $\sim$50% of the total OH maser emission seen on MERLIN scales (Montgomery & Cohen 1992). Most of the missing OH maser flux was found to be in a rotating OH maser ring using EVN+MERLIN observations (Pihlström *et al.* 2001). The previously known compact masers lie at the tangents of this ring. P01 explained both diffuse and compact masers using a mechanism based on a single phase of small OH masering clouds ($\sim$1 pc) amplifying background continuum. At the ring tangents multiple overlaps between clouds in space and velocity are likely due to the increased path length through the ring. These multiple cloud overlaps give rise to the bright compact maser features. Elsewhere, at the front and back of the ring where there are few such cloud overlaps, the emission consists of many weak maser spots. These spots are too weak to be detected individually in high resolution observations but in low resolution observations they are averaged together and give rise to an apparently diffuse emission.

In Parra *et al.* (2005) we investigate in greater depth the *clumpy ring model* proposed in Pihlström *et al.* (2001). We fully consider the spectral and kinematical properties of the maser clouds. We also model the continuum emission in a geometrically and physically consistent manner. Using numerical simulations we demonstrate that most of the available OH maser observations of IIIZw35 can be explained using an improved version of the Pihlström *et al.* (2001) model. Most of the input parameters for these simulations are constrained directly by the observations. Our modelling also illuminates general properties of maser ring geometries and clumpy maser media.

2. Description of the Model

The IIIZw35 OH MM observations are fitted by an inclined axisymmetric model in which both OH clouds and continuum emission coexist within circumnuclear rings (see Figure 1). In order to reconcile such a symmetric geometry with the observed East-West asymmetry in both line and continuum emission, our model also includes a bi-cone of free-free absorption which covers the eastern side of the source. This obscuration defines the maser ring orientation requiring the eastern side to be the most distant. Note that the existence of a free-free absorbing component is supported by independent evidence (Chapman *et al.* 1990). Physically the free-free absorbing cone could be the base of an outflowing superwind such as is often observed in energetic starbursts (Heckman 2003).

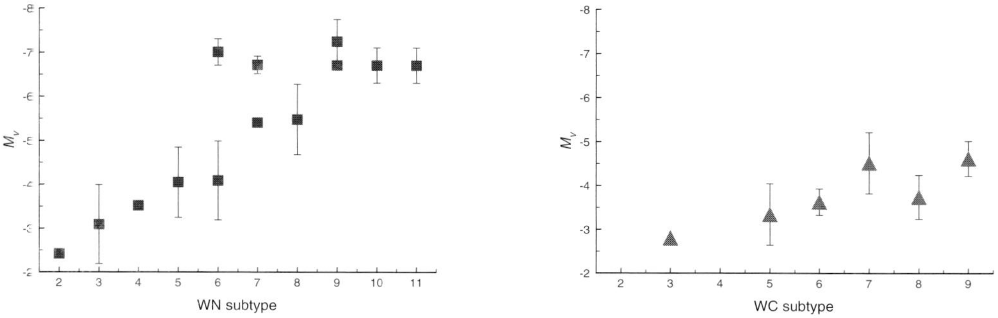

Figure 1. Cross section of the starburst ring plus outflow model proposed for IIIZw35. The proposed ring has internal radius of 22 pc, radial thickness of 3 pc and height of 6 pc. The rotation velocity is 57 km s^{-1} with the northern side being redshifted. The volume defined by the ring is filled with independently orbiting molecular clouds of diameter 0.7 pc, internal velocity dispersion of 20 km s^{-1} and mean density of $3\times10^3 \leqslant N_{\mathrm{H}_2} < 10^5$ cm^{-3}. These clouds are assumed to be the source of the OH maser emission. The mechanical energy released by supernova explosions occurring in the equatorial plane of the ring at a rate of $\nu_{\mathrm{SN}} \sim 0.9$ year^{-1}, pushes away from the ring plane the molecular clouds. These clouds are not confined but are instead freely expanding and outflowing away from the ring to eventually dissipate and form a cone of obscuring ionised gas. Although the OH molecules are expected to be in a more extended region than the ring, as suggested by the OH absorption features observed by Pihlström *et al.* (2001), the necessary conditions for maser action (e.g. pumping) are fulfilled only within the ring (see Parra *et al.* 2005).

Although the total maser emission is weaker on the eastern side, the LCR is about factor of 2 larger there than on the western side. This difference is explained if more of the seed continuum emission is background to the OH ring on the east side. This naturally occurs if *most* of the bright continuum emission comes from a larger radius than the OH masering gas. This is corroborated by the fact that the observed continuum radio source is larger than the megamaser source. The proposed geometry also predicts that the brightest masers do not occur exactly at the maser ring tangents but slightly to the east of the tangent points, just as found by the observations of Trotter *et al.* (1997) and Diamond *et al.* (1999).

The properties of the different components of the model are estimated in a detailed way from observations in Parra *et al.* (2005). These parameters are subsequently used in Monte-Carlo simulations to produce the results shown in Figure 2.

3. Conclusions

Observations of distinct regions of compact and apparently diffuse maser emission in OH MM have been used by other authors to argue for two distinct physical phases of OH masers (see Lonsdale *et al.* 1998; Diamond *et al.* 1999). In contrast, the mechanism discussed in Parra *et al.* (2005) explains both types of maser structures using a single phase of low opacity clouds within a thin circumnuclear ring.

The fact that the maser amplifying medium is composed of clouds is found *essential* to explain the range of maser brightness around the ring at low resolution. The same clouds also explain the bright maser spots seen only at the ring tangents in terms of multiple cloud overlaps in both space and velocity. The brightest features in our model arise from the alignment of 5 clouds giving a total maser optical depth of 7.5, well below the threshold of ~13 required for saturation (Elitzur 1992). It is also important to remember that saturation depends on the angle averaged intensity implying that very large maser

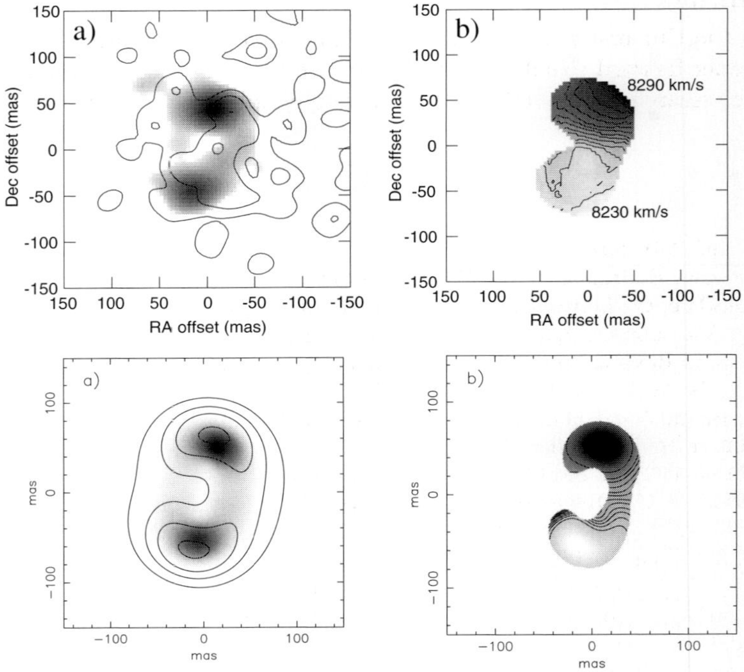

Figure 2. Comparison between observations and numerical simulations. The top row shows images reproduced from P01 showing observations of OH maser and continuum emission at EVN+MERLIN resolution **Top-Left** Greyscale shows the velocity integrated OH emission at a resolution of 34×29 mas. The contours show continuum emission at the same resolution. Contour levels are at -1,1,2, and 4 times the 3σ noise of 0.42mJy/beam. **Top-Right** Corresponding OH maser velocity centroid field. The greyscale is between 8224 (lightest grey) and 8300 km s^{-1}(black). Contours are from 8230 km s^{-1} increasing at 5 km s^{-1} intervals up to 8290 km s^{-1}. **Bottom-Left and Right** Same as top row but synthesized from a single realisation of our model (see Parra *et al.* 2005)

brightness temperatures can be achieved whenever the maser beaming angle is small. Such small angles arise naturally in our overlapping cloud model (see Parra *et al.* 2005).

The proposed model can explain the spectra of both the compact spots and the apparently diffuse areas of emission. The fact that the OH clouds are outflowing from the ring midplane explains the large velocity gradients observed amongst the compact maser spots. The model is also able to reproduce the LCRs in the bright spots and diffuse regions, and the large value of the 1667MHz:1665MHz line ratio in the compact spots.

We find that the OH masering ring is relatively narrow in radius which could be explained either in terms of a narrow circumnuclear ring of star formation, or due to the strong sensitivity of maser pumping to physical conditions which vary gradually with radius. The ring we find is narrower but qualitatively similar to those that have been produced in numerical simulations (Wada & Norman 2002; Wada & Tomisaka 2005).

Finally, the model seems to successfully apply to other sources like IRAS 17208–0014 (Momjian *et al.* 2006) or the eastern nucleus of Arp 220.

Acknowledgements

R. P. thanks the University of Kentucky for a most delightful month, which helped in bringing this paper forward to submission. J.C gratefully acknowledges support from the Swedish science research council (VR). M.E. gratefully acknowledges the partial support of NSF.

References

Baan, W. 1989, ApJ, 338, 804
Chapman, J., Staveley-Smith, L., Axon, D., *et al.* 1990, MNRAS, 244, 281
Diamond, P., Lonsdale, C., Lonsdale, C., & Smith, H. 1999, ApJ, 511, 178
Elitzur, M. 1992, Astronomical masers, Vol. 170 (Kluwer Academic Publishers ASSL), 365
Heckman, T. 2003, in Revista Mexicana de Astronomia y Astrofisica Conference Series, 47–55
Klöckner, H.-R. & Baan, W. 2004, A&A, 419, 887
Klöckner, H.-R., Baan, W., & Garrett, M. 2003, Nature, 421, 821
Lonsdale, C. 2002, in IAU Symposium, 413
Lonsdale, C., Diamond, P., & Smith, H. 1994, Nature, 370, 117
Lonsdale, C., Lonsdale, C., Diamond, P., & Smith, H. 1998, ApJ, 493, L13
Momjian, E., Romney, J., Carilli, C., & Troland, T. 2006, ApJ, 653, 1172
Montgomery, A. & Cohen, R. 1992, MNRAS, 254, 23P
Parra, R., Conway, J., Elitzur, M., & Pihlström, Y. 2005, A&A, 443, 383
Pihlström, Y., Conway, J., Booth, R., Diamond, P., & Polatidis, A. 2001, A&A, 377, 413
Smith, H., Lonsdale, C., Lonsdale, C., & Diamond, P. 1998, ApJ, 493, L17
Trotter, A., Moran, J., Greenhill, L., Zheng, X., & Gwinn, C. 1997, ApJ, 485, L79
Wada, K. & Norman, C. 2002, ApJ, 566, L21
Wada, K. & Tomisaka, K. 2005, ApJ, 619, 93

Astrophysical Masers and their Environments
Proceedings IAU Symposium No. 242, 2007
J.M. Chapman & W.A. Baan, eds.

doi:10.1017/S1743921307013609

Molecular properties of (U)LIRGs: CO, HCN, HNC and HCO^+

A. F. Loenen[1,2], W. A. Baan[2] and M. Spaans[1]

[1]Kapteyn Astronomical Institute, University of Groningen,
P.O. Box 800, 9700 AV Groningen, The Netherlands
email: loenen@astro.rug.nl, spaans@astro.rug.nl

[2]ASTRON, P.O. Box 2, 7990 AA Dwingeloo, The Netherlands
email: baan@astron.nl

Abstract. The observed molecular properties of a sample of FIR-luminous and OH megamaser (OH-MM) galaxies have been investigated. The ratio of high and low-density tracer lines is found to be determined by the progression of the star formation in the system. The HCO^+/HCN and HCO^+/HNC line ratios are good proxies for the density of the gas, and PDR and XDR sources can be distinguished using the HNC/HCN line ratio. The properties of the OH-MM sources in the sample can be explained by PDR chemistry in gas with densities higher than $10^{5.5}$ cm^{-3}, confirming the classical OH-MM model of IR pumped amplification with (variable) low gains.

Keywords. galaxies: nuclei, galaxies: ISM, galaxies: active, galaxies: starburst, ISM: molecules, ISM: evolution, radio lines: ISM

1. Introduction

High-density molecular gas plays an important role in the physics of (Ultra-)Luminous Infrared Galaxies ((U)LIRGs), giving rise to spectacular starbursts and possibly providing the fuel for an active galactic nucleus (AGN). The emission lines emanating from the nuclear gas provide information about the physical properties of the nuclear environment in these systems, e.g. the (column) density, temperature and chemical composition of the gas, and the type and strength of the central energy source. It can also provide us an insight into the processes influencing the (gas in the) nuclei: the star formation rate and history, fuelling of a possible central black hole and feedback processes.

Baan *et al.* (2007) present data of the CO, HCN, HNC, HCO^+, CN, and CS line emissions of a representative group of 37 FIR-luminous and OH megamaser (OH-MM) galaxies and 80 additional sources taken from the literature. In this work, the molecular characteristics of this sample are explained using several models. First, the properties of the different (density) components of the nuclear gas are explained in terms of starburst evolution (see Baan *et al.* 2007). Also the chemical properties of the high-density gas are analyzed, using chemical networks (Meijerink & Spaans 2005) and radiative transfer models Meijerink *et al.* (2007). Here only the J=1-0 transitions of the molecules are considered. A more detailed analysis will be presented in Loenen, Baan & Spaans (2007).

2. Starburst evolution

The gas in galaxies is build up from multiple components, each with different densities and temperatures. From our data we can derive information about these different components. The CO(1-0) line traces the large scale low-density (critical density $n_{cr}=3 \times 10^3$ cm^{-3}) component (LD), whereas the lines of HCN, HNC, and HCO^+

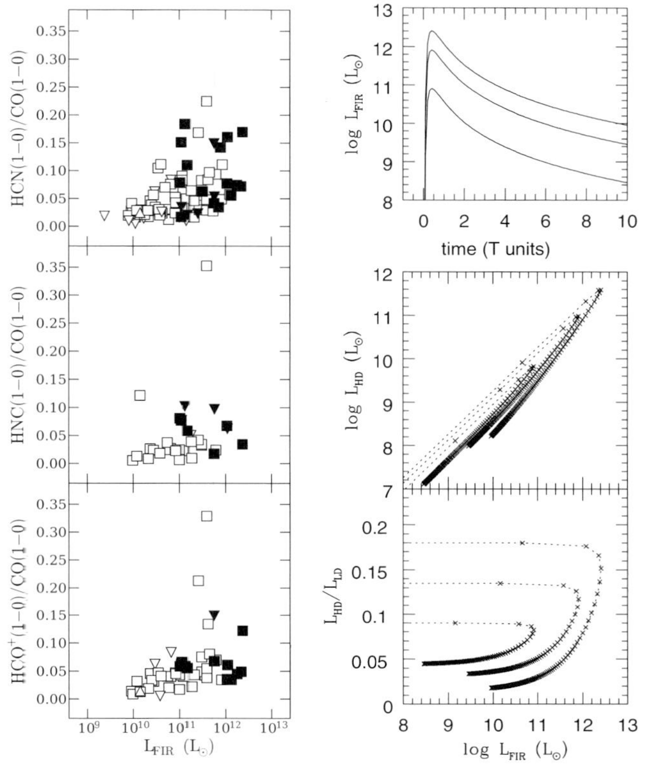

Figure 1. *left, from top to bottom:* Integrated line ratios HCN(1-0)/CO(1-0), HNC(1-0)/CO(1-0), and HCO^+(1-0)/CO(1-0) versus FIR luminosity. Squares represent reliable values and triangles upper or lower limits. Filled symbols are sources with OH-MM activity.
right, from top to bottom: Three FIR luminosity curves used in the simulations with different maximum luminosities, the luminosity of a high-density component, and the variation of the high- versus low-density ratio during the outburst starting at the upper part of the curve.

(all $n_{\rm cr}>10^5$ cm^{-3}) trace the high-density component (HD) which mostly resides in the cores of the galaxies, at the sites of the star-formation activity.

On the left hand side of Fig. 1, the relative contributions of these two density components are shown, by plotting the ratios of the high-density and CO(1-0) line strengths. The figure shows that the distribution of line ratios for all molecules increases with FIR luminosity, which gives an upwardly curved lower boundary for the distribution at higher $L_{\rm FIR}$. The highest values are found at $L_{\rm FIR}\geqslant 10^{10.5}$ L$_\odot$. The figure also shows that in general the OH-MM sources have a much larger spread in HD/CO(1-0).

This behavior can be explained as the result of ongoing star formation. The FIR luminosity of the ULIRG during the evolution of the outburst reflects energy generated by the star formation activity. The FIR luminosity integrated over the course of the outburst would reflect the amount of high-density molecular material consumed by the star formation process and destroyed or removed by feedback.

In the following simplified scenario (see Baan *et al.* 2007), we employ a model for the time-evolution of the high-density components in a galaxy during a FIR outburst. In the absence of a representative FIR light curve, we use a diffusion-like expression in time t as a response to a starburst starting at t=0 defined as:

$$L_{\rm FIR}(t) = 1.35 L_{\rm FIR}(0) \left(\frac{t}{T}\right)^{2.5} e^{-t/T}, \tag{2.1}$$

where $L_{FIR}(0)$ is the maximum luminosity of the burst and T is the timescale of the outburst. We note that a diffusion curve may not be the most appropriate representation of $L_{\rm FIR}$, but this curve does resemble the outcome of starburst-driven FIR evolution simulations (Loenen, Baan & Spaans 2006).

The high-density component HD can be defined as β LD, the low-density component. As a result, the high-low-density ratio varies with time during the FIR outburst as:

$$\frac{\rm HD(t)}{\rm LD} = \beta \left[1 - \gamma \frac{\int L_{\rm FIR}(t){\rm d}t}{L_{\rm FIR,int}}\right], \tag{2.2}$$

where γ is the fraction of the initial HD component that is consumed during the whole outburst, and $L_{\rm FIR,int}$ the FIR luminosity integrated over the whole course of the outburst. The large-scale low-density component LD is assumed to remain unchanged.

The results of these simulations have been presented on the right hand side in Fig. 1. The top panel shows the FIR light curve of the outburst for three peak luminosities. The middle panel shows the luminosity of a representative high-density component for the three $L_{\rm FIR}$ curves. The bottom panel shows the high-low-density ratio for these same FIR light curves. Combining the results of this simulation with the data shows that the HD/CO(1-0) data points in the left panels are a measure of the evolution of the starburst. This implies that the OH MM sources are galaxies in an early stage of star formation, which is consistent with OH MM sources being found in starburst-dominated galaxies (Genzel *et al.* 1998; Baan & Klöckner 2007).

3. Chemistry

The model presented in the previous section does well in explaining the evolution of the different gas components, but it makes no distinction between the different high-density tracers. Even though the emission of the different molecules originates in the same regions, the line strengths are influenced by the environmental properties like the (column) density, temperature, and the type and strength of the prevailing radiation field. In order to study the effects of these parameters on the emission characteristics of the sources, we remove the intrinsic difference in line strength between the galaxies in our sample and use line ratios to find diagnostic properties. Fig. 2 presents the ratios of the integrated lines of HCO^+/HCN, HNC/HCO^+ and HNC/HCN against each other.

In order to interpret the behavior of the sources in this diagram, we compare the data to the theoretical models that treat the chemistry and radiative transfer of molecular clouds, including all the relevant heating, cooling and chemical processes (see Meijerink & Spaans 2005; Meijerink *et al.* 2007, and Spaans *et al.*in these proceedings). A large grid of models was created by Meijerink *et al.* (2007), varying the strength of the radiation field, its type (UV and X-ray), the gas density and the column density. These results are compared to our observational data in Fig. 2 (note: not all models are shown, some fall out of the range of our figure).

3.1. *PDR models*

The results of the PDR models (UV radiation field) are shown in Fig. 2 with heavy lines, where the line styles indicate different gas densities (solid: n=$10^{4.5}$, dashed: n=$10^{5.0}$, and striped: n=$10^{5.5}$ cm^{-3}). The tracks vary as a function of column density, which ranges from N=10^{22} cm^{-2} (the column density below which the strength of the emission lines decreases rapidly) to N=10^{23}, N=$10^{23.5}$ and N=10^{24} cm^{-2} for the n=$10^{4.5}$, n=$10^{5.0}$, and n=$10^{5.5}$ cm^{-3} models, respectively (corresponding to a cloud size of 1 pc; indicated by the symbol at one end of the tracks). Two different radiation field strengths are shown: $F_{\rm UV}$=1.6 erg cm^{-2} s^{-1}, indicated by a plus symbol at the highest column density point, and $F_{\rm UV}$=160 erg cm^{-2} s^{-1}, indicated with a circle.

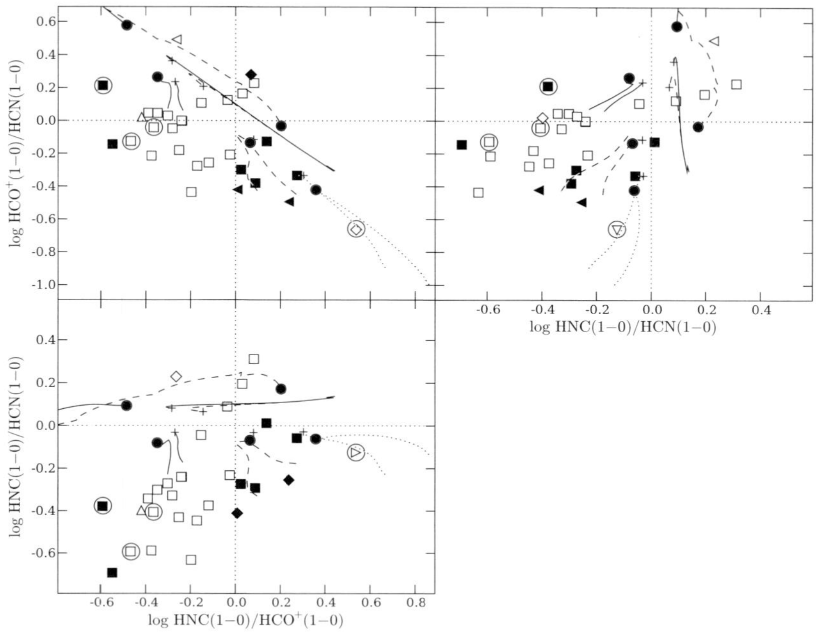

Figure 2. *top left* Integrated HCO^+(1-0)/HCN(1-0) versus HNC(1-0)/HCO^+(1-0) ratios. *top right* Integrated HCO^+(1-0)/HCN(1-0) versus HNC(1-0)/HCN(1-0) ratios. *bottom left* Integrated HNC(1-0)/HCN(1-0) versus HNC(1-0)/HCO^+(1-0) ratios. Explanation about the plot symbols and line styles is provided in Sect. 3.

Two observations can be made, when comparing the data and the models. First of all one can see that the models are separated based on the density (n) in the HCO^+/HCN and HCO^+/HNC line ratios. The HNC/HCN line ratio shows no differentiation. This can be explained in terms of the critical density of the individual molecules. HCO^+ has a critical density of about $3{\times}10^5$ cm^{-3}, whereas the critical density of HCN and HNC is around $3{\times}10^6$ cm^{-3}. Therefore the excitation of HCO^+ will differ from HCN and HNC for different densities. A second observation that can be made is that most of the OH MM sources cluster together in an area traced by PDR models that have a high-density ($n{\geqslant}10^{5.5}$ cm^{-3}), and a high column density $N{\geqslant}10^{22}$ cm^{-2}. This points to the classical OH MM model of IR (UV radiation reprocessed by the surrounding gas and dust) pumped, low (and variable) gain amplification (Baan 1989).

3.2. *XDR models*

The results of two XDR simulations are also shown in Fig. 2, using thin lines. Again, the line styles indicate different gas densities (solid: $n{=}10^{5.5}$, and dashed: $n{=}10^{6.0}$ cm^{-3}); column densities range from $N{=}10^{22}$ to $N{=}10^{24}$-$10^{24.5}$; and the radiation field strengths are $F_X{=}1.6$ (plus), and $F_X{=}160$ erg cm^{-2} s^{-1} (circle).

The XDR models are not as well differentiated as the PDR models. Because X-ray photons penetrate the molecular cloud much easier, the XDRs do not show the strong density dependency seen for the PDRs, making the distinction between different XDR models very difficult. The addition of higher transitions and other molecules (e.g. CN, CO^+, HOC^+) will most likely break this degeneracy. Another problem with trying to identify XDR sources is that they are in general smaller that PDR sources and thus are more affected by beam dilution, especially in single dish observations like ours. This will affect the detection rate of XDR sources and could make hybrid sources look like PDR sources, even if the XDRs are energetically more important than the PDRs.

Despite these drawbacks, the XDR models do provide diagnostics, since they clearly separate from the PDR models in their HNC/HCN line ratio. The PDR models approach, but never cross the HNC/HCN=1 line and the XDR models all have HNC/HCN>1. Therefore, the HNC/HCN line ratio is an excellent way to determine whether a system's chemistry is dominated by UV or X-ray photons.

3.3. *Terra Incognita*

Not all our observational data in Fig. 2 is covered by the models. The few sources in our sample with known H_2O MM activity (indicated by plot symbols surrounded by circles) are also located in this area, which is characterized by lower HNC and higher HCO^+ line strengths compared to HCN. The fact that this area is not covered by the models suggests that other processes influence the line ratios, such as strong shocks. Shocks are not treated in the models and can have profound effects on the chemistry in molecular clouds as they may selectively destroy HNC (Schilke *et al.* 1992) and enhance HCO^+ relative to HCN (Dickinson *et al.* 1980). A simultaneous increase in HCO^+ and a decrease in HNC shifts the PDR models to the uncovered region, implying that the H_2O MM sources in our sample are UV driven systems with strong shocks. This would suggest that these H_2O MM sources are similar to shock-induced Galactic H_2O maser spots (see other contributions in these proceedings).

4. Conclusions

Molecular line emissions of multiple species (and transitions) provide excellent diagnostics for understanding the status of the nuclear gas in extra-galactic sources. The different molecules trace different (density) components and the ratio of high and low-density tracer lines follows the star formation activity in the system. Comparing different high-density tracers tells a lot about physical characteristics of the gas. The HCO^+/HCN and HCO^+/HNC line ratios are good proxies for the density of the gas, due to the different critical densities of the species. PDR and XDR sources can be distinguished using the HNC/HCN line ratio: PDR sources all have ratios lower than unity and XDRs have ratios larger than 1. OH MM sources cluster in a particular part of the diagnostic diagram, which is only traced by PDR models with densities higher than $10^{5.5}$ cm^{-3}, confirming the classical OH MM model of FIR (UV radiation reprocessed by the surrounding gas and dust) pumped amplification with low but variable gains.

References

Baan, W. A. 1989, *ApJ* 338, 804
Baan, W. A., Henkel, C., Loenen, A. F., Baudry, A., & Wiklind, T. 2007, submitted to *A&A*
Baan, W. A., & Klöckner, H.-R. 2006, *A&A*, 449, 559
Dickinson, D. F., Dinger, A. S. C., Kuiper, T. B. H., & Rodriguez Kuiper, E. N. 1980, *ApJ*, 237, L43
Genzel, R., *et al.* 1998, *ApJ* 4498, 579
Loenen, A. F., Baan, W., A., & Spaans, M. 2006, *A&A* 458, 89
Loenen, A. F., Baan, W., A., & Spaans, M. 2007, in preparation
Meijerink, R., & Spaans, M. 2005, *A&A* 436, 397
Meijerink, R., Spaans, M., & Israel, F. P. 2007, *A&A* 461, 793
Schilke, P., Walmsley, C. M., Pineau Des Forets, G. *et al.* 1992, *A&A* 256, 595

Astrophysical Masers and their Environments
Proceedings IAU Symposium No. 242, 2007
J.M. Chapman & W.A. Baan, eds.

doi:10.1017/S1743921307013610

Zeeman splitting in OH megamasers

Timothy Robishaw, Carl Heiles, and Eliot Quataert
Department of Astronomy, University of California, Berkeley, CA 94720, USA

Abstract. We detected significant Zeeman splitting in the 1667 MHz OH megamaser emission from four ultraluminous galaxies. These detections, in addition to being the first extragalactic detection of the Zeeman effect in an emission line, suggest that OH megamasers are excellent extragalactic magnetometers.

Keywords. Masers, galaxies: magnetic fields, galaxies: starburst, ISM: molecules

1. Introduction

Ultraluminous infrared galaxies (ULIRGs) are a population of galaxies that emit far-infrared (FIR) radiation with energies comparable to those of the most luminous quasars ($\log(L_{\rm FIR}/L_\odot) > 12$). Nearly every ULIRG appears to have undergone a merger/interaction and contains massive star formation and/or an active galactic nucleus (AGN) induced by gravitational interactions. Diamond *et al.* (1999) conducted very long baseline interferometry (VLBI) observations of the 1667 MHz hydroxyl (OH) transition in the nuclear regions in ULIRGs and found multiple masering regions with $1 < \log(L_{\rm OH}/L_\odot) < 4$; these regions are known as OH megamasers (OH MMs). Each OHM has a spectral linewidth of between 50 and 150 km s^{-1}; when viewed by a single dish, these spectral components are superimposed. The 1667 MHz OHM flux is always a few to many times the flux of the 1665 MHz transition and in many cases the 1665 MHz line is absent (Darling & Giovanelli 2002); this is an interesting contrast to the case of OH masers in the Galaxy in which the 1665 MHz is usually the dominant transition (Reid & Moran 1988). The starbursts and AGNs in ULIRGs create strong FIR dust emission as well as a strong radio continuum; the OHMMs are generally believed to be pumped by the FIR radiation field (e.g., Randell *et al.* 1995) although collisional excitation may be important as well (e.g., Lonsdale *et al.* 1998). Given the conditions that exist in ULIRGs and considering that every OH maser in our Galaxy is associated with massive star-forming regions, it is not surprising that the entire OHM sample finds homes in LIRGs, strongly favoring the most FIR-luminous, the ULIRGs (Darling & Giovanelli 2002).

ULIRGs are natural locations to expect very strong magnetic fields given their high gas and energy densities. Much of the radio emission in ULIRGs is resolved on scales $\sim$ 100 pc with VLA observations (Condon *et al.* 1991). In Arp 220, high-resolution observations (Rovilos *et al.* 2003) show that the OH MMs arise in this region as well. With this size scale and the observed radio fluxes, minimum energy arguments suggest *volume averaged* field strengths of $\approx$ 1 mG (e.g., Condon *et al.* 1991, Thompson *et al.* 2006), significantly larger than the $\sim 10\mu$G fields measured in normal spirals. The field strengths in ULIRGs cannot be much below a mG or else inverse Compton cooling would dominate over synchrotron cooling, making it energetically difficult to explain the radio fluxes from ULIRGs and the fact that ULIRGs lie on the FIR-radio correlation. The field strengths could, however, in principle be larger than the minimum energy estimate if, as in our Galaxy, the magnetic energy density is in approximate equipartition with the total pressure (Thompson *et al.* 2006). This can be estimated from the observed surface density; CO observations of Arp 220 and several other systems reveal $\sim 10^9 M_\odot$ of molecular gas in the central ~ 100 pc (e.g., Downes & Solomon 1998) implying gas surface densities $\Sigma \sim 1 - 10\,\mathrm{g\,cm^{-2}}$, $10^3 - 10^4$ times larger than in the Milky Way. The equipartition field scales as $B \propto \Sigma$ implying that the mean field in ULIRGs could approach $\sim 10\,\mathrm{mG}$.

2. Observations

Motivated by the above considerations, we have used the 300 m Arecibo† telescope in full-Stokes mode to try to detect Zeeman splitting in megamaser emission of the 1667 MHz OH transition. The six strongest OH MMs visible from Arecibo were observed by spending equal time at on-source and off-source positions. As shown by all VLBA observations of OH MMs, the spatial extent of the OHM emission in any ULIRG is much smaller than Arecibo's beam; there are no instrumental polarization effects due to beam structure or sidelobes. The observations were made in full-Stokes mode and the Mueller matrix was calibrated using the standard Arecibo technique of observing the linearly-polarized continuum sources 3C138 and 3C286 (Heiles *et al.* 2001, Heiles & Troland 2004).

3. Results

Four of the six sources we observed exhibit significant Stokes V emission that can be interpreted as Zeeman splitting of the 1667 MHz line. We present the results for IRAS F12032+1707, one of only 3 known gigamasers, in Figure 1. The top panel shows the Stokes I spectrum‡ of the source as a function of frequency and optical heliocentric velocity (this source is at $z = 0.217$). It is clear that the OHM emission profile is a composite of many narrower components as the emission spans almost 2000 km s^{-1}. The 1665 and 1667 MHz lines are blended in this source. We decomposed the Stokes I profile into 13 Gaussian components; we attempted to use the fewest number of components to obtain reasonable residuals (which are plotted with an enhanced scale through the middle height of the Stokes I profile) while allowing enough components to reproduce multiple splittings in the Stokes V spectrum. The bottom panel shows the Stokes V spectrum and it is clear that the detectable circular polarization is restricted to the stronger of the two bright and narrow Stokes I peaks. The fitted field for this feature is $17.9 \pm 0.8\,\mathrm{mG}$.

A total of 15 independent field detections were made in multiple components within four extragalactic sources including Arp 220 and III Zw 35. The median absolute field strength is $\sim 3\,\mathrm{mG}$ and the direction of the field reverses within 3 of the galaxies. It is interesting to note that the standard deviation of the magnetic field distribution of the Galactic OH masers is also $\sim 3\,\mathrm{mG}$ (Fish *et al.* 2003).

4. OH megamasers: the new extragalactic magnetometers

The Zeeman detection rate for our sample of ULIRGs suggests that OH MMs are excellent extragalactic magnetometers. Since it is not possible to unambiguously associate

† The Arecibo Observatory is part of the National Astronomy and Ionosphere Center, which is operated by Cornell University under a cooperative agreement with the National Science Foundation.

‡ We use the classical definition of Stokes I, which is the sum (not the average) of two orthogonal polarizations. Thus, fluxes in Stokes I are twice the usual flux density given in catalogs.

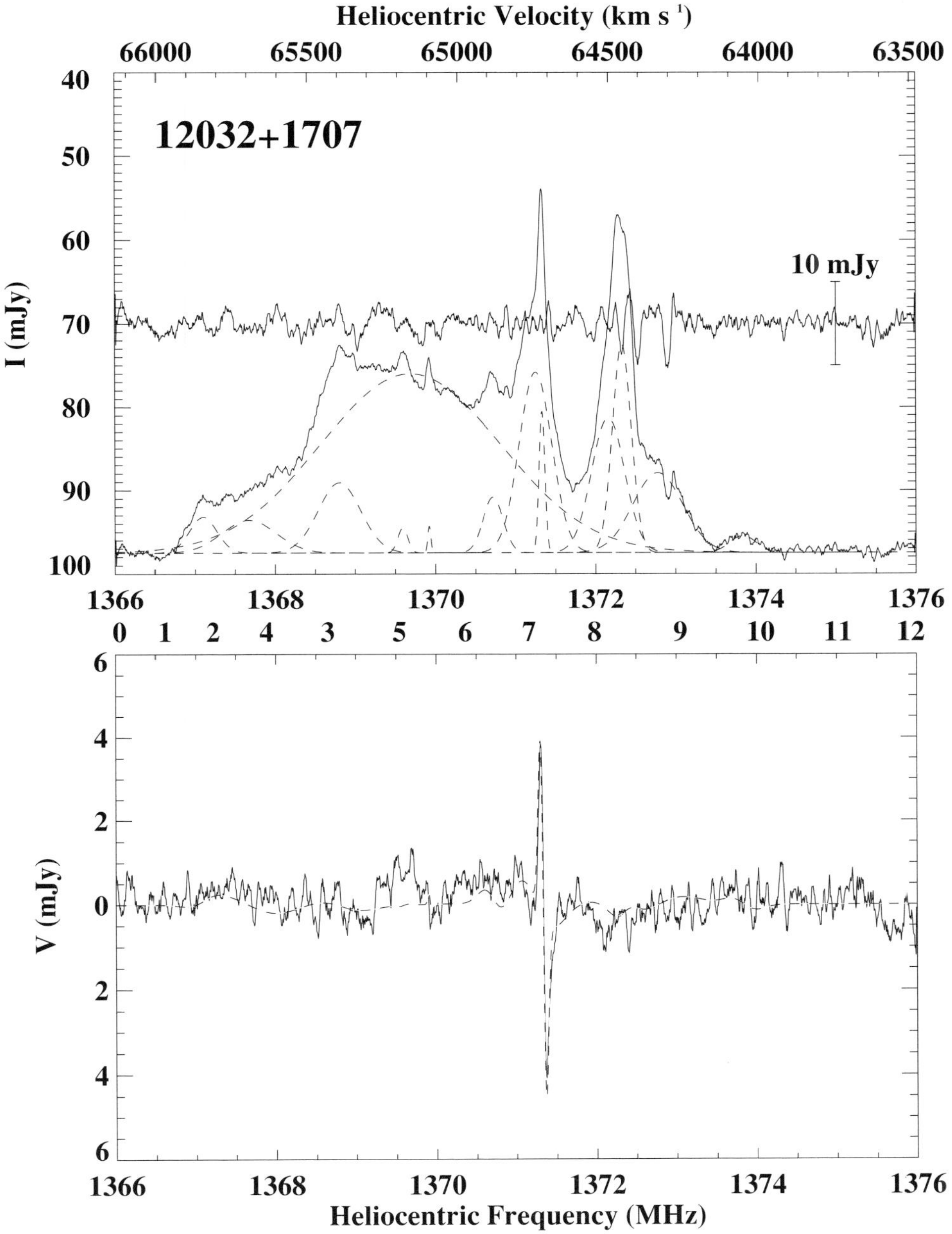

Figure 1. Stokes I spectrum (*top*) for IRAS F12032 + 1707. The residuals are plotted through the center of the Stokes I plot and are expanded by a factor of 2. The profile of each Gaussian component is plotted as a dashed line. The Stokes V (*bottom*) spectrum with its fit is plotted as a dashed line.

the VLBA spectra of the OH MM spots within any ULIRG with the Gaussian components in our single-dish spectra, we shall be observing Arp 220 and III Zw 35 using the High-Sensitivity Array configuration of the VLBA in order to attempt to detect Zeeman splitting directly in the OH MM spots. We shall also conduct a full-Stokes flux-limited Arecibo survey of OH MMs including the OH satellite lines.

Acknowledgements

This research was supported in part by NSF grant AST-0406987. Support for this work was also provided to TR through an IAU grant and by the NSF through award GSSP 06-0003 from the NRAO. TR would also like thank Vicki Drazenovic for her help.

References

Condon, J. J., Huang, Z.-P., Yin, Q. F., & Thuan, T. X. 1991, ApJ, 378, 65
Darling, J., & Giovanelli, R. 2002, AJ, 124, 100
Diamond, P. J., Lonsdale, C. J., Lonsdale, C. J., & Smith, H. E. 1999, ApJ, 511, 178
Downes, D., & Solomon, P. M. 1998, ApJ, 507, 615
Fish, V. L., Reid, M. J., Argon, A. L., & Menten, K. M. 2003, ApJ, 596, 328
Heiles, C., *et al.* 2001, PASP, 113, 1274
Heiles, C., & Troland, T. H. 2004, ApJS, 151, 271
Lonsdale, C. J., Lonsdale, C. J., Diamond, P. J., & Smith, H. E. 1998, ApJL, 493, L13+
Randell, J., Field, D., Jones, K. N., Yates, J. A., & Gray, M. D. 1995, A&A, 300, 659
Reid, M. J., & Moran, J. M. 1988, Astronomical masers (Galactic and Extragalactic Radio Astronomy), 255–294
Rovilos, E., Diamond, P. J., Lonsdale, C. J., Lonsdale, C. J., & Smith, H. E. 2003, MNRAS, 342, 373
Thompson, T. A., Quataert, E., Waxman, E., Murray, N., & Martin, C. L. 2006, ApJ, 645, 186

Astrophysical Masers and their Environments
Proceedings IAU Symposium No. 242, 2007
J.M. Chapman & W.A. Baan, eds.

doi:10.1017/S1743921307013622

Submillimeter and millimeter masers

E. M. L. Humphreys
Harvard-Smithsonian CfA, 60 Garden Street, Cambridge, MA 02138, USA

Abstract. Despite theoretical predictions of the existence of many submillimeter masers, and some pioneering observational discoveries over the past few decades, these lines have remained relatively unstudied due to (i) challenges associated with observing at shorter wavelength; and, (ii) the lack of possibility of high ($< 14''$ at 345 GHz) angular resolution observations. With the advent of the SMA, the first submillimeter imaging array capable of sub-arcsecond resolution, APEX, and the promise of ALMA, opportunities are opening for performing new science with millimeter/submillmeter masers. In this talk, I will review recent work in the field - including extragalactic H_2O millimeter masers, hydrogen recombination masers, submillimeter masers in star-forming regions, and in the envelopes of evolved stars - and discuss prospects for the future.

Keywords. masers, submillimeter

1. Introduction

Submillimeter masers exist in a wide range of astronomical environments, and provide the possibility to probe physical conditions, source dynamics and magnetic fields on small angular scales. They occur in several molecular and atomic species, including H_2O, SiO, H (recombination), CH_3OH, HCN, and SiS, and can be very strong (e.g., 8000 Jy for the 325 GHz H_2O masers in W49N; Menten *et al.* 1990a). However, lack of angular resolution at submillimeter wavelengths has, until recently, been a serious obstacle to realizing the potential of the masers. Relating cm-wave maser emission observed on, say $0\farcs001$ scales, with that of the submillimeter maser emission on $>10''$ scales (at 345 GHz), has made it difficult to constrain and test the radiative transfer models that we will need to use in the Atacama Large Millimeter Array (ALMA) era to map out precise source temperature and density distributions.

The Submillimeter Array (SMA) on Mauna Kea, operating from 0.3 to 2 mm, is the first instrument capable of imaging in the submillimeter on sub-arcsecond scales ($0\farcs25$ at 345 GHz), and ALMA will further transform maser science opportunities (see review by Wootten in these proceedings). In this review, I will discuss results for masers at wavelengths shorter than 1.6 mm ($\nu > 180$ GHz), and future prospects for their observation using e.g., ALMA, the *Herschel* satellite, the Stratospheric Observatory for Infra-Red Astronomy (SOFIA) and submillimeter Very Long Baseline Interferometry (VLBI).

2. (Sub)millimeter H_2O masers

The H_2O masers detected to date, from rotational transitions within the vibrational ground state and within the ν_2=1 bending mode, are listed in Table 1 and are marked on the energy level diagram in Figure 1. The most studied lines are those at 183, 321 and 325 GHz, despite the relatively low atmospheric transmission at 183 and 325 GHz due to their low energies above ground state (Figure 2). These masers are believed to be collisionally-pumped by a subset of the conditions that pump 22 GHz masers, for the

Table 1. H_2O masers

Freq. (GHz)	Transition J_{k_a,k_c} - J_{k_a,k_c}	Vib. State	Species[1]	E_u/k (K)	CSE[2]	SFR[2]	EXG[2]	Primary Reference
22.235	6_{16} - 5_{23}	G	O	644	Y	Y	Y	Cheung *et al.* (1969)
96.261	4_{40} - 5_{33}	ν_2=1	P	3065	Y			Menten & Melnick (1989)
183.308	3_{13} - 2_{20}	G	P	205	Y	Y	Y	Waters *et al.* (1980)
232.687	5_{50} - 6_{43}	ν_2=1	O	3463	Y			Menten & Melnick (1989)
293.439	6_{61} - 7_{52}	ν_2=1	O	3935	Y			Menten *et al.* (2006)
321.226	10_{29} - 9_{36}	G	O	1862	Y	Y		Menten *et al.* (1990a)
325.153	5_{15} - 4_{22}	G	P	470	Y	Y		Menten *et al.* (1990b
[3]336.228	5_{23} - 6_{16}	ν_2=1	O	2956	Y			Feldman *et al.* (1993)
354.885	17_{412} - 16_{710}	G	O	5782	Y			Feldman *et al.* (1991)
380.194	4_{14} - 3_{21}	G	O	324		Y		Phillips *et al.* (1980)
437.347	7_{53} - 6_{60}	G	P	1525	Y			Melnick *et al.* (1993)
439.151	6_{43} - 5_{50}	G	O	1089	Y	Y		Melnick *et al.* (1993)
470.889	6_{42} - 5_{51}	G	P	1091	Y	Y		Melnick *et al.* (1993)
658.007	1_{10} - 1_{01}	ν_2=1	O	2361	Y			Menten & Young (1995)

[1]O=ortho-H_2O (parallel hydrogen atom nuclear spins) and P=para-H_2O (anti-parallel hydrogen nuclear spins). In thermal equilibrium, the two forms are present in an O/P ratio of 3:1.
[2]CSE=Circumstellar Envelope; SFR=Star Forming Region; EXG=Extragalactic
[3]Quasi-maser (Feldman *et al.* 1993), or thermal (Menten *et al.* 2006), emission toward VY CMa.

parameter space investigated by Neufeld & Melnick (1991, hereafter NM91) and by Yates, Field & Gray (1997, hereafter YFG97). However, Cernicharo *et al.* (1994, 1999, 2006a, 2006b) find that the 183 and 325 GHz transitions can also be inverted in significantly lower temperature and density regimes of $T_k \sim 40$ K and n(H_2)=10^5-10^6 cm^{-3}. NM91 and YFG97 are in broad agreement, except that YFG97 find that the masers at 439 and 470 GHz are pumped by radiation from warm dust. Both NM91 and YFG97 make predictions for new H_2O masers (e.g., at 448, 1296, & 1322 GHz; also Deguchi 1977, Cooke & Elitzur 1985, Deguchi & Nguyen-Q-Rieu 1990, Humphreys *et al.* 2001), some of which could be observed using *Herschel*. NM91 and YFG97 do not include levels from the ν_2=1 vibrationally-excited state, see Deguchi (1977), Deguchi & Nguyen-Q-Rieu (1990) and Alcolea & Menten (1993) for ν_2=1 maser pumping models. Modelling of the ν_2=1 masers is severely hampered by lack of relevant collisional excitation rates.

2.1. *H_2O masers in evolved stars*

(Sub)millimeter masers at 183, 321 and 325 GHz are common in the circumstellar envelopes (CSEs) of evolved stars. 70% of the 22 GHz H_2O maser sources observed by Yates, Cohen & Hills (1995) also have H_2O maser emission at 321 and 325 GHz. Single-dish linewidths of 22 and 325 GHz masers have similar extents and peak flux densities, whereas 321 GHz maser line widths are narrower and weaker by a factor of a few (an exception is emission from R Aqr, a Mira variable in a symbiotic binary; Ivison *et al.* 1998). 321 GHz emission likely originates from a subset of the conditions that give rise to the 22 and 325 GHz emission, close to the central star. The 321 GHz line is generally more variable than the 22 and 325 GHz emission and variations in the 22, 321 and 325 GHz masers are not particularly well-correlated (in some cases they are completely anti-correlated; Yates *et al.* 1996). For 183 GHz H_2O masers, González-Alfonso *et al.* (1998) find that variability of the line profile and flux from one epoch to another is small in comparison with that of 22 GHz masers in a study of 23 evolved stars. As for 22 GHz H_2O masers, in stars of low mass-loss rates ($\dot{M}$) the 183 GHz emission peaks at a velocity similar to that of the star, whilst in stars with high $\dot{M}$ the emission peaks at velocities

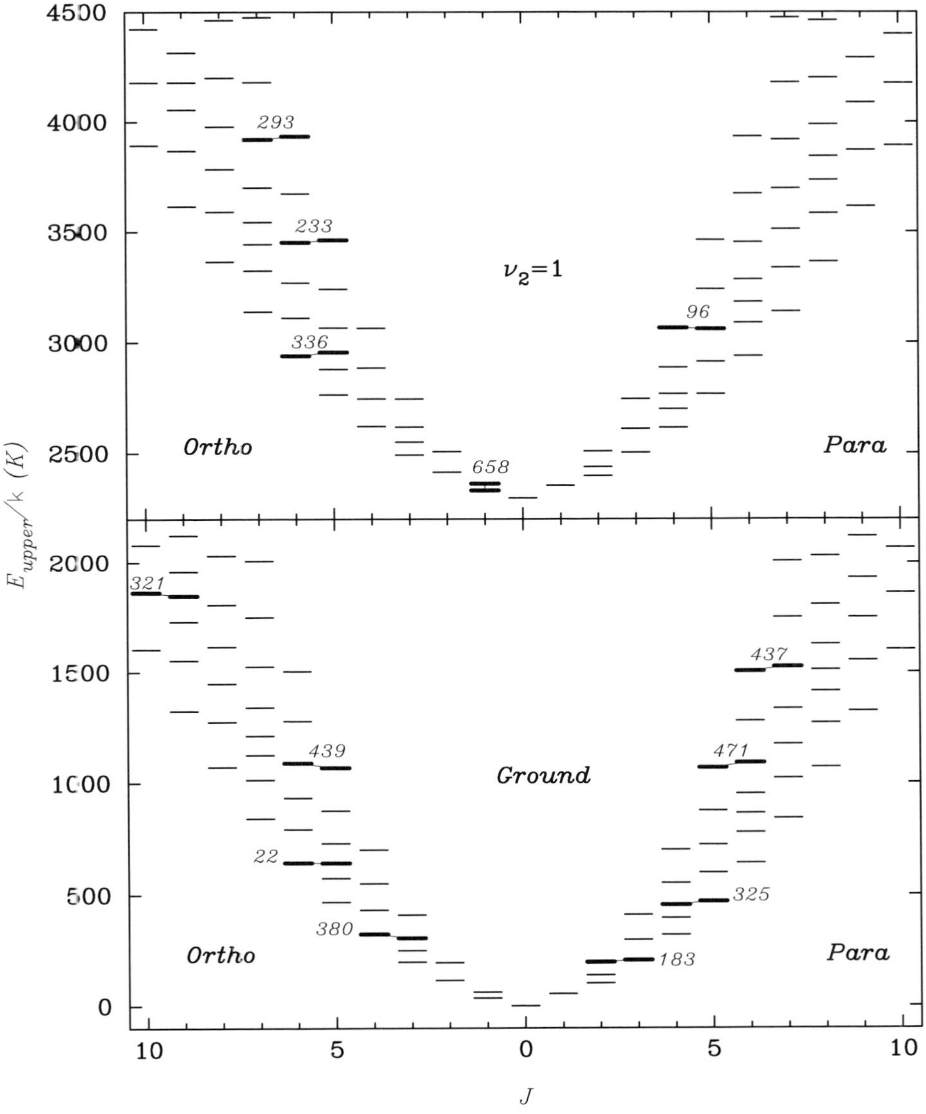

Figure 1. H_2O Energy Level Diagram. Rotational levels in the ground and vibrationally-excited ν_2 states are shown for energies between 0 to 2200 K, and 2200 to 4500 K respectively. Levels of maser transitions are plotted in bold, and labels are the transition frequencies in gigahertz. The maser at 355 GHz, at an energy of 5782 K above ground-state, is not shown on this plot. Ortho-H_2O plotted for values of the total molecular angular momentum J increasing to the left, para-H_2O to the right. Data are from the experimentally-derived energy levels of Tennyson *et al.* (2001), available on http://www.tampa.phys.ucl.ac.uk/ftp/astrodata/water/levels.

closer to the terminal velocity of the envelope (tangential vs. radial amplification as the envelope becomes denser at greater radii). Masers at 437, 439 and 471 GHz have all been detected in CSEs and the 437 GHz line has been found exclusively in this environment (Melnick *et al.* 1993; YFG97). Masers from the ν_2=1 state, at 96, 233, 293, (336) &

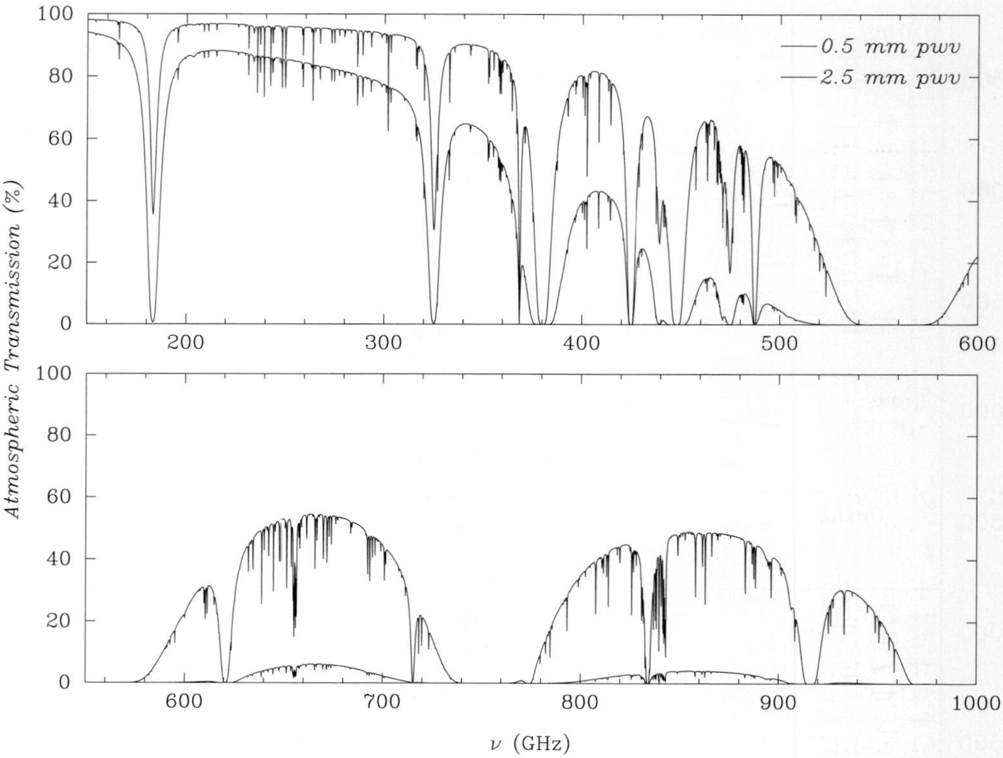

Figure 2. Zenith atmospheric transmission at Mauna Kea for column densities of 0.5 and 2.5 mm H_2O (τ_{225GHz}~0.04 and 0.13 respectively). Data are from the Caltech Submillimeter Observatory Atmospheric Transmission Interactive Plotter (http://www.submm.caltech.edu/cso/weather/atplot.shtml).

658 GHz, are only known to occur strongly from CSEs, although 658 GHz emission of undetermined nature is observed toward Orion-KL (Schilke *et al.* 2001). On the basis of excitation arguments, and similarity with SiO maser lineshapes in some cases, the ν_2=1 masers likely occur close (within a few R_*) to the central star. In recent Atacama Pathfinder Experiment (APEX) observations towards VY CMa, Menten *et al.* (2006) find weak maser emission from the 293 GHz line, a non-detection of emission at 297 GHz and thermal emission at 336 GHz. For a discussion of SMA observations of the 658 GHz masers, often particularly strong e.g., 3000 Jy toward VY CMa, see the review by Hunter in these proceedings.

2.2. *H_2O masers in star-forming regions*

Observations of the (sub)millimeter H_2O masers are summarized in Table 2. 183, 321, 325, 439 and 471 GHz masers have been observed towards high-mass star-forming regions, the 183 and 325 GHz lines have also been observed towards low-mass star-forming regions (HH7-11A, L1448IRS3, L1448-mm at 183 GHz; IRAS16293-2422 at 325 GHz). The velocity range covered by the 321 GHz maser is typically smaller than that observed at 22, 183, and 325 GHz. The 321 GHz emission is typically weakest of these four lines, and the 22 GHz is the strongest.

First arcsecond resolution observations of H_2O masers towards a star-forming region were performed at 325 GHz towards Orion-KL by Greenhill *et al.* (2007) using the compact configuration of the SMA, and followed up with a higher resolution (0.″65 circular),

Table 2. Some (Sub)millimeter H_2O observations towards star-forming regions

Freq. (GHz)	Sources	Telescope[1] (Beam)	Comments
183	Orion-KL	KAO (7′.5)	First 183 GHz detection (Waters *et al.* 1980)
	Orion-KL, Cep A, W49N, S252A, S158, HH7-11A, W3(H2O) NGC 7538S, RN013	IRAM 30-m (14″)	Established 183 GHz maser emission widespread (Cernicharo *et al.* 1990)
	Orion	IRAM 30-m	Spatially-extended emission; strong, narrow features at IRC2 (Cernicharo *et al.* 1994)
	W49N	IRAM 30-m	Spatially-extended; less time-variable than at 22 & 325 GHz (González-Alfonso *et al.* 1995)
	HH7-11A, L1448IRS3, L1448-mm	IRAM 30-m	183 GHz maser variability in low-mass star formation Cernicharo *et al.* (1996)
	Sgr B2	IRAM 30-m	Strong toward cores; moderate emission at Sgr B2 main condensations (Cernicharo *et al.* 2006a)
321	W3(OH), W49N, W51 IRS2 & Main	CSO (23″)	Strongest 22 GHz & 321 GHz features generally at similar velocities (Menten *et al.* 1990a)
	Cep A	SMA (0″.75)	22 & 321 GHz distributions perpendicular (cm & submm obs. ~1 mth apart) (Patel *et al.* 2007)
325	Orion-KL W49N, W51 Main IRAS 16293-2422	CSO (22″)	22 & 325 GHz cover similar velocity extents (Menten *et al.* 1990b)
	G34.3-0.2, W49N, Sgr B2	CSO	325, 439 & 470 GHz cover similar velocity extents (Melnick *et al.* 1993)
	Orion-KL	CSO	325 GHz emission much less extended than at 183 GHz (Cernicharo *et al.* 1999)
	Orion-KL	SMA (0″.65) Full Stokes	In high-mass protostar Source I outflow, 325 GHz emission more collimated than 22 GHz (cm & submm obs. ~5 yrs apart) (Greenhill *et al.* 2007)
439, 471	G34.3-0.2 W49N, Sgr B2	CSO (16″)	First detections: 325, 439 & 470 GHz cover similar velocity extents (Melnick *et al.* 1993)

[1]KAO = Kuiper Airborne Observatory; IRAM = Institut de Radioastronomie Millimétrique; CSO = Caltech Submillimeter Observatory; SMA = Submillimeter Array

full polarization epoch. In previous mapping of this region using the Caltech Submillimeter Observatory (CSO) with a 22″ beam, Cernicharo *et al.* (1999) concluded that the 325 GHz emission traces extended, low-density material of $n(H_2)\sim10^{5-6}$ cm^{-3}. However, Greenhill *et al.* (2007) find that it also arises from compact high-density clumps, much as the 22 GHz transition, although in the outflow of high-mass protostar Source I the 325 GHz emission appears more collimated. Line ratios of these H_2O transitions could therefore be valuable diagnostics for shocked material in protostellar outflows.

Using the SMA, Patel *et al.* (2007) imaged 321 GHz H_2O maser emission towards high-mass star-forming region Cepheus A with a resolution of 0″.75, in close time proximity to Very Large Array observations of the 22 GHz H_2O masers (43 days later). The majority of 321 GHz maser spots did not appear to be associated with those at 22 GHz, and the position angles of the roughly linear structures traced by the masers appeared perpendicular, perhaps tracing a jet and disk respectively. Patel *et al.* (2007) interpret the submillimeter masers in Cepheus A to be tracing significantly hotter regions (600-2000 K) than the centimeter masers, see the contribution by Patel in these proceedings for further details.

2.3. *Extragalactic H_2O masers*

There have been two recent detections of extragalactic H_2O masers at 183 GHz. Humphreys *et al.* (2005) detected emission toward the well-known 22 GHz H_2O megamaser galaxy NGC 3079 using the SMA. At a distance of 14 Mpc, NGC 3079 harbors an active galactic nucleus (AGN), and additionally has some starburst indicators. Spatially and kinematically the 183 GHz emission is associated with the AGN, with emission peaking at the same position as that of 22 GHz emission imaged by Kondratko *et al.* (2005) using VLBI. At 22 GHz, the emission has a time-variable peak flux density in the range 3-12 Jy, whereas at 183 GHz, the H_2O maser emission had a peak flux density of $\sim$0.5 Jy. Humphreys *et al.* (2005) also make a tentative detection of the 439 GHz maser using the JCMT.

Cernicharo, Pardo & Weiss (2006) detected a megamaser at 183.310 GHz in Arp 220 using the IRAM 30 m, with a line width of $\sim$350 km s^{-1} and total luminosity of $\sim 2.5 \times 10^8$ K km s^{-1} pc^2. This is very interesting since no emission at 22 GHz has been detected from Arp 220 (an OH megamaser source). This fact puts constraints on the physical conditions of the central region of Arp 220, which are further strengthened by observations of HCN and HNC $J = 3 - 2$ and $J = 1 - 0$, suggesting densities of n(H_2)=10^5 cm^{-3}. Cernicharo, Pardo & Weiss (2006) propose a scenario with $\sim 10^6$ star-forming cores similar to those found in Sgr B2 in the central kiloparsec of Arp 220. The 183 GHz line is therefore an additional tool to explore the physical conditions in starburst and AGN sources, with the potential for high angular resolution observations using ALMA.

3. (Sub)millimeter SiO masers in evolved stars

(Sub)millimeter ^{28}SiO masers have been detected from the J =5-4 $\approx$ 215 GHz ($v = 1$ & 2, Clemens & Lane 1983; $v = 3$ tentative detection from VX Sgr, Jewell *et al.* 1987; $v = 3$ & 4 from VY CMa, Cernicharo, Bujarrabal & Santaren 1993), $J = 6 - 5 \approx 258$ GHz (v =1, Jewell *et al.* 1987; v =2, VY CMa, Cernicharo, Bujarrabal & Santaren 1993), $J = 7 - 6 \approx 301$ GHz ($v = 1$ & 2, R Aqr, Gray *et al.* 1995), $J = 8 - 7 \approx 344$ GHz ($v = 1$, VY CMa, and tentative $v = 2$, Humphreys *et al.* 1997; $v = 2$, VY CMa, Gray, Humphreys & Yates 1999). The highly-rotationally excited masers are very rare from the $v = 3$ & 4 states (Pardo *et al.* 1998 and references therein) which lie at >5400 K above ground state. They are more common in the v =1 & 2 (Jewell *et al.* 1987; Cernicharo, Bujarrabal & Santaren 1993; Humphreys *et al.* 1997; Gray, Humphreys & Yates 1999) especially ing J =5-4 emission, but weaker than their lower frequency counterparts in the same vibrational states, and more time-variable. In a survey of 34 supergiant and long-period variable stars, Gray, Humphreys & Yates (1999) found that for Mira variables, emission from the high-frequency transitions is absent or weak from optical phase range $\phi \sim 0.4 - 0.7$ of the stellar pulsation cycle.

SiO maser emission at lower frequencies is well-known to display high degrees (tens of %) of linear polarisation e.g., $v = 1$ $J = 1-0$ (43 GHz) maser components can be $\sim$100% linearly polarized (e.g., Kemball & Diamond 1997). Using a partially-completed SMA, Shinnaga *et al.* (2004) imaged the $v = 1$, $J = 5-4$ SiO maser emission of supergiant VY CMa to investigate linear polarization properties at higher frequency. The majority of components showed significant degrees of linear polarization, with one at the 60% level, that Shinnaga *et al.* attribute to a radiative pumping process.

For the less abundant isotopomers ^{29}SiO and ^{30}SiO, Cernicharo & Bujarrabal (1992) detected maser emission from the $v = 0$ $J = 5 - 4$ transition for both species, the ^{29}SiO $v = 2$, $J = 6 - 5$ line, and the ^{30}SiO $v = 1$ $J = 6 - 5$ towards VY CMa. For ^{29}SiO, the $v = 3$ $J = 8 - 7$ at 335.9 GHz was detected toward TX Cam, R Leo and W Hya at optical stellar phases ϕ of 0.3, 0.15 and 0.25 respectively (González-Alfonso *et al.* 1996) and towards VY CMa (González-Alfonso *et al.* 1996; Menten *et al.* 2006 using APEX). Menten *et al.* (2006) also detected maser emission in the ^{30}SiO $v = 1$ $J = 8 - 7$ line towards VY CMa, whereas the ^{29}SiO $v = 0$ $J = 8 - 7$ transition appears thermal. Infrared line overlaps of the SiO isotopomers is believed to be important in the pump scheme of these masers (e.g., Herpin & Baudry 2000). For a detailed discussion of SiO masers in evolved stars, see the review by Bujarrabal in these proceedings.

4. (Sub)millimeter H recombination masers

Hydrogen recombination maser emission is known from two galactic peculiar stellar sources, MWC 349A (Martin-Pintado *et al.* 1989) and Eta Carinae (Cox *et al.* 1995). (Sub)-millimeter maser emission from MWC 349A has been detected from at least the H31α (210.5 GHz), H30α (231.9 GHz), H29α (256.302 GHz) (Martin-Pintado *et al.* 1989) from H26α (353.623 GHz; Thum *et al.* 1994a) and the H21α (662.405 GHz; 350 Jy; Thum *et al.* 1994b), H32β (366.6 GHz; Thum *et al.* 1995). Planesas, Martin-Pintado & Serabyn (1992) spatially resolved the double-peaked maser spectrum into two emitting regions, separated by 0."065, associated with the red and blue-shifted emission from a sub-arcsecond disk imaged in the near-infrared by Danchi, Tuthill & Monnier (2001). Weintroub *et al.* (2007) again detected H30α and H26α maser emission from the two regions using the SMA, but also found emission at positions between them with an accuracy of 0."01. The emission position-velocity diagram is consistent with that of an edge-on disk in approximate Keplerian rotation. However, Weintroub *et al.* (2007) argue that systematic deviation from Keplerian rotation may indicate the presence of spiral structure in the MWC 349A disk (see also these proceedings). From Zeeman observations of the H30α maser, Thum & Morris (1999) report a dynamically-important magnetic field associated with the corona of the circumstellar disk, possibly generated by a local disk dynamo. Pumping of the masers in MWC 349A has been explained by Strelnitski *et al.* (1996). Towards Eta Carinae, Cox *et al.* (1995) detected millimeter maser emission at H30α, H29α and H37β (240.021 GHz) (see also Abraham *et al.* (2002)).

Extragalactic H recombination maser emission from the H27α (316.416 GHz) transition has also been detected towards M82 (Seaquist *et al.* 1996). The emission is highly time-variable, and of peak flux density 1.5 Jy at the strongest epoch. We note that H recombination masers at lower frequency may also have been detected from starburst galaxies, see references in Seaquist *et al.* (1996), and that H recombination masers are predicted to probe the Epochs of Recombination and Reionisation (Spaans & Norman 1997).

5. (Sub)millimeter CH_3OH masers

In a survey of Galactic star-forming regions, Kalenskii, Slysh & Val'Tts (2002) detected maser emission from methanol 8_{-1} - 7_0 E at 229.8 GHz towards DR 21(OH) and DR 21 West, and toward two maser candidates, L 379IRS3 and NGC 6334I(N). The maser emission in DR21(OH) and DR 21 West indicates gas kinetic temperatures of $T_k \sim 50$ K and densities of $n(H_2) = 3 \times 10^4$ cm^{-3}. Towards 16 other sources, the emission detected from this line was thermal in nature. Sobolev *et al.* (2002) reported the detection of class II methanol emission at 216.9 GHz, and models by Cragg *et al.* (2005) predict the existence of many more (sub)millimeter Class II methanol masers.

6. (Sub)millimeter HCN & SiS masers in carbon stars

(Sub)millimeter HCN maser emission has been detected from carbon-rich circumstellar envelopes. Using the Caltech Submillimeter Observatory (CSO), Schilke, Mehringer & Menten (2000) and Schilke & Menten (2003) detected the $J = 9 - 8$ maser of the $(04^{0}0)$ vibrationally-excited state of HCN at a frequency of ≈804.751 GHz towards IRC+10216 (at two epochs of peak flux densities 1420 & 840 Jy) and CIT 6 (110 Jy). The lower level of the maser is at 4200 K above ground state, such that emission should originate from the innermost region of the CSEs (< 3.5 R_*). Schilke & Menten (2003) also detected the $(11^{1}0)$-$(04^{0}0)$, $J = 10 - 9$ maser at 890.761 GHz towards IRC+10216 (at four epochs with peak flux densities of 6120, 4430, 9230, 900 Jy), CIT 6 (1090 & 1150 Jy) and Y CVn (140 Jy). In surveys using the Heinrich-Hertz-Submillimeter Telescope, Bieging, Shaked & Gensheimer (2000) and Bieging (2001) discovered maser emission in the $J = 3 - 2$ (265.886 GHz) and 4−3 (354.505 GHz) transitions of the HCN $(01^{1c}0)$ vibrational bending mode toward five stars: R Scl, V384 Per, R Lep, Y CVn, and V Cyg (out of 12 observed). Submillimeter HCN masers at 964 and 968 GHz are also predicted by Schilke & Menten (2003), and could be detected using SOFIA.

SiS masers were first discovered by Henkel, Matthews & Morris (1983) from the $v = 0$, $J = 1 - 0$ transition at 18 GHz toward carbon-rich star IRC+10216. (Sub)millimeter SiS maser emission was also detected toward IRC+10216 from the $v = 0$, $J = 11 - 10$ (199.672 GHz), $J = 14 - 13$ (254.103 GHz) and $J = 15 - 14$ (272.243 GHz) transitions by Fonfría Expósito *et al.* (2006) using the IRAM 30-m. Line overlap is believed to be important in the pumping scheme of the highly-rotationally excited masers and they are thought to occupy ∼5 - 7 R_* in the CSE of IRC+10216. Future high-resolution observations of the HCN and SiS masers using ALMA will therefore yield new information on the dust formation zone of carbon stars.

7. Summary & future propects

Observations of submillimeter masers at high angular resolution provide new means of studying stellar evolution, star formation and AGN/starburst activity. Where different maser transitions trace the same gas, we will be able to place new constraints on radiative transfer models to determine small-scale source temperature and density distributions. Where maser lines trace different regions of sources, we will be able to map out more of source structures and dynamics than ever before. Submillimeter masers could be particularly important probes of regions in which longer wavelength maser emission is subject to obscuration e.g., due to free-free or synchrotron opacity.

The spatial resolution and sensitivity of ALMA will revolutionize submillimeter science. There have also been huge strides in submillimeter VLBI, with fringes obtained at 129, 147, and 230 GHz (see e.g., Krichbaum *et al.* (2007)) and with imaging of SiO $J = 3 - 2$ masers at 129 GHz in VY CMa and several AGB stars already achieved (Doeleman *et al.* 2005; Doeleman, private communication). Within the next decade, observations of submillimeter masers are likely to become very much more commonplace and, in conjunction with detailed modelling, will yield a wealth of new and exciting avenues of research.

Acknowledgements

EH thanks Lincoln Greenhill, Preethi Pratap, Andrej Sobolev, Vladimir Strelnitski and Jonathan Weintroub for providing unpublished results, and Jim Moran for helpful comments on this manuscript.

References

Abraham, Z., Damineli, A., Durouchoux, P., Nyman, L., & McAuliffe, F. 2002, in: V. Migenes & M. J. Reid (eds.), *Cosmic Masers: From Proto-Stars to Black Holes*, IAU Symposium, vol. 206, p. 234

Alcolea, J. & Menten, K. M. 1993, in: A. W. Clegg & G. E. Nedoluha (eds.), *Astrophysical Masers*, Lecture Notes in Physics (Berlin: Springer Verlag), vol. 412, p. 399

Bieging, J. H., Shaked, S., & Gensheimer, P. D. 2000, *ApJ*, 543, 897

Bieging, J. H. 2001, *ApJ*, 549, L125

Cernicharo, J., Thum, C., Hein, H., John, D., Garcia, P., & Mattioco, F. 1990, *A&A*, 231, L15

Cernicharo, J. & Bujarrabal, V. 1992, *ApJ*, 401, L109

Cernicharo, J., Bujarrabal, V., Santaren, J. L. 1993, *ApJ* 407, L33

Cernicharo, J., González-Alfonso, E., Alcolea, J., Bachiller, R., John, D. 1994, *ApJ*, 432, L59

Cernicharo, J., Bachiller, R., González-Alfonso, E. 1996, *A&A*, 305, L5

Cernicharo, J., Pardo, J. R., González-Alfonso, E., Serabyn, E., Phillips, T. G., Benford, D. J., Mehringer, D. 1999, *ApJ*, 520, L131

Cernicharo, J., Goicoechea, J. R., Pardo, J. R., Asensio-Ramos, A. 2006a, *ApJ*, 642, 940

Cernicharo, J., Pardo, J. R., & Weiss, A. 2006b, *ApJ*, 646, L49

Cheung, A. C., Rank, D. M., Townes, C. H., Thornton, D. D., Welch, W. J. 1969, *Nature*, 221, 626

Clemens, D. P. & Lane, A. P. 1983, *ApJ*, 266, L117

Cooke, B. & Elitzur, M. 1985, *ApJ*, 295, 175

Cox, P., Martin-Pintado, J., Bachiller, R., Bronfman, L., Cernicharo, J., Nyman, L.-A., Roelfsema, P. R. 1995, *A&A*, 295, L39

Cragg, D. M., Sobolev, A. M., Godfrey, P. D. 2005, *MNRAS*, 360, 533

Danchi, W. C., Tuthill, P. G., & Monnier, J. D. 2001, *ApJ*, 562, 440

Deguchi, S. 1977, *PASJ*, 29, 669

Deguchi, S., Nguyen-Q-Rieu 1990, *ApJ*, 360, L27

Doeleman, S. S., Phillips, R. B., Rogers, A. E. E., *et al.* 2005, in: J. Romney & M. Reid (eds.), *Future Directions in High Resolution Astronomy*, Astronomical Society of the Pacific Conference Series, vol. 340, p. 605

Feldman, P. A., Matthews, H. E., Cunningham, C. T., Hayward, R. H., Wade, J. D., Amano, T., Scappini, F. 1991, *JRASC*, 85, 191

Feldman, P. A., Matthews, H. E., Amano, T., Scappini, F., Lees, R. M. 1993, in: A. W. Clegg & G. E. Nedoluha (eds.), *Astrophysical Masers*, Lecture Notes in Physics (Berlin: Springer Verlag), vol. 412, p. 65

Fonfría Expósito, J. P., Agundez, M., Tercero, B., Pardo, J. R., Cernicharo, J. 2006, *ApJ*, 646, L127

González-Alfonso, E., Cernicharo, J., Bachiller, R., Fuente, A. 1995, *A&A*, 293, L9

González-Alfonso, E., Alcolea, J., Cernicharo, J. 1996, *A&A*, 313, L13

González-Alfonso, E., Cernicharo, J., Alcolea, J., & Orlandi, M. A. 1998, *A&A*, 334, 1016

Gray, M. D., Ivison, R. J., Yates, J. A., Humphreys, E. M. L., Hall, P. J., & Field, D. 1995, *MNRAS*, 277, L67

Gray, M. D., Humphreys, E. M. L., & Yates, J. A 1999, *MNRAS*, 304, 906

Greenhill, L. J., *et al.* 2007, in prep

Henkel, C., Matthews, H. E., & Morris, M. 1983, *ApJ*, 267, 184

Herpin, F. & Baudry, A. 2000, *A&A*, 359, 1117

Humphreys, E. M. L., Gray, M. D., Yates, J. A., Field, D. 1997, *MNRAS*, 287, 663

Humphreys, E. M. L., Yates, J. A., Gray, M. D., Field, D., Bowen, G. H. 2001, *A&A*, 379, 501

Humphreys, E. M. L., Greenhill, L. J., Reid, M. J., Beuther, H., Moran, J. M., Gurwell, M., Wilner, D. J., Kondratko, P. T. 2005, *ApJ*, 634, L133

Ivison, R. J., Yates, J. A., Hall, P. J. 1998, *MNRAS*, 295, 813

Jewell, P. R. and Dickinson, D. F. and Snyder, L. E. and Clemens, D. P. 1987, *ApJ*, 323, 749

Kalenskii, S. V., Slysh, V. I., Val'tts, I. E. 2002, *ARep*, 46, 49

Kemball, A. J., & Diamond, P. J. 1997, *ApJ*, 481, L111

Kondratko, P. T., Greenhill, L. J., & Moran, J. M. 2005, *ApJ*, 618, 618

Krichbaum, T. P., Graham, D. A., Witzel, A., *et al.* 2007, in: *Towards the Event Horizon: High Resolution VLBI Imaging of Nuclei of Active Galaxies*, Exploring the Cosmic Frontier, ESO Astrophysics Symposia European Southern Observatory, Springer, p. 189

Martin-Pintado, J., Bachiller, R., Thum, C., Walmsley, M. 1989, 215, L13

Menten, K. M. & Melnick, G. J. 1989, *ApJ*, 341, L91

Melnick, G. J., Menten, K. M., Phillips, T. G., Hunter, T. 1993, *ApJ*, 416, L37

Menten, K. M., Melnick, G. J., Phillips, T. G. 1990b, *ApJ*, 350, L41

Menten, K. M., Melnick, G. J., Phillips, T. G., Neufeld, D. A. 1990a, *ApJ*, 363, L27

Menten, K. M., & Young, K. 1995, *ApJ*, 450, L67

Menten, K. M., Philipp, S. D., Güsten, R., Alcolea, J., Polehampton, E. T., Brünken, S. 2006, *A&A*, 454, L107

Neufeld, D. A., Melnick, G. J. 1991, *ApJ*, 368, 215

Pardo, J. R., Cernicharo, J., Gonzalez-Alfonso, E. and Bujarrabal, V. 1998, A&A, 329, 219

Patel, N. A., Curiel, S., Zhang, Q., Sridharan, T. K., Ho, P. T. P., & Torrelles, J. M. 2007, *ApJ*, 658, L55

Phillips, T. G., Kwan, J., Huggins, P. J. 1980, in: B. H. Andrew (eds.), *Interstellar Molecules*, IAU Symposium, vol. 87, p. 21

Planesas, P., Martin-Pintado, J., Serabyn, E., 1992, *ApJ*, 386, L23

Schilke, P., Mehringer, D. M., & Menten, K. M. 2000, *ApJ*, 528, L37

Schilke, P., Benford, D. J., Hunter, T. R., Lis, D. C., Phillips, T. G. 2001, *ApJS*, 132, 281

Schilke, P., & Menten, K. M. 2003, *ApJ*, 583, 446

Seaquist, E. R., Carlstrom, J. E., Bryant, P. M., Bell, M. B. 1996, *ApJ*, 465, 691

Shinnaga, H., Moran, J. M., Young, K. H., Ho, P. T. P. 2004, *ApJ*, 616, L47

Sobolev, A. M., Ostrovskii, A. B., Malyshev, A. V., Cragg, D. M., Godfrey, P. D., Sutton, E. C., Watson, W. D., Ellingsen, S. P., Caswell, J. L. 2002, in: V. Migenes & M. J. Reid (eds.), *Cosmic Masers: From Proto-Stars to Black Holes*, IAU Symposium, vol. 206, p. 179

Spaans, M., & Norman, C. A. 1997, *ApJ*, 488, 27

Strelnitski, V. S., Ponomarev, V. O., Smith, H. A. 1996, *ApJ*, 470, 1118

Tennyson, J., Zobov, N. F., Williamson, R., Polyansky, O.L., Bernath, P. F. 2001, *J. Phys. Chem. Ref. Data*, 30, 735

Thum, C., Matthews, H. E., & Martin-Pintado, J., Serabyn, E., Planesas, P., Bachiller, R. 1994a, *A&A*, 283, 582

Thum, C., Matthews, H. E., Harris, A. I., Tacconi, L. J., Schuster, K. F., Martin-Pintado, J. 1994b, *A&A*, 288, L25

Thum, C., Strelnitski, V. S., Martin-Pintado, J., Matthews, H. E., Smith, H. A. 1995, *A&A*, 300, 843

Thum, C. & Morris, D. 1999, *A&A*, 344, 923

Waters, J. W., Kakar, R. K., Kuiper, T. B. H., Roscoe, H. K., Swanson, P. N., Rodriguez Kuiper, E. N., Kerr, A. R., Thaddeus, P., Gustincic, J. J. 1980, *ApJ*, 235, 57

Weintroub, J., *et al.* 2007, in prep

Yates, J. A., Cohen, R. J., Hills, R. E., 1995, *MNRAS*, 273, 529

Yates, J. A. & Cohen, R. J. 1996, *MNRAS*, 278, 655

Yates, J.A., Field, D., & Gray, M.D. 1997, *MNRAS*, 285, 303

Astrophysical Masers and their Environments
Proceedings IAU Symposium No. 242, 2007
J.M. Chapman & W.A. Baan, eds.

doi:10.1017/S1743921307013634

658 GHz vibrationally-excited water masers with the Submillimeter Array

T. R. Hunter[1], K. H. Young[2], R. D. Christensen[3] and M. A. Gurwell[2]

[1]NRAO, 520 Edgemont Rd, Charlottesville, VA 22903, USA
email: thunter@nrao.edu

[2]Harvard-Smithsonian Center for Astrophysics, 60 Garden St., Cambridge, MA 02138, USA

[3]Submillimeter Array, 645 North A'ohoku Place, Hilo, HI 96720, USA

Abstract. Discovered in 1995 at the Caltech Submillimeter Observatory (CSO), the vibrationally-excited water maser line at 658 GHz (455 micron) is seen in oxygen-rich giant and supergiant stars. Because this maser can be so strong (up to thousands of Janskys), it was very helpful during the commissioning phase of the highest frequency band (620-700 GHz) of the Submillimeter Array (SMA) interferometer. From late 2002 to early 2006, brief attempts were made to search for emission from additional sources beyond the original CSO survey. These efforts have expanded the source count from 10 to 16. The maser emission appears to be quite compact spatially, as expected from theoretical considerations; thus these objects can potentially be used as atmospheric phase calibrators. Many of these objects also exhibit maser emission in the vibrationally-excited SiO maser at 215 GHz. Because both maser lines likely originate from a similar physical region, these objects can be used to test techniques of phase transfer calibration between millimeter and submillimeter bands. The 658 GHz masers will be important beacons to assess the performance of the Atacama Large Millimeter Array (ALMA) in this challenging high-frequency band.

1. Introduction

Water is an asymmetric top molecule with three vibrational quantum numbers (Herzberg 1945). The presence of vibrationally-excited water has been detected at near-IR wavelengths in a variety of astrophysical environments such as the atmospheres of Mira variable stars (e.g. R Leo, Hinkle & Barnes 1979), semi-regular pulsating stars (e.g. W Hya, Justtanont *et al.* 2004), FU Orionis objects (Reipurth *et al.* 2007), Mars (Fouglet *et al.* 2006), and in the post-Deep Impact ejecta from Comet Temple 1 (Barber *et al.* 2006). Of the three vibrational modes, the $v_2 = 1$ bending mode has the lowest energy ground state (2297 K). Rotational levels for this mode are tabulated by Camy-Peyret *et al.* (1977) and a level diagram is presented by Alcolea & Menten (1993). Frequency measurements are given by Belov *et al.* (1987), Pearson *et al.*(1991), and Chen *et al.* (2000). Analogous frequencies for deuterated water are currently being measured (Brünken *et al.* 2005).

Table 1. Astronomically-detected (sub)millimeter transitions of vibrationally-excited water

Frequency (GHz)	Rotational Transition	Vibrational State	Species	$E_{\rm upper}$ (K)	Telescope	Discovery Reference
96.261	$4_{4,0} \rightarrow 5_{3,3}$	010	para	3066	IRAM 30m	Menten & Melnick 1989
232.686	$5_{5,0} \rightarrow 6_{4,3}$	010	ortho	3465	IRAM 30m	Menten & Melnick 1989
293.664	$6_{6,1} \rightarrow 7_{5,2}$	010	ortho	2941	APEX 12m	Menten *et al.* 2006
336.227	$5_{2,3} \rightarrow 6_{1,6}$	010	ortho	2958	APEX 12m	Menten *et al.* 2006
658.006	$1_{1,0} \rightarrow 1_{0,1}$	010	ortho	2362	CSO 10.4m	Menten & Young 1995

Emission from water vapor in the $v_2 = 1$ state has been detected in astronomical sources in several (sub)millimeter lines (see Table 1). Searches have been performed for three other $v_2 = 1$ transitions with negative results: para $6_{6,0} \rightarrow 7_{5,3}$ at 297.439 GHz (Menten *et al.* 2006), ortho $4_{2,3} \rightarrow 3_{3,0}$ at 12.008 GHz (Myers & Barrett 1982), and ortho $4_{1,4} \rightarrow 3_{2,1}$ at 67.704 GHz (Petuchowski & Bennett 1991). All of the lines in Table 1 have been detected in the hypergiant star VY CMa. With the exception of the 336 GHz line, all of the detected lines in this object show evidence for maser action, at least in some velocity components. The 232 GHz line is also seen in the semi-regular pulsating star W Hya. The 658 GHz transition to the ortho ground state is seen as a strong maser in both objects, along with eight other evolved stars including three semi-regular stars (R Crt, RT Vir, RX Boo), two hypergiants (VX Sgr, NML Cyg) and three Miras (R Leo, S CrB, U Her). Unlike the non-vibrational water lines at 22 GHz, 183 GHz (Cernicharo *et al.* 1990), 321 GHz (Menten *et al.* 1990a), 325 GHz (Menten *et al.* 1990b), 439 and 470 GHz (Melnick *et al.* 1993), the 658 GHz line is not detected as a maser in star-forming regions. However, the 658 GHz line is detected as a thermal line in the CSO Orion KL 600-720 GHz line survey (Schilke *et al.* 2001). In this paper, we describe the first $1'' - 2''$ resolution observations of the 658 GHz maser and present new detections from six additional Mira variables obtained by the Submillimeter Array (SMA)†.

2. Observations

Summarized in Table 2, the SMA observations were performed primarily during commissioning periods of the 600 GHz receivers and their associated IF path (Hunter *et al.* 2002). The SMA receivers are double sideband (DSB), fixed-tuned SIS mixers with an IF output band of 4-6 GHz. Further description of the SMA antennas and receivers can be found elsewhere (Blundell 2004). Prior to January 2005, SMA observations could only be obtained in one receiver band at a time. As of January 2005, simultaneous observations of the same target position with two receiver bands could be obtained: (200 GHz with 600 GHz) or (300 GHz with 600 GHz). During the observations, in most cases the 200 GHz band was tuned to the SiO v =1, J =5-4 maser line at 215 GHz in lower sideband (LSB). The 658 line was typically tuned in upper sideband (USB). The H_2O maser line was observed with a correlator channel spacing of 0.8125 MHz (0.4 km/s). The compact configuration was used, with baseline lengths ranging from 16 to 69 meters. Callisto was the flux calibrator.

Table 2. Log of significant SMA observations in the 658 GHz water maser line

Date	Log #	Antennas	$\tau_{225\mathrm{GHz,zenith}}$	New detections	Other detections‡
12 Dec 2002	4327	3	0.03 - 0.05	U Ori	
08 Apr 2004	7082	4	0.07	R Cas, R Aql	VY CMa, U Her, R Leo & W Hya
24 Jan 2005	8690	5	0.06		W Hya
28 Jan 2005	8726	5	0.06		R Leo, VY CMa
16 Feb 2005	8847	6	0.03 - 0.06		VY CMa
24 Aug 2005	9962	7	0.05 - 0.08		VX Sgr
14 Dec 2005	10802	7	0.028	TX Cam, NML Tau	R Cas
11 Jan 2006	10721	6	0.06	R Hya	VY CMa, R Leo

‡Non-detections: IRC+10011, Betelgeuse, IRC+70066, GX Mon, OH231.8+4.2, RS Cnc

† Located on Mauna Kea, Hawaii, the Submillimeter Array (SMA) is an eight-element interferometer built and operated as a collaborative project between the Smithsonian Astrophysical Observatory and the Academia Sinica Institute of Astronomy & Astrophysics of Taiwan.

3. Results

3.1. *New detections*

With the SMA, new detections were obtained of six Mira variables. As an example, calibrated uv spectra of R Cas and R Aql are shown in Fig. 1.

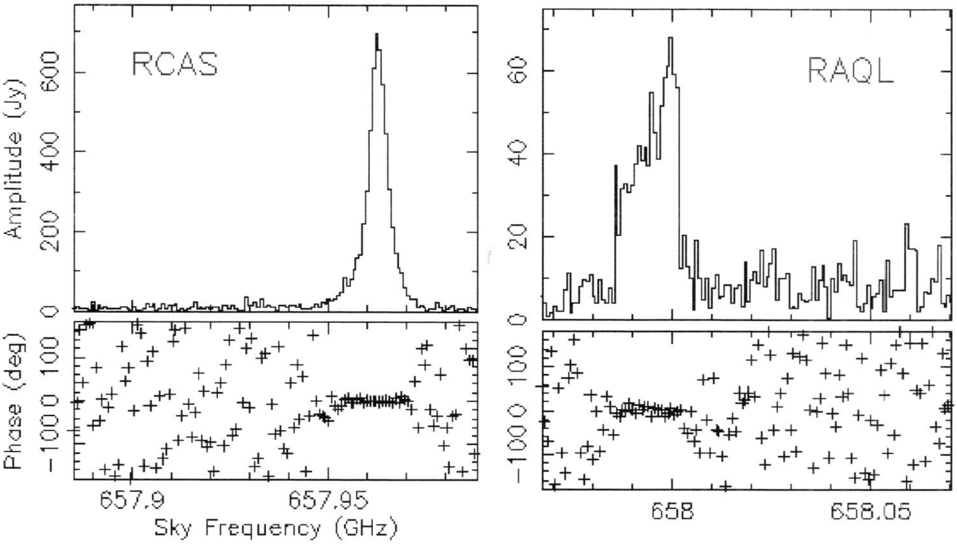

Figure 1. Calibrated SMA uv spectra of the Mira variables R Cas (left panel) and R Aql (right panel) in the 658 GHz H_2O maser line. The velocity range shown is 47 km/s (104 MHz).

These detections increase the total number of objects detected in this maser line either by CSO or SMA from 10 to 16. A complete listing of objects that have been detected in this line is given in Table 3.

Table 3. List of all objects detected in the 658 GHz water maser line (as of March 2007)

Object Class	Name	Position† α (J2000)	δ (J2000)	Spectral Type	Distance (pc)	Velocity km/s
semi-regular pulsating stars	R Crt	11 00 33.85	−18 19 29.6	M7III, SRb	186‡	8
	RT Vir	13 02 37.98	+05 11 08.4	M8III, SRb	220	15
	W Hya	13 49 02.00	−28 22 03.5	M7e, SRa	98	42
	RX Boo	14 24 11.63	+25 42 13.4	M7.5, SRb	141‡	1
hypergiants	VY CMa	07 22 58.33	−25 46 03.2	M3/M4II	1500¶	22
	VX Sgr	18 08 04.05	−22 13 26.6	M5/M6III, SRc	1400	7
	NML Cyg	20 46 25.46	+40 06 59.6	M6IIIe	2000	−1
Mira variables	NML Tau	03 53 28.84	+11 24 22.6	M6me	270	32
	TX Cam	05 00 50.39	+56 10 52.6	M8.5	380	9
	U Ori	05 55 49.17	+20 10 30.7	M8III	256	39
	R Leo	09 47 33.49	+11 25 43.6	M8IIIe	110	0
	R Hya	13 29 42.78	−23 16 52.8	M7IIIe	125	−11
	S CrB	15 21 23.96	+31 22 02.6	M7e	430	1
	U Her	16 25 47.47	+18 53 32.9	M7III	347‡	−15
	R Aql	19 06 22.25	+08 13 48.0	M7IIIev	220	48
	R Cas	23 58 24.87	+51 23 19.7	M7IIIe	184‡	26

†ICRS positions taken from SIMBAD ‡Colomer *et al.* (2000) ¶Lada & Reid (1978)

3.2. *Comparison of H_2O and SiO maser line widths*

A comparison of the simultaneous spectral observations of the 658 GHz water maser with the 215 GHz SiO $v = 1$ maser in four objects is shown in Fig. 2. The emission from these two species arises from a similar range of velocities, which has been interpreted as evidence for a common physical origin (Menten & Young 1995), particularly because the lower level of the SiO $v = 1$ maser lies at a similar energy above the ground state (1792 K) as does the 658 GHz line. A scatter plot of the line width (FWZI) of these two masers in all 16 objects is shown in Fig. 3.

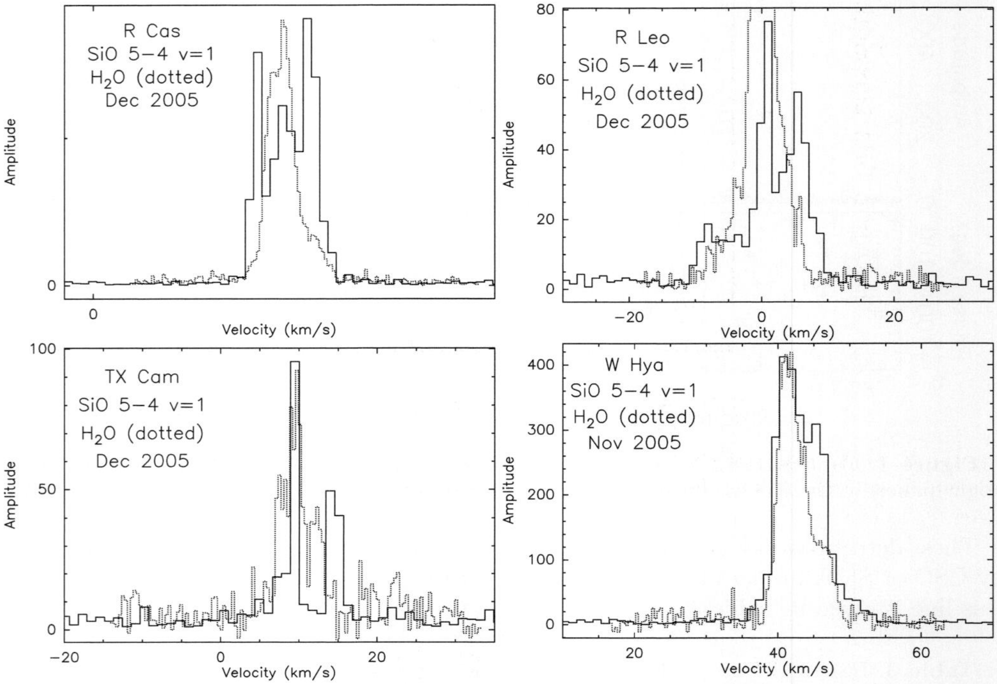

Figure 2. Simultaneous SMA uv spectra of SiO J =5-4, $v = 1$ (solid line) and 658 GHz H_2O (dotted line) in four evolved stars. The channel spacing was 3.25 MHz (4.5 km/s) for SiO, and 0.8125 MHz (0.4 km/s) for H_2O.

3.3. *Source size*

To investigate the source size of the 658 GHz masers, we have analyzed the data for VY CMa for which one of the highest signal-to-noise tracks of SMA data exists, amounting to 87 minutes on-source. The radio photosphere of this red, luminous (5×10^5 $L_\odot$ e.g. Humphreys 2006) hypergiant star was found to be unresolved with a 1″ beam at 22 GHz (Lipscy *et al.* 2005). In the mid-infrared, the same authors measured the size of the warm dust continuum emission to be 0.3″ at 17.9 μm. In the near-infrared, the K band size was measured to be 0.138″ × 0.205″ (Wittkowski *et al.* 1998). SMA observations at 230 GHz continuum with a 1.4″ beam find the emission to be unresolved (Muller *et al.* 2007). Proceeding with the likely assumption that the 658 GHz continuum emission is likewise compact with respect to the synthesized beam (1.6″ × 1.0″), we generated a continuum dataset using all the correlator chunks in USB except the one containing the maser line. We used that dataset to calibrate the antenna phases and amplitudes with a point source model, and applied the solutions to the spectral line data. A plot of the resulting amplitude vs. uv distance is shown in Fig. 4. There is no discernible drop in

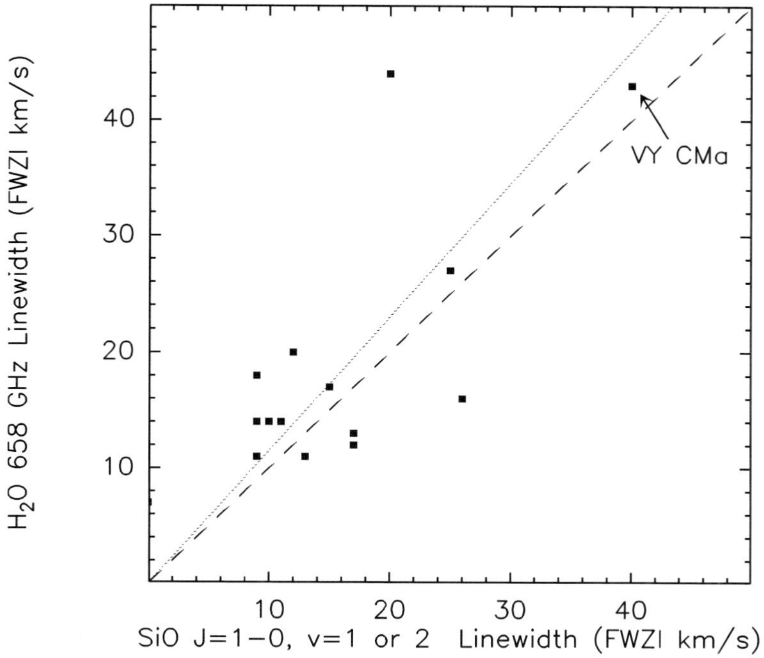

Figure 3. Scatter plot of FWZI of 658 GHz H_2O line vs FWZI of SiO $v = 1$, J=1-0 line (or in some cases J=2-1). The dashed line has a slope of unity, while the best fit dotted line has a slope of 1.15.

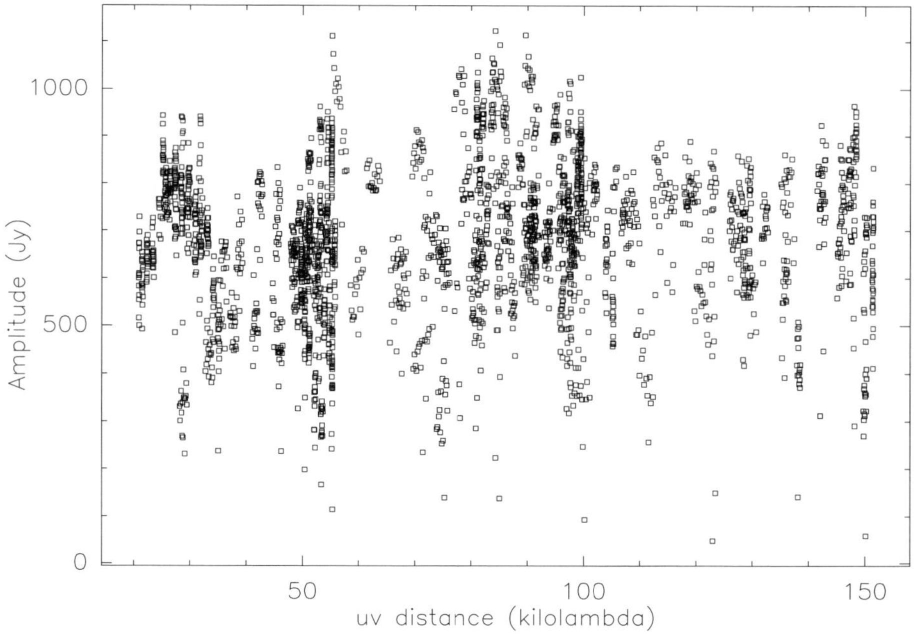

Figure 4. Calibrated amplitude vs. uv distance for the VY CMa 658 GHz maser observations of 16 Feb 2005 with 15 baselines.

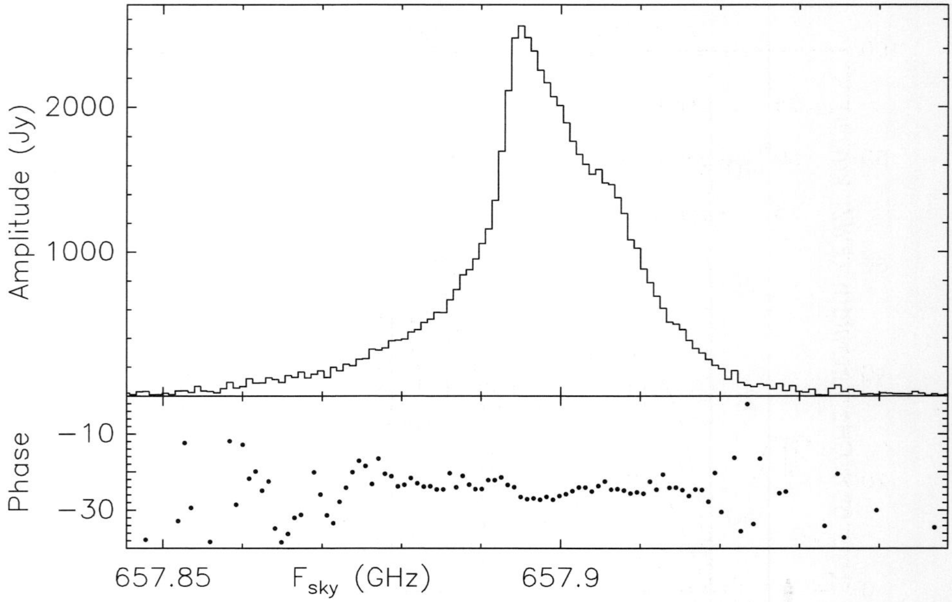

Figure 5. Calibrated phase in degrees (dots) and amplitude (histogram) vs. frequency for the VY CMa 658 GHz maser observations of 16 Feb 2005. This spectrum is a vector average of all 15 baselines.

amplitude with distance, suggesting that the source remains unresolved at these angular scales. Further evidence for this conclusion comes from the uv spectrum (Fig. 5) which shows only a very small phase gradient ($< 10°$) across the spectral line.

4. Phase calibration and phase transfer

Because they are bright celestial beacons, the 658 GHz water masers are helpful in debugging problems in the SMA frontends and backends. The 658 GHz maser line has also served as a successful phase calibrator for SMA observations of other sources. Chen *et al.* (2007) used VY CMa to calibrate observations of the ultracompact HII region G240.31+0.07, located less than 5° away. The measured phase of the water maser emission was used to calibrate the temporal changes in atmospheric phase plus instrumental phase time while the unresolved stellar continuum emission ($\sim$ 9 Jy) was used to calibrate the amplitudes. The 658 GHz maser line has also been used to test the effectiveness of phase transfer between the 200 and 600 GHz bands on the SMA. Using observations of the star W Hydra on 28 Jan 2005, the antenna-based phase solutions for the 215 GHz SiO maser emission and the 658 GHz H_2O maser emission were compared over a period of several hours and shown to change in a ratio close to that expected theoretically (Hunter *et al.* 2005). At SMA, the usage of this technique is limited due to some remaining instrumental thermal drifts and phase jumps currently under investigation (Kubo *et al.* 2006). However, the phase transfer technique is expected to be relied upon heavily by ALMA (Laing 2004), typically from Band 3 (84-116 GHz) where bright quasars are more numerous, to higher frequency bands where they are not (Butler 2003).

5. Future prospects

At the present time, the SMA has the potential to observe the 658 GHz maser sources with a beamsize as small as 0.15″, where it may begin to resolve the emission as a phase

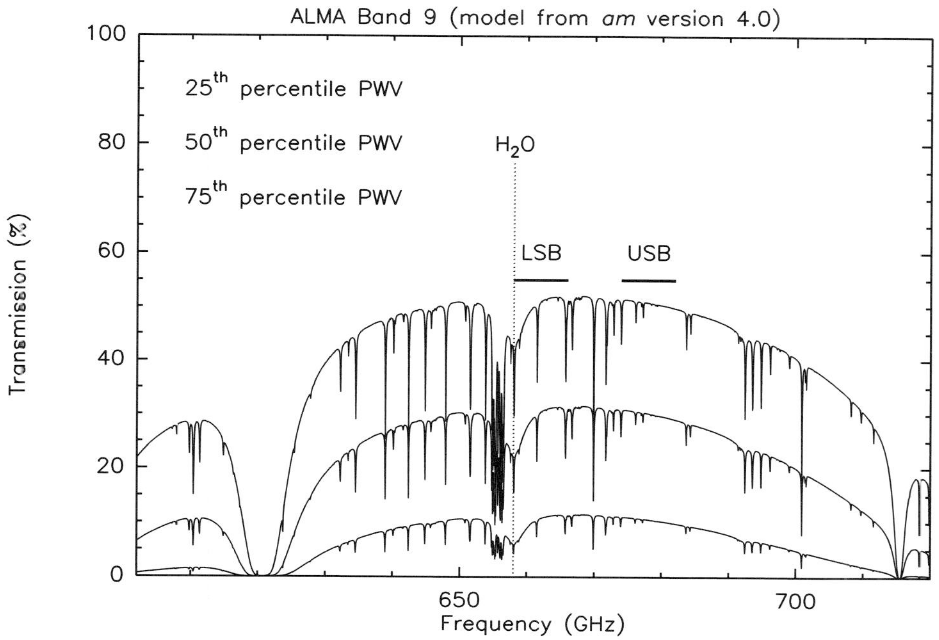

Figure 6. Proposed tuning for 660 GHz continuum observations with ALMA, including the 658 GHz line in the LSB passband in order to use an evolved star as a phase calibrator near the target source. The atmospheric model curves are from the *am* package (Paine 2006). The top curve is 25th %ile conditions, the middle curve is 50th %ile, and the bottom curve is 75th %ile.

gradient in velocity. For example, 43 GHz VLBI observations of TX Cam show a ring of SiO masers with a diameter of $\sim 0.04''$ (Diamond & Kemball 2003). After its launch in 2008, the Band 2 receiver (Teipen *et al.* 2005) of the Heterodyne Instrument for the Far-Infrared (HIFI) aboard the Herschel satellite (Poglitsch *et al.* 2006) will provide the opportunity to perform a sensitive survey of 658 GHz maser emission in a large number of stars. Beginning around 2010, ALMA will become the first telescope capable of observing all five known (sub)millimeter transitions of vibrationally-excited water (Wootten 2007). This capability will be important for obtaining near-simultaneous observations of all these lines in order to accurately determine the temperature and density of the emitting gas. As shown in Fig. 6, the Band 9 receivers of ALMA are DSB and can be tuned to provide the 658 GHz water maser line in the LSB and a 4 GHz region of continuum in the USB that is nearly free of atmospheric ozone absorption features.

References

Alcolea, J., Bujarrabal, V., & Gomez-Gonzalez, J. 1990, *A&A*, 231, 431

Alcolea, J., & Menten, K. M. 1993, LNP Vol. 412: Astrophysical Masers, 412, 399

Barber, R. J., Miller, S., Stallard, T. S., & Tennyson, J. 2006, ArXiv Astrophysics e-prints, arXiv:astro-ph/0610673

Belov, S. P., Kozin, I. N., Polyansky, O. L., Tret'yakov, M. Y., & Zobov, N. F. 1987, Journal of Molecular Spectroscopy, 126, 113

Blundell, R. 2004, Proceedings of the 15th International Symposium on Space Terahertz Technology, April 27-29, 2004, p. 3, ArXiv Astrophysics e-prints, arXiv:astro-ph/0508492

Boboltz, D. A., & Diamond, P. J. 2005, *ApJ*, 625, 978

Brünken, S., Müller, H. S. P., Endres, C. P., Giesen, T. F., Mäder, H., Pearson, J. C., & Drouin, B. J. 2005, Astrochemistry: Recent Successes and Current Challenges, 231, 97

Butler, B. 2003, ALMA Memo 478, "Distance to Possible Calibration Sources as a Function of Frequency for ALMA"

Camy-Peyret, C., Flaud, J. M., Maillard, J. P., & Guelachvili, G. 1977, Molecular Physics, 33, 1641

Cernicharo, J., Thum, C., Hein, H., John, D., Garcia, P., & Mattioco, F. 1990, *A&A*, 231, L15

Chen, P., Pearson, J. C., Pickett, H. M., Matsuura, S., & Blake, G. A. 2000, *ApJS*, 128, 371

Chen, X., Shen, Z.-Q., Imai, H., & Kamohara, R. 2006, *ApJ*, 640, 982

Chen, H.-R., *et al.* 2007, *ApJ*, 654, L87

Cho, S.-H., Kaifu, N., & Ukita, N. 1996, *A&AS*, 115, 117

Colomer, F., Reid, M. J., Menten, K. M., & Bujarrabal, V. 2000, *A&A*, 355, 979

Diamond, P. J., & Kemball, A. J. 2003, *ApJ*, 599, 1372

Herpin, F., Baudry, A., Thum, C., Morris, D., & Wiesemeyer, H. 2006, *A&A*, 450, 667

Herzberg, G. "Molecular spectra and molecular structure. Vol.2: Infrared and Raman spectra of polyatomic molecules", New York: Van Nostrand, Reinhold, 1945

Hinkle, K. H., & Barnes, T. G. 1979, *ApJ*, 227, 923

Humphreys, R. M. 2006, ArXiv Astrophysics e-prints, arXiv:astro-ph/0610433

Hunter, T. R., *et al.* 2002, Bulletin of the American Astronomical Society, 34, 1302, arXiv:astro-ph/0704.2641

Hunter, T. R., *et al.* 2005, ArXiv Astrophysics e-prints, arXiv:astro-ph/0509467

Jouglet, D., Poulet, F., Mustard, J. F., Milliken, R. E., Bibring, J. P., Langevin, Y., & Gondet, B. 2006, 37th Annual Lunar and Planetary Science Conference, 37, 1741

Justtanont, K., de Jong, T., Tielens, A. G. G. M., Feuchtgruber, H., & Waters, L. B. F. M. 2004, *A&A*, 417, 625

Kubo, D. Y., Hunter, T. R., Christensen, R. D., & Yamaguchi, P. I. 2006, *Proc. of the SPIE*, 6275, 63

Lada, C. J., & Reid, M. J. 1978, *ApJ*, 219, 95

Laing, R. 2004, ALMA Commissioning and Science Verification Plan, ALMA-90.00.00.00-007-B-PLA

Lipscy, S. J., Jura, M., & Reid, M. J. 2005, *ApJ*, 626, 439

Melnick, G. J., Menten, K. M., Phillips, T. G., & Hunter, T. 1993, *ApJ*, 416, L37

Menten, K. M., & Young, K. 1995, *ApJ*, 450, L67

Menten, K. M., & Melnick, G. J. 1989, *ApJ*, 341, L91

Menten, K. M., Melnick, G. J., Phillips, T. G., & Neufeld, D. A. 1990, *ApJ*, 363, L27

Menten, K. M., Melnick, G. J., & Phillips, T. G. 1990, *ApJ*, 350, L41

Menten, K. M., Philipp, S. D., Güsten, R., Alcolea, J., Polehampton, E. T., & Brünken, S. 2006, *A&A*, 454, L107

Muller, S., Dinh-V-Trung, Lim, J., Hirano, N., Muthu, C., & Kwok, S. 2007, *ApJ*, 656, 1109

Myers, P. C., & Barrett, A. H. 1982, *ApJ*, 263, 716

Paine, S. 2006, SMA Memo 152, rev. 3. "The *am* Atmospheric Model" (http://cfarx6.cfa.harvard.edu/am)

Pearson, J. C., De Lucia, F. C., Anderson, T., Herbst, E., & Helminger, P. 1991, *ApJ*, 379, L41

Petuchowski, S. J., & Bennett, C. L. 1991, *ApJ*, 367, 168

Poglitsch, A., et al. 2006, 36th COSPAR Scientific Assembly, 36, 215

Reipurth, B., Aspin, C., Beck, T., Brogan, C., Connelley, M. S., & Herbig, G. H. 2007, *AJ*, 133, 1000

Schilke, P., Benford, D. J., Hunter, T. R., Lis, D. C., & Phillips, T. G. 2001, *ApJS*, 132, 281

Teipen, R., Justen, M., Tils, T., Schultz, M., Glenz, S., Putz, P., Honingh, C.E., Jacobs K. 2005, 16th Int. Symp. on Space THz Technology; Göteborg, Sweden, May 2-4 2005

Wittkowski, M., Langer, N., & Weigelt, G. 1998, *A&A*, 340, L39

Wootten, A. 2007, ArXiv Astrophysics e-prints, arXiv:astro-ph/0702668

Yi, J., Booth, R. S., Conway, J. E., & Diamond, P. J. 2005, *A&A*, 432, 531

Astrophysical Masers and their Environments
Proceedings IAU Symposium No. 242, 2007
J.M. Chapman & W.A. Baan, eds.

doi:10.1017/S1743921307013646

Submillimeter Array observations of 321 GHz water maser emission in Cepheus A

Nimesh A. Patel[1], Salvador Curiel[2], Qizhou Zhang[1], T. K. Sridharan[1], Paul T. P. Ho[1,3] and José M. Torrelles[4]

[1]Harvard-Smithsonian Center for Astrophysics, Cambridge MA 02138, USA
email: npatel@cfa.harvard.edu

[2]Instituto de Astronomía, Universidad Nacional Autónoma de México (UNAM), Mexico

[3]Academia Sinica Institute of Astronomy and Astrophysics, Taipei, Taiwan

[4]Instituto de Ciencias del Espacio (CSIC)-IEEC, Facultat de Física, Universitat de Barcelona, Barcelona, Spain

Abstract. Using the Submillimeter Array (SMA) we have imaged for the first time the 321.226 GHz, $10_{29} - 9_{36}$ ortho-H_2O maser emission. This is also the first detection of this line in the Cepheus A high-mass star-forming region. The 22.235 GHz, $6_{16} - 5_{23}$ water masers were also observed with the Very Large Array 43 days following the SMA observations. Three of the nine detected submillimeter maser spots are associated with the centimeter masers spatially as well as kinematically, while there are 36 22 GHz maser spots without corresponding submillimeter masers. In the HW2 source, both the 321 GHz and 22 GHz masers occur within the region of $\sim 1''$ which includes the disk-jet system, but the position angles of the roughly linear structures traced by the masers indicate that the 321 GHz masers are along the jet while the 22 GHz masers are perpendicular to it. We interpret the submillimeter masers in Cepheus A to be tracing significantly hotter regions (600$\sim$2000 K) than the centimeter masers.

Keywords. ISM: jets and outflows, submillimeter, masers, stars: formation

1. Introduction

It is well known that water masers in star-forming regions are closely associated with highly energetic gas dynamics during an early stage of evolution of the protostar, as first pointed out by Strelńitskii and Sunyaev (1972). These masers arise in hot ($\geqslant 500$ K) and dense (10^9 cm^{-3}) gas in shocked regions associated with bipolar outflows (Elitzur *et al.* 1989). VLBI observations of the 22 GHz masers have been very helpful in revealing the kinematics of the gas around young stellar objects (YSO) over length scales of one to several hundreds of AU (e.g., Torrelles *et al.* 2001a; 2001b; Uscanga *et al.* 2007; Patel *et al.* 2000). From the observations of just this single $6_{16} - 5_{23}$ transition, however, it is difficult to infer the physical conditions of the masing regions very close to the YSO. With observations of multiple masing transitions, we can use the line ratios to check theoretical models of the excitation of these lines and infer the physical conditions in the masing regions. All the known masing transitions in H_2O, except for the $6_{16} - 5_{23}$ transition lie in the millimeter and submillimeter range of wavelengths (Neufeld and Melnick 1991). With the availability of the Submillimeter Array (SMA; Ho *et al.* 2004), interferometric observations of these lines are now possible (see review by Humphreys 2007). Here we briefly present results from our study of the $10_{29} - 9_{36}$ water maser emission at 321.226 GHz (along with the 22 GHz water masers) from the high-mass star-forming region Cepheus A, as reported in more detail in Patel *et al.* (2007).

2. 321 GHz water maser emission

The $10_{29} - 9_{36}$ ortho-H_2O transition at the frequency of 321.226 GHz lies about 1200 K above the well studied $6_{16} - 5_{23}$ transition at 22.235 GHz (which itself is about 640 K above the ground level). Hence, the excitation of the 321 GHz line would require relatively much higher temperature and these masers are expected to arise in highly shocked regions. The first astronomical detection of the 321 GHz maser emission was made by Menten *et al.* (1990) using the Caltech Submillimeter Observatory 10.4 m telescope toward the star-forming regions W3(OH), W49N, W51 IRS 2, and W51 Main and the supergiant star VY CMa. About one month after the submillimeter observations, Menten *et al.* obtained 22 GHz observations using the Haystack 37 m telescope. Their comparison of the 321 GHz and 22 GHz spectra (see Figure 3 of Menten *et al.* 1990) show the 321 GHz emission to be significantly weaker and having very different line profiles compared to the 22 GHz spectra. The angular resolution in these single dish observations was inadequate to check if the 22 GHz and 321 GHz masers were arising from the same location.

3. Cepheus A high-mass star-forming region

Cepheus A is a well-known high-mass star-forming region located at a distance of ~725 pc in the Cepheus OB3 complex of molecular clouds (Sargent 1977). Several compact objects in this region have been identified from radio continuum observations at centimeter wavelengths with the VLA (hereafter described as HW sources: Hughes & Wouterlout 1984; Garay *et al.* 1996). The 22 GHz water masers have been observed towards several of these HW sources (Torrelles *et al.* 1996, 2001a, 2001b; Vlemmings *et al.* 2006; Gallimore *et al.* 2004). HW2 appears to be the most dominant among these sources and the 1.3 cm and 3.6 cm thermal jet associated with HW2 is shown to have proper motions of ~500 km s^{-1} (Curiel *et al.* 2006). Perpendicular to this jet and nearly coincident with it, was found an elongated structure in 330 GHz continuum emission observed with the SMA. This structure was interpreted to be a circumstellar disk around HW2 (Patel *et al.* 2005). A different interpretation has been suggested by Brogan et al. (2007) who propose that the elongated structure is due to multiple sources within 1" of HW2. See also Comito *et al.* (2004), Martín-Pintado *et al.* (2005). VLBI observations of methanol masers in this region are reported by Sugiyama *et al.* (2007) and Torstensson & van Langevelde (2007).

4. SMA and VLA observations

We observed the Cepheus A region with the SMA on 2004 August 30 using seven antennas in the extended configuration with a maximum baseline length of 220 m. The phase center was $\alpha(2000) = 22^h 56^m 17.971^s, \delta(2000) = +62°01'49.''279$. We used a tuning of 321.226 GHz to center the water maser lines in the lower sideband (LSB). With uniform weighting, the synthesized beam was $0.''8 \times 0.''7$ with a position angle (P. A.) of -80°. The rms noise in the images was ~160 mJy/beam.

We also observed the 22.235 GHz water masers in Cepheus A with the VLA in the A configuration on 2004 October 12. The rms noise level in our VLA observations is about 3.3×10^{-3} Jy beam^{-1} in channels free of maser emission and $\sim 4.5 \times 10^{-1}$ Jy beam^{-1} in the channel with the strongest maser spot. The line and continuum data at 1.3 cm were obtained simultaneously and the astrometric registration between the maser and continuum emission is better than $0.''01$.

5. Results and discussion

The submillimeter maser emission was detected only from HW2 and HW3c in the Cepheus A region. The 1.3 cm continuum emission is shown as contours in Figure 1 for these two sources. The 22 GHz masers are shown as crosses and the positions of the 321 GHz masers are shown with error bars representing their 1σ uncertainties. The 321 GHz masers appear to lie along the thermal jet with velocities in agreement with the larger scale bipolar outflow (Gomez *et al.* 1999; Rodriguez *et al.* 1980). The 22 GHz masers on the other hand, appear to be along the perpendicular direction, presumably associated with the elongated structure seen in submillimeter continuum emission.

In HW2, only 2 out of the 7 submillimeter masers are found to be coincident with the 22 GHz masers (occurring within few tens of milliarcseconds and <3 km s^{-1}). In HW3c, the single submillimeter maser spot is found to be agreeing in position and velocity with the 22 GHz maser. There are several 22 GHz masers unaccompanied by the 321 GHz maser emission. The spectra of 321 GHz and 22 GHz masers (see Figure 2 of Patel *et al.* 2007) show that in HW2, the 321 GHz masers seem to prefer higher velocities w.r.t. the systemic velocity of HW2.

Shortly after the discovery of the 321 GHz masers, Neufeld & Melnick (1990), showed that the observed 22 and 321 GHz masers can be explained by collisional excitation within the same volume of gas. These authors presented models of the emissivity ratio R of 22 GHz/321 GHz maser luminosity as a function of ξ and T, where ξ is the "maser emission measure" defined by Elitzur *et al.* 1989 and is proportional to the square of the gas density. Yates *et al.* (1997) carried out similar calculations covering a wider range of physical conditions than in Neufeld & Melnick (1990). According to these models, the 321 GHz maser is strongly inverted under more restricted conditions than the 22 GHz line (see Fig. 3 of Neufeld & Melnick 1990 and Fig. 4a of Yates *et al.* 1997). Hence, there can be 22 GHz emission but the 321 GHz masers must always have associated 22 GHz masers.

In the three regions where the 321 and 22 GHz masers appear coincident, the observed emissivity ratio R (22/321 GHz) implies the kinetic temperature to be in the range of 500–2000K, depending on the value of ξ. According to Neufeld & Melnick (1990), the typical value of log ξ would be in the range of -1.25 to -0.25 for the shocked gas in star-forming regions and their Figure 3 would then imply a temperature of $\sim$ 600K to 2000K for a wide range of ξ.

With respect to the remaining six 321 GHz masers which do not appear to be associated with 22 GHz masers, we cannot apply any of the existing theoretical models to these $R \to 0$ cases that imply that the 22 GHz masers are quenched but the 321 GHz emission remains strong. According to Figure 3 of NM90, which has the lowest contour corresponding to $R = 1.5$, the regions with $R \to 0$ may imply temperatures greater than 2000K.

The 22 GHz masers are well known to be time-variable. Previous VLBA observations of Cepheus A (Torrelles *et al.* 2001a, 2001b) and IRAS 21391+5802 (Patel *et al.* 2000), show that several masing spots disappear over time-scales of 1 month. These masers however, tend to be relatively weaker. The theoretical models as in NM90 predict strong emission at 22 GHz corresponding to the 321 GHz masers. If these corresponding 22 GHz masers existed during the epoch of the submillimeter observations, they are unlikely to have been completely extinguished about a month later in the centimeter wavelength observations. See also Lekth *et al.* (1982) and Brand *et al.* (2007).

The 22 GHz masers that are not associated with the 321 GHz masers are likely to be arising in relatively cooler regions among the various HW sources in Cepheus A,

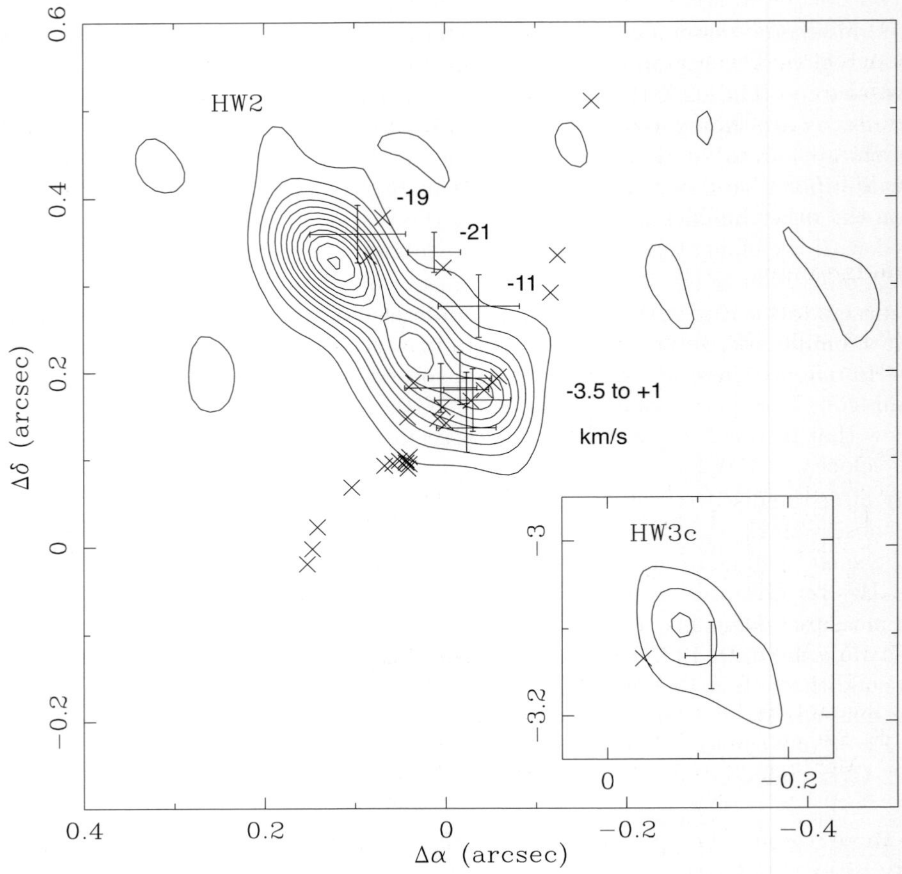

Figure 1. VLA map of 1.3 cm continuum emission from HW2 and HW3c (*inset*). The positions of submillimeter water masers observed with the SMA are shown with error bars representing the formal uncertainties. The position offsets were obtained by fitting two-dimensional elliptical Gaussians to integrated intensity maps over channels in which the maser emission appeared. The numbers denote the LSR velocities of the 321 GHz water masers. (The systemic velocity of the HW2 source is ~ -10 km s^{-1}.) The position offsets are with respect to the absolute coordinates: $\alpha(2000) = 22^h\,56^m\,17.971^s, \delta(2000) = +62°01'49.''279$. The crosses show the positions of the 22 GHz water masers observed with the VLA (43 days after the SMA observations). The errors in the positions of the 22 GHz water masers are smaller than the size of the symbols shown. The VLA beam is $0.''09 \times 0.''07$ (P.A. 14°). The SMA beam is $0.''8 \times 0.''7$ with a P.A. of -80°. Figure from Patel *et al.* 2007

compared to HW2 and HW3c. The fact that the submillimeter wavelength masers in HW2 are found along the major axis of the jet, lead us to propose that they are arising in hotter gas regions due to the impact of the jet.

6. Conclusions and future work

(1) The 321 GHz maser emission was detected in HW2 and HW3c in the Cepheus A high-mass star-forming region using the SMA. (2) 3 out of 9 submillimeter masers are found to be associated with the 22 GHz masers. (3) Submillimeter masers appear to

trace the outflow jet in HW2, presumably arising in hot gas of temperature about 1000–2000 K. (4) Multi-transition observations of water masers can be used to constrain theoretical models of their excitation, and when these models are improved, such observations can be useful to probe the physical conditions of the masing regions. We plan to observe other high-mass star-forming regions such as W75N, W3(OH), W3 IRS5 and NGC 6334N with the SMA and VLA, nearly simultaneously. We also look forward to the availability of the 400 GHz band receivers on the SMA to allow observations of water maser transitions in this band (Melnick *et al.* 1993).

Acknowledgements

We are very grateful to Jim Moran, Gary Melnick, David Neufeld, Elizabeth Humphreys and the SMA staff. NP would like to thank Jan Brand and Andrej Sobolev for helpful discussions on 22 GHz maser variability at this conference.

References

Brand, J., Felli, M., Cesaroni, R. *et al.* 2007, *this volume*
Brogan, C., Chandler, C., Hunter, T., *et al.* 2007 *to appear in ApJL, astro-ph/0703626*
Comito C., Schilke, P., Jiménez-Serra, I. & Martín-Pintado, J. 2005, *IAU 231 Symposium Astrochemistry throughout the Universe, eds: Lis, D. C., Blake, G. A. & Herbst, E., Cambridge University Press*
Curiel, S., *et al.* 2006, *ApJ*, 638, 878
Elitzur, M., Hollenbach, D. J., & McKee, C. F. 1989, *ApJ*, 246, 983
Gallimore, J. F., Cool, R. J., Thornley, M. D., & McMullin, J. 2003, *ApJ*, 586, 306
Garay, G., Ramírez, S., Rodríguez, L. F., Curiel, S., & Torrelles, J. M., 1996, *ApJ*, 459, 193
Gómez, J. F., Sargent, A. I., Torrelles, J. M., *et al.* 1999, *ApJ*, 514, 287
Ho, P. T. P., Moran, J. M. & Lo, K. Y., 2006, ApJ, 616, L1
Hughes, V. A. & Wouterloot, J. G. A. 1984, *ApJ*, 276, 204
Humphreys, E. 2007 *this volume*
Lekht, E., Pashchenko, M., Rudnitskii, G., & Sorochenko, R. 1982, *Soviet Astron.*, 26, 168
Martín-Pintado, J., Jiménez-Serra, I., Rodríguez-Franco, A., Martín, S., & Thum, C. 2005, ApJ, 628, L61
Menten, K. M., Melnick, G. J., & Phillips, T. G. 1990, *ApJ*, 350, L41
Melnick, G., J., Menten, K. M., Phillips, T. G. & Hunter, T., 1993, ApJ, 416, L37
Neufeld, D. A., & Melnick, G. J. 1990, *ApJ*, 352, L9
——— 1991, *ApJ*, 368, 215
Patel, N. A., Greenhill, L. J., Herrnstein, J., *et al.* 2000, *ApJ*, 538, 268
Patel, N. A, Curiel, S., Sridharan, T. K., *et al.* 2005, *Nature*, 437, 109
Patel, N. A, Curiel, S., Zhang, Q., *et al.* 2007, *ApJ*, 658, 55
Rodriguez, L., Moran, J. M., & Ho, P. T. P. 1980, *ApJ*, 240, 149.
Sargent, A. I. 1977, *ApJ*, 218, 736
Strelńitskii, V. S. & Sunyaev, R. A. 1972, *Soviet Astron.*, 16, 579.
Sugiayama, K., FujiSawa, K., Honma, M., *et al.* 2007, *this volume*
Torrelles, J. M., Gómez, J. F., Rodríguez, L. F., Curiel, S., Ho, P. T. P., & Garay, G., 1996, ApJ, 457, L107
Torrelles, J. M., *et al.* 2001a, ApJ, 560, 853
——— 2001b, Nature, 411, 277
Torstensson, K. & van Langevelde, H. 2007, *this volume*
Uscanga, L., Canto, J. & Raga, A. C. 2007, *this volume*
Vlemmings, W. H. T., Diamond, P. J., van Langevelde, H. J., & Torrelles, J. M. 2006, *A&A*, 448, 597.
Yates, J. A., Field, D., & Gray, M. D., 1997, *MNRAS*, 285, 303.

Astrophysical Masers and their Environments
Proceedings IAU Symposium No. 242, 2007
J.M. Chapman & W.A. Baan, eds.

doi:10.1017/S1743921307013658

SMA imaging of the masers in MWC349A

Jonathan Weintroub[1] †, James Moran[1], Ramprasad Rao[2], Hiroko Shinnaga[3], David Wilner[1] and Ken Young[1]

[1]Harvard-Smithsonian CfA, 60 Garden Street, MS78, Cambridge, MA 02138, USA
email: jweintroub@cfa.harvard.edu

[2]ASIAA, 645 North A'ohoku Place, Hilo, Hawaii 96720

[3]Caltech Submillimeter Observatory, 111 Nowelo Street, Hilo, HI 96720

Abstract. In September 2004 and September 2005 we used the Submillimeter Array (SMA) to map the distribution of hydrogen recombination line (HRL) maser spots that originate in the envelope of the peculiar star MWC349A. We measured positions in the line over a 100 km/s velocity range with 1 km/s spectral resolution. Relative angular position was estimated to an accuracy of about 0.01 arcseconds. The H30α maser components can be tracked continuously in frequency between the red- and blue-shifted peaks of the spectrum. The masers appear to trace a nearly edge-on disk-like structure. The position-velocity relationship is quite remarkable, and reminiscent of that of NCG4258. However, the "wings" in the PV relation are too steep for a simple Keplerian prescription to be correct. We present our data, and suggest a simple model to account for the steep wings.

Keywords. masers, instrumentation: high angular resolution, submillimeter, radio lines: stars

1. Introduction

The hydrogen recombination line emission from the peculiar B[e] star MWC349A is double-peaked at frequencies above 100 GHz, and is thought to be due to masering in a gaseous rotating circumstellar disk. The star also has an ionized stellar wind outflow. The outflow is suggested by VLA continuum maps [for example Tafoya *et al.* (2004)] showing an hourglass structure, as well as a characteristic spectral index of 0.6. Maser spot maps from OVRO [Planesas *et al.* (1992)], show two spots, one associated with each spectral peak, centered about 0.05 arcseconds apart with a position angle of about 107°. The position angle of the spots is aligned with that of the waist of the hourglass structure in the continuum map, suggesting that the maser spots are located in the waist. Figure 1 shows the VLA continuum data from White and Becker (1985) with the Keck aperture masking data of Danchi *et al.* (2001) and the OVRO 230 GHz spotmap of Planesas *et al.* (1992) overlaid.

2. Results

In this paper we present SMA images of the maser emission in the H30α line which show much more detail than the OVRO map. The upper panel of figure 2 shows the maser spot positions derived from the SMA data plotted as a spotmap. The position of the two spot clusters is consistent with the OVRO map in that they are centered approximately $0''.05$ apart, however there are many more spots. Spots are also present between the clusters in an elongated distribution. The lower panel shows radial velocity against RA.

† Present address: SAO, 60 Garden Street, MS-78, Cambridge, MA 02138

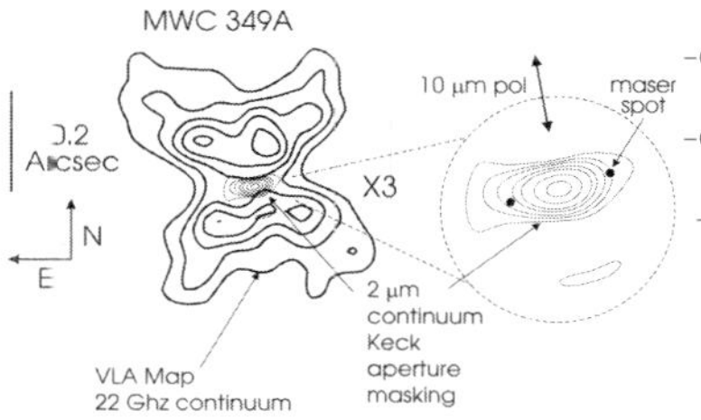

Figure 1. Taken from Danchi *et al.* (2001), this figure shows an overlay of representative continuum data and H30α interferometric data published on MWC349A prior to this work.

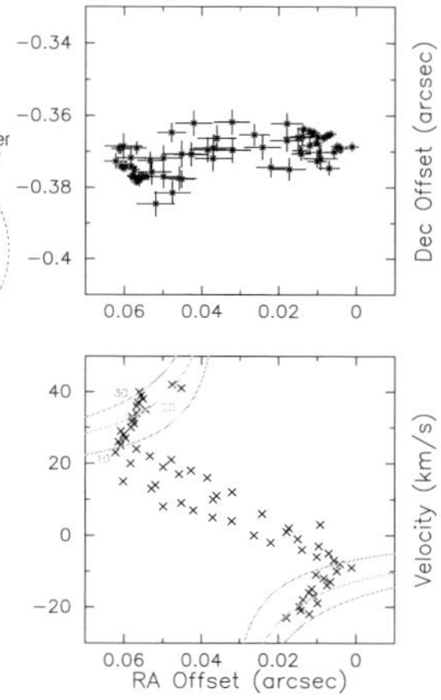

Figure 2. The SMA spotmaps and a PV diagram.

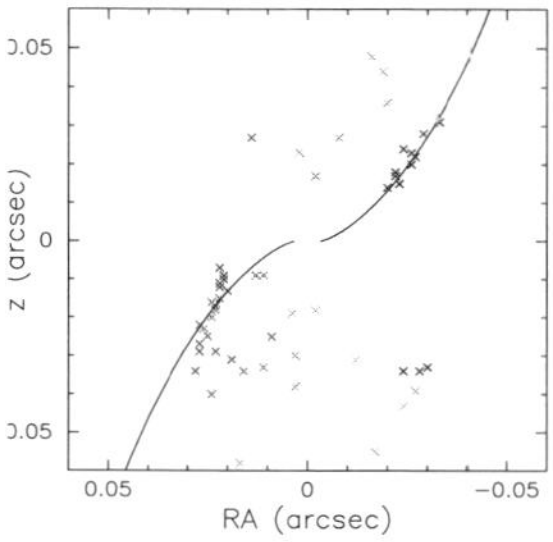

Figure 3. A deprojected SMA H30α spotmap made under the assumptions discussed in the text, with a possible spiral model for the maser spot distribution overlaid

This PV diagram shows a region of apparent solid body rotation with wings reminiscent of the Keplerian wings of NGC4258. However as the three overplotted Keplerian curves (10, 20 and 30 $M_{\odot}$) show, the wings in the measured data are in fact much steeper. An alternate explanation for the apparent solid body region is that the masers between the clusters might be concentrated in an annular ring.

Figure 3 shows a deprojected view of the maser distribution along an axis perpendicular to the line of sight and the axis of the maser distribution. The azimuth angle is determined by assuming Keplerian rotation. The red- and blue-shifted features are assumed to arise in front of and behind the midline, respectively. The rapid change in azimuth angle with velocity can explain the steepness of the PV plot, suggesting a spiral distribution of maser spots; a possible spiral model is overplotted.

3. Conclusions

With more baselines, greater sensitivity and a superior site, the SMA has allowed us to closely examine the dynamics of the stellar disk in MWC349A. We show that either the circumstellar disk is not in simple Keplerian rotation, or, if the disk rotation is Keplerian, we suggest that the high velocity masers may trace spiral density arms. We are working on more elaborate dynamical models which combine the disk rotation with the outflow.

Acknowledgements

The Submillimeter Array is a joint project between the Smithsonian Astrophysical Observatory and the Academia Sinica Institute of Astronomy and Astrophysics and is funded by the Smithsonian Institution and the Academia Sinica.

References

Danchi, W. C., Tuthill, P. G., & Monnier, J. D. 2001, ApJ, 562, 440
Planesas, P., Martín-Pintado, J., & Serabyn, E. 1992, ApJ, 386, L23
Tafoya, D., Gómez, Y., & Rodríguez, L. F. 2004, ApJ, 610, 827
White, R. L. & Becker, P. H. 1985, ApJ, 297, 677

Astrophysical Masers and their Environments
Proceedings IAU Symposium No. 242, 2007
J.M. Chapman & W.A. Baan, eds.

doi:10.1017/S174392130701366X

Exciting maser science with new instruments – the promise of the EVLA

Karl M. Menten

Max-Planck-Institut für Radioastronomie, Auf dem Hügel 69, D-53121, Germany

Abstract. In the near future, the Expanded Very Large Array (EVLA) will allow surveys for maser sources with unprecedented sensitivity, spectral coverage and spectroscopic capabilities. In particular, comprehensive surveys for many maser species with simultaneous sensitive continuum imaging and absorption studies will give a comprehensive radio picture of star formation in the Galactic plane and elsewhere. Very efficient EVLA surveys for H_2O megamasers in Active Galacic Nuclei will be possible to practically arbitrary redshifts.

EVLA and Atacama Large Millimeter Array (ALMA) studies of H_2O and SiO masers will serve as high resolution probes of the innermost envelopes of oxygen-rich evolved stars and HCN masers of carbon-rich stars.

Farther in the future, the Square Kilometer Array (SKA) promises the detection of OH gigamasers at all conceivable redshifts and maser astrometry with unprecedented accuracy.

Keywords. masers, instrumentation: high angular resolution, instrumentation: interferometers, surveys, stars: formation, Galaxy: structure

1. Introduction – maser surveys

Arguably, one of the main uses for astronomical masers is their signpost function. To give a few examples: Interstellar water vapor (H_2O) masers arise from high velocity outflows of accreting low- and high-mass protostars. Many class II methanol (CH_3OH) masers mark the positions of deeply embedded protostars that are difficult or frequently impossible to locate by any other phenomenon. 1612 MHz hydroxyl (OH) masers mark intermediate-mass stars during a short-lived phase of high mass loss. Surveys for maser emission thus help finding interesting sources and often have surprising results.

1.1. *Two types of surveys*

There are, essentially, two types of surveys - *Blind surveys* and *targeted surveys.*

1.1.1. *Blind surveys*

Blind surveys cover large areas of the sky in an unbiased way, integrating (possibly) uniformly to a pre-set flux density limit. The covered area may be the whole sky, like for the NRAO Very Large Array Sky Survey (NVSS), a 1.4 GHz continuum survey of all the sky north of $-40°$ declination (Condon *et al.* 1998), but for maser surveys they usually are not. Rather, most blind maser surveys cover selected molecular clouds or portions of the Galactic plane.

An important example of the latter is the unbiased OH maser survey of the southern Galactic plane conducted by Caswell and collaborators (Caswell & Haynes 1987 and references therein).

Surveys for H_2O masers in the spiral arms of external galaxies may cover whole galaxies, which for nearby systems such as M31 or M33 comprise many hundreds or even thousands of separate pointings. Blind surveys can be made either with multiple separate pointings, measuring, for full sampling, a grid of 0.5×FWHM beam-width spaced positions. Alternately and more efficiently, "on-the-fly" mapping may be employed, during which the telescope is constantly moved at a certain slew rate and data are taken continuously. In the past, most blind surveys were conducted using single dish telescopes. For newly found sources, in many cases improved positions were determined with follow-up observations with accuracies limited by telescope pointing errors (usually a few arcseconds) or much smaller ($\ll 1''$) with interferometers.

Clearly, full-interferometric surveys are highly preferable (see §2), delivering much better gain stability, position accuracy and direct information on the small-scale maser distributions. However, the limited correlator bandwidth of today's most effective interferometer, the NRAO Very Large Array (VLA), has been a severe problem in conducting efficient blind surveys at high frequencies, in particular for sources with a priori unknown velocities (see §2.1).

1.1.2. *Targeted surveys*

In contrast, *targeted surveys* consist of pointed observations toward sources that were identified as promising targets based on certain criteria. If one searches for masers in a certain species, one such criterion might be the existence of maser emission from another species at or in the vicinity of the target position. Examples for this are searches toward the positions of the OH masers found in the Caswell surveys (§1.1.1), for H_2O masers (Caswell *et al.* 1989 and references therein) and CH_3OH masers (Norris *et al.* 1987, Menten 1991, Caswell *et al.* 1995).

Another example is extragalactic H_2O masers, all of which were either found in targeted searches toward HII regions complexes (first in M33; Churchwell *et al.* 1977) or the nuclear regions of the target galaxies. The latter masers are located either in a parsec-scale accretion disk around an active galactic nucleus or interacting with a jet emerging from it (see, e.g., Lo 2005 and Greenhill, these proceedings) or from a starburst region surrounding the nucleus on a few hundred pc scale (e.g., Castangia *et al.*, these proceedings).

Extremely fruitful selection criteria for finding masers and the interesting sources that host them have been color criteria derived from mid/far-infrared surveys. Namely, the InfraRed Astronomy Satellite (IRAS) covered the whole sky at 12, 25, 60, and 100μm, which resulted in the IRAS Point Source Catalogue. A survey of a $\pm 5°$ latitude wide strip of the whole Galactic plane was made with the Midcourse Satellite Experiment, MSX, in several IR bands. Representative portions of the plane were covered by other surveys, e.g., "ISOGAL" with the Infrared Space Observatory, ISO (Omont *et al.* 2003, Schuller *et al.* 2003). New multi-IR band space-borne surveys of the Galactic plane and the whole sky with the Spitzer Space Observatory (GLIMPSE, Benjamin *et al.* 2003) and Akari (Murakami *et al.*2007), respectively, will yield many thousands of targets for maser surveys. We now give a few examples for IR color-selected maser surveys.

A series of successful 1612 MHz OH satellite line surveys with the Australia Telescope Compact Array (ATCA) and the VLA using IRAS colors resulted in the detection of more than a thousand OH/IR stars the asymptotic giant branch (AGB) lifetime (Sevenster *et al.* 2001 and references therein).

Deguchi and collaborators have highly effectively used IRAS colors to identify targets for very successful 43 GHz $v = 1$ and $2, J = 1 \rightarrow 0$ SiO and 22.2 cm H_2O maser surveys of practically all of the Galactic plane accessible with the Nobeyama 45 m telescope. Using infrared magnitudes and colors from the ISOGAL and MSX catalogues, Messineo *et al.* (2002) searched for SiO maser emission in 441 objects and detected 271 of them.

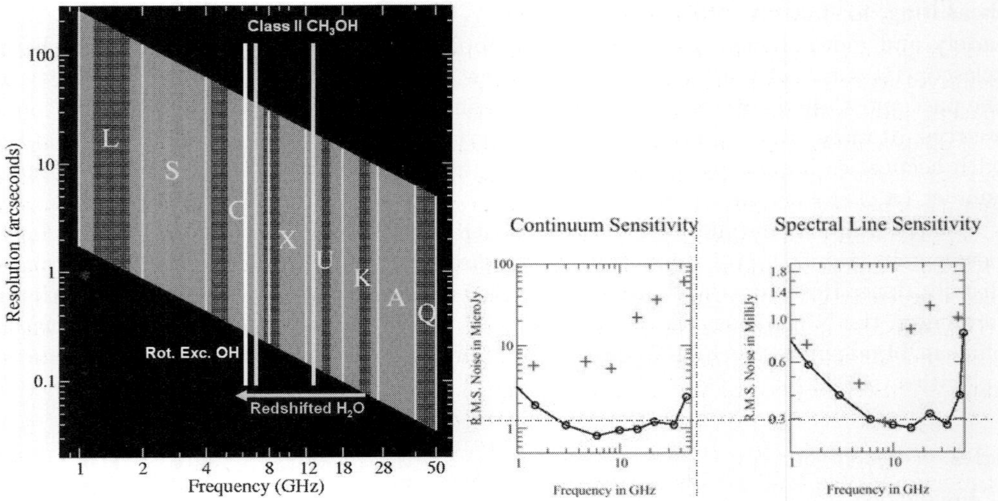

Figure 1. *Left*: Comparison of the frequency coverage of the present VLA (dark stripes) with that of the EVLA (dark plus lighter stripes, i.e. continuous coverage from 1 to 50 GHz). The frequencies of the 6.7 and 12.2 GHz CH_3OH and the 6.0 GHz OH lines are indicated. Also indicated is the EVLA's virtually limitless redshift coverage of the 22.2 GHz H_2O line. Note that the H_2O line from the $z = 0.660$ QSO SDS J0804+3607 is redshifted to 13.4 GHz (Barvainis & Antonucci 2005). *Right*: Comparison of the sensitivities of the present VLA (crosses) and the EVLA (continuous line). Shown is the 1σ rms noise level reached in 12 h of integration for continuum and spectral line observations in a 1 km s^{-1} wide velocity channel. Figures are taken from the EVLA proposal, which is available on the EVLA WWW site: http://www.aoc.nrao.edu/evla/pbook.shtml.

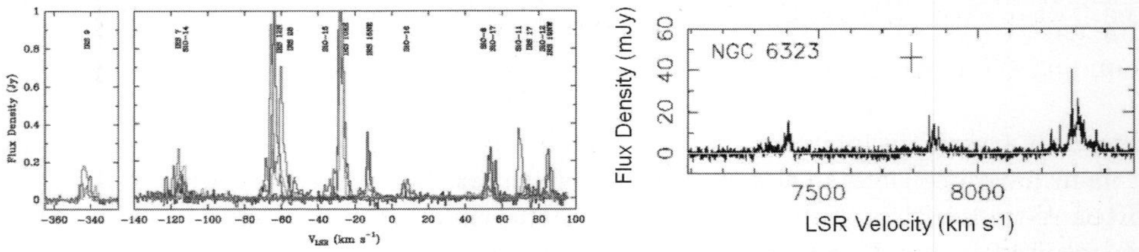

Figure 2. *Left*: Composite spectrum of the velocity range over which $v = 1, J = 1 \rightarrow 0$ SiO maser emission was found with the VLA toward the Galactic center. That this is a "mosaic" obtained with seven different frequency settings was necessitated by the present VLA correlator's limited capabilities (from Reid *et al.* 2007). *Right*: Spectrum of the 22.2 GHz H_2O line toward the active galaxy NGC 6323 taken with a wideband spectrometer at the Green Bank Telescope. The channel spacing is 0.33 km s^{-1} (from Braatz *et al.* 2004). To cover the 1100 km s^{-1} with emission at the same resolution with the current VLA correlator would require 37 settings.

While all of the above surveys found many hundreds of objects allowing statistical and kinematical studies (like characterizing the Galactic bar potential; see Habing *et al.* 2006 and Habing's contribution to these proceedings), they also resulted in the serendipitous discovery of a number of highly interesting objects. For example, 1612 MHz OH maser sources with extremely wide and/or peculiar velocity appearance could be classified as extreme mass-loss proto-planetary nebulae: important, short-lived and, therefore, difficult to find transition objects in the evolution from asymptotic giant branch (AGB) star to planetary nebula (e.g., Zijlstra *et al.* 2001, Deacon *et al.* 2007; see also Chapman's contribution to these proceedings).

2. The Expanded Very Large Array

Throughout its operation the VLA has made significant contributions to maser science, mainly through observations of ground-state hydroxyl (OH) and water vapor (H_2O) masers and, from the 1990s on, of SiO masers.

If phase calibration can be achieved, for maser surveys and time monitoring an interferometer is much preferred to a single dish telescope equipped with receivers of comparable quality. This is particularly true for the VLA, for the following reasons: (a) At K and Q band (H_2O and SiO masers), the VLA has a larger collecting area than any single dish (1.7 times that of a 100 m telescope). At longer wavelengths, the Arecibo 300m telescope has a much larger collecting area but can only observe a much smaller part of the sky. (b) Atmospheric disturbances do not correlate. Consequently, spectra produced from interferometer data have higher quality. (c) All of the VLA's receivers are available at all times. This is important for monitoring and not generally true for other observatories. (d) Importantly, even in its lowest resolution configuration (D-array) and, in particular at its maximum A-array resolutions around $1''$, $0.''08$, and $0.''04$ at 1.7, 22.2, and 43 GHz, respectively, the VLA yields good enough *relative* positions to get a detailed picture of the distribution of individual maser components and good enough *absolute* positions for VLBI correlators.

Representing 1970's technology, the VLA, despite its continuing success, has many shortcomings, preventing it from being the ideal spectral-line and in particular maser survey instrument. The Expanded Very Large Array (EVLA) will overcome many of the VLA's limitations. While using the existing VLA's infrastructure (antennas, railroad tracks, general logistics), it will be equipped with more sensitive receivers with much larger bandwidth, full spectral coverage from 1 to 50 GHz, an immensely more powerful correlator and much faster electronics.

Comprehensive, detailed information on all technical aspects of the EVLA can be found on the EVLA Project Book WWW site: http://www.aoc.nrao.edu/evla/pbook.shtml

2.1. *Frequency coverage*

Given the bandwidths of receivers available in the 1970s, the VLA has a number of discrete wavelength bands each of limited width (see Fig. 1). Two of these, the 18–21 cm L band and 1.3 cm K band, contain important maser lines from the OH and the H_2O molecules, respectively. The 7 mm Q band added in the mid-1990s gave access to SiO maser lines. The very widespread 6.7 GHz (4.5 cm) CH_3OH line, detected in 1991, and the 6.0 GHz (5 cm) OH hyperfine structure lines, excellent probes of the Zeeman effect, are not covered as aren't many other maser lines, e.g., from the NH_3 molecule (see Table 1).

A key EVLA requirement is continuous frequency coverage from 1 to 50 GHz. This will be met with 8 bands (see Fig. 1): Two existing (K, Q), four replaced (L, C, X, U), and two new (S, A) bands. The existing meter-wavelength bands, P (90 cm) and 4 (4 m) are retained with no changes. K and Q band will benefit from greatly improved receivers. In Fig. 1 the dark stripes show the frequency coverage of the present VLA while the dark plus light stripes show that of the EVLA. All bands will have significantly wider instantaneous width than in the current VLA. In the highest frequency (K and Q) bands the bandwidth will be 2×8 GHz each (RCP+LCP); compare to today's 2×100 MHz! While affording fantastic continuum sensitivities, all the maser lines listed in Table 1 will be observable. In Fig. 1 the frequencies of the 6.7 and 12.2 GHz CH_3OH and the 6.0 GHz OH lines are marked. Also indicated is the EVLA's virtually limitless redshift coverage of the 22.2 GHz H_2O line. Note that the (by far) highest redshift H_2O line yet detected (from the $z = 0.660$ QSO SDS J0804+3607) is shifted to 13.4 GHz (Barvainis & Antonucci 2005).

2.2. *The correlator*

Most higher frequency (K and Q band) maser observations with the VLA have been severely limited by its 1970s technology correlator. The VLA correlator has a maximum bandwidth of 50 MHz per intermediate frequency (IF) band. In unsplit mode (i.e. 1 IF) it has a maximum of 16 frequency channels, which reduces to 8 per band if we chose a 2 IF mode, e.g., right plus left circular polarization and 4 per band for a 2×(RCP+LCP) = 4 IF mode, where two IF pairs with different frequencies (which have to be within 500 MHz of each other) are employed. For simplicity's sake let us only consider the 1 IF (1 polarization) mode for the 22.2 GHz H_2O line: To adequately resolve individual narrow maser features (and to avoid "ringing" caused by poorly resolved strong lines), we require a velocity resolution of 0.2 km s^{-1}, corresponding to 14.8 kHz, which is close to a correlator configuration that affords 256 channels spaced by 12.2 km s^{-1}. Considering that, due to band edge roll-off, only the inner 70% of the bandpass are usable, the total velocity range that can be covered is 30 km s^{-1}, which is inadequate for many interstellar H_2O masers, which frequently have velocities spread over many tens of km s^{-1}, even if the systemic velocity of the source in question is known. Increasing the bandwidth by a factor of 2 results in half the channels and an inadequate four times worse resolution of 48.8 kHz (0.66 km s^{-1}) since we are dealing with a "recirculating" correlator. For SiO masers (Q band) the situation is even much worse, in particular for the kind of searches described in §1.1.2, in which the stars' velocities are a priori unknown.

Extragalactic H_2O megamaser spectra frequently consist of narrow features with a width comparable to that of Galactic maser features (≈ 1 km s^{-1}; see Fig. 2). For the most interesting H_2O megamaser sources (those with NGC 4258-like Keplerian rotation signatures; see Miyoshi *et al.* 1995) maser features are found around the systemic velocity as well as at velocities offset by up to ~ 1000 km s^{-1} of the latter. Consequently, when searching for new such sources one needs to scan a total range of ≈ 2000 km s^{-1}, a tedious job with 30 km s^{-1} increments. The situation is aggravated by the fact that these masers are variable and many of their high velocity components are weak, which requires repeated and deep observations. Fig. 2 shows an H_2O spectrum taken toward the AGN at the center of NGC 6323 taken with the Green Bank telescope equipped with 200 MHz wide spectrometer with a channel spacing corresponding to 0.33 km s^{-1}.

The EVLA's correlator will overcome all of the described problems.

2.3. *"Dead times"*

Every change in a VLA observation, be it a change in position or in the frequency setup, is afflicted by a "dead time" of 20 seconds, which derives mainly from limitations of the ancient electronics. This makes fast surveys in which integration times per position are small compared to these dead times extremely inefficient. As an example, the NVSS was conducted as discrete snapshot pointings with each pointing taking 30 s of which only 7.7 s was actual dwell time (Condon *et al.* 1998).

On-the-fly (OTF) observations (see §1.1.1) are impossible with the current VLA. A glimpse at the power of fast-scanning OTF interferometer surveys was recently provided by an 18 GHz continuum survey with a single baseline of the Australia Telescope Compact Array (ATCA). Ricci *et al.* (2004) covered 1216 square degrees with an effective dwell time of 0.2 seconds, resulting in an rms noise of 15 mJy, they obtained 221 confirmed detections > 60 mJy. For the EVLA very low overhead levels are a design feature.

Table 1. Astronomical Maser Transitions Observable with the EVLA

Species	Transition	Frequency (MHz)	Number/Type
OH $^2\Pi_{3/2}$	$J=\frac{3}{2}$, $F=1\rightarrow 2$	1612.2310(2)	*T* SFR/*H* O-CSE
	$F=1\rightarrow 1$	1665.4018(1)	*H* SFR/*T* O-CSE
	$F=2\rightarrow 2$	1667.3590(1)	*H* SFR/*T* O-CSE
	$F=2\rightarrow 1$	1720.5300(1)	*H* SFR
	$J=\frac{5}{2}$, $F=2\rightarrow 2$	6030.747(5)	*T* SFR
	$F=3\rightarrow 3$	6035.092(5)	*T* SFR
	$J=\frac{7}{2}$, $F=4\rightarrow 4$	13441.4173(2)	*S* SFR/*T* SNR
OH $^2\Pi_{1/2}$	$J=\frac{1}{2}$, $F=0\rightarrow 1$	4660.242(3)	*T* SFR
	$F=1\rightarrow 1$	4750.656(3)	*T* SFR
	$F=1\rightarrow 0$	4765.562(3)	*T* SFR
H_2O	$J_{K_a,K_c}=6_{16}\rightarrow 5_{23}$	22235.08(2)	*H* SFR/*H* O-CSE
CH_3OH[a]	$J_k=5_1\rightarrow 6_0A^+$	6668.5192(8)	*H* SFR/Cl II
	$9_{-1}\rightarrow 8_{-2}E$	9936.202(4)	*1* SFR/Cl I (W33-Met)
	$2_0\rightarrow 3_{-1}E$	12178.597(4)	*H* SFR/Cl II
	$2_1\rightarrow 3_0E$	19967.3961(2)	*S* SFR/Cl II
	$9_2\rightarrow 10_1A^+$	23121.0242(5)	*S* SFR/Cl II
	$6_2\rightarrow 6_1E^{b}$	25018.1225(4)	*T* SFR/Cl I
	$7_2\rightarrow 7_1E$	25124.8719(4)	*T* SFR/Cl I
	$8_2\rightarrow 9_1A^-$	28969.942(50)	*S* SFR/Cl I
	$4_{-1}\rightarrow 3_0E$	36169.265(30)	*T* SFR/Cl I
	$7_{-2}\rightarrow 8_{-1}E$	37703.700(30)	*S* SFR/Cl II
	$6_2\rightarrow 5_3A^+$	38293.268(50)	*S* SFR/Cl II
	$6_2\rightarrow 5_3A^-$	38452.677(50)	*S* SFR/Cl II
	$7_0\rightarrow 6_1A^+$	44069.410(10)	*T* SFR/Cl I
NH_3[c]	$(J,K)=$ (9,6)	18499.390(5)	*S* SFR
	(6,3)	19757.538(5)	*S* SFR
	(3,3)	23870.1296(1)	*S* SFR
HC_3N	$J=1\rightarrow 0$	9098.1152 (2)	*1* SFR
H_2CO	$J_{K_a,K_c}=1_{10}\rightarrow 1_{11}$	4829.6600	*S* SFR
CH $^2\Pi_{1/2}$	$J=\frac{1}{2}$, $F=0\rightarrow 1$	3263.795(3)	*T* SFR
	$F=1\rightarrow 1$	3335.481(2)	*T* SFR
	$F=1\rightarrow 0$	3349.194(3)	*T* SFR
SiO $v=0$	$J=1\rightarrow 0$	43423.858(10)	*1 SFR* (Orion-KL)/*S* O-CSE
SiO $v=1$	$J=1\rightarrow 0$	43122.079(21)	*3 SFR*/*H* O-CSE
SiO $v=2$	$J=1\rightarrow 0$	42820.582(23)	*3 SFR*/*H* O-CSE
SiO $v=3$	$J=1\rightarrow 0$	42519.373(27)	*S* O-CSE
^{29}SiO $v=0$	$J=1\rightarrow 0$	42879.916(10)	*1* SFR (Orion-KL)/*S* O-CSE
^{30}SiO $v=0$	$J=1\rightarrow 0$	42373.359(10)	*1* SFR (Orion-KL)/*S* O-CSE
SiS	$J=1\rightarrow 0$	18154.880(2)	*1* C-CSE (IRC+10216)

[a] A comprehensive compilation of CH_3OH maser frequencies has been published by Müller *et al.* 2004. [b] Emission and/or absorption from lines in the $J_2\rightarrow J_1E$ series of CH_3OH has been detected from $J=3, \nu=24928.70$ MHz to $J=17, \nu=30308.08$ MHz. Clear maser emission has been detected for the $J=5$ to $J=9$ lines. The $J=6$ and 7 lines listed here are usually the strongest maser lines of this series. [c] Weak maser emission from 6 other non-metastable ($J\neq K$) $^{14}NH_3$ and 3 $^{15}NH_3$ lines between 20.8 and 23.7 GHz has been reported in the literature.

General remarks: Except for the high excitation H_2O, NH_3, and vibrationally excited SiO lines, many lines may also be observable in (quasi)thermal emission and/or absorption in regions that are not conducive to maser excitation. See the Lovas WWW site for references to the laboratory measurements of the frequencies and for the first astronomical measurements of the lines (http://physics.nist.gov/cgi-bin/micro/table5/start.pl). The numbers in parentheses following the frequencies denote the measurement error in the last significant digit(s). Abbreviations for different maser environments are SFR for star-forming regions and O-CSE and C-CSE for oxygen-rich and carbon-rich circumstellar envelopes, respectively. Maser emission from the 1720 MHz OH line has also been found in supernova remnants (SNR). A leading "*H*" means that there are hundreds of sources known, a "*T*" means tens (< 100) and an "*S*" means some (< 10). A "*1*" means that only a single source has been found to show maser action and that source is listed. Apart from Orion-KL, SiO masers have been found in the *3* SFRs W51N and Sgr B2(M). Methanol masers are marked as Class (Cl) I or to II; see Menten 1991.

2.3.1. *The power of the EVLA: The ultimate Galactic plane star formation survey*

To demonstrate the awesome power of the EVLA let us consider a blind survey to find and characterize star-forming regions in the Galaxy. Let us first consider the 4–8 GHz EVLA C band. This band contains the frequency (6.7 GHz) of the by far most prominent class II CH_3OH maser line, the presence of which is a unique tracer of high-mass star formation. The power of sensitive CH_3OH maser surveys in this line has recently been demonstrated by a blind survey with the Arecibo 300m telescope: Pandian *et al.* 2007 (see also these proceedings) observed a $b = \pm 0\overset{\circ}{.}41$ wide strip from $l = 35\overset{\circ}{.}2$ to $53\overset{\circ}{.}7$ and found, at a completeness level of 0.27 Jy, a total of 86 sources, 48 of them new detections. A large scale single-dish survey covering all of the Galactic plane with an 8 channel array receiver on the Parkes and the Jodrell Bank telescopes is currently under way (Cohen *et al.* 2007).

A possible fully sampled C band Galactic plane survey would not only cover the 6.7 GHz methanol line but, in addition, all four hyperfine structure (hfs) lines of rotationally excited OH near 6.0 GHz and the three OH lines near 4.7 GHz (see Table 1). The two strongest of the former, at 6030 and 6035 MHz, are excellent lines to study Zeeman splitting and determining magnetic field strengths in ultracompact HII regions (UCHIIRs) where they frequently exhibit maser emission. In more developed, compact HII regions they are seen in absorption (see, e.g., Fish *et al.* 2006).

Most interestingly, one would also be able to detect continuum emission from extremely weak UCHIIRs. With the 4 GHz bandwidth one would reach in 10 sec an rms noise of 41 μJy beam^{-1}. Note that the famous dual frequency Wood & Churchwell 1989 targeted survey of 75 UCHIIRs reached a comparable noise level in 4 minutes!

Splitting the continuum band in, e.g., four parts one would be able to determine the spectral properties of the continuum emission between 4 and 8 GHz, a range in which compact and many ultracompact HII regions "turn over" from an optically thick spectrum with flux density $S \propto \nu^2$ to an optically thin spectrum ($S \propto \nu^{-0.1}$). The dynamics of the ionized gas could be studied from the velocities of recombination lines. Note also, that with 50% fractional bandwidth coverage, the *uv*-plane would be heavily filled, which should result in improved imaging of the continuum emission ("bandwidth synthesis").

Moreover, it would be possible to observe formaldehyde (H_2CO) absorption in the ubiquitous 4.8 GHz ground-state doublet line against the UCHIIRs' continuum emission and, thus, resolve the kinematic distance ambiguity for the detected star-forming regions. To emphasize, by setting spectral windows around the lines in question it will be possible to image all of the mentioned lines (and more) *and* the continuum *simultaneously*.

The ideal complement to this C band survey would be a K band (18–26 GHz) survey which would detect very low luminosity H_2O masers throughout the Galaxy. Note that in contrast to CH_3OH and OH masers, H_2O masers are found in high- *and* low-mass star-forming regions. Known masers in the nearby ρ Oph or Perseus low-mass star forming regions at distances D of 160 and 350 pc, respectively, have been found with flux densities of hundreds or even thousands of Jy (Claussen *et al.* 1996). Such masers placed at the distance of the Galactic center ($D = 8$ kpc) would have flux densities of a few tenths to a few Jy and be detectable in our survey. Such water masers would, in fact, be the only means to find evidence for active low-mass star formation at distances greater than a few kpc. †

† We note however, that the luminosity function of H_2O masers is poorly constrained at the lowest luminosity end (Palla *et al.* 1993). Therefore such detections would deliver little information for putting statistical or other constraints on low-mass star formation.

Apart from detecting H_2O masers, the K band survey would highly efficiently detect hypercompact HII regions (HCHIIRs; see Kurtz 2005). These represent highly compact (< 50 AU) radio continuum emission from embedded protostellar objects with early B to O star luminosities that have an optically thick spectrum up to the highest radio frequencies. With typical K band flux densities of a few mJy or even less (at a few kpc distance) the radio emission is thought to be "choked off" by accretion in these sources (Walmsley 1995) and its intensity cannot be predicted from their total luminosity. Representing the earliest (pre-UCHIIR) stages of high-mass star formation, observations of HCHIIRs are of central interest to constrain high-mass protostellar evolution schemes.

Also covered by the K-band survey would be simultaneous searches for all of the maser lines in that range listed in Table 1 and sensitive searches for absorption lines from ammonia and other molecules toward the continuum emission of stronger UCHIIRs.

An important by-product of the K band survey would be a list of high frequency calibrator candidates whose suitability, i.e., compactness, could be tested with multi-wavelength data, more extended configuration EVLA and/or VLBI observations. A dense net of calibrator sources will be highly useful for all high frequency EVLA observations and mandatory for ALMA, for which fast-switching (requiring a nearby calibrator) will be a standard observing procedure and not an exception as currently with the VLA.

If one conducted the C band survey in the B-configuration and the K band survey in the three times more compact C-configuration, both data sets could be restored with comparable synthesized beams around $1''$, which would be highly desirable for meaningful determinations of band-to-band spectral index distributions.

2.3.2. *What the EVLA will not do*

The EVLA will not provide any enhancements over the VLA for frequencies below 1 GHz. This means, for example, that searches for high-redshift 1667 MHz OH mega-masers will be restricted to the frequency range of the current P band system (306–340 MHz), corresponding to $z = 3.90$–4.45), which will be retained unchanged. While a VLA search for OH gigamaser emission in the $z = 3.911$ broad absorption line quasar APM08279+5255, a well-known source of redshifted dust and CO emission, was unsuccessful (Ivison 2006), we note that this was the *only* known such source to fall into this redshift band. Given this situation, the detection of high-z OH gigamasers at $z > 0.67$ may have to await the Square Kilometer Array (see §4).

Moreover, very unfortunately, EVLA Phase II, consisting of physically extending the VLA by adding additional antennas to bridge the gap from VLA to VLBA resolution ("The New Mexico Array") was not funded. For a long time MERLIN, will remain the only option to cover that interesting regime at radio wavelengths. MERLIN is currently undergoing an upgrade, resulting, a.o., in wider bandwidth receivers, that will enhance its sensitivity up to 30 fold ("e-MERLIN", Garrington *et al.* 2004).

Finally, we emphasize the importance of the ATCA, the only versatile operating radio wavelength interferometer in the southern hemisphere, which will greatly increase with the advent of ALMA. The successful recent 7 mm upgrade and the soon too be installed much wider band correlator are very significant milestones.

3. The Atacama Large Millimeter Array

ALMA in its highest resolution configurations at 345 GHz, a wavelength at which weather conditions should allow observations for most of the time, in current planning will have a synthesized beam width of 12 milliarcseconds (mas), ≈ 3 times better than the highest VLA resolution. For maser science with ALMA, see Wootten's contribution to these proceedings. A few additional thoughts follow.

3.1. *Methanol masers*

So far, a total of 19 class I and class II CH_3OH maser lines with frequencies between 84 and 230 GHz have been discovered in star-forming regions (Müller *et al.* 2004). Class II CH_3OH masers show mas structure, i.e., a few AU at a few kpc and are VLBI targets (see, e.g., the contributions by Reid *et al.* and Xu *et al.* to these proceedings). In contrast, class I CH_3OH masers are heavily resolved at the highest VLA resolutions: 25 GHz maser spots in the nearby (≈ 400 pc) Orion region have sizes between 20 and 60 AU, or 50 – 150 mas (Johnston *et al.* 1997) and VLBI observations of class I masers have not produced fringes (Lonsdale *et al.* 1998). With resolutions intermediate between VLA and VLBI values, ALMA observations of class I masers will reveal their morphology. Since existing observations suggest that they mark interface regions between protostellar outflows and dense ambient gas (Menten 1996) such observations will be very interesting.

3.2. *Water masers*

Apart from the most widespread 22.2 GHz $6_{16} \rightarrow 5_{23}$ transition, a number of (sub)millimeter lines have been discovered from the vibrational ground state as well as from the lowest vibrationally excited state (see the contribution of Hunter *et al.* to these proceedings). Excitation calculations predict temperature and density ranges in which one might expect inversion of specific lines (Neufeld & Melnick 1991). To exploit this, one has to ascertain that maser emission from different lines indeed arises from identical regions. Up to now this was not possible as virtually all submillimeter H_2O data came from single dish observations. Now, (quasi)simultaneous interferometric observations of the 22.2 and 321 GHz H_2O maser lines reported in these proceedings by Patel *et al.* indeed point to a more complicated picture. They find that in the Cepheus A star-forming region the H_2O maser emission from the much higher excitation 321 GHz line (1846 K above the ground-state) comes from different regions than the emission in the lower excitation 22.2 GHz (643 K) line. Needless to say, this is a wide field awaiting ALMA.

3.3. *HCN masers*

Maser emission from hydrogen cyanide (HCN) has been detected toward several mass-losing carbon stars. Most of the maser lines are emitted from the vibrationally excited ν_2 state. For all circumstellar (SiO, H_2O, and HCN) masers it should be noted that highest resolution ALMA observations will not only reveal the structure of the maser distributions but will also resolve the stellar submillimeter photospheres of their host AGB stars (Miras) out to several hundred pc and even of further away red supergiants.

4. The Square Kilometer Array

See Green & Baan's contribution to these proceedings for a description of maser science with the SKA. Again, a few more thoughts:

The usefulness of OH mega- (or rather giga-)masers as a probe of violent star formation at early epochs, complementary to (sub)millimeter dust and molecular and radio continuum studies, has been discussed in a number of studies (Briggs 1998, Townsend *et al.* 2001, Ivison 2006). The SKA will have the potential to detect OH gigamaser emission at all redshifts of potential interest.

The SKA will open new vistas for maser astrometry with microarcsecond accuracies (Fomalont & Reid 2004) and the collecting area necessary to image NGC 4248-like systems into the Hubble flow, allowing a high-accuracy determination the Hubble constant (see Greenhill s contribution to these proceedings).

Acknowledgements

I'm very grateful to Mark Reid for a thorough reading of the manuscript of this paper and to him and Rick Perley for useful suggestions.

References

Barvainis, R., & Antonucci, R. 2005, *ApJ*, 628, L89
Benjamin, R. A., *et al.* 2003, *PASP*, 115, 953
Braatz, J. A., Henkel, C., Greenhill, L. J., Moran, J. M., & Wilson, A. S. 2004, *ApJ*, 617, L29
Briggs, F. H. 1998, *A&A*, 336, 815
Caswell, J. L., & Haynes, R. F. 1987, *Austr. J. of Physics*, 40, 215
Caswell, J. L., Batchelor, R. A., Forster, J. R., & Wellington, K. J. 1989, *Austr. J. of Physics*, 42, 331
Caswell, J. L., Vaile, R. A., Ellingsen, S. P., Whiteoak, J. B., & Norris, R. P. 1995, *MNRAS*, 272, 96
Churchwell, E., Witzel, A., Huchtmeier, W., Pauliny-Toth, I., Roland, J., & Sieber, W. 1977, *A&A*, 54, 969
Claussen, M. J., Wilking, B. A., Benson, P. J., Wootten, A., Myers, P. C., & Terebey, S. 1996, *ApJS*, 106, 111
Cohen, R. J., *et al.* 2007, IAU Symposium, 237, 403
Condon, J. J., *et al.* 1998, *AN*, 115, 1693
Deacon, R. M., Chapman, J. M., Green, A. J., & Sevenster, M. N. 2007, *ApJ*, 658, 1096
Fish, V. L., Reid, M. J., Menten, K. M., & Pillai, T. 2006, *A&A*, 458, 485
Fomalont, E., & Reid, M. 2004, *New Astronomy Review*, 48, 1473
Garrington, S. T., *et al.* 2004, *Proc. SPIE*, 5489, 332
Habing, H. J., Sevenster, M. N., Messineo, M., van de Ven, G. & Kuijken, K. 2006, *A&A*, 458, 151
Ivison, R. J. 2006, *MNRAS*, 370, 495
Johnston, K. J., *et al.* 1997, *ApJ*, 490, 758
Kurtz, S. 2005, Massive Star Birth: A Crossroads of Astrophysics, 227, 111
Lo, K. Y. 2005, *ARA&A*, 43, 625
Lonsdale, C. J., *et al.* 1998, *Bulletin of the American Astronomical Society*, 30, 1355
Menten, K. M. 1991, *ApJ*, 380, L75
Menten, K. M. 1996, Molecules in Astrophysics: Probes & Processes, 178, 163
Messineo, M., Habing, H. J., Sjouwerman, L. O., Omont, A. & Menten, K. M. 2002, *A&A*, 393, 115
Miyoshi, M., *et al.* 1995, *Nature*, 373, 127
Müller, H. S. P., Menten, K. M., & Mäder, H. 2004, *A&A*, 428, 1019
Murakami, H., *et al.* 2007, ArXiv e-prints, 708, arXiv:0708.1796
Neufeld, D. A., & Melnick, G. J. 1991, *ApJ*, 368, 215
Norris, R. P., Caswell, J. L., Gardner, F. F., & Wellington, K. J. 1987, *ApJ*, 321, L159
Omont, A., *et al.* 2003, *A&A*, 403, 975
Palla, F., Cesaroni, R., Brand, J., Caselli, P., Comoretto, G., & Felli, M. 1993, *A&A*, 280, 599
Pandian, J. D., Goldsmith, P. F., & Deshpande, *A. A.* 2007, ApJ, 656, 255
Reid, M. J., Menten, K. M., Trippe, S., Ott, T., & Genzel, R. 2007, *ApJ*, 659, 378
Ricci, R., *et al.* 2004, *MNRAS*, 354, 305
Schuller, F., *et al.* 2003, *A&A*, 403, 955
Sevenster, M. N., van Langevelde, H. J., Moody, R. A., Chapman, J. M., Habing, H. J., & Killeen, N. E. B. 2001, *A&A*, 366, 481
Townsend, R. H. D., Ivison, R. J., Smail, I., Blain, A. W. & Frayer, D. T. 2001, *MNRAS*, 328, L17
Walmsley, M. 1995, Revista Mexicana de Astronomia y Astrofisica Conference Series, 1, 137
Wood, D. O. S., & Churchwell, E. 1989, *ApJS*, 69, 831
Zijlstra, A. A., *et al.*, 2001, *MNRAS*, 322, 280

Astrophysical Masers and their Environments
Proceedings IAU Symposium No. 242, 2007
J.M. Chapman & W.A. Baan, eds.

doi:10.1017/S1743921307013671

Opportunities for maser studies with the Square Kilometre Array

Anne J. Green[1] **and Willem A. Baan**[2]

[1]School of Physics, University of Sydney, NSW 2006, Australia
email: agreen@physics.usyd.edu.au
[2]ASTRON, 7991PD Dwingeloo, The Netherlands

Abstract. The Square Kilometre Array (SKA) is the radio telescope of the next generation, providing an increase in sensitivity and angular resolution of two orders of magnitude over existing telescopes. Currently, the SKA is expected to span the frequency range 0.1−25 GHz with capabilities including a wide field-of-view and measurement of polarised emission. Such a telescope has enormous potential for testing fundamental physical laws and producing transformational discoveries. Important science goals include using H_2O megamasers to make precise estimates of H_0, which will anchor the extragalactic distance scale, and to probe the central structures of accretion disks around supermassive black holes in AGNs, to study OH megamasers associated with extreme starburst activity in distant galaxies and to study with unprecedented precision molecular gas and star formation in our Galaxy.

1. Overview

The Square Kilometre Array (SKA) is a paradigm-shifting radio telescope for the next generation, providing an increase in sensitivity and angular resolution of two orders of magnitude over existing telescopes. It will be a truly global machine with an expected lifetime of at least 50 years. Such a telescope has enormous potential for testing fundamental physical laws and producing transformational discoveries. The aim of the telescope is to answer some of the "big" questions in astronomy, such as what is the nature of dark energy, are we alone in the Universe, how did galaxies and black holes form, what is the origin and evolution of cosmic magnetism, can pulsars be used to detect gravity waves?

The project is an international consortium of 50 institutions, spread over 17 countries. Present governance is through the International SKA Steering Committee, with 21 members, which oversees the International SKA Project Office and 6 targeted Working Groups. Engagement with inter-governmental agencies has begun to establish a governance model and funding strategies. Currently, there are several pathfinder and demonstrator projects in progress for both science and technology developments.

2. SKA Concept

The concept of this instrument is based on the following principles:

• A data network of sensors of the electromagnetic field are connected using a correlator to produce an interferometric array.

• The sensing antennas will be highly concentrated in a central core, with 20% of the collecting area within 1 km, 50% within 5 km and 75% within 150 km. Outlier stations will be distributed at distances up to at least 3000 km from the core.

• Antennas and stations will be connected via wide-band optic fibre links (data rates at 100 Gbits/sec) to the central processor, which will need to process 10 – 100 Pflops/sec.

• The telescope will be built in stages, with Phase 1 planned to be 10% of the total collecting area and able to undertake unique science.

3. Status at March 2007

Following a rigorous and objective assessment, two sites to host the SKA have been shortlisted for further evaluation and development. The locations encompass Australia and New Zealand, and South Africa with 7 partner countries. The key issues in the site selection process were a very low RFI environment, a large unencumbered site and low ionospheric and tropospheric turbulence. A site decision is expected about 2010 with the complete SKA operational in 2020.

A second milestone was the selection of a Reference Design, which was developed to focus engineering and science efforts, to provide the basis for detailed costing models and to provide a recognisable image for the SKA. The design is likely to evolve, but at present it comprises small dishes with smart feeds with aperture arrays for the lower frequencies.

4. Expected capabilities

The current SKA model has an estimated cost of about 1 Billion Euro for construction with an annual operating budget of about 70 Million Euro. This is based on the following intended capabilities and specifications:

• A sensitivity at least 50 times more than the EVLA. This will enable detection of atomic hydrogen and other molecules right to the edge of the Universe. The specifications are for a continuum sensitivity of 0.4 μJy in 1 hour and a spectral line sensitivity of 5 μJy/channel after 12 hours (both 5σ detections). To achieve this requires a *very large collecting area, $\sim$ 1 square kilometre.*

• A fast survey speed, up to 10,000 times better than currently possible. This requires a *very large field of view, projected to be 1 square degree at 1.4 GHz and 18 square arcminutes at 20 GHz.*

• A *wide frequency range of 0.1 – 25 GHz*, to handle the key science priorities.

• Moderately high angular resolution to make detailed images of structures including disks, outflows and planetary gaps. To do this requires *a large physical extent, at least 3000 km*, to produce beamsizes of $20/f_{GHz}$ mas. The size is limited by the Earth, if one assumes a real-time connected ground-based array.

• Good spectral resolution, with *more than 4000 dual polarization channels* to give velocity resolution of at least 0.2 km/sec.

5. Science goals for maser research

The strength of the SKA will be its great sensitivity with a wide field of view. Angular resolution is constrained by the size of the Earth. There are two main threads for the maser science projects, namely, one which focuses on the early Universe, dark energy and galaxy evolution, and one which will make discoveries in our Galaxy on star formation mechanisms and the interstellar medium.

5.1. *Masers in the early Universe*

Megamasers (MM) of the OH and H_2O molecules will be used to study the properties of prominent populations of active galaxies at cosmological distances. Extragalactic masering activity relies on amplification of radio continuum by foreground pumped molecular gas and the large pathlengths in galactic nuclei.

H_2O MMs are a signpost of AGN activity that may be used to study dark energy, make precise estimates of H_0 to anchor the extragalactic distance scale, and to probe the central structures of accretion disks around supermassive black holes (see Greenhill, these proceedings). Direct mapping of nuclear Keplerian disks such as found in the archetypal NGC 4258 (e.g. Claussen *et al.* 1984, Nakai *et al.* 1993, Haschick *et al.* 1994, Herrnstein *et al.* 1999) will enable determination of precision distance scales out to about 500 Mpc. For unresolved nuclear disks, observed velocity drifts and rotation velocities will extend this distance scale, albeit with less precision, well into the Hubble flow. More than a 1000 water masers may be detected to a flux limit of about 10 mJy (see Braatz *et al.*, these proceedings, for current catalog). The most distant H_2O maser found to date is in quasar SDSS J0804+3607 at a distance of 2.4 Gps (Barvainis & Antonucci 2005), which shows that molecular gas exists at very early epochs. A second class of H_2O MM probes the interaction between radio jets and encroaching molecular clouds away from the AGN, such as seen in nearby NGC 1052 (Claussen *et al.* 1998) or Mrk 34 at a distance of 205 Mpc (Henkel *et al.* 2005) or the FR-II radio galaxy 3C 403 (Tarchi *et al.* 2003).

Powerful OH MMs are associated with extreme starburst activity in (ultra-)luminous infrared galaxies (ULIRGs) resulting from mergers and interactions (see Darling, these proceedings). The redshift distribution of these dusty starburst galaxies reflects the galaxy merger history of the universe (Townsend *et al.* 2001, Briggs 1998). The SKA can probe this population of ULIRGs up and beyond its peak at redshifts between 2 or 3. Typical extended OH MM emission structures can be imaged up to redshifts of 0.6. Powerful OH Gigamasers in the most luminous ULIRGs have been detected and imaged out to redshifts of 0.265 (Baan *et al.* 1992, Darling & Giovanelli 2001, Pihlström *et al.* 2005). The FIR radiation field provides the pumping for the OH molecules and masering action increases with the FIR luminosity (Baan 1989, Henkel & Wilson 1990). The OH emission traces the filaments and cloud in the nuclear ISM and the toroidal structures of 60-100 pc that may surround the nucleus (Rovilos *et al.* 2003, Klöckner *et al.* 2003). The properties of the prototype OH MM Arp 220 are described elsewhere (Baan, these proceedings). The OH MM emission also probes starburst-related outflows and the surroundings of the population of supernovae and SNR in the nucleus (Rovilos *et al.* 2005, Lonsdale *et al.* 2006). The nuclear emission studies of OH MM complement similar studies with ALMA.

5.2. *Galactic and extragalactic masers of many flavours*

The SKA will be used to study with unprecedented precision molecular gas and star formation in our Galaxy and nearby galaxies. The increased sensitivity will make up for the absence of massive stars in the solar neighbourhood and enable detection of large numbers of protostellar Keplerian disks and mapping of outflows. Molecular masers are common in the vicinity of newly formed massive stars, with H_2O and CH_3OH being the signposts of massive star formation and some 70% of UCHIIs in the Galaxy are associated with H_2O masers (Churchwell *et al.* 1990, Walsh *et al.* 1998, Minier *et al.* 2000). Observations with multiple species of masers will significantly widen the studies of young stellar objects, ultra-compact HII regions, and evolved stars and will result in new and (more) accurate diagnostics of high-mass and low-mass stellar environments.

For later stages of stellar evolution, direct imaging of stellar photospheres and stellar winds in AGB stars will be possible. The study of multiple maser emissions and their phenomenology in stellar winds as well as the study of the surrounding ISM will be possible to much larger distances. In addition, the detection and study of (eclipsing) planets in these systems will become possible.

A key science objective will also be to make precision distance measurements in the Galaxy, using routine proper motions and parallax measurements for a large number of sources. This will determine peculiar motions in the spiral arms and resolve many apparent discrepancies in distances to particular objects (e.g. Xu *et al.* 2002).

Systematic and detailed studies of masers associated with evolved stars and star formation regions will also be possible in nearby galaxies with the sensitivity and resolution provided by SKA. In addition, the extended frequency range of SKA allows other molecular line studies of these active extragalactic regions. The H_2O maser emission in other galaxies may trace weak nuclear activity (as in H_2O MM) or massive star-forming regions similar to regions found in the Galaxies. NGC 2146 shows that the kilomaser H_2O emission from UCHII regions may be detected up to distances of 50 Mpc (Tarchi *et al.* 2002). Similarly, the properties of 1720 MHz OH masers measured in nearby spiral galaxies, such as M33 (expected source fluxes of 3 mJy at a distance of 1 Mpc), can be compared/correlated with those associated with supernova remnant (SNR)-molecular cloud interactions in our Galaxy (e.g. Frail *et al.* 1996, Yusef-Zadeh *et al.* 1996). The extension of the diagnostics for ongoing star formation and evolved stars to other galaxies, which may have more intense and more extended star-forming activity than is found in our Galaxy, will greatly benefit our ability to understand the important astrophysical processes that occur.

6. Conclusions

A summary of what the SKA is likely to produce for maser science include:

- A large increase in the number H_2O masers in sub-parsec disks around AGNS to be used to study structure and black hole properties. For resolved nuclear disks, precision distances and an improved measure of H_0 can be determined out to distances of about 500 Mpc.
- Probing the high-density ISM of high redshift galaxies using H_2O for parsec nuclear regions and jet interaction regions, and using OH for the nuclear ISM and torus structures in luminous (circum-)nuclear starbursts will be possible without a distance limit.
- Many more disks and outflows will be detected for star-forming regions and dusty proto-clusters in our Galaxy. Masers associated with SNRs and Compact HII regions can be studied in nearby galaxies.
- Proper motion and parallax studies will give a more precise picture of the structure and the peculiar motion of the spiral arms in our Galaxy.

A more detailed explanation can be found in *"Science with the Square Kilometre Array"*, edited by Carilli & Rawlings in New Astronomy Reviews (2004) volume 48.

References

Baan, W. A. 1989, *ApJ* 338, 804

Baan, W. A., Rhoads, J., Fisher, K., Altschuler, D. R. & Haschick, A., 1992, *ApJL*, 396, L99

Barvainis, R. & Antonucci, R. 2005, *ApJ* (Letters) 628, L89

Briggs, F. H. 1998, *A&A* 336, 815

Churchwell, E., Walmsley, C. M. & Cesaroni, R. 1990, *A&AS* 83, 119

Claussen, M. J., Heiligman, G. M., & Lo, K.Y 1984, *Nature* 310, 298

Claussen, M. J., Diamond, P. J., Braatz, J. A., Wilson, A. S. & Henkel, C. 1998, *ApJL* 500, L129

Darling, J. & Giovanelli, R., 2001, *AJ*, 121, 1278

Frail, D. A., Goss, W. M., Reynoso, E. M., Giacani, E. B., Green, A. J. & Otrupcek, R. 1996, *AJ* 111, 1651

Haschick, A. D., Baan, W. A. & Peng, E. W. 1994, *ApJL*, 437, L35
Henkel, C., Peck, A. B., Tarchi, A., Nagar, N. M. *et al.* 2005, *A&A* 436, 75
Henkel, C. & Wilson, T. L. 1990, *A&A* 229, 431
Herrnstein, J. R., Moran, J. M., Greenhill, L. J., Diamond, P. J., *et al.* 1999 *Nature* 400, 539
Klöckner, H.-R., Baan, W. A., & Garrett, M. A. 2003, *Nature*, 421, 821
Lonsdale, C. J.,Diamond, P. J., Thrall, H., Smith, H. E. & Lonsdale, C. J. 2006, *ApJ*, 647, 185
Minier, V., Booth, R. S. & Conway, J. 2000, *A&A* 362, 1093
Nakai, N., Inoue, M., & Miyoshi, M. 1993, *Nature* 361, 45
Pihlström, Y. M., Baan, W. A., Darling, J. Klöckner, H.-R. 2005, *ApJ*, 618, 705
Rovilos, E., Diamond, P. J., Lonsdale, C. J., *et al.* 2003, *MNRAS*, 342, 373
Rovilos, E., Diamond, P. J., Lonsdale, C. J. *et al.* 2005, *MNRAS*, 359, 827
Tarchi, A., Henkel, C., Peck, A. B., Menten, K. M. 2002, *A&A* 389, L39
Tarchi, A., Henkel, C., Chiaberge, M., Menten, K. M. 2003, *A&A* 407, L33
Townsend, R. H.D., Ivison, R. J., Smail, I., Blain, A. W., Frayer, D. T. 2001, *MNRAS* 328, L17
Walsh, A. J., Burton, M. G., Hyland, A. R. & Robinson, G. 1998, *MNRAS* 301, 640
Xu, Y., Reid, M. J., Zheng, X. W., Menten, K. M. 2006, *Science* 311, 54
Yusef-Zadeh, F., Roberts, D. A., Goss, W. M., Frail, D. A., Green, A. J. 1996, *ApJL* 466, L25

Astrophysical Masers and their Environments
Proceedings IAU Symposium No. 242, 2007
J.M. Chapman & W.A. Baan, eds.

doi:10.1017/S1743921307013683

ALMA's view of maser emission

Alwyn Wootten†

Joint ALMA Office, 40 El Golf, Piso 18, Las Condes, Santiago, Chile and NRAO, Charlottesville, Va, USA
email: awootten@nrao.edu

Abstract. ALMA, the Atacama Large Millimeter/submillimeter Array, provides a large collecting area at a location on a high dry site, endowing it with unparalleled potential for sensitive spectral line observations. Its wide frequency coverage, superb receivers and flexible backend will ensure that that potential is met. Although in the Southern Hemisphere, its tropical latitude ensures good coverage of the northern sky. Since the last meeting on astrophysical masers, the ALMA team has substantially enhanced its capability for line observations. Japan's entry into ALMA has provided increased sensitivity with the addition of the 16 antennas of the Atacama Compact Array, equivalent to eight additional 12m telescopes. The first four cartridges for the baseline ALMA receiver packages (to be augmented by three other bands owing to Japanese participation) have been accepted, with performance above the already-challenging specifications. At first light, ALMA will offer nearly complete frequency coverage of the millimeter and submillimeter spectral windows, ensuring coverage of a variety of masering transitions. ALMA's flexibility as a spectrometer has increased with the enhancement of the baseline correlator with tunable filter banks, and with the addition of a separate correlator for the ACA. As an example of the increased flexibility, ALMA is now capable of multi-spectral-region and multi-resolution modes. With the former, one might observe e.g. four separate transitions anywhere within one of four 2 GHz bands with a high resolution bandwidth. With the latter, one might simultaneously observe with low spectral resolution over a wide bandwidth and with high spectral resolution over a narrow bandwidth. Thus, one could simultaneously cover an extremely broad velocity range while providing high spectral resolution of groups of lines within that range. Several science examples will be presented illustrating ALMA's potential for transforming the millimeter and submillimeter study of masers.

1. Introduction

ALMA will observe the cosmic millimeter/submillimeter spectral region. Millimeter and submillimeter photons dominate the spectral energy distribution of the cosmos, and arise in two spectral components. The most luminous component is the 3K Cosmic Microwave Background (CMB). After the CMB, the strongest component is the submm/FIR component, which carries most of the remaining luminous energy in the Universe, and 40% of that in, for instance, the Milky Way Galaxy. ALMA's wavelength coverage, from 1cm to 0.3 mm, covers both components to the extent that the atmosphere of the Earth allows. In addition to dominating the spectrum of the distant Universe, millimeter/submillimeter spectral components dominate and characterize the spectrum of

† The Atacama Large Millimeter/submillimeter Array (ALMA) is an international astronomy facility. ALMA is a partnership between Europe, North America and Japan, in cooperation with the Republic of Chile. ALMA is funded in North America by the U.S. National Science Foundation (NSF) in cooperation with the National Research Council of Canada (NRC), and in Europe by the European Southern Observatory (ESO). ALMA construction and operations are led on behalf of North America by the National Radio Astronomy Observatory (NRAO), which is managed by Associated Universities, Inc. (AUI), on behalf of Europe by ESO, and on behalf of Japan by the National Astronomical Observatory of Japan.

planets, young stars, and many distant galaxies. Cool objects tend to be extended, hence ALMA's mandate to image with high sensitivity, recovering all of an objects emitted flux at the frequency of interest. Cool objects present an environment conducive to the existence of molecules. Most of the observed transitions of the 142 known interstellar molecules lie in the mm/submm spectral region – for example some 17,000 lines are seen in a small portion of the spectrum at 2mm. Unfortunately, molecules, particularly water, in the Earth′s atmosphere inhibit our study of many of these molecules. Furthermore, the long wavelength requires large aperture for high resolution, un-achievable from space. To explore the submillimeter spectrum, a telescope should be placed at the highest driest site on Earth. The 5000m Chajnantor site selected for ALMA allows complete spectral coverage using ten frequency bands coincident with atmospheric windows, eight of which will be populated upon the completion of construction.

2. ALMA science

The highest level document governing the ALMA Project is the Bilateral Agreement. Annex B of this agreement details ALMA's three highest level science goals.

•The ability to detect spectral line emission from CO or C+ in a normal galaxy like the Milky Way at a redshift of z = 3, in less than 24 hours of observation.

•The ability to image the gas kinematics in a solar-mass protostellar / protoplanetary disk at a distance of 150 pc (roughly, the distance of the star-forming clouds in Ophiuchus or Corona Australis), enabling one to study the physical, chemical, and magnetic field structure of the disk and to detect the tidal gaps created by planets undergoing formation.

•The ability to provide precise images at an angular resolution of 0.1". Here the term precise image means accurately representing the sky brightness at all points where the brightness is greater than 0.1% of the peak image brightness. This requirement applies to all sources visible to ALMA that transit at an elevation greater than 20 degrees. These requirements drive the technical specifications of ALMA.

While none of these goals specifically address maser science, they have guided the design of an instrument which directly addresses many issues facing maser astrophysicists. Some maser features which we may enumerate are their small size, their origin in multiple transitions of several molecules with emission across the millimeter/submillimeter spectrum, their association in some cases with weak thermal emission, their appearance in multiple velocity components, often widely separated in velocity, and, for stellar masers, their association with older stars found near the galactic center.

Several features of the ALMA design enable it to address key maser science. Several of these features represent enhancements to the design since the last maser conference, some enabled by Japan joining the project in 2004 September.

•ALMA provides complete frequency access from 31.3 GHz to 950 GHz, with coverage from 84 GHz to 950 GHz available upon completion. Since the previous maser conference, four bands have been added to the first light retinue: 2mm (125 – 169 GHz; provided by NAOJ), 0.6mm (385 – 500 GHz, provided by NAOJ), 1.5mm (163 – 211 GHz, six single polarization receivers, provided by the EU), and 0.35mm (a partial complement of receivers at 787 – 950 GHz, provided by NAOJ with the intention of completely populating the array eventually).

•ALMA has been enhanced by additional antennas, providing the equivalent of 8 12m antennas in collecting area, but comprised of 4 12m antennas and 12 7m antennas deployed in the Atacama Compact Array (ACA) for improved brightness temperature sensitivity, provided by NAOJ.

Table 1. ALMA antenna arrays and configurations

Array	**12m**	**Compact (ACA)**
Number of Antennas	$\leqslant 64^a$	16
Total Collecting Area	$\leqslant 7238\text{m}^2$	915^2
Array Configurations	*(dimension of filled area)*	
Compact filled	150 m	35 m
Largest Extent	18.5 km	
Total No of antenna stations	170	22
Antennas[b]		
Diameter	12 m	4 ×12-m + 12 × 7-m
Surface accuracy	25 μm	12-m:25μm 7-m:20μm

[a] Contracts are signed for provision of at least fifty, up to 64 antennas; two prototypes have already been procured and tested [b] Transportable by specially constructed vehicles with rubber tires

•A more versatile baseline correlator, in addition to a versatile new correlator dedicated to the ACA. Tunable filter bank elements developed by the University of Bordeaux in collaboration with ESO, have brought improved flexibility to the baseline correlator. The ACA correlator has been developed by NAOJ.

•ALMA is located just north of the Tropic of Capricorn, in the southern hemisphere but with good access to objects in the northern hemisphere; the galactic center transits nearly overhead.

•ALMA antennas may be deployed over baselines as long as 16 km on its extensive but high site, enabling it to achieve beamsizes as small as 5 milliarcseconds at its highest operating frequency, similar to the resolution achieved by very long baseline arrays at lower frequencies (Table 1).

•ALMA provides excellent brightness temperature sensitivity (Table 2). A line which is amplified only slightly can easily be identified and measured.

3. ALMA correlators

Of the two baseline correlators, the NRAO Baseline Correlator will be used with the longest baselines and most likely be most relevant to maser observations. There are four quadrants for 64 antennas processing 16 GHz of data for each antenna. The basic sampling is 2 bits but there are limited 3 or 4 bit sampling modes available. The first quadrant of this correlator has been operating for some time in the NRAO Technology Center, to be retrofitted with the Tunable Filter Bank enhancement and installed at the Array Operations Site (AOS) at 5000m in the Technical Building being completed within the next year. A detailed list of observational modes may be found in ALMA Memo No. 556. In addition, the ACA correlator, processing data from 16 antennas, will be installed in the building at about the same time.

Both correlators are very flexible. An observer may specify a set of disjoint or overlapping spectral regions for observation, each characterized by bandwidth ranging from 31.25 MHz to 2 GHz. Each of the eight available 2 GHz baseband inputs drives 32 tunable digital filters, which may be placed as desired within the baseband. A number of products are available: autocorrelation or correlation, several polarization products (XX or YY; XX and YY; XX, YY, XY and YX), and improved sensitivity (4x4 bit correlation or double Nyquist modes). The temporal resolution depends upon the mode and varies from 1 msec for autocorrelation and from 16 to 512 msec for cross correlation. The correlators offer simultaneous pseudo-continuum and spectral line operation.

Table 2. 'First Light' front ends on all antennas[a]

Band	λ mm	ν GHz	Type	Continuum Sensitivity (mJy)[b]
Band 3	3	84-116	SIS	0.04
Band 4	2	125-169	SIS	0.06
Band 5	1.5	163-211	SIS	[b]
Band 6	1.3	211-275	SIS	0.11
Band 7	0.9	275-373	SIS	0.19
Band 8	0.6	385-500	SIS	0.70
Band 9	0.5	602-720	SIS	1.0
Water Vapor Radiometer	1.6	183	Schottky	...

[a] Baseline ALMA plan; all are planned to have dual polarization with noise performance limited by atmosphere. [b] See the ALMA sensitivity calculator at http://www.eso.org/projects/alma/science/bin/sensitivity.html; these numbers are for nominal atmospheric conditions and one minute of integration at band center.

Table 3. Brightness temperature sensitivity of ALMA, 14.7 km baseline

Band	λ mm	Beam	ΔT (Line Sensitivity) (1 km s^{-1}, K)[a]
Band 3	3	0.038	482
Band 4	2	0.030	495
Band 5	1.5	0.022	9500[b]
Band 6	1.3	0.018	709
Band 7	0.9	0.012	1128
Band 8	0.6	0.010	1569
Band 9	0.5	0.006	4305

[a] See the ALMA sensitivity calculator at http://www.eso.org/projects/alma/science/bin/sensitivity.html; these numbers are for nominal atmospheric conditions and one minute of integration at band center. [b] Only a partial receiver complement is available at first light, single polarization.

The observer might set up multiple spectral windows within any of the 2 GHz basebands for modes with total bandwidth from 125 MHz to 1 GHz. This will prove useful for high spectral resolution observations, which can be provided over a large bandwidth for sources such as masers and which can show complex narrow emission over a vast velocity range. One may also have multi-resolution modes, which may prove useful for sources with different characteristic linewidths such as planetary atmospheres or star-forming cores containing outflows, rotating disks and quiescent material. As one ALMA goal is ease of use, the ALMA Observing Tool will guide the observer through the maze of possibilities.

4. Example

There are many molecules known with transitions having maser action within the ALMA bands. A summary is given in Table 3. Of particular interest is the abundant and simple water molecule, which emits in all of the ALMA bands. Owing to its abundance, water is an excellent tracer of star-forming regions, where it is seen in shocks and warm, chemically active regions. ALMA's high resolution, excellent brightness temperature

Table 4. ALMA bands with known masering lines

Array	**12m** and **Compact (ACA)**
Species	Bands
H_2O	B3, B5, B6, B7, B8, B9
CH_3OH	B1, B3, B4, B6
SiO	B1, B2, B3, B4, B5, B6, B7
HCN	B3, B5, B6, B7, B9
Etc: Hnα, SiS	B3, B4, B6, B7, B9

sensitivity, spectral coverage, polarization capability and flexibility will prove particularly useful in these regions. These features will also serve science projects which target objects as diverse as evolved stars and galactic nuclei.

Water is perhaps the most observed of these as it incorporates two of the three most abundant elements in the Universe, has a myriad of transitions and has a singular importance in cooling interstellar clouds. Owing to its very high brightness temperatures and ubiquity, it has been used as a probe of interstellar shocks, including measuring their motion across the sky, and for measuring parallactic distance, normally through Very Long Baseline (VLB) techniques using its 22 GHz line. Owing to its exceptionally dry location high on the Chilean Altiplano, ALMA can observe many additional lines from both higher and lower energy levels. Some of these transitions occur at sufficiently high frequency that ALMA can achieve near-VLB resolution across its 15 km baselines. Since the ALMA brightness temperature sensitivity is so high, one can literally watch chemistry taking place behind the shocks. Water masers are also sent coursing through the atmospheres of evolved stars as they pulsate. In both these environments, high magnetic fields can cause observable Zeeman splitting in the lines, leading to magnetic field measurements. ALMA will also be able to constrain conditions in masers spiraling inward toward black holes in galactic nuclei by simultaneously observing several lines at once; the enormous bandwidth of the receivers and correlators will be particularly useful for this task.

The power of the instrument is perhaps best demonstrated by an example. We will illustrate one receiver/correlator setting at 321-337 GHz, which encompasses three lines of water emission at different excitations, two lines of water isotopes as well as a selection of other lines of shock tracers such as methanol, SO, SiO and SO_2. A main target of this setting is the para-H_2O $J_{K_a K_c}$ 5_{15} - 4_{22} line at 325.15 GHz, which lies at an energy 470 K above ground. At such a low energy, this line is strong in the Earth's atmosphere and only very limited observations have been made (see review by Humphreys in this volume (Humphrey 2007) since its discovery by Menten *et al.* (1990). A line from the ^{18}O isotopomer of this same water line lies at 322.97 GHz, in the center of the band. At the lower end of the same 4 GHz lower sideband, just fitting in, lies the higher energy (1861 K) ortho-H_2O $J_{K_a K_c}$ 10_{29} - 9_{36} line at 321.23 GHz. In the upper sideband, 8 GHz higher in frequency, lie two water lines, the HDO $J_{K_a K_c}$ 3_{31} - 4_{22} line at 335.40 GHz, 319 K above ground, and the ortho-H_2O $J_{K_a K_c}$ 5_{23} - 6_{16} $\nu_2 = 1$ line at 336.23 GHz, 2939 K above ground. Sprinkled throughout the band are the additional lines of SO, SO_2 and CH_3OH. For the largest configuration, the ALMA beam will be 14 milliarcsec. In 8 hours of integration, during best octile weather (0.52 mm of precipitable water vapor) the resulting spectrum should have an rms in the center of the atmospheric line of $\Delta T = 52$ K, or about 0.8 mJy. For 5s integrations, the 30Mbytes/s data rate exceeds the average ALMA rate of 6 Mbyte/s but fits comfortably within the allowed maximum, producing a dataset of 860 MB size.

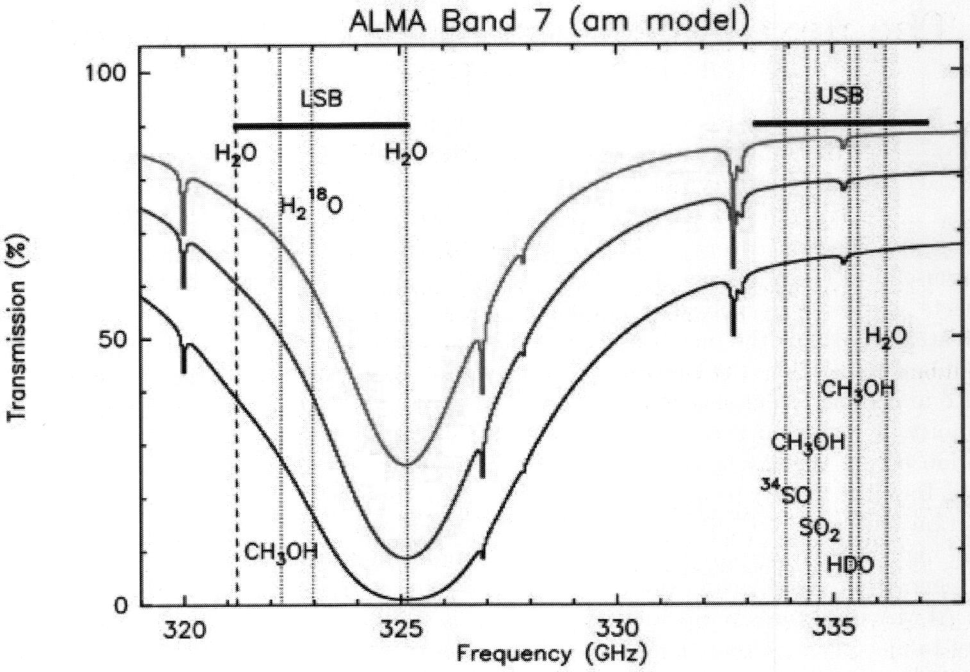

Figure 1. Atmospheric transparency for the ALMA site during the 25% weather quartiles is plotted against the receiver sideband setup with the targeted spectral lines.

Water is also of interest in stars, though the other molecules in the setup are of only occasional interest. However, some isotopic and highly excited vibrational states of SiO lie within the band.

5. Conclusions

ALMA's complete frequency access across atmospherically accessible bands, its large collecting area and excellent brightness temperature sensitivity, its high altitude equatorial location, and its high resolution make ALMA an ideal instrument for study of the astrophysics of masers.

References

Humphreys, E. M. L. 2007, ArXiv e-prints, 705, arXiv:0705.4456
Menten, K. M., Melnick, G. J., & Phillips, T. G. 1990, ApJL, 35
Menten, K. M., Melnick, G. J., Phillips, T. G., & Neufeld, D. A. 1990, ApJL, 363, L27

Astrophysical Masers and their Environments
Proceedings IAU Symposium No. 242, 2007
J.M. Chapman & W.A. Baan, eds.

doi:10.1017/S1743921307013695

The next generation space VLBI project, VSOP-2

Yasuhiro Murata[1], Nanako Mochizuki[1], Hirofumi Saito[1], Hisashi Hirabayashi[1], Makoto Inoue[2], Hideyuki Kobayashi[2], Philip G. Edwards[3] and the Next Generation Space VLBI Working Group

[1]Institute of Space and Astronautical Science, Japan Aerospace Exploration Agency, 3-1-1 Yoshinodai, Sagamihara, Kanagawa 229-8510, Japan.
email:murata@vsop.isas.jaxa.jp

[2]National Astronomical Observatory of Japan, 2-21-1 Osawa, Mitaka, Tokyo 181-8588, Japan

[3] CSIRO, Australia Telescope National Facility, Locked Bag 194, Narrabri, NSW 2390, Australia

Abstract. The first dedicated space-VLBI project, the VLBI Space Observatory Programme (VSOP), commenced with the successful launch of radio-astronomical satellite HALCA in 1997. Plans for a second generation space-VLBI project have been made by a working group over a number of years. This project, VSOP-2, has now been approved by Japan's space agency, JAXA, as the ASTRO-G project. It is planned for the spacecraft to observe in the 8, 22 and 43 GHz bands with cooled receivers for the two higher bands, which include important maser lines. It will have a maximum angular resolution at 43 GHz (7 mm) of about 40 micro-arcseconds. Although the VSOP project mainly observed continuum emission from active galactic nuclei (AGN), VSOP-2/ASTRO-G is expected to enable a variety of high angular resolution maser line observations.

Keywords. space vehicles, instrumentation: high angular resolution, telescopes

1. VSOP-1/HALCA project

The Institute of Space and Astronautical Science (ISAS) started the VLBI Space Observatory Programme (VSOP) in 1989, and launched the first space VLBI satellite, HALCA, in 1997. Launch was on the first flight of the M-V rocket (Hirabayashi *et al.* 1999, Hirabayashi *et al.* 2000)

HALCA carried an 8 m diameter center-fed Cassegrain mesh antenna. The on-board radio astronomy subsystem was composed of low-noise amplifiers for three frequency bands, 1.60–1.73 GHz, 4.7–5.0 GHz and 22.0–22.3 GHz. The 22 GHz system temperature was found to be unexpectedly high, and we could get space-ground fringes only for a bursting Orion-KL maser.

Scientific observations were undertaken at 1.6 and 5 GHz, which provided the highest angular resolution images ever obtained in these frequency bands (0.36 mas at 5 GHz, and 1.1 mas at 1.6 GHz). VSOP observations were mostly devoted to AGN, although several observations were made of non-AGN sources, such as hydroxyl (OH) masers, pulsars, and X-ray binaries. VSOP observations yielded new results on topics such as the internal structure and motion of AGN jets, the highest brightness temperature sources, and the fine structure of the plasma shadow of free-free absorption.

Figure 1. Schematic view of the ASTRO-G (VSOP-2) satellite.

In total, 750 VSOP observations were successfully made. The HALCA satellite lost attitude in October, 2003, and satellite operations were finally ended in November 2005, on the 3213th day after launch.

2. VSOP-2/ASTRO-G project

A next-generation space VLBI mission Working Group, which consists of the space VLBI research community, was established by ISAS in May 1997, shortly after fringes were first found to HALCA. The working group submitted a VSOP-2 mission proposal to ISAS in September 2005. The VSOP-2 mission (Hirabayashi *et al.* 2000) was selected as the 25th scientific mission of JAXA in May 2006 and the VSOP-2 satellite given the developmental name ASTRO-G. VSOP/HALCA was formally an ISAS engineering test mission, but VSOP-2 has been selected as a ISAS/JAXA science mission. To support the ambitious science goals, the ASTRO-G spacecraft requires improvements in both sensitivity and angular resolution when compared with HALCA.

The VSOP-2 spacecraft, ASTRO-G (figure 1), will have a deployable 9-m off-axis paraboloid antenna with an uncooled receiver operating in the range 8.0–8.8 GHz, and cryogenically cooled receivers operating from 20.6–22.6 GHz and 41–45 GHz in both LHCP and RHCP.

To achieve an order of magnitude higher sensitivity for continuum sources, VLBI data will be down-linked in real-time at 1 gigabit per second using the 37–38 GHz band. The on-board system is locked to a reference phase, derived from a H-maser at one of 3∼4 tracking stations, and uplinked as a tone at 40 GHz. ASTRO-G has 2 IF channels with 2 sampling modes. One uses 256 MHz bandwidth, 1-bit sampled channels, and the other has 128 MHz, 2- bit sampled channels.

The satellite will be placed in an elliptical orbit with an apogee height of 25,000 km above the Earth's surface and a perigee height of 1,000 km, resulting in a period of 7.5 hours.

With an apogee height of 25,000 km, 43 GHz observations can achieve an angular resolution of 38 micro-arcseconds. Furthermore, a phase-referencing capability is being actively considered which will not only increase the number of observable sources but will also allow state-of-the-art astrometric measurements to be undertaken. ASTRO-G should achieve an order of magnitude higher sensitivity than HALCA with these new capabilities. Table 1 shows the target sensitivities for line and continuum sources.

Table 1. Target VSOP-2 observational parameters

Observing Frequency	8 GHz	22 GHz	43 GHz
Resolution	205 mas	75 mas	38 mas
Target SEFD	4080 Jy	2200 Jy	3170 Jy
Target Tsys	60 K	30K	40K
Line sensitivity			
Target Line	—	H_2O	SiO
7 σ Fringe detection sensitivity (1km/s line width)			
with VLBA 25m	—	2.9 Jy	4.5 Jy
with phased-VLA	—	0.66 Jy	0.93 Jy
Phase referencing sensitivity			
90 min., VLBA 25m	—	0.43 Jy	0.47 Jy
Continuum sensitivities			
7 σ Fringe detection sensitivity			
with VLBA 25m	25 mJy	20 mJy	110 mJy
with phased-VLA	5 mJy	11 mJy	22 mJy
Phase referencing sensitivity			
90 min., VLBA 25m	6 mJy	8 mJy	11 mJy Jy
Image sensitivity (12hour)	0.034 mJy/beam	0.064 mJy/beam	0.100 mJy/ beam
Detectable Brightness temperature	6.8×10^7 K	1.3×10^8 K	2.1×10^8 K

The design of the VSOP-2 instruments is intended to realize the following science goals: (i) The structures and magnetic field configurations of accretion disks in nearby active galactic nuclei (AGNs), (ii) The mechanism of jet acceleration and collimation, (iii) The motion of masers in galactic star forming regions, (iv) The study of proto-stellar magnetospheres, and (v) The structures and magnetic fields of accretion disks in active galactic nuclei.

As a result, VSOP-2 will allow studies of regions where extreme physical conditions are encountered. Consequently the high-resolution imaging capability of VSOP-2 will enable new science in fundamental astrophysics to be undertaken.

3. VSOP-2 maser observations

Extragalactic water masers are found within a few parsecs of several active galactic nuclei. H_2O masers in the spiral galaxy NGC 4258 are located 40,000 Schwarzschild radii from the central engine in a rapidly rotating accretion disk. Monitoring observations to determine the three dimensional motions of these maser spots were used to prove the existence of a supermassive black hole in the center of the galaxy. The motions of such spots allows not only the supermassive black hole mass in the AGN center to be obtained but also the temperature and density of the emitting gas to be probed, revealing the physical conditions of the accretion disk. At present about 50 extragalactic water 'megamaser' sources are known. An advantage of VSOP-2 observations for megamasers is again, the ability to measure the proper motions of the maser spots in a shorter time than required for ground-based observations.

H_2O and SiO masers observations are a main target of the VSOP-2 project. Candidate objects to observe with those maser lines are galaxies, low and high mass star forming regions, and late-type stars (Mira type stars, OH/IR stars and planetary nebulae) in our Galaxy. Masers in the nuclei of low redshift (z <0.05) galaxies, and emissions from Local group galaxies, such as the LMC, SMC, and M33, can also be studied.

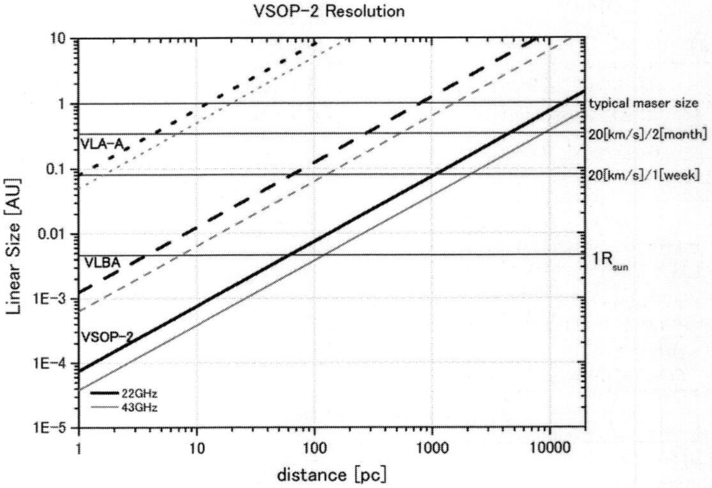

Figure 2. Diagram indicating how the VSOP-2 angular resolution compares with sources in our Galaxy. The horizontal axis shows the distance from the sun, and the vertical axis shows the absolute size of the sources.

By combining phase referencing VLBI or Space-VLBI observations with radial velocity information derived from the maser spot Doppler shift, it is possible to reconstruct the the 3-D motions of the maser spots. Such 3-D motions are required for determining the gravitational field in which gas is moving and for the study of other environmental conditions.

In star forming regions, water maser motions appear as an outflow from the protostellar disk which allows important constraints to be placed on star formation mechanisms. Compared to ground-VLBI observations, VSOP-2 observations have the distinct advantage of being able to measure proper motions over a shorter period of time, and, as a result, will enable shorter-lifetime maser spot motions to be determined. We show how small scale we can measure with VSOP-2 on figure 2.

H_2O and SiO masers associated with late type stars probe an important stage in stellar evolution. H_2O and several SiO maser transitions fall in the observing band of VSOP-2. Multi-transitional observations of these SiO masers excited in different inversion schemes will give new information on the local physical conditions of the maser emission, thereby tracing the mass loss process.

One of the problems when we observe Galactic maser emission with VSOP-2 is its spot size. Larger size spots cannot be detected on the longest baselines as the sources are resolved out. From figure 3 of Gwinn *et al.* (1988), the size of nearby/neighborhood sources are large, but more distant sources are smaller with the exception of sources near the Galactic center direction because of the scatter broadening. The VSOP pre-launch maser survey of Migenes *et al.* (1999) pointed out there is a large range of angular sizes of maser spots from 10 mas to less than 0.1 mas. Though it is difficult to observe larger size spots by space VLBI, there are still many targets with smaller size spots. We think new pre-launch survey of maser spots will be useful for VSOP-2 maser observations.

4. Current status of VSOP-2/ASTRO-G

VSOP-2 received a new start at the beginning of Japanese 2007 fiscal year and funding is currently in place for satellite development. A schedule to develop ASTRO-G is shown

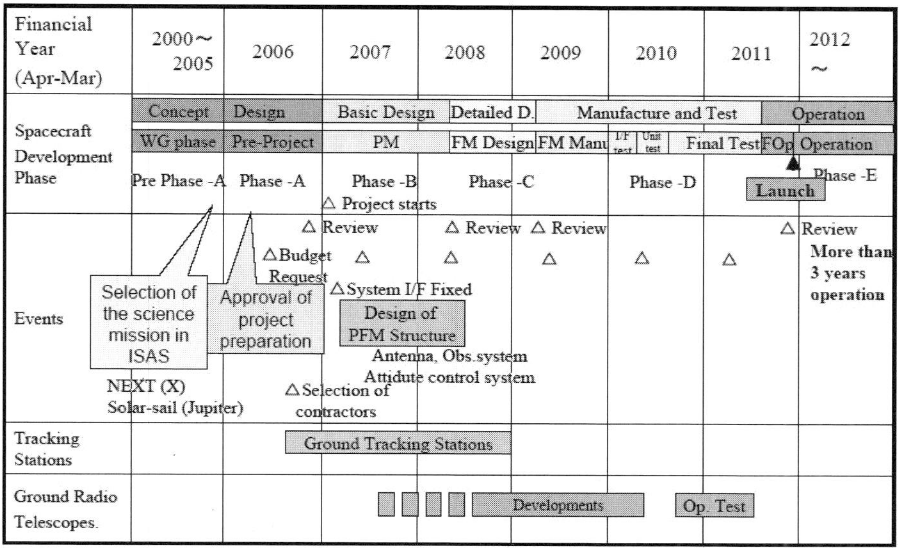

Figure 3. Current development schedule of ASTRO-G/VSOP-2 satellite.

in Figure 3. The current launch date is early 2012. A mission lifetime of more than 3 years is planned, and hopefully more than the 6 years of HALCA. We have now almost finished its concept design, and will move to the more detailed design of the satellite, such as the location of the instruments, and defining the mechanical and electrical interfaces. We also start to configure the international collaboration in basically the same manner as for VSOP-1.

5. Summary

The budget for the VSOP-2 (ASTRO-G) project has been approved with the target launch date of 2012. Reviews to approve the start of the Phase B stage (engineering-model development) were held in March/April 2007. Extragalactic masers and compact Galactic maser sources are important targets of VSOP-2.

Acknowledgements

We would like to acknowledge all those who are working on, or have contributed to, the VSOP-2 project.

References

Gwinn, C. R., Moran, J. M., Reid, M. J., & Schneps, M. H. 1988, *ApJ* 330, 817

Migenes, V., Horiuchi, S., Slysh, V. I., Val'tts, I. E., Golubev, V.V, Edwards, P. G., Fomalont, E. B., Okayasu, R., Diamond, P. J., Umemoto, T., Shibata K. M., & Inoue, M. 1999, *ApJS* 123, 487

Hirabayashi *et al.* 1998, *Science* 281, 1825 (erratum 282, p. 1995, 1998)

Hirabayashi *et al.* 2000, *PASJ* 52, 955

Hirabayashi, H. Murata, Y., Asaki, Y., Edwards, P. G., Mochizuki, N., Natori, M. C., Inoue, M., Umemoto, T., Kameno, S. Kono, Y., Tsuboi, M., Kasuga, T., 2004, *Optical, Infrared, and Millimeter Space Telescopes. Edited by Mather, John C. Proceedings of the SPIE* 5487, 1646.

Astrophysical Masers and their Environments
Proceedings IAU Symposium No. 242, 2007
J.M Chapman & W.A. Baan, eds.

doi:10.1017/S1743921307013701

IAU 242 closing summary

M. J. Reid

Harvard-Smithsonian CfA, Cambridge, MA, USA
email: reid@cfa.harvard.edu

Abstract. This paper focuses on highlights of the symposium and attempts to synthesize recent results in terms of key advances and unsolved problems related to maser research.

Keywords.

1. Introduction

In the six years since the second international meeting devoted to astronomical masers (held near Rio de Janeiro, Brazil), there have been many advances in observations and understanding of masers. In this summary, I will attempt to consolidate, from the roughly 130 talks and poster papers, where the field stands. As this is an opportunity to explore astrophysical problems without the constraints and formalities of refereed journals, I will attempt to be speculative and provocative. Likely some of these speculations will prove wrong, but hopefully they will stimulate new ideas for research.

Before starting, I would like to note the incredible complexity associated with maser emission. In order to understand observations fully, we must know the "chemistry," amplification process, and lifetimes of the masers. Chemical effects that control molecular maser emission include not only temperature (T) and total density (n_{tot}), but also ionization, dissociation, and sublimation of dust. Amplification in masers is very complicated and may involve details of the clumping of dense gas, the alignment of clumps, velocity and magnetic field gradients, beaming and saturation effects, and, of course pumping.

There are many examples of complicated phenomena that need to be explained. (1) While the crude nature of maser pumps (ie, whether dominated by radiative, collisional or chemical processes) is agreed upon for most molecules and classes of maser emission, we now suspect that, for example, many pumping routes are involved in the inversion of interstellar OH masers. (2) We heard that essentially all masers are variable, new data suggests that interstellar OH maser line wings can vary on timescales of 1000 seconds while the line centers are much more stable. These results are quite puzzling and seem to require disturbances propagating at the speed of light over one maser e-folding gain length. (3) The lack of many, if any, clear cases of pure π-component emission (seen as 100% linearly polarized) from interstellar OH masers is puzzling. The reason for this might involve the amplification in one cloud of a pi-component followed by further amplification in a second cloud by sigma-components (that are appropriately shifted in velocity to compensate for Zeeman splitting). (4) Any of the above mentioned complexities can be further "enhanced" by the dynamic process of cloud formation and dissipation.

As if our problems were not difficult enough, they are often compounded by the fact that astronomical images reveal only the 2-dimensional projection of the 3-dimensional structure. Also, when placing the masers in context, via complementary observations across the electromagnetic spectrum, the effects of absorption and scattering of radiation can be considerable.

2 Pumps

After decades of study, details of many maser pumps remain unknown. There is some general agreement on the broadest categorization of maser pumping: ie, are they dominated by collisional or radiative processes? For example, Class-I methanol masers are inverted predominantly by collisions, whereas Class-II methanol masers and interstellar hydroxyl masers are inverted mostly by radiative processes. Beyond these cases, it is becoming clear that population inversion can be a very complex process. Many levels and pumping routes seem to contribute to the inversions. However it is unclear to me whether the intrinsic complexity of pumping or our ignorance of atomic and molecular cross-sections is the main reason for our difficulties.

3 Millimeter and sub-millimeter wavelength masers

At this meeting we heard about large numbers of newly discovered sources of high-frequency masers. Even though (as V. Strelnitski pointed at in the Brazil meeting) it gets progressively harder to invert populations the higher energy levels involved, Nature manages somehow. There is hope that newly discovered extragalactic water masers in the sub-millimeter band may provide key data that constrain, for example, the physical conditions in AGN accretion disks.

4 Interstellar masers

4.1. *Characteristics*

The spatial distributions of interstellar masers can provide strong clues to their origins. The 2-point spatial correlation function, which is proportional to the probability of finding neighboring maser spots, is a power law for water masers. There is no clustering scale to give a physical scale length for the condensations or their amplification lengths. Rather, the structures are fractal in nature and can best be explained as owing their existence to turbulent processes. The outer scale length, where energy from e.g. strong proto-stellar outflows is deposited, seems to be ~ 0.1 pc. The inner scale for the dissipation of this energy appears to be associated with the scale of individual maser spots (~ 1 AU).

Hydroxyl masers, in contrast to water masers, do not follow a power-law spatial correlation pattern. They appear to cluster on a scale of ~ 100 AU. This probably indicates a true cloud formation size, which may find an origin in Rayleigh-Taylor instabilities between cool and dense material supported by a hot and light wind or plasma associated with a young star. More complex structures, such as lines or arcs of spots on the sky or "worms" on position-velocity diagrams have now been well documented but have yet to be explained.

The spatial correlation properties of methanol masers have yet to be fully explored. One might expect Class-II sources to display properties similar to that of interstellar hydroxyl masers, and Class-I sources to show more fractal structure similar to water masers. However, perhaps they will reveal interesting surprises.

4.2. *Variability*

The report of quasi-periodic variability of the 6.7 GHz methanol masers in G9.6 was one highlight of the Brazil maser meeting. While the evidence then was quite strong, there was only one source with well documented periodicity and some of us did not know what to make of this phenomenon. Perhaps a rare chance meeting of a star-forming region with an old Mira variable happened in the G9.6 case, and the IR variations of the Mira light were affecting the population inversion in a periodic manner.

We have learned in this meeting that the phenomenon is not unique to G9.6, but occurs in roughly 10% of sources surveyed. VLBA maps show that individual spots pulsate, while the overall form and distribution of the maser emission persists. The variability period among sources ranges between about 130 and 500 days, which is reminiscent of Mira variables. However, the chances of Miras wandering into so many star forming regions is vanishingly small. Perhaps embedded, pulsating, pre-main-sequence stars are responsible for the periodic maser emission. (VY CMa might be a high-mass example of such an object.) Should this speculation be correct, the variability statistics would imply that the duration of pre-main-sequence pulsations would be $\sim$ 10% of the maser lifetime.

4.3. *Young stellar objects: disks & jets*

Maser emission is a common phenomenon in young stellar objects (YSOs). Most known interstellar masers seem to come from outflows in YSOs. Both fast and slow outflows are observed, probably related to high speed jets and the interaction (entrainment) of dense material with these jets. Water masers are the best tracer of fast outflows. Indeed motions of hundreds of km s^{-1} have been detected in many sources. These motions are usually not highly collimated and may proceed into solid angles of order a steradian. Slow outflows may be traced by many masering species, including water, CH_3OH, OH and occasionally SiO. It is unclear if fast and slow outflows reflect the same general processes, possibly differing in the amount of momentum transferred to the dense masering material from faster, lighter jets.

YSO outflows are thought to be related to disk-based accretion and may be important for solving the angular momentum problem in star formation. While evidence for outflows is abundant, clear evidence for disks is much harder to come by. The H-recombination line masers in MWC 349 appear to trace the limbs of an edge-on YSO disk. The young star Orion I (in the BN/KL region) provides what may be the best picture of a high-mass stellar accretion disk (seen in radio continuum) and its associated outflows (traced by SiO maser emission). Few other examples exist and they usually have less clear interpretations.

It is often hard to differentiate between an edge-on disk and its bipolar outflows. Indeed, this observational ambiguity is striking for the case of Orion I, where convincing claims have been made for orthogonal disk-jet orientations. Often one must rely on indirect indications of jet orientations. These include the H_2 2 μm emission, thermal SiO lines or other molecular emission associated with jets.

In addition to masers associated with (at least partially) collimated jets, some masers form in the relatively gentle expansion surrounding hyper- and ultra-compact H II regions. Interstellar OH and Class-II CH_3OH masers are the classic examples, although both species can maser in other circumstances. For example, weak OH maser action is known to accompany strong water masers in outflows.

Finally, interstellar masers are also found far from young stars in regions where shocks have propagated and probably dissipated enough energy to prepare dense material to maser. Class-I CH_3OH masers seem to fall into this category. Also, 1720 MHz OH masers are found at the interface of SNR shocks and dense molecular material.

4.4. *Magnetic fields*

Magnetic fields (B) derived from maser polarization (Zeeman) measurements are consistent with, and greatly extend, the correlation of B with total molecular density (n). Above a critical density of ≈ 10 cm^{-3}, $B \propto n^{1/2}$.

The most straight forward Zeeman results are for OH masers, where the Zeeman splitting is often larger than the line widths. This makes estimation of B fairly simple, and even gives the *full magnitude* of the B-field (as apposed to the line-of-sight component often measured in the small splitting regime). For decades OH maser B-field estimates have been ranged between 1 and 10 mG, associated with densities between 10^5 and 10^7, respectively. A natural explanation for the observed range is that for densities below 10^5 insufficient maser gain occurs, while for densities above 10^7 the populations thermalize.

Recent observations pose a challenge. Very large Zeeman splitting requiring magnetic fields of up to 40 mG have now been discovered in some OH masers. These values of magnetic field are reminiscent of the tens to hundreds of mG typical for interstellar water masers (which are thought to operate at significantly higher densities than OH masers). If physical conditions, other than density, are the same for all OH masers, this suggests total densities exceeding $\sim 10^8$ cm^{-3}. At such densities it is unclear how the OH masers could operate. It is unlikely that the $B \propto n^{1/2}$ relation breaks down at high densities, since water masers observations have extended the relation to even higher densities and magnetic fields. It is unclear what is going on in the high B-field OH masers.

One other result of Zeeman observations of interstellar OH maser remains quite robust. The magnetic field directions (toward or away from the observer) are highly ordered across entire sources (ie, ~ 0.1 pc). In nearly every source where high quality VLBI data are available, there are either 0 or 1 magnetic field reversals. When there is a reversal, one can "draw a smooth line" on the sky separating the two regions with differing B-field directions. The origin of such organized fields is unclear, but it is probably an important clue to the interaction of magnetic fields and material in massive star forming regions.

4.5. *Chronologies*

One of the most interesting, yet vexing, research goals is to establish a chronology for the formation and evolution of masers associated with star formation. Generally, we have thought that water are found in the youngest sources, followed by CH_3OH, and then OH masers. However, there is probably considerable overlap in time, and sometimes in space, for these masering species. At the Brazil maser meeting, I attempted to depict this general understanding in a figure, similar to Fig. 1.

Based on results presented at this meeting (or published since the Brazil meeting), the general chronology seems to stand the test of observations. The existence of molecular cores (eg, NH_3 cores) without masers suggests that a true protostar is needed for most strong masers (other than Class-I methanol masers).

Water masers have been found associated with mm-wavelength-only cores, which are thought to be among the youngest of star forming sources. Also, water masers may be associated with the first generation of star formation in a given region, followed by Class-II methanol masers which may appear only in the 2nd generation stars.

While OH masers, especially those associated with UCHII regions, are clearly around "old objects" (eg, main sequence OB stars), some weak OH masers have been found projected toward water-maser outflows, which are clearly quite young. The observation that water maser sources are nearly twice as numerous as OH masers, suggests that water masers are longer lived and/or are extend to lower mass stars than OH masers.

Class-I CH_3OH masers remain poorly understood. Clearly they are very young in the sense that they do not form around a YSO. However, it is unclear if there will be any significant star formation at the sites of these masers. They may simply occur at the interfaces of shocks from relatively distant forming stars and high density clumps of molecular cloud material.

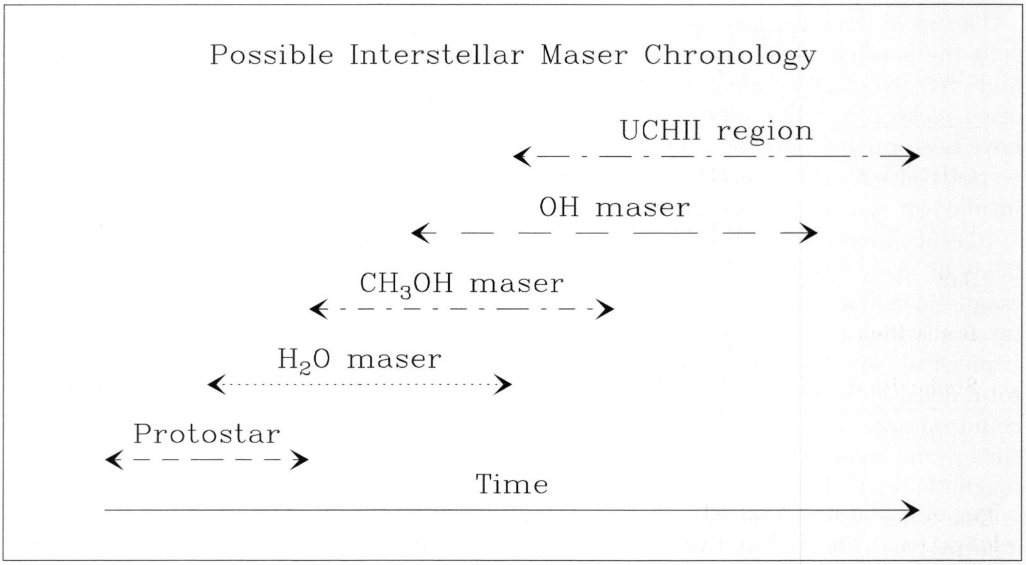

Figure 1. Possible chronology for masers associated with regions of massive star formation. Time increases toward the right. Note the considerable time overlap of various masering species and phenomena.

Finally, it is worth noting that attempting to infer age from spatial distributions, (projected) associations, and numbers is a risky enterprise. Factors other than age that may influence the observed properties are stellar mass, external ionizing sources, and shocks from nearby YSOs and even SNRs.

5. Stellar masers

5.1. *Standard model*

Evolved stars (eg, long period variables, Miras, AGB stars, red super-giants) display rich structure, and masers contribute a great deal to our understanding. A typical AGB star with circumstellar masers has a "surface" temperature below 3000 K, where the surface (at $R_* \approx 1 - 2AU$) is defined in line-free regions of the infrared spectrum. At this low temperature, hydrogen is predominantly atomic and free electrons (important for radio frequency opacity) come from the low ionization potential metals (eg, Na and K). Radio frequency opacity is dominated by H^- free-free interactions, which create a "radio photosphere" extending to about $2R_*$. The radio photosphere approximately matches a "molecular photosphere" seen in strong metallic oxide lines optically and in water and carbon monoxide lines in the infrared.

At about 2–$3R_*$($\approx$ 4 AU) in oxygen-rich stars, SiO masers form, expand outwards, and dissipate in roughly a stellar cycle ($\approx$ 1 yr). Beyond the SiO masers, dust forms copiously. Thus, SiO masers exist between a "rock (dust formation) and a hard place (strong continuum opacity)." Water masers are found at $\sim$ 15 AU and OH masers at $\sim$ 40 AU. Thus, masers trace most of the inner circumstellar envelope and provide excellent probes of physical and chemical conditions.

5.2. *Post-AGB stars*

Understanding the changes in appearance from AGB stars to post-AGB stars to planetary nebulae is one of the great challenges of modern astrophysics. The central stars show some

Table 1. Magnetic Fields and Pressures in Stellar Masers

Maser	R (AU)	B (G)	n (cm^{-3})	T (K)	V_{exp} (km/s)	$B^2/8\pi$ (dyne/cm^2)	nkT (dyne/cm^2)	ρV_{exp}^2 (dyne/cm^2)
OH(1665/7)	40	0.003	10^6	300	10	$10^{-6.4}$	$10^{-7.4}$	$10^{-5.9}$
H_2O	15	0.3	10^8	500	8	$10^{-2.4}$	$10^{-5.2}$	$10^{-4.1}$
SiO	4	3	10^{10}	1300	5	$10^{-0.4}$	$10^{-2.7}$	$10^{-2.5}$
Radio Phot.	3	?	10^{12}	1500	5	?	$10^{-0.7}$	$10^{-0.5}$
Stellar Phot.	1.5	?	10^{14}	2500	15	?	$10^{+1.5}$	$10^{+2.4}$

signs of asphericity and VLA images of the radio photospheres indicate some significant elongations. The SiO and water masers appear as partially filled rings and show modest asymmetries. However, during the post-AGB phase, strong departures from sphericity are observed. These include evidence for disks (seen in SiO maser emission) and jets (seen in radio continuum and water masers). Understanding the transition from the AGB to the post-AGB is critical to understanding the beautiful and highly aspherical appearance of planetary nebulae. Clearly, masers can play a significant role in this field.

5.3. *Magnetic fields*

Magnetic fields have been estimated from maser polarization studies and these observations pose significant theoretical challenges. For Mira-like variables, OH masers indicate magnetic field strengths of a few mG at radii of $\sim$ 40 AU; however water and SiO masers show evidence of fields up to a few Gauss. Table 1 summarizes characteristic values for physical and dynamical variables as a function of radius, as well as rough estimates of the implied magnetic, thermal, and ram pressures. The factors of 100 or 1000 jump in magnetic field strength between radii of $\sim$ 40 AU (OH masers) and $\sim$ 15 AU (water masers) or $\sim$ 4 AU (SiO masers), respectively, stand out. Taken at face value, they would require the magnetic field strength to decrease as a high power of radius (eg, $B \propto R^{-4}$, between the water and OH masering radii. If the water and SiO magnetic field estimates are correct, then it appears that magnetic pressure dominates in these regions. Explaining how this happens may be both fascinating and challenging. Clearly, further observations and theoretical/computational work is required.

5.4. *Super- & hyper-giant stars*

VY CMa is one of the most luminous and interesting supergiants in the Galaxy. It is associated with a region of star formation, but there are no lithium lines in its spectrum, indicating that it is indeed a post-main-sequence star. Its luminosity has been estimated to be $\approx 5 \times 10^5$ $L_\odot$, which is near the maximum allowed for a red supergiant. However, recent trigonometric parallax measurements made with the VLBA, using its SiO masers as astrometric targets, indicate that rather than being at its commonly assumed distance of 1.5 kpc, VY CMa is $\approx$ 25% nearer. This reduces estimates of its luminosity by about a factor of two.

There is evidence that over the past $\sim 10^3$ years VY CMa has been losing mass extremely rapidly and possibly asymmetrically. Complex asymmetric structures are seen in all maser species and in IR images, but a quasi-symmetry axis of $\approx 50°$ (E of N) appears in many images.

In the supergiant S Per, the water and OH masers appear distributed similarly, raising the question are they co-spatial or is this a projection effect. One clue is that Zeeman measurements indicate ~ 3 mG fields for OH masers and ~ 150 mG fields for water masers. This suggests that the masers are not co-spatial.

6. Galactic structure

The spiral nature of the Milky Way remains a large unknown, even after decades of research. Indeed, the most widely used models of the spiral structure of the Milky Way can be traced to the work of Georgelin & Georgelin in the 1970's. This situation may change soon. At this meeting we heard of about a dozen trigonometric parallaxes to masers in massive star forming regions measured with VLBI techniques. These parallaxes have accuracies as good as ± 10 μas. In the future the VLBA and VERA will provide ~ 100 high quality parallaxes that should serve to trace the number and the location of spiral arms in the Milky Way.

One side benefit of these parallax observations are proper motions of massive star-forming regions accurate to ~ 1 km s^{-1}. These motions will yield valuable data on the rotation curve and distribution of mass, especially dark matter, in the Milky Way. Also, the fundamental constants of the Milky Way, R_0 and Θ_0, should be directly measured with high accuracy soon.

7. AGN water masers

We now have a clear understanding that water masers populate both the sub-parsec scale accretion disks and jets associated with super-massive black holes in the centers of active galaxies. A wide variety of structures are inferred from observations, including thin and thick disks, narrow jets and wide angle winds. Water masers are proving exceptional, if not unique, probes of AGN disks, yielding images of warps, indications of spiral structures, and measurements of magnetic fields, densities and temperatures. Of course, they also supply by far the best measurements of super-massive black hole masses.

NGC 4258 remains the "Rosetta Stone" of AGN water masers. It has a remarkably thin disk (< 10 μas) which, if in hydrostatic equilibrium, suggests a temperature of 600 K. There is a clear warping. The (spatial) periodicity in the locations of the high velocity masers has been confirmed and is probably a result of spiral structures. Indeed, kinematic signatures of spiral structure might already have been detected.

Water masers in AGN can be precisely modeled and a combination of VLBA maser maps with maser accelerations measured from a time series of spectra yield a geometric distance estimate. NGC 4258 observations have yielded the most accurate, direct, distance for a galaxy; this helps anchor the extragalactic distance scale and improves the accuracy of estimates of H_0. The NRAO/CfA/MPIfR Maser Cosmology Project seeks to expand on the number and accuracy of AGN water maser distances, with a goal of measuring H_0 to $\pm 3\%$ accuracy, by directly measuring distances to galaxies in the Hubble Flow. Independent knowledge of the value of H_0 is crucial for braking cosmological parameter degeneracies, for example to establish the equation of state of dark energy. This project has already discovered and mapped the "NGC 4258-like" water masers in UGC 3789, a galaxy at about 50 Mpc distance.

8. OH megamasers

The physical picture of OH megamasers has become clear over the past years. These masers are associated with starbursts, which are the result of major mergers; they are not associated with AGN. Very dense gas ($n_H \sim 10^6$ cm^{-3}) is required to support these masers, and recent measurements of large Zeeman splitting (large B-fields) suggests even higher densities may be involved, if the relation between magnetic field and density seen in Milky Way star formation regions holds.

The model of a torus filled with clouds (each with low maser gain) that occasionally align both along the line-of-sight and in velocity, can possibly explain both high and low gain maser emission. Aligned, well-separated (along the line-of-sight) clouds can produce very highly beamed emission that is likely to be unsaturated. However, such a model does not explain why 1667 MHz masers are compact, while 1665 MHz masers are extended in Arp 220.

Other observed features of OH megamasers remain unexplained. For example, what leads to the origin of strong velocity gradients associated with arcs of maser spots in Arp 220? If gravitation in origin, the arcs enclose very large masses of $\sim 10^5$ M$_\odot$.

1720 MHz OH masers and weak formaldehyde masers have also been discovered in megamasers. Interestingly, a 183 GHz water maser has been discovered in Arp 220, even though no 22 GHz water maser is detected in this galaxy. So perhaps the 183 GHz water line can trace mergers and starbursts.

9. The future

In the next ten years we can expect many great discoveries in the field of maser research. I suspect that millimeter and sub-millimeter wavelength maser observations will be as common as centimeter wavelength observations are now. Blind surveys (of at least the Galactic Plane) should be completed for all major classes of masers. The Milky Way will be mapped with $\sim$ 100 trigonometric parallaxes and proper motions for star forming regions.

At this point it seems that masers may soon yield the best estimate of the Hubble constant: a 3% uncertainty is within reach. This result, coupled with other cosmological probes, may establish that $w = 1$, and with nothing left to solve but details, astrophysics will come to an end ;-). The constancy of fundamental constants should be further tested. But, hopefully, something quite unusual, such as the detection of hydrogen recombination-line masers from the epoch of reionization, will be discovered.

There have now been three international meetings on masers. The first was in Washington, D.C. in 1992 and the second near Rio De Janeiro in 2001. This meeting marks the third maser conference held on a third different continent: Australia. That suggests the next three maser conferences should be held in Africa, Asia, and Europe. If these meetings are held every 5 years, the 7th international maser conference should be held in 2027 in the Antarctic.

Author Index